MW01284523

OPTIMAL ESTIMATION of DYNAMIC SYSTEMS

CHAPMAN & HALL/CRC APPLIED MATHEMATICS
AND NONLINEAR SCIENCE SERIES
Series Editors *Goong Chen and Thomas J. Bridges*

Published Titles

Mathematical Methods in Physics and Engineering with Mathematica,
Ferdinand F. Cap

Optimal Estimation of Dynamic Systems, John L. Crassidis and John L. Junkins

Forthcoming Titles

An Introduction to Partial Differential Equations with MATLAB,
Matthew P. Coleman

Mathematical Theory of Quantum Computation, Goong Chen and Zijian Diao

CHAPMAN & HALL/CRC APPLIED MATHEMATICS
AND NONLINEAR SCIENCE SERIES

OPTIMAL ESTIMATION of DYNAMIC SYSTEMS

John L. Crassidis and John L. Junkins

A CRC Press Company
Boca Raton London New York Washington, D.C.

Library of Congress Cataloging-in-Publication Data

Crassidis, John L.
Optimal estimation of dynamic systems / John L. Crassidis, John L. Junkins.
p. cm. — (Chapman & Hall/CRC applied mathematics and nonlinear science series; 2)
Includes bibliographical references and index.
ISBN1-58488-391-X (alk. paper)
1. System analysis. 2. Estimation theory. I. Junkins, John L. II. Title. III. Series.

QA402.C73 2004
003—dc22 2004043817

Direct all inquiries to CRC Press LLC, 2000 N.W. Corporate Blvd., Boca Raton, Florida 33431.

Visit the CRC Press Web site at www.crcpress.com

International Standard Book Number 1-58488-391-X
Library of Congress Card Number 2004043817
1 2 3 4 5 6 7 8 9 0
Printed on acid-free paper

Preface

THIS text is designed to introduce the fundamentals of estimation to engineers, scientists, and applied mathematicians. This text is the follow-on to the first estimation book written by the second author in 1978. The current text expands upon the past treatment to provide more comprehensive developments and updates, including new theoretical results in the area. The level of the presentation should be accessible to senior undergraduate and first-year graduate students, and should prove especially well-suited as a self study guide for practicing professionals. The primary motivation of this text is to make a significant contribution toward minimizing the painful process most newcomers must go through in digesting and applying the theory. By stressing the interrelationships between estimation and modelling of dynamic systems, it is hoped that this new and unique perspective will be of perennial interest to other students, scholars, and employees in engineering disciplines.

This work is the outgrowth of the authors' multiple encounters with the subject while motivated by practical problems with spacecraft attitude determination and control, aircraft navigation and tracking, orbit determination, powered rocket trajectories, photogrammetry applications, and identification of vibratory systems. The text has evolved from lecture notes for short courses and seminars given to professionals at various private laboratories and government agencies, and in conjunction with courses taught at the University at Buffalo and Texas A&M University.

To motivate the reader's thinking, the structure of a typical estimation problem often assumes the following form:

- Given a dynamical system, a mathematical model is hypothesized based upon the experience of the investigator, which is consistent with whatever physical laws known to govern the system's behavior, the number and nature of the available measurements, and the degree of accuracy desired. Such mathematical models almost invariably embody a number of poorly known parameters.

- Determine "best" estimates of all poorly known parameters so that the mathematical model provides an "optimal estimate" of the system's actual behavior.

Any systematic method which seeks to solve a problem of the above structure should generally be referred to as an estimation process. Depending upon the nature of the mathematical model of the system and the statistical properties of the measurement errors, the degree of difficulty associated with solution of such problems ranges from near-trivial to impossible.

In writing this text, we have kept in mind three principal objectives:

1. Document the development of the central concepts and methods of optimal estimation theory in a manner accessible to engineering students, applied mathematicians, and practicing engineers.

2. Illustrate the application of the methods to problems having varying degrees of analytical and numerical difficulty. Where applicable, compare competitive approaches to help the reader develop a feel for the absolute and relative utility of various methods.

3. Present prototype algorithms, giving sufficient detail and discussion to stimulate development of efficient computer programs, as well as intelligent use of programs.

Consistent with the first objective, the major results are developed initially by the route requiring minimum reliance upon the reader's mathematical skills and *a priori* knowledge. This is shown by the first chapter, which introduces least squares methods without the requirement of probability and statistics knowledge. We have decided to include the required prerequisites (such as matrix properties, probability and statistics, and optimization methods) as appendices, so that this information can be made accessible to the readers at their own leisure. Our approach should give the reader an immediate sense of the usefulness of estimation concepts from first principles, while later chapters provide more rigorous developments that use higher-level mathematics and knowledge. In many cases, subsequent developments re-establish the same "end results" by alternative logical/mathematical processes (e.g., the derivation of the continuous-time Kalman filter in Chapter 5). These developments should provide fresh insight and greater appreciation of the underlying theory.

The set of problems selected to accomplish the second objective are typically idealized versions of real-world engineering problems. We believe that bridging the gap between theory and application is important. Several examples are given in each chapter to illustrate the methods of that chapter. The main focus of the text is to stress actual dynamical models. The methods shown are applicable to "block box" representations, but it is hoped that the expanded dynamical models will more clearly illustrate the importance of the theoretical methods in estimation. Chapter 3 provides a review of dynamical systems, which spans the central core of the subject matter and provides a reasonable foundation for immediate application of estimation concepts to a significant class of problems. Chapters 4 and 7 use this subject matter to provide realistic examples, thereby giving the reader a deep understanding of the value of estimation concepts in actual engineering practice. In the applications of Chapters 4 and 7, the methods of the remaining chapters are applied; often with two or more estimation strategies compared and two or more prototype models of the system considered (e.g., the comparison of GPS position determination using nonlinear least squares in §4.1 versus a Kalman filter approach in §7.1).

In adopting the last objective, the authors remain sensitive to the pitfalls of "cookbooks" for a subject as diverse as estimation. The problem solutions and algorithms are not put forth as optimal implementations of the various facets of the theory, nor will the methods succeed in solving every problem to which they formally apply.

Nonetheless, it is felt that the example algorithms will prove useful, if accepted in the spirit that they are offered; namely as implementations which have proven successful in previous applications. Also, general computer software and coded scripts have deliberately not been included with this text. Instead, a website with computer programs for all the examples shown in the text can be accessed by the reader (see Appendix D). Although computer routines can provide some insights to the subject, we feel that they may hinder rigorous theoretical studies that are required to properly comprehend the material. Therefore, we strongly encourage students to program their own computer routines, using the codes provided from the website for verification purposes only. Most of the general algorithms are summarized in flowchart or table form, which should be adequate for the mechanization of computer routines.

Our philosophy involves rigorous theoretical derivations along with a significant amount of qualitative discussion and judgments. The text is written to enhance student learning by including several practical examples and projects taken from experiences gained by the authors. One of our purposes is to illustrate the importance of both physical and numerical modelling in solving dynamics-based estimation problems found in engineering systems. To encourage student learning we have incorporated both analytical and computer-based problems at the end of each chapter. This promotes working problems from first principles. Furthermore, advanced topics are placed in the chapters for the purpose of engaging the interest of students for further study. These advanced topics also give the practicing engineer a preview of important research issues and current methods. Finally, we have included many qualitative comments where such seems appropriate, and have also provided insights to the practical applications of the methods gained from years of intimate experience with the systems described in the book.

We are indebted to numerous colleagues and students for contributions to various aspects of this work. Many students have provided excellent insights and recommendations to enhance the pedagogical value, as well as developing new problems which are used as exercises. Although there are far too many students to name individually here, our heartfelt thanks and appreciation go out to them. We do wish to acknowledge the significant contributions and discussions on the subject matter to the following individuals: Terry Alfriend, Roberto Alonso, Mark Balas, Itzhack Bar-Itzhack, Russell Carpenter, Yang Cheng, Agamemnon Crassidis, Glenn Creamer, Chris Hall, Jer-Nan Juang, Kok-Lam Lai, E. Glenn Lightsey, F. Landis Markley, Paul Mason, Tom Meyer, D. Joseph Mook, Daniele Mortari, Yaakov Oshman, Tom Pollock, Mark Psiaki, Hanspeter Schaub, Malcolm Shuster, Tarun Singh, Debo Sun, S. Rao Vadali, John Valasek, and Bong Wie. Also, many thanks are due to several people at CRC Press, including: Bob Stern, Jessica Vakili, Nishith Arora, and Helena Redshaw. Finally, our deepest and most sincere appreciation must be expressed to our families for their patience and understanding throughout the years while we prepared this text. This text was produced using LaTeX 2_ε (thanks Yaakov and HP!). Any corrections are welcome via email to *johnc@buffalo.edu* or *junkins@tamu.edu*.

John L. Crassidis
John L. Junkins

To Pam and Lucas, and in memory of Lucas G.J. Crassidis
and
To Elouise, Stephen and Kathryn

Contents

1

Least Squares Approximation

Theory attracts practice as the magnet attracts iron. Gauss, Karl Friedrich

THE celebrated concept of least squares approximation is introduced in this chapter. Least squares can be used in a wide variety of categorical applications, including: curve fitting of data, parameter identification, and system model realization. Many examples from diverse fields fall under these categories, for instance determining the damping properties of a fluid-filled damper as a function of temperature, identification of aircraft dynamic and static aerodynamic coefficients, orbit and attitude determination, position determination using triangulation, and modal identification of vibratory systems. Even modern control strategies, for instance certain adaptive controllers, use the least squares approximation to update model parameters in the control system. The broad utility implicit in the aforementioned examples strongly confirm that the least squares approximation is worthy of study.

Before we begin analytical and mathematical discussions, let us first define some common quantities used throughout this chapter and the text. For any variable or parameter in estimation, there are three quantities of interest: the true value, the measured value, and the estimated value. The true value (or "truth") is usually unknown in practice. This represents the actual value sought of the quantity being approximated by the estimator. Unadorned symbols are used to represent the true values. The measured value denotes the quantity which is directly determined from a sensor. For example, in orbit determination a radar is often used to obtain a measure of the range to a vehicle. In actuality, this is not a totally accurate statement since the truly measured quantity given by the radar is not the range. Radars work by "shining" a beam of energy (usually microwaves) at an object and analyzing the spectral content of the energy that gets reflected back. Signal processing of the measured return energy can yield estimates of range (or range rate). For navigation purposes, we often assume that the measured quantity is the computed range, because this is a direct function of the truly measured quantity, which is the reflected energy received by the radar. Measurements are never perfect, since they will always contain errors. Thus, measurements are usually modelled using a function of the true values plus some error. The measured values of the truth x are typically denoted by $\tilde{x}$. Estimated values of x are determined from the estimation process itself, and are found using a combination of a static/dynamic model and the measurements. These values are denoted by $\hat{x}$. Other quantities used commonly in estimation are the measurement error (measurement value minus true value), and the residual error (measurement value minus estimated value). Thus, for a measurable quantity x, the following two equations hold:

measured value	=	true value	+	measurement error
$\tilde{x}$	=	x	+	v

and

measured value	=	estimated value	+	residual error
$\tilde{x}$	=	$\hat{x}$	+	e

The actual measurement error (v), like the true value, is never known in practice. However, the errors in the mechanism that physically generate this error are usually approximated by some known process (often by a zero-mean Gaussian noise process with known variance). These assumed known statistical properties of the measurement errors are often employed to weight the relative importance of various measurements used in the estimation scheme. Unlike the measurement error, the residual error is known explicitly and is easily computed once an estimated value has been found. The residual error is often used to drive the estimator itself. It should be evident that both measurement errors and residual errors play important roles in the theoretical and computational aspects of estimation.

1.1 A Curve Fitting Example

To explore Gauss' connection between theory and practice, we introduce the concept of least squares by considering a simple example that will be used to motivate the theoretical developments of this chapter. Displayed in Figure 1.1 are measurements of some process $y(t)$. At this point we do not consider the physical connotations of the particular process, but it may be useful to think of $y(t)$ as a stock quote history for a particular company. You want to determine a mathematical model for $y(t)$ in order to predict future prospects for the company. Measurements (e.g., closing stock price) of $y(t)$, denoted by $\tilde{y}(t)$, are given for a 6-month time frame. In order to insure an accurate model fit, you have been informed that the residual errors (i.e., between the measured values and estimated values) must have an absolute mean of ≤ 0.0075, and a standard deviation of ≤ 0.125. With a large number of samples (m), the sample mean (μ) and sample standard deviation (σ) for the residual error can be computed using[1] (we will derive these later)

$$\mu = \frac{1}{m}\sum_{i=1}^{m}\left[\tilde{y}(t_i) - \hat{y}(t_i)\right] \tag{1.1}$$

$$\sigma^2 = \frac{1}{m-1}\sum_{i=1}^{m}\left\{\left[\tilde{y}(t_i) - \hat{y}(t_i)\right] - \mu\right\}^2 \tag{1.2}$$

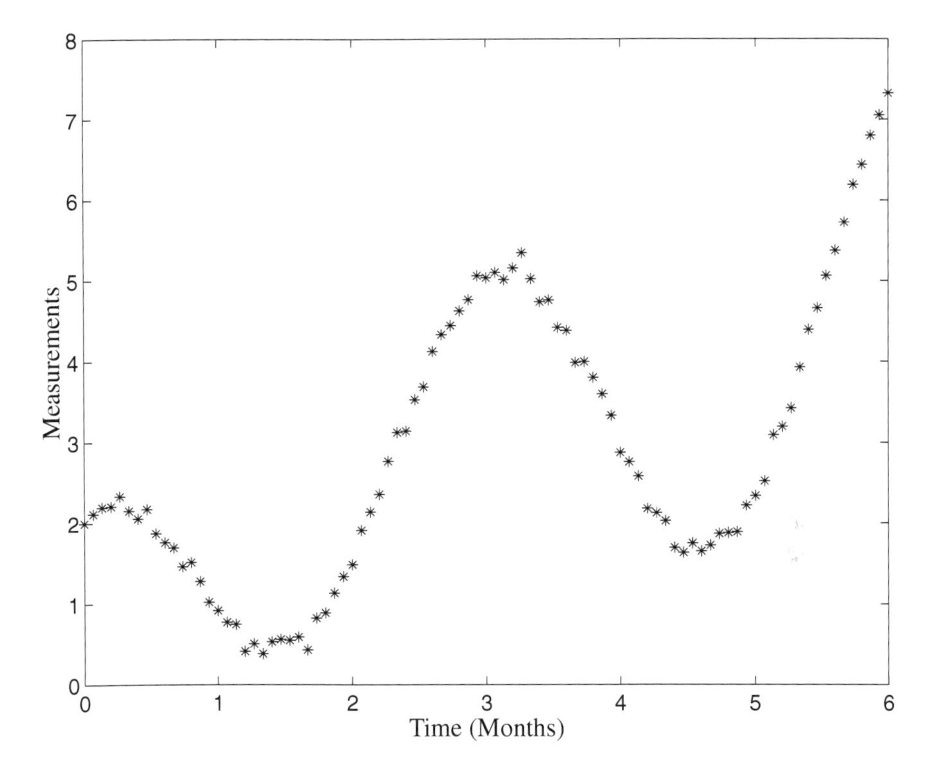

Figure 1.1: Measurements of $y(t)$

where $\hat{y}(t)$ denotes the estimate of $y(t)$.

Now in your quest to establish a model which predicts the behavior of $y(t)$, you might naturally attempt evaluation of some previously developed models. After some research you have found two models, given by

$$\text{Model 1}: \quad y_1(t) = c_1 t + c_2 \sin(t) + c_3 \cos(2t) \tag{1.3}$$

$$\text{Model 2}: \quad y_2(t) = d_1(t+2) + d_2 t^2 + d_3 t^3 \tag{1.4}$$

where t is given in months, and c_1, c_2, c_3 and d_1, d_2, d_3 are constants. The next step is to evaluate "how well" each of these models predicts the measurements with "optimum" values of c_i and d_i. The process of fitting curves, such as Models 1 and 2, to measured data is known in statistics as *regression*.

For the moment, continuing the discussion of the hypothetical problem solving situation, let us assume that you have read and digested the discussion that will come later in §1.2.1 on the method of *linear least squares*. Also, you have employed a least squares algorithm to determine the coefficients in the two models, and found that the "optimum" coefficients are

$$(\hat{c}_1, \hat{c}_2, \hat{c}_3) = (0.9967, 0.9556, 2.0030) \tag{1.5}$$

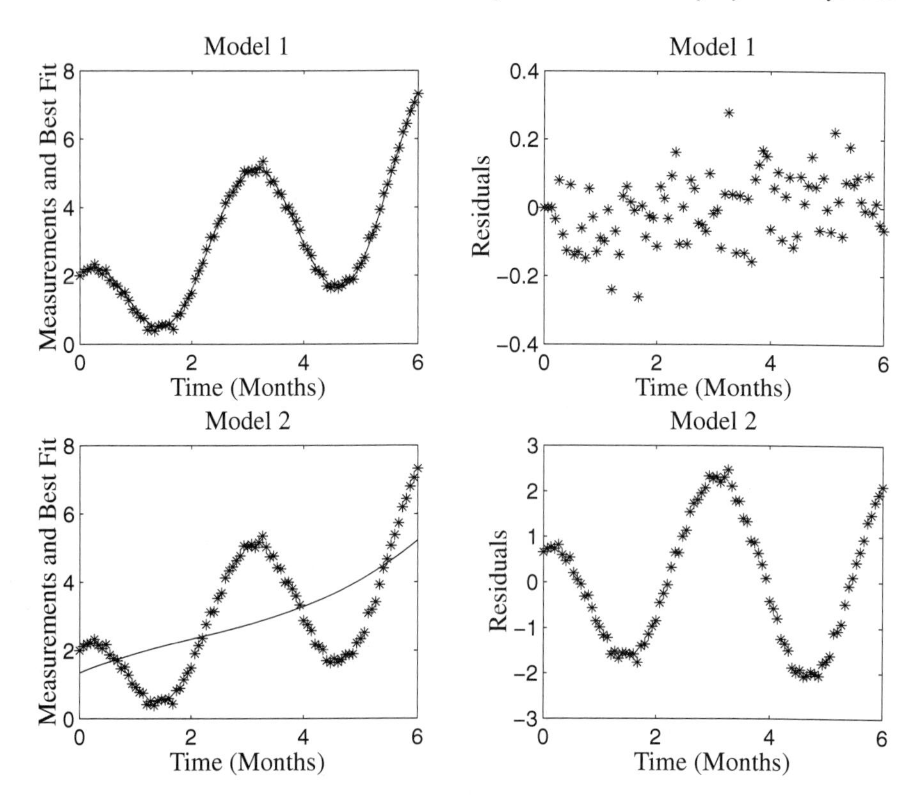

Figure 1.2: Best Fit and Residual Errors for Both Models

$$(\hat{d}_1, \hat{d}_2, \hat{d}_3) = (0.6721, -0.1303, 0.0210) \tag{1.6}$$

Plots of each model's fit superposed on the measured data, and residual errors are shown in Figure 1.2. As is clearly evident, Model 1 is able to obtain the best fit with the determined coefficients. This can also be seen by comparing the sample mean and sample standard deviation of both fits using eqns. (1.1) and (1.2). For Model 1 the sample mean is 1×10^{-5} and the sample standard deviation is 0.0921. For Model 2 the sample mean is 1×10^{-5} and the sample standard deviation is 1.3856. This shows that Model 1 meets both minimum requirements for a good fit, while Model 2 does not.

From the above analysis, you make the qualitative observation that Model 1 is a much better representation of $y(t)$'s behavior than is Model 2. From Figure 1.2, you observe that Model 1's residual errors are "random" in appearance, while Model 2's best fit failed to predict significant trends in the data. Having no reason to suspect that systematic errors are present in the measurements or in Model 1, you conclude that Model 1 can be used to provide an accurate assessment of $y(t)$'s behavior.

Since Model 1 was used to fit the measured data accurately, you might now make the logical hypothesis that this model can be used to *predict* future values for $y(t)$.

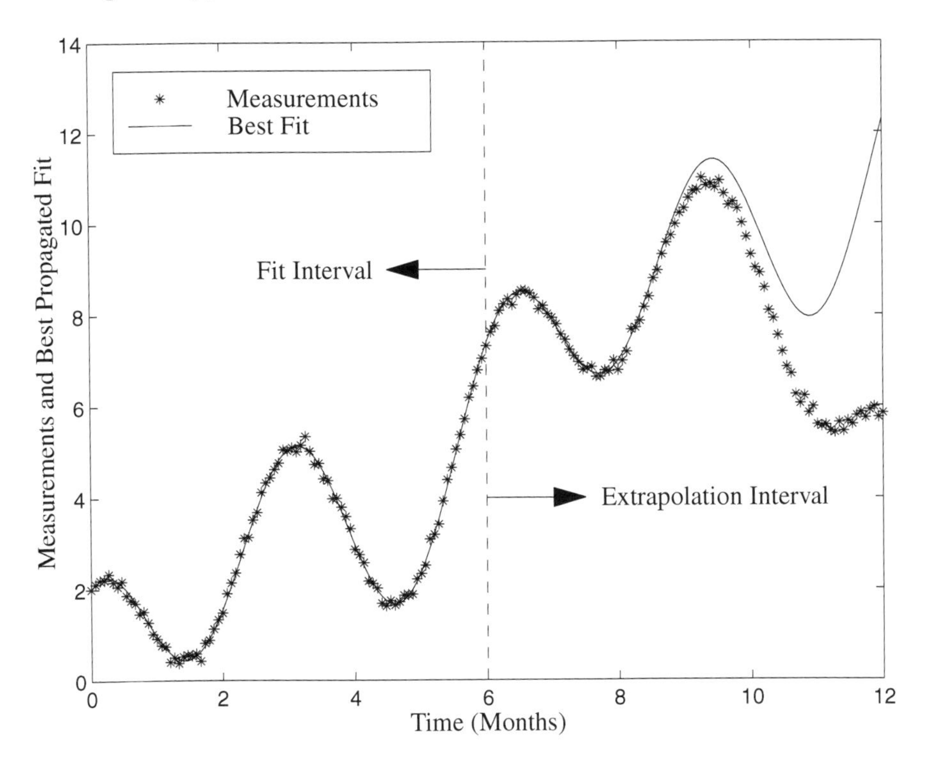

Figure 1.3: Best Fit for $y(t)$ Propagated to 12 Months

The trends in the data of the fit interval, and therefore our model, indicate that the stock prices will continue an upward trend and will more than double in 12 months. Putting your trust in this "get rich quick" scheme, suppose you invest a great amount of money in the stock. But, as is often true in many "get rich quick" schemes, this dangerous extrapolation failed. A plot of Model 1's predictions, with coefficients given in eqn. (1.5), superimposed on the measured data over a twelve month period is shown in Figure 1.3. This shows that you have actually lost money in the stock if you invest after 6 months and hold it until 12 months.

In reality, the synthetic measurements of Figure 1.1 were calculated using the following equation:

$$\tilde{y}(t) = t + \sin(t) + 2\cos(2t) - \frac{0.4e^t}{1 \times 10^4} + v(t) \tag{1.7}$$

where the simulated measurement errors $v(t)$ were calculated by a zero-mean Gaussian noise generator with a standard deviation given by $\sigma = 0.1$. In the above example, Model 1 clearly can be used to "estimate" $y(t)$ for the first 6 months where the estimate is "supported" by many measurements, but does a poor job predicting future values. This is due to the fact that the unmodelled exponential term in eqn. (1.7) begins to dominate the other terms after time $t = 10$. To further illustrate this, let us consider the following model:

$$\text{Model 3}: \quad y_3(t) = x_1 t + x_2 \sin(t) + x_3 \cos(2t) + x_4 e^t \tag{1.8}$$

We observe that this model is in fact the correct model, in the absence of measurement errors. Upon applying the method of least squares using the first 6 months of measurements in Figure 1.1, we find the optimal estimates of the coefficients $\hat{x}_i$ are

$$(\hat{x}_1, \hat{x}_2, \hat{x}_3, \hat{x}_4) = (0.9958, 0.9979, 2.0117, -4.232 \times 10^{-5}) \tag{1.9}$$

We note these estimates differ slightly from the true values $(1, 1, 2, -4 \times 10^{-5})$ as a consequence of the small errors contained in the measurements. It is also of interest to ask the question: "How well can we predict the future when we use the correct model?" This question is answered by repeating the calculation underlying Figure 1.3, using the correct model (1.8) and best estimates (1.9) derived over the first 6 months of data. These results are shown in Figure 1.4. Comparing Figures 1.3 and 1.4, it is evident that using the correct model (1.8) vastly improves the 6-month extrapolation accuracy. The extrapolation still diverges slowly from the subsequent measurements over months 10 to 12. This is because the coefficient estimates derived from any finite set of measurements can be expected to contain estimation errors even when the model structure is perfect. We will develop full insight into the issue: "How do measurement errors propagate into errors of the estimated parameters?"

The above contrived example demonstrates many important issues in estimation theory. First, a challenging facet of practical estimation applications is correctly specifying the system's mathematical model. Also, the first two models contain a t term, but the corresponding numerical estimates of the t coefficient are drastically different in the two best fits. In many real-world problems, dominant terms in a mathematical model will have a correct mathematical structure, but higher-order effects may be poorly understood. Finally, unknown higher order effects and parameter estimation errors can produce erroneous results, especially outside of the measurement domain considered, as shown in Figure 1.3.

Model development is the least tractable aspect of the problem setup and solution, insofar as employing universally applicable procedures. It is unlikely, indeed, that mathematically complicated physical phenomena can be correctly modelled *a priori* by anyone unfamiliar with the basic principles underlying the phenomena. In short, intelligent formulation and application of estimation algorithms require intimate knowledge of the field in which the estimation problem is embedded. In numerous cases, decisions regarding which variable should be measured, the frequency with which data should be collected, the necessary measurement accuracy, and the best mathematical model can be inferred directly from theoretical analysis of the system. *Estimation theory can be developed apart from considering a particular dynamic system, but successful applications almost invariably rely jointly upon understanding estimation theory and the principles governing the system under consideration.*

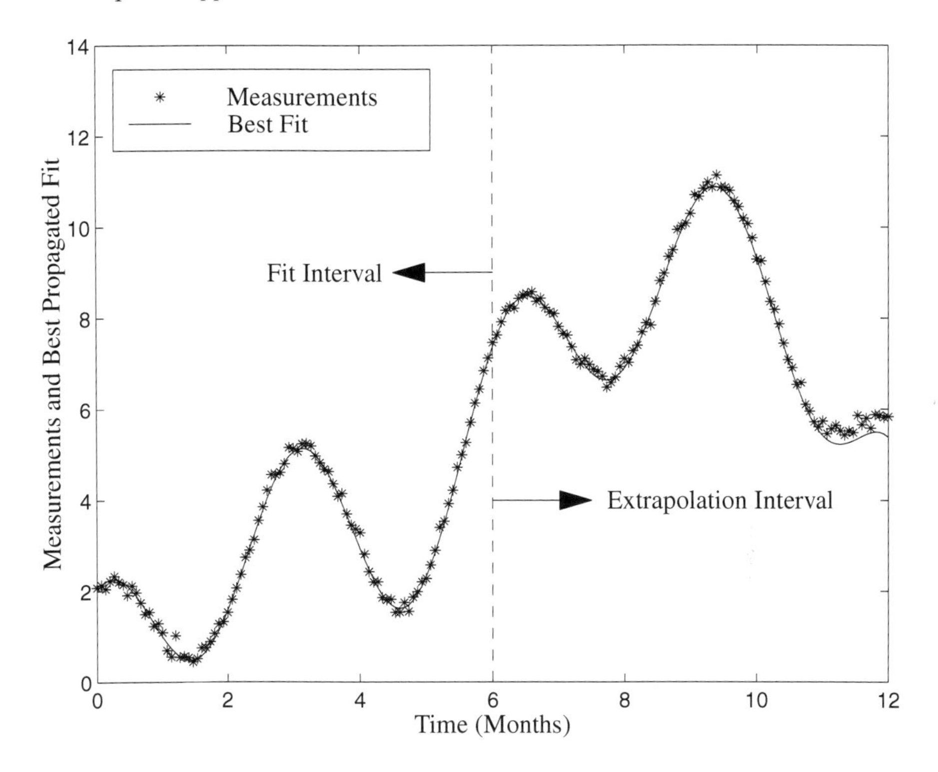

Figure 1.4: Best Fit for $y(t)$ Propagated to 12 Months

1.2 Linear Batch Estimation

In this section we formally introduce Gauss' principle of linear least squares. This principle will be found to be central to the solution of a large family of estimation problems. Suppose that you have in hand a set (or a "batch") of measured values, $\tilde{y}_j$, of a process $y(t)$, taken at known discrete instants of time t_j:

$$\{\tilde{y}_1, t_1;\ \tilde{y}_2, t_2;\ \ldots;\ \tilde{y}_m, t_m\} \tag{1.10}$$

and a proposed mathematical model of the form

$$y(t) = \sum_{i=1}^{n} x_i\, h_i(t), \quad m \geq n \tag{1.11}$$

where

$$h_i(t) \in \{h_1(t), h_2(t), \ldots, h_n(t)\} \tag{1.12}$$

are a set of independent specified *basis* functions. For example, eqns. (1.3) and (1.4) each contain three basis functions in our previous work in §1.1. The x_i are a set of constants whose numerical values are unknown. From eqn. (1.11) it follows that the variables x and y are related according to a simple linear regression model. It seems altogether reasonable to select the optimum x-values based upon a measure of "how well" the proposed model (1.11) predicts the measurements (1.10). Toward this end, we seek a set of estimates, denoted by $\{\hat{x}_1, \hat{x}_2, \ldots, \hat{x}_n\}$, which can be used in eqn. (1.11) to predict $y(t)$. Errors, however, can arise between the "true" value $y(t)$ and the predicted (estimated) value $\hat{y}(t)$ from a number of sources, including:

- measurement errors
- incorrect choice of x-values
- modelling errors, i.e., the actual process being observed may not be accurately modelled by eqn. (1.11).

In virtually every application, some combination of these error sources is present.

We first formally relate the measurements $\tilde{y}_j$ and the estimated output $\hat{y}_j$ to the true and estimated x-values using the mathematical model of eqn. (1.11):

$$\tilde{y}_j \equiv \tilde{y}(t_j) = \sum_{i=1}^{n} x_i\, h_i(t_j) + v_j, \quad j = 1, 2, \ldots, m \tag{1.13}$$

$$\hat{y}_j \equiv \hat{y}(t_j) = \sum_{i=1}^{n} \hat{x}_i\, h_i(t_j), \quad j = 1, 2, \ldots, m \tag{1.14}$$

where v_j is the measurement error. At this point of the discussion, we consider the measurement error to be some unknown process that may include random as well as deterministic characteristics (in the next chapter, we will elaborate more on v_j). It is important to remember that $\tilde{y}_j$ is a *measured* quantity (i.e., it is the output of the measurement process). We have assumed that the measurement process is *modelled* by eqn. (1.13). Next, consider the following identity:

$$\tilde{y}_j = \sum_{i=1}^{n} \hat{x}_i\, h_i(t_j) + e_j, \quad j = 1, 2, \ldots, m \tag{1.15}$$

where the *residual error* e_j is defined by

$$e_j \equiv \tilde{y}_j - \hat{y}_j \tag{1.16}$$

Equation (1.15) can be rewritten in compact matrix form as

$$\tilde{\mathbf{y}} = H\hat{\mathbf{x}} + \mathbf{e} \tag{1.17}$$

where

$$\begin{aligned}\tilde{\mathbf{y}} &= \begin{bmatrix}\tilde{y}_1 \; \tilde{y}_2 \; \cdots \; \tilde{y}_m\end{bmatrix}^T = \textit{measured } y\text{-values}\\ \mathbf{e} &= \begin{bmatrix}e_1 \; e_2 \; \cdots \; e_m\end{bmatrix}^T = \textit{residual errors}\\ \hat{\mathbf{x}} &= \begin{bmatrix}\hat{x}_1 \; \hat{x}_2 \; \cdots \; \hat{x}_n\end{bmatrix}^T = \textit{estimated } x\text{-values}\end{aligned}$$

$$H = \begin{bmatrix} h_1(t_1) & h_2(t_1) & \cdots & h_n(t_1) \\ h_1(t_2) & h_2(t_2) & \cdots & h_n(t_2) \\ \vdots & \vdots & & \vdots \\ h_1(t_m) & h_2(t_m) & \cdots & h_n(t_m) \end{bmatrix}$$

and the superscript T denotes the matrix transpose operation. In a similar manner, eqns. (1.13) and (1.14) can also be written in compact form as

$$\tilde{\mathbf{y}} = H\mathbf{x} + \mathbf{v} \tag{1.18}$$

$$\hat{\mathbf{y}} = H\hat{\mathbf{x}} \tag{1.19}$$

where

$$\begin{aligned}\mathbf{x} &= \begin{bmatrix}x_1 \; x_2 \; \cdots \; x_n\end{bmatrix}^T = \textit{true } x\text{-values}\\ \mathbf{v} &= \begin{bmatrix}v_1 \; v_2 \; \cdots \; v_m\end{bmatrix}^T = \textit{measurement errors}\\ \hat{\mathbf{y}} &= \begin{bmatrix}\hat{y}_1 \; \hat{y}_2 \; \cdots \; \hat{y}_m\end{bmatrix}^T = \textit{estimated } y\text{-values}\\ \tilde{\mathbf{y}} &= \begin{bmatrix}\tilde{y}_1 \; \tilde{y}_2 \; \cdots \; \tilde{y}_m\end{bmatrix}^T = \textit{measured } y\text{-values}\end{aligned}$$

Equations (1.17) and (1.18) are identical, of course, if $\hat{\mathbf{x}} = \mathbf{x}$, and if the assumption of zero model errors is valid. Both of these equations, (1.17) and (1.18), are commonly referred to as the "observation equations."

1.2.1 Linear Least Squares

Gauss's celebrated *principle of least squares*[2] selects, as an optimum choice for the unknown parameters, the particular $\hat{\mathbf{x}}$ that minimizes the sum square of the residual errors, given by

$$J = \frac{1}{2}\mathbf{e}^T\mathbf{e} \tag{1.20}$$

Substituting eqn. (1.17) for $\mathbf{e}$ into eqn. (1.20) and using the fact that a scalar equals its transpose yields

$$J = J(\hat{\mathbf{x}}) = \frac{1}{2}(\tilde{\mathbf{y}}^T\tilde{\mathbf{y}} - 2\tilde{\mathbf{y}}^T H\hat{\mathbf{x}} + \hat{\mathbf{x}}^T H^T H\hat{\mathbf{x}}) \tag{1.21}$$

The $1/2$ multiplier of J does have a statistical significance, as will be shown in Chapter 2. We seek to find the $\hat{\mathbf{x}}$ that minimizes J. Using the matrix calculus differentiation rules developed in §A.5, it follows that for a global minimum of the quadratic function of eqn. (1.21) we have the following requirements:

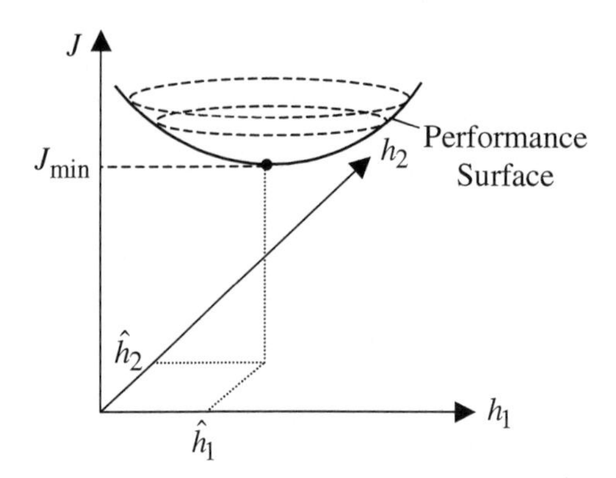

Figure 1.5: Convex Performance Surface for Order $n = 2$ Problem

necessary condition

$$\nabla_{\hat{\mathbf{x}}} J \equiv \begin{bmatrix} \dfrac{\partial J}{\partial \hat{x}_1} \\ \vdots \\ \dfrac{\partial J}{\partial \hat{x}_n} \end{bmatrix} = H^T H \hat{\mathbf{x}} - H^T \tilde{\mathbf{y}} = \mathbf{0} \tag{1.22}$$

sufficient condition

$$\nabla^2_{\hat{\mathbf{x}}} J \equiv \frac{\partial^2 J}{\partial \hat{\mathbf{x}}\, \partial \hat{\mathbf{x}}^T} = H^T H \text{ must be positive definite} \tag{1.23}$$

where $\nabla_{\hat{\mathbf{x}}} J$ is the *Jacobian* and $\nabla^2_{\hat{\mathbf{x}}} J$ is the *Hessian* (see Appendix A). Consider the sufficient condition first. Any matrix B such that

$$\mathbf{x}^T B \mathbf{x} \geq 0 \tag{1.24}$$

for all $\mathbf{x} \neq \mathbf{0}$ is called positive semi-definite. By setting $\mathbf{h} = H\mathbf{x}$ and squaring, we easily obtain the scalar $h^2 = \mathbf{h}^T \mathbf{h} \geq 0$, so, $H^T H$ is always positive semi-definite. It becomes positive definite when H is of maximum rank (n).

The function J is a performance surface in $n+1$-dimensional space.[3] This performance surface has a convex shape of an n-dimensional parabola with one *distinct* minimum. An example of this performance surface for $n = 2$ is the three-dimensional bowl-shaped surface shown in Figure 1.5.

From the necessary conditions of eqn. (1.22), we now have the "normal equations"

$$(H^T H)\hat{\mathbf{x}} = H^T \tilde{\mathbf{y}} \tag{1.25}$$

If the rank of H is n (i.e., there are at least n independent observation equations), then $H^T H$ is *strictly* positive definite and can be inverted to obtain the explicit solution for the optimal estimate:

$$\boxed{\hat{\mathbf{x}} = (H^T H)^{-1} H^T \tilde{\mathbf{y}}} \tag{1.26}$$

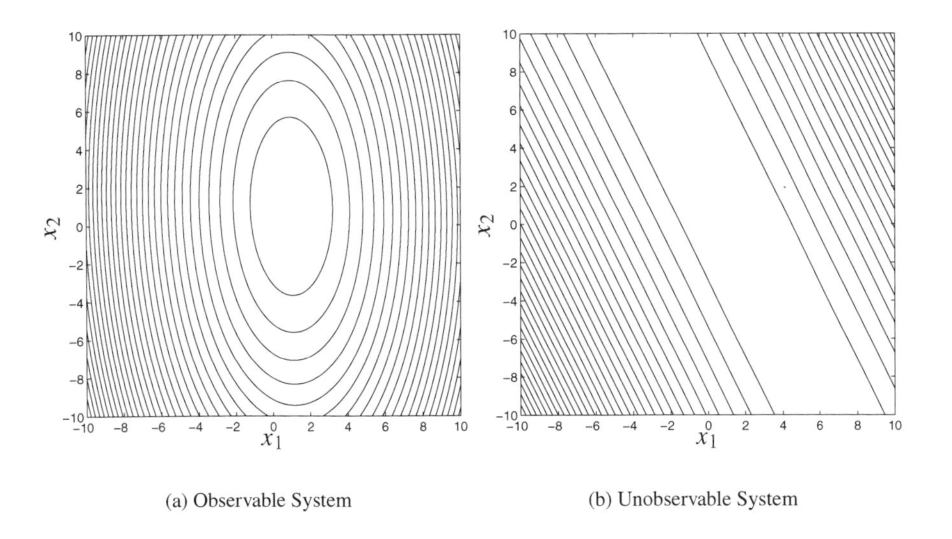

(a) Observable System (b) Unobservable System

Figure 1.6: Contour Plots for an Observable and Unobservable System

Equation (1.17) is the matrix equivalent of Gauss' original "equations of condition" which he wrote in index/summation notation.[2] Equation (1.26) serves as the most common basis for algorithms that solve simple least squares problems.

The inverse of $H^T H$ is required to determine $\hat{\mathbf{x}}$. This inverse exists only if the number of linearly independent observations is equal to or greater than the number of unknown x_i. To show this concept consider a simple least squares problem with $\mathbf{x} = \begin{bmatrix} 1 & 1 \end{bmatrix}^T$, and two basis functions given by $H_1 = \begin{bmatrix} \sin t & 2\cos t \end{bmatrix}$ and $H_2 = \begin{bmatrix} \sin t & 2\sin t \end{bmatrix}$. Clearly, H_1 provides a linearly independent set of basis functions, while H_2 does not because the second column of H_2 is twice the first column. A plot of the contour lines using H_1 is shown in Figure 1.6(a), which clearly shows a minimum at the true value for $\mathbf{x} = \begin{bmatrix} 1 & 1 \end{bmatrix}^T$. A plot of the contour lines using H_2 is shown in Figure 1.6(b), which shows that an infinite number of solutions are possible. More details on observability for dynamic systems is discussed in §3.4.

One of the implicit advantages of least squares is that the order of the matrix inverse is equal to the number of *unknowns*, not the number of measurement observations. The explicit solution (1.26) can be seen to play a role similar to $\mathbf{x} = H^{-1}\mathbf{y}$ in solving $\mathbf{y} = H\mathbf{x}$ for the $m = n$ case. We note that Gauss introduced his method of Gaussian elimination to solve the normal equations (1.25), by reducing $(H^T H)$ to upper triangular form, then solving for $\hat{\mathbf{x}}$ by back substitution (see Appendix A).

Example 1.1: Let us illustrate the basic concept of using linear least squares to estimate the parameters of a simple dynamic system. Consider the following dynamic system:

$$\dot{y} = ay + bu, \quad \dot{(\,)} \equiv \frac{d}{dt}(\,)$$

where u is an exogenous (i.e., externally specified) input, and a and b are constants. The system can also be represented in discrete-time with constant sampling interval Δt by (see §3.5)

$$y_{k+1} = \Phi y_k + \Gamma u_k$$

where the integer k is the sample index, and

$$\Phi = e^{a\Delta t}$$

$$\Gamma = \int_0^{\Delta t} b e^{at}\, dt = \frac{b}{a}(e^{a\Delta t} - 1)$$

The goal of this problem is to determine the constants Φ and Γ given a discrete set of measurements $\tilde{y}_k$ and inputs u_k. For the particular problem in which it is known that u is given by an impulse input with magnitude 100 (i.e., $u_1 = 100$ and $u_k = 0$ for $k \geq 2$), a total of 101 discrete measurements of the system are given with $\Delta t = 0.1$, and are shown in Figure 1.7. In order to set up the least squares problem, we construct the following basis function matrix:

$$H = \begin{bmatrix} \tilde{y}_1 & u_1 \\ \tilde{y}_2 & u_2 \\ \vdots & \vdots \\ \tilde{y}_{100} & u_{100} \end{bmatrix}$$

so

$$\begin{bmatrix} \tilde{y}_2 \\ \tilde{y}_3 \\ \vdots \\ \tilde{y}_{101} \end{bmatrix} = H \begin{bmatrix} \hat{\Phi} \\ \hat{\Gamma} \end{bmatrix} + \begin{bmatrix} e_2 \\ e_3 \\ \vdots \\ e_{101} \end{bmatrix}$$

Now, estimates for Φ and Γ can be determined using eqn. (1.26) directly:

$$\begin{bmatrix} \hat{\Phi} \\ \hat{\Gamma} \end{bmatrix} = (H^T H)^{-1} H^T \begin{bmatrix} \tilde{y}_2 \ \tilde{y}_3 \ \ldots \ \tilde{y}_{101} \end{bmatrix}^T$$

Using the measurements shown in Figure 1.7 the computed estimates are found to be

$$\begin{bmatrix} \hat{\Phi} \\ \hat{\Gamma} \end{bmatrix} = \begin{bmatrix} 0.9048 \\ 0.0950 \end{bmatrix}$$

In reality, the synthetic measurements of Figure 1.7 were generated using the following true values:

$$\begin{bmatrix} \Phi \\ \Gamma \end{bmatrix} = \begin{bmatrix} 0.9048 \\ 0.0952 \end{bmatrix}$$

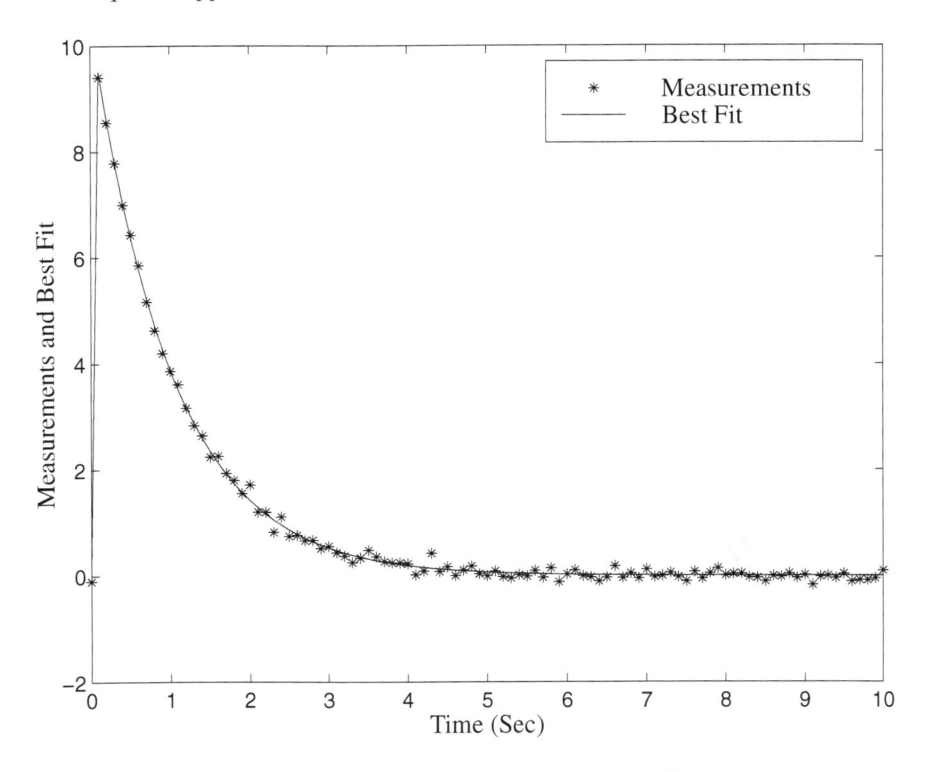

Figure 1.7: Measurements of $y(t)$ and Best Fit

with simulated measurement errors calculated using a zero-mean Gaussian noise generator with a standard deviation given by $\sigma = 0.08$.

The above example clearly involves a *dynamic* system; however, even though this system is modelled using a linear differential equation with constant coefficients, we are still able to bring the relationship (between measured quantities and constants which determine the model) to a linear algebraic equation, and therefore, we can use the principle of linear least squares. Also, the basis functions involve the measurements themselves, which is perhaps counterintuitive, but still is a valid approach. This is due to the fact that one of the sought parameters, Φ, multiplies y_k in the assumed model (the other parameter multiplies the input). This example clearly shows the power of least squares for dynamic model *identification*. We note in passing that the multi-dimensional generalization and sophistication of this example leads to the Eigensystem Realization Algorithm (ERA).[4] This algorithm is presented in Chapter 4.

1.2.2 Weighted Least Squares

The least squares criterion in eqn. (1.20), minimized to determine $\hat{\mathbf{x}}$, implicitly places equal emphasis on each measurement $\tilde{y}_j$. For the common event that the measurements are made with unequal precision, this "equal weight" approach seems logically unsound. Thus, the question arises as to how to select proper weights. One might intuitively select weights for each measurement that are inversely proportional to the measurement's estimated precision (i.e., a measurement with zero error should be weighted infinitely, while a measurement with infinite error should be weighted zero). Additionally, we shall see in Chapter 2 that a statistically optimal ("maximum likelihood") choice for the weights is the reciprocal of the measurement error variance. In order to incorporate appropriate weighting, we set up a least squares criterion of the form

$$J = \frac{1}{2}\mathbf{e}^T W \mathbf{e} \tag{1.27}$$

We now seek to determine $\hat{\mathbf{x}}$ that minimizes J, where W is an $m \times m$ symmetric matrix (it is symmetric because the terms $e_i e_j$, $i \neq j$, are always weighted equally with the corresponding $e_j e_i$ terms). In order that $\hat{\mathbf{x}}$ yield a minimum of eqn. (1.27), we have the requirements:

necessary condition

$$\nabla_{\hat{\mathbf{x}}} J = H^T W H \hat{\mathbf{x}} - H^T W \tilde{\mathbf{y}} = \mathbf{0} \tag{1.28}$$

sufficient condition

$$\nabla^2_{\hat{\mathbf{x}}} J = H^T W H \text{ must be positive definite.} \tag{1.29}$$

From the necessary condition in eqn. (1.28), we obtain the solution for $\hat{\mathbf{x}}$ given by

$$\boxed{\hat{\mathbf{x}} = (H^T W H)^{-1} H^T W \tilde{\mathbf{y}}} \tag{1.30}$$

Also, eqn. (1.29) clearly shows that W must be positive definite.

Example 1.2: To illustrate the power of weighted least squares, we will employ a subset of 31 measurements from the 91 measurements shown in Figure 1.1. Also, the first three measurements are known to contain less measurement errors than the remaining measurements. Toward this end, the structure of the weighting matrix now becomes

$$W = \text{diag}\begin{bmatrix} w & w & w & 1 & \cdots & 1 \end{bmatrix}$$

where diag[ ] denotes a diagonal matrix. Using Model 1 in eqn. (1.3) and the subset of 31 measurements with $w = 1$ (i.e., reduces to standard least squares) yields the following estimates:

$$(\hat{c}_1, \hat{c}_2, \hat{c}_3) = (1.0278, 0.8750, 1.9884)$$

Observe the unsurprising fact that the estimates are further from their true values $(1, 1, 2)$ than the estimates (1.5) resulting from all 91 measurements. However, since we know that the first three measurements are better than the remaining measurements, we can improve the estimates using weighted least squares. A summary of the solutions for $\hat{\mathbf{x}}$ with various values of w is shown below.

w	$\hat{\mathbf{x}}$	constraint residual norm
1×10^0	$(1.0278, 0.8750, 1.9884)$	3.21×10^{-2}
1×10^1	$(1.0388, 0.8675, 2.0018)$	1.17×10^{-2}
1×10^2	$(1.0258, 0.8923, 2.0049)$	7.87×10^{-3}
1×10^5	$(0.9047, 1.0949, 2.0000)$	5.91×10^{-5}
1×10^7	$(0.9060, 1.0943, 2.0000)$	1.10×10^{-5}
1×10^{10}	$(0.9932, 1.0068, 2.0000)$	4.55×10^{-7}
1×10^{15}	$(0.9970, 1.0030, 2.0000)$	0.97×10^{-9}

One can see that the residual constraint error (i.e., the computed norm of the measurements minus the estimates for the first three observations) decreases as more weight is used. However, this does not generally guarantee that the estimates ($\hat{\mathbf{x}}$) are closer to their true values. The interaction of the basis function therefore plays an important role in weighted least squares. Still, if the weight is sufficiently large, the estimates are indeed closer to their true values, as expected. In this simulation, the first three measurements were obtained with no measurement errors. However, perfect estimates (with zero associated model error) cannot be achieved since the exponential term in eqn. (1.7) is still present, which is not in the assumed model. Weighted least squares can improve the estimates if some knowledge of the relative accuracy of the measurements is known, and can obviously be used to approximately impose constraints on an estimation process.

1.2.3 Constrained Least Squares

Minimization of the weighted least squares criterion (1.27) allows relative emphasis to be placed upon the model agreeing with certain measurements more closely than others. Consider the limiting case of a perfect measurement where the corresponding diagonal element of the weight matrix should be ∞. This can often be accomplished in a practical situation by replacing ∞ with a "sufficiently large" number to obtain satisfactory approximations. However, we might be motivated to seek a rigorous means for imposing equality constraints in estimation problems.[5]

Suppose the original observations in eqn. (1.17) partition naturally into the subsystems $\tilde{\mathbf{y}}_1$ and $\tilde{\mathbf{y}}_2$ as

$$\begin{bmatrix} \tilde{\mathbf{y}}_1 \\ \cdot\cdot \\ \tilde{\mathbf{y}}_2 \end{bmatrix} = \begin{bmatrix} H_1 \\ \cdots \\ H_2 \end{bmatrix} \hat{\mathbf{x}} + \begin{bmatrix} \mathbf{e}_1 \\ \cdot\cdot \\ \mathbf{0} \end{bmatrix} \tag{1.31}$$

or

$$\tilde{\mathbf{y}}_1 = H_1\hat{\mathbf{x}} + \mathbf{e}_1 \tag{1.32}$$

and

$$\tilde{\mathbf{y}}_2 = H_2\hat{\mathbf{x}} \tag{1.33}$$

where

$\tilde{\mathbf{y}}_1 =$ an $m_1 \times 1$ vector of measured y-values
$H_1 =$ an $m_1 \times n$ basis function matrix corresponding with the measured y-values
$\mathbf{e}_1 =$ an $m_1 \times 1$ vector of residual errors
$\tilde{\mathbf{y}}_2 =$ an $m_2 \times 1$ vector of perfectly measured y-values
$H_2 =$ an $m_2 \times n$ basis function matrix corresponding with the perfectly measured y-values

and further assume that the dimensions satisfy

$$n \geq m_2$$
$$n \leq m_1$$

The absence of the residual error matrix $\mathbf{e}_2$ in eqns. (1.31) and (1.33) reflects the fact that $H_2\hat{\mathbf{x}}$ is required to equal $\tilde{\mathbf{y}}_2$ *exactly*. Thus we can formulate the problem as a constrained minimization problem of the type discussed in Appendix C. We seek a vector $\hat{\mathbf{x}}$ that minimizes

$$J = \frac{1}{2}\mathbf{e}_1^T W_1 \mathbf{e}_1 = \frac{1}{2}(\tilde{\mathbf{y}}_1 - H_1\hat{\mathbf{x}})^T W_1 (\tilde{\mathbf{y}}_1 - H_1\hat{\mathbf{x}}) \tag{1.34}$$

subject to the satisfaction of the equality constraint

$$\tilde{\mathbf{y}}_2 - H_2\hat{\mathbf{x}} = \mathbf{0} \tag{1.35}$$

Using the method of Lagrange multipliers (Appendix C), the necessary conditions are found by minimizing the augmented function

$$J = \frac{1}{2}\left[\tilde{\mathbf{y}}_1^T W_1 \tilde{\mathbf{y}}_1 - 2\tilde{\mathbf{y}}_1^T W_1 H_1 \hat{\mathbf{x}} + \hat{\mathbf{x}}^T (H_1^T W_1 H_1)\hat{\mathbf{x}}\right] + \boldsymbol{\lambda}^T(\tilde{\mathbf{y}}_2 - H_2\hat{\mathbf{x}}) \tag{1.36}$$

where

$$\boldsymbol{\lambda} = \begin{bmatrix}\lambda_1 & \lambda_2 & \cdots & \lambda_{m_2}\end{bmatrix}^T \tag{1.37}$$

is a vector of Lagrange multipliers. As necessary conditions for constrained minimization of J, we have the requirements:

$$\nabla_{\hat{\mathbf{x}}} J = -H_1^T W_1 \tilde{\mathbf{y}}_1 + (H_1^T W_1 H_1)\hat{\mathbf{x}} - H_2^T \boldsymbol{\lambda} = \mathbf{0} \tag{1.38}$$

and

$$\nabla_{\boldsymbol{\lambda}} J = \tilde{\mathbf{y}}_2 - H_2\hat{\mathbf{x}} = \mathbf{0}, \quad \rightarrow \tilde{\mathbf{y}}_2 = H_2\hat{\mathbf{x}} \tag{1.39}$$

Solving eqn. (1.38) for $\hat{\mathbf{x}}$ yields

$$\hat{\mathbf{x}} = (H_1^T W_1 H_1)^{-1} H_1^T W_1 \tilde{\mathbf{y}}_1 + (H_1^T W_1 H_1)^{-1} H_2^T \boldsymbol{\lambda} \tag{1.40}$$

Substituting eqn. (1.40) into eqn. (1.39) allows for solution of the Lagrange multipliers as

$$\boldsymbol{\lambda} = \left[H_2(H_1^T W_1 H_1)^{-1} H_2^T\right]^{-1} \left[\tilde{\mathbf{y}}_2 - H_2(H_1^T W_1 H_1)^{-1} H_1^T W_1 \tilde{\mathbf{y}}_1\right] \tag{1.41}$$

Finally, substituting eqn. (1.41) into eqn. (1.40) allows for elimination of $\boldsymbol{\lambda}$, yielding an explicit solution for the equality constrained least squares coefficient estimates as

$$\boxed{\hat{\mathbf{x}} = \bar{\mathbf{x}} + K(\tilde{\mathbf{y}}_2 - H_2\bar{\mathbf{x}})} \tag{1.42}$$

where

$$\boxed{K = (H_1^T W_1 H_1)^{-1} H_2^T \left[H_2(H_1^T W_1 H_1)^{-1} H_2^T\right]^{-1}} \tag{1.43}$$

and

$$\boxed{\bar{\mathbf{x}} = (H_1^T W_1 H_1)^{-1} H_1^T W_1 \tilde{\mathbf{y}}_1} \tag{1.44}$$

Observe that $\bar{\mathbf{x}}$, the first term of eqn. (1.42), is the least squares estimate of $\mathbf{x}$ in the absence of the constraint equations (1.33). The second term is an additive correction in which an optimal "gain matrix" K multiplies the constraint residual $(\tilde{\mathbf{y}}_2 - H_2\bar{\mathbf{x}})$ prior to the correction. This general "update form" (1.42) is seen often in estimation theory and is therefore an important result.

Due to the more complicated structure of eqns. (1.42), (1.43), and (1.44), in comparison to algorithms for solution of the weighted least squares problem, it often proves more expedient to simply use a least squares solution with a large weight on the constraint equation. However, if the number m_2 of constraint equations is small, the number of arithmetic operations in eqns. (1.42) and (1.43) can be much less than eqn. (1.30). In the limit, of $m_2 = 1$ constraint, then the matrix inverse in eqn. (1.43) simplifies to a scalar division.

As another important special case, consider $m_2 = n$. In this case H_2 is a square matrix, so eqn. (1.43) reduces to

$$K = H_2^{-1} \tag{1.45}$$

Thus, the constrained least squares estimate becomes

$$\hat{\mathbf{x}} = H_2^{-1}\tilde{\mathbf{y}}_2 \tag{1.46}$$

This shows that the solution is dependent on the perfectly measured values and H_2 only, which is the same result obtained using a square H matrix in the standard least squares solution. Thus if $m_2 = n$ perfect measurements are available, the solution is unaffected by an arbitrary number m of erroneous measurements.

Example 1.3: In example 1.2, weighted least squares was used to improve the estimates by incorporating knowledge of the perfectly known measurements. This result can also be obtained using constrained least squares. Again, a subset of 31 measurements is used. Three cases have been examined for the equality constraint, summarized by

$$\begin{aligned}
&\text{case 1,} \quad \tilde{\mathbf{y}}_1 = \left[\tilde{y}_2\ \tilde{y}_3\ \cdots\ \tilde{y}_{31}\right]^T, \quad \tilde{\mathbf{y}}_2 = y_1 \\
&\text{case 2,} \quad \tilde{\mathbf{y}}_1 = \left[\tilde{y}_3\ \tilde{y}_4\ \cdots\ \tilde{y}_{31}\right]^T, \quad \tilde{\mathbf{y}}_2 = \left[y_1\ y_2\right]^T \\
&\text{case 3,} \quad \tilde{\mathbf{y}}_1 = \left[\tilde{y}_4\ \tilde{y}_5\ \cdots\ \tilde{y}_{31}\right]^T, \quad \tilde{\mathbf{y}}_2 = \left[y_1\ y_2\ y_3\right]^T
\end{aligned}$$

Results using constrained least squares for $\bar{\mathbf{x}}$ and $\hat{\mathbf{x}}$ are summarized for each case below

case	$\bar{\mathbf{x}}$	$\hat{\mathbf{x}}$
1	$(1.0261, 0.8766, 1.9869)$	$(1.0406, 0.8629, 2.0000)$
2	$(1.0233, 0.8789, 1.9840)$	$(0.9039, 1.0901, 2.0000)$
3	$(1.0192, 0.8820, 1.9793)$	$(0.9970, 1.0030, 2.0000)$

We see that when one perfect measurement is used (case 1), the solution is not substantially improved over conventional least squares since $\bar{\mathbf{x}} \approx \hat{\mathbf{x}}$. However, when two perfect measurements are used (case 2), the estimates are closer to their true values. When three perfect measurements are used (case 3), which implies that $n = m_2$, the estimates are even closer to their true values. In fact, the estimates are identical within several significant digits to the case of $w = 1 \times 10^{15}$ in example 1.2. Were it not for the unaccounted error term $-0.4e^t/1 \times 10^4$, these would be found to agree exactly with the true coefficients $(1, 1, 2)$.

The theoretical equivalence of an infinitely weighted measurement to an equality constraint, from the viewpoint that eqns. (1.30) and (1.42) for this limiting case, is algebraically difficult to establish. It is possible, however, and is an intuitively pleasing truth. In practical applications, one can often obtain satisfactory solutions of constrained least squares problems in a fashion analogous to this example.

1.3 Linear Sequential Estimation

In the developments of the previous section, an implicit assumption is present, namely that all measurements are available for simultaneous ("batch") processing. In numerous real-world applications, the measurements become available sequentially in subsets and, immediately upon receipt of a new data subset, it may be desirable to determine new estimates based upon all previous measurements (including the current subset). To simplify the initial discussion, consider only two subsets:

$$\tilde{\mathbf{y}}_1 = \left[\tilde{y}_{11}\ \tilde{y}_{12}\ \cdots\ \tilde{y}_{1m_1}\right]^T = \text{an } m_1 \times 1 \text{ vector of measurements} \tag{1.47a}$$

$$\tilde{\mathbf{y}}_2 = \left[\tilde{y}_{21}\ \tilde{y}_{22}\ \cdots\ \tilde{y}_{2m_2}\right]^T = \text{an } m_2 \times 1 \text{ vector of measurements} \tag{1.47b}$$

and the associated observation equations

$$\tilde{\mathbf{y}}_1 = H_1\mathbf{x} + \mathbf{v}_1 \tag{1.48a}$$

$$\tilde{\mathbf{y}}_2 = H_2\mathbf{x} + \mathbf{v}_2 \tag{1.48b}$$

where

$H_1 =$ an $m_1 \times n$ known coefficient matrix of maximum rank $n \le m_1$
$H_2 =$ an $m_2 \times n$ known coefficient matrix
$\mathbf{v}_1, \mathbf{v}_2 =$ vectors of measurement errors
$\mathbf{x} =$ the $n \times 1$ vector of unknown parameters

The least squares estimate, $\hat{\mathbf{x}}$, of $\mathbf{x}$ based upon the *first* measurement subset (1.47a) follows from eqn. (1.30) as

$$\hat{\mathbf{x}}_1 = (H_1^T W_1 H_1)^{-1} H_1^T W_1 \tilde{\mathbf{y}}_1 \tag{1.49}$$

where W_1 is an $m_1 \times m_1$ symmetric, positive definite matrix associated with measurements $\tilde{\mathbf{y}}_1$. It is possible to consider $\tilde{\mathbf{y}}_1$ and $\tilde{\mathbf{y}}_2$ *simultaneously* and determine an estimate $\hat{\mathbf{x}}_2$ of $\mathbf{x}$ based upon *both* measurement subsets (1.47a) and (1.47b). Toward this end, we form the *merged* observation equations

$$\tilde{\mathbf{y}} = H\mathbf{x} + \mathbf{v} \tag{1.50}$$

where

$$\tilde{\mathbf{y}} = \begin{bmatrix} \tilde{\mathbf{y}}_1 \\ \cdot\cdot \\ \tilde{\mathbf{y}}_2 \end{bmatrix}, \quad H = \begin{bmatrix} H_1 \\ \cdots \\ H_2 \end{bmatrix}, \quad \mathbf{v} = \begin{bmatrix} \mathbf{v}_1 \\ \cdot\cdot \\ \mathbf{v}_2 \end{bmatrix} \tag{1.51}$$

Next, we assume that the merged weight matrix is in block diagonal structure, so that*

$$W = \begin{bmatrix} W_1 & \vdots & 0 \\ \cdots & & \cdots \\ 0 & \vdots & W_2 \end{bmatrix} \tag{1.52}$$

Then, the optimal least squares estimate based upon the first two measurement subsets follows from eqn. (1.30) as

$$\hat{\mathbf{x}}_2 = (H^T W H)^{-1} H^T W \tilde{\mathbf{y}} \tag{1.53}$$

*In Chapter 2 and Appendix B, we will see that an implicit assumption here is that measurement errors can be *correlated* only to other measurements belonging to the same subset.

Now, since W is block diagonal, eqn. (1.53) can be expanded as

$$\hat{\mathbf{x}}_2 = [H_1^T W_1 H_1 + H_2^T W_2 H_2]^{-1}(H_1^T W_1 \tilde{\mathbf{y}}_1 + H_2^T W_2 \tilde{\mathbf{y}}_2) \tag{1.54}$$

It is clearly possible, in principle, to continue forming merged normal equations using the above procedure (upon receipt of each data subset) and solving for new optimal estimates as in eqn. (1.54). However, the above route does not take efficient advantage of the calculations done in processing the previous subsets of data. The essence of the *sequential* approach to the least squares problem is to simply arrange calculations for the new estimate (e.g., $\hat{\mathbf{x}}_2$) to make efficient use of previous estimates and the associated side calculations. We begin the derivation of this approach by defining the following variables:

$$P_1 \equiv [H_1^T W_1 H_1]^{-1} \tag{1.55}$$

$$P_2 \equiv [H_1^T W_1 H_1 + H_2^T W_2 H_2]^{-1} \tag{1.56}$$

From these definitions it immediately follows that (assuming that both P_1^{-1} and P_2^{-1} exist)

$$P_2^{-1} = P_1^{-1} + H_2^T W_2 H_2 \tag{1.57}$$

We now rewrite eqns. (1.49) and (1.54) using the definitions in eqns. (1.55) and (1.56) as

$$\hat{\mathbf{x}}_1 = P_1 H_1^T W_1 \tilde{\mathbf{y}}_1 \tag{1.58}$$

$$\hat{\mathbf{x}}_2 = P_2(H_1^T W_1 \tilde{\mathbf{y}}_1 + H_2^T W_2 \tilde{\mathbf{y}}_2) \tag{1.59}$$

Pre-multiplying eqn. (1.58) by P_1^{-1} yields

$$P_1^{-1}\hat{\mathbf{x}}_1 = H_1^T W_1 \tilde{\mathbf{y}}_1 \tag{1.60}$$

Next, from eqn. (1.57) we have

$$P_1^{-1} = P_2^{-1} - H_2^T W_2 H_2 \tag{1.61}$$

Substituting eqn. (1.61) into eqn. (1.60) leads to

$$H_1^T W_1 \tilde{\mathbf{y}}_1 = P_2^{-1}\hat{\mathbf{x}}_1 - H_2^T W_2 H_2 \hat{\mathbf{x}}_1 \tag{1.62}$$

Finally, substituting eqn. (1.62) into eqn. (1.59) and collecting terms gives

$$\hat{\mathbf{x}}_2 = \hat{\mathbf{x}}_1 + K_2(\tilde{\mathbf{y}}_2 - H_2\hat{\mathbf{x}}_1) \tag{1.63}$$

where

$$K_2 \equiv P_2 H_2^T W_2 \tag{1.64}$$

We now have a mechanism to *sequentially* provide an updated estimate, $\hat{\mathbf{x}}_2$, based upon the previous estimate, $\hat{\mathbf{x}}_1$, and associated side calculations. We can easily generalize eqns. (1.63) and (1.64) to use the kth estimate to determine estimate at $k+1$

from the $k+1$ subset of measurements, which leads to a most important result in sequential estimation theory:

$$\hat{\mathbf{x}}_{k+1} = \hat{\mathbf{x}}_k + K_{k+1}(\tilde{\mathbf{y}}_{k+1} - H_{k+1}\hat{\mathbf{x}}_k) \tag{1.65}$$

where

$$K_{k+1} = P_{k+1} H_{k+1}^T W_{k+1} \tag{1.66}$$

$$P_{k+1}^{-1} = P_k^{-1} + H_{k+1}^T W_{k+1} H_{k+1} \tag{1.67}$$

Equation (1.65) modifies the previous best correction $\hat{\mathbf{x}}_k$ by an additional correction to account for the information contained in the $k+1$ measurement subset. This equation is a *Kalman update equation*[6] for computing the improved estimate $\hat{\mathbf{x}}_{k+1}$. Also, notice the similarity between eqn. (1.65) and eqn. (1.42). Equation (1.66) is the correction term, known as the *Kalman gain matrix*. The sequential least squares algorithm plays an important role for linear (and nonlinear) dynamic *state* estimation, as will be seen in the Kalman filter in §5.3. Equation (1.65) is in fact a linear difference equation, commonly found in digital control analysis. This equation may be rearranged as

$$\hat{\mathbf{x}}_{k+1} = \left[I - K_{k+1} H_{k+1}\right]\hat{\mathbf{x}}_k + K_{k+1}\tilde{\mathbf{y}}_{k+1} \tag{1.68}$$

which clearly is in the form of a time-varying dynamical system. Therefore, linear tools can be used to check stability, dynamic response times, etc.

The specific form for P^{-1} in eqn. (1.67) is known as the *information matrix recursion.*† The current approach for computing P_{k+1} involves computing the inverse of eqn. (1.67) which offers no advantage over inverting the normal equations in their original *batch* processing in eqn. (1.53). This is due to the fact that an $n \times n$ inverse must still be performed. We might wonder if there is an easier way to compute P_{k+1} given that we have computed P_k previously. As it turns out, when the number of measurements m in the new data subset is small compared to n (as is usually the case), a *small rank adjustment* to the already computed P_k can be calculated efficiently using the Sherman-Morrison-Woodbury *matrix inversion lemma.*[7] Let

$$F = [A + B\,C\,D]^{-1} \tag{1.69}$$

where

$$\begin{aligned} F &= \text{an arbitrary } n \times n \text{ matrix} \\ A &= \text{an arbitrary } n \times n \text{ matrix} \\ B &= \text{an arbitrary } n \times m \text{ matrix} \\ C &= \text{an arbitrary } m \times m \text{ matrix} \\ D &= \text{an arbitrary } m \times n \text{ matrix} \end{aligned}$$

†As is evident in Chapter 2, the interpretation of P^{-1} as the *information matrix* (and P as the *covariance matrix*) hinges upon several assumptions, most notably, that W_k is the inverse of the measurement error covariance.

Then, assuming all inverses exist

$$F = A^{-1} - A^{-1}B\left(D\,A^{-1}B + C^{-1}\right)^{-1} D\,A^{-1} \tag{1.70}$$

The matrix inversion lemma can be proved by showing that $F^{-1}F = I$. Brute force calculation of $F^{-1}F$ gives

$$\begin{aligned} F^{-1}F = I - B\Big[&\left(D\,A^{-1}B + C^{-1}\right)^{-1} - C \\ &+ C\,D\,A^{-1}B\left(D\,A^{-1}B + C^{-1}\right)^{-1}\Big]D\,A^{-1} \end{aligned} \tag{1.71}$$

To prove the matrix inversion lemma, it is enough to show that the quantity inside the square brackets of eqn. (1.71) is identically zero. Therefore, we need to prove that

$$\left(D\,A^{-1}B + C^{-1}\right)^{-1} = C - C\,D\,A^{-1}B\left(D\,A^{-1}B + C^{-1}\right)^{-1} \tag{1.72}$$

Right multiplying both sides of eqn. (1.72) by $\left(D\,A^{-1}B + C^{-1}\right)$ reduces eqn. (1.72) to

$$I = C\left(D\,A^{-1}B + C^{-1}\right) - C\,D\,A^{-1}B = I \tag{1.73}$$

This completes the proof.

Our next step is to apply the matrix inversion lemma to eqn. (1.67). The "judicious choices" for F, A, B, C, and D are

$$F = P_{k+1} \tag{1.74a}$$
$$A = P_k^{-1} \tag{1.74b}$$
$$B = H_{k+1}^T \tag{1.74c}$$
$$C = W_{k+1} \tag{1.74d}$$
$$D = H_{k+1} \tag{1.74e}$$

The matrix information recursion now becomes

$$P_{k+1} = P_k - P_k H_{k+1}^T\left(H_{k+1}P_k H_{k+1}^T + W_{k+1}^{-1}\right)^{-1} H_{k+1}P_k \tag{1.75}$$

Thus, P_{k+1}, which is used in eqn. (1.66), can be obtained by "updating" P_k, and the update process usually requires inverting a matrix with rank less than n. A large number of successive applications of the recursion (1.75) occasionally introduces arithmetic errors which can invalidate the estimates (1.65). In connection with the applications of Chapter 4, alternatives to (1.75) which are numerically superior are presented.

The "update equation" (1.65) can also be rearranged in several alternate forms. One of the more common is obtained by substituting eqn. (1.75) into eqn. (1.66) to obtain

$$K_{k+1} = \left[P_k - P_k H_{k+1}^T \left(H_{k+1} P_k H_{k+1}^T + W_{k+1}^{-1}\right)^{-1} H_{k+1} P_k\right] \times H_{k+1}^T W_{k+1} \tag{1.76a}$$

$$= P_k H_{k+1}^T \left[I - \left(H_{k+1} P_k H_{k+1}^T + W_{k+1}^{-1}\right)^{-1} H_{k+1} P_k H_{k+1}^T\right] W_{k+1} \tag{1.76b}$$

Now, factoring $\left(H_{k+1} P_k H_{k+1}^T + W_{k+1}^{-1}\right)^{-1}$ outside of the square brackets leads directly to

$$K_{k+1} = P_k H_{k+1}^T \left(H_{k+1} P_k H_{k+1}^T + W_{k+1}^{-1}\right)^{-1} \times \left[W_{k+1}^{-1} + H_{k+1} P_k H_{k+1}^T - H_{k+1} P_k H_{k+1}^T\right] W_{k+1} \tag{1.77}$$

This leads to the *covariance recursion form*, given by

$$\boxed{\hat{\mathbf{x}}_{k+1} = \hat{\mathbf{x}}_k + K_{k+1}(\tilde{\mathbf{y}}_{k+1} - H_{k+1}\hat{\mathbf{x}}_k)} \tag{1.78}$$

where

$$\boxed{K_{k+1} = P_k H_{k+1}^T \left[H_{k+1} P_k H_{k+1}^T + W_{k+1}^{-1}\right]^{-1}} \tag{1.79}$$

$$\boxed{P_{k+1} = \left[I - K_{k+1} H_{k+1}\right] P_k} \tag{1.80}$$

The covariance form of sequential least squares is most commonly used in practice, because it is more computationally efficient. However, the information form may be numerically superior in the initialization stage. The process may be initiated at any step by an *a priori* estimate, $\hat{\mathbf{x}}_1$, and covariance estimate P_1. If *a priori* estimates are not available, then the first data subset can be used for initialization by using a batch least squares to determine $\hat{\mathbf{x}}_q$ and P_q, where $q \geq n$. Then the sequential least squares algorithm can be invoked for $k \geq q$. However, sequential least squares can still be used for $k = 1, 2, \ldots, q-1$ if one uses

$$P_1 = \left[\frac{1}{\alpha^2} I + H_1^T W_1 H_1\right]^{-1} \tag{1.81}$$

$$\hat{\mathbf{x}}_1 = P_1 \left[\frac{1}{\alpha}\boldsymbol{\beta} + H_1^T W_1 \tilde{\mathbf{y}}_1\right] \tag{1.82}$$

where α is a very "large" number and $\boldsymbol{\beta}$ is a vector of very "small" numbers. It can be shown that the resulting recursive least squares values of P_n and $\hat{\mathbf{x}}_n$ are very close to the corresponding batch values at time t_n.

If the model is in fact linear and if there is no correlation between measurement errors of different measurement subsets (so that the assumed block structure of W is strictly valid), then the sequential solution for $\hat{\mathbf{x}}$ in eqn. (1.65) will agree exactly with the batch solution in eqn. (1.30), to within arithmetic errors. This is because eqn. (1.65) is simply an algebraic rearrangement of the normal equations (1.30).

Example 1.4: In example 1.1, we used a batch least squares process to estimate the parameters of a simple dynamic system. We now will use this same system to determine the parameters sequentially using recursive least squares with one measurement $\tilde{y}_k$ at a time. In order to initialize the routine, we will use eqns. (1.81) and (1.82) with $\alpha = 1 \times 10^3$ and $\beta = \left[1 \times 10^{-2}\ 1 \times 10^{-2}\right]^T$. As mentioned in example 1.1, the measurement errors were simulated using a zero-mean Gaussian noise generator with a standard deviation given by $\sigma = 0.08$. We will see in Chapter 2 that an "optimal" choice for W_k is given by $W_k = \sigma^{-2}$. The calculated initial values for P_1 and $\hat{\mathbf{x}}_1$ are given by

$$P_1 = \begin{bmatrix} 1.000 \times 10^6 & 1.038 \times 10^3 \\ 1.038 \times 10^3 & 1.077 \times 10^0 \end{bmatrix}$$

$$\hat{\mathbf{x}}_1 = \begin{bmatrix} 10.010 \\ 0.014 \end{bmatrix}$$

Plots of the estimates $\hat{\mathbf{x}}_k$ and diagonal elements of P_k are shown in Figure 1.8. As can be seen from these plots, convergence is reached very quickly for this example. This is not the case in all systems, but is typical for well-conditioned linear systems. The sequential estimates at the final time agree exactly with the batch estimates in example 1.1. The diagonal elements of P_k actually have a physical meaning, as shown in Chapter 2, which can be used to develop a suitable stopping criterion. This example clearly shows the power of sequential least squares to identify the parameters of a dynamic system in *real time*.

1.4 Nonlinear Least Squares Estimation

It is a fact of life that most real-world estimation problems are nonlinear. The preceding developments of this chapter apply rigorously to only a small subset of problems encountered in practice. Fortunately, most nonlinear estimation problems can be accurately solved by a judiciously chosen successive approximation procedure. In this section we develop the most widely used successive approximation procedure, *nonlinear least squares*; otherwise known as *Gaussian least squares differential correction*. This method was originally developed by Gauss and employed

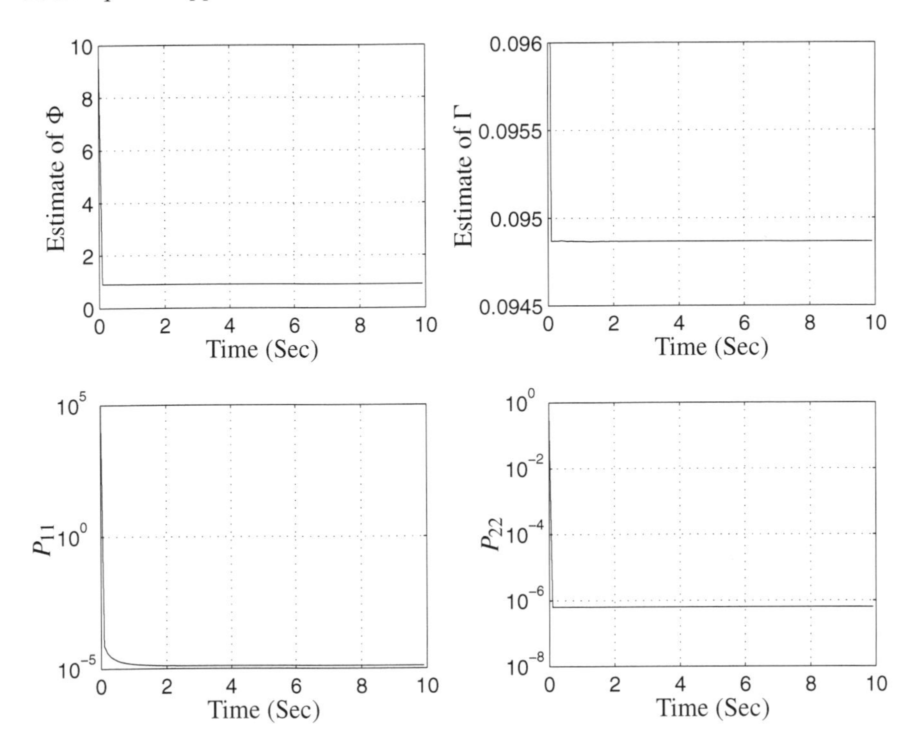

Figure 1.8: Estimates and Diagonal Elements of P_k

to determine planetary orbits (during the early 1800s) from telescope measurements of the "line of sight angles" to the planets.[2]

The method to be developed here is an $m \times n$ generalization of Newton's root solving method[8] for finding x-values satisfying $y - f(x) = 0$. As with Newton's method, convergence of the multi-dimensional generalization is guaranteed only under rather strict requirements on the functions and their first two partial derivatives as well as on the closeness of the starting estimates. Let us not be concerned with convergence at this stage (although be informed, convergence difficulties do occasionally occur!). Rather, let us proceed with formulating the method and look at typical applications.

Assume m observable quantities modelled as

$$y_j = f_j(x_1, x_2, \ldots, x_n); \qquad j = 1, 2, \ldots, m; \qquad m \geq n \tag{1.83}$$

where the $f_j(x_1, x_2, \ldots, x_n)$ are m arbitrary independent functions of the unknown parameters x_i. These should be interpreted as "functions" in the general sense, as specifying "whatever process one must go through" to compute the y_j given the x_i (including, for example, numerical solution of differential equations). We do require that $f_j(x_1, x_2, \ldots, x_n)$ and at least its first partial derivatives be single-valued, continuous and at least once differentiable. Additionally, suppose that a set of observed values of the variables y_j are available:

$$y_j \in \{y_1, y_2, \ldots, y_m\} \tag{1.84}$$

As done in §1.2, we can rewrite the measurement model with eqn. (1.84) in compact form as

$$\tilde{\mathbf{y}} = \mathbf{f}(\mathbf{x}) + \mathbf{v} \tag{1.85}$$

where

$$\begin{aligned}
\tilde{\mathbf{y}} &= \begin{bmatrix} \tilde{y}_1 & \tilde{y}_2 & \cdots & \tilde{y}_m \end{bmatrix}^T = \textit{measured } y\text{-values} \\
\mathbf{f}(\mathbf{x}) &= \begin{bmatrix} f_1 & f_2 & \cdots & f_m \end{bmatrix}^T = \textit{independent functions} \\
\mathbf{x} &= \begin{bmatrix} x_1 & x_2 & \cdots & x_n \end{bmatrix}^T = \textit{true } x\text{-values} \\
\mathbf{v} &= \begin{bmatrix} v_1 & v_2 & \cdots & v_m \end{bmatrix}^T = \textit{measurement errors}
\end{aligned}$$

Likewise, the estimated y-values, denoted by $\hat{y}_j$ and residual errors $e_j = \tilde{y}_j - \hat{y}_j$, can also be written in compact form as

$$\hat{\mathbf{y}} = \mathbf{f}(\hat{\mathbf{x}}) \tag{1.86}$$

$$\mathbf{e} = \tilde{\mathbf{y}} - \hat{\mathbf{y}} \equiv \Delta\mathbf{y} \tag{1.87}$$

where

$$\begin{aligned}
\hat{\mathbf{y}} &= \begin{bmatrix} \hat{y}_1 & \hat{y}_2 & \cdots & \hat{y}_m \end{bmatrix}^T = \textit{estimated } y\text{-values} \\
\mathbf{e} &= \begin{bmatrix} e_1 & e_2 & \cdots & e_m \end{bmatrix}^T = \textit{residual errors} \\
\hat{\mathbf{x}} &= \begin{bmatrix} \hat{x}_1 & \hat{x}_2 & \cdots & \hat{x}_n \end{bmatrix}^T = \textit{estimated } x\text{-values}
\end{aligned}$$

The measurement model in eqn. (1.86) can again be written using the residual errors $\mathbf{e}$ as

$$\tilde{\mathbf{y}} = \mathbf{f}(\hat{\mathbf{x}}) + \mathbf{e} \tag{1.88}$$

As done in §1.2 we seek an estimate ($\hat{\mathbf{x}}$) for $\mathbf{x}$ that minimizes

$$J = \frac{1}{2}\mathbf{e}^T W \mathbf{e} = \frac{1}{2}[\tilde{\mathbf{y}} - \mathbf{f}(\hat{\mathbf{x}})]^T W [\tilde{\mathbf{y}} - \mathbf{f}(\hat{\mathbf{x}})] \tag{1.89}$$

where W is an $m \times m$ weighting matrix again used to weight the relative importance of each measurement.

In most practical problems, J cannot be directly minimized by application of ordinary calculus to eqn. (1.89), in the sense that explicit closed form solutions for $\hat{\mathbf{x}}$ result. The case where $\mathbf{f}(\hat{\mathbf{x}}) = H\hat{\mathbf{x}}$ reduces to the standard linear least squares solution; however, general nonlinear functions for $\mathbf{f}(\hat{\mathbf{x}})$ typically make the solution difficult to find explicitly. For this reason, attention is directed to construction of a successive approximation procedure due to Gauss, that is designed to converge to accurate least squares estimates, given approximate starting values (the determination of sufficiently close starting estimates is a problem that cannot be dealt with in general, but can usually be overcome, as seen in applications of Chapter 4 and in §1.6.3).

Assume that the *current* estimates of the unknown $\mathbf{x}$-values are available, denoted by

$$\mathbf{x}_c = \begin{bmatrix} x_{1c} \; x_{2c} \; \cdots \; x_{nc} \end{bmatrix}^T \tag{1.90}$$

Whatever the unknown objective $\mathbf{x}$*-values* $\hat{\mathbf{x}}$ *are*, we assume that they are related to their respective current estimates, $\mathbf{x}_c$, by an also unknown set of corrections, $\Delta\mathbf{x}$, as

$$\hat{\mathbf{x}} = \mathbf{x}_c + \Delta\mathbf{x} \tag{1.91}$$

If the components of $\Delta\mathbf{x}$ are sufficiently small, it may be possible to solve for approximations to them and thereby update $\mathbf{x}_c$ with an improved estimate of $\mathbf{x}$ from eqn. (1.91). With this assumption, we may *linearize* $\mathbf{f}(\hat{\mathbf{x}})$ in eqn. (1.86) about $\mathbf{x}_c$ using a first-order Taylor series expansion as

$$\mathbf{f}(\hat{\mathbf{x}}) \approx \mathbf{f}(\mathbf{x}_c) + H\,\Delta\mathbf{x} \tag{1.92}$$

where

$$H \equiv \left.\frac{\partial \mathbf{f}}{\partial \mathbf{x}}\right|_{\mathbf{x}_c} \tag{1.93}$$

The gradient matrix H is known as a *Jacobian* matrix (see Appendix A). The measurement residual "after the correction" can now be linearly approximated as

$$\Delta\mathbf{y} \equiv \tilde{\mathbf{y}} - \mathbf{f}(\hat{\mathbf{x}}) \approx \tilde{\mathbf{y}} - \mathbf{f}(\mathbf{x}_c) - H\,\Delta\mathbf{x} = \Delta\mathbf{y}_c - H\,\Delta\mathbf{x} \tag{1.94}$$

where the residual "before the correction" is

$$\Delta\mathbf{y}_c \equiv \tilde{\mathbf{y}} - \mathbf{f}(\mathbf{x}_c) \tag{1.95}$$

Recall that the objective is to minimize the weighted sum squares, J, given by eqn. (1.89). The local strategy for determining the approximate corrections ("differential corrections") in $\Delta\mathbf{x}$ is to select the particular corrections that lead to the *minimum sum of squares of the linearly predicted residuals* J_p:

$$J = \frac{1}{2}\Delta\mathbf{y}^T W \Delta\mathbf{y} \approx J_p \equiv \frac{1}{2}(\Delta\mathbf{y}_c - H\,\Delta\mathbf{x})^T W (\Delta\mathbf{y}_c - H\,\Delta\mathbf{x}) \tag{1.96}$$

Before carrying out the minimization, we note (to the approximation that the linearization implicit in the prediction (1.92) is valid) that the minimization of J_p in eqn. (1.96) is equivalent to the minimization of J in eqn. (1.89). If the process is convergent, then $\Delta\mathbf{x}$ determined by minimizing eqn. (1.96) would be expected to decrease on successive iterations until (on the final iteration) the linearization is an extremely good approximation.

Observe that the minimization of eqn. (1.96) is completely analogous to the previously minimized quadratic form (1.27). Thus any algorithm for solving the weighted least squares problem directly applies to solving for $\Delta\mathbf{x}$ in eqn. (1.96). Therefore,

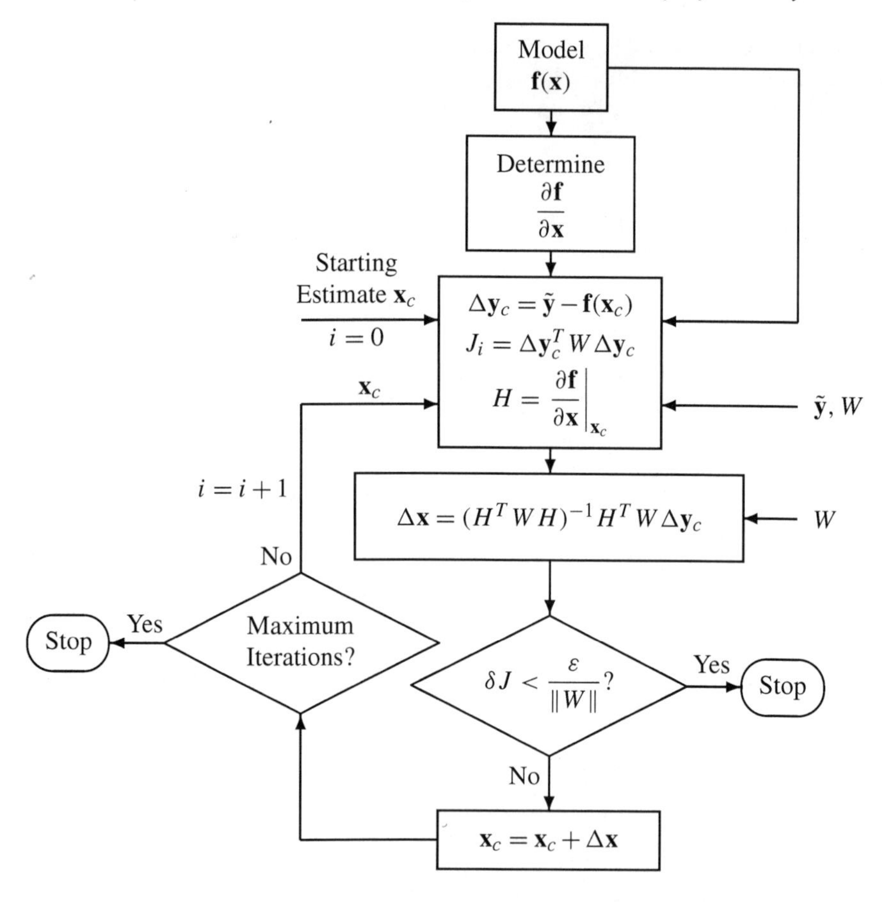

Figure 1.9: Nonlinear Least Squares Algorithm

the appropriate version of the normal equations follows as in the development of eqns. (1.28)-(1.30), as

$$\boxed{\Delta \mathbf{x} = (H^T W H)^{-1} H^T W \Delta \mathbf{y}_c} \tag{1.97}$$

The complete nonlinear least squares algorithm is summarized in Figure 1.9. An initial guess $\mathbf{x}_c$ is required to begin the algorithm. Equation (1.97) is then calculated using the residual measurements ($\Delta \mathbf{y}_c$), Jacobian matrix (H), and weighting matrix (W), so that current estimate can be updated. A stopping condition with an accuracy dependent tolerance for the minimization of J is given by

$$\delta J \equiv \frac{|J_i - J_{i-1}|}{J_i} < \frac{\varepsilon}{\|W\|} \tag{1.98}$$

where ε is a prescribed small value. If eqn. (1.98) is not satisfied, then the update procedure is iterated with the new estimate as the current estimate until the process converges, or unsatisfactory convergence progress is evident (e.g., a maximum allowed number of iterations is exceeded, or J increases on successive iterations).

The above least squares differential correction process, while far from fail-safe, has been successfully applied to an extremely wide variety of nonlinear estimation problems. Convergence difficulties usually stem from one of the following sources: (1) the initial $\mathbf{x}$-estimate is too far from the minimizing $\hat{\mathbf{x}}$ (for the nonlinearity of the particular application), resulting in the implicit local linearity assumption being invalid; (2) numerical difficulties are encountered in solving for the corrections, $\Delta\mathbf{x}$, due to $(2a)$ arithmetic errors corrupting the particular algorithm used to calculate the $\Delta\mathbf{x}$, or $(2b)$ the H matrix having fewer than n linearly independent rows or columns (i.e., rank deficient). The difficulties (1) and $(2a)$ can usually be overcome by a resourceful analyst; however, the least squares criterion does not uniquely define $\Delta\mathbf{x}$ in the $(2b)$ case, and therefore some other criterion must be employed to select $\Delta\mathbf{x}$. The initial estimate convergence difficulty can also be overcome by using the Levenberg-Marquardt algorithm shown in §1.6.3, which combines the least squares differential correction process with a gradient search.

Example 1.5: In this simple example, we consider the 1×1 special case of nonlinear least squares with $m = n = 1$. Suppose we have the following model:

$$y = x^3 + 6x^2 + 11x + 6 = 0$$

For this model, we can assume that

$$\begin{aligned} \mathbf{y} &= y = 0 \\ \mathbf{f}(\mathbf{x}) &= f(x) = x^3 + 6x^2 + 11x + 6 \end{aligned}$$

For this case, eqn. (1.97) becomes simply

$$x = x_c - \left[\left. \frac{\partial f}{\partial x} \right|_{x_c} \right]^{-1} f(x_c)$$

where

$$\frac{\partial f}{\partial x} = 3x^2 + 12x + 11$$

As seen in the above equations, this special scalar case reduces to the classical Newton root solving method. Therefore, eqn. (1.97) actually represents an $m \times n$ generalization of Newton's root solver. Seven iterations for three different starting values of x are given below.

iteration	x	x	x
0	0.0000	−1.6000	−5.0000
1	−0.5455	−2.2462	−4.0769
2	−0.8490	−1.9635	−3.5006
3	−0.9747	−2.0001	−3.1742
4	−0.9991	−2.0000	−3.0324
5	−1.0000	−2.0000	−3.0015
6	−1.0000	−2.0000	−3.0000
7	−1.0000	−2.0000	−3.0000

This clearly shows that different solutions are possible for various starting conditions. In this case, we know this to be true since we are solving a cubic equation, which has three possible solutions, and obviously, we have converged to all three roots. More generally, complex algebra would have to be used to find complex roots.

Example 1.6: In example 1.1, we used linear least squares to estimate the parameters of a simple dynamic system. Recall that the system is given by

$$y_{k+1} = \left[e^{a\Delta t}\right] y_k + \left[\frac{b}{a}(e^{a\Delta t} - 1)\right] u_k$$

Suppose that we now wish to determine a and b directly from the above equation. To accomplish this task, we must now use nonlinear least squares, with

$$\mathbf{x} = \left[a \; b\right]^T$$
$$\tilde{\mathbf{y}} = \left[\tilde{y}_2 \; \tilde{y}_3 \; \cdots \; \tilde{y}_{101}\right]^T$$
$$f_k = \left[e^{a\Delta t}\right] y_k + \left[\frac{b}{a}(e^{a\Delta t} - 1)\right] u_k$$

The appropriate partials are given by

$$\frac{\partial f_k}{\partial a} = \Delta t\left[e^{a\Delta t}\right] y_k + \left[\frac{b}{a^2}(1 - e^{a\Delta t}) + \frac{b}{a}\Delta t e^{a\Delta t}\right] u_k$$

$$\frac{\partial f_k}{\partial b} = \frac{1}{a}(e^{a\Delta t} - 1) u_k$$

Then, the H matrix is given by

$$H = \begin{bmatrix} \Delta t\left[e^{a\Delta t}\right]\tilde{y}_1 + \left[\frac{b}{a^2}(1-e^{a\Delta t}) + \frac{b}{a}\Delta t e^{a\Delta t}\right]u_1 & \frac{1}{a}(e^{a\Delta t}-1)u_1 \\ \Delta t\left[e^{a\Delta t}\right]\tilde{y}_2 + \left[\frac{b}{a^2}(1-e^{a\Delta t}) + \frac{b}{a}\Delta t e^{a\Delta t}\right]u_2 & \frac{1}{a}(e^{a\Delta t}-1)u_2 \\ \vdots & \vdots \\ \Delta t\left[e^{a\Delta t}\right]\tilde{y}_{100} + \left[\frac{b}{a^2}(1-e^{a\Delta t}) + \frac{b}{a}\Delta t e^{a\Delta t}\right]u_{100} & \frac{1}{a}(e^{a\Delta t}-1)u_{100} \end{bmatrix}$$

The nonlinear least squares algorithm in Figure 1.9 can now be used to determine a and b. The starting guess for the iteration is given by

$$\mathbf{x}_c = [5\ 5]^T$$

Also, the stopping criterion is given by $\varepsilon = 1 \times 10^{-8}$. Results are tabulated below.

iteration	$\hat{a}$	$\hat{b}$
0	5.0000	5.0000
1	0.4876	1.9540
2	−0.8954	1.0634
3	−1.0003	0.9988
4	−1.0009	0.9985
5	−1.0009	0.9985
6	−1.0009	0.9985

If we convert the final values for $\hat{a}$ and $\hat{b}$ into their discrete time equivalents, we see that $\hat{\Phi} = 0.9048$ and $\hat{\Gamma} = 0.0950$, which agree with the results obtained in example 1.1. This example clearly shows that the *form* of the model chosen can have a highly significant impact on the complexity of the required estimator. If we choose to determine Φ and Γ directly, then *linear* least squares may be employed. However, if we choose to determine a and b, then nonlinear least squares must be used. Clearly, by using creative system model choices, one can greatly simplify the overall solution process. This point is further explored in §1.5 and in Chapter 4.

Example 1.7: Under certain approximations, the pitch (θ) and yaw (ψ) attitude dynamics of an inertially and aerodynamically symmetric projectile can be modelled via a pair of equations

$$\begin{aligned} \theta(t) &= k_1 e^{\lambda_1 t}\cos(\omega_1 t + \delta_1) + k_2 e^{\lambda_2 t}\cos(\omega_2 t + \delta_2) \\ &\quad + k_3 e^{\lambda_3 t}\cos(\omega_3 t + \delta_3) + k_4 \end{aligned}$$

$$\begin{aligned} \psi(t) &= k_1 e^{\lambda_1 t}\sin(\omega_1 t + \delta_1) + k_2 e^{\lambda_2 t}\sin(\omega_2 t + \delta_2) \\ &\quad + k_3 e^{\lambda_3 t}\sin(\omega_3 t + \delta_3) + k_5 \end{aligned}$$

where k_1, k_2, k_3, k_4, k_5, λ_1, λ_2, λ_3, ω_1, ω_2, ω_3, δ_1, δ_2, δ_3 are 14 constants which can be related to the aerodynamic and mass characteristics of the projectile and to the initial motion conditions. These constants are often estimated by nonlinear least squares to "best fit" measured pitch and yaw histories modelled by the above equations.

As an example of such a data reduction process, consider the simulated measurements of $\theta(t)$ and $\psi(t)$ with the measurement error generated by using a zero-mean Gaussian noise process with a standard deviation given by $\sigma = 0.0002$. The measurements are sampled at 1 sec intervals, shown in Figure 1.10. The *a priori* constant estimates and true values are given by

Constant Parameter	Start Value	True Value
k_1	0.5000	0.2000
k_2	0.2500	0.1000
k_3	0.1250	0.0500
k_4	0.0000	0.0001
k_5	0.0000	0.0001
λ_1	−0.1500	−0.1000
λ_2	−0.0600	−0.0500
λ_3	−0.0300	−0.0250
ω_1	0.2600	0.2500
ω_2	0.5500	0.5000
ω_3	0.9500	1.0000
δ_1	0.0100	0.0000
δ_2	0.0100	0.0000
δ_3	0.0100	0.0000

For the problem at hand the necessary conditions in eqn. (1.97) are defined as

$$\overset{(14\times 1)}{\mathbf{x}} = \begin{bmatrix} k_1 & k_2 & k_3 & k_4 & k_5 & \lambda_1 & \lambda_2 & \lambda_3 & \omega_1 & \omega_2 & \omega_3 & \delta_1 & \delta_2 & \delta_3 \end{bmatrix}^T$$

$$\overset{(52\times 1)}{\tilde{\mathbf{y}}} = \begin{bmatrix} \tilde{\theta}(0) & \tilde{\psi}(0) & \tilde{\theta}(1) & \tilde{\psi}(1) & \cdots & \tilde{\theta}(25) & \tilde{\psi}(25) \end{bmatrix}^T$$

$$\overset{(52\times 14)}{H} = \begin{bmatrix} \left.\dfrac{\partial\theta(0)}{\partial x_1}\right|_{\mathbf{x}_c} & \cdots & \left.\dfrac{\partial\theta(0)}{\partial x_{14}}\right|_{\mathbf{x}_c} \\ \left.\dfrac{\partial\psi(0)}{\partial x_1}\right|_{\mathbf{x}_c} & \cdots & \left.\dfrac{\partial\psi(0)}{\partial x_{14}}\right|_{\mathbf{x}_c} \\ \vdots & & \vdots \\ \left.\dfrac{\partial\theta(25)}{\partial x_1}\right|_{\mathbf{x}_c} & \cdots & \left.\dfrac{\partial\theta(25)}{\partial x_{14}}\right|_{\mathbf{x}_c} \\ \left.\dfrac{\partial\psi(25)}{\partial x_1}\right|_{\mathbf{x}_c} & \cdots & \left.\dfrac{\partial\psi(25)}{\partial x_{14}}\right|_{\mathbf{x}_c} \end{bmatrix}$$

$$\overset{(52\times 52)}{W} = 10^8 \begin{bmatrix} 0.25 & & 0 \\ & 0.25 & \\ & & \ddots & \\ 0 & & & 0.25 \end{bmatrix}$$

and the 28 partial derivative expressions (needed to fill the H-matrix) are given by

$$\frac{\partial\theta(t_j)}{\partial k_i} = e^{\lambda_i t_j}\cos(\omega_i t_j + \delta_i), \qquad i = 1,2,3$$
$$\frac{\partial\psi(t_j)}{\partial k_i} = e^{\lambda_i t_j}\sin(\omega_i t_j + \delta_i), \qquad i = 1,2,3$$

$$\frac{\partial\theta(t_j)}{\partial k_4} = 1, \quad \frac{\partial\psi(t_j)}{\partial k_4} = 0, \quad \frac{\partial\theta(t_j)}{\partial k_5} = 0, \quad \frac{\partial\psi(t_j)}{\partial k_5} = 1$$

$$\frac{\partial\theta(t_j)}{\partial \lambda_i} = t_j k_i e^{\lambda_i t_j}\cos(\omega_i t_j + \delta_i), \qquad i = 1,2,3$$
$$\frac{\partial\psi(t_j)}{\partial \lambda_i} = t_j k_i e^{\lambda_i t_j}\sin(\omega_i t_j + \delta_i), \qquad i = 1,2,3$$

$$\frac{\partial\theta(t_j)}{\partial \omega_i} = -t_j k_i e^{\lambda_i t_j}\sin(\omega_i t_j + \delta_i), \qquad i = 1,2,3$$
$$\frac{\partial\psi(t_j)}{\partial \omega_i} = t_j k_i e^{\lambda_i t_j}\cos(\omega_i t_j + \delta_i), \qquad i = 1,2,3$$

$$\frac{\partial\theta(t_j)}{\partial \delta_i} = -k_i e^{\lambda_i t_j}\sin(\omega_i t_j + \delta_i), \qquad i = 1,2,3$$
$$\frac{\partial\psi(t_j)}{\partial \delta_i} = k_i e^{\lambda_i t_j}\cos(\omega_i t_j + \delta_i), \qquad i = 1,2,3$$

Results in the convergence history are summarized below.

Parameter	Iteration Number					σ
	0	1	2	$\cdots$	5	
k_1	0.5000	0.1852	0.1975		0.1999	0.0006
k_2	0.2500	0.1075	0.1012		0.0997	0.0005
k_3	0.1250	0.0567	0.0505		0.0500	0.0001
k_4	0.0000	−0.0006	0.0001		0.0002	0.0001
k_5	0.0000	−0.0018	−0.0005		0.0001	0.0001
λ_1	−0.1500	−0.1234	−0.0954		−0.0998	0.0004
λ_2	−0.0600	−0.0661	−0.0585		−0.0497	0.0004
λ_3	−0.0300	−0.0398	−0.0338		−0.0250	0.0002
ω_1	0.2600	0.2490	0.2471		0.2500	0.0004
ω_2	0.5500	0.5300	0.4955		0.4999	0.0004
ω_3	0.9500	0.9697	1.0068		0.9998	0.0002
δ_1	0.0100	0.0344	0.0143		0.0010	0.0031
δ_2	0.0100	−0.0447	0.0051		0.0001	0.0048
δ_3	0.0100	0.0024	−0.0570		−0.0001	0.0024

Observe the rather dramatic convergence progress shown in the results. The right-most column is obtained by taking the square root of the 14 diagonal elements of $(H^T W H)^{-1}$ on the final iteration. We prove this interpretation of $(H^T W H)^{-1}$ in Chapter 2. Thus, a by-product of the least squares algorithm is an uncertainty measure of the answer! Note that the convergence errors are comparable in size to the corresponding σ. Also, for this example the weighted sum square of residuals (i.e., the value of J) at each iteration is given by

Cost	Iteration Number				
	0	1	2	$\cdots$	5
J	1.08×10^7	2.51×10^5	1.17×10^4		1.93×10^1

Clearly, the dramatic convergence is evidenced by the decrease of the weighted sum square of the residuals by six orders of magnitude in five iterations. Also, observe that the final converged values of the fifth iteration are in reasonable agreement with their respective true values.

1.5 Basis Functions

This section gives an overview of some common basis functions used in least squares. Although the discussion here is not exhaustive, it will serve to introduce

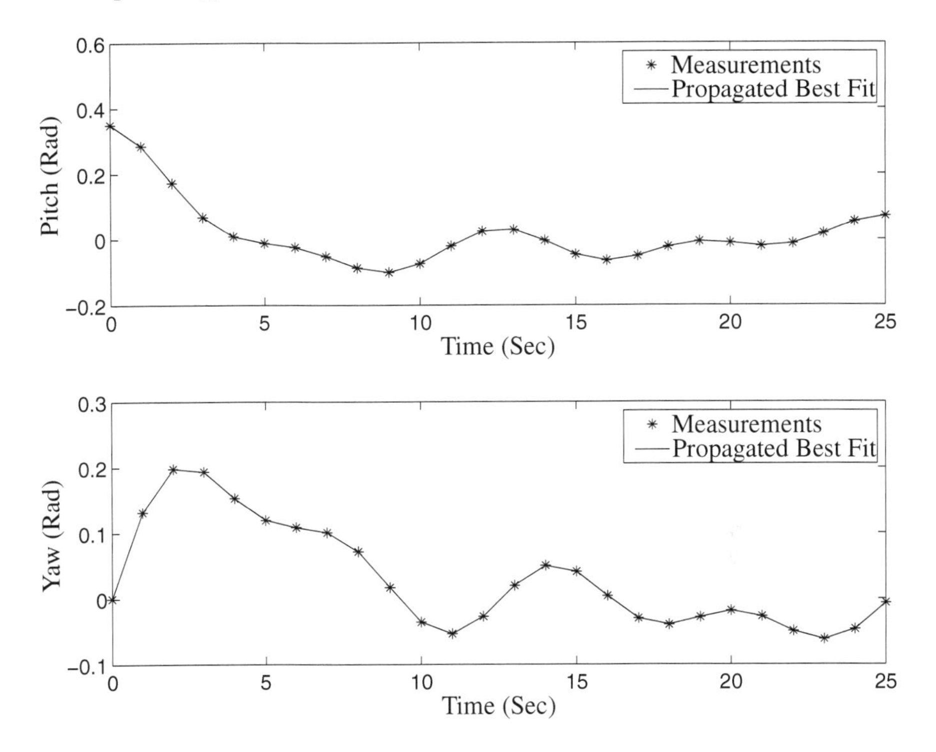

Figure 1.10: Simulated Pitch and Yaw Measurements and Best Fits

the subject matter. As seen in previous examples from this chapter, various basis functions have been used to identify system parameters. How to choose these basis functions usually comes from experience and knowledge of the particular dynamical system under investigation. Still, some commonly used basis functions can be used for a wide variety of systems. A very common choice for the linearly independent basis functions (1.12) are the powers of t:

$$\left\{1, t, t^2, t^3, \ldots\right\} \tag{1.99}$$

in which case the model (1.11) is a power series polynomial

$$y(t) = x_1 + x_2 t + x_3 t^2 + \cdots = \sum_{i=1}^{n} x_i t^{i-1} \tag{1.100}$$

The least squares coefficients estimates then follow from eqn. (1.26) with the coefficient matrix

$$H = \begin{bmatrix} 1 & t_1 & t_1^2 & \cdots & t_1^{n-1} \\ 1 & t_2 & t_2^2 & \cdots & t_2^{n-1} \\ \vdots & \vdots & \vdots & & \vdots \\ 1 & t_m & t_m^2 & \cdots & t_m^{n-1} \end{bmatrix} \tag{1.101}$$

Table 1.1: Change of Variables into Powers of t

Basis Function	New Form	Change of Variables
$y = x_1 + \frac{x_2}{a} + \frac{x_3}{a^2} + \cdots$	$y = x_1 + x_2 t + x_3 t^2 + \cdots$	$t = \frac{1}{a}$
$y = Be^{at}$	$z = x_1 + x_2 t$	$z = \ln y$ $x_1 = \ln B$ $x_2 = a$
$y = x_1 w^{-m} + x_2 w^n$	$z = x_1 + x_2 t$	$z = y\, w^m$ $t = w^{m+n}$
$y = B \exp\left[-\frac{(1-at)^2}{2\sigma^2}\right]$	$z = x_1 + x_2 t + x_3 t^2$	$z = \ln y$ $x_1 = \ln B - \frac{\ln e}{2\sigma^2}$ $x_2 = \frac{a \ln e}{\sigma^2}$ $x_3 = -\frac{\ln e}{2\sigma^2} a^2$

known as the *Vandermonde matrix*.[7, 9] Often, one encounters a nonlinear system where the basis functions are not polynomials. However, through a change of variables, one may be able to transform the original basis functions into powers of t.[10] Examples of such change are given in Table 1.1.

Therefore, linear least squares may often be used to determine the parameters that *appear* to be nonlinear in nature. Through judicious change of variables, a linear solution is now possible. Note that the Vandermonde matrix may have numerical problems if $n > 10$, but these may be partially overcome by using least squares matrix decompositions, which are discussed in §1.6.1.

Another common choice for the linearly independent basis functions (1.12) are harmonic series, which can be used to approximate y:

$$\begin{aligned} y_j &= a_0 + a_1 \cos(\omega t_j) + b_1 \sin(\omega t_j) + \ldots \\ &\quad + a_n \cos(n\omega t_j) + b_n \sin(n\omega t_j), \\ j &= 1, \ldots, m; \quad m \geq 2n+1 \end{aligned} \tag{1.102}$$

where the amplitudes (a_i, b_i) are the sought parameters. Suppose we are given $\tilde{y}_j$, t_j, $W = (W_{ij})$, and $\omega = 2\pi / T$, where T is the period under consideration. Then the desired least squares estimate $(\hat{a}_i, \hat{b}_i)$ is computable as

$$\hat{\mathbf{x}} = \begin{bmatrix} \hat{a}_0 \\ \hat{a}_1 \\ \hat{b}_1 \\ \vdots \\ \hat{a}_n \\ \hat{b}_n \end{bmatrix} = (H^T W H)^{-1} H^T W \tilde{\mathbf{y}} \tag{1.103}$$

where

$$H = \begin{bmatrix} 1 & \cos(\omega t_1) & \sin(\omega t_1) & \cdots & \cos(n\omega t_1) & \sin(n\omega t_1) \\ 1 & \cos(\omega t_2) & \sin(\omega t_2) & \cdots & \cos(n\omega t_2) & \sin(n\omega t_2) \\ \vdots & \vdots & \vdots & & \vdots & \vdots \\ 1 & \cos(\omega t_m) & \sin(\omega t_m) & \cdots & \cos(n\omega t_m) & \sin(n\omega t_m) \end{bmatrix} \tag{1.104}$$

In the case above, if W is chosen as an identity matrix and the sample points $\{t_1, t_2, \ldots\}$ are symmetric about $T/2$, then the matrix inverse becomes trivial because $(H^T W H)$ will be found to be a diagonal matrix. This leads to a simple solution, given by

$$\hat{x}_i = \left[\sum_{j=1}^{m} h_i^2(t_j) \right]^{-1} \sum_{j=1}^{m} h_i(t_j)\tilde{y}_j, \quad i = 1, 2, \ldots, n \tag{1.105}$$

where

$$\begin{aligned} \mathbf{h}(t) &\equiv \left[h_1(t) \; h_2(t) \; h_3(t) \; \cdots \right]^T \\ &= \left[1 \; \cos(\omega t) \; \sin(\omega t) \; \cdots \; \cos(n\omega t) \; \sin(n\omega t) \right]^T \end{aligned} \tag{1.106}$$

A significant advantage of the uncoupled solution for the coefficients in eqn. (1.105) is that adding another $(n+1)$ basis function (which has the same form as any of the first n) does affect the first n solutions for $\hat{x}_i$.

The least squares estimate for the coefficients has a strong connection to the continuous approximation for $\tilde{y}(t)$. Before we formally prove this, let us review the concept of an *orthogonal* set of functions.[11, 12] An infinite system of real functions

$$\{\varphi_1(t), \varphi_2(t), \varphi_3(t), \ldots, \varphi_n(t), \ldots\} \tag{1.107}$$

is said to be orthogonal on the interval $[\alpha, \beta]$ if

$$\int_{\alpha}^{\beta} \varphi_p(t)\varphi_q(t)\,dt = 0 \quad (p \neq q, \; p, q = 1, 2, 3, \ldots) \tag{1.108}$$

and

$$\int_{\alpha}^{\beta} \varphi_p^2(t)\,dt \equiv c_p \neq 0 \quad (p = 1, 2, 3, \ldots) \tag{1.109}$$

The series given in eqn. (1.106) can be shown to be orthogonal over any interval centered on $t = T/2$. We further note the distinction between the continuous orthogonality conditions of eqns. (1.108) and the corresponding discrete orthogonality conditions

$$\sum_{j=1}^{m} \varphi_p(t_j)\varphi_q(t_j) = c_p\,\delta_{pq} \tag{1.110}$$

where the Kronecker delta δ_{pq} is defined as

$$\begin{aligned} \delta_{pq} &= 0 \quad \text{if } p \neq q \\ &= 1 \quad \text{if } p = q \end{aligned} \tag{1.111}$$

For the discrete orthogonality case, a specific pattern of sample points underlies this condition. In particular, the discrete orthogonality condition for the columns of H in eqn. (1.104) holds for any set of points $\{t_1, t_2, \ldots, t_m\}$ that are symmetric about $t = T/2$. We also mention that the most general forms of the continuous and discrete orthogonality conditions are

$$\int_{\alpha}^{\beta} w(t)\varphi_p(t)\varphi_q(t)\,dt = c_p\,\delta_{pq} \tag{1.112}$$

and

$$\sum_{j=1}^{m} w(t_j)\varphi_p(t_j)\varphi_q(t_j) = c_p\,\delta_{pq} \tag{1.113}$$

where $w(t)$ is an associated weight function.

The orthogonality condition on the individual integrals of the terms $\sin(2\pi pt/T)$ and $\cos(2\pi pt/T)$ are trivial to prove on the interval $[0, T]$. A slightly more complex case involves the integral of $\sin(ct)\sin(dt)$ for any $c \neq d$ on the interval $[0, T]$:

$$\begin{aligned} \int_0^T \sin(ct)\sin(dt)\,dt &= \frac{1}{2}\int_0^T [\cos(ct-dt) - \cos(ct+dt)]\,dt \\ &= \left[\frac{\sin(ct-dt)}{2(c-d)} - \frac{\sin(ct+dt)}{2(c+d)}\right]\Bigg|_0^T \end{aligned} \tag{1.114}$$

If we let $c = 2\pi p/T$ and $d = 2\pi q/T$, then it is easy to see that eqn. (1.114) is identically zero for any $p \neq q$. Therefore, this system is orthogonal with the associated weight function $w(t) = 1$. It can also be shown that all integrals of any combinations of the functions in eqn. (1.106) are orthogonal on the interval $[0, T]$. Of course, we may also replace the integral with a summation; for symmetrically located samples, we have discrete orthogonality and this leads directly to the solution in eqn. (1.105).

The *Fourier series* of a function is a harmonic expansion of sines and cosines, given by

$$y(t) = a_0 + \sum_{n=1}^{\infty} a_n \cos(n\omega t) + \sum_{n=1}^{\infty} b_n \sin(n\omega t) \tag{1.115}$$

To compute a coefficient such as a_1, multiply both sides of eqn. (1.115) by $\cos(\omega t)$ and integrate from 0 to T (the function y is given on this interval). This leads to

$$\int_0^T y(t)\cos(\omega t)\,dt = a_0 \int_0^T \cos(\omega t)\,dt + a_1 \int_0^T [\cos(\omega t)]^2\,dt + \cdots + \\ + b_1 \int_0^T \cos(\omega t)\sin(\omega t)\,dt + \cdots \tag{1.116}$$

Every integral on the right side of eqn. (1.116) is zero (since the sines and cosines are mutually orthogonal) except the one in which $\cos(\omega t)$ multiplies itself. Therefore, a_1 is given by

$$a_1 = \frac{\int_0^T y(t)\cos(\omega t)\,dt}{\int_0^T [\cos(\omega t)]^2\,dt} \tag{1.117}$$

The coefficient b_1 would have $\sin(\omega t)$ in place of $\cos(\omega t)$, and b_2 would use $\sin(2\omega t)$, and so on. Evaluating the integral in the denominator of eqn. (1.117), and likewise for the other coefficients leads to the *Fourier coefficients*,[13, 14] given by

$$a_0 = \frac{1}{T}\int_0^T y(t)\,dt \tag{1.118a}$$

$$a_n = \frac{2}{T}\int_0^T y(t)\cos(n\omega t)\,dt \tag{1.118b}$$

$$b_n = \frac{2}{T}\int_0^T y(t)\sin(n\omega t)\,dt \tag{1.118c}$$

The Fourier coefficients can also be determined using linear least squares, and in the process, we establish that the determined coefficients are simply a special case of least squares approximation. For this development we will assume that our measurement model, $\tilde{y}(t)$, is given by eqn. (1.115), so that $\tilde{y}(t) = y(t)$. Consider minimizing the following function:

$$J = \frac{1}{2}\int_0^T [y(t) - \hat{\mathbf{x}}^T\mathbf{h}(t)]^T[y(t) - \hat{\mathbf{x}}^T\mathbf{h}(t)]\,dt \tag{1.119}$$

or

$$J = \frac{1}{2}\int_0^T [y(t)]^2\,dt - \left[\int_0^T y(t)\mathbf{h}^T(t)\,dt\right]\hat{\mathbf{x}} \\ + \frac{1}{2}\hat{\mathbf{x}}^T\left[\int_0^T \mathbf{h}(t)\mathbf{h}^T(t)\,dt\right]\hat{\mathbf{x}} \tag{1.120}$$

The necessary condition $\nabla_{\hat{\mathbf{x}}} J = \mathbf{0}$ leads to

$$\hat{\mathbf{x}} = \left[\int_0^T \mathbf{h}(t)\mathbf{h}^T(t)\,dt\right]^{-1}\left[\int_0^T y(t)\mathbf{h}(t)\,dt\right] \tag{1.121}$$

Since $\mathbf{h}(t)$ represents a set of orthogonal functions on the interval $[0, T]$, i.e., the functions satisfy eqn. (1.108), so that $\mathbf{h}(t)\mathbf{h}^T(t)$ is a diagonal matrix, then the individual components of $\hat{\mathbf{x}}$ are simply given by the uncoupled equations

$$\hat{x}_i = \frac{\int_0^T y(t)h_i(t)\,dt}{\int_0^T [h_i(t)]^2\,dt}, \quad i = 1, 2, \ldots, n \tag{1.122}$$

This is identical to the solution shown in eqn. (1.118). Therefore, the Fourier coefficients are just "least square" estimates using the particular orthogonal basis function in eqn. (1.106). On several occasions herein, we will make use of orthogonal basis functions; however, this subject is not treated comprehensively within the scope of this text. Most standard mathematical handbooks, such as Abramowitz and Stegun,[15] and Ledermann,[16] summarize a large family of orthogonal polynomials and discuss their use in approximation.

1.6 Advanced Topics

In this section we will show some advanced topics used in least squares. Although an exhaustive treatment is beyond the scope of this text, we hope that the subjects presented herein will motivate the interested reader to pursue them in the referenced literature.

1.6.1 Matrix Decompositions in Least Squares

The core component of any least squares algorithm is $(H^T H)^{-1}$. As an alternative to direct computation of this inverse, it is common to decompose H in some way which simplifies the calculations and/or is more robust with respect to near singularity conditions. A more detailed mathematical development of some of the topics presented here is provided in §A.4.

A particularly useful decomposition of the matrix H is the QR decomposition. Before we discuss this decomposition, let us first review the definition and properties of orthogonal vectors and matrices. Two vectors, $\mathbf{u}$ and $\mathbf{v}$, are *orthogonal* if the angle between them is $\pi/2$. This can be true if and only if $\mathbf{u}^T\mathbf{v} = 0$. An *orthogonal matrix*[7, 17] Q is a square matrix with *orthonormal* column vectors. Orthonormal vectors are orthogonal vectors each with unit lengths. Since the columns of an orthogonal matrix Q are orthonormal, then $Q^T Q = I$ (where $Q^T Q$ is a matrix of vector-space inner-products) and $Q^T = Q^{-1}$. This clearly shows that the inverse of an orthogonal matrix is given by its transpose!

An example of an orthogonal matrix in dynamic systems is the *rotation* matrix.

For example, let

$$Q = \begin{bmatrix} 1 & 0 & 0 \\ 0 & \cos\phi & \sin\phi \\ 0 & -\sin\phi & \cos\phi \end{bmatrix} \tag{1.123}$$

This matrix is clearly orthogonal, since the column vectors are orthonormal.

The QR decomposition factors a full rank matrix H as the product of an orthogonal matrix Q and an upper-triangular matrix R, given by

$$H = QR \tag{1.124}$$

where Q is an $m \times n$ matrix with $Q^T Q = I$, and R is an upper triangular $n \times n$ matrix with all elements $R_{ij} = 0$ for $i > j$. The QR decomposition can be accomplished using the *modified Gram-Schmidt* algorithm (see §A.4). The advantage of the QR decomposition is that it greatly simplifies the least squares problem. The term $H^T H$ in the normal equations is easier to invert since

$$H^T H = R^T Q^T Q R = R^T R \tag{1.125}$$

Therefore, the normal equations (1.26) simplify to

$$R^T R\hat{\mathbf{x}} = R^T Q^T \tilde{\mathbf{y}} \tag{1.126}$$

or

$$\boxed{R\hat{\mathbf{x}} = Q^T \tilde{\mathbf{y}}} \tag{1.127}$$

The solution to eqn. (1.127) can easily be accomplished since R is upper triangular (see Appendix A). The real cost is in the $2mn^2$ operations in the modified Gram-Schmidt algorithm, which are required to compute Q and R. The QR decomposition can also be used in linear least squares to improve an approximate solution using iterative refinement.[18] Notice it is not necessary to square H (i.e., form $H^T H$); the QR algorithm operates directly on H. If H is poorly conditioned, it is easy to verify that $H^T H$ is much more poorly conditioned than H itself.

Another decomposition of the matrix H is the *singular-value decomposition*,[7, 17] which decomposes a matrix into a diagonal matrix and two orthogonal matrices:

$$H = U\,S\,V^T \tag{1.128}$$

where U is the $m \times n$ matrix with orthonormal columns, $\mathcal{S}$ is an $n \times n$ diagonal matrix such that $\mathcal{S}_{ij} = 0$ for $i \neq j$, and V is an $n \times n$ orthogonal matrix. Note that $U^T U = I$, but it is no longer possible to make the same statement for $U U^T$. Now, substitute eqn. (1.128) into eqn. (1.25):

$$(H^T H)\hat{\mathbf{x}} = H^T \tilde{\mathbf{y}} \tag{1.129a}$$

$$(V S U^T U S V^T)\hat{\mathbf{x}} = V S U^T \tilde{\mathbf{y}} \tag{1.129b}$$

$$(V S S V^T)\hat{\mathbf{x}} = V S U^T \tilde{\mathbf{y}} \tag{1.129c}$$

$$(S V^T)\hat{\mathbf{x}} = U^T \tilde{\mathbf{y}} \tag{1.129d}$$

Therefore, the solution for $\hat{\mathbf{x}}$ is simply given by

$$\boxed{\hat{\mathbf{x}} = V S^{-1} U^T \tilde{\mathbf{y}}} \tag{1.130}$$

Note that the inverse of S is easy to compute since it is a diagonal matrix (i.e., $S = \text{diag}\begin{bmatrix} s_1 & \cdots & s_n \end{bmatrix}$). The elements of S are known as the *singular values* of H.

The singular value decomposition can also be used to perform a least squares minimization subject to a spherical (ball) constraint on $\hat{\mathbf{x}}$.[7] Consider the minimization of

$$J = \frac{1}{2}(\tilde{\mathbf{y}} - H\hat{\mathbf{x}})^T(\tilde{\mathbf{y}} - H\hat{\mathbf{x}}) \tag{1.131}$$

subject to the following constraint:

$$\sqrt{\hat{\mathbf{x}}^T\hat{\mathbf{x}}} \le \gamma \tag{1.132}$$

where γ is some known constant. Equation (1.132) constrains $\hat{\mathbf{x}}$ to lie within or on a sphere. The solution to this problem can be given using a singular value decomposition as follows[7]

$$H = U S V^T \tag{1.133a}$$

$$\begin{bmatrix} \mathbf{v}_1, & \ldots, & \mathbf{v}_n \end{bmatrix} = V \tag{1.133b}$$

$$\mathbf{z} = U^T \tilde{\mathbf{y}} \tag{1.133c}$$

$$r = \text{rank}(H) \tag{1.133d}$$

If the following inequality is true:

$$\sum_{i=1}^{r} \left(\frac{z_i}{s_i}\right)^2 > \gamma^2 \tag{1.134}$$

then find λ^* such that

$$\sum_{i=1}^{r} \left(\frac{s_i z_i}{s_i^2 + \lambda^*}\right)^2 = \gamma^2 \tag{1.135}$$

and the optimal estimate is given by

$$\hat{\mathbf{x}} = \sum_{i=1}^{r} \left(\frac{s_i z_i}{s_i^2 + \lambda^*}\right) \mathbf{v}_i \tag{1.136}$$

If the inequality in eqn. (1.134) is not satisfied, then the optimal estimate is given by

$$\hat{\mathbf{x}} = \sum_{i=1}^{r} \left(\frac{z_i}{s_i}\right) \mathbf{v}_i \tag{1.137}$$

It can be shown that there exists a unique positive solution for λ^*, which can be found using Newton's root solving method. A more general case of the quadratic inequality constraint can be found in Golub.[7]

Example 1.8: Consider the following model:

$$y = x_1 + x_2 t + x_3 t^2$$

Given a set of 101 measurements, shown in Figure 1.11, we are asked to determine $\hat{\mathbf{x}}$ such that $\hat{\mathbf{x}}^T \hat{\mathbf{x}} \leq 14$. After forming the H matrix, we determine that the rank of H is $r = 3$, and the singular values are given by

$$S = \text{diag}\left[456.3604 \;\; 15.5895 \;\; 3.1619\right]$$

The singular values clearly show that this least squares problem is well posed since the condition number is given by $456.36/3.16 = 144.33$. Forming the $\mathbf{z}$ vector, and with $\gamma^2 = 14$, we see that the inequality in eqn. (1.134) is satisfied with the given measurements. The optimal value for λ^* in eqn. (1.135) was determined using Newton's root solving with a starting value of 0, and converged to a value of $\lambda^* = 0.245$. The optimal estimate in eqn. (1.136) is given by

$$\hat{\mathbf{x}} = \begin{bmatrix} 3.0209 \\ 1.9655 \\ 1.0054 \end{bmatrix}$$

The inequality constraint in eqn. (1.132) is clearly satisfied since $\hat{\mathbf{x}}^T \hat{\mathbf{x}} = 14$ (in this case the equality condition is actually satisfied). It is interesting to note that the solution using standard least squares in eqn. (1.26) is given by

$$\hat{\mathbf{x}}_{ls} = \begin{bmatrix} 3.0686 \\ 1.9445 \\ 1.0067 \end{bmatrix}$$

We can see that the solutions are nearly identical; however, the standard least squares solution violates the inequality constraint since $\hat{\mathbf{x}}_{ls}^T \hat{\mathbf{x}}_{ls} = 14.2109 \geq 14$. Also, since the standard least squares solution gives a condition that violates the constraint, we expect that the optimal solution should give estimates that lie on the surface of the sphere (i.e., on the equality constraint).

This section has introduced some popular matrix decompositions used in linear least squares. Choosing which decomposition to use is primarily dependent upon the particular application, numerical concerns, and desired level of accuracy. For example, the singular value decomposition is one of the most robust algorithms to compute the least squares estimates. However, it is also one of the most computationally expensive algorithms. The decompositions presented in this section do not represent an exhaustive treatise of the subject. For the interested reader, the many references cited throughout this section give more thorough treatments of the subject matter. In particular, both the QR and singular-value decomposition algorithms can be generalized to include the case that H is either row or column rank deficient.[18]

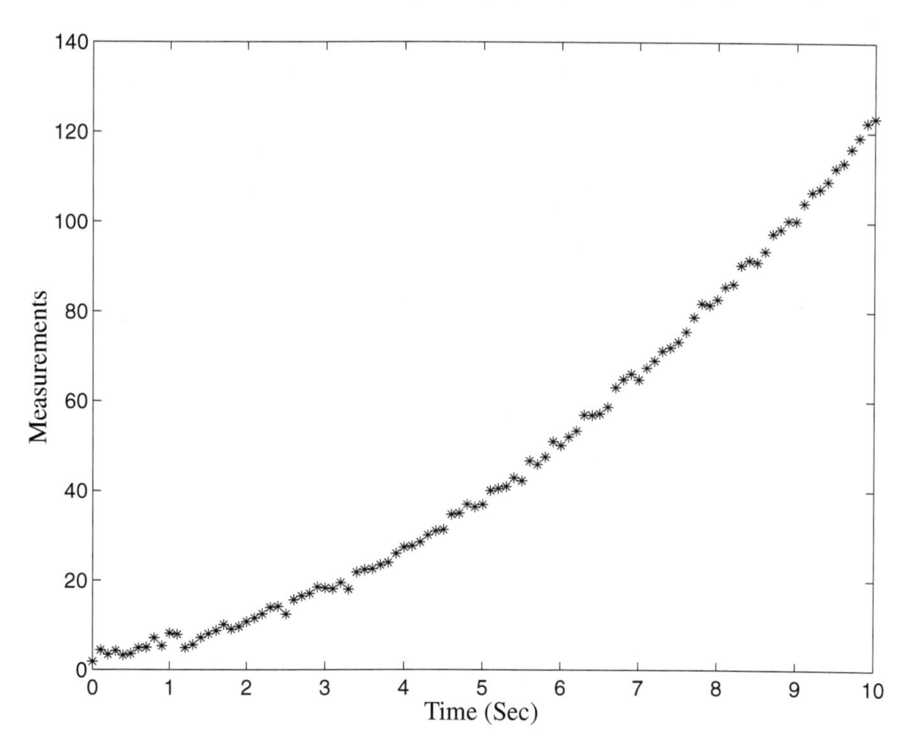

Figure 1.11: Measurements of $y(t)$

1.6.2 Kronecker Factorization and Least Squares

The SVD approach of §1.6.1 can be used to improve the numerical accuracy of the solution over the equivalent standard least squares solution. However, this comes at a significant computational cost. In this section another approach based on the Kronecker factorization[19] is shown that can be used to improve the accuracy and reduce the computational costs for a certain class of problems. The Kronecker product is defined as

$$H = A \otimes B \equiv \begin{bmatrix} a_{11}B & a_{12}B & \cdots & a_{1\beta}B \\ a_{21}B & a_{22}B & \cdots & a_{2\beta}B \\ \vdots & \vdots & \ddots & \vdots \\ a_{\alpha 1}B & a_{\alpha 2}B & \cdots & a_{\alpha\beta}B \end{bmatrix} \tag{1.138}$$

where H is an $M \times N$ dimension matrix, A is an $\alpha \times \beta$ matrix, and B is a $\gamma \times \delta$ matrix. The Kronecker product is only valid when $M = \alpha\gamma$ and $N = \beta\delta$. The key results for least squares problems is that if $H = A \otimes B$, then eqn. (1.26) reduces down to

$$\hat{\mathbf{x}} = \left\{[(A^T A)^{-1} A^T] \otimes [(B^T B)^{-1} B^T]\right\} \tilde{\mathbf{y}} \tag{1.139}$$

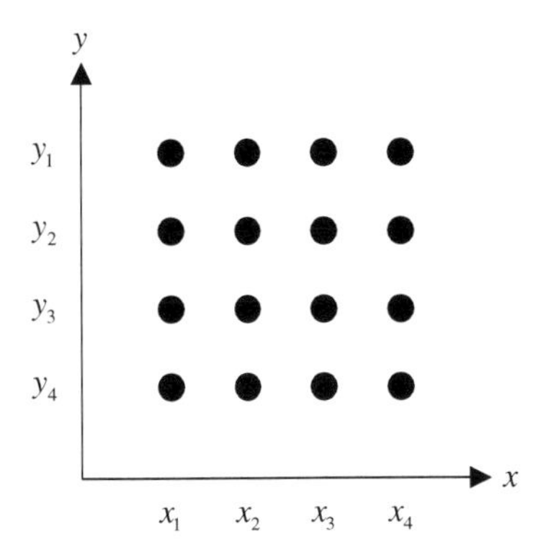

Figure 1.12: Gridded Data

In essence the Kronecker product takes the square root of the matrix dimensions in regards to the computational difficulty.

A key question now arises: "Under what conditions can a matrix be factored as a Kronecker product of smaller matrices?" This is a difficult question to answer, but fortunately it is easy to show that some important curve fitting problems lead to a Kronecker factorization, such as the case of gridded data depicted in Figure 1.12. We first consider the case of fitting a two-variable polynomial to data on an x-y grid:

$$z = f(x, y) = \sum_{p=0}^{M} \sum_{q=0}^{M} c_{pq} x^p y^q \tag{1.140}$$

where the measurements are now defined by

$$\tilde{z}_{ij} = f(x_i, y_i) + v_{ij} \tag{1.141}$$

for $i = 1, 2, \ldots, n_x$ and $j = 1, 2, \ldots, n_y$. Now consider the special case of $M = 2$, $N = 1$, $n_x = 4$, and $n_y = 3$. The quantity z in eqn. (1.140) is given by

$$z = c_{00} + c_{01} y + c_{10} x + c_{11} x\, y + c_{20} x^2 + c_{21} x^2 y \tag{1.142}$$

The least squares measurement model is now given by

$$\begin{bmatrix} \tilde{z}_{11} \\ \tilde{z}_{12} \\ \tilde{z}_{13} \\ \vdots \\ \tilde{z}_{41} \\ \tilde{z}_{42} \\ \tilde{z}_{43} \end{bmatrix} = \begin{bmatrix} 1 & y_1 & x_1 & x_1 y_1 & x_1^2 & x_1^2 y_1 \\ 1 & y_2 & x_1 & x_1 y_2 & x_1^2 & x_1^2 y_2 \\ 1 & y_3 & x_1 & x_1 y_3 & x_1^2 & x_1^2 y_3 \\ \vdots & \vdots & \vdots & \vdots & \vdots & \vdots \\ 1 & y_1 & x_4 & x_4 y_1 & x_4^2 & x_4^2 y_1 \\ 1 & y_2 & x_4 & x_4 y_2 & x_4^2 & x_4^2 y_2 \\ 1 & y_3 & x_4 & x_4 y_3 & x_4^2 & x_4^2 y_3 \end{bmatrix} \begin{bmatrix} c_{00} \\ c_{01} \\ c_{10} \\ c_{11} \\ c_{20} \\ c_{21} \end{bmatrix} + \begin{bmatrix} v_{11} \\ v_{12} \\ v_{13} \\ \vdots \\ v_{41} \\ v_{42} \\ v_{43} \end{bmatrix} \equiv H\mathbf{c} + \mathbf{v} \tag{1.143}$$

where H, $\mathbf{c}$, and $\mathbf{v}$ have dimensions of 12×6, 6×1, and 12×1, respectively. We can now easily verify that the matrix H has a Kronecker factorization given by

$$H = \begin{bmatrix} 1 & x_1 & x_1^2 \\ 1 & x_2 & x_2^2 \\ 1 & x_3 & x_3^2 \\ 1 & x_4 & x_4^2 \end{bmatrix} \otimes \begin{bmatrix} 1 & y_1 \\ 1 & y_2 \\ 1 & y_3 \end{bmatrix} \equiv H_x \otimes H_y \tag{1.144}$$

where H_x and H_y have dimensions of 4×3 and 3×2, respectively. Thus, perhaps, it is not surprising that the two-variable Vandermonde matrix can be produced by the Kronecker product of the corresponding one-variable Vandermonde matrices. The consequences in the least squares solution are enormous, since the estimate for the coefficient vector, $\mathbf{c}$, can be computed by

$$\hat{\mathbf{c}} = (H^T H)^{-1} H^T \tilde{\mathbf{z}} = \left\{ [(H_x^T H_x)^{-1} H_x^T] \otimes [(H_y^T H_y)^{-1} H_y^T] \right\} \tilde{\mathbf{z}} \tag{1.145}$$

Hence, only inverses of 3×3 and 2×2 matrices need to be computed instead of an inverse of a 6×6 matrix. In general, for H of dimension $M \times N$, and H_x and H_y of dimensions about $\sqrt{M}/2$ and $\sqrt{N}/2$, respectively, the least squares computational burden is reduced from an order of n^3 operations to an order of $(\sqrt{n})^3$ operations! Furthermore, as will be shown in example 1.9, the accuracy of the solution is also vastly improved.

The previous Kronecker factorization solution in the least squares problem can be expanded to the n-dimensional case, where data are at the vertices of an n-dimensional grid:

$$z = f(x_1, x_2, \ldots, x_n) = \sum_{i_1=1}^{N_1} \sum_{i_2=1}^{N_2} \cdots \sum_{i_n=1}^{N_n} c_{i_1 i_2 \cdots i_n} \phi_{i_1}(x_1) \phi_{i_2}(x_2) \cdots \phi_{i_n}(x_n) \tag{1.146}$$

where $\phi_{i_j}(x_j)$ are basis functions. The measurements now follow

$$\tilde{z}_{j_1 j_2 \cdots j_n} \quad \text{at} \quad (x_{1_{j_1}}, x_{2_{j_2}}, \ldots, x_{n_{j_n}}) \tag{1.147}$$

for $j_1 = 1, 2, \ldots, M_1$ through $j_n = 1, 2, \ldots, M_n$. The vectors $\tilde{\mathbf{z}}$ and $\mathbf{c}$ are now denoted by

$$\tilde{\mathbf{z}} = \left[\tilde{z}_{11\cdots 11} \; \cdots \; \tilde{z}_{11\cdots 1M_n} \; \cdots \; \tilde{z}_{M_1 M_2 \cdots M_{n-1} 1} \; \cdots \; \tilde{z}_{M_1 M_2 \cdots M_{n-1} M_n} \right]^T \tag{1.148a}$$

$$\mathbf{c} = \left[c_{11\cdots 11} \; \cdots \; c_{11\cdots 1N_1} \; \cdots \; c_{N_1 N_2 \cdots N_{n-1} 1} \; \cdots \; c_{N_1 N_2 \cdots N_{n-1} N_n} \right]^T \tag{1.148b}$$

The matrix H is given by

$$H = H_1 \otimes H_2 \otimes \cdots \otimes H_N \tag{1.149}$$

with

$$H_i = \begin{bmatrix} \Phi_1(x_{i_1}) & \Phi_2(x_{i_1}) & \cdots & \Phi_{N_i}(x_{i_1}) \\ \vdots & \vdots & \ddots & \vdots \\ \Phi_1(x_{i_{M_i}}) & \Phi_2(x_{i_{M_i}}) & \cdots & \Phi_{N_i}(x_{i_{M_i}}) \end{bmatrix}, \quad i = 1, 2, \ldots, N \tag{1.150}$$

where the Φ's are sub-matrices composed of the basis functions $\phi_{i_1}(x_1)$ through $\phi_{i_n}(x_n)$. The estimate for the coefficient vector, $\mathbf{c}$, can be computed by

$$\boxed{\hat{\mathbf{c}} = \left\{[(H_1^T H_1)^{-1} H_1^T] \otimes \cdots \otimes [(H_N^T H_N)^{-1} H_N^T]\right\} \tilde{\mathbf{z}}} \tag{1.151}$$

Therefore, the least squares solution is given by a Kronecker product of sub-matrices with much smaller dimension than the original problem.

Example 1.9: In this simple example, the power of the Kronecker product in least squares problems is illustrated. We consider a 21×21 grid over the intervals $-2 \leq x \leq 2$ and $-2 \leq y \leq 2$ with functions given by

$$\begin{bmatrix} 1 \; x \; x^2 \; x^3 \; x^4 \; x^5 \end{bmatrix}$$
$$\begin{bmatrix} 1 \; y \; y^2 \; y^3 \; y^4 \; y^5 \end{bmatrix}$$

The 21×6 matrices H_x and H_y are given by

$$H_x = \begin{bmatrix} 1 & x_1 & x_1^2 & x_1^3 & x_1^4 & x_1^5 \\ 1 & x_2 & x_2^2 & x_2^3 & x_2^4 & x_2^5 \\ \vdots & \vdots & \vdots & \vdots & \vdots & \vdots \\ 1 & x_{21} & x_{21}^2 & x_{21}^3 & x_{21}^4 & x_{21}^5 \end{bmatrix}, \quad H_y = \begin{bmatrix} 1 & y_1 & y_1^2 & y_1^3 & y_1^4 & y_1^5 \\ 1 & y_2 & y_2^2 & y_2^3 & y_2^4 & y_2^5 \\ \vdots & \vdots & \vdots & \vdots & \vdots & \vdots \\ 1 & y_{21} & y_{21}^2 & y_{21}^3 & y_{21}^4 & y_{21}^5 \end{bmatrix}$$

The 441×36 matrix H is just the Kronecker product of H_x and H_y, so that $H = H_x \otimes H_y$. The true coefficient vector, $\mathbf{c}$, has elements simply given by 1 in this formulation. As shown previously the Kronecker factorization gives a substantial savings in numerical computations. We also wish to investigate the accuracy of this approach. To accomplish this task, no noise is added to form the 441×1 vector of measurements, which is simply given by $\tilde{\mathbf{z}} = H\,\mathbf{c}$.

The numerical accuracy is shown by computing $\epsilon \equiv ||\hat{\mathbf{c}} - \mathbf{c}||$, which is ideally zero. Using the standard least squares solution of §1.2.1, which takes the inverse of a 36×36 matrix, gives $\epsilon = 7.15 \times 10^{-10}$. Using the SVD solution of §1.6.1 gives $\epsilon = 1.15 \times 10^{-12}$, which provides more accuracy but at a price of a substantial computational cost over the standard least squares solution. Using the Kronecker factorization gives $\epsilon = 1.66 \times 10^{-13}$, which provides even better accuracy than the SVD solution, but is more computationally efficient than the standard least squares solution. An SVD solution for each inverse in the Kronecker factorization can also be used instead of the standard inverse. This approach gives $\epsilon = 1.20 \times 10^{-13}$, which provides the most accurate solution with only a modest increase in computational cost over the standard Kronecker factorization solution. This example clearly shows the power of the Kronecker factorization for curve fitting problems with gridded data.

This section summarized a powerful solution to the curve fitting problem involving gridded data. The Kronecker factorization leads to substantial computational savings, while improving the numerical accuracy of the solution, over the standard least squares solution. This is especially significant for systems involving polynomial models, which have a tendency to be ill conditioned. This approach has substantial advantages for applications in many systems, such as satellite imagery, terrain modelling, and photogrammetry. More details on the usefulness of the Kronecker factorization in least squares applications can be found in Ref. [19].

1.6.3 Levenberg-Marquardt Method

The differential correction algorithm in §1.4 may not be suitable for some nonlinear problems since convergence cannot be guaranteed, unless the *a priori* estimate is close to a minimum in the loss function. This difficulty may be overcome by using the *method of steepest descent* (see Appendix C). This method adjusts the current estimate so that the most favorable direction is given (i.e., the direction of steepest descent), which is along the negative gradient of J. The method of steepest descent often converges rapidly for the first few iterations, but has difficulty converging to a solution because the slope becomes more and more shallow as the number of iterations increases.

The Levenberg-Marquardt algorithm[20] overcomes both the difficulties of the standard differential correction approach when an accurate initial estimate is not given, and the slow convergence problems of the method of steepest descent when the solution is close to minimizing the nonlinear least squares loss function (1.89). The paper by Marquardt develops the entire algorithm; however a significant acknowledgment is given to Levenberg.[21] Hence, the algorithm is usually referred to by both authors. This algorithm performs an optimum interpolation between the differential correction, which approximates a second-order Taylor series expansion of J, and the method of steepest descent, which uses a first-order approximation of local J behavior.

We first derive an expression for the gradient correction. Consider the loss function given by eqn. (1.96):

$$J = \frac{1}{2}\Delta\mathbf{y}^T W \Delta\mathbf{y} \tag{1.152}$$

The gradient of eqn. (1.152) is given by

$$\nabla_{\hat{\mathbf{x}}} J = -H^T W \Delta\mathbf{y}_c \tag{1.153}$$

where

$$H \equiv \left.\frac{\partial \mathbf{f}}{\partial \mathbf{x}}\right|_{\hat{\mathbf{x}}} \tag{1.154}$$

The method of gradients seeks corrections down the gradient:

$$\Delta\mathbf{x} = -\frac{1}{\eta}\nabla_{\hat{\mathbf{x}}} J = \frac{1}{\eta} H^T W \Delta\mathbf{y}_c \tag{1.155}$$

where $1/\eta$ is a scalar which controls the step size. The poor terminal convergence of the first-order gradient and the less reliable early convergence of the second-order differential correction algorithm can be compromised, as in the Levenberg-Marquardt algorithm with the modified normal equations:

$$\boxed{\Delta\mathbf{x} = (H^T W H + \eta\mathcal{H})^{-1} H^T W \Delta\mathbf{y}_c} \tag{1.156}$$

where $\mathcal{H}$ is a diagonal matrix with entries given by the diagonal elements of $H^T W H$ or in some cases simply the identity matrix. By using the algorithm in eqn. (1.156) the search direction is an intermediate between the steepest descent and the differential correction direction. As $\eta \rightarrow 0$, eqn. (1.156) is equivalent to the differential correction method; however, as $\eta \rightarrow \infty$, if $\mathcal{H} = I$, eqn. (1.156) reduces to a steepest descent search along the negative gradient of J.

Controlling η (and therefore both the magnitude and direction of $\Delta\mathbf{x}$) is a heuristic art form that can be tuned by the user. Generally η is large in early iterations and should definitely be reduced toward zero in the region near the minimum. To capture the spirit of the approach, here is a typical recipe for implementing the Levenberg-Marquardt algorithm:

1. Compute eqn. (1.89) using an initial estimate for $\hat{\mathbf{x}}$, denoted by $\mathbf{x}_c$.

2. Use eqns. (1.156) and (1.91) to update the current estimate with a large value for η (usually much larger than the norm of $H^T W H$, typically 10 to 100 times the norm).

3. Recompute eqn. (1.89) with the new estimate. If the new value for eqn. (1.89) is $\geq$ the value computed in step 1, then the new estimate is disregarded and η is replaced by $f\eta$, where f is a fixed positive constant, usually between 1 and 10 (we suggest a default of 5). Otherwise, retain the estimate, and replace η with η/f.

4. After each subsequent iteration, compare the new value of eqn. (1.89) with its value using the previous estimate and replace η with $f\eta$ or η/f as in step 3. The estimate $\hat{\mathbf{x}}$ is retained if J in eqn. (1.89) continues to decrease and discarded if (1.89) increases.

This procedure continues until the difference in eqn. (1.89) between two consecutive iterations is small. The Levenberg-Marquardt method is heuristic, seeking to find the middle ground between the method of steepest descent and the Gaussian differential correction, tending toward the Gaussian differential correction in the terminal corrections. However, a little effort in tuning this algorithm often leads to a significantly enhanced domain of convergence.

Example 1.10: In example 1.7, we used nonlinear least squares to determine the parameters of an inertially and aerodynamically symmetric projectile. In this example we begin with the same start values, except that the start value for λ_1 is equal to

−0.8500 instead of −0.1500. For this initial value, the standard least squares solution diverges rapidly with each iteration. Therefore, we must use a different starting set or, in this case we choose to use the Levenberg-Marquardt algorithm. For this algorithm, we set the initial value for η to 1×10^6. Results in the convergence history are summarized below.

Parameter	Iteration Number				
	0	10	15	···	20
k_1	0.5000	0.3601	0.0844		0.1999
k_2	0.2500	0.1946	0.2099		0.0997
k_3	0.1250	0.0905	0.0620		0.0500
k_4	0.0000	−0.0062	0.0111		0.0002
k_5	0.0000	−0.0047	−0.0004		0.0001
λ_1	−0.8500	−0.7977	−0.0436		−0.0998
λ_2	−0.0600	−0.0760	−0.1270		−0.0497
λ_3	−0.0300	−0.0418	−0.0436		−0.0250
ω_1	0.2600	0.1094	0.1621		0.2500
ω_2	0.5500	0.5505	0.4950		0.4999
ω_3	0.9500	0.9582	0.9874		0.9998
δ_1	0.0100	0.0060	0.5068		0.0010
δ_2	0.0100	−0.1234	−0.3482		0.0001
δ_3	0.0100	0.1225	0.1918		−0.0001
η	10^6	0.5120	0.0041		10^{-6}

Clearly, the Levenberg-Marquardt algorithm converges to the correct estimates for this case, where the classical Gaussian differential correction fails.

1.6.4 Projections in Least Squares

In this section we give a geometrical interpretation of least squares. The term "normal" in Normal Equations implies that there is a geometrical interpretation to least squares. In fact, we will show that the least squares solution for $\hat{\mathbf{x}}$ provides the *orthogonal projection*, hence normal, of $\tilde{\mathbf{y}}$ onto a *subspace* which is spanned by columns of the matrix H. Let us illustrate this concept using the simple scalar case of least squares. Say we wish to determine $\hat{x}$ which minimizes

$$J = \frac{1}{2}(\tilde{\mathbf{y}} - \hat{x}\mathbf{h})^T(\tilde{\mathbf{y}} - \hat{x}\mathbf{h}) \tag{1.157}$$

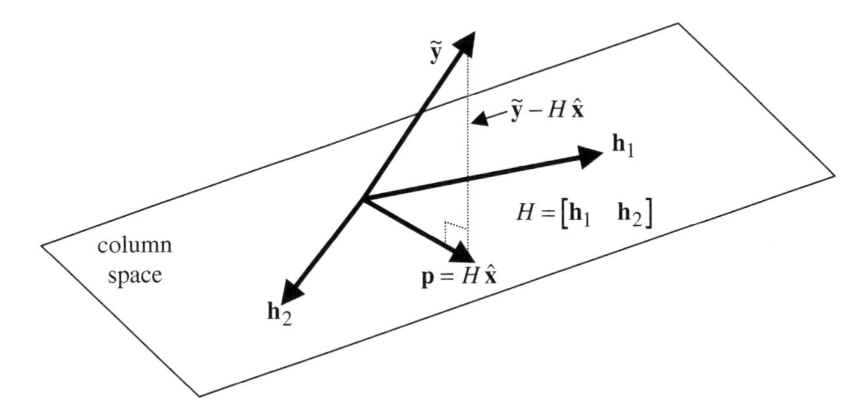

Figure 1.13: Projection onto the Column Space of a 3×2 Matrix

where $\mathbf{h}$ is the basis function vector. The necessary conditions yield the following simple solution:

$$\hat{x} = \frac{\mathbf{h}^T \tilde{\mathbf{y}}}{\mathbf{h}^T \mathbf{h}} \tag{1.158}$$

The residual error is given by

$$\mathbf{e} = (\tilde{\mathbf{y}} - \hat{x}\mathbf{h}) \tag{1.159}$$

Now, left multiply the residual error by $\mathbf{h}^T$ in eqn. (1.159) and substitute eqn. (1.158) into eqn. (1.159). This yields

$$\begin{aligned}
\mathbf{h}^T \mathbf{e} &= \mathbf{h}^T (\tilde{\mathbf{y}} - \hat{x}\mathbf{h}) \\
&= \mathbf{h}^T (\tilde{\mathbf{y}} - \frac{\mathbf{h}^T \tilde{\mathbf{y}}}{\mathbf{h}^T \mathbf{h}}\mathbf{h}) \\
&= \mathbf{h}^T \tilde{\mathbf{y}} - \frac{\mathbf{h}^T \tilde{\mathbf{y}}}{\mathbf{h}^T \mathbf{h}}\mathbf{h}^T \mathbf{h} \\
&= 0
\end{aligned} \tag{1.160}$$

This shows that the angle between $\mathbf{h}$ and $\mathbf{e}$ is 90 degrees, so that the line connecting $\tilde{\mathbf{y}}$ to $\hat{x}\mathbf{h}$ must be *perpendicular* to $\mathbf{h}$.

The aforementioned scalar case is easily expanded to the multi-dimensional case where $\tilde{\mathbf{y}}$ is *projected* onto a subspace rather than just onto a line. In this case, the vector $\mathbf{p} \equiv H\hat{\mathbf{x}}$ must be the projection of $\tilde{\mathbf{y}}$ onto the column space of H, and the residual error $\mathbf{e}$ must be perpendicular to that space.[22] This is illustrated for a simple 3×2 case in Figure 1.13. In other words, the residual error must be perpendicular to every column ($\mathbf{h}_i$) of H, so that

$$\begin{aligned}
\mathbf{h}_1^T (\tilde{\mathbf{y}} - H\hat{x}) &= 0 \\
\mathbf{h}_2^T (\tilde{\mathbf{y}} - H\hat{x}) &= 0 \\
&\vdots \\
\mathbf{h}_n^T (\tilde{\mathbf{y}} - H\hat{x}) &= 0
\end{aligned} \tag{1.161}$$

or

$$H^T(\tilde{\mathbf{y}} - H\hat{x}) = 0 \tag{1.162}$$

which gives the normal equations again. The projection of $\tilde{\mathbf{y}}$ onto the column space is therefore given by

$$\mathbf{p} = H(H^T H)^{-1} H^T \tilde{\mathbf{y}} \tag{1.163}$$

Geometrically, this means that the closest point to $\tilde{\mathbf{y}}$ on the column space of H is $\mathbf{p}$. Equation (1.163) expresses in matrix terms the construction of a perpendicular line from $\tilde{\mathbf{y}}$ to the column space of H.[22] The *projection matrix* is given by

$$\mathcal{P} = H(H^T H)^{-1} H^T \tag{1.164}$$

The projection matrix $\mathcal{P}$ can readily be seen to be symmetric. More importantly, the projection matrix has another property known of as *idempotence*, which states

$$\mathcal{P}\tilde{\mathbf{y}} = [\mathcal{P}\mathcal{P}\ldots\mathcal{P}]\tilde{\mathbf{y}} \tag{1.165}$$

The idempotence property shows that once a vector has been obtained as the projection onto a subspace using $\mathcal{P}$, it can never be modified by any further application of $\mathcal{P}$.[3] The corresponding prediction error, $\mathbf{e}_{min}$, once the solution for $\hat{\mathbf{x}}$ has been found, is given by

$$\mathbf{e}_{min} = (I - \mathcal{P})\tilde{\mathbf{y}} \tag{1.166}$$

where the matrix $(I - \mathcal{P})$ is the *orthogonal complement* of $\mathcal{P}$. It is easy to show that $(I - \mathcal{P})$ must also be a projection matrix, since it projects $\tilde{\mathbf{y}}$ onto the orthogonal complement.

1.7 Summary

With some reluctance, the curve fitting example of §1.1 was presented prior to discussion of the methods of §1.2 necessary to carry out the calculations. On several subsequent occasions herein, theoretical development of *methods* follow typical *results*, to provide motivation and to allow some *a priori* evaluation by the reader of the role played by the methodology under development.

The results developed in §1.2 are among the most important in estimation theory. Indeed, the bulk of estimation theory could be viewed as extensions, modifications, or generalizations of these basic results that address a wider variety of mathematical models and measurement strategies. We shall see, however, that the results of §1.2 can be placed upon a more rigorous foundation and several important new insights gained through study of the developments of Chapter 2 and Appendices A and B.

The sequential estimation results in §1.3 are the simplest version of a class of procedures known as *Kalman Filter* algorithms. Indeed, with the advancement of computer technology in today's age, sequential algorithms have found their way into mainstream applications in a wide variety of areas. Numerous investigators have extended/applied these algorithms since the most fundamental results were published by Kalman.[6] The constrained least squares solution[5] in eqn. (1.42) is closely related to the sequential estimation solution in eqn. (1.78), and can in fact be obtained from it by limiting arguments (allowing the weight of the constraint "observation" equations to approach infinity). A substantial portion of the present text deals with sequential estimation methodology and applications thereof.

The differential correction procedures documented in §1.4 are most fundamental whenever estimation methods must be applied to a nonlinear problem. It is interesting to note that the original estimation problem motivating Gauss (i.e., determination of the planetary orbits from telescope/sextant observations) was nonlinear, and his methods (essentially §1.4) have survived as a standard operating procedure to this day. Other *mathematical programming* methods (Appendix C), such as the gradient method, can also be employed in minimizing the sum square residuals.

A summary of the key formulas presented in this chapter is given below.

- Linear Least Squares

$$\begin{aligned}\tilde{\mathbf{y}} &= H\mathbf{x}+\mathbf{v}\\ \hat{\mathbf{x}} &= (H^T H)^{-1} H^T \tilde{\mathbf{y}}\end{aligned}$$

- Weighted Least Squares

$$\begin{aligned}\tilde{\mathbf{y}} &= H\mathbf{x}+\mathbf{v}\\ \hat{\mathbf{x}} &= (H^T W H)^{-1} H^T W \tilde{\mathbf{y}}\end{aligned}$$

- Constrained Least Squares

$$\begin{aligned}\tilde{\mathbf{y}}_1 &= H_1\mathbf{x}+\mathbf{v}\\ \tilde{\mathbf{y}}_2 &= H_2\hat{\mathbf{x}}\\ \hat{\mathbf{x}} &= \bar{\mathbf{x}}+K(\tilde{\mathbf{y}}_2 - H_2\bar{\mathbf{x}})\\ K &= (H_1^T W_1 H_1)^{-1} H_2^T \left[H_2(H_1^T W_1 H_1)^{-1} H_2^T\right]^{-1}\\ \bar{\mathbf{x}} &= (H_1^T W_1 H_1)^{-1} H_1^T W_1 \tilde{\mathbf{y}}_1\end{aligned}$$

- Sequential Least Squares

$$\begin{aligned}\hat{\mathbf{x}}_{k+1} &= \hat{\mathbf{x}}_k + K_{k+1}(\tilde{\mathbf{y}}_{k+1} - H_{k+1}\hat{\mathbf{x}}_k)\\ K_{k+1} &= P_k H_{k+1}^T \left[H_{k+1} P_k H_{k+1}^T + W_{k+1}^{-1}\right]^{-1}\\ P_{k+1} &= \left[I - K_{k+1} H_{k+1}\right] P_k\end{aligned}$$

- Nonlinear Least Squares (see Figure 1.9)

$$\tilde{\mathbf{y}} = \mathbf{f}(\mathbf{x}) + \mathbf{v}$$
$$H \equiv \left.\frac{\partial \mathbf{f}}{\partial \mathbf{x}}\right|_{\mathbf{x}_c}$$
$$\Delta\mathbf{y} \equiv \tilde{\mathbf{y}} - \mathbf{f}(\mathbf{x}_c)$$
$$\Delta\mathbf{x} = (H^T W H)^{-1} H^T W \Delta\mathbf{y}$$
$$\hat{\mathbf{x}} = \mathbf{x}_c + \Delta\mathbf{x}$$

- QR Decomposition

$$H = QR$$
$$R\hat{\mathbf{x}} = Q^T \tilde{\mathbf{y}}$$

- Singular Value Decomposition

$$H = USV^T$$
$$\hat{\mathbf{x}} = VS^{-1}U^T \tilde{\mathbf{y}}$$

- Kronecker Factorization

$$\hat{\mathbf{c}} = \left\{[(H_1^T H_1)^{-1} H_1^T] \otimes \cdots \otimes [(H_N^T H_N)^{-1} H_N^T]\right\} \tilde{\mathbf{z}}$$

- The Levenberg-Marquardt Algorithm

$$\Delta\mathbf{x} = (H^T W H + \eta\mathcal{H})^{-1} H^T W \Delta\mathbf{y}_c$$
$$\mathcal{H} = \text{diag}[H^T W H]$$

- Projection Matrix and Idempotence

$$\mathcal{P} = H(H^T H)^{-1} H^T$$
$$\mathcal{P}\tilde{\mathbf{y}} = [\mathcal{P}\mathcal{P} \ldots \mathcal{P}]\tilde{\mathbf{y}}$$

Exercises

1.1 Prove that $H^T H$ is a symmetric matrix.

1.2 Prove that if W is a symmetric positive definite matrix, then $H^T W H$ will always be positive semi-definite (hint: any positive definite matrix W can be factored into $W = R^T R$, where R is an upper triangular matrix, known as the Cholesky Decomposition).

1.3 Following the notation of §1.2 consider the m dimensional observation equation

$$\tilde{\mathbf{y}} = Hx + \mathbf{v}$$
$$\tilde{\mathbf{y}} = H\hat{x} + \mathbf{e}$$

with

$$H = \begin{bmatrix} 1 & 1 & \dots & 1 \end{bmatrix}^T$$

These observation equations hold for the simplest situation in which an unknown scalar parameter x is *directly* measured m times (assume that the measurements errors have zero mean and known, equal variances). From the normal equations (1.26), establish the well known truth that the optimum least squares estimate $\hat{x}$ of x is the sample mean

$$\hat{x} = \frac{1}{m}\sum_{i=1}^{m} \tilde{y}_i$$

1.4 Suppose that $\mathbf{v}$ in exercise 1.3 is a constant vector (i.e., a *bias error*). Evaluate the loss function (1.21) in terms of v_i only and discuss how the value of the loss function changes with a bias error in the measurements instead of a zero mean assumption.

1.5 Show that the mean of the linear least squares residuals, given by eqn. (1.1), vanishes identically if *one* of the linearly independent basis functions is a constant.

1.6 In this problem we will consider a simple linear *regression* model. The vertical deviation of a point (z_j, y_j) from the line $y = a + bz$ is $e_j = y_j - (a + bz_j)$. Determine closed-form least squares estimates of a and b given measurements sets for z_j and y_j.

1.7 Using the simple model

$$y = x_1 + x_2 \sin 10t + x_3 e^{2t^2}$$

with $x_1 = x_2 = x_3 = 1.0$, generate four sets of "synthetic data" at the instants $t = 0, 0.1, 0.2, 0.3, \dots, 1.0$ by truncating each y value after $6, 4, 2,$ and 1 significant figures, respectively, to simulate (crudely) measurement errors. Use the normal equations (1.26) to process the measurements and derive $\hat{x}_i$ estimates for each of the four cases. Compare the estimates with the true values $(1, 1, 1)$ in each case.

1.8 Use the sequential estimation algorithm (1.78) to (1.80) to process the first three measurements of exercise 1.7 as a single measurement subset and then consider the remaining measurements to become available one at a time, for each of the four synthetic data sets of exercise 1.7.

1.9 Consider the following partitioned matrix (assume that $|A_{11}| \neq 0$ and $|A_{22}| \neq 0$):

$$A = \begin{bmatrix} A_{11} & A_{12} \\ A_{21} & A_{22} \end{bmatrix}$$

Prove that the following matrices are all valid inverses:

$$A^{-1} = \begin{bmatrix} A_{11}^{-1} + A_{11}^{-1} A_{12} B_{22}^{-1} A_{21} A_{11}^{-1} & -A_{11}^{-1} A_{12} B_{22}^{-1} \\ -B_{22}^{-1} A_{21} A_{11}^{-1} & B_{22}^{-1} \end{bmatrix}$$

$$A^{-1} = \begin{bmatrix} B_{11}^{-1} & -B_{11}^{-1} A_{12} A_{22}^{-1} \\ -A_{22}^{-1} A_{21} B_{11}^{-1} & A_{22}^{-1} + A_{22}^{-1} A_{21} B_{11}^{-1} A_{12} A_{22}^{-1} \end{bmatrix}$$

$$A^{-1} = \begin{bmatrix} B_{11}^{-1} & -A_{11}^{-1} A_{12} B_{22}^{-1} \\ -A_{22}^{-1} A_{21} B_{11}^{-1} & B_{22}^{-1} \end{bmatrix}$$

where B_{ii} is the *Schur complement* of A_{ii}, given by

$$B_{11} = A_{11} - A_{12} A_{22}^{-1} A_{21}$$
$$B_{22} = A_{22} - A_{21} A_{11}^{-1} A_{12}$$

Also, prove the matrix inversion lemma from these matrix inverses.

1.10 Create 101 synthetic measurements $\tilde{y}$ at 0.1 second intervals of the following:

$$\tilde{y}_j = a \sin t_j - b \cos t_j + v_j$$

where $a = b = 1$, and v is a zero-mean Gaussian noise process with standard deviation given by 0.01. Determine the unweighted least squares estimates for a and b. Using the same measurements, find a value of $\tilde{\mathbf{y}}$ that is near zero (near time $\pi/4$), and set that "measurement" value to 1. Compute the unweighted least squares solution, and compare it to the original solution. Then, use weighted least squares to "deweight" the measurement.

1.11 In the derivation of the weighted least squares estimator of §1.2.2, the weight matrix W is assumed to be symmetric. How does the solution change if W is no longer symmetric (but still positive definite)?

1.12 Using the method of Lagrange multipliers, find all solutions $\mathbf{x}$ of the first necessary conditions for extremals of the function

$$J(\mathbf{x}) = (\mathbf{x} - \mathbf{a})^T W (\mathbf{x} - \mathbf{a})$$
$$\text{subject to } \mathbf{b}^T \mathbf{x} = c$$

where $\mathbf{a}$ and $\mathbf{b}$ are constant vectors, c is a scalar, and W is a symmetric, positive definite matrix.

1.13 Consider the following dynamic model:

$$y_k = \sum_{i=1}^{n} \phi_i y_{k-i} + \sum_{i=1}^{p} \gamma_i u_{k-i}$$

where u_i is a known input. This ARMA (AutoRegressive Moving Average) model extends the simple scalar model given in example 1.1. Given measurements of y_i and the known inputs u_i recast the above model into least squares form and determine estimates for ϕ_i and γ_i.

1.14 Program a sequential estimation algorithm to determine in real-time the parameters of the ARMA model shown in exercise 1.13. Develop some synthetic data with various system models, and verify your algorithm.

1.15 One of the most important mathematical equations in history is given by Kepler's equation, which provides powerful geometrical insights into orbiting bodies. This equation is given by

$$M = E - e \sin E$$

where M and E are known as the mean anomaly and eccentric anomaly, respectively, both given in radians, and e is the eccentricity of the orbit. For elliptical orbits $0 < e < 1$. To date, no one has found a closed-form solution for E in terms of M and e. Pick various values for M and e and use nonlinear least squares, which reduces to Newton's method for this equation, to determine E.

1.16 Consider the following dynamic model:

$$\begin{bmatrix} z_1 \\ z_2 \end{bmatrix}_{k+1} = \begin{bmatrix} 1 & 0 \\ 0 & 1 \end{bmatrix} \begin{bmatrix} z_1 \\ z_2 \end{bmatrix}_k$$

and measurement model

$$\tilde{y}_k = \begin{bmatrix} \sin(\omega_0 \, \Delta t \, k) & \cos(\omega_0 \, \Delta t \, k) \end{bmatrix} \begin{bmatrix} z_1 \\ z_2 \end{bmatrix}_k + v_k$$

where ω_0 is the harmonic frequency, and Δt is the sampling interval. Create synthetic measurements of the above process with $\omega_0 = 0.4\pi$ rad/sec and $\Delta t = 0.1$ seconds. Also, create different synthetic measurement sets using various values for the standard deviation of v in the measurements errors. Use nonlinear least squares to find an estimate for ω_0 for each synthetic measurement set.

1.17 A measurement process used in three-axis magnetometers for low-Earth attitude determination involves the following measurement model:

$$\mathbf{b}_j = A_j \mathbf{r}_j + \mathbf{c} + \epsilon_j$$

where $\mathbf{b}_j$ is the measurement of the magnetic field (more exactly, magnetic induction) by the magnetometer at time t_j, $\mathbf{r}_j$ is the corresponding value of the geomagnetic field with respect to some reference coordinate system, A_j is the orthogonal attitude matrix (see §3.7.1), $\mathbf{c}$ is the magnetometer bias, and ϵ_j is the measurement error. We can eliminate the dependence on the attitude by transposing terms and computing the square, and can define an effective measurement by

$$\tilde{y}_j = \mathbf{b}_j^T \mathbf{b}_j - \mathbf{r}_j^T \mathbf{r}_j$$

which can be rewritten to form the following measurement model:

$$\tilde{y}_j = 2\mathbf{b}_j^T \mathbf{c} - \mathbf{c}^T \mathbf{c} + \upsilon_j$$

where υ_j is the effective measurement error, whose closed-form expression is not required for this problem. For this exercise assume that

$$A\mathbf{r} = \begin{bmatrix} 10\sin(0.001t) \\ 5\sin(0.002t) \\ 10\cos(0.001t) \end{bmatrix}, \quad \mathbf{c} = \begin{bmatrix} 0.5 \\ 0.3 \\ 0.6 \end{bmatrix}$$

Also, assume that ϵ is given by a zero-mean Gaussian-noise process with standard deviation given by 0.05 in each component. Using the above values create 1001 synthetic measurements of $\mathbf{b}$ and $\tilde{y}$ at 5-second intervals. The estimated output is computed from

$$\hat{y}_j = 2\mathbf{b}_j^T \hat{\mathbf{c}} - \hat{\mathbf{c}}^T \hat{\mathbf{c}}$$

where $\hat{\mathbf{c}}$ is the estimated solution from the nonlinear least square iterations. Use nonlinear least squares to determine $\hat{\mathbf{c}}$ for a starting value of $\mathbf{x}_c = [0\ 0\ 0]^T$. Also, try various starting values to check convergence. Note: $\mathbf{r}^T\mathbf{r} = \mathbf{r}^T A^T A\mathbf{r}$, since $A^T A = I$.

1.18 An approximate linear solution to exercise 1.17 is possible. The original loss function is quartic in $\hat{\mathbf{c}}$. But this can be approximated by a quadratic loss function using a process known as *centering*.[23] The linearized solution proceeds as follows. First, compute the following averaged values:

$$\bar{y} = \frac{1}{m}\sum_{j=1}^{m} \tilde{y}_j$$

$$\bar{\mathbf{b}} = \frac{1}{m}\sum_{j=1}^{m} \mathbf{b}_j$$

where m is the total number of measurements, which is equal to 1001 from exercise 1.17. Next, define the following variables:

$$\breve{y}_j = \tilde{y}_j - \bar{y}$$

$$\breve{\mathbf{b}}_j = \mathbf{b}_j - \bar{\mathbf{b}}$$

The centered estimate now minimizes the following loss function:

$$\bar{J}(\hat{\mathbf{c}}) = \frac{1}{2}\sum_{j=1}^{m}\left(\breve{y}_j - 2\breve{\mathbf{b}}_j^T \hat{\mathbf{c}}\right)^2$$

Minimizing this function yields

$$\hat{\mathbf{c}} = P\sum_{j=1}^{m} 2\breve{y}_j \breve{\mathbf{b}}_j$$

where

$$P \equiv \left[\sum_{i=1}^{m} 4\breve{\mathbf{b}}_j \breve{\mathbf{b}}_j^T \right]^{-1}$$

Using the parameters described in exercise 1.17, compare the linear solution described here to the solution obtained by nonlinear least squares. Furthermore, find solutions for $\hat{\mathbf{c}}$ using both approaches with the following trajectory for $A\mathbf{r}$:

$$A\mathbf{r} = \begin{bmatrix} 10\sin(0.001t) \\ 5 \\ 10\cos(0.001t) \end{bmatrix}$$

Discuss the performance of the linear solution using this assumed trajectory for $A\mathbf{r}$.

1.19 ♣ Convert the linear batch solution shown in exercise 1.18 to a sequential form (hint: use the matrix inversion lemma in eqn. (1.69) to find a sequential form for P). Perform a simulation using the parameters in exercise 1.17 to test your algorithm.

1.20 Consider the following measurement model:

$$\tilde{y}_j = B \exp\left[-\frac{(1-at)^2}{2\sigma^2} \right] + v_j$$

with $a = 1$, $B = 2$, $\sigma = 3$, and let v be represented by a zero-mean Gaussian noise process with standard deviation given by 0.001. Create 101 synthetic measurements at 0.1-second intervals. Use the change of variables in Table 1.1 to determine *linear* least squares estimates for a, B, and σ.

1.21 Analytically expand $y = |\sin t|$ in a Fourier series. Compute the Fourier coefficients using least squares with the basis functions in eqn. (1.104) for $n = 10$ and compare the numerical solutions to the analytically derived solutions.

1.22 Consider the following matrix commonly used to describe attitude motion:

$$A = \begin{bmatrix} \cos\theta & \sin\theta & 0 \\ -\sin\theta & \cos\theta & 0 \\ 0 & 0 & 1 \end{bmatrix}$$

Prove that the columns of the A are orthonormal.

1.23 Show that the vector $(\mathbf{x}-\mathbf{y})$ is orthogonal to the vector $(\mathbf{x}+\mathbf{y})$ if and only if $\|\mathbf{x}\| = \|\mathbf{y}\|$.

1.24 Prove that the Kronecker product in eqn. (1.144) is indeed equivalent to the matrix H given in eqn. (1.143).

1.25 Reproduce the results of example 1.9. Try some higher-order polynomials to further show the importance of the solution using the Kronecker factorization.

1.26 Find starting values in exercise 1.17 that cause the standard nonlinear least squares problem to diverge using the following trajectory for $A\mathbf{r}$:

$$A\mathbf{r} = \begin{bmatrix} 10\sin(0.001t) \\ 5 \\ 10\cos(0.001t) \end{bmatrix}$$

For example, try starting values of $\mathbf{x}_c = \begin{bmatrix} 10 & 10 & 10 \end{bmatrix}^T$. Program the Levenberg-Marquardt method, and check convergence for this starting condition as well as various other starting conditions. Also, check the performance of the Levenberg-Marquardt method for various values of η and f (start with $\eta = 10||H^T H||$ and $f = 5$).

1.27 Consider the projection onto the θ-direction in the $x - y$ plane. Find the projection matrix for the line through $\mathbf{h} = \begin{bmatrix} \cos\theta & \sin\theta \end{bmatrix}^T$. Is this matrix invertible? Explain.

1.28 Prove that $(I - \mathcal{P})$, with $\mathcal{P}$ given by eqn. (1.164), has the idempotence property.

References

[1] Devore, J.L., *Probability and Statistics for Engineering and Sciences*, Duxbury Press, Pacific Grove, CA, 1995.

[2] Gauss, K.F., *Theory of the Motion of the Heavenly Bodies Moving about the Sun in Conic Sections, A Translation of Theoria Motus*, Dover Publications, New York, NY, 1963.

[3] Strobach, P., *Linear Prediction Theory*, Springer-Verlag, Berlin, 1990.

[4] Juang, J.N. and Pappa, R.S., "An Eigensystem Realization Algorithm for Modal Parameter Identification and Model Reduction," *Journal of Guidance, Control, and Dynamics*, Vol. 8, No. 5, Sept.-Oct. 1985, pp. 620–627.

[5] Junkins, J.L., "On the Optimization and Estimation of Powered Rocket Trajectories Using Parametric Differential Correction Processes," Tech. Rep. SM G1793, McDonnell Douglas Astronautics Co., 1969.

[6] Kalman, R.E. and Bucy, R.S., "New Results in Linear Filtering and Prediction Theory," *Journal of Basic Engineering*, March 1961, pp. 95–108.

[7] Golub, G.H. and Van Loan, C.F., *Matrix Computations*, The Johns Hopkins University Press, Baltimore, MD, 3rd ed., 1996.

[8] Saaty, T.L., *Modern Nonlinear Equations*, Dover Publications, New York, NY, 1981.

[9] Mirsky, L., *An Introduction to Linear Algebra*, Dover Publications, New York, NY, 1990.

[10] Sveshnikov, A.A., *Problems in Probability Theory, Mathematical Statistics and Theory of Random Functions*, Dover Publications, New York, NY, 1978.

[11] Chihara, T.S., *An Introduction to Orthogonal Polynomials*, Gordan and Breach Science Publishers, New York, NY, 1978.

[12] Datta, K.B. and Mohan, B.M., *Orthogonal Functions in Systems and Control*, World Scientific, Singapore, 1995.

[13] Tolstov, G.P., *Fourier Series*, Dover Publications, New York, NY, 1972.

[14] Gasquet, C. and Witomski, P., *Fourier Analysis and Applications: Filtering, Numerical Computations, Wavelets*, Springer-Verlag, New York, NY, 1978.

[15] Abramowitz, M. and Stengun, I.A., *Handbook of Mathematical Functions with Formulas, Graphs and Mathematical Tables*, Applied Mathematics Series - 55, National Bureau of Standards, Washington, D.C., 1964.

[16] Ledermann, W., *Handbook of Applicable Mathematics: Analysis*, Vol. 4, John Wiley & Sons, New York, NY, 1982.

[17] Horn, R.A. and Johnson, C.R., *Matrix Analysis*, Cambridge University Press, Cambridge, MA, 1985.

[18] Stewart, G.W., *Introduction to Matrix Computations*, Academic Press, New York, NY, 1973.

[19] Snay, R.A., "Applicability of Array Algebra," *Reviews of Geophysics and Space Physics*, Vol. 16, No. 3, Aug. 1978, pp. 459–464.

[20] Marquardt, D.W., "An Algorithm for Least-Squares Estimation of Nonlinear Parameters," *Journal of the Society for Industrial and Applied Mathematics*, Vol. 11, No. 2, June 1963, pp. 431–441.

[21] Levenberg, K., "A Method for the Solution of Certain Nonlinear Problems in Least Squares," *Quarterly of Applied Mathematics*, Vol. 2, 1944, pp. 164–168.

[22] Strang, G., *Linear Algebra and its Applications*, Saunders College Publishing, Fort Worth, TX, 1988.

[23] Alonso, R. and Shuster, M.D., "A New Algorithm for Attitude-Independent Magnetometer Calibration," *Proceedings of the Flight Mechanics/Estimation Theory Symposium*, NASA-Goddard Space Flight Center, Greenbelt, MD, May 1994, pp. 513–527.

2

Probability Concepts in Least Squares

The excitement that a gambler feels when making a bet is equal to the amount he might win times the probability of winning it. Pascal, Blaise

THE intuitively reasonable *principle of least squares* was put forth in §1.2 and employed as the starting point for all developments of Chapter 1. In the present chapter, several alternative paths are followed to essentially the same mathematical conclusions as Chapter 1. The primary function of the present chapter is to place the results of Chapter 1 upon a more rigorous (or at least, a better understood) foundation. A number of new and computationally most useful extensions of the estimation results of Chapter 1 come from the developments shown herein. In particular, minimal variance estimation and maximum likelihood estimation will be explored, and a connection to the least squares problem will be shown. Using these estimation techniques, the elusive weight matrix will be rigorously identified as the inverse of the measurement-error covariance matrix, and some most important *nonuniqueness properties* developed in §2.5. Methods for rigorously accounting for *a priori* parameter estimates and their uncertainty will also be developed. Finally, many other useful concepts will be explored, including: unbiased estimates and the Cramér-Rao inequality; other advanced topics such as Bayesian estimation, analysis of covariance errors, and ridge estimation are introduced as well. These concepts are useful for the analysis of least squares estimation by incorporating probabilistic approaches.

Familiarity with basic concepts in probability is necessary for comprehension of the material in the present chapter. Should the reader anticipate or encounter difficulty in the following developments, Appendix B provides an adequate review of the concepts needed herein.

2.1 Minimum Variance Estimation

Here we introduce one of the most important and useful concepts in estimation. Minimum variance estimation can give the "best way" (in a probabilistic sense) to find the optimal estimates. First, a minimum variance estimator is derived without *a priori* estimates. Then these results are extended to the case where *a priori* estimates are given.

2.1.1 Estimation without *a priori* State Estimates

As in Chapter 1, we assume a linear observation model

$$\underset{(m\times 1)}{\tilde{\mathbf{y}}} = \underset{(m\times n)}{H}\,\underset{(n\times 1)}{\mathbf{x}} + \underset{(m\times 1)}{\mathbf{v}} \tag{2.1}$$

We desire to estimate $\mathbf{x}$ as a linear combination of the measurements $\tilde{\mathbf{y}}$ as

$$\underset{(n\times 1)}{\hat{\mathbf{x}}} = \underset{(n\times m)}{M}\,\underset{(m\times 1)}{\tilde{\mathbf{y}}} + \underset{(n\times 1)}{\mathbf{n}} \tag{2.2}$$

An "optimum" choice of the quantities M and $\mathbf{n}$ is sought. The minimum variance definition of "optimum" M and $\mathbf{n}$ is that the variance of *all* n estimates, $\hat{x}_i$, from their respective "true" values is minimized:*

$$J_i = \frac{1}{2}E\left\{\left(\hat{x}_i - x_i\right)^2\right\}, \quad i = 1, 2, \ldots, n \tag{2.3}$$

This clearly requires n minimizations depending upon the same M and $\mathbf{n}$; it may not be clear at this point that the problem is well-defined and whether or not M and $\mathbf{n}$ exist (or can be found if they do exist) to accomplish these n minimizations.

If the linear model (2.1) is strictly valid, then, for the special case of perfect measurements $\mathbf{v} = 0$ the model (2.1) should be exactly satisfied by the perfect measurements $\mathbf{y}$ and the true state $\mathbf{x}$ as

$$\tilde{\mathbf{y}} \equiv \mathbf{y} = H\mathbf{x} \tag{2.4}$$

An obvious requirement upon the desired estimator (2.2) is that perfect measurements should result (if a solution is possible) when $\hat{\mathbf{x}} = \mathbf{x} =$ true state. Thus, this requirement can be written by substituting $\hat{\mathbf{x}} = \mathbf{x}$ and $\tilde{\mathbf{y}} = H\mathbf{x}$ into eqn. (2.2) as

$$\mathbf{x} = MH\mathbf{x} + \mathbf{n} \tag{2.5}$$

We conclude that M and $\mathbf{n}$ satisfy the constraints

$$\mathbf{n} = \mathbf{0} \tag{2.6}$$

and

$$MH = I \tag{2.7a}$$

$$H^T M^T = I \tag{2.7b}$$

Equation (2.6) is certainly useful information! The desired estimator then has the form

$$\hat{\mathbf{x}} = M\tilde{\mathbf{y}} \tag{2.8}$$

We are now concerned with determining the optimum choice of M which accomplishes the n minimizations of (2.3), subject to the constraint (2.7).

---

*$E\{\ \}$ denotes "*expected value*" of $\{\ \}$, see Appendix B.

Subsequent manipulations will be greatly facilitated by partitioning the various matrices as follows: The unknown M-matrix is partitioned by rows as

$$M = \begin{bmatrix} M_1 \\ M_2 \\ \vdots \\ M_n \end{bmatrix}, \quad M_i \equiv \{M_{i1}\ M_{i2} \cdots M_{im}\} \tag{2.9}$$

or

$$M^T = \begin{bmatrix} M_1^T\ M_2^T \ \cdots\ M_n^T \end{bmatrix} \tag{2.10}$$

The identity matrix can be partitioned by rows and columns as

$$I = \begin{bmatrix} I_1^r \\ I_2^r \\ \vdots \\ I_n^r \end{bmatrix} = \begin{bmatrix} I_1^c\ I_2^c \ \cdots\ I_n^c \end{bmatrix}, \quad \text{note } I_i^r = (I_i^c)^T \tag{2.11}$$

The constraint in eqn. (2.7) can now be written as

$$H^T M_i^T = I_i^c, \quad i = 1, 2, \ldots, n \tag{2.12a}$$

$$M_i H = I_i^r, \quad i = 1, 2, \ldots, n \tag{2.12b}$$

and the ith element of $\hat{\mathbf{x}}$ from eqn. (2.8) can be written as

$$\hat{x}_i = M_i \tilde{\mathbf{y}}, \quad i = 1, 2, \ldots, n \tag{2.13}$$

A glance at eqn. (2.13) reveals that $\hat{x}_i$ depends *only* upon the elements of M contained in the ith row. A similar statement holds for the constraint equations (2.12); the elements of the ith row are independently constrained. This "uncoupled" nature of eqns. (2.12) and (2.13) is the key feature which allows one to carry out the n "separate" minimizations of eqn. (2.3).

The ith variance (2.3) to be minimized, upon substituting eqn. (2.13), can be written as

$$J_i = \frac{1}{2} E\left\{\left(M_i \tilde{\mathbf{y}} - x_i\right)^2\right\}, \quad i = 1, 2, \ldots, n \tag{2.14}$$

Substituting the observation from eqn. (2.1) into eqn. (2.14) yields

$$J_i = \frac{1}{2} E\left\{(M_i H \mathbf{x} + M_i \mathbf{v} - x_i)^2\right\}, \quad i = 1, 2, \ldots, n \tag{2.15}$$

Incorporating the constraint equations from eqn. (2.12) into eqn. (2.15) yields

$$J_i = \frac{1}{2} E\left\{\left(I_i^r \mathbf{x} + M_i \mathbf{v} - x_i\right)^2\right\}, \quad i = 1, 2, \ldots, n \tag{2.16}$$

But $I_i^r \mathbf{x} = x_i$, so that eqn. (2.16) reduces to

$$J_i = \frac{1}{2} E\left\{(M_i \mathbf{v})^2\right\}, \quad i = 1, 2, \ldots, n \tag{2.17}$$

which can be rewritten as

$$J_i = \frac{1}{2} E\left\{M_i \left(\mathbf{v}\mathbf{v}^T\right) M_i^T\right\}, \quad i = 1, 2, \ldots, n \tag{2.18}$$

But the only random variable on the right-hand side of eqn. (2.18) is $\mathbf{v}$; introducing the covariance matrix of measurement errors (assuming that $\mathbf{v}$ has zero mean, i.e., $E\{\mathbf{v}\} = 0$),

$$\text{cov}\{\mathbf{v}\} \equiv R = E\left\{\mathbf{v}\mathbf{v}^T\right\} \tag{2.19}$$

then eqn. (2.18) reduces to

$$J_i = \frac{1}{2} M_i R M_i^T, \quad i = 1, 2, \ldots, n \tag{2.20}$$

The ith constrained minimization problem can now be stated as: Minimize each of equations (2.20) subject to the corresponding constraint in eqn. (2.12). Using the method of Lagrange multipliers (Appendix C), the ith augmented function is introduced as

$$J_i = \frac{1}{2} M_i R M_i^T + \boldsymbol{\lambda}_i^T \left(I_i^c - H^T M_i^T\right), \quad i = 1, 2, \ldots, n \tag{2.21}$$

where

$$\boldsymbol{\lambda}_i^T = \left\{\lambda_{1_i}, \lambda_{2_i}, \ldots, \lambda_{n_i}\right\} \tag{2.22}$$

are n vectors of Lagrange multipliers.

The necessary conditions for eqn. (2.21) to be minimized are then

$$\nabla_{M_i^T} J_i = R M_i^T - H \boldsymbol{\lambda}_i = \mathbf{0}, \quad i = 1, 2, \ldots, n \tag{2.23}$$

$$\nabla_{\boldsymbol{\lambda}_i} J_i = I_i^c - H^T M_i^T = \mathbf{0}, \text{ or } M_i H = I_i^r, \quad i = 1, 2, \ldots, n \tag{2.24}$$

From eqn. (2.23), we obtain

$$M_i = \boldsymbol{\lambda}_i^T H^T R^{-1}, \quad i = 1, 2, \ldots, n \tag{2.25}$$

Substituting eqn. (2.25) into the second equation of eqn. (2.24) yields

$$\boldsymbol{\lambda}_i^T = I_i^r \left(H^T R^{-1} H\right)^{-1} \tag{2.26}$$

Therefore, substituting eqn. (2.26) into eqn. (2.25), the n rows of M are given by

$$M_i = I_i^r \left(H^T R^{-1} H\right)^{-1} H^T R^{-1}, \quad i = 1, 2, \ldots, n \tag{2.27}$$

It then follows that

$$M = \left(H^T R^{-1} H\right)^{-1} H^T R^{-1} \tag{2.28}$$

and the desired estimator (2.8) then has the final form

$$\boxed{\hat{\mathbf{x}} = \left(H^T R^{-1} H\right)^{-1} H^T R^{-1} \tilde{\mathbf{y}}} \tag{2.29}$$

which is referred to as the *Gauss-Markov Theorem.*

The minimal variance estimator (2.29) is identical to the least squares estimator (1.30), *provided that the weight matrix is identified as the inverse of the observation error covariance.* Also, the "sequential least squares estimation" results of §1.3 are seen to embody a special case "sequential minimal variance estimation;" it is simply necessary to employ R^{-1} as W in the sequential least squares formulation, but we still require R^{-1} to have the block diagonal structure assumed for W.

The previous derivation can also be shown in compact form, but requires using vector matrix differentiation. This is shown for completeness. We will see in §2.2 that the condition $MH = I$ gives an *unbiased* estimate of $\mathbf{x}$. Let us first define the error covariance matrix for an unbiased estimator, given by (see Appendix B for details)

$$P = E\left\{(\hat{\mathbf{x}} - \mathbf{x})(\hat{\mathbf{x}} - \mathbf{x})^T\right\} \tag{2.30}$$

We wish to determine M that minimizes eqn. (2.30) in some way. We will choose to minimize the trace of P since this is a common choice and intuitively makes sense. Therefore, applying this choice with the constraint $MH = I$ gives the following loss function to be minimized:

$$J = \frac{1}{2}\mathrm{Tr}\left[E\left\{(\hat{\mathbf{x}} - \mathbf{x})(\hat{\mathbf{x}} - \mathbf{x})^T\right\}\right] + \mathrm{Tr}[\Lambda(I - MH)] \tag{2.31}$$

where Tr denotes the trace operator, and Λ is an $n \times n$ matrix of Lagrange multipliers. We can also make use of the parallel axis theorem[1†] for an unbiased estimate (i.e., $MH = I$), which states that

$$E\left\{(\hat{\mathbf{x}} - \mathbf{x})(\hat{\mathbf{x}} - \mathbf{x})^T\right\} = E\left\{\hat{\mathbf{x}}\hat{\mathbf{x}}^T\right\} - E\{\mathbf{x}\}\,E\{\mathbf{x}\}^T \tag{2.32}$$

Substituting eqn. (2.1) into eqn. (2.8) leads to

$$\begin{aligned} \hat{\mathbf{x}} &= M\tilde{\mathbf{y}} \\ &= MH\mathbf{x} + M\mathbf{v} \end{aligned} \tag{2.33}$$

Next, taking the expectation of both sides of eqn. (2.33) and using $E\{\mathbf{v}\} = 0$ gives (note, $\mathbf{x}$ on the right-hand side of eqn. (2.33) is treated as a deterministic quantity)

$$E\left\{\hat{\mathbf{x}}\right\} = MH\mathbf{x} \tag{2.34}$$

†This terminology is actually more commonly used in analytical dynamics to determine the moment of inertia about some arbitrary axis, related by a parallel axis through the center of mass.[2, 3] However, in statistics the form of the equation is identical when taking second moments about an arbitrary random variable.

In a similar fashion, using $E\{\mathbf{v}\mathbf{v}^T\} = R$ and also assuming that $E\{\mathbf{x}\mathbf{v}^T\} = 0$ and $E\{\mathbf{v}\mathbf{x}^T\} = 0$ (i.e., $\mathbf{x}$ and $\mathbf{v}$ are assumed to be *uncorrelated*), we obtain

$$E\left\{\hat{\mathbf{x}}\hat{\mathbf{x}}^T\right\} = MH\mathbf{x}\mathbf{x}^T H^T M^T + MRM^T \tag{2.35}$$

Therefore, the loss function in eqn. (2.31) becomes

$$J = \frac{1}{2}\text{Tr}(MRM^T) + \text{Tr}[\Lambda(I - MH)] \tag{2.36}$$

Next, we will make use of the following useful trace identities (see Appendix A):

$$\frac{\partial}{\partial A}\text{Tr}(BAC) = B^T C^T \tag{2.37a}$$

$$\frac{\partial}{\partial A}\text{Tr}(ABA^T) = A(B + B^T) \tag{2.37b}$$

Thus, we have the following necessary conditions:

$$\nabla_M J = MR - \Lambda^T H^T = 0 \tag{2.38}$$

$$\nabla_\Lambda J = I - MH = 0 \tag{2.39}$$

Solving eqn. (2.38) for M yields

$$M = \Lambda^T H^T R^{-1} \tag{2.40}$$

Substituting eqn. (2.40) into eqn. (2.39), and solving for Λ^T gives

$$\Lambda^T = (H^T R^{-1} H)^{-1} \tag{2.41}$$

Finally, substituting eqn. (2.41) into eqn. (2.40) yields

$$M = (H^T R^{-1} H)^{-1} H^T R^{-1} \tag{2.42}$$

This is identical to the solution given by eqn. (2.28).

2.1.2 Estimation with *a priori* State Estimates

The preceding results will now be extended to allow rigorous incorporation of *a priori* estimates, $\hat{\mathbf{x}}_a$, of the state and associated *a priori* error covariance matrix Q. We again assume the linear observation model

$$\tilde{\mathbf{y}} = H\mathbf{x} + \mathbf{v} \tag{2.43}$$

and associated (assumed known) measurement-error covariance matrix

$$R = E\left\{\mathbf{v}\mathbf{v}^T\right\} \tag{2.44}$$

Suppose that the variable $\mathbf{x}$ is also unknown (i.e., it is now treated as a *random variable*). The *a priori* state estimates are given as the sum of the true state $\mathbf{x}$ and the errors in the *a priori* estimates $\mathbf{w}$, so that

$$\hat{\mathbf{x}}_a = \mathbf{x} + \mathbf{w} \tag{2.45}$$

with associated (assumed known) *a priori* error covariance matrix

$$\text{cov}\{\mathbf{w}\} \equiv Q = E\left\{\mathbf{w}\mathbf{w}^T\right\} \tag{2.46}$$

where we assume that $\mathbf{w}$ has zero mean. We also assume that the measurement errors and *a priori* errors are uncorrelated so that $E\left\{\mathbf{w}\mathbf{v}^T\right\} = 0$.

We desire to estimate $\mathbf{x}$ as a linear combination of the measurements $\tilde{\mathbf{y}}$ and *a priori* state estimates $\hat{\mathbf{x}}_a$ as

$$\hat{\mathbf{x}} = M\tilde{\mathbf{y}} + N\hat{\mathbf{x}}_a + \mathbf{n} \tag{2.47}$$

An "optimum" choice of the M $(n \times m)$, N $(n \times n)$, and $\mathbf{n}$ $(n \times 1)$ matrices is desired. As before, we adopt the minimal variance definition of "optimum" to determine M, N, and $\mathbf{n}$ for which the variances of all n estimates, $\hat{x}_i$, from their respective true values, x_i, are minimized:

$$J_i = \frac{1}{2}E\left\{\left(\hat{x}_i - x_i\right)^2\right\}, \quad i = 1, 2, \ldots, n \tag{2.48}$$

If the linear model (2.43) is strictly valid, then for the special case of perfect measurements ($\mathbf{v} = \mathbf{0}$), the measurements $\mathbf{y}$ and the true state $\mathbf{x}$ should satisfy eqn. (2.43) exactly as

$$\mathbf{y} = H\mathbf{x} \tag{2.49}$$

If, in addition, the *a priori* state estimates are also perfect ($\hat{\mathbf{x}}_a = \mathbf{x}$, $\mathbf{w} = \mathbf{0}$), an obvious requirement upon the estimator in eqn. (2.47) is that it yields the true state as

$$\mathbf{x} = MH\mathbf{x} + N\mathbf{x} + \mathbf{n} \tag{2.50}$$

or

$$\mathbf{x} = (MH + N)\mathbf{x} + \mathbf{n} \tag{2.51}$$

Equation (2.51) indicates that M, N, and $\mathbf{n}$ must satisfy the constraints

$$\mathbf{n} = \mathbf{0} \tag{2.52}$$

and

$$MH + N = I \text{ or } H^T M^T + N^T = I \tag{2.53}$$

Because of eqn. (2.52), the desired estimator (2.47) has the form

$$\hat{\mathbf{x}} = M\tilde{\mathbf{y}} + N\hat{\mathbf{x}}_a \tag{2.54}$$

It is useful in subsequent developments to partition M, N, and I as

$$M = \begin{bmatrix} M_1 \\ M_2 \\ \vdots \\ M_n \end{bmatrix}, \quad M^T = \begin{bmatrix} M_1^T \ M_2^T \ \cdots \ M_n^T \end{bmatrix} \tag{2.55}$$

$$N = \begin{bmatrix} N_1 \\ N_2 \\ \vdots \\ N_n \end{bmatrix}, \quad N^T = \begin{bmatrix} N_1^T \ N_2^T \ \cdots \ N_n^T \end{bmatrix} \tag{2.56}$$

and

$$I = \begin{bmatrix} I_1^r \\ I_2^r \\ \vdots \\ I_n^r \end{bmatrix} = \begin{bmatrix} I_1^c \ I_2^c \ \cdots \ I_n^c \end{bmatrix}, \quad I_i^r = (I_i^c)^T \tag{2.57}$$

Using eqns. (2.55), (2.56), and (2.57), the constraint equation (2.53) can be written as n independent constraints as

$$H^T M_i^T + N_i^T = I_i^c, \quad i = 1, 2, \ldots, n \tag{2.58a}$$

$$M_i H + N_i = I_i^r, \quad i = 1, 2, \ldots, n \tag{2.58b}$$

The ith element of $\hat{\mathbf{x}}$, from eqn. (2.54), is

$$\hat{x}_i = M_i \tilde{\mathbf{y}} + N_i \hat{\mathbf{x}}_a, \quad i = 1, 2, \ldots, n \tag{2.59}$$

Note that both eqns. (2.58) and (2.59) depend *only* upon the elements of the ith row, M_i, of M and the ith row, N_i, of N. Thus the ith variance (2.48) to be minimized is a function of the same $n+m$ unknowns (the elements of M_i and N_i) as is the ith constraint, eqn. (2.58a) or eqn. (2.58b).

Substituting eqn. (2.59) into eqn. (2.48) yields

$$J_i = \frac{1}{2} E\left\{ \left(M_i \tilde{\mathbf{y}} + N_i \hat{\mathbf{x}}_a - x_i \right)^2 \right\}, \quad i = 1, 2, \ldots, n \tag{2.60}$$

Substituting eqns. (2.43) and (2.45) into eqn. (2.60) yields

$$J_i = \frac{1}{2} E\left\{ [(M_i H + N_i)\mathbf{x} + M_i \mathbf{v} + N_i \mathbf{w} - x_i]^2 \right\}, \quad i = 1, 2, \ldots, n \tag{2.61}$$

Making use of eqn. (2.58a), eqn. (2.61) becomes

$$J_i = \frac{1}{2} E\left\{ \left(I_i^r \mathbf{x} + M_i \mathbf{v} + N_i \mathbf{w} - x_i \right)^2 \right\}, \quad i = 1, 2, \ldots, n \tag{2.62}$$

Since $I_i^r \mathbf{x} = x_i$, eqn. (2.62) reduces to

$$J_i = \frac{1}{2} E\left\{(M_i \mathbf{v} + N_i \mathbf{w})^2\right\}, \quad i = 1, 2, \ldots, n \tag{2.63}$$

or

$$J_i = \frac{1}{2} E\left\{(M_i \mathbf{v})^2 + 2(M_i \mathbf{v})(N_i \mathbf{w}) + (N_i \mathbf{w})^2\right\}, \quad i = 1, 2, \ldots, n \tag{2.64}$$

which can be written as

$$\begin{aligned} J_i = \frac{1}{2} E\Big\{ & M_i \left(\mathbf{v}\,\mathbf{v}^T\right) M_i^T + 2 M_i \left(\mathbf{v}\,\mathbf{w}^T\right) N_i^T \\ & + N_i \left(\mathbf{w}\,\mathbf{w}^T\right) N_i^T \Big\}, \quad i = 1, 2, \ldots, n \end{aligned} \tag{2.65}$$

Therefore, using the defined covariances in eqns. (2.44) and (2.46), and since we have assumed that $E\left\{\mathbf{v}\,\mathbf{w}^T\right\} = 0$ (i.e., the errors are uncorrelated), eqn. (2.65) becomes

$$J_i = \frac{1}{2}\left[M_i R M_i^T + N_i Q N_i^T\right], \quad i = 1, 2, \ldots, n \tag{2.66}$$

The ith minimization problem can then be restated as: Determine the M_i and N_i to minimize the ith equation (2.66) subject to the constraint equation (2.53).

Using the method of Lagrange multipliers (Appendix C), the augmented functions are defined as

$$\begin{aligned} J_i = & \frac{1}{2}\left[M_i R M_i^T + N_i Q N_i^T\right] \\ & + \boldsymbol{\lambda}_i^T \left(I_i^c - H^T M_i^T - N_i^T\right), \quad i = 1, 2, \ldots, n \end{aligned} \tag{2.67}$$

where

$$\boldsymbol{\lambda}_i^T = \left\{\lambda_{1_i}, \lambda_{2_i}, \ldots, \lambda_{n_i}\right\} \tag{2.68}$$

is the ith matrix of n Lagrange multipliers.

The necessary conditions for a minimum of eqn. (2.67) are

$$\nabla_{M_i^T} J_i = R M_i^T - H \boldsymbol{\lambda}_i = \mathbf{0}, \quad i = 1, 2, \ldots, n \tag{2.69}$$

$$\nabla_{N_i^T} J_i = Q N_i^T - \boldsymbol{\lambda}_i = \mathbf{0}, \quad i = 1, 2, \ldots, n \tag{2.70}$$

and

$$\nabla_{\boldsymbol{\lambda}_i} J_i = I_i^c - H^T M_i^T - N_i^T = \mathbf{0}, \quad i = 1, 2, \ldots, n \tag{2.71}$$

From eqns. (2.69) and (2.70), we obtain

$$M_i = \boldsymbol{\lambda}_i^T H^T R^{-1}, \; M_i^T = R^{-1} H \boldsymbol{\lambda}_i, \quad i = 1, 2, \ldots, n \tag{2.72}$$

and

$$N_i = \boldsymbol{\lambda}_i^T Q^{-1}, \; N_i^T = Q^{-1} \boldsymbol{\lambda}_i, \quad i = 1, 2, \ldots, n \tag{2.73}$$

Substituting eqns. (2.72) and (2.73) into (2.71) allows immediate solution for $\boldsymbol{\lambda}_i^T$ as

$$\boldsymbol{\lambda}_i^T = I_i^r \left(H^T R^{-1} H + Q^{-1}\right)^{-1}, \quad i = 1, 2, \ldots, n \tag{2.74}$$

Then substituting eqn. (2.74) into eqns. (2.72) and (2.73), the rows of M and N are

$$M_i = I_i^r \left(H^T R^{-1} H + Q^{-1}\right)^{-1} H^T R^{-1}, \quad i = 1, 2, \ldots, n \tag{2.75}$$

$$N_i = I_i^r \left(H^T R^{-1} H + Q^{-1}\right)^{-1} Q^{-1}, \quad i = 1, 2, \ldots, n \tag{2.76}$$

Therefore, the M and N matrices are

$$M = \left(H^T R^{-1} H + Q^{-1}\right)^{-1} H^T R^{-1} \tag{2.77}$$

$$N = \left(H^T R^{-1} H + Q^{-1}\right)^{-1} Q^{-1} \tag{2.78}$$

Finally, substituting eqns. (2.77) and (2.78) into eqn. (2.54) yields the minimum variance estimator

$$\boxed{\hat{\mathbf{x}} = \left(H^T R^{-1} H + Q^{-1}\right)^{-1} \left(H^T R^{-1} \tilde{\mathbf{y}} + Q^{-1} \hat{\mathbf{x}}_a\right)} \tag{2.79}$$

which allows rigorous processing of *a priori* state estimates $\hat{\mathbf{x}}_a$ and associated covariance matrices Q.

Notice the following limiting cases:

1. *A priori* knowledge very poor

   $$\left(R \text{ finite}, \; Q \to \infty, \; Q^{-1} \to 0\right)$$

   then eqn. (2.79) reduces immediately to the standard minimal variance estimator (2.29).

2. Measurements very poor

   $$\left(Q \text{ finite}, \; R^{-1} \to 0\right)$$

   then eqn. (2.79) yields $\hat{\mathbf{x}} = \hat{\mathbf{x}}_a$, an intuitively pleasing result!

Notice also that eqn. (2.79) can be obtained from the sequential least squares formulation of §1.3 by processing the *a priori* state information as a subset of the "observation" as follows: In eqns. (1.53) and (1.54) of the sequential estimation developments:

1. Set $\tilde{\mathbf{y}}_2 = \hat{\mathbf{x}}_a$, $H_2 = I$ (note: the dimension of $\tilde{\mathbf{y}}_2$ is n in this case), and $W_1 = R^{-1}$ and $W_2 = Q^{-1}$.

2. Ignore the "1" and "2" subscripts.

Then one immediately obtains eqn. (2.79).

We thus conclude that the minimal variance estimate (2.79) is in all respects consistent with the sequential estimation results of §1.3; to start the sequential process, one would probably employ the *a priori* estimates as

$$\hat{\mathbf{x}}_1 = \hat{\mathbf{x}}_a$$
$$P_1 = Q$$

and process subsequent measurement subsets $\{\tilde{\mathbf{y}}_k,\, H_k,\, W_k\}$ with $W_k = R^{-1}$ for the minimal variance estimates of $\mathbf{x}$.

As in the case of estimation without *a priori* estimates, the previous derivation can also be shown in compact form. The following loss function to be minimized is

$$J = \frac{1}{2}\mathrm{Tr}\left[E\left\{(\hat{\mathbf{x}}-\mathbf{x})(\hat{\mathbf{x}}-\mathbf{x})^T\right\}\right] + \mathrm{Tr}[\Lambda(I - MH - N)] \tag{2.80}$$

Substituting eqns. (2.43) and (2.45) into eqn. (2.54) leads to

$$\begin{aligned} \hat{\mathbf{x}} &= M\tilde{\mathbf{y}} + N\hat{\mathbf{x}}_a \\ &= (MH+N)\mathbf{x} + M\mathbf{v} + N\mathbf{w} \end{aligned} \tag{2.81}$$

Next, as before we assume that the true state $\mathbf{x}$ and error terms $\mathbf{v}$ and $\mathbf{w}$ are uncorrelated with each other. Using eqns. (2.44) and (2.46) with the uncorrelated assumption leads to

$$J = \frac{1}{2}\mathrm{Tr}(MRM^T + NQN^T) + \mathrm{Tr}[\Lambda(I - MH - N)] \tag{2.82}$$

Therefore, we have the following necessary conditions:

$$\nabla_M J = MR - \Lambda^T H^T = 0 \tag{2.83}$$

$$\nabla_N J = NQ - \Lambda^T = 0 \tag{2.84}$$

$$\nabla_\Lambda J = I - MH - N = 0 \tag{2.85}$$

Solving eqn. (2.83) for M yields

$$M = \Lambda^T H^T R^{-1} \tag{2.86}$$

Solving eqn. (2.84) for N yields

$$N = \Lambda^T Q^{-1} \tag{2.87}$$

Substituting eqns. (2.86) and (2.87) into eqn. (2.85), and solving for Λ^T gives

$$\Lambda^T = (H^T R^{-1} H + Q^{-1})^{-1} \tag{2.88}$$

Finally, substituting eqn. (2.88) into eqns. (2.86) and (2.87) yields

$$M = \left(H^T R^{-1} H + Q^{-1}\right)^{-1} H^T R^{-1} \tag{2.89}$$

$$N = \left(H^T R^{-1} H + Q^{-1}\right)^{-1} Q^{-1} \tag{2.90}$$

This is identical to the solutions given by eqns. (2.77) and (2.78).

2.2 Unbiased Estimates

The structure of eqn. (2.8) can also be used to prove that the minimal variance estimator is "unbiased." An estimator $\hat{\mathbf{x}}(\tilde{\mathbf{y}})$ is said to be an "unbiased estimator" of $\mathbf{x}$ if $E\left\{\hat{\mathbf{x}}(\tilde{\mathbf{y}})\right\} = \mathbf{x}$ for every possible value of $\mathbf{x}$.[4‡] If $\hat{\mathbf{x}}$ is biased, the difference $E\left\{\hat{\mathbf{x}}(\tilde{\mathbf{y}})\right\} - \mathbf{x}$ is called the "bias" of $\hat{\mathbf{x}}$. For the minimum variance estimate $\hat{\mathbf{x}}$, given by eqn. (2.29), to be unbiased M must satisfy the following condition:

$$MH = I \tag{2.91}$$

The proof of the unbiased condition is given by first substituting eqn. (2.1) into eqn. (2.13), leading to

$$\begin{aligned} \hat{\mathbf{x}} &= M\tilde{\mathbf{y}} \\ &= MH\mathbf{x} + M\mathbf{v} \end{aligned} \tag{2.92}$$

Next, taking the expectation of both sides of (2.92) and using $E\{\mathbf{v}\} = 0$ gives (again $\mathbf{x}$ on the right-hand side of eqn. (2.92) is treated as a deterministic quantity)

$$E\left\{\hat{\mathbf{x}}\right\} = MH\mathbf{x} \tag{2.93}$$

which gives the condition in eqn. (2.91). Substituting eqn. (2.28) into eqn. (2.91) shows that the estimator clearly produces an unbiased estimate of $\hat{\mathbf{x}}$.

The sequential least squares estimator can also be shown to produce an unbiased estimate. A more general definition for an unbiased estimator is given by the following:

$$\boxed{E\left\{\hat{\mathbf{x}}_k(\tilde{\mathbf{y}})\right\} = \mathbf{x} \quad \text{for all } k} \tag{2.94}$$

Similar to the batch estimator, it is desired to estimate $\hat{\mathbf{x}}_{k+1}$ as a linear combination of the previous estimate $\hat{\mathbf{x}}_k$ and measurements $\tilde{\mathbf{y}}_{k+1}$ as

$$\hat{\mathbf{x}}_{k+1} = G_{k+1}\hat{\mathbf{x}}_k + K_{k+1}\tilde{\mathbf{y}}_{k+1} \tag{2.95}$$

where G_{k+1} and K_{k+1} are deterministic matrices. To determine the conditions for an unbiased estimator, we begin by assuming that the (sequential) measurement is modelled by

$$\tilde{\mathbf{y}}_{k+1} = H_{k+1}\mathbf{x}_{k+1} + \mathbf{v}_{k+1} \tag{2.96}$$

Substituting eqn. (2.96) into the estimator equation (2.95) gives

$$\hat{\mathbf{x}}_{k+1} = G_{k+1}\hat{\mathbf{x}}_k + K_{k+1}H_{k+1}\mathbf{x}_{k+1} + K_{k+1}\mathbf{v}_{k+1} \tag{2.97}$$

Taking the expectation of both sides of eqn. (2.97) and using eqn. (2.94) gives the following condition for an unbiased estimate:

$$G_{k+1} = I - K_{k+1}H_{k+1} \tag{2.98}$$

‡This implies that the estimate is a *function* of the measurements.

Substituting eqn. (2.98) into eqn. (2.95) yields

$$\hat{\mathbf{x}}_{k+1} = \hat{\mathbf{x}}_k + K_{k+1}(\tilde{\mathbf{y}}_{k+1} - H_{k+1}\hat{\mathbf{x}}_k) \tag{2.99}$$

which clearly has the structure of the sequential estimator in eqn. (1.65). Therefore, the sequential least squares estimator also produces an unbiased estimate. The case for the unbiased estimator with *a priori* estimates is left as an exercise for the reader.

Example 2.1: In this example we will show that the sample variance in eqn. (1.2) produces an unbiased estimate of $\hat{\sigma}^2$. For random data $\{\tilde{y}(t_1), \tilde{y}(t_2), \ldots, \tilde{y}(t_m)\}$ the sample variance is given by

$$\hat{\sigma}^2 = \frac{1}{m-1}\sum_{i=1}^{m}\left[\tilde{y}(t_i) - \hat{\mu}\right]^2$$

For any random variable z, the variance is given by $\text{var}\{z\} = E\{z^2\} - E\{z\}^2$, which is derived from the parallel axis theorem. Defining $E\{\hat{\sigma}^2\} \equiv S^2$, and applying this to the sample variance equation with the definition of the sample mean gives

$$\begin{aligned}
S^2 &= \frac{1}{m-1}\left[\sum_{i=1}^{m} E\left\{[\tilde{y}(t_i)]^2\right\} - \frac{1}{m}E\left\{\left[\sum_{i=1}^{m}\tilde{y}(t_i)\right]^2\right\}\right] \\
&= \frac{1}{m-1}\left[\sum_{i=1}^{m}\left(\sigma^2+\mu^2\right) - \frac{1}{m}\left\{\text{var}\left[\sum_{i=1}^{m}\tilde{y}(t_i)\right] + \left[E\left\{\sum_{i=1}^{m}\tilde{y}(t_i)\right\}\right]^2\right\}\right] \\
&= \frac{1}{m-1}\left[m\sigma^2 + m\mu^2 - \frac{1}{m}m\sigma^2 - \frac{1}{m}m^2\mu^2\right] \\
&= \frac{1}{m-1}\left[m\sigma^2 - \sigma^2\right] \\
&= \sigma^2
\end{aligned}$$

Therefore, this estimator is unbiased. However, the sample variance shown in this example does not give an estimate with the smallest mean-square-error for Gaussian (normal) distributions.[1]

2.3 Maximum Likelihood Estimation

We have seen that minimum variance estimation provides a powerful method to determine least squares estimates through rigorous proof of the relationship between the weight matrix and measurement-error covariance matrix. In this section another powerful method, known as *maximum likelihood estimation* is shown. This method was first introduced by R.A. Fisher, a geneticist and statistician in the 1920s. Maximum likelihood yields estimates for the unknown quantities which maximize the probability of obtaining the observed set of data. Although fundamentally different than minimum variance we will show that under the assumption of zero-mean Gaussian noise measurement-error process, both maximum likelihood and minimum variance estimation yield the same exact results for the least squares estimates.

We begin the topic of maximum likelihood by first considering a probability density function (see Appendix B) which is a function of the measurements and unknown parameters, denoted by $f(\tilde{\mathbf{y}};\mathbf{x})$. For motivational purposes, let $\tilde{\mathbf{y}}$ be a random sample from a simple Gaussian distribution. The density function is given by (see Appendix B)

$$f(\tilde{\mathbf{y}};\mathbf{x}) = \left(\frac{1}{2\pi\sigma^2}\right)^{m/2} e^{\left[-\sum\limits_{i=1}^{m}(\tilde{y}_i-\mu)^2\Big/(2\sigma^2)\right]} \tag{2.100}$$

Clearly, the Gaussian distribution is a monotonic exponential function for the mean (μ) and variance (σ^2). Due to the monotonic aspect of the function, this fit can be accomplished by also taking the natural logarithm of eqn. (2.100), which yields

$$\ln\left[f(\tilde{\mathbf{y}};\mathbf{x})\right] = -\frac{m}{2}\ln\left(2\pi\sigma^2\right) - \frac{1}{2\sigma^2}\sum_{i=1}^{m}(\tilde{y}_i-\mu)^2 \tag{2.101}$$

Now the fit leads immediately to an equivalent quadratic optimization problem to maximize the function in eqn. (2.101). This leads to the concept of maximum likelihood estimation, which is stated as follows. Given a measurement $\tilde{\mathbf{y}}$, the maximum-likelihood estimate $\hat{\mathbf{x}}$ is the value of $\mathbf{x}$ which maximizes $f(\tilde{\mathbf{y}};\mathbf{x})$, which is the likelihood that $\mathbf{x}$ resulted in the measured $\tilde{\mathbf{y}}$.

The *likelihood function* $L(\tilde{\mathbf{y}};\mathbf{x})$ is also a probability density function, given by

$$L(\tilde{\mathbf{y}};\mathbf{x}) = \prod_{i=1}^{p} f_i(\tilde{\mathbf{y}};\mathbf{x}) \tag{2.102}$$

where p is the total number of density functions (a product of a number of density functions, known as a joint density, is also a density function in itself). The goal of the method of maximum likelihood is to choose as our estimate of the unknown parameters $\mathbf{x}$ that value for which the *probability* of obtaining the observations $\tilde{\mathbf{y}}$ is maximized. Many likelihood functions contain exponential terms, which can complicate the mathematics involved in obtaining a solution. However, since $\ln\left[L(\tilde{\mathbf{y}};\mathbf{x})\right]$ is a monotonic function of $L(\tilde{\mathbf{y}};\mathbf{x})$, finding $\mathbf{x}$ to maximize $\ln\left[L(\tilde{\mathbf{y}};\mathbf{x})\right]$ is equivalent to maximizing $L(\tilde{\mathbf{y}};\mathbf{x})$.[§] It follows that for a maximum we have the following:

[§]Also, taking the natural logarithm changes a product to a sum which often simplifies the problem to be solved.

necessary condition

$$\boxed{\left\{\frac{\partial}{\partial \mathbf{x}} \ln\left[L(\tilde{\mathbf{y}};\mathbf{x})\right]\right\}\bigg|_{\hat{\mathbf{x}}} = \mathbf{0}} \tag{2.103}$$

sufficient condition

$$\frac{\partial^2}{\partial \mathbf{x}\,\partial \mathbf{x}^T} \ln\left[L(\tilde{\mathbf{y}};\mathbf{x})\right] \text{ must be negative definite.} \tag{2.104}$$

Equation (2.103) is often called the *likelihood equation*.[5, 6] Let us demonstrate this method by a few simple examples.

Example 2.2: Let $\tilde{\mathbf{y}}$ be a random sample from a Gaussian distribution. We desire to determine estimates for the mean (μ) and variance (σ^2), so that $\mathbf{x}^T = \left[\mu\ \sigma^2\right]^T$. For this case the likelihood function is given by eqn. (2.100):

$$L(\tilde{\mathbf{y}};\mathbf{x}) = \left(\frac{1}{2\pi\sigma^2}\right)^{m/2} e^{\left[-\sum\limits_{i=1}^{m}(\tilde{y}_i-\mu)^2/(2\sigma^2)\right]}$$

The log likelihood function is given by

$$\ln\left[L(\tilde{\mathbf{y}};\mathbf{x})\right] = -\frac{m}{2}\ln\left(2\pi\sigma^2\right) - \frac{1}{2\sigma^2}\sum_{i=1}^{m}(\tilde{y}_i-\mu)^2$$

To determine the maximizing μ we take the partial derivative of $\ln\left[L(\tilde{\mathbf{y}};\mathbf{x})\right]$ with respect to μ, evaluated at $\hat{\mu}$, and equate the resultant to zero, giving

$$\left\{\frac{\partial}{\partial \mu} \ln\left[L(\tilde{\mathbf{y}};\hat{\mathbf{x}})\right]\right\}\bigg|_{\hat{\mu}} = \frac{1}{\sigma^2}\sum_{i=1}^{m}(\tilde{y}_i-\hat{\mu}) = 0$$

Solving for $\hat{\mu}$ yields

$$\hat{\mu} = \frac{1}{m}\sum_{i=1}^{m}\tilde{y}_i$$

which is the well known sample mean. To determine the maximizing σ^2 we take the partial derivative of $\ln\left[L(\tilde{\mathbf{y}};\hat{\mathbf{x}})\right]$ with respect to σ^2, evaluated at $\hat{\sigma}^2$, and equate the resultant to zero, giving

$$\left\{\frac{\partial}{\partial \sigma^2} \ln\left[L(\tilde{\mathbf{y}};\hat{\mathbf{x}})\right]\right\}\bigg|_{\hat{\sigma}^2} = -\frac{m}{2\hat{\sigma}^2} + \frac{1}{2\hat{\sigma}^4}\sum_{i=1}^{m}(\tilde{y}_i-\mu)^2 = 0$$

Solving for $\hat{\sigma}^2$ yields

$$\hat{\sigma}^2 = \frac{1}{m}\sum_{i=1}^{m}(\tilde{y}_i-\mu)^2$$

which is the sample variance. It is easy to show that this estimate for σ^2 is biased, whereas the estimate shown in example 2.1 is unbiased. Thus, two different principles of estimation (unbiased estimator and maximum likelihood) give two different estimators.

Example 2.3: An advantage of using maximum likelihood is that we are not limited to Gaussian distributions. For example, suppose we wish to determine the probability of obtaining a certain number of heads in multiple flips of a coin. We are given $\tilde{y}$ "successes" in n trials, and wish to estimate the probability of success x of the *binomial* distribution.[7] The likelihood function is given by

$$L(\tilde{y};x) = \binom{n}{\tilde{y}} x^{\tilde{y}}(1-x)^{n-\tilde{y}}$$

The log likelihood function is given by

$$\ln\left[L(\tilde{y};x)\right] = \ln\binom{n}{\tilde{y}} + \tilde{y}\ln(x) + (n-\tilde{y})\ln(1-x)$$

To determine the maximizing x we take the partial derivative of $\ln\left[L(\tilde{y};x)\right]$ with respect to x, evaluated at $\hat{x}$, and equate the resultant to zero, giving

$$\left\{\frac{\partial}{\partial x}\ln\left[L(\tilde{y};x)\right]\right\}\bigg|_{\hat{x}} = \frac{\tilde{y}}{\hat{x}} - \frac{n-\tilde{y}}{1-\hat{x}} = 0$$

Therefore, the likelihood function has a maximum at

$$\hat{x} = \frac{\tilde{y}}{n}$$

This intuitively makes sense for our coin toss example, since we expect to obtain a probability of $1/2$ in n flips (for a balanced coin).

Maximum likelihood has many desirable properties. The first is the *invariance principle*,[5] which is stated as follows: Let $\hat{\mathbf{x}}$ be the maximum likelihood estimate of $\mathbf{x}$. Then the maximum likelihood estimate of any function $g(\mathbf{x})$ of these parameters is the function $g(\hat{\mathbf{x}})$ of the maximum likelihood estimate. This is a powerful tool since we do not have to take more partial derivatives to determine the maximum likelihood estimate! A simple example involves estimating the standard deviation, σ, in example 2.2. Using the invariance principle the solution is simply given by $\hat{\sigma} = \sqrt{\hat{\sigma}^2}$. Another property is that maximum likelihood is an *asymptotically efficient* estimator. This means that if the sample size m is large, the maximum likelihood estimate is approximately unbiased and has a variance that approaches the smallest that can be achieved by any estimator.[4] We see that this property is true in example 2.2, since as m becomes large the maximum likelihood estimate for the variance approaches the unbiased estimate asymptotically. Finally, the estimation errors in the maximum likelihood estimate can be shown to be *asymptotically Gaussian* no matter what density function is used in the likelihood function. Proofs of these properties can be found in Sorenson.[5]

We now turn our attention to the least squares problem. We will again use the linear observation model from eqn. (2.1), but we assume that $\mathbf{v}$ has a zero mean Gaussian distribution with covariance given by eqn. (2.19). To compute the maximum likelihood estimate of $\mathbf{x}$ we need the probability density function of $\tilde{\mathbf{y}}$, which we know is Gaussian since measurements of a linear system, such as eqn. (2.1), driven by Gaussian noise are also Gaussian (see Appendix B). To determine the mean of the observation model, we take the expectation of both sides of eqn. (2.1)

$$\boldsymbol{\mu} \equiv E\left\{\tilde{\mathbf{y}}\right\} = E\left\{H\mathbf{x}\right\} + E\left\{\mathbf{v}\right\} \tag{2.105}$$

Since both H and $\mathbf{x}$ are *deterministic* quantities and since $\mathbf{v}$ has zero mean (so that $E\left\{\mathbf{v}\right\} = \mathbf{0}$), eqn. (2.105) reduces to

$$\boldsymbol{\mu} = H\mathbf{x} \tag{2.106}$$

Next, we determine the covariance of the observation model, which is given by

$$\text{cov}\left\{\tilde{\mathbf{y}}\right\} \equiv E\left\{\left(\tilde{\mathbf{y}} - \boldsymbol{\mu}\right)\left(\tilde{\mathbf{y}} - \boldsymbol{\mu}\right)^T\right\} \tag{2.107}$$

Substituting eqns. (2.1) and (2.106) into (2.107) gives

$$\text{cov}\left\{\tilde{\mathbf{y}}\right\} = R \tag{2.108}$$

In shorthand notation it is common to use $\tilde{\mathbf{y}} \sim \mathcal{N}(\boldsymbol{\mu}, R)$ to represent a Gaussian (normal) noise process with mean $\boldsymbol{\mu}$ and covariance R. Next, from Appendix B, we use the *multidimensional* or *multivariate normal distribution* for the likelihood function, and from eqns. (2.106) and (2.108) we have

$$L(\tilde{\mathbf{y}};\mathbf{x}) = \frac{1}{(2\pi)^{m/2}\left[\det(R)\right]^{1/2}} \exp\left\{-\frac{1}{2}\left[\tilde{\mathbf{y}} - H\mathbf{x}\right]^T R^{-1}\left[\tilde{\mathbf{y}} - H\mathbf{x}\right]\right\} \tag{2.109}$$

The log likelihood function is given by

$$\ln\left[L(\tilde{\mathbf{y}};\mathbf{x})\right] = -\frac{1}{2}\left[\tilde{\mathbf{y}} - H\mathbf{x}\right]^T R^{-1}\left[\tilde{\mathbf{y}} - H\mathbf{x}\right] - \frac{m}{2}\ln(2\pi) - \frac{1}{2}\ln\left[\det(R)\right] \tag{2.110}$$

We can ignore the last two terms of the right-hand side of eqn. (2.110) since they are independent of $\mathbf{x}$. Also, if we take the negative of eqn. (2.110), then maximizing the log likelihood function to determine the optimal estimate $\hat{\mathbf{x}}$ is equivalent to *minimizing*

$$J(\hat{\mathbf{x}}) = \frac{1}{2}\left[\tilde{\mathbf{y}} - H\hat{\mathbf{x}}\right]^T R^{-1}\left[\tilde{\mathbf{y}} - H\hat{\mathbf{x}}\right] \tag{2.111}$$

The optimal estimate for $\mathbf{x}$ found by minimizing eqn. (2.111) is exactly equivalent to the minimum variance solution given in eqn. (2.29)! Therefore, for the case of Gaussian measurement errors the minimum variance and maximum likelihood estimates are identical to the least squares solution with the weight replaced with the inverse measurement-error covariance. The term $\frac{1}{2}$ in the loss function comes directly from maximum likelihood, which also helps simplify the mathematics when taking partials.

Example 2.4: In example 2.2 we estimated the variance using a random measurement sample from a normal distribution. In this example we will expand upon this to estimate the covariance from a multivariate normal distribution given a set of observations:

$$\{\tilde{\mathbf{y}}_1, \tilde{\mathbf{y}}_2, \ldots, \tilde{\mathbf{y}}_p\}$$

The likelihood function in this case is the joint density function, given by

$$L(R) = \prod_{i=1}^{p} \frac{1}{(2\pi)^{m/2}\left[\det(R)\right]^{1/2}} \exp\left\{-\frac{1}{2}\left[\tilde{\mathbf{y}}_i - \boldsymbol{\mu}\right]^T R^{-1}\left[\tilde{\mathbf{y}}_i - \boldsymbol{\mu}\right]\right\}$$

The log likelihood function is given by

$$\ln[L(R)] = \sum_{i=1}^{p}\left\{-\frac{1}{2}\left[\tilde{\mathbf{y}}_i - \boldsymbol{\mu}\right]^T R^{-1}\left[\tilde{\mathbf{y}}_i - \boldsymbol{\mu}\right] - \frac{m}{2}\ln(2\pi) - \frac{1}{2}\ln\left[\det(R)\right]\right\}$$

To determine an estimate of R we need to take the partial of $\ln[L(R)]$ with respect to R and set the resultant to zero. In order to accomplish this task, we will need to review some matrix calculus differentiating rules. For any given matrices R and G we have

$$\frac{\partial \ln\left[\det(R)\right]}{\partial R} = (R^T)^{-1}$$

and

$$\frac{\partial\, \mathrm{Tr}(R^{-1}G)}{\partial R} = -(R^T)^{-1}G(R^T)^{-1}$$

where Tr denotes the trace operator. It can also be shown through simple matrix manipulations that

$$\sum_{i=1}^{p}\left[\tilde{\mathbf{y}}_i - \boldsymbol{\mu}\right]^T R^{-1}\left[\tilde{\mathbf{y}}_i - \boldsymbol{\mu}\right] = \mathrm{Tr}(R^{-1}G)$$

where

$$G = \sum_{i=1}^{p}\left[\tilde{\mathbf{y}}_i - \boldsymbol{\mu}\right]\left[\tilde{\mathbf{y}}_i - \boldsymbol{\mu}\right]^T$$

Now, since R is symmetric we have

$$\frac{\partial \ln[L(R)]}{\partial R} = -\frac{p}{2}R^{-1} + \frac{1}{2}R^{-1}GR^{-1}$$

Therefore, the maximum likelihood estimate for the covariance is given by

$$\hat{R} = \frac{1}{p}\sum_{i=1}^{p}[\tilde{\mathbf{y}}_i - \mu][\tilde{\mathbf{y}}_i - \mu]^T$$

It can also be shown that this estimate is biased.

2.4 Cramér-Rao Inequality

This section describes one of the most useful and important concepts in estimation theory. The Cramér-Rao inequality[8] can be used to give us a lower bound on the expected errors between the estimated quantities and the *true* values from the known statistical properties of the measurement errors. The theory was proved independently by Cramér and Rao, although it was found earlier by Fisher[9] for the special case of a Gaussian distribution. Let $f(\tilde{\mathbf{y}};\mathbf{x})$ be the probability density function of the sample $\tilde{\mathbf{y}}$. The Cramér-Rao inequality for an unbiased estimate $\hat{\mathbf{x}}$ is given by[¶]

$$\boxed{P \equiv E\left\{(\hat{\mathbf{x}}-\mathbf{x})(\hat{\mathbf{x}}-\mathbf{x})^T\right\} \geq F^{-1}} \tag{2.112}$$

where the *Fisher information matrix*, F, is given by

$$F = E\left\{\left[\frac{\partial}{\partial \mathbf{x}}\ln f(\tilde{\mathbf{y}};\mathbf{x})\right]\left[\frac{\partial}{\partial \mathbf{x}}\ln f(\tilde{\mathbf{y}};\mathbf{x})\right]^T\right\} \tag{2.113}$$

It can be shown that the Fisher information matrix[10] can also be computed using the Hessian matrix, given by

$$F = -E\left\{\frac{\partial^2}{\partial \mathbf{x}\,\partial \mathbf{x}^T}\ln f(\tilde{\mathbf{y}};\mathbf{x})\right\} \tag{2.114}$$

The first- and second-order partial derivatives are assumed to exist and to be absolutely integrable. A formal proof of the Cramér-Rao inequality requires using the "conditions of regularity."[1] However, a slightly different approach is taken here. We begin the proof by using the definition of a probability density function

$$\int_{-\infty}^{\infty}\int_{-\infty}^{\infty}\cdots\int_{-\infty}^{\infty} f(\tilde{\mathbf{y}};\mathbf{x})\,d\tilde{y}_1\,d\tilde{y}_2\cdots d\tilde{y}_m = 1 \tag{2.115}$$

¶For a definition of what it means for one matrix to be greater than another matrix see Appendix A.

In short-hand notation, we write eqn. (2.115) as

$$\int_{-\infty}^{\infty} f(\tilde{\mathbf{y}};\mathbf{x})\, d\tilde{\mathbf{y}} = 1 \tag{2.116}$$

Taking the partial of eqn. (2.116) with respect to $\mathbf{x}$ gives

$$\frac{\partial}{\partial \mathbf{x}} \int_{-\infty}^{\infty} f(\tilde{\mathbf{y}};\mathbf{x})\, d\tilde{\mathbf{y}} = \int_{-\infty}^{\infty} \left[\frac{\partial f(\tilde{\mathbf{y}};\mathbf{x})}{\partial \mathbf{x}}\right]^T d\tilde{\mathbf{y}} = \mathbf{0} \tag{2.117}$$

Next, since $\hat{\mathbf{x}}$ is assumed to be unbiased, we have

$$E\left\{\hat{\mathbf{x}} - \mathbf{x}\right\} = \int_{-\infty}^{\infty} \left(\hat{\mathbf{x}} - \mathbf{x}\right) f\left(\tilde{\mathbf{y}};\mathbf{x}\right) d\tilde{\mathbf{y}} = \mathbf{0} \tag{2.118}$$

Differentiating both sides of eqn. (2.118) with respect to $\mathbf{x}$ gives

$$\int_{-\infty}^{\infty} \left(\hat{\mathbf{x}} - \mathbf{x}\right) \left[\frac{\partial f(\tilde{\mathbf{y}};\mathbf{x})}{\partial \mathbf{x}}\right]^T d\tilde{\mathbf{y}} - I = 0 \tag{2.119}$$

The identity matrix in eqn. (2.119) is obtained since a probability density function always satisfies eqn. (2.116). Next, we use the following logarithmic differentiation rule:[11]

$$\frac{\partial f(\tilde{\mathbf{y}};\mathbf{x})}{\partial \mathbf{x}} = \left[\frac{\partial}{\partial \mathbf{x}} \ln f(\tilde{\mathbf{y}};\mathbf{x})\right] f(\tilde{\mathbf{y}};\mathbf{x}) \tag{2.120}$$

Substituting eqn. (2.120) into eqn. (2.119) leads to

$$I = \int_{-\infty}^{\infty} \left(\mathbf{a}\,\mathbf{b}^T\right) d\tilde{\mathbf{y}} \tag{2.121}$$

where

$$\mathbf{a} \equiv f(\tilde{\mathbf{y}};\mathbf{x})^{1/2} \left(\hat{\mathbf{x}} - \mathbf{x}\right) \tag{2.122a}$$

$$\mathbf{b} \equiv f(\tilde{\mathbf{y}};\mathbf{x})^{1/2} \left[\frac{\partial}{\partial \mathbf{x}} \ln f(\tilde{\mathbf{y}};\mathbf{x})\right] \tag{2.122b}$$

The error-covariance expression in eqn. (2.112) can be rewritten using the definition in eqn. (2.122a) by

$$P = \int_{-\infty}^{\infty} \left(\mathbf{a}\,\mathbf{a}^T\right) d\tilde{\mathbf{y}} \tag{2.123}$$

Also, the Fisher information matrix can be rewritten as

$$F = \int_{-\infty}^{\infty} \left(\mathbf{b}\,\mathbf{b}^T\right) d\tilde{\mathbf{y}} \tag{2.124}$$

Now, multiply eqn. (2.121) on the left by an arbitrary row vector $\boldsymbol{\alpha}^T$ and on the right by an arbitrary column vector $\boldsymbol{\beta}$, so that

$$\boldsymbol{\alpha}^T \boldsymbol{\beta} = \int_{-\infty}^{\infty} \boldsymbol{\alpha}^T \left(\mathbf{a}\,\mathbf{b}^T\right) \boldsymbol{\beta}\, d\tilde{\mathbf{y}} \tag{2.125}$$

Next, we make use of the *Schwartz inequality* (see §A.2), which is given by[‖]

$$\left[\int_{-\infty}^{\infty} g(\tilde{\mathbf{y}};\mathbf{x})\, h(\tilde{\mathbf{y}};\mathbf{x})\, d\tilde{\mathbf{y}}\right]^2 \leq \int_{-\infty}^{\infty} g^2(\tilde{\mathbf{y}};\mathbf{x})\, d\tilde{\mathbf{y}} \int_{-\infty}^{\infty} h^2(\tilde{\mathbf{y}};\mathbf{x})\, d\tilde{\mathbf{y}} \tag{2.126}$$

If we let $g(\tilde{\mathbf{y}};\mathbf{x}) = \boldsymbol{\alpha}^T\mathbf{a}$ and $h(\tilde{\mathbf{y}};\mathbf{x}) = \mathbf{b}^T\boldsymbol{\beta}$, then eqn. (2.126) becomes

$$\left[\int_{-\infty}^{\infty} \boldsymbol{\alpha}^T(\mathbf{a}\mathbf{b}^T)\boldsymbol{\beta}\, d\tilde{\mathbf{y}}\right]^2 \leq \int_{-\infty}^{\infty} \boldsymbol{\alpha}^T(\mathbf{a}\,\mathbf{a}^T)\boldsymbol{\alpha}\, d\tilde{\mathbf{y}} \int_{-\infty}^{\infty} \boldsymbol{\beta}^T(\mathbf{b}\,\mathbf{b}^T)\boldsymbol{\beta}\, d\tilde{\mathbf{y}} \tag{2.127}$$

Using the definitions in eqns. (2.123) and (2.124), and assuming that $\boldsymbol{\alpha}$ and $\boldsymbol{\beta}$ are independent of $\tilde{\mathbf{y}}$ gives

$$\left(\boldsymbol{\alpha}^T\boldsymbol{\beta}\right)^2 \leq \left(\boldsymbol{\alpha}^T P \boldsymbol{\alpha}\right)\left(\boldsymbol{\beta}^T F \boldsymbol{\beta}\right) \tag{2.128}$$

Finally, choosing the particular choice $\boldsymbol{\beta} = F^{-1}\boldsymbol{\alpha}$ gives

$$\boldsymbol{\alpha}^T(P - F^{-1})\boldsymbol{\alpha} \geq 0 \tag{2.129}$$

Since $\boldsymbol{\alpha}$ is arbitrary then $P \geq F^{-1}$, which proves the Cramér-Rao inequality.

The Cramér-Rao inequality gives a *lower* bound on the expected errors. When the equality in eqn. (2.112) is satisfied, then the estimator is said to be *efficient*. This can be useful for the investigation of the quality of a particular estimator. Therefore, the Cramér-Rao inequality is certainly useful information! It should be stressed that the Cramér-Rao inequality gives a lower bound on the expected errors only for the case of unbiased estimates.

Let us now turn our attention to the Gauss-Markov Theorem in eqn. (2.29). The negative of the log likelihood function for this system is given by $J(\mathbf{x})$ plus terms independent of $\mathbf{x}$. Therefore, to compute the Fisher information matrix we can directly use the loss function in eqn. (2.111) so that

$$F = \frac{\partial^2}{\partial\mathbf{x}\,\partial\mathbf{x}^T} J(\mathbf{x}) \tag{2.130}$$

An expectation is not required in eqn. (2.130) since the loss function is quadratic in $\mathbf{x}$. Carrying out this computation leads to

$$F = (H^T R^{-1} H) \tag{2.131}$$

Hence, the Cramér-Rao inequality is given by

$$P \geq (H^T R^{-1} H)^{-1} \tag{2.132}$$

Let us now find an expression for the estimate covariance P. Using eqns. (2.29) and (2.1) leads to

$$\hat{\mathbf{x}} - \mathbf{x} = (H^T R^{-1} H)^{-1} H^T R^{-1}\mathbf{v} \tag{2.133}$$

[‖]If $\int_{-\infty}^{\infty} a(\mathbf{x})b(\mathbf{x})\, d\mathbf{x} = 1$ then $\int_{-\infty}^{\infty} a^2(\mathbf{x})\, d\mathbf{x} \int_{-\infty}^{\infty} b^2(\mathbf{x})\, d\mathbf{x} \geq 1$; the equality holds if $a(\mathbf{x}) = c\, b(\mathbf{x})$ where c is not a function of $\mathbf{x}$.

Using $E\{\mathbf{v}\mathbf{v}^T\} = R$ leads to the following estimate covariance:

$$P = (H^T R^{-1} H)^{-1} \tag{2.134}$$

Therefore, the *equality* in eqn. (2.132) is satisfied, so, the least squares estimate from the Gauss-Markov Theorem is the most efficient possible estimate!

Example 2.5: In this example we will show how the covariance expression in eqn. (2.134) can be used to provide boundaries on the expected errors. For this example a set of 1001 measurement points sampled at 0.01-second intervals was taken using the following observation model:

$$y(t) = \cos(t) + 2\sin(t) + \cos(2t) + 2\sin(3t) + v(t)$$

where $v(t)$ is a zero-mean Gaussian noise process with variance given by $R = 0.01$. The least squares estimator from eqn. (2.29) was used to estimate the coefficients of the transcendental functions. In this example the basis functions used in the estimator are equivalent to the functions in the observation model. Estimates were found from 1000 trial runs using a different random number seed between runs. Statistical conclusions can be made if the least squares solution is performed many times using different measurement sets. This approach is known as *Monte Carlo simulation.* A plot of the actual errors for each estimate and associated 3σ boundaries (found from taking the square root of the diagonal elements of P and multiplying the result by 3) is shown in Figure 2.1. From probability theory, for a Gaussian distribution, there is a 0.9974 probability that the estimate error will be inside of the 3σ boundary. We see that the estimate errors in Figure 2.1 agree with this assessment, since for 1000 trial runs we expect about 3 estimates to be outside of the 3σ boundary. This example clearly shows the power of the estimate covariance and Cramér-Rao lower bound. It is important to note that in this example the estimate covariance, P, can be computed *without* any measurement information, since it only depends on H and R. This powerful tool allows one to use probabilistic concepts to compute estimate error boundaries, and subsequently analyze the expected performance in a dynamical system. This is demonstrated further in Chapter 4.

Example 2.6: In this example we will show the usefulness of the Cramér-Rao inequality for parameter estimation. Suppose we wish to estimate a nonlinear appearing parameter, $a > 0$, of the following exponential model:

$$\tilde{y}_k = B\, e^{a t_k} + v_k, \quad k = 1, 2 \ldots, m$$

where v_k is a zero-mean Gaussian white-noise process with variance given by σ^2. We can choose to employ nonlinear least squares to iteratively determine the parameter a, given measurements y_k and a known $B > 0$ coefficient. If this approach is taken,

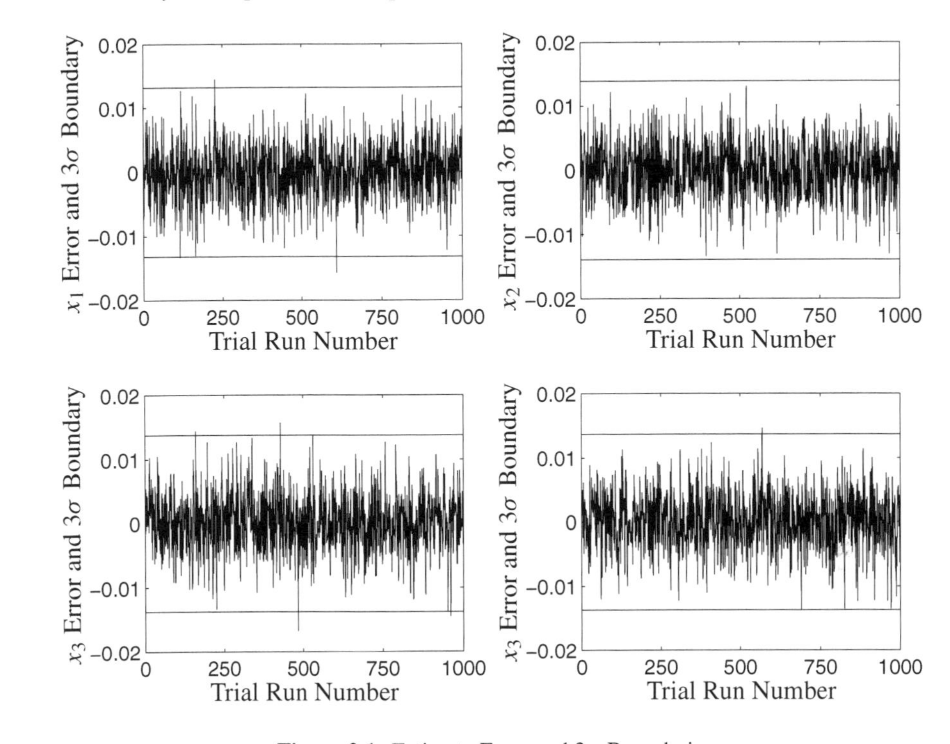

Figure 2.1: Estimate Error and 3σ Boundaries

then the covariance of the estimate error is given by

$$P = \sigma^2 (H^T H)^{-1}$$

where

$$H = \left[B\,t_1\,e^{a\,t_1} \;\; B\,t_2\,e^{a\,t_2} \;\; \cdots \;\; B\,t_m\,e^{a\,t_m} \right]^T \tag{2.135}$$

The matrix P is also equivalent to the Cramér-Rao lower bound. Suppose instead we wish to simplify the estimation process by defining $\tilde{z}_k \equiv \ln \tilde{y}_k$, using the change of variables approach shown in Table 1.1. Then, linear squares can be applied to determine a. But how optimal is this solution? It is desired to study the effects of applying this linear approach because the logarithmic function also affects the Gaussian noise. Expanding $\tilde{z}_k$ in a first-order series gives

$$\ln \tilde{y}_k - \ln B \approx a\,t_k + \frac{2\,v_k}{2\,B\,e^{a\,t_k} + v_k}$$

The linear least squares "H matrix," denoted by $\mathcal{H}$, is now simply given by

$$\mathcal{H} = \left[t_1 \; t_2 \; \cdots \; t_m \right]^T \tag{2.136}$$

However, the new measurement noise will certainly not be Gaussian anymore. A first-order expansion of the new measurement noise is given by

$$\varepsilon_k \equiv 2\, v_k (2\, B\, e^{a\, t_k} + v_k)^{-1} \approx \frac{v_k}{B\, e^{a\, t_k}} \left(1 - \frac{v_k}{2\, B\, e^{a\, t_k}}\right)$$

The variance of ε_k, denoted by ς_k^2, is derived from

$$\begin{aligned} \varsigma_k^2 &= E\{\varepsilon_k^2\} - E\{\varepsilon_k\}^2 \\ &= E\left\{\left(\frac{v_k}{B\, e^{a\, t_k}} - \frac{v_k^2}{2\, B^2 e^{2a\, t_k}}\right)^2\right\} - \frac{\sigma^4}{4\, B^2 e^{4a\, t_k}} \end{aligned}$$

This leads to (which is left as an exercise for the reader)

$$\varsigma_k^2 = \frac{\sigma^2}{B^2 e^{2a\, t_k}} + \frac{\sigma^4}{2\, B^4 e^{4a\, t_k}}$$

Note that ε_k contains both Gaussian and χ^2 components (see Appendix B). Therefore, the covariance of the linear approach, denoted by $\mathcal{P}$, is given by

$$\mathcal{P} = \left(\mathcal{H}^T \operatorname{diag}\left[\varsigma_1^{-2}\ \varsigma_2^{-2}\ \cdots\ \varsigma_m^{-2}\right] \mathcal{H}\right)^{-1}$$

Notice that $\mathcal{P}$ is equivalent to P if $\sigma^4/(2\, B^4 e^{4a\, t_k})$ is negligible. If this is not the case, then the Cramér-Rao lower bound is not achieved and the linear approach does not lead to an efficient estimator. This clearly shows how the Cramér-Rao inequality can be particularly useful to help quantify the errors introduced by using an approximate solution instead of the optimal approach. A more practical application of the usefulness of the Cramér-Rao lower bound is given in Ref. [12] and exercise 4.15.

2.5 Nonuniqueness of the Weight Matrix

Here we study the truth that more than one weight matrix in the normal equations can yield identical $\mathbf{x}$ estimates. Actually two classes of weight matrices (which preserve $\hat{\mathbf{x}}$) exist; the first is rather well known, the second is less known and its implications are more subtle.

We first consider the class of weight matrices which is formed by multiplying all elements of W by some scalar α as

$$W' = \alpha W \tag{2.137}$$

The $\mathbf{x}$ estimate corresponding to W' follows from eqn. (1.30) as

$$\hat{\mathbf{x}}' = \frac{1}{\alpha}(H^T W H)^{-1} H^T (\alpha W)\tilde{\mathbf{y}} = (H^T W H)^{-1} H^T W \tilde{\mathbf{y}} \tag{2.138}$$

so that

$$\hat{\mathbf{x}}' \equiv \hat{\mathbf{x}} \tag{2.139}$$

Therefore, scaling all elements of W does not (formally) affect the estimate solution $\hat{\mathbf{x}}$. Numerically, possible significant errors may result if extremely small or extremely large values of α are used, due to computed truncation errors.

We now consider a second class of weight matrices obtained by adding a nonzero $(m \times m)$ matrix ΔW to W as

$$W'' = W + \Delta W \tag{2.140}$$

Then the estimate solution $\hat{\mathbf{x}}''$ corresponding to W'' is obtained from eqn. (1.30) as

$$\hat{\mathbf{x}}'' = (H^T W'' H)^{-1} H^T W'' \tilde{\mathbf{y}} \tag{2.141}$$

Substituting eqn. (2.140) into eqn. (2.141) yields

$$\hat{\mathbf{x}}'' = \left[H^T W H + (H^T \Delta W) H\right]^{-1} \left[H^T W \tilde{\mathbf{y}} + (H^T \Delta W)\tilde{\mathbf{y}}\right] \tag{2.142}$$

If $\Delta W \neq 0$ exists such that

$$H^T \Delta W = 0 \tag{2.143}$$

then eqn. (2.142) clearly reduces to

$$\hat{\mathbf{x}}'' = (H^T W H)^{-1} H^T W \tilde{\mathbf{y}} \equiv \hat{\mathbf{x}} \tag{2.144}$$

There are, in fact, an infinity of matrices ΔW satisfying the *orthogonality constraint* in eqn. (2.143). To see this, assume that all elements of ΔW except those in the first column are zero, then eqn. (2.143) becomes

$$H^T \Delta W = \begin{bmatrix} h_{11} & h_{21} & \cdots & h_{m1} \\ h_{12} & h_{22} & \cdots & h_{m2} \\ \vdots & \vdots & \ddots & \vdots \\ h_{1n} & h_{2n} & \cdots & h_{mn} \end{bmatrix} \begin{bmatrix} \Delta W_{11} & 0 & \cdots & 0 \\ \Delta W_{21} & 0 & \cdots & 0 \\ \vdots & \vdots & \ddots & \vdots \\ \Delta W_{m1} & 0 & \cdots & 0 \end{bmatrix} = 0 \tag{2.145}$$

which yields the scalar equations

$$h_{1i}\Delta W_{11} + h_{2i}\Delta W_{21} + \ldots + h_{mi}\Delta W_{m1} = 0, \quad i = 1, 2, \ldots, n \tag{2.146}$$

Equation (2.146) provides n equations to be satisfied by the m unspecified ΔW_{j1}'s. Since any n of the ΔW_{j1}'s can be determined to satisfy eqns. (2.146), while the remaining $(m-n)$ ΔW_{j1}'s can be given arbitrary values, it follows that an infinity of ΔW matrices satisfy eqn. (2.145) and therefore eqn. (2.143).

The fact that more than one weight matrix yields the same estimates for $\mathbf{x}$ is no cause for alarm though. Interpreting the covariance matrix as the inverse of the measurement-error covariance matrix associated with a specific $\tilde{\mathbf{y}}$ of measurements, the above results imply that one can obtain the same $\mathbf{x}$-estimate from the given measured $\mathbf{y}$-values, for a variety of measurement weights, according to eqn. (2.137) or eqns. (2.140) and (2.143). A most interesting question can be asked regarding the covariance matrix of the estimated parameters. From eqn. (2.134), we established that the estimate covariance is

$$P = (H^T W H)^{-1}, \quad W = R^{-1} \tag{2.147}$$

For the first class of weight matrices $W' = \alpha W$ note that

$$P' = \frac{1}{\alpha}(H^T W H)^{-1} = \frac{1}{\alpha}(H^T R^{-1} H)^{-1} \tag{2.148}$$

or

$$P' = \frac{1}{\alpha} P \tag{2.149}$$

Thus linear scaling of the observation weight matrix results in reciprocal linear scaling of the estimate covariance matrix, an intuitively reasonable result.

Considering now the second class of error covariance matrices $W'' = W + \Delta W$, with $H^T \Delta W = 0$, it follows from eqn. (2.147) that

$$P'' = (H^T W H + H^T \Delta W H)^{-1} = (H^T W H)^{-1} \tag{2.150}$$

or

$$P'' = P \tag{2.151}$$

Thus, the additive class of observation weight matrices preserves not only the $\mathbf{x}$-estimates, but also the associated estimate covariance matrix. It may prove possible, in some applications, to exploit this truth since a family of measurement-error covariances can result in the same estimates and associated uncertainties.

Example 2.7: Given the following linear system:

$$\tilde{\mathbf{y}} = H\mathbf{x}$$

with

$$\tilde{\mathbf{y}} = \begin{bmatrix} 2 \\ 1 \\ 3 \end{bmatrix}, \quad H = \begin{bmatrix} 1 & 3 \\ 2 & 2 \\ 3 & 4 \end{bmatrix}$$

For each of the three weight matrices

$$W = I, \quad W' = 3W, \quad W'' = W + \begin{bmatrix} 1/4 & 5/8 & -1/2 \\ 5/8 & 25/16 & -5/4 \\ -1/2 & -5/4 & 1 \end{bmatrix}$$

determine the least squares estimates

$$\hat{\mathbf{x}} = (H^T W H)^{-1} H^T W \tilde{\mathbf{y}}$$
$$\hat{\mathbf{x}}' = (H^T W' H)^{-1} H^T W' \tilde{\mathbf{y}}$$
$$\hat{\mathbf{x}}'' = (H^T W'' H)^{-1} H^T W'' \tilde{\mathbf{y}}$$

and corresponding error-covariance matrices

$$P = (H^T W H)^{-1}$$
$$P' = (H^T W' H)^{-1}$$
$$P'' = (H^T W'' H)^{-1}$$

The reader can verify the numerical results

$$\hat{\mathbf{x}} = \hat{\mathbf{x}}' = \hat{\mathbf{x}}'' = \begin{bmatrix} -1/15 \\ 11/15 \end{bmatrix}$$

and

$$P = P'' = \begin{bmatrix} 29/45 & -19/45 \\ -19/45 & 14/45 \end{bmatrix}$$
$$P' = \frac{1}{3} P = \begin{bmatrix} 29/135 & -19/135 \\ -19/135 & 14/135 \end{bmatrix}$$

These results are consistent with eqns. (2.139), (2.144), (2.149), and (2.151).

2.6 Bayesian Estimation

The parameters that we have estimated in this chapter have been assumed to be unknown constants. In Bayesian estimation, we consider that these parameters are random variables with some *a priori* distribution. Bayesian estimation combines this *a priori* information with the measurements through a conditional density function of $\mathbf{x}$ given the measurements $\tilde{\mathbf{y}}$. This conditional function is known as the *a posteriori distribution* of $\mathbf{x}$. Therefore, Bayesian estimation requires the probability density functions of both the measurement noise and unknown parameters. The posterior density function $f(\mathbf{x}|\tilde{\mathbf{y}})$ for $\mathbf{x}$ (taking the measurements $\tilde{\mathbf{y}}$ into account) is given by *Bayes rule* (see Appendix B for details):

$$\boxed{f(\mathbf{x}|\tilde{\mathbf{y}}) = \frac{f(\tilde{\mathbf{y}}|\mathbf{x})\, f(\mathbf{x})}{f(\tilde{\mathbf{y}})}} \tag{2.152}$$

Note since $\tilde{\mathbf{y}}$ is treated as a set of known quantities, then $f(\tilde{\mathbf{y}})$ is just a normalization factor to ensure that $f(\mathbf{x}|\tilde{\mathbf{y}})$ is a probability density function. Therefore,

$$f(\tilde{\mathbf{y}}) = \int_{-\infty}^{\infty} f(\tilde{\mathbf{y}}|\mathbf{x})\, f(\mathbf{x})\, d\mathbf{x} \tag{2.153}$$

If the integral in eqn. (2.153) exists then the posterior function $f(\mathbf{x}|\tilde{\mathbf{y}})$ is said to be *proper*; if it does not exist then $f(\mathbf{x}|\tilde{\mathbf{y}})$ is *improper*, in which case we let $f(\tilde{\mathbf{y}}) = 1$ (see [13] for sufficient conditions).

Maximum *a posteriori* (MAP) estimation finds an estimate for $\mathbf{x}$ that maximizes eqn. (2.152).[6] Since $f(\tilde{\mathbf{y}})$ does not depend on $\mathbf{x}$, this is equivalent to maximizing $f(\tilde{\mathbf{y}}|\mathbf{x})\, f(\mathbf{x})$. We can again use the natural logarithm (as shown in §2.3) to simplify the problem by maximizing

$$J_{\text{MAP}}(\hat{\mathbf{x}}) = \ln\left[f(\tilde{\mathbf{y}}|\hat{\mathbf{x}})\right] + \ln\left[f(\hat{\mathbf{x}})\right] \tag{2.154}$$

The first term in the sum is actually the log-likelihood function, and the second term gives the *a priori* information on the to-be-determined parameters. Therefore, the MAP estimator maximizes

$$\boxed{J_{\text{MAP}}(\hat{\mathbf{x}}) = \ln\left[L(\tilde{\mathbf{y}}|\hat{\mathbf{x}})\right] + \ln\left[f(\hat{\mathbf{x}})\right]} \tag{2.155}$$

Maximum *a posteriori* estimation has the following properties: (1) if the *a priori* distribution $f(\hat{\mathbf{x}})$ is uniform, then MAP estimation is equivalent to maximum likelihood estimation, (2) MAP estimation shares the asymptotic consistency and efficiency properties of maximum likelihood estimation, (3) the MAP estimator converges to the maximum likelihood estimator for large samples, and (4) the MAP estimator also obeys the invariance principle.

Example 2.8: Suppose we wish to estimate the mean μ of a Gaussian variable from a sample of m independent measurements known to have a standard deviation of $\sigma_{\tilde{y}}$. We have been given that the *a priori* density function of μ is also Gaussian with zero mean and standard deviation σ_μ. The density functions are therefore given by

$$f(\tilde{y}_i|\mu) = \frac{1}{\sigma_{\tilde{y}}\sqrt{2\pi}} \exp\left\{-\frac{1}{2}\frac{(\tilde{y}_i - \mu)^2}{\sigma_{\tilde{y}}^2}\right\}, \quad i = 1, 2, \ldots, m$$

and

$$f(\mu) = \frac{1}{\sigma_\mu\sqrt{2\pi}} \exp\left\{-\frac{\mu^2}{2\sigma_\mu^2}\right\}$$

Since the measurements are independent we can write

$$f(\tilde{\mathbf{y}}|\mu) = \frac{1}{(\sigma_{\tilde{y}}\sqrt{2\pi})^m} \exp\left\{-\frac{1}{2}\sum_{i=1}^{m} \frac{(\tilde{y}_i - \mu)^2}{\sigma_{\tilde{y}}^2}\right\}$$

Using eqn. (2.154) and ignoring terms independent of μ we now seek to maximize

$$J_{\text{MAP}}(\hat{\mu}) = -\frac{1}{2}\left[\sum_{i=1}^{m}\frac{(\tilde{y}_i - \hat{\mu})^2}{\sigma_{\tilde{y}}^2} + \frac{\hat{\mu}^2}{\sigma_{\mu}^2}\right]$$

Taking the partial of this equation with respect to $\hat{\mu}$ and equating the resultant to zero gives

$$\sum_{i=1}^{m}\frac{(\tilde{y}_i - \hat{\mu})}{\sigma_{\tilde{y}}^2} - \frac{\hat{\mu}}{\sigma_{\mu}^2} = 0$$

Recall that the maximum likelihood estimate for the mean from example 2.2 is given by

$$\hat{\mu}_{\text{ML}} = \frac{1}{m}\sum_{i=1}^{m}\tilde{y}_i$$

Therefore we can write the maximum *a posteriori* estimate for the mean as

$$\hat{\mu} = \frac{\sigma_{\mu}^2}{\frac{1}{m}\sigma_{\tilde{y}}^2 + \sigma_{\mu}^2}\hat{\mu}_{\text{ML}}$$

Notice that $\hat{\mu} \to \hat{\mu}_{\text{ML}}$ as either $\sigma_{\mu}^2 \to \infty$ or as $m \to \infty$. This is consistent with the properties discussed previously of a maximum *a posteriori* estimator.

Maximum *a posteriori* estimation can also be used to find an optimal estimator for the case with *a priori* estimates, modelled using eqns. (2.43) through (2.46). The assumed probability density functions for this case are given by

$$L(\tilde{\mathbf{y}}|\hat{\mathbf{x}}) = f(\tilde{\mathbf{y}}|\hat{\mathbf{x}}) = \frac{1}{(2\pi)^{m/2}\left[\det(R)\right]^{1/2}}\exp\left\{-\frac{1}{2}\left[\tilde{\mathbf{y}} - H\hat{\mathbf{x}}\right]^T R^{-1}\left[\tilde{\mathbf{y}} - H\hat{\mathbf{x}}\right]\right\} \tag{2.156}$$

$$f(\hat{\mathbf{x}}) = \frac{1}{(2\pi)^{n/2}\left[\det(Q)\right]^{1/2}}\exp\left\{-\frac{1}{2}\left[\hat{\mathbf{x}}_a - \hat{\mathbf{x}}\right]^T Q^{-1}\left[\hat{\mathbf{x}}_a - \hat{\mathbf{x}}\right]\right\} \tag{2.157}$$

Maximizing eqn. (2.155) leads to the following estimator:

$$\boxed{\hat{\mathbf{x}} = \left(H^T R^{-1} H + Q^{-1}\right)^{-1}\left(H^T R^{-1}\tilde{\mathbf{y}} + Q^{-1}\hat{\mathbf{x}}_a\right)} \tag{2.158}$$

which is the same result obtained through minimum variance. However, the solution using MAP estimation is much simpler since we do not need to solve a constrained minimization problem using Lagrange multipliers.

The Cramér-Rao inequality can be extended for a Bayesian estimator. The Cramér-Rao inequality for the case of *a priori* information is given by[5, 14]

$$\boxed{\begin{aligned} P &\equiv E\left\{(\hat{\mathbf{x}}-\mathbf{x})(\hat{\mathbf{x}}-\mathbf{x})^T\right\} \\ &\geq \left[F + E\left\{\left[\frac{\partial}{\partial \mathbf{x}}\ln f(\mathbf{x})\right]\left[\frac{\partial}{\partial \mathbf{x}}\ln f(\mathbf{x})\right]^T\right\}\right]^{-1} \end{aligned}} \tag{2.159}$$

This can be used to test the efficiency of the MAP estimator. The Fisher information matrix has been computed in eqn. (2.131) as

$$F = \left(H^T R^{-1} H\right) \tag{2.160}$$

Using the *a priori* density function in eqn. (2.157) leads to

$$\begin{aligned} E\left\{\left[\frac{\partial}{\partial \mathbf{x}}\ln f(\mathbf{x})\right]\left[\frac{\partial}{\partial \mathbf{x}}\ln f(\mathbf{x})\right]^T\right\} &= Q^{-1}E\left\{(\hat{\mathbf{x}}_a-\mathbf{x})^T(\hat{\mathbf{x}}_a-\mathbf{x})\right\}Q^{-1} \\ &= Q^{-1}E\left\{\mathbf{w}\,\mathbf{w}^T\right\}Q^{-1} = Q^{-1} \end{aligned} \tag{2.161}$$

Next, we need to compute the covariance matrix P. From eqn. (2.81) and using $MH + N = I$, the estimate can be written as

$$\hat{\mathbf{x}} = \mathbf{x} + M\mathbf{v} + N\mathbf{w} \tag{2.162}$$

Using the definitions in eqns. (2.44) and (2.46), and assuming that $E\left\{\mathbf{v}\,\mathbf{w}^T\right\} = 0$ and $E\left\{\mathbf{w}\,\mathbf{v}^T\right\} = 0$, the covariance matrix can be written as

$$P = MRM^T + NQN^T \tag{2.163}$$

From the solutions for M and N in eqns. (2.77) and (2.78), the covariance matrix becomes

$$P = \left(H^T R^{-1} H + Q^{-1}\right)^{-1} \tag{2.164}$$

Therefore, the lower bound in the Cramér-Rao inequality is achieved, and thus the estimator (2.158) is efficient. Equation (2.164) can be alternatively written using the matrix inversion lemma, shown by eqns. (1.69) and (1.70), as

$$P = Q - QH^T\left(R + HQH^T\right)^{-1}HQ \tag{2.165}$$

Equation (2.165) may be preferred over eqn. (2.164) if the dimension of R is less than the dimension of Q.

Another approach for Bayesian estimation is a minimum risk (MR) estimator.[14, 15] In practical engineering problems, we are often faced with making a decision in the face of uncertainty. An example involves finding the best value for an aircraft model parameter given wind tunnel data in the face of measurement error uncertainty. Bayesian estimation chooses a *course of action* that has the largest expectation of gain (or smallest expectation of loss). This approach assumes the existence (or at least a guess) of the *a priori* probability function. Minimum risk estimators also use this information to find the best estimate based on decision theory, which assigns a cost to any loss suffered due to errors in the estimate. Our goal is to evaluate the cost $c(\mathbf{x}^*|\mathbf{x})$ of believing that the value of the estimate is $\mathbf{x}^*$ when it is actually $\mathbf{x}$. Since $\mathbf{x}$ is unknown, the actual cost cannot be evaluated; however, we usually assume that $\mathbf{x}$ is distributed by the *a posteriori* function. This approach minimizes the risk, defined as the mean of the cost over all possible values of $\mathbf{x}$, given a set of observations $\tilde{\mathbf{y}}$. The risk function is given by

$$J_{\text{MR}}(\mathbf{x}^*) = \int_{-\infty}^{\infty} c(\mathbf{x}^*|\mathbf{x})\, f(\mathbf{x}|\tilde{\mathbf{y}})\, d\mathbf{x} \tag{2.166}$$

Using Bayes rule we can rewrite the risk as

$$\boxed{J_{\text{MR}}(\mathbf{x}^*) = \int_{-\infty}^{\infty} c(\mathbf{x}^*|\mathbf{x}) \frac{f(\tilde{\mathbf{y}}|\mathbf{x})\, f(\mathbf{x})}{f(\tilde{\mathbf{y}})}\, d\mathbf{x}} \tag{2.167}$$

The *minimum risk* estimate is defined as the value of $\mathbf{x}^*$ that minimizes the loss function in eqn. (2.167).

A common choice for the cost $c(\mathbf{x}^*|\mathbf{x})$ is a quadratic function taking the form

$$c(\mathbf{x}|\mathbf{x}^*) = \frac{1}{2}(\mathbf{x}^* - \mathbf{x})^T S (\mathbf{x}^* - \mathbf{x}) \tag{2.168}$$

where S is a positive definite weighting matrix. The risk is now given by

$$J_{\text{MR}}(\mathbf{x}^*) = \frac{1}{2}\int_{-\infty}^{\infty} (\mathbf{x}^* - \mathbf{x})^T S (\mathbf{x}^* - \mathbf{x})\, f(\mathbf{x}|\tilde{\mathbf{y}})\, d\mathbf{x} \tag{2.169}$$

To determine the minimum risk estimate we take the partial of eqn. (2.169) with respect to $\mathbf{x}^*$, evaluated at $\hat{\mathbf{x}}$, and set the resultant to zero:

$$\left.\frac{\partial J_{\text{MR}}(\mathbf{x}^*)}{\partial \mathbf{x}}\right|_{\hat{\mathbf{x}}} = \mathbf{0} = S \int_{-\infty}^{\infty} (\hat{\mathbf{x}} - \mathbf{x})\, f(\mathbf{x}|\tilde{\mathbf{y}})\, d\mathbf{x} \tag{2.170}$$

Since S is invertible eqn. (2.170) simply reduces down to

$$\hat{\mathbf{x}} \int_{-\infty}^{\infty} f(\mathbf{x}|\tilde{\mathbf{y}})\, d\mathbf{x} = \int_{-\infty}^{\infty} \mathbf{x}\, f(\mathbf{x}|\tilde{\mathbf{y}})\, d\mathbf{x} \tag{2.171}$$

The integral on the left-hand side of eqn. (2.171) is clearly unity, so that

$$\hat{\mathbf{x}} = \int_{-\infty}^{\infty} \mathbf{x}\, f(\mathbf{x}|\tilde{\mathbf{y}})\, d\mathbf{x} \equiv E\left\{\mathbf{x}|\tilde{\mathbf{y}}\right\} \tag{2.172}$$

Notice that the minimum risk estimator is independent of S in this case. Additionally, the optimal estimate is seen to be the expected value (i.e., the mean) of $\mathbf{x}$ given the measurements $\tilde{\mathbf{y}}$. From Bayes rule we can rewrite eqn. (2.172) as

$$\hat{\mathbf{x}} = \int_{-\infty}^{\infty} \mathbf{x} \frac{f(\tilde{\mathbf{y}}|\mathbf{x})\, f(\mathbf{x})}{f(\tilde{\mathbf{y}})} d\mathbf{x} \tag{2.173}$$

We will now use the minimum risk approach to determine an optimal estimate with *a priori* information. Recall from §2.1.2 that we have the following models:

$$\tilde{\mathbf{y}} = H\mathbf{x} + \mathbf{v} \tag{2.174a}$$

$$\hat{\mathbf{x}}_a = \mathbf{x} + \mathbf{w} \tag{2.174b}$$

with associated known expectations and covariances

$$E\{\mathbf{v}\} = \mathbf{0} \tag{2.175a}$$

$$\text{cov}\{\mathbf{v}\} = E\left\{\mathbf{v}\mathbf{v}^T\right\} = R \tag{2.175b}$$

and

$$E\{\mathbf{w}\} = \mathbf{0} \tag{2.176a}$$

$$\text{cov}\{\mathbf{w}\} = E\left\{\mathbf{w}\mathbf{w}^T\right\} = Q \tag{2.176b}$$

Also, recall that $\mathbf{x}$ is now a random variable with associated expectation and covariance

$$E\{\mathbf{x}\} = \hat{\mathbf{x}}_a \tag{2.177a}$$

$$\text{cov}\{\mathbf{x}\} = E\left\{\mathbf{x}\mathbf{x}^T\right\} - E\{\mathbf{x}\}\,E\{\mathbf{x}\}^T = Q \tag{2.177b}$$

The probability functions for $f(\tilde{\mathbf{y}}, \mathbf{x})$ and $f(\mathbf{x})$ are given by

$$f(\tilde{\mathbf{y}}|\mathbf{x}) = \frac{1}{(2\pi)^{m/2}\,[\det(R)]^{1/2}} \exp\left\{-\frac{1}{2}\left[\tilde{\mathbf{y}} - H\mathbf{x}\right]^T R^{-1}\left[\tilde{\mathbf{y}} - H\mathbf{x}\right]\right\} \tag{2.178}$$

$$f(\mathbf{x}) = \frac{1}{(2\pi)^{n/2}\,[\det(Q)]^{1/2}} \exp\left\{-\frac{1}{2}\left[\hat{\mathbf{x}}_a - \mathbf{x}\right]^T Q^{-1}\left[\hat{\mathbf{x}}_a - \mathbf{x}\right]\right\} \tag{2.179}$$

We now need to determine the density function $f(\tilde{\mathbf{y}})$. Since a sum of Gaussian random variables is itself a Gaussian random variable, then we know that $f(\tilde{\mathbf{y}})$ must also be Gaussian. The mean of $\tilde{\mathbf{y}}$ is simply

$$E\left\{\tilde{\mathbf{y}}\right\} = E\{H\mathbf{x}\} = H\hat{\mathbf{x}}_a \tag{2.180}$$

Assuming that $\mathbf{x}$, $\mathbf{v}$, and $\mathbf{w}$ are uncorrelated with each other, the covariance of $\tilde{\mathbf{y}}$ is given by

$$\begin{aligned}\text{cov}\left\{\tilde{\mathbf{y}}\right\} &= E\left\{\tilde{\mathbf{y}}\tilde{\mathbf{y}}^T\right\} - E\left\{\tilde{\mathbf{y}}\right\} E\left\{\tilde{\mathbf{y}}\right\}^T \\ &= E\left\{H\mathbf{w}\mathbf{w}^T H^T\right\} + E\left\{\mathbf{v}\mathbf{v}^T\right\}\end{aligned} \tag{2.181}$$

Therefore, using eqns. (2.175) and (2.176), then eqn. (2.181) can be written as

$$\text{cov}\left\{\tilde{\mathbf{y}}\right\} = HQH^T + R \equiv D \tag{2.182}$$

Hence, $f(\tilde{\mathbf{y}})$ is given by

$$f(\tilde{\mathbf{y}}) = \frac{1}{(2\pi)^{m/2}\left[\det(D)\right]^{1/2}} \exp\left\{-\frac{1}{2}\left[\tilde{\mathbf{y}} - H\hat{\mathbf{x}}_a\right]^T D^{-1}\left[\tilde{\mathbf{y}} - H\hat{\mathbf{x}}_a\right]\right\} \tag{2.183}$$

Using Bayes rule and the matrix inversion lemma shown by eqns. (1.69) and (1.70), it can be shown that $f(\mathbf{x}|\tilde{\mathbf{y}})$ is given by

$$\begin{aligned} f(\mathbf{x}|\tilde{\mathbf{y}}) &= \frac{\left[\det\left(HQH^T + R\right)\right]^{1/2}}{(2\pi)^{n/2}\left[\det(R)\right]^{1/2}\left[\det(Q)\right]^{1/2}} \\ &\times \exp\left\{-\frac{1}{2}\left[\mathbf{x} - H\mathbf{p}\right]^T (H^T R^{-1} H + Q^{-1})\left[\mathbf{x} - H\mathbf{p}\right]\right\} \end{aligned} \tag{2.184}$$

where

$$\mathbf{p} = \left(H^T R^{-1} H + Q^{-1}\right)^{-1}\left(H^T R^{-1}\tilde{\mathbf{y}} + Q^{-1}\hat{\mathbf{x}}_a\right) \tag{2.185}$$

Clearly, since eqn. (2.172) is $E\left\{\mathbf{x}|\tilde{\mathbf{y}}\right\}$, then the minimum risk estimate is given by

$$\hat{\mathbf{x}} = \mathbf{p} = \left(H^T R^{-1} H + Q^{-1}\right)^{-1}\left(H^T R^{-1}\tilde{\mathbf{y}} + Q^{-1}\hat{\mathbf{x}}_a\right) \tag{2.186}$$

which is equivalent to the estimate found by minimum variance and maximum *a posteriori*.

The minimum risk approach can be useful since it incorporates a decision-based means to determine the optimal estimate. However, there are many practical disadvantages. Although an analytical solution for the minimum risk using Gaussian distributions can be found in many cases, the evaluation of the integral in eqn. (2.173) may be impractical for general distributions. Also, the minimum risk estimator does not (in general) converge to the maximum likelihood estimate for uniform *a priori* distributions. Finally, unlike maximum likelihood, the minimum risk estimator is not invariant under reparameterization. For these reasons, minimum risk approaches are often avoided in practical estimation problems.

Some important properties of the *a priori* estimator in eqn. (2.186) are given by the following:

$$E\left\{(\mathbf{x} - \hat{\mathbf{x}})\tilde{\mathbf{y}}^T\right\} = 0 \tag{2.187}$$

$$E\left\{(\mathbf{x} - \hat{\mathbf{x}})\hat{\mathbf{x}}^T\right\} = 0 \tag{2.188}$$

The proof of these relations now follows. We first substitute $\hat{\mathbf{x}}$ from eqn. (2.186) into eqn. (2.187), with use of the model given in eqn. (2.174a). Then taking the expectation of the resultant, with $E\{\mathbf{v}\mathbf{x}^T\} = E\{\mathbf{x}\mathbf{v}^T\} = 0$, and using eqn. (2.177a) gives

$$E\left\{(\mathbf{x}-\hat{\mathbf{x}})\tilde{\mathbf{y}}^T\right\} = (I - KH^TR^{-1}H)E\left\{\mathbf{x}\mathbf{x}^T\right\}H^T - KQ^{-1}\hat{\mathbf{x}}_a\hat{\mathbf{x}}_a^T H^T - KH^T \tag{2.189}$$

where

$$K \equiv \left(H^TR^{-1}H + Q^{-1}\right)^{-1} \tag{2.190}$$

Next, using the following identity:

$$(I - KH^TR^{-1}H) = KQ^{-1} \tag{2.191}$$

yields

$$E\left\{(\mathbf{x}-\hat{\mathbf{x}})\tilde{\mathbf{y}}^T\right\} = K\left(Q^{-1}E\left\{\mathbf{x}\mathbf{x}^T\right\}H^T - Q^{-1}\hat{\mathbf{x}}_a\hat{\mathbf{x}}_a^T H^T - H^T\right) \tag{2.192}$$

Finally, using eqn. (2.177b) in eqn. (2.192) leads to

$$E\left\{(\mathbf{x}-\hat{\mathbf{x}})\tilde{\mathbf{y}}^T\right\} = 0 \tag{2.193}$$

To prove eqn. (2.188), we substitute eqn. (2.186) into eqn. (2.188), again with use of the model given in eqn. (2.174a). Taking the appropriate expectations leads to

$$\begin{aligned} E\left\{(\mathbf{x}-\hat{\mathbf{x}})\hat{\mathbf{x}}^T\right\} &= E\left\{\mathbf{x}\mathbf{x}^T\right\}H^TR^{-1}HK + \hat{\mathbf{x}}_a\hat{\mathbf{x}}_a^T Q^{-1}K \\ &\quad - KH^TR^{-1}HE\left\{\mathbf{x}\mathbf{x}^T\right\}H^TR^{-1}HK \\ &\quad - KH^TR^{-1}H\hat{\mathbf{x}}_a\hat{\mathbf{x}}_a^T Q^{-1}K - KH^TR^{-1}HK \\ &\quad - KQ^{-1}\hat{\mathbf{x}}_a\hat{\mathbf{x}}_a^T H^TR^{-1}HK - KQ^{-1}\hat{\mathbf{x}}_a\hat{\mathbf{x}}_a^T Q^{-1}K \end{aligned} \tag{2.194}$$

Next, using eqn. (2.177b) and the identity in eqn. (2.191) leads to

$$E\left\{(\mathbf{x}-\hat{\mathbf{x}})\hat{\mathbf{x}}^T\right\} = 0 \tag{2.195}$$

Equations (2.187) and (2.188) show that the residual error is orthogonal to both the measurements and the estimates. Therefore, the concepts shown in §1.6.4 also apply to the *a priori* estimator.

2.7 Advanced Topics

In this section we will show some advanced topics used in probabilistic estimation. As in Chapter 1 we encourage the interested reader to pursue these topics further in the references provided.

2.7.1 Analysis of Covariance Errors

In §2.5 an analysis was shown for simple errors in the measurement-error covariance matrix. In this section we expand upon these results to the case of general errors in the assumed measurement-error covariance matrix. Say that the assumed measurement-error covariance is denoted by $\tilde{R}$ and the actual covariance is denoted by R. The least squares estimate with the assumed covariance matrix is given by

$$\hat{\mathbf{x}} = (H^T \tilde{R}^{-1} H)^{-1} H^T \tilde{R}^{-1} \tilde{\mathbf{y}} \tag{2.196}$$

Using the measurement model in eqn. (2.1) leads to the following residual error:

$$\hat{\mathbf{x}} - \mathbf{x} = (H^T \tilde{R}^{-1} H)^{-1} H^T \tilde{R}^{-1} \mathbf{v} \tag{2.197}$$

The estimate $\hat{\mathbf{x}}$ is unbiased since $E\{\mathbf{v}\} = \mathbf{0}$. Using $E\left\{\mathbf{v}\mathbf{v}^T\right\} = R$, the estimate covariance is given by

$$\tilde{P} = (H^T \tilde{R}^{-1} H)^{-1} H^T \tilde{R}^{-1} R \tilde{R}^{-1} H (H^T \tilde{R}^{-1} H)^{-1} \tag{2.198}$$

Clearly $\tilde{P}$ reduces to $(H^T R^{-1} H)^{-1}$ when $\tilde{R} = R$ or when H is square (i.e., $m = n$). Next, we define the following relative inefficiency parameter e, which gives a measure of the error induced by the incorrect measurement-error covariance:

$$\boxed{e = \frac{\det\left[(H^T \tilde{R}^{-1} H)^{-1} H^T \tilde{R}^{-1} R \tilde{R}^{-1} H (H^T \tilde{R}^{-1} H)^{-1}\right]}{\det\left[(H^T R^{-1} H)^{-1}\right]}} \tag{2.199}$$

We will now prove that $e \geq 1$. Since for any invertible matrix A, $\det(A^{-1}) = 1/\det(A)$, eqn. (2.199) reduces to

$$e = \frac{\det(H^T \tilde{R}^{-1} R \tilde{R}^{-1} H)\det(H^T R^{-1} H)}{\det(H^T \tilde{R}^{-1} H)^2} \tag{2.200}$$

Performing a singular value decomposition of the matrix $\tilde{R}^{1/2} H$ gives

$$\tilde{R}^{1/2} H = X S Y^T \tag{2.201}$$

where X and Y are orthogonal matrices.[16] Also, define the following matrix:

$$D \equiv X^T \tilde{R}^{-1/2} R \tilde{R}^{-1/2} X \tag{2.202}$$

Using the definitions in eqns. (2.201) and (2.202), then eqn. (2.200) can be written as

$$e = \frac{\det(Y S^T D S Y^T)\det(Y S^T D^{-1} S Y^T)}{\det(Y S^T S Y^T)} \tag{2.203}$$

This can easily be reduced to give

$$e = \frac{\det(S^T D S)\det(S^T D^{-1} S)}{\det(S^T S)^2} \tag{2.204}$$

Next, we partition the $m \times n$ matrix S into an $n \times n$ matrix S_1 and an $(m-n) \times n$ matrix of zeros so that

$$S = \begin{bmatrix} S_1 \\ 0 \end{bmatrix} \tag{2.205}$$

where S_1 is a diagonal matrix of the singular values. Also, partition D as

$$D = \begin{bmatrix} D_1 & F \\ F^T & D_2 \end{bmatrix} \tag{2.206}$$

where D_1 is a square matrix with the same dimension as S_1 and D_2 is also square. The inverse of D is given by (see Appendix A)

$$D^{-1} = \begin{bmatrix} (D_1 - F\,D_2^{-1}F^T)^{-1} & G \\ G^T & (D_2 - F^T D_1^{-1} F)^{-1} \end{bmatrix} \tag{2.207}$$

where the closed-form expression for G is not required in this development. Substituting eqns. (2.205), (2.206), and (2.207) into eqn. (2.204) leads to

$$e = \frac{\det(D_1)}{\det(D_1 - F\,D_2^{-1}F^T)} \tag{2.208}$$

Next, we use the following identity (see Appendix A):

$$\det(D) = \det(D_2)\det(D_1 - F\,D_2^{-1}F^T) \tag{2.209}$$

which reduces eqn. (2.208) to

$$e = \frac{\det(D_1)\det(D_2)}{\det(D)} \tag{2.210}$$

By Fischer's inequality[16] $e \geq 1$. The specific value of e gives an indication of the inefficiency of the estimator, and can be used to perform a sensitivity analysis given bounds on matrix R. A larger value for e means that the estimates are further (in a statistical sense) from their true values.

Example 2.9: In this simple example we consider a two measurement case with the true covariance given by the identity matrix. The assumed covariance $\tilde{R}$ and H matrices are given by

$$\tilde{R} = \begin{bmatrix} 1+\alpha & 0 \\ 0 & 1+\beta \end{bmatrix}, \quad H = \begin{bmatrix} 1 \\ 1 \end{bmatrix}$$

where α and β can vary from -0.99 to 1. A three-dimensional plot of the inefficiency in eqn. (2.199) for varying α and β is shown in Figure 2.2. The minimum value (1) is given when $\alpha = \beta = 0$ as expected. Also, the values for e are significantly lower when both α and β are greater than 1 (the average value for e in this

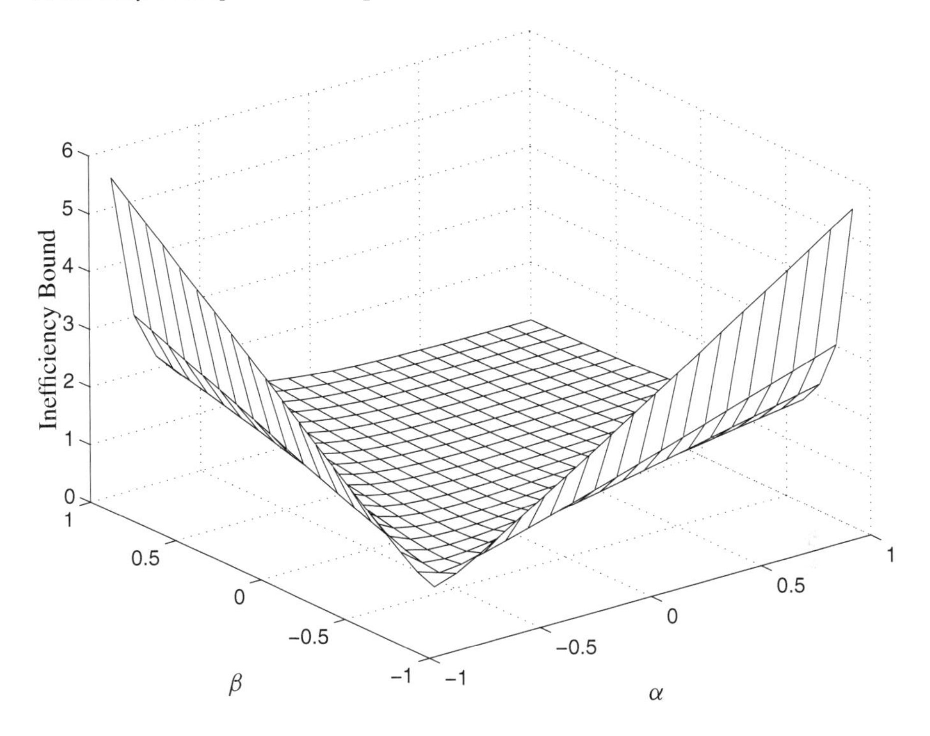

Figure 2.2: Measurement-Error Covariance Inefficiency Plot

case is 1.1681), as compared to when both are less than 1 (the average value for e in this case is 1.0175). This states that the estimate errors are worse when the assumed measurement-error covariance matrix is lower than the true covariance. This example clearly shows the influence of the measurement-error covariance on the performance characteristics of the estimates.

2.7.2 Ridge Estimation

As mentioned in §1.2.1 the inverse of $H^T H$ exists only if the number of linearly independent observations is equal to or greater than the number of unknowns, and if independent basis functions are used to form H. If the matrix $H^T H$ is close to being ill-conditioned, then the model is known as *weak multicollinear*. We can clearly see that weak multicollinearity may produce a large covariance in the estimated parameters. A strong multicollinearity exists if there are exact linear relations among the observations so that the rank of H equals n.[17, 18] This corresponds to the case of having linearly dependent rows in H. Another situation for $H^T H$ ill-conditioning is due to H having linearly independent columns, which occurs when the basis functions themselves are not independent of each other (e.g., choosing t, t^2 and $at+bt^2$, where a and b are constants, as basis functions leads to an ill-conditioned H matrix). Hoerl and Kennard[19] have proposed a class of estimators, called *ridge regression* estimators, that have a less total mean error than ordinary least squares (which is useful for the case of weak multicollinearity). However, as will be shown, the estimates are biased. Ridge estimation involves adding a positive constant, ϕ, to each diagonal element of $H^T H$, so that

$$\boxed{\hat{\mathbf{x}} = (H^T H + \phi I)^{-1} H^T \tilde{\mathbf{y}}} \tag{2.211}$$

Note the similarity between the ridge estimator and the Levenberg-Marquardt method in §1.6.3. Also note that even though the ridge estimator is a heuristic step motivated by numerical issues, comparing eqn. (2.79) to eqn. (2.211) leads to an equivalent relationship of formally treating $\hat{\mathbf{x}}_a = \mathbf{0}$ as an *a priori* estimate with associated covariance $Q = (1/\phi)I$. More generally, we may desire to use $\hat{\mathbf{x}}_a \neq \mathbf{0}$ and Q equal to some best estimate of the covariance of the errors in $\hat{\mathbf{x}}_a$.

We will first show that the ridge estimator produces biased estimates. Substituting eqn. (2.1) into eqn. (2.211) and taking the expectation leads to

$$E\{\hat{\mathbf{x}}\} = (H^T H + \phi I)^{-1} H^T H \mathbf{x} \tag{2.212}$$

Therefore, the bias is given by

$$\mathbf{b} \equiv E\{\hat{\mathbf{x}}\} - \mathbf{x} = \left[(H^T H + \phi I)^{-1} H^T H - I\right] \mathbf{x} \tag{2.213}$$

This can be simplified to yield

$$\mathbf{b} = -\phi (H^T H + \phi I)^{-1} \mathbf{x} \tag{2.214}$$

We clearly see that the ridge estimates are unbiased only when $\phi = 0$, which reduces to the standard least squares estimator.

Let us compute the covariance of the ridge estimator. Recall that the covariance is defined as

$$P \equiv E\left\{\hat{\mathbf{x}}\hat{\mathbf{x}}^T\right\} - E\{\hat{\mathbf{x}}\} E\{\hat{\mathbf{x}}\}^T \tag{2.215}$$

Assuming that $\mathbf{v}$ and $\mathbf{x}$ are uncorrelated leads to

$$P_{\text{ridge}} = (H^T H + \phi I)^{-1} H^T R H (H^T H + \phi I)^{-1} \tag{2.216}$$

Clearly, as ϕ increases the ridge covariance decreases, but at a price! The estimate becomes more biased, as seen in eqn. (2.214). We wish to find ϕ that minimizes the error $\hat{\mathbf{x}} - \mathbf{x}$, so that the estimate is as close to the truth as possible. A natural choice is to investigate the characteristics of the following matrix:

$$\Upsilon \equiv E\left\{(\hat{\mathbf{x}} - \mathbf{x})(\hat{\mathbf{x}} - \mathbf{x})^T\right\} \tag{2.217}$$

Note, this is *not* the covariance of the ridge estimate since $E\{\hat{\mathbf{x}}\} \neq \mathbf{x}$ in this case (therefore, the parallel axis theorem cannot be used). First, define

$$\Gamma \equiv (H^T H + \phi I)^{-1} \tag{2.218}$$

The following expectations can readily be derived

$$E\left\{\hat{\mathbf{x}}\hat{\mathbf{x}}^T\right\} = \Gamma\left(H^T R H + H^T H \mathbf{x}\mathbf{x}^T H^T H\right)\Gamma \tag{2.219}$$

$$E\left\{\mathbf{x}\hat{\mathbf{x}}^T\right\} = \mathbf{x}\mathbf{x}^T H^T H \Gamma \tag{2.220}$$

$$E\left\{\hat{\mathbf{x}}\mathbf{x}^T\right\} = \Gamma H^T H \mathbf{x}\mathbf{x}^T \tag{2.221}$$

Next, we make use of the following identities:

$$I - \Gamma H^T H = \phi\Gamma \tag{2.222}$$

and

$$\Gamma^{-1} - H^T H = \phi I \tag{2.223}$$

Hence, eqn. (2.217) becomes

$$\Upsilon = \Gamma\left(H^T R H + \phi^2 \mathbf{x}\mathbf{x}^T\right)\Gamma \tag{2.224}$$

We now wish to investigate the possibility of finding a range of ϕ that produces a lower Υ than the standard least squares covariance. In this analysis we will assume isotropic measurement errors so that $R = \sigma^2 I$. The least squares covariance can be manipulated using eqn. (2.218) to yield

$$\begin{aligned}
P_{\text{ls}} &= \sigma^2 (H^T H)^{-1} \\
&= \sigma^2 \Gamma\left[\Gamma^{-1}(H^T H)^{-1}\Gamma^{-1}\right]\Gamma \\
&= \sigma^2 \Gamma\left[I + \phi(H^T H)^{-1}\right]\left[H^T H + \phi I\right]\Gamma \\
&= \sigma^2 \Gamma\left[\phi^2 (H^T H)^{-1} + 2\phi I + H^T H\right]\Gamma
\end{aligned} \tag{2.225}$$

Using eqns. (2.216), (2.218), and (2.225), the condition for $P_{\text{ls}} - \Upsilon \geq 0$ is given by

$$\phi\Gamma\left\{\sigma^2\left[2I + \phi(H^T H)^{-1}\right] - \phi\mathbf{x}\mathbf{x}^T\right\}\Gamma \geq 0 \tag{2.226}$$

A sufficient condition for this inequality to hold true is $\phi \geq 0$ and

$$2\sigma^2 I - \phi\mathbf{x}\mathbf{x}^T \geq 0 \tag{2.227}$$

Left multiplying eqn. (2.227) by $\mathbf{x}^T$ and right multiplying the resulting expression by $\mathbf{x}$ leads to the following condition:

$$0 \leq \phi \leq \frac{2\sigma^2}{\mathbf{x}^T\mathbf{x}} \tag{2.228}$$

This guarantees that the inequality is satisfied; however, it is only a sufficient condition since we ignored the term $(H^T H)^{-1}$ in eqn. (2.226).

We can also choose to minimize the trace of Υ as well, which reduces the residual errors. Without loss in generality we can replace $H^T H$ with Λ, which is a diagonal matrix with elements given by the eigenvalues of $H^T H$. The trace of Υ is given by

$$\text{Tr}(\Upsilon) = \text{Tr}\left[(\Lambda+\phi I)^{-1}(\sigma^2\Lambda+\phi^2\mathbf{x}\mathbf{x}^T)(\Lambda+\phi I)^{-1}\right] \tag{2.229}$$

Therefore, we can now express the trace of Υ simply by

$$\text{Tr}(\Upsilon) = \sum_{i=1}^{n} \frac{\sigma^2\lambda_i+\phi^2 x_i^2}{(\lambda_i+\phi)^2} \tag{2.230}$$

where λ_i is the ith diagonal element of Λ. Minimizing eqn. (2.230) with respect to ϕ yields the following condition:

$$2\phi\sum_{i=1}^{n}\frac{\lambda_i x_i^2}{(\lambda_i+\phi)^3} - 2\sigma^2\sum_{i=1}^{n}\frac{\lambda_i}{(\lambda_i+\phi)^3} = 0 \tag{2.231}$$

Since $\mathbf{x}$ is unknown, the optimal ϕ cannot be determined *a priori*.[20] One possible procedure to determine ϕ involves plotting each component of $\hat{\mathbf{x}}$ against ϕ, which is called a *ridge trace*. The estimates will stabilize at a certain value of ϕ. Also, the residual sum squares should be checked so that the condition in eqn. (2.228) is met.

Example 2.10: As an example of the performance tradeoffs in ridge estimation, we will consider a simple case with $x = 1.5$, $\sigma^2 = 2$, and $\lambda = 2$. A plot of the ridge variance, the least squares variance, the ridge residual sum squares, and the bias-squared quantities as a function of the ridge parameter ϕ is shown in Figure 2.3. From eqn. (2.231), using the given parameters, the optimal value for ϕ is 0.89. This is verified in Figure 2.3. From eqn. (2.228), the region where the residual sum squares is less than the least squares residual is given by $0 \leq \phi \leq 1.778$, which is again verified in Figure 2.3. As mentioned previously, this is a conservative condition (the actual upper bound is 3.200). From Figure 2.3, we also see that the ridge variance is always less than the least squares variance; however, the bias increases as ϕ increases.

Ridge estimation provides a powerful tool that can produce estimates that have smaller residual errors than traditional least squares. It is especially useful when $H^T H$ is close to being singular. However, in practical engineering applications involving dynamic systems biases are usually not tolerated, and thus the advantage of ridge estimation is diminished. In short, careful attention needs to be placed by the design engineer in order to weigh the possible advantages with the inevitable biased estimates in the analysis of the system. Alternatively, it may be possible to justify a

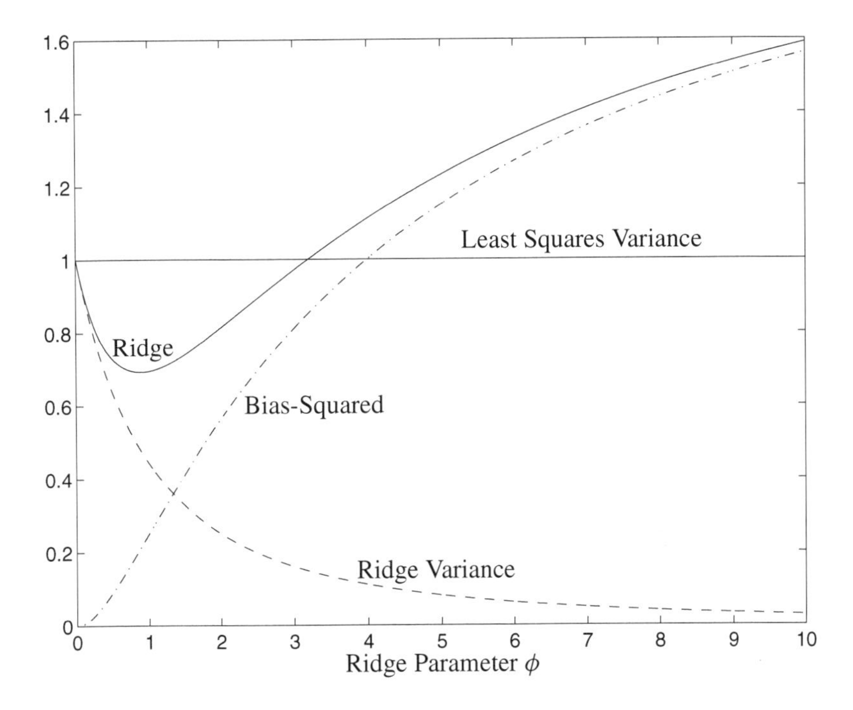

Figure 2.3: Ridge Estimation for a Scalar Case

particular ridge estimation process by using eqn. (2.79) for the case that a rigorous covariance Q is available for an *a priori* estimate $\hat{\mathbf{x}}_a$. Of course, in this theoretical setting, eqn. (2.79) is an unbiased estimator.

2.7.3 Total Least Squares

The standard least squares model in eqn. (2.1) assumes that there are no errors in the H matrix. Although this situation occurs in many systems, this assumption may not be always true. The least squares formulation in example 1.1 uses the measurements themselves in H, which contain random measurement errors. These "errors" were ignored in the least squares solution. Total least squares[21, 22] addresses errors in the H matrix, and can provide higher accuracy than ordinary least squares. In order to introduce this subject we begin by considering estimating a scalar parameter x:[22]

$$\tilde{\mathbf{y}} = \tilde{\mathbf{h}}x \tag{2.232}$$

with

$$\tilde{y}_i = y_i + v_i, \quad i = 1, 2, \ldots, m \tag{2.233a}$$

$$\tilde{h}_i = h_i + u_i, \quad i = 1, 2, \ldots, m \tag{2.233b}$$

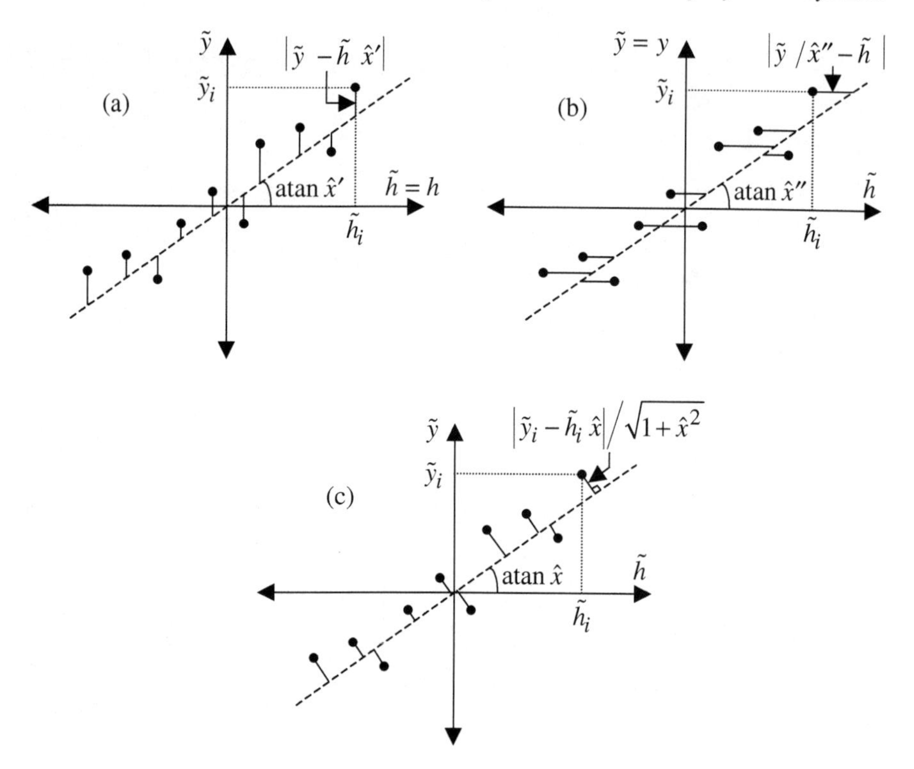

Figure 2.4: Geometric Interpretation of Total Least Squares

where v_i and u_i represent errors to the true values y_i and h_i, respectively.

When $u_i = 0$ then the estimate for x, denoted by $\hat{x}'$, is found by minimizing:

$$J(\hat{x}') = \sum_{i=1}^{m} (\tilde{y}_i - h_i\,\hat{x}')^2 \tag{2.234}$$

which yields

$$\hat{x}' = \left[\sum_{i=1}^{m} h_i^2\right]^{-1} \sum_{i=1}^{m} h_i\,\tilde{y}_i \tag{2.235}$$

The geometric interpretation of this result is shown by Case (a) in Figure 2.4. The residual is perpendicular to the $\tilde{h}$ axis. When $v_i = 0$ then the estimate for x, denoted by $\hat{x}''$, is found by the minimizing:

$$J(\hat{x}'') = \sum_{i=1}^{m} (y_i/\hat{x}'' - \tilde{h}_i)^2 \tag{2.236}$$

which yields

$$\hat{x}'' = \left[\sum_{i=1}^{m} \tilde{h}_i y_i\right]^{-1} \sum_{i=1}^{m} y_i^2 \tag{2.237}$$

The geometric interpretation of this result is shown by Case (b) in Figure 2.4. The residual is perpendicular to the $\tilde{y}$ axis. If the errors in both y_i and h_i have zero mean and have the same variance, then the total least squares estimate for x, denoted $\hat{x}$, is found by minimizing the sum of squared distances of the measurement points from the fitted line:

$$J(\hat{x}) = \sum_{i=1}^{m} (\tilde{y}_i - h_i \hat{x}')^2 / (1 + \hat{x}^2) \tag{2.238}$$

The solution for this minimization problem will be shown later. The geometric interpretation of this result is shown by Case (c) in Figure 2.4. The residual is now perpendicular to the fitted line. This geometric interpretation leads to the *orthogonal regression* approach in the total least squares problem.

For the general problem, the total least squares model is given by

$$\tilde{\mathbf{y}} = (H + U)\mathbf{x} + \mathbf{v} \tag{2.239}$$

where U represents the error to the model H. We assume $E\{\mathbf{v}\} = \mathbf{0}$ and $E\{\mathbf{v}\mathbf{v}^T\} = \sigma^2 I$ (i.e., the errors are isotropic). Furthermore we assume that the rows of U have zero mean with the same variance as the measurement error $(\sigma^2 I)$. If this is not a valid assumption, then the matrix $\left[H \; \tilde{\mathbf{y}}\right]$ can be multiplied by an appropriate $m \times m$ matrix D such that the assumption is valid.[23]

The total least squares problem seeks an optimal estimate of $\mathbf{x}$ that minimizes

$$J = \left|\left|\left[\tilde{H} \; \tilde{\mathbf{y}}\right] - \left[\hat{H} \; \hat{\mathbf{y}}\right]\right|\right|_F \tag{2.240}$$

where $\tilde{H} = H + U$ and $||\cdot||_F$ denotes the Frobenius norm (see §A.3). The solution for the total least squares problem is given by taking a singular value decomposition of the following augmented matrix:

$$\left[\tilde{H} \; \tilde{\mathbf{y}}\right] = U S V^T \tag{2.241}$$

with $S = \text{diag}\left[s_1 \; \cdots \; s_{n+1}\right]$. The total least squares solution is then given by[24]

$$\boxed{\hat{\mathbf{x}}_{\text{TLS}} = (\tilde{H}^T \tilde{H} - s_{n+1}^2 I)^{-1} \tilde{H}^T \tilde{\mathbf{y}}} \tag{2.242}$$

Notice the resemblance to ridge estimation in §2.7.2, but here the positive multiple is subtracted from $\tilde{H}^T \tilde{H}$. Therefore, the total least squares problem is a *deregularization* of the least squares problem, which means that it is always worse conditioned than the ordinary least squares problem.

Total least squares has been shown to provide parameter error accuracy gains of 10 to 15 percent in typical applications.[24] In order to quantify the bounds on the difference between total least squares and ordinary least squares we begin by using the following identity:

$$(\tilde{H}^T \tilde{H} - s_{n+1}^2 I)\hat{\mathbf{x}}_{\text{LS}} = \tilde{H}^T \tilde{\mathbf{y}} - s_{n+1}^2 \hat{\mathbf{x}}_{\text{LS}} \tag{2.243}$$

Subtracting eqn. (2.243) from eqn. (2.242) leads to

$$\hat{\mathbf{x}}_{\text{TLS}} - \hat{\mathbf{x}}_{\text{LS}} = s_{n+1}^2 (\tilde{H}^T \tilde{H} - s_{n+1}^2 I)^{-1} \hat{\mathbf{x}}_{\text{LS}} \tag{2.244}$$

Using the norm inequality now leads to:

$$\frac{||\hat{\mathbf{x}}_{\text{TLS}} - \hat{\mathbf{x}}_{\text{LS}}||}{||\hat{\mathbf{x}}_{\text{LS}}||} \leq \frac{s_{n+1}^2}{\bar{s}_n^2 - s_{n+1}^2} \tag{2.245}$$

where $\bar{s}_n$ is the smallest singular value of $\tilde{H}$ and the assumption $\bar{s}_n > s_{n+1}$ must be valid. The accuracy of total least squares will be more pronounced when the ratio of the singular values $\bar{s}_n$ and s_{n+1} is large. The "errors-in-variables" estimator shown in Ref. [25] coincides with the total least squares solution. This indicates that the total least squares estimate is a strongly consistent estimate for large samples, which leads to an asymptotic unbiasedness property. Ordinary least squares with errors in H produces biased estimates as the sample size increases. However, the covariance of total least squares is larger than the ordinary least squares covariance, but by increasing the noise in the measurements the bias of ordinary least squares becomes more important and even the dominating term.[22] Several aspects and properties of the total least squares problem can be found in the references cited in this section.

Example 2.11: We will show the advantages of total least squares by re-considering the problem of estimating the parameters of a simple dynamic system shown in example 1.1. To compare the accuracy of total least squares with ordinary least squares we will use the square root of the diagonal elements of mean-squared-error (MSE) matrix, defined as

$$\begin{aligned}
\text{MSE} &= E\left\{(\hat{\mathbf{x}} - \mathbf{x})(\hat{\mathbf{x}} - \mathbf{x})^T\right\} \\
&= E\left\{(\hat{\mathbf{x}} - E\{\hat{\mathbf{x}}\})(\hat{\mathbf{x}} - E\{\hat{\mathbf{x}}\})^T\right\} + \left\{(E\{\hat{\mathbf{x}}\} - \mathbf{x})(E\{\hat{\mathbf{x}}\} - \mathbf{x})^T\right\} \\
&= \text{cov}\{\hat{\mathbf{x}}\} + \text{squared bias}\{\hat{\mathbf{x}}\}
\end{aligned}$$

For this particular problem it is known that u is given by an impulse input with magnitude $10/\Delta t$ (i.e., $u_1 = 10/\Delta t$ and $u_k = 0$ for $k \geq 2$). A total of 10 seconds is considered with sampling intervals ranging from $\Delta t = 2$ seconds down to $\Delta t = 0.001$ seconds. Synthetic measurements are again generated with $\sigma = 0.08$. This example tests the accuracy of both approaches for various measurement sample lengths (i.e., from 5 samples when $\Delta t = 2$ to 10,000 samples when $\Delta t = 0.001$). For each simulation 1,000 runs were performed each with different random number seeds. Results for $\hat{\Phi}$ are given in the following table:

Δt	$\text{bias}\{\hat{\Phi}\}_{LS}$	$\text{bias}\{\hat{\Phi}\}_{TLS}$	$\sqrt{\text{MSE}\{\hat{\Phi}\}_{LS}}$	$\sqrt{\text{MSE}\{\hat{\Phi}\}_{TLS}}$
2	3.12×10^{-4}	3.89×10^{-4}	1.82×10^{-2}	1.83×10^{-2}
1	5.52×10^{-4}	2.43×10^{-4}	1.12×10^{-2}	1.12×10^{-2}
0.5	1.03×10^{-3}	3.67×10^{-4}	6.36×10^{-3}	6.28×10^{-3}
0.1	1.24×10^{-3}	9.68×10^{-5}	1.99×10^{-3}	1.54×10^{-3}
0.05	1.23×10^{-3}	2.30×10^{-5}	1.47×10^{-3}	7.90×10^{-4}
0.01	1.26×10^{-3}	7.08×10^{-6}	1.28×10^{-3}	1.62×10^{-4}
0.005	1.27×10^{-3}	3.48×10^{-6}	1.27×10^{-3}	8.26×10^{-5}
0.001	1.28×10^{-3}	5.32×10^{-7}	1.27×10^{-3}	1.60×10^{-5}

Results for $\hat{\Gamma}$ are given in the following table:

Δt	$\text{bias}\{\hat{\Gamma}\}_{LS}$	$\text{bias}\{\hat{\Gamma}\}_{TLS}$	$\sqrt{\text{MSE}\{\hat{\Gamma}\}_{LS}}$	$\sqrt{\text{MSE}\{\hat{\Gamma}\}_{TLS}}$
2	1.37×10^{-4}	1.11×10^{-4}	8.37×10^{-3}	8.78×10^{-3}
1	1.32×10^{-4}	6.24×10^{-5}	6.64×10^{-3}	6.71×10^{-3}
0.5	1.29×10^{-4}	2.25×10^{-5}	4.76×10^{-3}	4.76×10^{-3}
0.1	1.52×10^{-5}	2.11×10^{-5}	1.07×10^{-3}	1.07×10^{-3}
0.05	2.71×10^{-5}	2.87×10^{-5}	5.61×10^{-4}	5.62×10^{-4}
0.01	7.04×10^{-6}	7.10×10^{-6}	1.12×10^{-4}	1.13×10^{-4}
0.05	2.02×10^{-6}	2.00×10^{-6}	5.90×10^{-5}	5.91×10^{-5}
0.001	1.79×10^{-7}	2.78×10^{-7}	1.10×10^{-5}	1.11×10^{-5}

These tables indicate that when using a small sample size ordinary least squares and total least squares have the same accuracy. However, as the sampling interval decreases (i.e., giving more measurements) the bias in $\hat{\Phi}$ increases using ordinary least squares, but substantially decreases using total least squares. Also, the bias is the dominating term in the MSE when the sample size is large. Results for $\hat{\Gamma}$ indicate that the ordinary least squares estimate is comparable to the total least squares estimate. This is due to the fact that u contains no errors. Nevertheless, this example clearly shows that improvements can be made using total least squares.

2.8 Summary

In this chapter we have presented several approaches to establish a class of linear estimation algorithms, and we have developed certain important properties of the weighting matrix used in weighted least squares. The end products of the developments for minimum variance estimation in §2.1.1 and maximum likelihood estimation in §2.3 are seen to be equivalent for Gaussian measurement errors to the linear weighted least squares results of §1.2.2, with interpretation of the weight matrix as the measurement-error covariance matrix. An interesting result is that several different theoretical/conceptual estimation approaches give the same estimator. In particular, when weighing the advantages and disadvantages of each approach one realizes that maximum likelihood provides a solution more directly than minimum variance, since a constrained optimization problem is not required. Therefore, in practice, maximum likelihood estimation is usually preferred over minimum variance. Several useful properties were also derived in this chapter, including unbiased estimates and the Cramér-Rao inequality. In estimation of dynamic systems, an unbiased estimate is always preferred, if obtainable, over a biased estimate. Also, an efficient estimator, which is achieved if the equality in the Cramér-Rao inequality is satisfied, gives the lowest estimation error possible from a statistical point of view. This allows the design engineer to quantify the performance of an estimation algorithm using a covariance analysis on the expected performance.

The interpretation of the *a priori* estimates in §2.1.2 is given as a measurement subset in the sequential least squares developments of §1.3. Several other approaches, such as maximum *a posteriori* estimation and minimum risk estimation of §2.6 were shown to be equivalent to the minimum variance solution of §2.1.2. Each of these approaches provides certain illuminations and useful insights. Maximum *a posteriori* estimation is usually preferred over the other approaches since it follows many of the same principles and properties of maximum likelihood estimation, and in fact reduces to the maximum likelihood estimate if the *a priori* distribution is uniform or for large samples. The Cramér-Rao bound for *a priori* estimation was also shown, which again provides a lower bound on the estimation error.

In §2.5 a discussion on the nonuniqueness of the weight matrix was given. It should be noted that specification and calculations involving the weight matrices are the source of most practical difficulties encountered in applications. Additionally, an analysis of errors in the assumed measurement-error covariance matrix was shown in §2.7.1. This analysis can be useful to quantify the expected performance of the estimate in the face of an incorrectly defined measurement-error covariance matrix. Ridge estimation, shown in §2.7.2, is useful for the case of weak multicollinear systems. This case involves the near ill-conditioning of the matrix to be inverted in the least squares solutions. It has also been established that the ridge estimate covariance is less than the least squares estimate covariance. However, if the least squares solution is well posed, then the advantage of a lower covariance is strongly outweighed by the inevitable biased estimate in ridge estimation. Also, a connection between ridge estimation and *a priori* state estimation has been established by noting that resemblance of the ridge parameter to the *a priori* covariance. Finally, total least squares, shown in §2.7.3, can give significant improvements in the accuracy of the estimates over ordinary least squares if errors are present in the model matrix. This approach synthesizes an optimal methodology for solving a variety of problems in many dynamic system applications.

A summary of the key formulas presented in this chapter is given below.

- Gauss-Markov Theorem

$$\tilde{\mathbf{y}} = H\mathbf{x} + \mathbf{v}$$
$$E\{\mathbf{v}\} = \mathbf{0}, \quad E\left\{\mathbf{v}\,\mathbf{v}^T\right\} = R$$
$$\hat{\mathbf{x}} = (H^T R^{-1} H)^{-1} H^T R^{-1} \tilde{\mathbf{y}}$$

- *A priori* Estimation

$$\tilde{\mathbf{y}} = H\mathbf{x} + \mathbf{v}$$
$$E\{\mathbf{v}\} = \mathbf{0}, \quad E\left\{\mathbf{v}\,\mathbf{v}^T\right\} = R$$
$$\hat{\mathbf{x}}_a = \mathbf{x} + \mathbf{w}$$
$$E\{\mathbf{w}\} = \mathbf{0}, \quad E\left\{\mathbf{w}\,\mathbf{w}^T\right\} = Q$$
$$\hat{\mathbf{x}} = \left(H^T R^{-1} H + Q^{-1}\right)^{-1} \left(H^T R^{-1} \tilde{\mathbf{y}} + Q^{-1} \hat{\mathbf{x}}_a\right)$$

- Unbiased Estimates

$$E\left\{\hat{\mathbf{x}}_k(\tilde{\mathbf{y}})\right\} = \mathbf{x} \quad \text{for all } k$$

- Maximum Likelihood Estimation

$$L(\tilde{\mathbf{y}}; \mathbf{x}) = \prod_{i=1}^{p} f_i(\tilde{\mathbf{y}}; \mathbf{x})$$
$$\left\{\frac{\partial}{\partial \mathbf{x}} \ln\left[L(\tilde{\mathbf{y}}; \mathbf{x})\right]\right\}\Bigg|_{\hat{\mathbf{x}}} = \mathbf{0}$$

- Cramér-Rao Inequality

$$P \equiv E\left\{(\hat{\mathbf{x}} - \mathbf{x})(\hat{\mathbf{x}} - \mathbf{x})^T\right\} \geq F^{-1}$$
$$F = -E\left\{\frac{\partial^2}{\partial \mathbf{x}\,\partial \mathbf{x}^T} \ln f(\tilde{\mathbf{y}}; \mathbf{x})\right\}$$

- Bayes Rule

$$f(\mathbf{x}|\tilde{\mathbf{y}}) = \frac{f(\tilde{\mathbf{y}}|\mathbf{x})\, f(\mathbf{x})}{f(\tilde{\mathbf{y}})}$$

- Maximum *A Posteriori* Estimation

$$J_{\text{MAP}}(\hat{\mathbf{x}}) = \ln\left[L(\tilde{\mathbf{y}}|\hat{\mathbf{x}})\right] + \ln\left[f(\hat{\mathbf{x}})\right]$$

- Cramér-Rao Inequality for Bayesian Estimators

$$P \equiv E\left\{(\hat{\mathbf{x}} - \mathbf{x})(\hat{\mathbf{x}} - \mathbf{x})^T\right\}$$
$$\geq \left[F + E\left\{\left[\frac{\partial}{\partial \mathbf{x}} \ln f(\mathbf{x})\right]\left[\frac{\partial}{\partial \mathbf{x}} \ln f(\mathbf{x})\right]^T\right\}\right]^{-1}$$

- Minimum Risk Estimation

$$J_{\text{MR}}(\mathbf{x}^*) = \int_{-\infty}^{\infty} c(\mathbf{x}^*|\mathbf{x}) \frac{f(\tilde{\mathbf{y}}|\mathbf{x}) f(\mathbf{x})}{f(\tilde{\mathbf{y}})} d\mathbf{x}$$

$$c(\mathbf{x}|\mathbf{x}^*) = \frac{1}{2}(\mathbf{x}^* - \mathbf{x})^T S(\mathbf{x}^* - \mathbf{x})$$

$$\hat{\mathbf{x}} = \int_{-\infty}^{\infty} \mathbf{x} \frac{f(\tilde{\mathbf{y}}|\mathbf{x}) f(\mathbf{x})}{f(\tilde{\mathbf{y}})} d\mathbf{x}$$

- Inefficiency for Covariance Errors

$$e = \frac{\det\left[(H^T \tilde{R}^{-1} H)^{-1} H^T \tilde{R}^{-1} R \tilde{R}^{-1} H (H^T \tilde{R}^{-1} H)^{-1}\right]}{\det\left[(H^T R^{-1} H)^{-1}\right]}$$

- Ridge Estimation

$$\hat{\mathbf{x}} = (H^T H + \phi I)^{-1} H^T \tilde{\mathbf{y}}$$

- Total Least Squares

$$\tilde{\mathbf{y}} = \tilde{H}\mathbf{x} + \mathbf{v}$$
$$\left[\tilde{H}\ \tilde{\mathbf{y}}\right] = U S V^T$$
$$S = \text{diag}\left[s_1 \ \cdots \ s_{n+1}\right]$$
$$\hat{\mathbf{x}}_{\text{TLS}} = (\tilde{H}^T \tilde{H} - s_{n+1}^2 I)^{-1} \tilde{H}^T \tilde{\mathbf{y}}$$

Exercises

2.1 Consider estimating a constant unknown variable x, which is measured twice with some error

$$\tilde{y}_1 = x + v_1$$
$$\tilde{y}_2 = x + v_2$$

where the random errors have the following properties:

$$E\{v_1\} = E\{v_2\} = E\{v_1 v_2\} = 0$$
$$E\left\{v_1^2\right\} = 1$$
$$E\left\{v_2^2\right\} = 4$$

Perform a weighted least squares solution with $H = \begin{bmatrix} 1 & 1 \end{bmatrix}^T$ for the following two cases:

$$W = \frac{1}{2}\begin{bmatrix} 1 & 0 \\ 0 & 1 \end{bmatrix}$$

and

$$W = \frac{1}{4}\begin{bmatrix} 4 & 0 \\ 0 & 1 \end{bmatrix}$$

Compute the variance of the estimation error (i.e. $E\left\{(x - \hat{x})^2\right\}$) and compare the results.

2.2 Write a simple computer program to simulate measurements of some discretely measured process

$$\tilde{y}_j = x_1 + x_2 \sin(10t_j) + x_3 e^{2t_j^2} + v_j, \quad j = 1, 2, \ldots, 11$$

with t_j sampled every 0.1 seconds. The true values (x_1, x_2, x_3) are $(1, 1, 1)$ and the measurement errors are synthetic Gaussian random variables with zero mean. The measurement-error covariance matrix is diagonal with

$$R = E\left\{\mathbf{v}\mathbf{v}^T\right\} = \text{diag}\left[\sigma_1^2 \ \sigma_2^2 \ \cdots \ \sigma_{11}^2\right]$$

where

$$\begin{array}{llll} \sigma_1 = 0.001 & \sigma_2 = 0.002 & \sigma_3 = 0.005 & \sigma_4 = 0.010 \\ \sigma_5 = 0.008 & \sigma_6 = 0.002 & \sigma_7 = 0.010 & \sigma_8 = 0.007 \\ \sigma_9 = 0.020 & \sigma_{10} = 0.006 & \sigma_{11} = 0.001 & \end{array}$$

You are also given the *a priori* $\mathbf{x}$-estimates

$$\hat{\mathbf{x}}_a^T = (1.01, 0.98, 0.99)$$

and associated *a priori* covariance matrix

$$Q = \begin{bmatrix} 0.001 & 0 & 0 \\ 0 & 0.001 & 0 \\ 0 & 0 & 0.001 \end{bmatrix}$$

Your tasks are as follows:

(A) Use the minimal variance estimation version of the normal equations

$$\hat{\mathbf{x}} = P\left(H^T R^{-1} \tilde{\mathbf{y}} + Q^{-1} \hat{\mathbf{x}}_a\right)$$

to compute the parameter estimates and estimate covariance matrix

$$P = \left(H^T R^{-1} H + Q^{-1}\right)^{-1}$$

with the jth row of H given by $\begin{bmatrix} 1 & \sin(10t_j) & e^{2t_j^2} \end{bmatrix}$. Calculate the mean and standard deviation of the residual

$$r_j = \tilde{y}_j - \left(\hat{x}_1 + \hat{x}_2 \sin(10t_j) + \hat{x}_3 e^{2t_j^2}\right)$$

as

$$r = \frac{1}{11} \sum_{j=1}^{11} r_j$$

$$\sigma_r = \left[\frac{1}{10} \sum_{j=1}^{11} r_j^2\right]^{\frac{1}{2}}$$

(B) Do a parametric study in which you hold the *a priori* estimate covariance Q fixed, but vary the measurement-error covariance according to

$$R' = \alpha R$$

with $\alpha = 10^{-3}$, 10^{-2}, 10^{-1}, 10, 10^2, 10^3. Study the behavior of the calculated results for the estimates $\hat{\mathbf{x}}$, the estimate covariance matrix P, and mean r and standard deviation σ_r of the residual.

(C) Do a parametric study in which R is held fixed, but Q is varied according to

$$Q' = \alpha Q$$

with α taking the same values as in (B). Compare the results for the estimates $\hat{\mathbf{x}}$, the estimate covariance matrix P, and mean r and standard deviation σ_r of the residual with those of part (B).

2.3 Suppose that $\mathbf{v}$ in exercise 1.3 is a constant vector (i.e., a *bias error*). Evaluate the loss function (2.111) in terms of v_i only and discuss how the value of the loss function changes with a bias error in the measurements instead of a zero mean assumption.

2.4 Consider the following constrained minimization problem:

$$\begin{aligned} &\text{minimize} \quad J(\hat{\mathbf{x}}) = (\tilde{\mathbf{y}} - H\hat{\mathbf{x}})^T R^{-1} (\tilde{\mathbf{y}} - H\hat{\mathbf{x}}) \\ &\text{subject to} \quad S\hat{\mathbf{x}} = \mathbf{z} \end{aligned}$$

where $\mathbf{z}$ and S are known (the matrix S has dimensions less than or equal to the dimension of the vector $\hat{\mathbf{x}}$). You are to perform the following tasks:

(A) Determine the least squares estimate $\hat{\mathbf{x}}$.

(B) Check to see if the estimator is unbiased.

(C) Derive the estimate covariance matrix.

(D) Determine the estimate when S is square with rank n.

2.5 A "Monte Carlo" approach to calculating covariance matrices is often necessary for nonlinear problems. The algorithm has the following structure: Given a functional dependence of two sets of random variables in the form

$$z_i = F_i(y_1, y_2, \ldots, y_m), \quad i = 1, 2, \ldots, n$$

where the y_j are random variables whose joint probability density function is known and the F_i are generally nonlinear functions. The Monte Carlo approach requires that the probability density function of y_j be sampled many times to calculate corresponding samples of the z_i joint distribution. Thus if the kth particular sample ("simulated measurement") of the y_j values is denoted as

$$(\tilde{y}_{1k}, \tilde{y}_{2k}, \ldots, \tilde{y}_{mk}), \quad k = 1, 2, \ldots, M$$

then the corresponding z_i sample is calculated as

$$z_{ik} = F_i(\tilde{y}_{1k}, \tilde{y}_{2k}, \ldots, \tilde{y}_{mk}), \quad k = 1, 2, \ldots, M$$

The first two moments of z_i's joint density function are then approximated by

$$\mu_i = E\{z_{ik}\} \simeq \frac{1}{M} \sum_{k=1}^{M} z_{ik}$$

and

$$\hat{R} = E\left\{(\mathbf{z} - \boldsymbol{\mu})(\mathbf{z} - \boldsymbol{\mu})^T\right\} \simeq \frac{1}{M-1} \sum_{k=1}^{M} [\mathbf{z}_k - \boldsymbol{\mu}][\mathbf{z}_k - \boldsymbol{\mu}]^T$$

where

$$\mathbf{z}_k^T \equiv (z_{1k}, z_{2k}, \ldots, z_{nk})$$
$$\boldsymbol{\mu}^T \equiv (\mu_1, \mu_2, \ldots, \mu_n)$$

The Monte Carlo approach can be used to experimentally verify the interpretation of $P = (H^T R^{-1} H)^{-1}$ as the $\hat{\mathbf{x}}$ covariance matrix in the minimal variance estimate

$$\hat{\mathbf{x}} = P H^T R^{-1} \tilde{\mathbf{y}}$$

To carry out this experiment, use the model in exercise 2.2 to simulate $M = 100$ sets of y-measurements. For each set (e.g., the kth) of the measurements, the corresponding $\hat{\mathbf{x}}$ follows as

$$\hat{\mathbf{x}}_k = P H^T R^{-1} \tilde{\mathbf{y}}_k$$

Then the $\hat{\mathbf{x}}$ mean and covariance matrices can be approximated by

$$\boldsymbol{\mu}_x = E\{\hat{\mathbf{x}}\} \simeq \frac{1}{M} \sum_{k=1}^{M} \hat{\mathbf{x}}_k$$

and

$$\hat{R}_{xx} = E\left\{(\hat{\mathbf{x}} - \boldsymbol{\mu}_x)(\hat{\mathbf{x}} - \boldsymbol{\mu}_x)^T\right\} \simeq \frac{1}{M-1} \sum_{k=1}^{M} [\hat{\mathbf{x}}_k - \boldsymbol{\mu}_x][\hat{\mathbf{x}}_k - \boldsymbol{\mu}_x]^T$$

In your simulation $\hat{R}_{xx}$ should be compared element-by-element with the covariance $P = (H^T R^{-1} H)^{-1}$, whereas μ_x should compare favorably with the true values $\mathbf{x}^T = (1, 1, 1)$.

2.6 Let $\tilde{\mathbf{y}} \sim \mathcal{N}(\mu, R)$. Show that

$$\hat{\mu} = \frac{1}{m} \sum_{k=1}^{M} \tilde{\mathbf{y}}_i$$

is an efficient estimator for the mean.

2.7 Consider estimating a constant unknown variable x, which is measured twice with some error

$$\tilde{y}_1 = x + v_1$$
$$\tilde{y}_2 = x + v_2$$

where the random errors have the following properties:

$$E\{v_1\} = E\{v_2\} = 0$$
$$E\left\{v_1^2\right\} = \sigma_1^2$$
$$E\left\{v_2^2\right\} = \sigma_2^2$$

The errors follow a bivariate normal distribution with joint density function given by

$$f(v_1, v_2) = \frac{1}{2\pi\sigma_1\sigma_2(1-\rho^2)} \exp\left[-\frac{1}{2(1-\rho^2)}\left(\frac{v_1^2}{\sigma_1^2} - \frac{2\rho v_1 v_2}{\sigma_1\sigma_2} + \frac{v_2^2}{\sigma_2^2}\right)\right]$$

where the correlation coefficient, ρ, is defined as

$$\rho \equiv \frac{E\{v_1 v_2\}}{\sigma_1\sigma_2}$$

Derive the maximum likelihood estimate for x. Also, how does the estimate change when $\rho = 0$?

2.8 Suppose that z_1 is the mean of a random sample of size m from a normal distributed system with mean μ and variance σ_1^2, and z_2 is the mean of a random sample of size m from a normal distributed system with mean μ and variance σ_2^2. Show that $\hat{\mu} = \alpha z_1 + (1-\alpha) z_2$, where $0 \le \alpha \le 1$, is an unbiased estimate of μ. Also, show that the variance of the estimate is minimum when $\alpha = \sigma_2^2(\sigma_1^2 + \sigma_2^2)^{-1}$.

2.9 Show that if $\hat{x}$ is an unbiased estimate of x and $\text{var}\{\hat{x}\}$ does not equal 0, then $\hat{x}^2$ is not an unbiased estimate of x^2.

2.10 If $\hat{x}$ is an estimate of x, its bias is $b = E\{\hat{x}\} - x$. Show that $E\left\{(\hat{x} - x)^2\right\} = \text{var}\{\hat{x}\} + b^2$.

2.11 Prove that the *a priori* estimator given in eqn. (2.47) is unbiased when $MH + N = I$ and $\mathbf{n} = \mathbf{0}$.

2.12 Prove that the Cramér-Rao inequality given by eqn. (2.112) achieves the equality if and only if

$$\left[\frac{\partial}{\partial \mathbf{x}} \ln f(\tilde{\mathbf{y}}; \mathbf{x})\right] = c\,(\mathbf{x} - \hat{\mathbf{x}})$$

where c is independent of $\mathbf{x}$ and $\tilde{\mathbf{y}}$.

2.13 Suppose that an estimator of a non-random scalar x is biased, with bias denoted by $b(x)$. Show that a lower bound on the variance of the estimate $\hat{x}$ is given by

$$\text{var}(\hat{x} - x) \geq \left(1 - \frac{db}{dx}\right)^2 J^{-1}$$

where

$$J = E\left\{\left[\frac{\partial}{\partial x} \ln f(\tilde{\mathbf{y}}; x)\right]^2\right\}$$

and

$$b(x) \equiv \int_{-\infty}^{\infty} (x - \hat{x})\, f(\tilde{\mathbf{y}}; x)\, d\tilde{\mathbf{y}}$$

2.14 Prove that the estimate for the covariance in example 2.4 is biased. Also, what is the unbiased estimate?

2.15 Prove that eqn. (2.114) is equivalent to eqn. (2.113).

2.16 Perform a simulation of the parameter identification problem shown in example 2.6 with $B = 10$ and varying σ for the measurement noise. Compare the nonlinear least squares solution to the linear approach for various noise levels. Also, check the performance of the two approaches by comparing P with $\mathcal{P}$. At what measurement noise level does the linear solution begin to degrade from the nonlinear least squares solution?

2.17 ♣ In example 2.6 an expression for the variance of the new measurement noise, denoted by ϵ_k, is derived. Prove the following expression:

$$E\left\{\left(\frac{v_k}{B\,e^{a\,t_k}} - \frac{v_k^2}{2\,B^2 e^{2a\,t_k}}\right)^2\right\} = \frac{\sigma^2}{B^2 e^{2a\,t_k}} + \frac{3\,\sigma^4}{4\,B^4 e^{4a\,t_k}}$$

Hint: use the theory behind χ^2 distributions.

2.18 ♣ Prove the inequality in eqn. (2.159).

2.19 The parallel axis theorem was used several times in this chapter to derive the covariance expression, e.g., in eqn. (2.181). Prove the following identity:

$$E\left\{(\mathbf{x} - E\{\mathbf{x}\})\,(\mathbf{x} - E\{\mathbf{x}\})^T\right\} = E\left\{\mathbf{x}\,\mathbf{x}^T\right\} - E\{\mathbf{x}\}\,E\{\mathbf{x}\}^T$$

2.20 Fully derive the density function given in eqn. (2.183).

2.21 Show that $\mathbf{e}^T R^{-1}\mathbf{e}$ is equivalent to $\text{Tr}\left(R^{-1}E\right)$ with $E = \mathbf{e}\,\mathbf{e}^T$.

2.22 Prove that $E\left\{\mathbf{x}^T A\mathbf{x}\right\} = \boldsymbol{\mu}^T A\boldsymbol{\mu} + \text{Tr}(A\,\Xi)$, where $E\{\mathbf{x}\} = \boldsymbol{\mu}$ and $\text{cov}(\mathbf{x}) = \Xi$.

2.23 Prove the following results for the *a priori* estimator in eqn. (2.186):

$$E\left\{\mathbf{x}\hat{\mathbf{x}}^T\right\} = E\left\{\hat{\mathbf{x}}\hat{\mathbf{x}}^T\right\}$$

$$\left(H^T R^{-1} H + Q^{-1}\right)^{-1} = E\left\{\mathbf{x}\mathbf{x}^T\right\} - E\left\{\hat{\mathbf{x}}\hat{\mathbf{x}}^T\right\}$$

$$E\left\{\mathbf{x}\mathbf{x}^T\right\} \geq E\left\{\hat{\mathbf{x}}\hat{\mathbf{x}}^T\right\}$$

2.24 Consider the 2×2 case for $\tilde{R}$ and R in eqn. (2.199). Verify that the inefficiency e in eqn. (2.210) is bounded by

$$1 \leq e \leq \frac{(\lambda_{\max} + \lambda_{\min})^2}{4\lambda_{\max}\lambda_{\min}}$$

where $\lambda_{\max}$ and $\lambda_{\min}$ are the maximum and minimum eigenvalues of the matrix $\tilde{R}^{-1/2} R\,\tilde{R}^{-1/2}$. Note, this inequality does not generalize to the case where $m \geq 3$.

2.25 ♣ An alternative to minimizing the trace of Υ in §2.7.2 is to minimize the generalized cross-validation (GRV) error prediction,[26] given by

$$\hat{\sigma}^2 = \frac{m\,\tilde{\mathbf{y}}^T \mathcal{P}^2 \tilde{\mathbf{y}}}{\text{Tr}(P)^2}$$

where m is the dimension of the vector $\tilde{y}$ and $\mathcal{P}$ is a projection matrix, given by

$$\mathcal{P} = I - H(H^T H + \phi I)^{-1} H^T$$

Determine the minimum of the GRV error, as a function of the ridge parameter ϕ. Also, prove that $\mathcal{P}$ is a projection matrix.

2.26 Consider the following model:

$$y = x_1 + x_2 t + x_3 t^2$$

Create a set of 101 noise-free observations at 0.01-second intervals with $x_1 = 3$, $x_2 = 2$, and $x_3 = 1$. Form the H matrix to be used in least squares with basis functions given by $\left\{1,\, t,\, t^2,\, 2t + 3t^2\right\}$. Show that H is rank deficient. Use the ridge estimator in eqn. (2.211) to determine the parameter estimates with the aforementioned basis functions. How does varying ϕ affect the solution?

2.27 Write a computer program to reproduce the total least squares results shown in example 2.11.

2.28 ♣ Let the last column (denoted by $\mathbf{v}$) of the matrix of V in eqn. (2.241) be partitioned into the $n \times 1$ vector ($\mathbf{g}$) and the remaining scalar component (γ), such that $\mathbf{v} = \left[\mathbf{g}^T \ \gamma\right]^T$. Show that the total least squares estimate in eqn. (2.242) is equivalent to $\hat{\mathbf{x}}_{\text{TLS}} = -\gamma^{-1}\mathbf{g}$.

References

[1] Berry, D.A. and Lingren, B.W., *Statistics, Theory and Methods*, Brooks/Cole Publishing Company, Pacific Grove, CA, 1990.

[2] Goldstein, H., *Classical Mechanics*, Addison-Wesley Publishing Company, Reading, MA, 2nd ed., 1980.

[3] Baruh, H., *Analytical Dynamics*, McGraw-Hill, Boston, MA, 1999.

[4] Devore, J.L., *Probability and Statistics for Engineering and Sciences*, Duxbury Press, Pacific Grove, CA, 1995.

[5] Sorenson, H.W., *Parameter Estimation, Principles and Problems*, Marcel Dekker, New York, NY, 1980.

[6] Sage, A.P. and Melsa, J.L., *Estimation Theory with Applications to Communications and Control*, McGraw-Hill Book Company, New York, NY, 1971.

[7] Freund, J.E. and Walpole, R.E., *Mathematical Statistics*, Prentice Hall, Englewood Cliffs, NJ, 4th ed., 1987.

[8] Cramér, H., *Mathematical Methods of Statistics*, Princeton University Press, Princeton, NJ, 1946.

[9] Fisher, R.A., *Contributions to Mathematical Statistics (collection of papers published 1920-1943)*, Wiley, New York, NY, 1950.

[10] Fisher, R.A., *Statistical Methods and Scientific Inference*, Hafner Press, New York, NY, 3rd ed., 1973.

[11] Stein, S.K., *Calculus and Analytic Geometry*, McGraw-Hill Book Company, New York, NY, 3rd ed., 1982.

[12] Crassidis, J.L. and Markley, F.L., "New Algorithm for Attitude Determination Using Global Positioning System Signals," *Journal of Guidance, Control, and Dynamics*, Vol. 20, No. 5, Sept.-Oct. 1997, pp. 891–896.

[13] Bard, Y., *Nonlinear Parameter Estimation*, Academic Press, New York, NY, 1974.

[14] Walter, E. and Pronzato, L., *Identification of Parametric Models from Experimental Data*, Springer Press, Paris, France, 1994.

[15] Schoukens, J. and Pintelon, R., *Identification of Linear Systems, A Practical Guide to Accurate Modeling*, Pergamon Press, Oxford, Great Britain, 1991.

[16] Horn, R.A. and Johnson, C.R., *Matrix Analysis*, Cambridge University Press, Cambridge, MA, 1985.

[17] Toutenburg, H., *Prior Information in Linear Models*, John Wiley & Sons, New York, NY, 1982.

[18] Magnus, J.R., *Matrix Differential Calculus with Applications in Statistics and Econometrics*, John Wiley & Sons, New York, NY, 1997.

[19] Hoerl, A.E. and Kennard, R.W., "Ridge Regression: Biased Estimation for Nonorthogonal Problems," *Technometrics*, Vol. 12, No. 1, Feb. 1970, pp. 55–67.

[20] Vinod, H.D., "A Survey of Ridge Regression and Related Techniques for Improvements Over Ordinary Least Squares," *The Review of Economics and Statistics*, Vol. 60, No. 1, Feb. 1978, pp. 121–131.

[21] Golub, G.H. and Van Loan, C.F., "An Analysis of the Total Least Squares Problem," *SIAM Journal on Numerical Analysis*, Vol. 17, No. 6, Dec. 1980, pp. 883–893.

[22] Huffel, S.V. and Vandewalle, J., "On the Accuracy of Total Least Squares and Least Squares Techniques in the Presence of Errors on all Data," *Automatica*, Vol. 25, No. 5, Sept. 1989, pp. 765–769.

[23] Björck, Å., *Numerical Methods for Least Squares Problems*, Society for Industial and Applied Mathematics, Philadelphia, PA, 1996.

[24] Huffel, S.V. and Vandewalle, J., *The Total Least Squares Problem: Computational Aspects and Analysis*, Society for Industial and Applied Mathematics, Philadelphia, PA, 1991.

[25] Gleser, L.J., "Estimation in a Multivariate Errors-in-Variables Regression Model: Large Sample Results," *Annals of Statistics*, Vol. 9, No. 1, Jan. 1981, pp. 24–44.

[26] Golub, G.H., Heath, M., and Wahba, G., "Generalized Cross-Validation as a Method for Choosing a Good Ridge Parameter," *Technometrics*, Vol. 21, No. 2, May 1979, pp. 215–223.

3

Review of Dynamical Systems

All the effects of nature are only the mathematical consequences of a small number of immutable laws. Laplace, Pierre-Simon

THIS chapter serves to provide a review of the equations and concepts of dynamical systems. These equations will subsequently be used in later chapters to illustrate the importance of estimation for actual applications in dynamical systems. In particular several systems will be reviewed in this chapter; including spacecraft dynamics, orbital mechanics, aircraft flight dynamics, and vibrational systems. A thorough treatise of these subjects is not possible, and only the fundamental equations and concepts will be reviewed here. The interested reader can pursue these subjects in more depth by studying the many references cited in this chapter.

The mathematical models of most physical processes are embodied by one or more differential equations. A large fraction of practical problems is included if we restrict our attention to the case in which the state is the solution of a system of ordinary differential equations (ODEs). The differential equations usually arise quite naturally from application of fundamental principles (e.g., Newton's laws of motion) known to govern the particular dynamical system's behavior. In a significant fraction of the applications, it is possible to obtain explicit algebraic solutions of the system of differential equations; when this is possible, the results of the first two chapters may be immediately employed (e.g., see example 1.1). If simple algebraic analytical solutions of the differential equations cannot be found, one need not (necessarily!) despair, as will be demonstrated in later chapters. We begin the present chapter with an overview of the analytical and numerical methods for solving differential equations.

3.1 Linear System Theory

We first consider linear ODEs, which can be used to describe the behavior of a large class of dynamical systems. A linear system follows the *superposition principle*,[1] which states that a linear combination of inputs produces an output that is the superposition (linear combination) of the outputs if the outputs of each input term were applied separately. Mathematically expressed, a system is linear if the following holds true:

$$
\begin{aligned}
y = f(a\,x_1 + b\,x_2) &= a f(x_1) + b f(x_2) \\
&= a\,y_1 + b\,y_2
\end{aligned}
\tag{3.1}
$$

where $y_1 = f(x_1)$ and $y_2 = f(x_2)$, and a and b are constants.

Example 3.1: We wish to investigate the linearity of the following functions:

1. $y = m\,x$
2. $y = x^2$
3. $y = 3\ddot{x} + 4\dot{x}$

The first equation is clearly linear since $y = m(a\,x_1 + b\,x_2) = a\,y_1 + b\,y_2$, with $y_1 = m\,x_1$ and $y_2 = m\,x_2$. The second equation is not linear since $y = (a\,x_1 + b\,x_2)^2 \neq a\,x_1^2 + b\,x_2^2$. The third equation is linear even though it involves a differential equation since $y = 3(a\,\ddot{x}_1 + b\,\ddot{x}_2) + 4(a\,\dot{x}_1 + b\,\dot{x}_2) \equiv a\,y_1 + b\,y_2$. Superposition is a powerful tool for solving linear ODEs since the homogeneous and forced response can be found individually, and then summed to form the entire solution.

3.1.1 The State Space Approach

The state space approach is extremely useful for many reasons, including: the approach reduces an nth-order linear ODE to n first-order ODEs, matrix analysis tools can easily be used, and it provides a convenient representation for multi-input-multi-output (MIMO) systems. We begin this topic by considering a simple single-input-single-output (SISO) n^{th}-order linear ODE, given by

$$
\frac{d^n y}{dt^n} + a_{n-1}\frac{d^{n-1} y}{dt^{n-1}} + \cdots + a_1\frac{dy}{dt} + a_0\,y = u
\tag{3.2}
$$

where y is the output variable and u is the input variable. In order to convert the ODE into first-order form, consider the following variable change:

$$
\begin{aligned}
x_1 &= y \\
x_2 &= \frac{dy}{dt} \\
&\vdots \\
x_n &= \frac{d^{n-1} y}{dt^{n-1}}
\end{aligned}
\tag{3.3}
$$

This leads to the following equivalent system of n first-order equations:

$$\begin{aligned} \dot{x}_1 &= x_2 \\ \dot{x}_2 &= x_3 \\ &\vdots \\ \dot{x}_n &= -a_0 x_1 - a_1 x_2 - \cdots - a_{n-1} x_n + u \end{aligned} \tag{3.4}$$

which can be represented in matrix form by

$$\dot{\mathbf{x}}(t) = F\,\mathbf{x} + B\,u(t) \tag{3.5}$$

where the vector $\mathbf{x}$ contains the *state variables*:

$$\mathbf{x} = \begin{bmatrix} x_1 & x_2 & \cdots & x_n \end{bmatrix}^T \tag{3.6}$$

and the matrices F and B are given by

$$F = \begin{bmatrix} 0 & 1 & 0 & \cdots & 0 \\ 0 & 0 & 1 & \cdots & 0 \\ \vdots & \vdots & \vdots & \ddots & \vdots \\ 0 & 0 & 0 & \cdots & 1 \\ -a_0 & -a_1 & -a_2 & \cdots & -a_{n-1} \end{bmatrix} \tag{3.7a}$$

$$B = \begin{bmatrix} 0 \; 0 \; \cdots \; 1 \end{bmatrix}^T \tag{3.7b}$$

The general SISO n^{th}-order linear ODE is given by

$$\begin{aligned} &\frac{d^n y}{dt^n} + a_{n-1}\frac{d^{n-1} y}{dt^{n-1}} + \cdots + a_1 \frac{dy}{dt} + a_0\, y \\ &= b_n \frac{d^n u}{dt^n} + b_{n-1}\frac{d^{n-1} u}{dt^{n-1}} + \cdots + b_1 \frac{du}{dt} + b_0\, u \end{aligned} \tag{3.8}$$

In order to convert the ODE into first-order form we first rewrite eqn. (3.8) into an equivalent form involving two ODEs, given by

$$y = b_n \frac{d^n x}{dt^n} + b_{n-1} d^{n-1} x dt^{n-1} + \cdots + b_1 \frac{dx}{dt} + b_0\, x \tag{3.9a}$$

$$u = \frac{d^n x}{dt^n} + a_{n-1}\frac{d^{n-1} x}{dt^{n-1}} + \cdots + a_1 \frac{dx}{dt} + a_0\, x \tag{3.9b}$$

where x is an intermediate variable. Now, consider the following variable change:

$$\begin{aligned} x_1 &= x \\ x_2 &= \frac{dx}{dt} \\ &\vdots \\ x_n &= \frac{d^{n-1} x}{dt^{n-1}} \end{aligned} \tag{3.10}$$

This leads to the following equivalent system of n first-order equations, given in matrix form by

$$\boxed{\begin{aligned}\dot{\mathbf{x}}(t) &= F\,\mathbf{x}(t) + B\,u(t) \\ y(t) &= H\,\mathbf{x}(t) + D\,u(t)\end{aligned}} \qquad \begin{aligned}&(3.11\text{a})\\&(3.11\text{b})\end{aligned}$$

where the matrices F and B are given by eqn. (3.7), and H and D are given by

$$H = \left[(b_0 - b_n\,a_0)\ (b_1 - b_n\,a_1)\ \cdots\ (b_{n-1} - b_n\,a_{n-1})\right] \tag{3.12a}$$

$$D = b_n \tag{3.12b}$$

Clearly, if $b_0 = 1$ and the remaining coefficients $b_i = 0$, $i = 1, 2, \ldots, n$, then the intermediate variable $x = y$, which reduces the general case in eqn. (3.8) to the simple case in eqn. (3.2).

The matrix representation in eqn. (3.11) is the basis for modern estimation and controls. The matrix F is known as the *state matrix* and defines the *stability* of the overall system. The matrix representation is useful for MIMO systems as well since additional inputs can be added simply by using additional columns in the B matrix (likewise, additional outputs can be added by using additional rows in the H matrix). Among developers of computer software for solution of ODEs, there is now a universal adoption of the standardized form of eqn. (3.11). Thus, in theoretical developments whose end products are likely to be implemented on a computer, adherence to this convention is justified on practical grounds. We mention that the particular transformations of eqns. (3.9) and (3.10) represent only one of many linear transformations that brings eqn. (3.8) to the form of eqn. (3.11). Each such transformation, leading to the associated $(F,\ B,\ H,\ D)$, is called a *realization*. Other aspects of the state space representation (such as transmission zeroes, internal and external descriptions, geometric visualization, balanced realizations, etc.), can be found in Refs. [1]-[9].

The MIMO version of the system in eqn. (3.11) can be represented in *transfer function* form by taking the Laplace transform[10] of both sides with zero initial conditions:

$$s\,\mathbf{X}(s) = F\,\mathbf{X}(s) + B\,\mathbf{U}(s) \tag{3.13a}$$

$$\mathbf{Y}(s) = H\,\mathbf{X}(s) + D\,\mathbf{U}(s) \tag{3.13b}$$

where s is the Laplace variable. Solving for $\mathbf{X}(s)$ in eqn. (3.13a) and substituting the resulting expression into eqn. (3.13b) yields

$$\mathbf{Y}(s) = \left\{H\,[s\,I - F]^{-1}\,B + D\right\}\mathbf{U}(s) \tag{3.14}$$

Since the inverse of $[s\,I - F]$ is given by its adjoint divided by its determinant, then the determinant of $[s\,I - F]$ gives the *poles* of the transfer function. Also, the eigenvalues of F are equivalent to the roots of the denominator of the transfer function. The transfer function representation can be useful; however, it becomes impractical for large order systems.

3.1.2 Homogeneous Linear Dynamical Systems

Consider the homogeneous matrix differential equation

$$\dot{\mathbf{x}}(t) = F(t)\,\mathbf{x}(t), \quad \mathbf{x}(t_0) \text{ known} \tag{3.15}$$

The standard approach for solving equations of the form (3.15) is to determine the "fundamental" or "state transition" matrix $\Phi(t, t_0)$ which "maps" the initial state into the current state as

$$\mathbf{x}(t) = \Phi(t, t_0)\,\mathbf{x}(t_0) \tag{3.16}$$

Before developing means for determining $\Phi(t, t_0)$, three important group properties of the transition matrix which follow from inspection of eqn. (3.16) are stated as

$$\Phi(t_0, t_0) = I \tag{3.17a}$$

$$\Phi(t_0, t) = \Phi^{-1}(t, t_0) \tag{3.17b}$$

$$\Phi(t_2, t_0) = \Phi(t_2, t_1)\Phi(t_1, t_0) \tag{3.17c}$$

A differential equation for determining $\Phi(t, t_0)$ can be developed by substituting eqn. (3.16) into the right-hand side of eqn. (3.15) and the derivative of eqn. (3.16) into the left-hand side of eqn. (3.15) to obtain

$$\dot{\Phi}(t, t_0)\,\mathbf{x}(t_0) = F(t)\,\Phi(t, t_0)\,\mathbf{x}(t_0) \tag{3.18}$$

from which we conclude that the transition matrix satisfies the differential equation

$$\dot{\Phi}(t, t_0) = F(t)\,\Phi(t, t_0) \tag{3.19}$$

with the identity matrix in eqn. (3.17a) as the initial condition. Only under ideal circumstances can a practical analytical solution of eqn. (3.19) be obtained; otherwise, numerical techniques must be employed to compute $\Phi(t, t_0)$. We now consider several standard approaches for extracting analytical or approximate solutions for $\Phi(t, t_0)$.

To develop one approach for solving eqn. (3.19), we rewrite it in integral form as

$$\Phi(t, t_0) = I + \int_{t_0}^{t} F(\tau_1)\,\Phi(\tau_1, t_0)\,d\tau_1 \tag{3.20}$$

which is a "matrix Volterra integral equation." We "casually note" that the integrand of eqn. (3.20) contains the left side; so it does not appear that any progress has been made writing eqn. (3.19) in integral form. One "might consider the wisdom" of substituting eqn. (3.20) *into its own integrand*; while this process may appear not only obscene, but futile, it does turn out to be profitable! For $\Phi(\tau_1, t_0)$ in the integrand of eqn. (3.20), we substitute from eqn. (3.20)

$$\Phi(\tau_1, t_0) = I + \int_{t_0}^{\tau_1} F(\tau_2)\,\Phi(\tau_2, t_0)\,d\tau_2 \tag{3.21}$$

to obtain

$$\Phi(t,t_0) = I + \int_{t_0}^{t} F(\tau_1)\, d\tau_1 + \int_{t_0}^{t} F(\tau_1) \int_{t_0}^{\tau_1} F(\tau_2)\, \Phi(\tau_2,t_0)\, d\tau_2 d\tau_1 \tag{3.22}$$

One can now re-use eqn. (3.20) to write

$$\Phi(\tau_2,t_0) = I + \int_{t_0}^{\tau_2} F(\tau_3)\, \Phi(\tau_3,t_0)\, d\tau_3 \tag{3.23}$$

which, when substituted into the final integrand of eqn. (3.22) yields

$$\begin{aligned}\Phi(t,t_0) = I &+ \int_{t_0}^{t} F(\tau_1)\, d\tau_1 \\ &+ \int_{t_0}^{t} F(\tau_1) \int_{t_0}^{\tau_1} F(\tau_2)\, d\tau_2 d\tau_1 \\ &+ \int_{t_0}^{t} F(\tau_1) \int_{t_0}^{\tau_1} F(\tau_2) \int_{t_0}^{\tau_2} F(\tau_3)\, d\tau_3\, d\tau_2\, d\tau_1 \\ &+ \cdots\end{aligned} \tag{3.24}$$

This procedure is known as the Peano-Baker Method; as is shown by Ince (1926),[11] uniform and absolute convergence is guaranteed. Whether or not this process is practical depends, of course, upon how difficult the elements of the $F(t)$ are to integrate, and how quickly convergence occurs.

Considering an important special case that F equals a constant matrix; F can be brought from under all integrands of eqn. (3.24), we immediately find

$$\Phi(t,t_0) = I + F(t-t_0) + \frac{1}{2!}F^2(t-t_0)^2 + \cdots + \frac{1}{n!}F^n(t-t_0)^n + \cdots \tag{3.25}$$

which is recognized to be the e^x series with the matrix $F(t-t_0)$ as the argument. For notational compactness, eqn. (3.25) is often written compactly as

$$\Phi(t,t_0) = e^{F(t-t_0)}, \quad \text{for } F = \text{constant} \tag{3.26}$$

Thus, returning to eqn. (3.16), we see that the solution for constant F is

$$\boxed{\mathbf{x}(t) = e^{F(t-t_0)}\mathbf{x}(t_0)} \tag{3.27}$$

Consider the analogy of the matrix differential equation (3.15) with the scalar differential equation

$$\dot{x}(t) = f(t)\, x(t), \quad x(t_0) \text{ known} \tag{3.28}$$

For the special case that f equals a constant, then the solution of eqn. (3.28) is

$$x(t) = x(t_0)e^{f(t-t_0)} \tag{3.29}$$

Thus, except for the constrained order of multiplication, the matrix solution (3.27) of eqn. (3.15) is completely analogous to the scalar solution (3.29) of eqn. (3.28) for constant coefficient matrices.

For the general case that f does not equal a constant, the general solution of eqn. (3.28) is

$$x(t) = x(t_0)e^{\int_{t_0}^{t} f(\tau)\, d\tau} \tag{3.30}$$

One might naturally conjecture that the general solution of eqn. (3.15) is

$$\mathbf{x}(t) = \Phi(t, t_0)\,\mathbf{x}(t_0) = \left[e^{\int_{t_0}^{t} F(\tau)\, d\tau}\right]\mathbf{x}(t_0) \tag{3.31}$$

This conjecture turns out to be false, in general. To see under what conditions eqn. (3.31) *is* a correct solution of eqn. (3.15), note

$$\Phi = e^{\int_{t_0}^{t} F\, d\tau} = I + \left[\int_{t_0}^{t} F\, d\tau\right] + \frac{1}{2!}\left[\int_{t_0}^{t} F\, d\tau\right]^2 + \frac{1}{3!}\left[\int_{t_0}^{t} F\, d\tau\right]^3 + \cdots \tag{3.32}$$

$$\begin{aligned}\dot{\Phi} = {} & 0 + F + \frac{1}{2!}F\left[\int_{t_0}^{t} F\, d\tau\right] + \frac{1}{2!}\left[\int_{t_0}^{t} F\, d\tau\right]F \\ & + \frac{1}{3!}F\left[\int_{t_0}^{t} F\, d\tau\right]^2 + \frac{1}{3!}\left[\int_{t_0}^{t} F\, d\tau\right]F\left[\int_{t_0}^{t} F\, d\tau\right] \\ & + \frac{1}{3!}\left[\int_{t_0}^{t} F\, d\tau\right]^2 F + \cdots\end{aligned} \tag{3.33}$$

and

$$F\Phi = F + F\left[\int_{t_0}^{t} F\, d\tau\right] + \frac{1}{2!}F\left[\int_{t_0}^{t} F\, d\tau\right]^2 + \frac{1}{3!}F\left[\int_{t_0}^{t} F\, d\tau\right]^3 \tag{3.34}$$

Clearly, for eqns. (3.33) and (3.34) to be equal {which they must if eqn. (3.31) is a solution of eqn. (3.15)} then it is necessary that the following "commutivity property" be satisfied:

$$F(t)\left[\int_{t_0}^{t} F(\tau)\, d\tau\right] = \left[\int_{t_0}^{t} F(\tau)\, d\tau\right]F(t) \tag{3.35}$$

This property defines only a very special class of matrices!

The conclusion is that the analogy between solutions of eqn. (3.16) and its scalar analog is not complete. The Peano-Baker solution (3.24) can be written in shorthand notation as

$$\Phi(t, t_0) = I + \sum_{i=1}^{\infty} l_i(t) \tag{3.36}$$

where the integrals are defined as

$$l_1(t) = \int_{t_0}^{t} F(\tau_1) d\tau_1 \tag{3.37a}$$

$$l_2(t) = \int_{t_0}^{t} F(\tau_1) \int_{t_0}^{\tau_1} F(\tau_2)\, d\tau_2 d\tau_1 = \int_{t_0}^{t} F(\tau_1) l_1(\tau_1)\, d\tau_1 \tag{3.37b}$$

$$l_3(t) = \int_{t_0}^{t} F(\tau_1) l_2(\tau_1)\, d\tau_1 \tag{3.37c}$$

or

$$l_n(t) = \int_{t_0}^{t} F(\tau_1) l_{n-1}(\tau_1)\, d\tau_1, \quad \text{for } n \geq 2 \tag{3.38}$$

As an alternative to the Peano-Baker solution, consider the Taylor's Series

$$\Phi(t, t_0) = I + \sum_{j=1}^{\infty} \frac{(t-t_0)^j}{j!} \left. \frac{d^j \Phi}{dt^j} \right|_{t=t_0} \tag{3.39}$$

where the necessary partial derivatives are evaluated sequentially from the following equations:

In General	Evaluated Initially
$\dfrac{d\Phi}{dt} = F\,\Phi$	$\left.\dfrac{d\Phi}{dt}\right\|_{t=t_0} = F(t_0)$
$\dfrac{d^2\Phi}{dt^2} = \dfrac{dF}{dt}\Phi + F\dfrac{d\Phi}{dt}$	$\left.\dfrac{d^2\Phi}{dt^2}\right\|_{t=t_0} = \left.\dfrac{dF}{dt}\right\|_{t=t_0} + F^2(t_0)$
$\vdots$	$\vdots$

In particular, if F is constant, then

$$\left. \frac{d^j \Phi}{dt^j} \right|_{t=t_0} = F^j \tag{3.40}$$

and eqn. (3.39) becomes

$$\Phi(t, t_0) = I + \sum_{j=1}^{\infty} \frac{(t-t_0)^j}{j!} F^j \equiv e^{F(t-t_0)} \tag{3.41}$$

In practice, if F is not constant, and the Peano-Baker or Taylor's Series prove too cumbersome (due to slow convergence or algebraic difficulties), then one must resort to a numerical solution of eqn. (3.18) or eqn. (3.15).

3.1.3 Forced Linear Dynamical Systems

We now direct our attention to the multi-input inhomogeneous differential equation

$$\dot{\mathbf{x}}(t) = F(t)\,\mathbf{x}(t) + B(t)\,\mathbf{u}(t) \tag{3.42}$$

Using Lagrange's method of *variation of parameters*, a solution of eqn. (3.42) having the following form is assumed:

$$\mathbf{x}(t) = \Phi(t,t_0)\,\mathbf{g}(t), \quad \mathbf{g}(t_0) = \mathbf{x}(t_0) \tag{3.43}$$

where $\mathbf{g}(t)$ is an $n \times 1$ vector of unknown functions and $\Phi(t,t_0)$ is the homogeneous transition matrix. Differentiating eqn. (3.43), we obtain

$$\dot{\mathbf{x}}(t) = \Phi(t,t_0)\,\dot{\mathbf{g}}(t) + \dot{\Phi}(t,t_0)\,\mathbf{g}(t) \tag{3.44}$$

which, upon substitution of eqn. (3.18) for $\dot{\Phi}(t,t_0)$, becomes

$$\dot{\mathbf{x}}(t) = \Phi(t,t_0)\,\dot{\mathbf{g}}(t) + F(t)\,\Phi(t,t_0)\,\mathbf{g}(t) \tag{3.45}$$

Substituting eqns. (3.43) and (3.45) into eqn. (3.42) yields

$$\Phi(t,t_0)\,\mathbf{g}(t) + F(t)\,\Phi(t,t_0)\,\dot{\mathbf{g}}(t) = F(t)\,\Phi(t,t_0)\,\mathbf{g}(t) + B(t)\,\mathbf{u}(t) \tag{3.46}$$

Therefore

$$\dot{\mathbf{g}}(t) = \Phi^{-1}(t,t_0)\,B(t)\,\mathbf{u}(t) \tag{3.47}$$

which we integrate to obtain {noting $\mathbf{g}(t_0) = \mathbf{x}(t_0)$}

$$\mathbf{g}(t) = \mathbf{x}(t_0) + \int_{t_0}^{t} \Phi^{-1}(\tau,t_0)\,B(\tau)\,\mathbf{u}(\tau)\,d\tau \tag{3.48}$$

Therefore, the general solution of eqn. (3.42) is

$$\mathbf{x}(t) = \Phi(t,t_0)\,\mathbf{x}(t_0) + \Phi(t,t_0)\int_{t_0}^{t} \Phi^{-1}(\tau,t_0)\,B(\tau)\,\mathbf{u}(\tau)\,d\tau \tag{3.49}$$

Application of eqn. (3.17b) allows the integrand to be written as

$$\Phi^{-1}(\tau,t_0) = \Phi(t_0,\tau) \tag{3.50}$$

Using eqn. (3.17c) gives

$$\Phi^{-1}(\tau,t_0) = \Phi(t_0,t)\,\Phi(t,\tau) \tag{3.51}$$

or

$$\Phi^{-1}(\tau,t_0) = \Phi^{-1}(t,t_0)\,\Phi(t,\tau) \tag{3.52}$$

which, when substituted into eqn. (3.49) yields

$$\boxed{\mathbf{x}(t) = \Phi(t,t_0)\,\mathbf{x}(t_0) + \int_{t_0}^{t} \Phi(t,\tau)\,B(\tau)\,\mathbf{u}(\tau)\,d\tau} \tag{3.53}$$

as the final form of the solution of eqn. (3.42) for arbitrary $F(t)$, $B(t)$, and $\mathbf{u}(t)$. Equation (3.53) must typically be solved numerically.

Example 3.2: Consider the motion of a projectile in a constant gravity field. The equations of motion are

$$\begin{aligned}\ddot{x} &= 0\\ \ddot{y} &= 0\\ \ddot{z} &= -g\end{aligned}$$

which integrate immediately to give

$$\begin{aligned}\dot{x} &= \dot{x}_0\\ \dot{y} &= \dot{y}_0\\ \dot{z} &= \dot{z}_0 - g\,(t-t_0)\end{aligned}$$

and

$$\begin{aligned}x &= x_0 + \dot{x}_0\,(t-t_0)\\ y &= y_0 + \dot{y}_0\,(t-t_0)\\ z &= z_0 + \dot{z}_0\,(t-t_0) - 1/2\,g\,(t-t_0)^2\end{aligned}$$

where g is the gravity constant, and (x_0, y_0, z_0) and $(\dot{x}_0, \dot{y}_0, \dot{z}_0)$ are the initial positions and velocities, respectively.

Alternatively, we could have employed the variable change

$$x_1 = x,\ x_2 = y,\ x_3 = z,\ x_4 = \dot{x},\ x_5 = \dot{y},\ x_6 = \dot{z},\ u = -g$$

so that the following state space form can be written:

$$\begin{bmatrix}\dot{x}_1\\ \dot{x}_2\\ \dot{x}_3\\ \dot{x}_4\\ \dot{x}_5\\ \dot{x}_6\end{bmatrix} = \begin{bmatrix}0&0&0&1&0&0\\ 0&0&0&0&1&0\\ 0&0&0&0&0&1\\ 0&0&0&0&0&0\\ 0&0&0&0&0&0\\ 0&0&0&0&0&0\end{bmatrix}\begin{bmatrix}x_1\\ x_2\\ x_3\\ x_4\\ x_5\\ x_6\end{bmatrix} - \begin{bmatrix}0\\0\\0\\0\\0\\1\end{bmatrix} g$$

Notice, by inspection of the analytical position and velocity solutions that the state transition matrix is

$$\Phi(t,t_0) = \begin{bmatrix}1&0&0&(t-t_0)&0&0\\ 0&1&0&0&(t-t_0)&0\\ 0&0&1&0&0&(t-t_0)\\ 0&0&0&1&0&0\\ 0&0&0&0&1&0\\ 0&0&0&0&0&1\end{bmatrix}$$

Therefore, the "forced" solution (including gravity) follows from eqn. (3.53), given by

$$\begin{bmatrix} x_1(t) \\ x_2(t) \\ x_3(t) \\ x_4(t) \\ x_5(t) \\ x_6(t) \end{bmatrix} = \begin{bmatrix} 1\,0\,0 & (t-t_0) & 0 & 0 \\ 0\,1\,0 & 0 & (t-t_0) & 0 \\ 0\,0\,1 & 0 & 0 & (t-t_0) \\ 0\,0\,0 & 1 & 0 & 0 \\ 0\,0\,0 & 0 & 1 & 0 \\ 0\,0\,0 & 0 & 0 & 1 \end{bmatrix} \begin{bmatrix} x_1(t_0) \\ x_2(t_0) \\ x_3(t_0) \\ x_4(t_0) \\ x_5(t_0) \\ x_6(t_0) \end{bmatrix}$$
$$- \int_{t_0}^{t} \left[0\;0\;(\tau - t_0)\;0\;0\;1\right]^T g\, d\tau$$

or

$$\begin{bmatrix} x_1(t) \\ x_2(t) \\ x_3(t) \\ x_4(t) \\ x_5(t) \\ x_6(t) \end{bmatrix} = \begin{bmatrix} x_1(t_0) + x_4(t_0)(t-t_0) \\ x_2(t_0) + x_5(t_0)(t-t_0) \\ x_3(t_0) + x_6(t_0)(t-t_0) - 1/2\, g\,(t-t_0)^2 \\ x_4(t_0) \\ x_5(t_0) \\ x_6(t_0) - g\,(t-t_0) \end{bmatrix}$$

which verify (again) the previous results and demonstrates the equivalence between the preceding results of this section and conventional integration of the differential equations.

3.1.4 Linear State Variable Transformations

The matrix exponential for an arbitrary constant matrix is expensive to compute if one requires a large number of terms in eqn. (3.25). Often, one can carry out a coordinate transformation which "blasts this problem into trivia." Consider the introduction of a new state vector $\mathbf{z}$ which is linearly related to $\mathbf{x}$ via

$$\mathbf{x} = T\,\mathbf{z} \tag{3.54}$$

where T is a constant $n \times n$ matrix. Taking the time derivative of eqn. (3.54) and solving for $\dot{\mathbf{z}}$ yields

$$\dot{\mathbf{z}} = T^{-1}\dot{\mathbf{x}} \tag{3.55}$$

Now, substitution of the $\mathbf{x}$-differential equation (3.15) yields

$$\dot{\mathbf{z}} = T^{-1} F\,\mathbf{x} \tag{3.56}$$

and substitution of eqn. (3.54) then yields the differential equation for $\mathbf{z}$ as

$$\dot{\mathbf{z}} = \Lambda\,\mathbf{z} \tag{3.57}$$

where the new coefficient matrix is given by the similarity transformation

$$\Lambda = T^{-1} F T \tag{3.58}$$

Now the unspecified T-matrix can often be judiciously chosen so that Λ is diagonal; more generally, Λ can be brought to a block diagonal form (the "Jordan Canonical Form"). If Λ is in fact diagonal, it is clear that the solution is trivial since eqn. (3.57) can be written as

$$\begin{bmatrix} \dot{\mathbf{z}}_1 \\ \dot{\mathbf{z}}_2 \\ \vdots \\ \dot{\mathbf{z}}_n \end{bmatrix} = \begin{bmatrix} \lambda_1 & 0 & \cdots & 0 \\ 0 & \lambda_2 & \cdots & 0 \\ \vdots & \vdots & \ddots & \vdots \\ 0 & 0 & \cdots & \lambda_n \end{bmatrix} \begin{bmatrix} \mathbf{z}_1 \\ \mathbf{z}_2 \\ \vdots \\ \mathbf{z}_n \end{bmatrix} \tag{3.59}$$

or

$$\dot{z}_i = \lambda_i z_i, \quad i = 1, 2, \ldots, n \tag{3.60}$$

and the solution is simply

$$z_i(t) = z_i(t_0) e^{\lambda_i (t - t_0)}, \quad i = 1, 2, \ldots, n \tag{3.61}$$

The solution in eqn. (3.61) can be written in state transition matrix form as

$$\mathbf{z}(t) = \Psi(t, t_0) \mathbf{z}(t_0) \tag{3.62}$$

where

$$\Psi(t, t_0) \equiv \begin{bmatrix} e^{\lambda_1 (t - t_0)} & 0 & \cdots & 0 \\ 0 & e^{\lambda_2 (t - t_0)} & \cdots & 0 \\ \vdots & \vdots & \ddots & \vdots \\ 0 & 0 & \cdots & e^{\lambda_n (t - t_0)} \end{bmatrix} \tag{3.63}$$

Now substituting eqn. (3.62) into eqn. (3.54) and using $\mathbf{z}(t_0) = T^{-1}\mathbf{x}(t_0)$ yields

$$\mathbf{x}(t) = T\, \Psi(t, t_0)\, T^{-1} \mathbf{x}(t_0) \tag{3.64}$$

The state transition matrix for $\mathbf{x}$ is then clearly identified as

$$\Phi(t, t_0) \equiv T\, \Psi(t, t_0)\, T^{-1} \tag{3.65}$$

Let us now see how to construct the elements of the T and Λ matrices. We require that the similarity transformation yields a diagonal Λ matrix as

$$\Lambda = T^{-1} F T \tag{3.66}$$

or

$$T \Lambda = F T \tag{3.67}$$

In detail, the equations (3.67) are

$$\begin{bmatrix} t_{11} & \cdots & t_{1n} \\ \vdots & \ddots & \vdots \\ t_{n1} & \cdots & t_{nn} \end{bmatrix} \begin{bmatrix} \lambda_1 & \cdots & 0 \\ \vdots & \ddots & \vdots \\ 0 & \cdots & \lambda_n \end{bmatrix} = \begin{bmatrix} f_{11} & \cdots & f_{1n} \\ \vdots & \ddots & \vdots \\ f_{n1} & \cdots & f_{nn} \end{bmatrix} \begin{bmatrix} t_{11} & \cdots & t_{1n} \\ \vdots & \ddots & \vdots \\ t_{n1} & \cdots & t_{nn} \end{bmatrix} \tag{3.68}$$

Equating the ith column resulting from the matrix product on the left-hand side of eqn. (3.68) to the ith column on the right-hand side yields

$$\lambda_i \begin{bmatrix} t_{1i} \\ t_{2i} \\ \vdots \\ t_{ni} \end{bmatrix} = F \begin{bmatrix} t_{1i} \\ t_{2i} \\ \vdots \\ t_{ni} \end{bmatrix}, \quad i = 1, 2, \ldots, n \tag{3.69}$$

Thus, the conclusion is that the diagonal elements of Λ are the *eigenvalues* of F, and the columns of the required matrix T are the corresponding *eigenvectors* of F. The λ's are the n roots of the characteristic equation

$$\det(\lambda I - F) = 0 \rightarrow \lambda_1, \lambda_2, \ldots, \lambda_n \tag{3.70}$$

Upon determining λ_i's from eqn. (3.70), the t_{ij}'s are determined (to within an arbitrary multiplicative constant for each column) from eqn. (3.69). For the most common case that the n λ's satisfying eqn. (3.70) are distinct, the independent columns of T can always be found to satisfy eqn. (3.70). For the case that eqn. (3.70) has multiple roots, it is not always possible to find independent columns of T from eqn. (3.69) which will guarantee eqn. (3.58) to be diagonal. The difficulties encountered for repeated eigenvalues are not always trivial to resolve; see Ref. [12] for a more detailed treatment of this subject.

We can easily prove that a transformation of state does not alter the transfer function of a system. Taking the Laplace transform of eqn. (3.54) and substituting the resultant into eqn. (3.13) gives

$$s\,\mathbf{Z}(s) = T^{-1}FT\,\mathbf{Z}(s) + T^{-1}B\,\mathbf{U}(s) \tag{3.71a}$$

$$\mathbf{Y}(s) = HT\,\mathbf{Z}(s) + D\,\mathbf{U}(s) \tag{3.71b}$$

The transfer function from $\mathbf{U}(s)$ to $\mathbf{Y}(s)$ is given by

$$\begin{aligned} \mathbf{Y}(s) &= \left\{ HT\left[sI - T^{-1}FT\right]^{-1} T^{-1}B + D \right\} \mathbf{U}(s) \\ &= \left\{ HT\left[T^{-1}(sI - F)\,T\right]^{-1} T^{-1}B + D \right\} \mathbf{U}(s) \\ &= \left\{ HTT^{-1}[sI - F]^{-1}\,TT^{-1}B + D \right\} \mathbf{U}(s) \\ &= \left\{ H[sI - F]^{-1}B + D \right\} \mathbf{U}(s) \end{aligned} \tag{3.72}$$

Therefore, the overall transfer function is unaffected. Clearly, there are an infinity number of state-space representations that yield identical transfer functions.

3.2 Nonlinear Dynamical Systems

We now consider the circumstance in which the original system of differential equations is nonlinear and can be brought to the standard form

$$\boxed{\begin{aligned}\dot{\mathbf{x}} &= \mathbf{f}(t, \mathbf{x}, \mathbf{u}) \\ \mathbf{y} &= \mathbf{h}(t, \mathbf{x}, \mathbf{u})\end{aligned}} \tag{3.73a, 3.73b}$$

Some of the nonlinear systems of differential equations encountered in applications can be solved for an exact analytical solution (e.g., as will be demonstrated for the elliptic two-body problem in §3.8.2). Unfortunately, only a minority of these systems have known analytical solutions and no standardized methods exist for finding *exact analytical solutions*. In many cases a reference motion may be known, which is "close" to the actual state history. In these cases the *departure* of the actual state history from a known reference motion may be adequately described by eqn. (3.11). The nominal reference ($\mathbf{x}_N$) trajectory's integration is formally indicated as

$$\mathbf{x}_N(t) = \mathbf{x}_N(t_0) + \int_0^t \mathbf{f}(\tau, \mathbf{x}_N, \mathbf{u}_N)\, d\tau \tag{3.74a}$$

$$\mathbf{y}(t) = \mathbf{h}(t, \mathbf{x}_N, \mathbf{u}_N) \tag{3.74b}$$

Now, we assume that the actual quantities are given by the nominal quantities plus a perturbation:

$$\mathbf{x}(t) = \mathbf{x}_N(t) + \delta\mathbf{x}(t) \tag{3.75a}$$

$$\mathbf{u}(t) = \mathbf{u}_N(t) + \delta\mathbf{u}(t) \tag{3.75b}$$

$$\mathbf{y}(t) = \mathbf{y}_N(t) + \delta\mathbf{y}(t) \tag{3.75c}$$

where $\delta\mathbf{x}(t)$, $\delta\mathbf{u}(t)$, and $\delta\mathbf{y}(t)$ are state, input, and output perturbations, respectively. Results from a first-order Taylor series expansion for $\mathbf{f}(t, \mathbf{x}, \mathbf{u})$ and $\mathbf{h}(t, \mathbf{x}, \mathbf{u})$ yield

$$\delta\dot{\mathbf{x}}(t) = F(t)\,\delta\mathbf{x}(t) + B(t)\,\delta\mathbf{u}(t) \tag{3.76a}$$

$$\delta\mathbf{y}(t) = H(t)\,\delta\mathbf{x}(t) + D(t)\,\delta\mathbf{u}(t) \tag{3.76b}$$

where

$$F(t) = \left.\frac{\partial \mathbf{f}}{\partial \mathbf{x}}\right|_{\mathbf{x}_N, \mathbf{u}_N}, \quad B(t) = \left.\frac{\partial \mathbf{f}}{\partial \mathbf{u}}\right|_{\mathbf{x}_N, \mathbf{u}_N} \tag{3.77a}$$

$$H(t) = \left.\frac{\partial \mathbf{h}}{\partial \mathbf{x}}\right|_{\mathbf{x}_N, \mathbf{u}_N}, \quad D(t) = \left.\frac{\partial \mathbf{h}}{\partial \mathbf{u}}\right|_{\mathbf{x}_N, \mathbf{u}_N} \tag{3.77b}$$

Equation (3.76) can be integrated and then employed in eqns. (3.75a) and (3.75c) to approximate trajectories in a sufficiently small neighborhood of eqn. (3.74). Errors arise when the "departure" from the nominal reference trajectory is not small (i.e., when the higher-order expansion terms in Taylor's series are not negligible).

For the *perturbation class* of nonlinear system whose differential equations can be brought to the form

$$\dot{\mathbf{x}} = \mathbf{f}(t, \mathbf{x}, \mathbf{u}) + \delta\mathbf{f}(t, \mathbf{x}, \mathbf{u}) \tag{3.78}$$

in which the "unperturbed" *generating system*

$$\dot{\mathbf{z}} = \mathbf{f}(t, \mathbf{x}, \mathbf{u}) \tag{3.79}$$

has a known analytical solution, and the perturbation $\delta\mathbf{f}(t, \mathbf{x}, \mathbf{u})$ follows

$$||\delta\mathbf{f}(t, \mathbf{x}, \mathbf{u})|| \ll ||\mathbf{f}(t, \mathbf{x}, \mathbf{u})|| \tag{3.80}$$

for all t and $\mathbf{x}$ of interest, numerous methods are available for construction of *approximate* analytical solutions. The interested reader is referred to Refs. [13] and [14] for development of basic perturbation methods which are not developed herein due to space limitations. Let us remark, however, that the perturbation approach suffers from one fundamental drawback; for each specification of the functions $\mathbf{f}$ and $\delta\mathbf{f}$ in eqn. (3.78), lengthy algebraic developments must be carried through to obtain an *approximate* solution. In many cases the practical constraints imposed by "having but one life to give" and the desirability of constructing general-purpose algorithms make the analytical perturbation approach unattractive. On the other hand, general purpose numerical methods exist which are routinely employed to solve a wide variety of highly nonlinear systems of the form (3.73) with excellent, near arbitrary control over precision of the solution (e.g., Runge-Kutta methods).

Estimation theory based upon a linear differential equation of the form (3.76) is seen to be applicable (at least approximately) to a wide class of dynamical systems. In any given application to nonlinear problems, of course, one must realistically face the problems of choosing suitable nominal trajectories to linearize about, and analyzing the effects of errors introduced through the linearization. Many of the available tools for linear systems (such as superposition, Laplace transforms, Bode plots, observability, etc.[1−9]) are not directly applicable to nonlinear systems. Still, the linearized system in eqn. (3.76) can be used to prove local stability and analyze the nonlinear system near an equilibrium point by using Lyapunov's linearization method. Also, Lyapunov's direct method can be used to prove global stability (whether the system is linear or nonlinear) by examining the variation of a single *scalar* function, which is often the total energy of the dynamical system.[15] These concepts are demonstrated in §3.6.

Example 3.3: In this example the linear perturbation technique described previously is used to study the behavior of a highly maneuverable aircraft which exhibits nonlinear behavior. This behavior occurs when the aircraft operates at high angles of attack, in which the lift coefficient cannot be accurately represented as a linear function of angle of attack. Using the coefficients for an F-8 aircraft and normalizing with respect to trim values yields the following nonlinear differential equations for

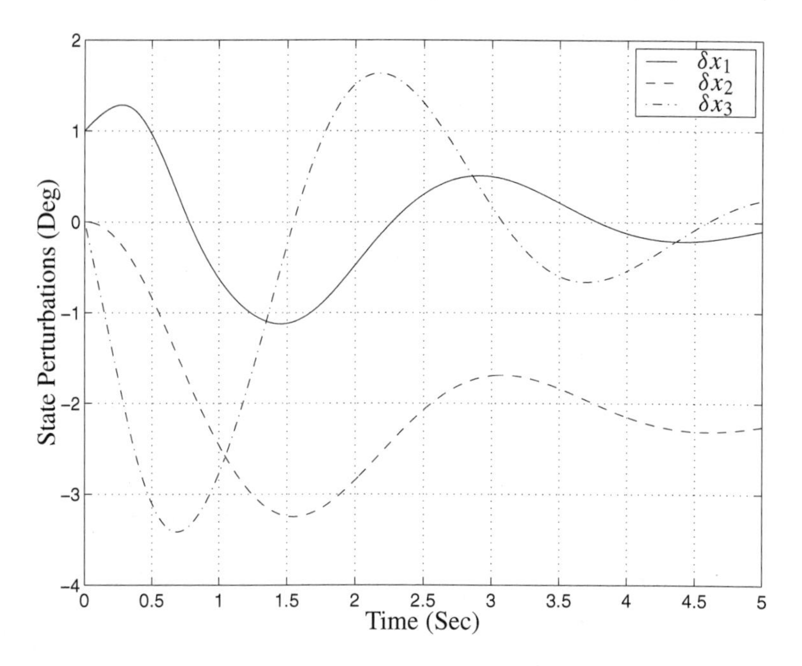

Figure 3.1: State Perturbation Trajectories

the longitudinal motion:[16]

$$\dot{\alpha} = \dot{\theta} - \alpha^2\dot{\theta} - 0.09\alpha\dot{\theta} - 0.88\alpha + 0.47\alpha^2 + 3.85\alpha^3 - 0.02\theta^2$$
$$\ddot{\theta} = -0.396\dot{\theta} - 4.208\alpha - 0.47\alpha^2 - 3.564\alpha^3$$

where α is the angle of attack and θ is the pitch angle (see §3.9). The state vector is chosen as $\mathbf{x} = \begin{bmatrix}\alpha \ \theta \ \dot{\theta}\end{bmatrix}^T$. Therefore, the linearized state matrix is

$$F = \begin{bmatrix} f_{11} & f_{12} & f_{13} \\ 0 & 0 & 1 \\ f_{31} & 0 & f_{33} \end{bmatrix}$$

where

$$f_{11} = -2x_1x_3 - 0.09x_3 - 0.88 + 0.94x_1 + 11.55x_1^2$$
$$f_{12} = -0.04x_2$$
$$f_{13} = 1 - x_1^2 - 0.09x_1$$
$$f_{31} = -4.208 - 0.94x_1 - 10.692x_1^2$$
$$f_{33} = -0.396$$

For the actual system the initial angle of attack is 25 degrees and the pitch and pitch rate are both zero. The nominal state quantities are found by integrating the nonlinear equations with initial conditions given by 24 degrees for the angle of attack and

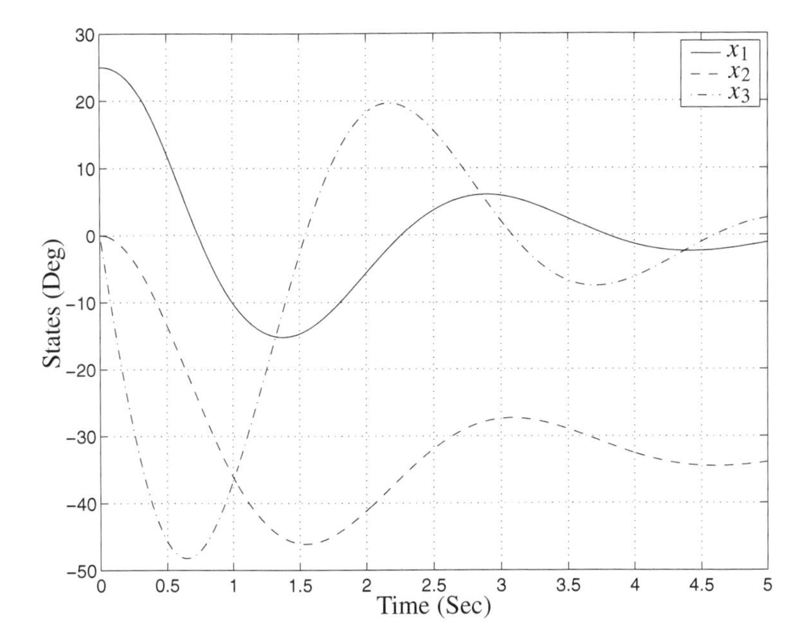

Figure 3.2: State Trajectories

zero for both the pitch and pitch rate. Then the linearized system is integrated with initial conditions given by $\boldsymbol{\delta}\mathbf{x}(t_0) = \left[\pi/180\ 0\ 0\right]^T$. A plot of the state perturbations is shown in Figure 3.1. As shown by this plot the perturbation trajectories are small compared to the large initial condition for the angle of attack. These perturbations are then added to the nominal quantities to form the state trajectories, shown in Figure 3.2. These trajectories closely match the actual state trajectories. Although the nominal trajectory typically involves the integration of the full nonlinear equations, the exercise of performing the linearization still remains useful, as will be demonstrated in the extended Kalman filter of §5.6.

3.3 Parametric Differentiation

Estimation or optimization algorithms are often applied to systems whose state is governed by a system of equations of the form

$$\dot{\mathbf{x}} = \mathbf{f}(t, \mathbf{x}, \mathbf{p}) \tag{3.81}$$

where

$$\mathbf{p} = \begin{bmatrix} p_1 & p_2 & \cdots & p_q \end{bmatrix}^T \tag{3.82}$$

is a set of q *model constants* which appear in the system's differential equations. In many applications, the initial conditions $\mathbf{x}(t_0)$ of eqn. (3.81) will be poorly known, as well as one or more elements of the model parameter vector $\mathbf{p}$. Thus it may be necessary to estimate both $\mathbf{x}(t_0)$ and $\mathbf{p}$ based upon measurements of $\mathbf{x}(t)$ or a function thereof. As will be seen in the applications of Chapter 4, conventional estimation will require the partial derivative matrices

$$\Phi(t, t_0) = \frac{\partial \mathbf{x}(t)}{\partial \mathbf{x}(t_0)} \tag{3.83}$$

and

$$\Psi(t, t_0) = \frac{\partial \mathbf{x}(t)}{\partial \mathbf{p}} \tag{3.84}$$

We now investigate methods for calculating these derivative matrices.

Equation (3.81) can be written in integral form as

$$\mathbf{x}(t) = \mathbf{x}(t_0) + \int_{t_0}^{t} \mathbf{f}(\tau, \mathbf{x}, \mathbf{p})\, d\tau \tag{3.85}$$

from which it follows

$$\Phi(t, t_0) = I + \int_{t_0}^{t} \frac{\partial \mathbf{f}(\tau, \mathbf{x}, \mathbf{p})}{\partial \mathbf{x}(\tau)} \frac{\partial \mathbf{x}(\tau)}{\partial \mathbf{x}(t_0)}\, d\tau \tag{3.86}$$

and

$$\Psi(t, t_0) = \int_{t_0}^{t} \left(\frac{\partial \mathbf{f}(\tau, \mathbf{x}, \mathbf{p})}{\partial \mathbf{p}} + \frac{\partial \mathbf{f}(\tau, \mathbf{x}, \mathbf{p})}{\partial \mathbf{x}(\tau)} \frac{\partial \mathbf{x}(\tau)}{\partial \mathbf{p}} \right) d\tau \tag{3.87}$$

Taking the time derivative of eqns. (3.86) and (3.87), it follows that the desired derivative matrices satisfy the first-order linear differential equations

$$\dot{\Phi}(t, t_0) = F(t)\Phi(t, t_0), \quad \Phi(t_0, t_0) = I \tag{3.88}$$

and

$$\dot{\Psi}(t, t_0) = F(t)\Psi(t, t_0) + \frac{\partial \mathbf{f}(t, \mathbf{x}, \mathbf{p})}{\partial \mathbf{p}}, \quad \Psi(t_0, t_0) = 0 \tag{3.89}$$

where

$$F(t) \equiv \frac{\partial \mathbf{f}(t, \mathbf{x}, \mathbf{p})}{\partial \mathbf{x}(t)} \tag{3.90}$$

Observe that $F(t)$ and $\partial \mathbf{f}(t, \mathbf{x}, \mathbf{p})/\partial \mathbf{p}$ in eqns. (3.88) and (3.89) depend on $\mathbf{x}(t)$. If numerical methods are required to solve the differential equations (3.81) for $\mathbf{x}(t)$, it is usually convenient to employ the same numerical process to simultaneously integrate eqns. (3.88) and (3.89) to obtain $\Phi(t, t_0)$ and $\Psi(t, t_0)$. Clearly, if the original system can be solved analytically for $\mathbf{x}(t)$, then the partial derivatives can be taken formally and analytical solutions can be determined for $\Phi(t, t_0)$ and $\Psi(t, t_0)$.

As is evident by comparison of eqns. (3.88) and (3.76a), the derivative matrix has the interpretation

$$\delta\mathbf{x}(t) = \Phi(t,t_0)\,\delta\mathbf{x}(t_0) \tag{3.91}$$

where $\delta\mathbf{x}$ are small variations about a reference solution of eqn. (3.81). One important conclusion of the above is that if the original nonlinear system can be solved analytically, then the linear variational equations (3.76a), (3.88), and (3.89) can be solved analytically (i.e., their solution is reduced to a process of formal partial differentiation). This approach will be demonstrated in the orbit determination problem given in §4.3.

The above developments can be derived via a different path that is illuminating. Consider using the following augmented system:

$$\dot{\mathbf{x}} = \mathbf{f}(t,\mathbf{x},\mathbf{p}) \tag{3.92a}$$

$$\dot{\mathbf{p}} = \mathbf{0} \tag{3.92b}$$

Equation (3.92) can be rewritten in compact form as

$$\dot{\mathbf{z}} = \mathbf{g}(t,\mathbf{z}) \tag{3.93}$$

where $\mathbf{z} \equiv \begin{bmatrix}\mathbf{x}^T & \mathbf{p}^T\end{bmatrix}^T$ and $\mathbf{g}(t,\mathbf{z}) \equiv \begin{bmatrix}\mathbf{f}^T & \mathbf{0}^T\end{bmatrix}^T$. We now seek the following augmented matrix:

$$\Gamma(t,t_0) \equiv \frac{\partial\mathbf{z}(t)}{\partial\mathbf{z}(t_0)} = \begin{bmatrix}\Phi(t,t_0) & \Psi(t,t_0)\\ 0 & I\end{bmatrix} \tag{3.94}$$

We know the augmented state transition matrix satisfies

$$\dot{\Gamma}(t,t_0) = \frac{\partial\mathbf{g}(t,\mathbf{z})}{\partial\mathbf{z}(t)}\Gamma(t,t_0), \quad \Gamma(t_0,t_0) = I \tag{3.95}$$

where

$$\frac{\partial\mathbf{g}(t,\mathbf{z})}{\partial\mathbf{z}(t)} = \begin{bmatrix} F(t) & \dfrac{\partial\mathbf{f}(t,\mathbf{x},\mathbf{p})}{\partial\mathbf{p}} \\ 0 & 0\end{bmatrix} \tag{3.96}$$

Making use of eqns. (3.94) and (3.96) in eqn. (3.95) immediately verifies eqns. (3.88) and (3.89). Thus augmenting the state vector as in eqns. (3.92) and (3.93) and computing the augmented state transition matrix as in eqns. (3.94) and (3.95) is theoretically equivalent to the sensitivities computed from eqns. (3.88) and (3.89).

3.4 Observability

This section presents one of the most useful concepts in estimation. Observability gives us an indication of the state quantities that can be monitored ("observed") from the measurements. An observable state-space form is given by the observer canonical form:

$$\dot{\mathbf{x}}_o = F_o\,\mathbf{x}_o + B_o\,u \tag{3.97a}$$

$$y_o = H_o\,\mathbf{x}_o + D_o\,u \tag{3.97b}$$

where the matrices F_o, B_o, H_o, and D_o are given by

$$F_o = \begin{bmatrix} 0 & 0 & \cdots & 0 & -a_0 \\ 1 & 0 & \cdots & 0 & -a_1 \\ 0 & 1 & \cdots & 0 & -a_2 \\ \vdots & \vdots & \ddots & \vdots & \vdots \\ 0 & 0 & \cdots & 1 & -a_{n-1} \end{bmatrix} \tag{3.98a}$$

$$B_o = \begin{bmatrix}(b_0 - b_n\,a_0) & (b_1 - b_n\,a_1) & \cdots & (b_{n-1} - b_n\,a_{n-1})\end{bmatrix}^T \tag{3.98b}$$

$$H_o = \begin{bmatrix}0 & 0 & \cdots & 1\end{bmatrix} \tag{3.98c}$$

$$D_o = b_n \tag{3.98d}$$

Clearly, since all states are "coupled" together in the F_o matrix, we only need to monitor one state (given as the last state by H_o) to observe *all* states. The matrix F_o is called the *right companion matrix* to the characteristic equation since the coefficients of eqn. (3.9b) appear on the right side of the matrix.

A general single-output system $(F,\,B,\,H,\,D)$ is "fully observable" if it can be converted into observer canonical form. This is achieved via a transformation of state shown in §3.1.4:

$$F_o = T^{-1}\,F\,T \tag{3.99}$$

where T is a nonsingular constant matrix. To demonstrate the general form for T, we begin by considering the third-order case (the extension to the general case will be clear from this development). Left multiplying both sides of eqn. (3.99) by T gives

$$T\,F_o = F\,T \tag{3.100}$$

For the third-order case let T be partitioned into column vectors so that $T = \begin{bmatrix}\mathbf{t}_1 & \mathbf{t}_2 & \mathbf{t}_3\end{bmatrix}$. This leads directly to

$$\begin{bmatrix}\mathbf{t}_1 & \mathbf{t}_2 & \mathbf{t}_3\end{bmatrix} \begin{bmatrix} 0 & 0 & -a_0 \\ 1 & 0 & -a_1 \\ 0 & 1 & -a_2 \end{bmatrix} = F \begin{bmatrix}\mathbf{t}_1 & \mathbf{t}_2 & \mathbf{t}_3\end{bmatrix} \tag{3.101}$$

Next, solving for $\mathbf{t}_2$ and $\mathbf{t}_3$ gives

$$\mathbf{t}_2 = F\,\mathbf{t}_1 \tag{3.102a}$$

$$\mathbf{t}_3 = F\,\mathbf{t}_2 \tag{3.102b}$$

Since $\mathbf{x} = T\,\mathbf{x}_o$, then $H\,T = H_o$, which gives the following three equations:

$$H\,\mathbf{t}_1 = 0 \tag{3.103a}$$

$$H\,\mathbf{t}_2 = 0 \tag{3.103b}$$

$$H\,\mathbf{t}_3 = 1 \tag{3.103c}$$

Substituting eqn. (3.102) into eqn. (3.103) leads to

$$\mathbf{t}_1 = \begin{bmatrix} H \\ HF \\ HF^2 \end{bmatrix}^{-1} \begin{bmatrix} 0 \\ 0 \\ 1 \end{bmatrix} \tag{3.104}$$

Clearly, the original system can only be transformed into observer canonical form if the matrix inverse in eqn. (3.104) exists. The extension to higher-order systems is given by the following $n \times n$ *observability matrix*:

$$\boxed{\mathcal{O} = \begin{bmatrix} H \\ HF \\ HF^2 \\ \vdots \\ HF^{n-1} \end{bmatrix}} \tag{3.105}$$

For a system to be fully observable, the observability matrix $\mathcal{O}$ must be non-singular. Also, a multi-output system is observable if the rank of the $mn \times n$ matrix (where m is the number of outputs) is equal to n.

Example 3.4: In this simple example we consider only a second-order system, with state matrices given by

$$F = \begin{bmatrix} 0 & 1 \\ -2 & -f_{22} \end{bmatrix}, \quad B = \begin{bmatrix} b_{11} \\ b_{21} \end{bmatrix}, \quad H = \begin{bmatrix} 1 & 1 \end{bmatrix}$$

where f_{22}, b_{11}, and b_{21} are real numbers. Computing the observability matrix in eqn. (3.105) with $n = 2$ gives

$$\mathcal{O} = \begin{bmatrix} 1 & 1 \\ -2 & 1 - f_{22} \end{bmatrix}$$

Clearly, the system is observable unless $f_{22} = 3$. Let us compute the transfer function using eqn. (3.14) to gain some physical insight for the case when $f_{22} = 3$:

$$\frac{Y(s)}{U(s)} = \frac{(b_{11} + b_{21})(s+1)}{(s+1)(s+3)}$$

This clearly indicates that a "pole-zero cancellation" has occurred (i.e., one of the roots of the numerator polynomial cancels one of the roots of the denominator polynomial). Therefore, we cannot observe the state associated with $s + 1 = 0$.

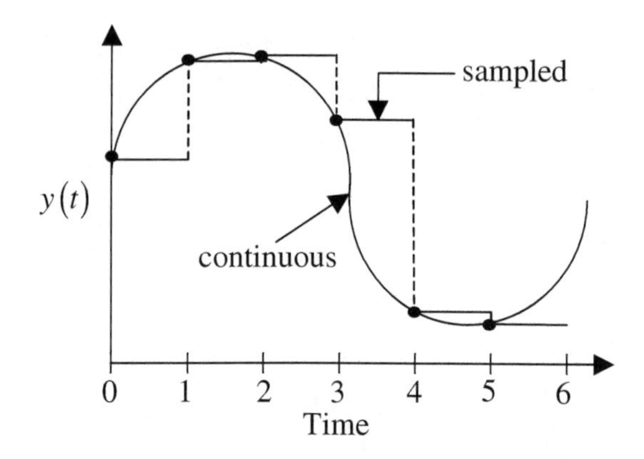

Figure 3.3: Continuous Signal and Sampled Zero-Order Hold

Observability is a powerful tool for state estimation. If a system is not fully observable then all is not lost. A singular value decomposition of the observability matrix can give us insight as to what states are observable. If the observed states are adequate for the dynamical system's requirements (e.g., for control requirements), then a fully observable system may not be necessary. Finally, an extension of observability to nonlinear systems is possible; however, for most nonlinear dynamical systems only local observability can be proven mathematically.[17, 18]

3.5 Discrete-Time Systems

All of the concepts shown in §3.1 extend to discrete-time systems. Discrete-time systems have now become standard in most dynamical applications with the advent of digital computers, which are used to process sampled-data systems for estimation and control purposes. The mechanism that acts on the sensor output and supplies numbers to the digital computer is the analog-to-digital (A/D) converter. Then, the numbers are processed through numerical subroutines and sent to the dynamical system input through the digital-to-analog (D/A) converter. This allows the use of software driven systems to accommodate the estimation/control aspect of a dynamical system, which can be modified simply by uploading new subroutines to the computer.

We shall only consider the most common sampled-type system given by a "zero-order hold" which holds the sampled point to a constant value throughout the interval. Figure 3.3 shows a sampled signal using a zero-order hold. Obviously as the sample interval decreases the sampled signal more closely approximates the continuous signal. Consider the case where time is set to the first sample interval, denoted by Δt, and $F(t)$ and $B(t)$ are constants in eqn. (3.42). Then eqn. (3.53) reduces to

$$\mathbf{x}(\Delta t) = e^{F\Delta t}\mathbf{x}(0) + \left[\int_0^{\Delta t} e^{F(\Delta t - \tau)}\, d\tau\right] B\,\mathbf{u}(0) \tag{3.106}$$

The integral on the right-hand side of eqn. (3.106) can be simplified by defining $\zeta = \Delta t - \tau$, which leads to

$$\int_0^{\Delta t} e^{F(\Delta t - \tau)}\, d\tau = -\int_{\Delta t}^{0} e^{F\zeta}\, d\zeta = \int_0^{\Delta t} e^{F\zeta}\, d\zeta \tag{3.107}$$

Therefore, eqn. (3.106) becomes

$$\mathbf{x}(\Delta t) = \Phi\,\mathbf{x}(0) + \Gamma\,\mathbf{u}(0) \tag{3.108}$$

where

$$\Phi \equiv e^{F\Delta t} \tag{3.109a}$$

$$\Gamma \equiv \left[\int_0^{\Delta t} e^{Ft}\, dt\right] B \tag{3.109b}$$

Expanding (3.108) for $k+1$ samples gives

$$\mathbf{x}[(k+1)\Delta t] = \Phi\,\mathbf{x}(k\,\Delta t) + \Gamma\,\mathbf{u}(k\,\Delta t) \tag{3.110}$$

It is common convention to drop Δt notation from eqn. (3.110) so that the entire discrete state-space representation is given by

$$\boxed{\begin{aligned} \mathbf{x}_{k+1} &= \Phi\,\mathbf{x}_k + \Gamma\,\mathbf{u}_k \qquad (3.111\text{a}) \\ \mathbf{y}_k &= H\,\mathbf{x}_k + D\,\mathbf{u}_k \qquad (3.111\text{b}) \end{aligned}}$$

Notice that the output system matrices H and D are unaffected by the conversion to a discrete-time system. The system can be shown to be stable if all eigenvalues of Φ lie within the unit circle.[3]

Example 3.5: In this example we will perform a conversion from the continuous-time domain to the discrete-time domain for a second-order system, given by

$$F = \begin{bmatrix} -1 & 0 \\ 1 & 0 \end{bmatrix}, \quad B = \begin{bmatrix} 1 \\ 0 \end{bmatrix}$$

To compute Φ we will enlist the help of Laplace transforms, with

$$\begin{aligned} \Phi = e^{F\Delta t} &= \left.\left\{\mathcal{L}^{-1}[sI - F]^{-1}\right\}\right|_{\Delta t} = \left.\left\{\mathcal{L}^{-1}\begin{bmatrix} \dfrac{1}{s+1} & 0 \\ \dfrac{1}{s(s+1)} & \dfrac{1}{s} \end{bmatrix}\right\}\right|_{\Delta t} \\ &= \begin{bmatrix} e^{-\Delta t} & 0 \\ 1 - e^{-\Delta t} & 1 \end{bmatrix} \end{aligned}$$

where $\mathcal{L}^{-1}$ denotes the inverse Laplace transform. The matrix Γ is computed using eqn. (3.109b):

$$\Gamma = \int_0^{\Delta t} \begin{bmatrix} e^{-t} \\ 1-e^{-t} \end{bmatrix} dt = \begin{bmatrix} 1-e^{-\Delta t} \\ \Delta t + e^{-\Delta t} - 1 \end{bmatrix}$$

If the sampling interval is chosen to be $\Delta t = 0.1$ seconds, then Φ and Γ become

$$\Phi = \begin{bmatrix} 0.9048 & 0 \\ 0.0952 & 1 \end{bmatrix}, \quad \Gamma = \begin{bmatrix} 0.0952 \\ 0.0048 \end{bmatrix}$$

Determining analytical expressions for Φ and Γ can be tedious and difficult for large order systems. Fortunately, several numerical approaches exist for computing these matrices.[19] A computationally efficient and accurate approach involves a series expansion:

$$\boxed{\Phi = I + F\Delta t + \frac{1}{2!}F^2\Delta t^2 + \frac{1}{3!}F^3\Delta t^3 + \cdots} \tag{3.112}$$

The matrix Γ is obtained from integration of eqn. (3.112):

$$\boxed{\Gamma = \left[I\Delta t + \frac{1}{2!}F\Delta t^2 + \frac{1}{3!}F^2\Delta t^3 + \cdots \right] B} \tag{3.113}$$

Adequate results can be obtained in most cases using only a few of the terms in the series expansion. For the matrices in example 3.5, using three terms in the series expansion yields

$$\Phi = \begin{bmatrix} 0.9048 & 0 \\ 0.0952 & 1 \end{bmatrix}, \quad \Gamma = \begin{bmatrix} 0.0952 \\ 0.0048 \end{bmatrix} \tag{3.114}$$

The series results for Φ and Γ are accurate to within four significant digits. Results vary with sampling interval. As a general rule of thumb, if the sampling interval is below Nyquist's upper limit, then three to four terms in the series expansion gives accurate results.[20]

The concept of observability can be extended to discrete-time systems. The discrete system is observable if there exists a finite k such that knowledge of the outputs to $k-1$ is sufficient to determine the initial state of the system.[21] Expanding eqn. (3.111), for single output with $\mathbf{u}_k = \mathbf{0}$, to $n-1$ points to obtain n equations for the n unknown initial condition gives

$$\begin{aligned} y_0 &= H\mathbf{x}_0 \\ y_1 &= H\mathbf{x}_1 = H\Phi\mathbf{x}_0 \\ y_2 &= H\mathbf{x}_2 = H\Phi^2\mathbf{x}_0 \\ &\vdots \\ y_{n-1} &= H\mathbf{x}_{n-1} = H\Phi^{n-1}\mathbf{x}_0 \end{aligned} \tag{3.115}$$

Solving eqn. (3.115) for $\mathbf{x}_0$ yields

$$\mathbf{x}_0 = \begin{bmatrix} H \\ H\Phi \\ H\Phi^2 \\ \vdots \\ H\Phi^{n-1} \end{bmatrix}^{-1} \begin{bmatrix} y_0 \\ y_1 \\ y_2 \\ \vdots \\ y_{n-1} \end{bmatrix} \tag{3.116}$$

Clearly, the initial state $\mathbf{x}_0$ can be obtained only if the following observability matrix is nonsingular:

$$\boxed{\mathcal{O}_d = \begin{bmatrix} H \\ H\Phi \\ H\Phi^2 \\ \vdots \\ H\Phi^{n-1} \end{bmatrix}} \tag{3.117}$$

If multiple outputs are given, then for the system to be fully observable $\mathcal{O}_d$ must have rank n.

The main difference in the analysis tools for discrete-time versus continuous-time systems is in the sampling interval. The sampling interval can adversely affect the system's response, but it can also be actually used as another design parameter in a dynamical system to achieve a desired response characteristic. Available tools for discrete-time systems include: z-transforms, bilinear transformations, stability, etc. These concepts are beyond the scope of the present text, since only the required basic fundamentals have been presented. The interested reader can pursue these subjects in more depth by studying the references cited in this section.

3.6 Stability of Linear and Nonlinear Systems

Stability of linear and nonlinear systems is extremely important in both control and estimation algorithms. In estimation the stability of a sequential process is a stringent requirement so that the estimated quantities remain within a bounded region. The general definition of stability begins with Bounded-Input-Bounded-Output (BIBO) stability. Before providing the definition of BIBO stability, we first must describe a *relaxed system*. A system is said to be relaxed at time t_0 if and only if the output $\mathbf{y}_{[t_o,\infty)}$ is solely and uniquely excited by $\mathbf{u}_{[t_o,\infty)}$.[12] For linear systems the relaxed condition follows $\mathbf{y}(t) = H\mathbf{u}_{(-\infty,t_0)} = \mathbf{0}$ for all $t \geq t_0$. A relaxed system is said to be BIBO stable if and only if for any bounded input, the output is bounded.

Let us consider the linear time-invariant model of eqn. (3.14). Since we assume that the input is bounded, we have

$$||\mathbf{u}(t)|| \leq \alpha < \infty \quad \text{for all } t \geq 0 \tag{3.118}$$

where α is a positive constant. The solution for $\mathbf{y}(t)$ assuming a relaxed condition (i.e., $\mathbf{x}(t_0) = \mathbf{0}$) is given by eqn. (3.53):

$$\mathbf{y}(t) = H \int_{t_0}^{t} \Phi(t,\tau)\, B\, \mathbf{u}(\tau)\, d\tau \tag{3.119}$$

Since BIBO stability must be valid for all time, we can allow $t \to \infty$. Next, making use of the convolution integral for $\mathbf{y}(t)$ as $t \to \infty$ gives

$$\mathbf{y}(\infty) = H \int_{t_0}^{\infty} \Phi(\tau)\, B\, \mathbf{u}(t-\tau)\, d\tau \tag{3.120}$$

Taking the norm of both sides of eqn. (3.120) and using eqn. (3.118) yields

$$||\mathbf{y}(\infty)|| \le \alpha \left\| H \int_{t_0}^{\infty} \Phi(\tau)\, B\, d\tau \right\| \tag{3.121}$$

Therefore, the system is bounded if

$$\left\| H \int_{t_0}^{\infty} \Phi(\tau)\, B\, d\tau \right\| < \infty \tag{3.122}$$

which can only be true if

$$\lim_{t\to\infty} ||\Phi(t,t_0)|| = 0 \tag{3.123}$$

From eqn. (3.26) the condition in eqn. (3.123) is satisfied if and only if all the eigenvalues of F have negative real parts.

BIBO stability for nonlinear systems is much more difficult to prove. Fortunately, Lyapunov methods can be applied to show BIBO stability for both nonlinear and linear systems. Two methods for stability were introduced by Lyapunov. The first is given by Lyapunov's linearization method. Before proceeding with this method we must first define an equilibrium point. An equilibrium is defined as a point where the system states remain indefinitely, so that $\dot{\mathbf{x}} = \mathbf{0}$. For linear systems there is usually only one equilibrium point given at $\mathbf{x} = \mathbf{0}$, although there are exceptions (see exercise 3.9). In Lyapunov's linearization method each equilibrium point is considered and evaluated in the linearized model of eqn. (3.76). The equilibrium point is said to be Lyapunov stable if we can select a bound on initial conditions that results in trajectories that remain with a chosen finite limit. Furthermore, the equilibrium point is asymptotically stable if the state also approaches zero as time approaches infinity. Lyapunov's linearization method gives the following stability conditions:[15]

- The equilibrium point is asymptotically stable for the actual nonlinear system if the linearized system is strictly stable, with all eigenvalues of F strictly in the left-hand plane.
- The equilibrium point is unstable for the actual nonlinear system if the linearized system is strictly unstable, with at least one eigenvalue strictly on the right-hand plane.

- Nothing can be concluded if the linearized system is marginally stable, with at least one eigenvalue of F on the imaginary axis and the remainder in the left-hand plane (the equilibrium point may be stable or unstable for the nonlinear system).

Lyapunov's linearization method provides a powerful approach to help qualify the stability of a system if a control (or estimation) scheme is designed to remain within a linear region, but does not give a thorough understanding of the nonlinear system in many cases.

Lyapunov's direct method gives a global stability condition for the general nonlinear system. This concept is closely related to the energy of a system, which is a *scalar* function. The scalar function must in general be continuous and have continuous derivatives with respect to all components of the state vector. Lyapunov showed that if the total energy of a system is dissipated, then the state is confined to a volume bounded by a surface of constant energy, so that the system must eventually settle to an equilibrium point. This concept is valid for both linear and nonlinear systems. Lyapunov stability is given if a chosen scalar function $V(\mathbf{x})$ satisfies the following conditions:

- $V(\mathbf{0}) = 0$
- $V(\mathbf{x}) > 0$ for $\mathbf{x} \neq \mathbf{0}$
- $\dot{V}(\mathbf{x}) \leq 0$

If these conditions are met, then $V(\mathbf{x})$ is a *Lyapunov function*. Furthermore, if $\dot{V}(\mathbf{x}) < 0$ for $\mathbf{x} \neq \mathbf{0}$ then the system is asymptotically stable.

Example 3.6: Consider the following spring-mass-damper system with nonlinear spring and damper components:

$$m\ddot{x} + c\dot{x}|\dot{x}| + k_1 x + k_2 x^3 = 0$$

where m, c, k_1, and k_2 have positive values. The system can be represented in first-order form by defining the following state vector $\mathbf{x} = \begin{bmatrix} x \ \dot{x} \end{bmatrix}^T$:

$$\begin{aligned} \dot{x}_1 &= x_2 \\ \dot{x}_2 &= -(k_1/m)x_1 - (k_2/m)x_1^3 - (c/m)x_2|x_2| \end{aligned}$$

The system has only one equilibrium point at $\mathbf{x} = \begin{bmatrix} 0 \ 0 \end{bmatrix}^T$ that is physically correct (the other one is complex). We wish to investigate the stability of this nonlinear system using Lyapunov's direct method. Intuitively, we choose a candidate Lyapunov function that is given by the total mechanical energy of the system, which is the sum of its kinetic and potential energies:

$$V(\mathbf{x}) = \frac{1}{2}m\dot{x}^2 + \int_0^x (k_1 x + k_2 x^3)\, dx$$

Evaluating this integral yields

$$V(\mathbf{x}) = \frac{1}{2}m\dot{x}^2 + \frac{1}{2}k_1x^2 + \frac{1}{4}k_2x^4$$

Note that zero energy corresponds to the equilibrium point ($\mathbf{x}=\mathbf{0}$), which satisfies the first condition for a valid Lyapunov function. Also, the second condition, $V(\mathbf{x}) > 0$ for $\mathbf{x} \neq \mathbf{0}$, is clearly satisfied. Taking the time derivative of $V(\mathbf{x})$ gives

$$\dot{V}(\mathbf{x}) = m\ddot{x}\dot{x} + (k_1x + k_2x^3)\dot{x}$$

Solving the original system equation for $m\ddot{x}$, and substituting the resulting expression into the equation for $\dot{V}(\mathbf{x})$ yields

$$\dot{V}(\mathbf{x}) = -c|\dot{x}|^3$$

Clearly this expression meets the final condition for a valid Lyapunov function since $\dot{V}(\mathbf{x}) \leq 0$ for all nonzero values of x and $\dot{x}$. Therefore, since $V(\mathbf{x})$ is indeed a Lyapunov function the system is asymptotically stable. This example shows how an "energy-like" function can be used to find a Lyapunov function, since the energy of this system is dissipated by the damper until the mass settles down. More details on Lyapunov methods for stability can be found in Ref. [15].

Lyapunov's global method also is valid for linear time-invariant systems with $\dot{\mathbf{x}} = F\mathbf{x}$. Consider the function $V(\mathbf{x}) = \mathbf{x}^T P\mathbf{x}$, where P is a positive definite symmetric matrix. Clearly, $V(\mathbf{x}) > 0$ for all $\mathbf{x} \neq \mathbf{0}$. The time derivative of $V(\mathbf{x})$ is given by

$$\dot{V}(\mathbf{x}) = \dot{\mathbf{x}}^T P\mathbf{x} + \mathbf{x}^T P\dot{\mathbf{x}} \tag{3.124a}$$

$$= \mathbf{x}^T(F^T P + PF)\mathbf{x} \tag{3.124b}$$

Next, define the following *matrix Lyapunov equation*:

$$\boxed{F^T P + PF = -Q} \tag{3.125}$$

If Q is strictly positive definite then the system is asymptotically stable. Lyapunov showed that this condition is true if and only if all eigenvalues of F are strictly in the left-hand plane. The proof begins by using $\mathbf{x}(t) = e^{Ft}\mathbf{x}_0$ and setting

$$P = \int_0^\infty e^{F^T t} Q e^{Ft}\, dt \tag{3.126}$$

where Q is assumed to be strictly positive definite. Then $F^T P + PF$ is given by

$$F^T P + PF = \int_0^\infty \left(F^T e^{F^T t} Q e^{Ft} + e^{F^T t} Q e^{Ft} F\right) dt \tag{3.127}$$

Next, we use the time derivative of e^{Ft}, which is given by

$$\frac{d}{dt}e^{Ft} = F e^{Ft} = e^{Ft} F \tag{3.128}$$

The second equality in eqn. (3.128) is due to the fact that F and e^{Ft} commute (see Appendix A). Then, the quantity within the integral of eqn. (3.127) can be written as

$$\frac{d}{dt}\left(e^{F^T t} Q e^{Ft}\right) = F^T e^{F^T t} Q e^{Ft} + e^{F^T t} Q e^{Ft} F \tag{3.129}$$

Therefore, we have

$$\begin{aligned} F^T P + P F &= \int_0^\infty \frac{d}{dt}\left(e^{F^T t} Q e^{Ft}\right) dt \\ &= e^{F^T t} Q e^{Ft}\Big|_0^\infty \end{aligned} \tag{3.130}$$

If all eigenvalues of F have negative real parts then the integral in eqn. (3.130) is given by

$$e^{F^T t} Q e^{Ft}\Big|_0^\infty = -Q \tag{3.131}$$

which gives the original matrix Lyapunov equation. Since eqn. (3.126) actually shows the existence of a solution P for *any* square matrix Q, then for any Q the solution for P is unique.[15] A simple choice of Q is given by the identity matrix.

Example 3.7: Given the following state matrix:

$$F = \begin{bmatrix} -a & b \\ -b & -a \end{bmatrix}$$

we wish to determine the ranges for a and b that yield a stable response using Lyapunov's direct method. Choosing $Q = I$, Lyapunov's matrix equation leads to the following three algebraic equations:

$$\begin{aligned} -a\,p_{11} - b\,p_{12} - a\,p_{11} - b\,p_{12} &= -1 \\ -a\,p_{12} - b\,p_{22} - a\,p_{12} + b\,p_{11} &= 0 \\ -a\,p_{22} + b\,p_{12} - a\,p_{22} + b\,p_{12} &= -1 \end{aligned}$$

where p_{11}, p_{22}, and p_{12} are the elements of the P matrix. The solutions for these elements are straightforward and are given by $p_{11} = p_{22} = 1/2a$ and $p_{12} = 0$, so that

$$P = \begin{bmatrix} \frac{1}{2a} & 0 \\ 0 & \frac{1}{2a} \end{bmatrix}$$

The matrix P is positive definite when $a > 0$, which gives the range for stability of the overall system matrix. This is easily confirmed by computing the eigenvalues of F, which are found from the roots of the following characteristic equation:

$$s^2 + 2a\,s + a^2 + b^2 = 0$$

This again shows that the real parts of s are negative when $a > 0$. Note that b may take any value.

Lyapunov's linearization and direct methods can also be applied to discrete-time systems. A nonlinear discrete system with no forcing input is represented by

$$\mathbf{x}_{k+1} = \mathbf{f}(\mathbf{x}_k) \tag{3.132}$$

Equilibrium points are determined by allowing $k+1 \rightarrow k$. Lypaunov's linearization method involves evaluating the equilibrium points using the following linearized model:

$$\Phi = \frac{\partial \mathbf{f}}{\partial \mathbf{x}} \tag{3.133}$$

The stability conditions are exactly the same as in the case for the continuous system. All eigenvalues of Φ must be within the unit circle for the equilibrium point to be stable. If at least one eigenvalue of Φ is on the unit circle then nothing can be concluding for the linearization method. The theory for the discrete-time case of Lyapunov's direct method has been presented by Kalman and Bertram.[22] For Lyapunov's direct method the discrete-time system is stable if the following conditions are satisfied for a chosen scalar function $V(\mathbf{x})$:[3]

- $V(\mathbf{0}) = 0$
- $V(\mathbf{x}) > 0$ for $\mathbf{x} \neq \mathbf{0}$
- $\Delta V(\mathbf{x}) = V[\mathbf{f}(\mathbf{x})] - V(\mathbf{x}) \leq 0$

When these conditions are satisfied then $V(\mathbf{x})$ is a discrete Lyapunov function. Furthermore, if $\Delta V(\mathbf{x}) < 0$ for $\mathbf{x} \neq \mathbf{0}$ then the system is asymptotically stable.

Lyapunov's global method also is valid for linear time-invariant systems with $\mathbf{x}_{k+1} = \Phi \mathbf{x}_k$. Consider the function $V(\mathbf{x}) = \mathbf{x}^T P \mathbf{x}$, where P is a positive definite symmetric matrix. Clearly, $V(\mathbf{x}) > 0$ for all $\mathbf{x} \neq \mathbf{0}$. The increment of $V(\mathbf{x})$ is given by

$$\Delta V(\mathbf{x}) = V(\Phi \mathbf{x}) - V(\mathbf{x}) \tag{3.134a}$$

$$= \mathbf{x}^T (\Phi^T P \Phi - P) \mathbf{x} \tag{3.134b}$$

Next, define the following matrix Lyapunov equation:

$$\boxed{\Phi^T P \Phi - P = -Q} \tag{3.135}$$

If Q is strictly positive definite then the system is asymptotically stable. This condition is true if and only if all eigenvalues of Φ are within the unit circle.

We shall now prove that the linear sequential estimator given by eqns. (1.77) to (1.80) is asymptotically stable. For this proof (assuming a bounded input $\mathbf{y}$) we can ignore the measurements and only treat the following recursion:

$$\hat{\mathbf{x}}_{k+1} = [I - K_{k+1}H_{k+1}]\hat{\mathbf{x}}_k \tag{3.136}$$

Next, we consider the following candidate Lyapunov function!

$$V(\hat{\mathbf{x}}) = \hat{\mathbf{x}}^T P^{-1}\hat{\mathbf{x}} \tag{3.137}$$

The increment of $V(\hat{\mathbf{x}})$ is given by

$$\Delta V(\hat{\mathbf{x}}) = \hat{\mathbf{x}}_{k+1}^T P_{k+1}^{-1}\hat{\mathbf{x}}_{k+1} - \hat{\mathbf{x}}_k^T P_k^{-1}\hat{\mathbf{x}}_k \tag{3.138}$$

Substituting eqn. (3.136) and the inverse of eqn. (1.80) into eqn. (3.138), and simplifying yields

$$\Delta V(\hat{\mathbf{x}}) = -\hat{\mathbf{x}}_k^T H_{k+1}^T K_{k+1}^T P_k^{-1}\hat{\mathbf{x}}_k \tag{3.139}$$

Finally, substituting the transpose of eqn. (1.79) into eqn. (3.139) gives

$$\Delta V(\hat{\mathbf{x}}) = -\hat{\mathbf{x}}_k^T H_{k+1}^T [H_{k+1}P_k H_{k+1}^T + W_{k+1}^{-1}]^{-1} H_{k+1}\hat{\mathbf{x}}_k \tag{3.140}$$

Therefore, since $H_{k+1}^T [H_{k+1}P_k H_{k+1}^T + W_{k+1}^{-1}]^{-1} H_{k+1}$ is positive definite, then we have $\Delta V(\hat{\mathbf{x}}) < 0$, and the sequential estimator is asymptotically stable. Further details on Lyapunov stability can be found in the references cited in this section.

3.7 Attitude Kinematics and Rigid Body Dynamics

This section reviews the equations and concepts of rotational attitude kinematics and dynamics. These equations form the basis for spacecraft, aircraft, and robotic dynamical systems. Only a brief review of the concepts are presented in this chapter.

3.7.1 Attitude Kinematics

The attitude of a vehicle is defined as its orientation with respect to some reference frame. If the reference frame is non-moving, then it is commonly referred to as an *inertial* frame. To describe the attitude two coordinate systems are usually defined: one on the vehicle body and one on the reference frame. For most dynamical applications these coordinate systems have orthogonal unit vectors that follow the right-hand rule. The *attitude matrix* (A), often referred to as the direction cosine matrix or rotation matrix, maps one frame to another (for spacecraft and aircraft kinematics this mapping is usually from the reference frame to the vehicle body frame). A graphical representation of this concept is shown in Figure 3.4.

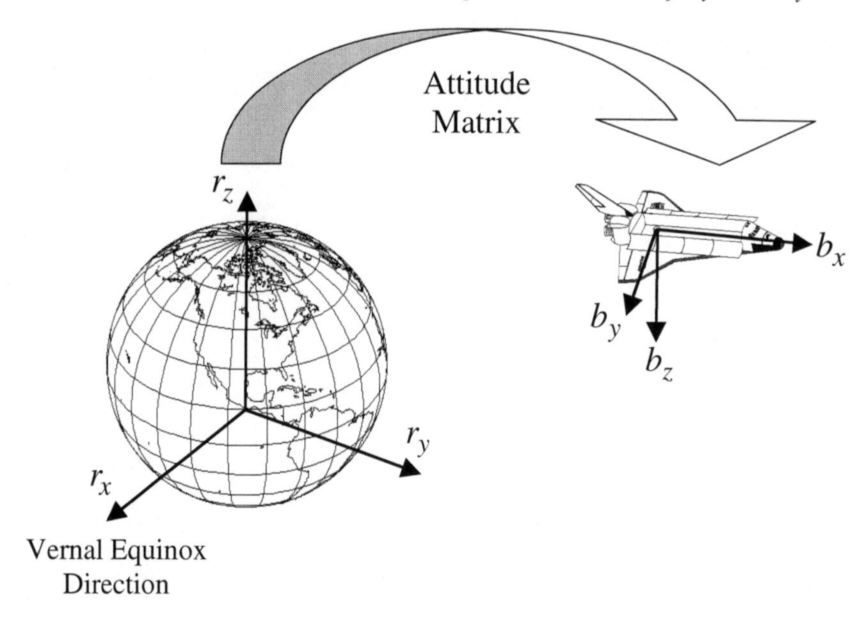

Figure 3.4: Relationship between Reference and Body Frames

Mathematically, the mapping from the reference frame to the body frame is given by

$$\boxed{\mathbf{b} = A\mathbf{r}} \tag{3.141}$$

where $\mathbf{b} = \begin{bmatrix} b_x & b_y & b_z \end{bmatrix}^T$ is the body-frame vector and $\mathbf{r} = \begin{bmatrix} r_x & r_y & r_z \end{bmatrix}^T$ is the reference-frame vector. These vectors are sometimes given by a sum of unit vectors, with orthonormal bases:

$$\mathbf{b} = b_x \hat{\mathbf{b}}_1 + b_y \hat{\mathbf{b}}_2 + b_z \hat{\mathbf{b}}_3 \tag{3.142a}$$

$$\mathbf{r} = r_x \hat{\mathbf{r}}_1 + r_y \hat{\mathbf{r}}_2 + r_z \hat{\mathbf{r}}_3 \tag{3.142b}$$

As an aside, we note the projections of the $\hat{\mathbf{b}}_i$ unit vectors onto the $\hat{\mathbf{r}}_i$ unit vectors are accomplished by the same matrix as

$$\begin{Bmatrix} \hat{\mathbf{b}}_1 \\ \hat{\mathbf{b}}_2 \\ \hat{\mathbf{b}}_3 \end{Bmatrix} = A \begin{Bmatrix} \hat{\mathbf{r}}_1 \\ \hat{\mathbf{r}}_2 \\ \hat{\mathbf{r}}_3 \end{Bmatrix} \tag{3.143}$$

where *vectrix* notation is used in eqn. (3.143) (see Ref. [23] for details). The matrix A is in fact an *orthogonal* matrix since its inverse is given by its transpose. Also, for right-handed systems the determinant of A is given by $+1$.[24] In other words, the attitude matrix is a *proper real orthogonal* matrix. Many parameterizations exist for the attitude matrix, including: the Euler angles, Euler axis/angle, the quaternion, Cayley-Klein parameters, Gibb's vector, modified Rodrigues parameters, etc.[25]

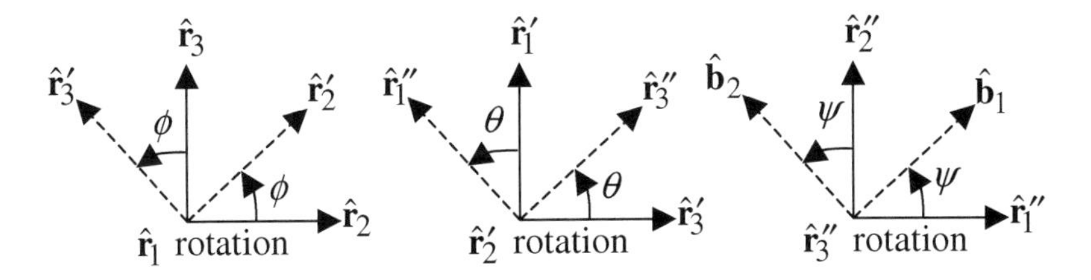

Figure 3.5: Euler Angles for a 1-2-3 Rotation Sequence

Euler angles are commonly used to parameterize the attitude matrix since they give a physical representation. The classical Euler angles are denoted by the roll (ϕ), pitch (θ), and yaw (ψ) angles. Consider a 1-2-3 Euler angle sequence, as shown by Figure 3.5. This sequence performs a rotation from the reference vector ($\mathbf{r}$) to the body vector ($\mathbf{b}$) through a rotation about the $\hat{\mathbf{r}}_1$ vector (the 1-axis rotation) first, with

$$\mathbf{r}' = \begin{bmatrix} 1 & 0 & 0 \\ 0 & \cos\phi & \sin\phi \\ 0 & -\sin\phi & \cos\phi \end{bmatrix} \mathbf{r} \tag{3.144}$$

Then a rotation about the $\hat{\mathbf{r}}_2'$ vector is performed (the 2-axis rotation), with

$$\mathbf{r}'' = \begin{bmatrix} \cos\theta & 0 & -\sin\theta \\ 0 & 1 & 0 \\ \sin\theta & 0 & \cos\theta \end{bmatrix} \mathbf{r}' \tag{3.145}$$

Finally a rotation about the $\hat{\mathbf{r}}_3'$ vector is performed (the 3-axis rotation), with

$$\mathbf{b} = \begin{bmatrix} \cos\psi & \sin\psi & 0 \\ -\sin\psi & \cos\psi & 0 \\ 0 & 0 & 1 \end{bmatrix} \mathbf{r}'' \tag{3.146}$$

Substituting eqn. (3.144) into eqn. (3.145), and substituting the resulting equation into eqn. (3.146) leads to the following form for the attitude matrix:

$$A = \begin{bmatrix} c\psi\, c\theta & s\psi\, c\phi + c\psi\, s\theta\, s\phi & s\psi\, s\phi - c\psi\, s\theta\, c\phi \\ -s\psi\, c\theta & c\psi\, c\phi - s\psi\, s\theta\, s\phi & c\psi\, s\phi + s\psi\, s\theta\, c\phi \\ s\theta & -c\theta\, s\phi & c\theta\, c\phi \end{bmatrix} \tag{3.147}$$

where $c\psi \equiv \cos\psi$, $s\phi \equiv \sin\phi$, etc. There are in fact twelve possible rotation sequences: six asymmetric (1-2-3, 1-3-2, 2-1-3, 2-3-1, 3-1-2, 3-2-1) and six symmetric (1-2-1, 1-3-1, 2-1-2, 2-3-2, 3-1-3, 3-2-3). An interesting case for the attitude matrix occurs when the Euler angles are small so that the cosine of the angle is approximately one and the sine of the angle is approximately the angle. In this case the attitude matrix is adequately approximated by

$$A \approx \begin{bmatrix} 1 & \psi & -\theta \\ -\psi & 1 & \phi \\ \theta & -\phi & 1 \end{bmatrix} = I_{3\times 3} - [\boldsymbol{\alpha}\times] \tag{3.148}$$

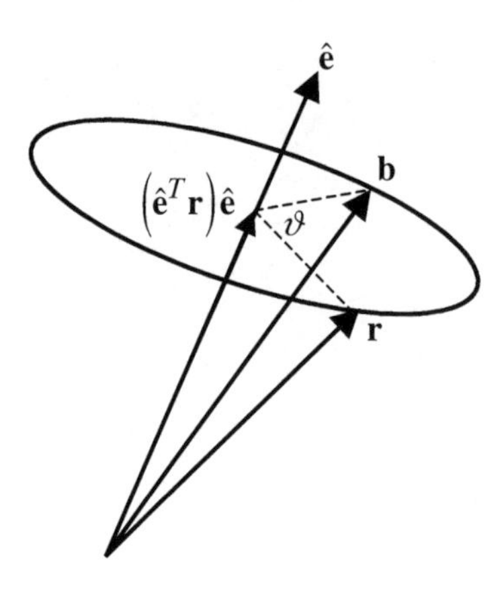

Figure 3.6: Euler Axis and Angle

where $\boldsymbol{\alpha} \equiv \left[\phi\ \theta\ \psi\right]^T$, $I_{3\times 3}$ is a 3×3 identity matrix, and $[\boldsymbol{\alpha}\times]$ is referred to as a cross product matrix because $\boldsymbol{\alpha}\times\boldsymbol{\beta} = [\boldsymbol{\alpha}\times]\boldsymbol{\beta}$, with

$$[\boldsymbol{\alpha}\times] \equiv \begin{bmatrix} 0 & -\alpha_3 & \alpha_2 \\ \alpha_3 & 0 & -\alpha_1 \\ -\alpha_2 & \alpha_1 & 0 \end{bmatrix} \tag{3.149}$$

Another attitude parameterization is given by the Euler axis $\hat{\mathbf{e}}$ and angle ϑ. Euler's theorem states that the most general motion of a rigid body with one point fixed is a rotation about some axis. This is represented by Figure 3.6, and can mathematically be written as

$$\mathbf{b} = \left(\hat{\mathbf{e}}^T\mathbf{r}\right)\hat{\mathbf{e}} + \cos\vartheta\left[\mathbf{r} - \left(\hat{\mathbf{e}}^T\mathbf{r}\right)\hat{\mathbf{e}}\right] - \sin\vartheta\left(\hat{\mathbf{e}}\times\mathbf{r}\right) \tag{3.150}$$

Comparing eqn. (3.150) with eqn. (3.141) gives the following attitude matrix:

$$\boxed{A = (\cos\vartheta)I_{3\times 3} + (1-\cos\vartheta)\hat{\mathbf{e}}\hat{\mathbf{e}}^T - \sin\vartheta[\hat{\mathbf{e}}\times]} \tag{3.151}$$

We also note that the Euler axis $\hat{\mathbf{e}}$ is unchanged by the attitude matrix, so that $A\hat{\mathbf{e}} = \hat{\mathbf{e}}$. This is true since any proper orthogonal 3×3 matrix has at least one eigenvector with unity eigenvalue.[24]

One of the most useful attitude parameterization is given by the *quaternion*.[26] Like the Euler axis/angle parameterization, the quaternion is also a four-dimensional vector, defined as

$$\mathbf{q} \equiv \begin{bmatrix} \boldsymbol{\varrho} \\ q_4 \end{bmatrix} \tag{3.152}$$

with

$$\boldsymbol{\varrho} \equiv \left[q_1\ q_2\ q_3\right]^T = \hat{\mathbf{e}}\sin(\vartheta/2) \tag{3.153a}$$

$$q_4 = \cos(\vartheta/2) \tag{3.153b}$$

Since a four-dimensional vector is used to describe three dimensions, the quaternion components cannot be independent of each other. The quaternion satisfies a single constraint given by $\mathbf{q}^T\mathbf{q} = 1$, which is analogous to requiring that $\hat{\mathbf{e}}$ be a unit vector in the Euler axis/angle parameterization. The attitude matrix is related to the quaternion by

$$\boxed{A(\mathbf{q}) = \Xi^T(\mathbf{q})\Psi(\mathbf{q})} \tag{3.154}$$

with

$$\Xi(\mathbf{q}) \equiv \begin{bmatrix} q_4 I_{3\times 3} + [\boldsymbol{\varrho}\times] \\ -\boldsymbol{\varrho}^T \end{bmatrix} \tag{3.155a}$$

$$\Psi(\mathbf{q}) \equiv \begin{bmatrix} q_4 I_{3\times 3} - [\boldsymbol{\varrho}\times] \\ -\boldsymbol{\varrho}^T \end{bmatrix} \tag{3.155b}$$

An advantage to using quaternions, which will be exploited in Chapter 4, is that the attitude matrix is quadratic in the parameters and also does not involve transcendental functions. For small angles the vector part of the quaternion is approximately equal to half angles so that $\boldsymbol{\varrho} \approx \boldsymbol{\alpha}/2$ and $q_4 \approx 1$.

The attitude kinematics equation can be derived by considering a state transition matrix $\Phi(t+\Delta t, t)$ that maps the attitude from one time to the next:

$$A(t+\Delta t) = \Phi(t+\Delta t, t)A(t) \tag{3.156}$$

Obviously $\Phi(t+\Delta t, t)$ must also be an attitude matrix, which can be given by eqn. (3.148) plus higher-order terms. Then, from the definition of the derivative we have

$$\lim_{\Delta t\to 0}\left\{\frac{A(t+\Delta t) - A(t)}{\Delta t}\right\} = -\lim_{\Delta t\to 0}\left\{\frac{1}{\Delta t}[\boldsymbol{\alpha}(t)\times]\right\}A(t) \tag{3.157}$$

where the higher-order terms vanish in the limit. Hence, the following kinematics equation can be derived:

$$\boxed{\dot{A} = -[\boldsymbol{\omega}\times]A} \tag{3.158}$$

where $\boldsymbol{\omega}$ is the angular velocity vector of the body frame relative to the reference frame. The Euler angle kinematics equation is given by substituting eqn. (3.147) into eqn. (3.158), leading to

$$\begin{bmatrix} \dot{\phi} \\ \dot{\theta} \\ \dot{\psi} \end{bmatrix} = \frac{1}{\cos\theta}\begin{bmatrix} \cos\psi & -\sin\psi & 0 \\ \cos\theta\sin\psi & \cos\theta\cos\psi & 0 \\ -\sin\theta\cos\psi & \sin\theta\sin\psi & \cos\theta \end{bmatrix}\boldsymbol{\omega} \tag{3.159}$$

We clearly see that the Euler angle kinematics become singular when θ is either 90 or 270 degrees. In fact all three-dimensional (minimal) parameterizations have a singularity, which can cause difficulties in a particular application. The inverse kinematics are given by

$$\boldsymbol{\omega} = \begin{bmatrix} \cos\theta\cos\psi & \sin\psi & 0 \\ -\cos\theta\sin\psi & \cos\psi & 0 \\ \sin\theta & 0 & 1 \end{bmatrix}\begin{bmatrix} \dot{\phi} \\ \dot{\theta} \\ \dot{\psi} \end{bmatrix} \tag{3.160}$$

The quaternion kinematics equation are given by

$$\boxed{\dot{\mathbf{q}} = \frac{1}{2}\Xi(\mathbf{q})\boldsymbol{\omega} = \frac{1}{2}\Omega(\boldsymbol{\omega})\mathbf{q}} \tag{3.161}$$

where

$$\Omega(\boldsymbol{\omega}) \equiv \begin{bmatrix} -[\boldsymbol{\omega}\times] & \boldsymbol{\omega} \\ -\boldsymbol{\omega}^T & 0 \end{bmatrix} \tag{3.162}$$

The matrix $\Xi(\mathbf{q})$ obeys the following helpful relations:

$$\Xi^T(\mathbf{q})\Xi(\mathbf{q}) = (\mathbf{q}^T\mathbf{q})I_{3\times 3} \tag{3.163a}$$

$$\Xi(\mathbf{q})\Xi^T(\mathbf{q}) = (\mathbf{q}^T\mathbf{q})I_{4\times 4} - \mathbf{q}\mathbf{q}^T \tag{3.163b}$$

$$\Xi^T(\mathbf{q})\mathbf{q} = \mathbf{0}_{3\times 1} \tag{3.163c}$$

$$\Xi^T(\mathbf{q})\boldsymbol{\lambda} = -\Xi^T(\boldsymbol{\lambda})\mathbf{q} \quad \text{for any } \boldsymbol{\lambda}_{4\times 1} \tag{3.163d}$$

Also, another useful identity is given by

$$\Psi(\mathbf{q})\boldsymbol{\omega} = \Gamma(\boldsymbol{\omega})\mathbf{q} \tag{3.164}$$

where

$$\Gamma(\boldsymbol{\omega}) \equiv \begin{bmatrix} [\boldsymbol{\omega}\times] & \boldsymbol{\omega} \\ -\boldsymbol{\omega}^T & 0 \end{bmatrix} \tag{3.165}$$

The inverse kinematics are given by multiplying eqn. (3.161) by $\Xi^T(\mathbf{q})$, and using the identity in eqn. (3.163a), leading to

$$\boldsymbol{\omega} = 2\,\Xi^T(\mathbf{q})\dot{\mathbf{q}} \tag{3.166}$$

A major advantage of using quaternions is that the kinematics equation is linear in the quaternion and is also free of singularities. Another advantage of quaternions is that successive rotations can be accomplished using quaternion multiplication. Here we adopt the convention of Lefferts, Markley, and Shuster[27] who multiply the quaternions in the same order as the attitude matrix multiplication (in contrast to the usual convention established by Hamiliton[26]). Suppose we wish to perform a successive rotation. This can be written using

$$A(\mathbf{q}')A(\mathbf{q}) = A(\mathbf{q}'\otimes\mathbf{q}) \tag{3.167}$$

The composition of the quaternions is bilinear, with

$$\mathbf{q}'\otimes\mathbf{q} = \begin{bmatrix}\Psi(\mathbf{q}') & \mathbf{q}'\end{bmatrix}\mathbf{q} = \begin{bmatrix}\Xi(\mathbf{q}) & \mathbf{q}\end{bmatrix}\mathbf{q}' \tag{3.168}$$

Also, the inverse quaternion is defined by

$$\mathbf{q}^{-1} \equiv \begin{bmatrix} -\boldsymbol{\varrho} \\ q_4 \end{bmatrix} \tag{3.169}$$

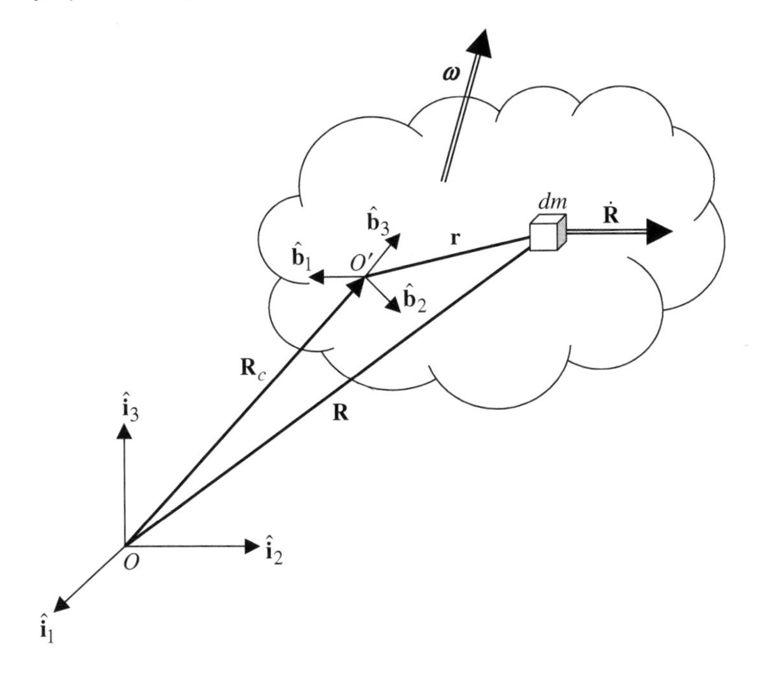

Figure 3.7: General Rigid Body Motion

Note that $\mathbf{q} \otimes \mathbf{q}^{-1} = \begin{bmatrix} 0 & 0 & 0 & 1 \end{bmatrix}^T$, which is the identity quaternion. A computationally efficient algorithm to extract the quaternion from the attitude matrix is given in Ref. [28]. A more thorough review of the attitude representations shown in this section, as well as others, can be found in the excellent survey paper by Shuster[25] and in the book by Kuipers.[29]

3.7.2 Rigid Body Dynamics

The rigid body equations of motion of a vehicle in both translation and rotation with respect to some inertial frame are obtained from Newton's second law. We first consider the angular momentum $\mathbf{H}_{\text{tot}}$ of a body defined as an integral over a continuous mass density (see Figure 3.7):

$$\mathbf{H}_{\text{tot}} = \int_B \mathbf{R} \times \dot{\mathbf{R}}\, dm \tag{3.170}$$

From Figure 3.7 the following vector relation is given:

$$\mathbf{R} = \mathbf{R}_c + \mathbf{r} \tag{3.171}$$

In order to determine the derivative of eqn. (3.171), since the velocity vector of $\mathbf{r}$ is defined to be an inertial derivative, we must employ the *transport theorem*:[23]

$$\dot{\mathbf{r}} \equiv \frac{{}^{\mathcal{N}}d}{dt}(\mathbf{r}) = \frac{{}^{\mathcal{B}}d}{dt}(\mathbf{r}) + \boldsymbol{\omega} \times \mathbf{r} \tag{3.172}$$

where ${}^{\mathcal{N}}d/dt$ denotes the derivative with respect to the inertial frame, ${}^{\mathcal{B}}d/dt$ denotes the derivative with respect to the body frame, and ω is the angular velocity of the body relative to the inertial frame. Since we have assumed that the body is rigid, then ${}^{\mathcal{B}}d/dt(\mathbf{r})$ is zero. Therefore, the derivative of eqn. (3.171) with respect to the inertial frame is given by

$$\dot{\mathbf{R}} = \dot{\mathbf{R}}_c + \omega \times \mathbf{r} \tag{3.173}$$

Substituting eqns. (3.171) and (3.173) into eqn. (3.170), and assuming that the point O' is the center of mass (so that $\int_B \mathbf{r}\, dm = 0$) leads to

$$\mathbf{H}_{\text{tot}} = \mathbf{H} + m\mathbf{R}_c \times \dot{\mathbf{R}}_c \tag{3.174}$$

where the contribution of the mass relative to the center of mass is defined by

$$\mathbf{H} \equiv \int_B \mathbf{r} \times (\omega \times \mathbf{r})\, dm = \left(\int_B -[\mathbf{r}\times][\mathbf{r}\times]\, dm \right) \omega \tag{3.175}$$

This is most often written in compact form:

$$\mathbf{H} = J\,\omega \tag{3.176}$$

with

$$J \equiv \int_B -[\mathbf{r}\times][\mathbf{r}\times]\, dm \tag{3.177}$$

The matrix J is called the *moment of inertia* or simply *inertia matrix*, which is a positive definite, symmetric matrix (with three orthogonal eigenvectors). The off-diagonal terms are sometimes referred to as *products of inertia*. The moment of inertia about some given axis is related simply to the moment about a parallel axis through the center of mass, which can be computed using the *parallel axis theorem*.[24, 30]

The rate of change of the angular momentum with respect to the inertial frame is equal to the applied torque $\mathbf{L}$:

$$\dot{\mathbf{H}} = \mathbf{L} \tag{3.178}$$

Using the transport theorem on eqn. (3.178) gives

$$\dot{\mathbf{H}} = \frac{{}^{\mathcal{B}}d}{dt}(\mathbf{H}) + \omega \times \mathbf{H} = \mathbf{L} \tag{3.179}$$

Substituting eqn. (3.176) into eqn. (3.179) gives *Euler's equations of motion*:

$$\boxed{J\dot{\omega} = -[\omega\times]J\omega + \mathbf{L}} \tag{3.180}$$

Equation (3.180) represents a set of three coupled, first-order, nonlinear differential equations. Closed-form solutions exist for special cases only.[31]

The linear force of a body relative to a body's center of mass is given by Newton's law:

$$\mathbf{F} = m\dot{\mathbf{v}} \tag{3.181}$$

where $\mathbf{F}$ is the total external force acting on the rigid body and $\mathbf{v}$ is the absolute velocity of the center of mass. In order to determine the acceleration in the body frame, the transport theorem must be again used:[32]

$$\boxed{\dot{\mathbf{v}} = \frac{^{\mathcal{B}}d}{dt}(\mathbf{v}) + \boldsymbol{\omega} \times \mathbf{v}} \tag{3.182}$$

Substituting eqn. (3.182) into eqn. (3.181) leads to the following scalar equations for the force:

$$f_1 = m(\dot{v}_1 + v_3\omega_2 - v_2\omega_3) \tag{3.183a}$$

$$f_2 = m(\dot{v}_2 + v_1\omega_3 - v_3\omega_1) \tag{3.183b}$$

$$f_3 = m(\dot{v}_3 + v_2\omega_1 - v_1\omega_2) \tag{3.183c}$$

The components of $\boldsymbol{\omega}$ can be obtained from the solution of eqn. (3.180). The force equations have been derived for a frame fixed to the body. In order to determine the position of the body a transformation of the velocity components v_1, v_2, and v_3 to the reference frame must be made using the attitude matrix, which are then integrated to obtain the absolute position.

3.8 Spacecraft Dynamics and Orbital Mechanics

This section reviews the basic equations for spacecraft dynamics and orbital mechanics. The equations are fairly straightforward, but carry deep meaning and revolutionary concepts, as attested to by the numerous publications in these areas since their conception. We only present the equations necessary to demonstrate the basics of attitude estimation and orbit determination of vehicles.

3.8.1 Spacecraft Dynamics

To fully describe the rotational motion of a rigid spacecraft, a kinematic and a dynamic equation of motion are required. For most modern spacecraft applications the quaternion kinematics equation is preferred. Therefore, the following equations are used:

$$\dot{\mathbf{q}} = \frac{1}{2}\Omega(\boldsymbol{\omega})\mathbf{q} \tag{3.184a}$$

$$J\dot{\boldsymbol{\omega}} = -[\boldsymbol{\omega}\times]J\boldsymbol{\omega} + \mathbf{L} \tag{3.184b}$$

If a spacecraft is equipped with reaction wheels (which are common on most spacecraft) the angular momentum can be modified as[31]

$$\mathbf{H} = J\boldsymbol{\omega} + \mathbf{h} \tag{3.185}$$

where $\mathbf{h}$ is the angular momentum due to the rotation of the wheels relative to the spacecraft, and the inertia J now contains the mass of the wheels. Using eqn. (3.185) in eqn. (3.179) gives

$$\dot{\mathbf{H}} = -[J^{-1}(\mathbf{H}-\mathbf{h})\times]\mathbf{H}+\mathbf{L} \tag{3.186}$$

Equation (3.186) can also be rewritten in Euler's form as

$$J\dot{\boldsymbol{\omega}} = -[\boldsymbol{\omega}\times](J\boldsymbol{\omega}+\mathbf{h})+\mathbf{L}-\dot{\mathbf{h}} \tag{3.187}$$

Equation (3.186) is often preferred since it does not involve the derivative of the wheel momentum.

An interesting and useful case of Euler's rotational equations of motion is given by defining the body coordinate system to coincide with the principal axes (i.e., along the eigenvectors of J). In this case the inertia matrix J is diagonal with elements denoted by J_1, J_2, and J_3 (i.e., the eigenvalues of J). Euler's equations then become:

$$J_1\dot{\omega}_1 = (J_2 - J_3)\omega_2\omega_3 + L_1 \tag{3.188a}$$

$$J_2\dot{\omega}_2 = (J_3 - J_1)\omega_3\omega_1 + L_2 \tag{3.188b}$$

$$J_3\dot{\omega}_3 = (J_1 - J_2)\omega_1\omega_2 + L_3 \tag{3.188c}$$

The stability of rotation about the principal axes can be shown by assuming a constant rotation about one of the axes, e.g., axis 3, and allowing a small perturbation. This indicates that the motion is stable if J_3 is the largest or smallest principal moment of inertia.[33]

We now consider the torque-free case (i.e., $\mathbf{L} = \mathbf{0}$) with two of the principal moments of inertia equal (say $J_1 = J_2 \equiv J_T$), which is the *axially symmetric* case. Euler's equations become

$$J_T\dot{\omega}_1 = -(J_3 - J_T)\omega_2\omega_3 \tag{3.189a}$$

$$J_T\dot{\omega}_2 = (J_3 - J_T)\omega_3\omega_1 \tag{3.189b}$$

$$J_3\dot{\omega}_3 = 0 \tag{3.189c}$$

Equation (3.189c) clearly indicates that ω_3 is constant, with $\omega_3(t) = \omega_3(t_0)$. Next we impose that $\omega_3 > 0$, which can be accomplished by choosing the proper sense of the third principal axis. This leads to the following equations for ω_1 and ω_2:

$$\dot{\omega}_1 - \omega_n\omega_2 = 0 \tag{3.190a}$$

$$\dot{\omega}_2 + \omega_n\omega_1 = 0 \tag{3.190b}$$

where $\omega_n = (1 - J_3/J_T)\omega_3(t_0)$ is a constant. The solutions for ω_1 and ω_2 are given by

$$\omega_1(t) = \omega_1(t_0)\cos\omega_n t + \omega_2(t_0)\sin\omega_n t \tag{3.191a}$$

$$\omega_2(t) = \omega_2(t_0)\cos\omega_n t - \omega_1(t_0)\sin\omega_n t \tag{3.191b}$$

This indicates that the system is *marginally stable*.[2] The constant ω_n is known as the *body nutation rate*. Also, the magnitude of the angular momentum can be shown to be given by

$$||\mathbf{H}|| = \left\{ J_T^2[\omega_1^2(t_0) + \omega_2^2(t_0)] + J_3^2\omega_3^2(t_0) \right\}^{1/2} \tag{3.192}$$

which is constant and inertially fixed along the third principal axis. This also indicates that energy is conserved. The angular momentum in body coordinates can be computed using the attitude matrix:

$$^{\mathcal{B}}\begin{bmatrix} H_1 \\ H_2 \\ H_3 \end{bmatrix} = A \, ^{\mathcal{I}}\begin{bmatrix} 0 \\ 0 \\ ||\mathbf{H}|| \end{bmatrix} \tag{3.193}$$

Since the body is spinning about its axis of symmetry (the third axis) a convenient parameterization of the attitude matrix is the 3-1-3 sequence. This leads to

$$H_1 = J_T\omega_1 = ||\mathbf{H}|| \sin\theta \sin\psi \tag{3.194a}$$

$$H_2 = J_T\omega_2 = ||\mathbf{H}|| \sin\theta \cos\psi \tag{3.194b}$$

$$H_3 = J_3\omega_3 = ||\mathbf{H}|| \cos\theta \tag{3.194c}$$

Since H_3 and $||\mathbf{H}||$ are constants then $\theta = \cos^{-1}(H_3/||\mathbf{H}||)$ is constant as well. This angle is known as the *nutation angle*. The solution for the yaw angle ψ is given by $\psi = \tan^{-1}(H_1/H_2)$. The solution for the roll angle ϕ is given from the 3-1-3 kinematics equation and can be shown to be given by $\dot{\phi} = ||\mathbf{H}||/J_T$. The asymmetric case with $J_1 \neq J_2$ can be solved in closed-form using *Jacobian elliptic functions*.[31]

As mentioned previously, a thorough treatise of spacecraft dynamics would entail significant effort. Other topics, such as dual-spin spacecraft, kinetic-energy and angular momentum ellipsoids, variable mass, passive and active control techniques, attitude torque disturbances, etc., can be found in the references in this section. Other reference includes works by Kane, Likens, and Levinson,[34] Hughes,[35] Kaplan,[36] Wiesel,[37] and Junkins and Turner.[38]

3.8.2 Orbital Mechanics

The study of bodies in orbit has attracted the world's greatest mathematicians in the past, and still is a flourishing subject area in the present. In fact many useful mathematical concepts, such as Bessel functions and nonlinear least squares, can be directly traced back to the study of orbital motion. As with spacecraft dynamics, a thorough treatise of orbital mechanics is not possible in the present text. We again only treat the basic equations and concepts that are required to demonstrate orbit determination.

An unperturbed orbiting body follows Kepler's three laws, originally given by

1. The orbit of each planet is an ellipse, with the Sun at a focus.

2. The line joining the planet to the sun sweeps out equal areas in equal times.

3. The square of the period of a planet is proportional to the cube of its mean distance from the sun.

These powerful statements define the shape of planetary orbits, the velocity at which planets travel around the sun, and the time required from a planet to complete an orbit. These laws can be proven mathematically from Newton's universal law of gravitation, which states: any two bodies with mass M and m attract each other by a force that is proportional to the product of their masses and inversely proportional to the square of the distance r between them. Mathematically, this statement is given by

$$F_g = \frac{GMm}{r^2} \tag{3.195}$$

where G is the *universal gravitation constant.*[39, 40] Consider the two bodies in Figure 3.8. The axes $\hat{\mathbf{i}}_1$, $\hat{\mathbf{i}}_2$, and $\hat{\mathbf{i}}_3$ are an inertial frame, and the axes $\hat{\mathbf{b}}_1$, $\hat{\mathbf{b}}_2$, and $\hat{\mathbf{b}}_3$ are a non-rotating frame with origin coincident with the center of mass. Applying Newton's law in the inertial frame for each body we obtain

$$M\ddot{\mathbf{r}}_M = \frac{GMm}{||\mathbf{r}||^3}\mathbf{r} \tag{3.196a}$$

$$m\ddot{\mathbf{r}}_m = -\frac{GMm}{||\mathbf{r}||^3}\mathbf{r} \tag{3.196b}$$

The negative sign in eqn. (3.196a) is due to the opposite direction of the force. Since, as shown in Figure 3.8, $\mathbf{r} = \mathbf{r}_m - \mathbf{r}_M$ then from eqn. (3.196) we obtain

$$\ddot{\mathbf{r}} = -\frac{G(M+m)}{||\mathbf{r}||^3}\mathbf{r} \tag{3.197}$$

If the mass m is much smaller than M (which is a very accurate assumption for orbiting spacecraft) then we can effectively ignore m so that

$$\boxed{\ddot{\mathbf{r}} = -\frac{\mu}{||\mathbf{r}||^3}\mathbf{r}} \tag{3.198}$$

where $\mu \equiv GM$ is called the *gravitational parameter*. The gravitational parameter is more commonly used in orbital mechanics of spacecraft since it can be measured to high precision, unlike the mass M.

Equation (3.198) is the most fundamental equation used in orbital mechanics, and can be used to prove Kepler's laws. In particular one can show that mechanical energy and angular momentum are conserved. The conservation of mechanical energy gives rise to the *vis-viva integral.*[39] Since angular momentum is related to $\mathbf{r} \times \dot{\mathbf{r}}$, which is constant, then the spacecraft's motion must be confined to a plane inertially fixed in space. The two-body relative equations represent a coupled nonlinear set of differential equations. Fortunately, analytical solutions to this set of equations exist. Herrick[41] establishes the solution of eqn. (3.198), given initial conditions $\mathbf{r}(t_0)$ and

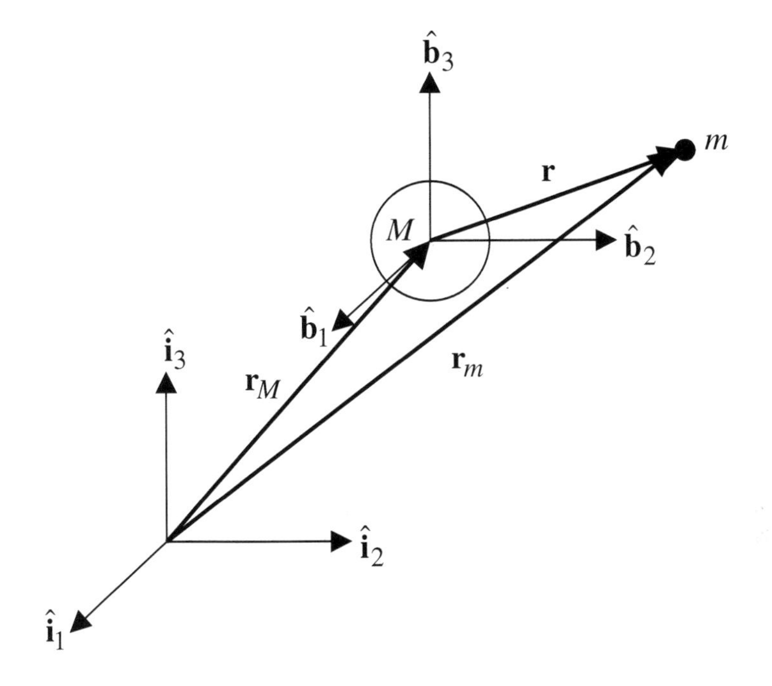

Figure 3.8: Relative Motion of Two Bodies

$\dot{\mathbf{r}}(t_0)$. First we compute the semimajor axis (a) using the vis-viva integral:

$$a = \left(\frac{2}{||\mathbf{r}(t_0)||} - \frac{||\dot{\mathbf{r}}(t_0)||^2}{\mu} \right)^{-1} \tag{3.199}$$

Then, given the current time of interest (t), we solve the following equation for ϕ (using Newton's method):*

$$t - t_0 = \frac{a^{3/2}}{\mu^{1/2}} \left[\phi - \left(1 - \frac{||\mathbf{r}(t_0)||}{a} \right) \sin\phi + \frac{\mathbf{r}^T(t_0)\dot{\mathbf{r}}(t_0)}{(\mu a)^{1/2}} (1 - \cos\phi) \right] \tag{3.200}$$

Next, compute the following variables:

$$f = 1 - a(1 - \cos\phi)/||\mathbf{r}(t_0)|| \tag{3.201a}$$

$$g = (t - t_0) - a^{3/2}(\phi - \sin\phi)/\mu^{1/2} \tag{3.201b}$$

$$r = a[1 - (1 - ||\mathbf{r}(t_0)||/a)\cos\phi] + \mathbf{r}^T(t_0)\dot{\mathbf{r}}(t_0)(a/\mu)^{1/2} \sin\phi \tag{3.201c}$$

$$\dot{f} = -(r||\mathbf{r}(t_0)||)^{-1}(\mu a)^{1/2} \sin\phi \tag{3.201d}$$

$$\dot{g} = 1 - a(1 - \cos\phi)/r \tag{3.201e}$$

*We note ϕ has the geometric interpretation as the change in eccentric anomaly.

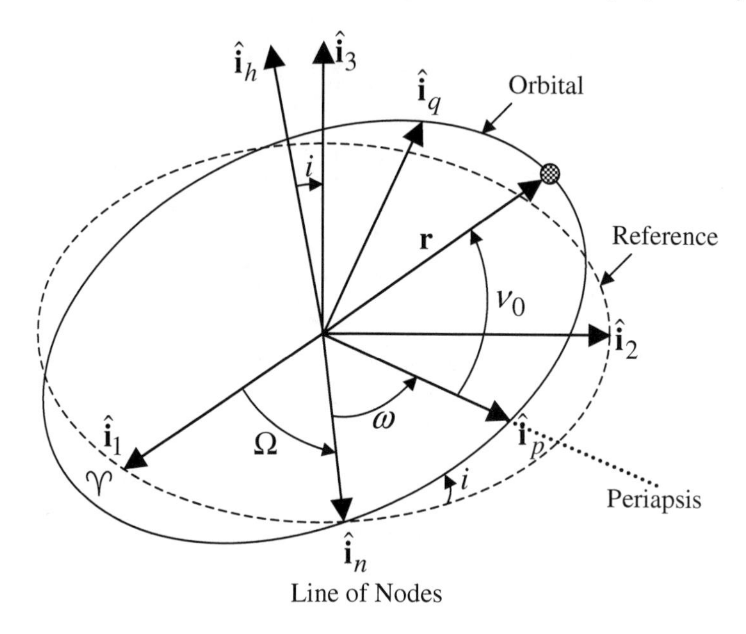

Figure 3.9: Coordinate System Geometry and Orbital Elements

Then, the solution to eqn. (3.198) is given by

$$\mathbf{r}(t) = f\,\mathbf{r}(t_0) + g\,\dot{\mathbf{r}}(t_0) \tag{3.202a}$$

$$\dot{\mathbf{r}}(t) = \dot{f}\,\mathbf{r}(t_0) + \dot{g}\,\dot{\mathbf{r}}(t_0) \tag{3.202b}$$

Unfortunately knowing $\mathbf{r}(t_0)$ and $\dot{\mathbf{r}}(t_0)$ does not provide a physical meaning of the orbit. To characterize an orbit six classical Keplerian orbital elements are given in the place of $\mathbf{r}(t_0)$ and $\dot{\mathbf{r}}(t_0)$, which do provide a physical meaning. Figure 3.9 shows the orbit system geometry and orbital elements. The dimensional elements are given by

- a = semimajor axis (size of the orbit)
- e = eccentricity (shape of the orbit)
- τ = time reference of periapsis or perigee

The orientation elements are given by

- i = inclination (angle between orbit plane and reference plane)
- Ω = right ascension of the ascending node (angle between vernal equinox direction and the line of nodes)
- ω = argument of periapsis or perigee (angle between the ascending node direction and periapsis or perigee direction)

The line of nodes vector is given by the intersection of the reference plane (e.g., the Earth's equatorial plane) and the orbital plane.

From these classical elements it is possible to determine $\mathbf{r}(t_0)$ and $\dot{\mathbf{r}}(t_0)$. Before we state the solution of this problem, we first define some other well known orbital quantities. The *mean motion* is defined by

$$n = \sqrt{\frac{\mu}{a^3}} \tag{3.203}$$

The *mean anomaly* is given by

$$M = n(t - \tau) \tag{3.204}$$

where M is not to be confused with the mass M, as defined previously. Note that M often replaces τ for one of the classical elements (e.g., see §7.1.1). To determine the position vector $\mathbf{r}(t_0)$ the initial *true anomaly*, ν_0, must be first determined. From Figure 3.9 the initial true anomaly is defined as the angle between the periapsis direction and the position vector. Unfortunately, this quantity cannot be determined in a straightforward manner. To facilitate this task Kepler used an intermediate step. First, given M and e, Kepler's equation is solved for the *eccentric anomaly* E:

$$\boxed{M = E - e\sin E} \tag{3.205}$$

The eccentric anomaly can be determined using Newton's method (see exercise 1.15). A series expansion of eqn. (3.205) gives the following approximation for E, which is accurate up to third-order in the eccentricity:[39]

$$E = M + \frac{e\sin M}{1 - e\cos M} - \frac{1}{2}\left(\frac{e\sin M}{1 - e\cos M}\right)^3 + \cdots \tag{3.206}$$

Equation (3.206) can be used as the starting guess in Newton's method. The true anomaly is then given by

$$\nu_0 = \text{atan2}\left[\frac{\sqrt{1-e^2}\sin E}{1 - e\cos E}, \frac{\cos E - e}{1 - e\cos E}\right] \tag{3.207}$$

where atan2 is a four quadrant inverse tangent function. Next, the *semilatus rectum* is computed by

$$p = a(1 - e^2) \tag{3.208}$$

Also, the magnitude of the momentum vector is given by

$$||\mathbf{H}|| = \sqrt{\mu\, p} \tag{3.209}$$

Then, using the equation of an ellipse in polar coordinates, the magnitude of the position vector is given by

$$||\mathbf{r}(t_0)|| = \frac{p}{1 + e\cos\nu_0} \tag{3.210}$$

Finally, the initial position and velocity vectors are determined using a coordinate transformation,[39] given by

$$\mathbf{r}(t_0) = ||\mathbf{r}(t_0)|| \begin{bmatrix} \cos\Omega\cos\theta - \sin\Omega\sin\theta\cos i \\ \sin\Omega\cos\theta + \cos\Omega\sin\theta\cos i \\ \sin\theta\sin i \end{bmatrix} \tag{3.211}$$

and

$$\dot{\mathbf{r}}(t_0) = -\frac{\mu}{||\mathbf{H}||} \begin{bmatrix} (\sin\theta + e\sin\omega)\cos\Omega + (\cos\theta + e\cos\omega)\sin\Omega\cos i \\ (\sin\theta + e\sin\omega)\sin\Omega - (\cos\theta + e\cos\omega)\cos\Omega\cos i \\ (\cos\theta + e\cos\omega)\sin i \end{bmatrix} \tag{3.212}$$

where $\theta = \omega + \nu_0$.

The orbital equations of motion described herein are sufficient to demonstrate the basic concepts of orbit determination and estimation. The two-body problem can also be extended to the n-body problem. The analysis of even the two- and three-body problem provides a wealth of information, which will not be addressed in the present text. Also, perturbation methods discussed in §3.2 can be used for both the problem of determining precision orbits and the problem of ensuring that a spacecraft in orbit will meet certain boundary conditions.[39] The interested reader is encouraged to pursue the vast knowledge base and developments on orbital mechanics in the open literature and texts such as Battin.[39]

3.9 Aircraft Flight Dynamics

This section presents a summary of the equations of motion of aircraft. Once again, we only introduce the fundamentals required within the scope of the present text. Aircraft flight dynamics is only one of three disciplines which encompass flight mechanics; the other two being performance and aeroelasticity.[42] Performance deals with determining various quantities (such as climb rate, range, etc.) that give an indication of the basic characteristics of a particular aircraft. Aeroelasticity involves the structural flexibility of modern aircraft. We will cover the basics of flexibility in §3.10.

We begin our discussion of flight dynamics by defining a number of various aircraft angles (see Figure 3.10): angle of attack (α), sideslip angle (β), flight path angle (γ), and pitch angle (θ). Referring to Figure 3.10, the angle of attack is the angle between the $\hat{\mathbf{b}}_1$ body axis and the projected free-stream velocity vector ($\mathbf{v}_{p_1}$) onto the $\hat{\mathbf{b}}_1$-$\hat{\mathbf{b}}_3$ (body axis) plane. The sideslip angle is the angle between the $\hat{\mathbf{b}}_1$ body axis and the projected free-stream velocity vector ($\mathbf{v}_{p_2}$) onto the $\hat{\mathbf{b}}_1$-$\hat{\mathbf{b}}_2$ (body axis) plane. The flight path angle is the angle between the horizon (which is assumed to be inertial) and the $\mathbf{v}_{p_1}$ axis. The pitch angle is the angle between the horizon and

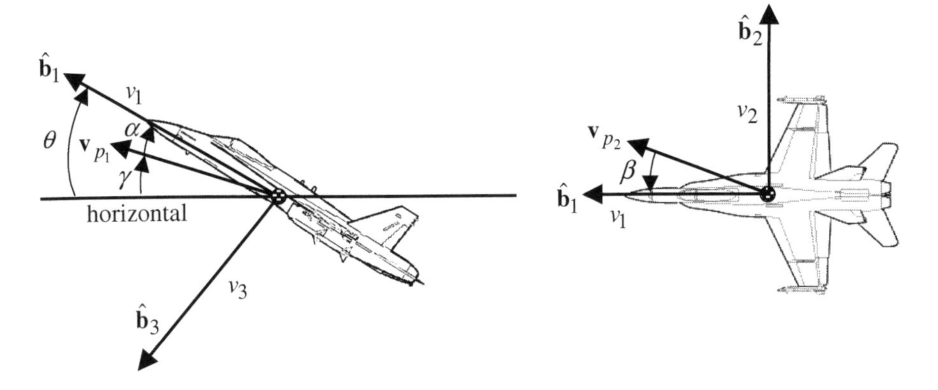

Figure 3.10: Definition of Various Aircraft Angles (Positive Senses Shown)

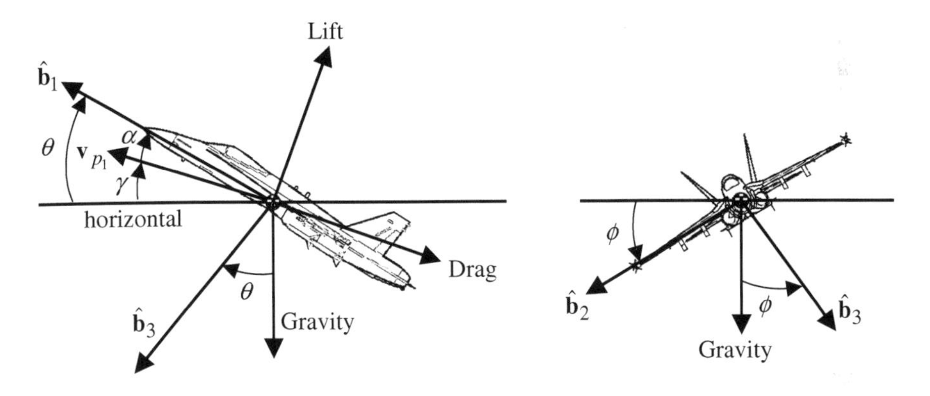

Figure 3.11: Aircraft Forces

the $\hat{\mathbf{b}}_1$ body axis, which is also given by

$$\theta = \alpha + \gamma \tag{3.213}$$

The equations for α and β are given by

$$\alpha = \tan^{-1} \frac{v_3}{v_1} \tag{3.214a}$$

$$\beta = \sin^{-1} \frac{v_2}{||\mathbf{v}||} \tag{3.214b}$$

where v_1, v_2, and v_3 are the free-stream velocity components along the $\hat{\mathbf{b}}_1$, $\hat{\mathbf{b}}_2$, and $\hat{\mathbf{b}}_3$ axes, respectively, and $||\mathbf{v}||$ is the free-stream velocity magnitude, given by

$$||\mathbf{v}|| = (v_1^2 + v_2^2 + v_3^2)^{1/2} \tag{3.215}$$

The rigid body equations of motion of an aircraft can be derived from Newton's second law, as described in §3.7.2. Figure 3.11 shows the forces acting on an aircraft.

The roll angle ϕ is defined as the angle between the horizon and the $\hat{\mathbf{b}}_2$ body axis. It is important to realize that drag is opposite the velocity vector, not the body axis vector (also, lift is perpendicular to the velocity vector). The force equations are derived from eqn. (3.183), with the addition of gravity, aerodynamic forces, and thrust forces. These equations are given by

$$T_1 - D\cos\alpha + L\sin\alpha - mg\sin\theta = m(\dot{v}_1 + v_3\omega_2 - v_2\omega_3) \tag{3.216a}$$

$$Y + mg\cos\theta\sin\phi = m(\dot{v}_2 + v_1\omega_3 - v_3\omega_1) \tag{3.216b}$$

$$T_3 - D\sin\alpha - L\cos\alpha + mg\cos\theta\cos\phi = m(\dot{v}_3 + v_2\omega_1 - v_1\omega_2) \tag{3.216c}$$

where D is the drag force, Y is the side force due to rudder, L is the lift force, and T_1 and T_3 are the thrust components along $\hat{\mathbf{b}}_1$ and $\hat{\mathbf{b}}_3$, respectively. The total drag equation , side-force equation, and lift equation are given by

$$D = C_D\,\bar{q}\,S \tag{3.217a}$$

$$Y = C_Y\,\bar{q}\,S \tag{3.217b}$$

$$L = C_L\,\bar{q}\,S \tag{3.217c}$$

where C_D, C_Y, and C_L are the total drag, side-force, and lift coefficients, respectively, S is the known reference area, and $\bar{q}$ is the dynamic pressure which is a function of the known air density (ρ) and velocity magnitude:

$$\bar{q} = \frac{1}{2}\rho||\mathbf{v}||^2 \tag{3.218}$$

The aerodynamic coefficients are given by

$$C_D = C_{D_0} + C_{D_\alpha}\alpha + C_{D_{\delta_E}}\delta_E \tag{3.219a}$$

$$C_Y = C_{Y_0} + C_{Y_\beta}\beta + C_{Y_{\delta_R}}\delta_R + C_{Y_{\delta_A}}\delta_A \tag{3.219b}$$

$$C_L = C_{L_0} + C_{L_\alpha}\alpha + C_{L_{\delta_E}}\delta_E \tag{3.219c}$$

where δ_E, δ_R, and δ_A are the elevator (or stabilizer), rudder, and aileron angle deflections. The other terms in eqn. (3.219) are the known aerodynamic coefficients (defined by the particular aircraft of interest). These reflect the contributions of the individual quantities (e.g., C_{D_α} is the drag coefficient contribution due to angle of attack, C_{D_0} is the drag coefficient for $\alpha = \delta_E = 0$, etc.). Note, the aerodynamic coefficients are first-order Taylor series with an infinite number of terms (we have chosen to show these with only a few of the most basic terms). Also, instead of eqn. (3.219a), the *drag polar*[43] is often used to approximate the drag coefficient.

The aircraft rotational equations of motion are given by eqn. (3.180). For conventional aircraft configurations the $\mathbf{b}_1$-$\mathbf{b}_3$ plane is usually a plane of symmetry so that $J_{23} = J_{12} = 0$. Therefore, Euler's equations in component form are given by

$$J_{11}\dot{\omega}_1 - J_{13}\dot{\omega}_3 - J_{13}\omega_1\omega_2 + (J_{33} - J_{22})\omega_2\omega_3 = L_{A_1} + L_{T_1} \tag{3.220a}$$

$$J_{22}\dot{\omega}_2 + (J_{11} - J_{33})\omega_1\omega_3 + J_{13}(\omega_1^2 - \omega_3^2) = L_{A_2} + L_{T_2} \tag{3.220b}$$

$$J_{33}\dot{\omega}_3 - J_{13}\dot{\omega}_1 + J_{13}\omega_2\omega_3 + (J_{22} - J_{11})\omega_1\omega_2 = L_{A_3} + L_{T_3} \tag{3.220c}$$

where L_{A_1}, L_{A_2}, and L_{A_3} are the aerodynamic torques, and L_{T_1}, L_{T_2}, and L_{T_3} are the known thrust torques. The aerodynamic torque equations are given by

$$L_{A_1} = C_l \, \bar{q} \, S b \tag{3.221a}$$

$$L_{A_2} = C_m \, \bar{q} \, S \bar{c} \tag{3.221b}$$

$$L_{A_3} = C_n \, \bar{q} \, S b \tag{3.221c}$$

where C_l, C_m, and C_n are the rolling, pitching, and yawing torque coefficients, respectively, b is the known wing span, and $\bar{c}$ is the known mean geometric chord.[43] The torque coefficients are given by

$$C_l = C_{l_0} + C_{l_\beta}\beta + C_{l_{\delta_R}}\delta_R + C_{l_{\delta_A}}\delta_A + C_{l_p}\frac{\omega_1 b}{2||\mathbf{v}||^2} + C_{l_r}\frac{\omega_3 b}{2||\mathbf{v}||^2} \tag{3.222a}$$

$$C_m = C_{m_0} + C_{m_\alpha}\alpha + C_{m_{\delta_E}}\delta_E + C_{m_q}\frac{\omega_2 \bar{c}}{2||\mathbf{v}||^2} \tag{3.222b}$$

$$C_n = C_{n_0} + C_{n_\beta}\beta + C_{n_{\delta_R}}\delta_R + C_{n_{\delta_A}}\delta_A + C_{n_p}\frac{\omega_1 b}{2||\mathbf{v}||^2} + C_{n_r}\frac{\omega_3 b}{2||\mathbf{v}||^2} \tag{3.222c}$$

By integrating eqns. (3.216) and (3.220) the body linear velocities and angular velocities can be determined. To determine the linear velocities with respect to the reference frame we utilize the inverse attitude matrix, which is usually defined by the 3-2-1 sequence, so that

$$\begin{bmatrix} \dot{x} \\ \dot{y} \\ \dot{z} \end{bmatrix} = \begin{bmatrix} c\theta\, c\psi & s\phi\, s\theta\, c\psi - c\phi\, s\psi & c\phi\, s\theta\, c\psi + s\phi\, s\psi \\ c\theta\, s\psi & s\phi\, s\theta\, s\psi + c\phi\, c\psi & c\phi\, s\theta\, s\psi - s\phi\, c\psi \\ -s\theta & s\phi\, c\theta & c\phi\, c\theta \end{bmatrix} \begin{bmatrix} v_1 \\ v_2 \\ v_3 \end{bmatrix} \tag{3.223}$$

where $\dot{x}$, $\dot{y}$, and $\dot{z}$ are the velocity components with respect to the reference frame. The aircraft's position relative to the reference frame can be determined by integrating eqn. (3.223). In a similar fashion the Euler rates can be expressed using the 3-2-1 kinematics equations:

$$\begin{bmatrix} \dot{\phi} \\ \dot{\theta} \\ \dot{\psi} \end{bmatrix} = \begin{bmatrix} 1 & \sin\phi \tan\theta & \cos\phi \tan\theta \\ 0 & \cos\phi & -\sin\phi \\ 0 & \sin\phi \sec\theta & \cos\phi \sec\theta \end{bmatrix} \begin{bmatrix} \omega_1 \\ \omega_2 \\ \omega_3 \end{bmatrix} \tag{3.224}$$

The roll (ϕ), pitch (θ), and yaw (ψ) angles can be determined by integrating the set in eqn. (3.224).

The equations presented in this section allow one to simulate the basic motion of an aircraft. As is the case with spacecraft dynamics, a thorough treatise of aircraft flight dynamics would entail significant effort which is beyond the scope of this text. Other important topics such as small-disturbance theory, atmospheric inputs, flying qualities, etc., can be found in Nelson[42] and Roskam.[43]

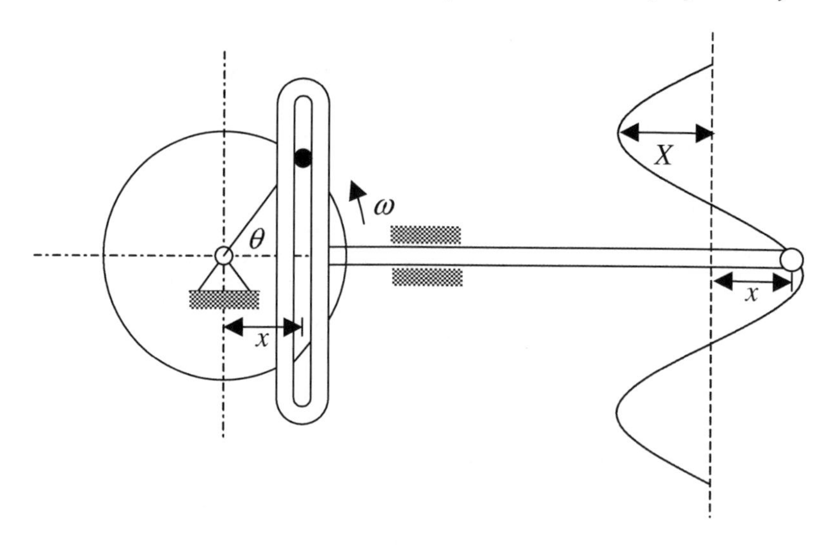

Figure 3.12: Harmonic Motion in a Yoke

3.10 Vibration

Vibration is a kind of motion where an object oscillates with respect to some reference frame. Any body that possesses mass and elasticity, such as flexible structures, aircraft wings, bridges, buildings, strings, etc., can vibrate. Vibration thus covers a wide range of disciplines, which still has a thriving research thrust to this day (especially in the control of vibratory systems). Many devastating failures have resulted when the effects of vibration on structures have not been adequately investigated (e.g., the infamous Tacoma Narrows bridge collapse due to wind-induced vibration[44]).

In order to introduce the concepts involved with vibration, we begin our discussion with the simplest form of periodic motion, known as *harmonic motion*. To illustrate this motion, we consider a simple mechanism called a yoke,[45] shown in Figure 3.12. A pin is attached to a wheel, which can slide freely in a slot attached to a stem. The stem then moves in a periodic manner, which can be expressed by the equation

$$x = X\cos\theta = X\cos\omega t \tag{3.225}$$

where X is the radius of the wheel, and ω is the angular velocity. Taking two time derivatives of eqn. (3.225) and back substituting yields

$$\ddot{x} + \omega^2 x = 0 \tag{3.226}$$

Therefore, in harmonic motion the acceleration is proportional to the displacement.

Harmonic motion can be related to Newton's second law of motion, which states that acceleration is proportional to force. Consider the spring-mass-damper system in Figure 3.13. From Newton's law we have:

$$m\ddot{x} + c\dot{x} + kx = F \tag{3.227}$$

We now consider the free response case only with $F = 0$, and assume an exponential solution for x, given by $x = Ae^{st}$. Taking time derivatives of x and substituting the resultants into eqn. (3.227) leads to

$$(ms^2 + cs + k)Ae^{st} = 0 \tag{3.228}$$

Since Ae^{st} is never zero, eqn. (3.228) holds true if and only if

$$ms^2 + cs + k = 0 \tag{3.229}$$

Equation (3.229) is called the *characteristic equation* of the system. The same equation can also be derived by taking the Laplace transform of eqn. (3.227), with $F = 0$ again. The roots of this equation are clearly given by

$$s_{1,2} = \frac{-c \pm \sqrt{c^2 - 4mk}}{2m} \tag{3.230}$$

Three possibilities for $s_{1,2}$ exist: 1) the roots are real and unequal for $c^2 - 4mk > 0$; 2) the roots are real and repeated for $c^2 - 4mk = 0$; and 3) the roots are complex conjugates for $c^2 - 4mk < 0$. The solution for each of these cases is given by

$$\text{real and unequal} \quad x = A_1 e^{s_1 t} + A_2 e^{s_2 t} \tag{3.231a}$$

$$\text{real and repeated} \quad x = A_1 e^{s_1 t} + t\, A_2 e^{s_1 t} \tag{3.231b}$$

$$\text{complex conjugates} \quad x = Be^{-at}\sin(bt + \phi) \tag{3.231c}$$

where $a = c/2m$ and $b = \sqrt{4mk - c^2}/2m$. The constants A_1, A_2, ϕ, and B are determined from initial conditions $x(t_0)$ and $\dot{x}(t_0)$:[2]

$$\text{real and unequal} \quad A_1 = \frac{\dot{x}(t_0) - s_2 x(t_0)}{s_1 - s_2}, \quad A_2 = x(t_0) - A_1 \tag{3.232a}$$

$$\text{real and repeated} \quad A_1 = x(t_0), \quad A_2 = \dot{x}(t_0) - s_1 x(t_0) \tag{3.232b}$$

$$\text{complex conjugates} \quad \phi = \text{atan2}[bx(t_0), \dot{x}(t_0) + ax(t_0)], \quad B = \frac{x(t_0)}{\sin\phi} \tag{3.232c}$$

Another way to represent the characteristic equation is given by

$$\boxed{s^2 + 2\zeta\omega_n s + w_n^2 = 0} \tag{3.233}$$

where the *damping ratio* ζ and *natural frequency* ω_n are defined as

$$\zeta = \frac{c}{2\sqrt{mk}} \tag{3.234a}$$

$$\omega_n = \sqrt{\frac{k}{m}} \tag{3.234b}$$

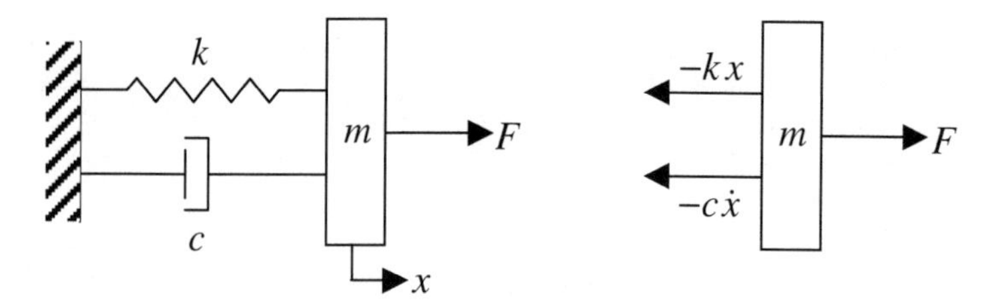

Figure 3.13: Simple Spring-Mass-Damper System

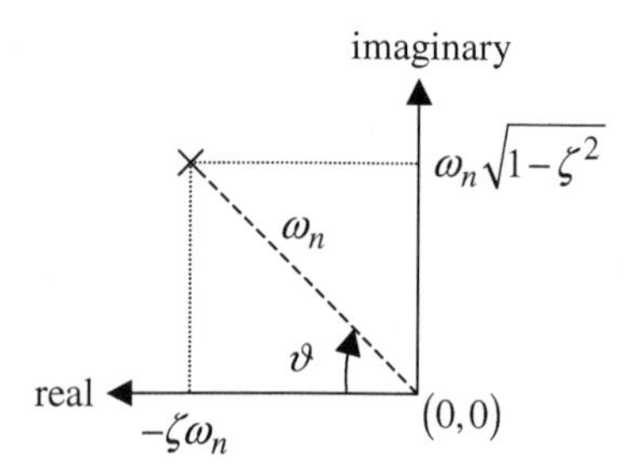

Figure 3.14: Root Location in the Complex Plane

The roots of the characteristic equation are now given by

$$s_{1,2} = -\zeta\,\omega_n \pm \omega_n\sqrt{\zeta^2 - 1} \tag{3.235}$$

The three cases shown in eqn. (3.231) depend on three variables (m, c, and k). The convenient notation in eqn. (3.234a) allows us to represent these three cases from the characteristic value of ζ only: 1) the roots are real and unequal for $\zeta > 1$; 2) the roots are real and repeated for $\zeta = 1$; and 3) the roots are complex conjugates for $0 \leq \zeta < 1$. A graphical representation of case 3 is shown in Figure 3.14. Since the natural frequency is the magnitude from the origin to the root, all roots with the same natural frequency must lie on a circle centered at the origin. The damping ratio is given by $\zeta = \cos\vartheta$, where ϑ is the angle between the natural frequency line and the negative real axis. If $\zeta = 0$ then the system reduces to the simple harmonic oscillator in eqn. (3.226) with $\omega_n = \omega$. Also, the *damped natural frequency* is defined by $\omega_d \equiv \omega_n\sqrt{1-\zeta^2}$, which is equivalent to b (the frequency of oscillation) in eqn. (3.231c).

Newton's law can easily be extended for a system of particles. In this text we consider a *lumped parameter system*,[46] where each mass corresponds to one degree of freedom. Many systems, such as bridges, trusses, aircraft structures, etc., can be sufficiently modelled using the lumped parameter concept. In order to demonstrate a lump parameter system with multiple springs, masses, and dampers we first consider the system shown in Figure 3.15. This system has two degrees of freedom (with

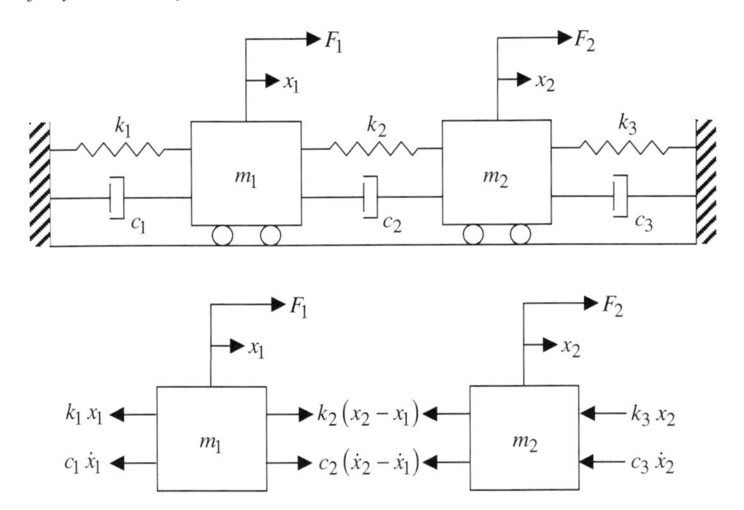

Figure 3.15: Multiple Spring-Mass-Damper System

mass positions given by x_1 and x_2). Applying Newton's law to this system yields

$$\begin{bmatrix} m_1 & 0 \\ 0 & m_2 \end{bmatrix} \begin{bmatrix} \ddot{x}_1 \\ \ddot{x}_2 \end{bmatrix} + \begin{bmatrix} (c_1+c_2) & -c_2 \\ -c_2 & (c_2+c_3) \end{bmatrix} \begin{bmatrix} \dot{x}_1 \\ \dot{x}_2 \end{bmatrix} + \begin{bmatrix} (k_1+k_2) & -k_2 \\ -k_2 & (k_2+k_3) \end{bmatrix} \begin{bmatrix} x_1 \\ x_2 \end{bmatrix} = \begin{bmatrix} F_1 \\ F_2 \end{bmatrix} \tag{3.236}$$

Equation (3.236) can be put into compact form using matrix notation:

$$\boxed{M\ddot{\mathbf{x}} + C\dot{\mathbf{x}} + K\mathbf{x} = \mathbf{F}} \tag{3.237}$$

with obvious definitions of M (the mass matrix), C (the damping matrix), K (the stiffness matrix), $\mathbf{x}$, and $\mathbf{F}$. The matrices M, C, and K are symmetric and must be positive definite to ensure stability.

In order to investigate the properties of a lump parameter system we first consider an undamped system (i.e., $C = 0$) with no forced input:

$$M\ddot{\mathbf{x}} + K\mathbf{x} = \mathbf{0} \tag{3.238}$$

subject to the given initial conditions $\mathbf{x}(t_0)$ and $\dot{\mathbf{x}}(t_0)$. An exponential solution to eqn. (3.238) is assumed with[47]

$$\mathbf{x}(t) = e^{st}\mathbf{u} \tag{3.239}$$

where s and $\mathbf{u}$ are constants. Taking two time derivatives of eqn. (3.239) and substituting the resultant into eqn. (3.238) leads to

$$(K - \lambda M)\mathbf{u} = \mathbf{0} \tag{3.240}$$

where $\lambda = -s^2$. Equation (3.240) corresponds to eigenvalue/eigenvector problem with $s = \pm\lambda j$. We seek to find a physical solution that does not entail complex numbers. It is common to perform a linear state transformation using $\mathbf{x} = M^{-1/2}\mathbf{z}$, which leads to the following differential equation:

$$\ddot{\mathbf{z}} + M^{-1/2} K M^{-1/2} \mathbf{z} = \mathbf{0} \tag{3.241}$$

This transformation is performed since the matrix $M^{-1/2} K M^{-1/2}$ is a symmetric matrix, whereas $M^{-1}K$ is generally not symmetric. As shown in §3.1.4 the eigenvalues of the system are invariant to this transformation. The eigenvectors of $M^{-1/2} K M^{-1/2}$ are denoted $\mathbf{v}_i$ for $i = 1, 2, \ldots, p$, where p is the number of degrees of freedom. The solution for $\mathbf{z}$ is given by[48]

$$\mathbf{z}(t) = \sum_{i=1}^{p} a_i \sin(\omega_i t + \phi_i)\mathbf{v}_i \tag{3.242}$$

where the natural frequencies are given by $\omega_i = \sqrt{\lambda_i}$, and the constants ϕ_i and a_i are given by

$$\phi_i = \tan^{-1}\left[\frac{\omega_i \mathbf{v}_i^T \mathbf{z}(t_0)}{\mathbf{v}_i^T \dot{\mathbf{z}}(t_0)}\right] \tag{3.243}$$

$$a_i = \frac{\mathbf{v}_i^T \mathbf{z}(t_0)}{\sin\phi_i} \tag{3.244}$$

The vectors $\mathbf{v}_i$ are called the *mode shapes* since they give an indication of the "shape" of the vibration for each mass, and the constants a_i are the *modal participation factors* since their value indicates how each mode influences the overall response. Once $\mathbf{z}(t)$ has been determined then $\mathbf{x}(t)$ can be found by simply using $\mathbf{x}(t) = M^{-1/2}\mathbf{z}(t)$.

Analytical solutions for the full system in eqn. (3.237) with $\mathbf{F} = \mathbf{0}$ cannot be found in general. However, special cases do exist where the equations of motion decouple. These cases exist if any of the following conditions exist:[48, 49]

1. $C = \alpha M + \beta K$, where α and β are any real scalars.

2. $C = \sum_{i=1}^{p} \gamma_{i-1} K^{i-1}$, where γ_i are real scalars.

3. $C M^{-1} K = K M^{-1} C$.

If any of these conditions holds true then the eigenvectors of eqn. (3.237) are the same as the eigenvectors with $D = 0$. Such systems are known as *normal mode systems*. These systems can be decoupled by the eigenvector matrix of K. Let V be the matrix of eigenvectors of $M^{-1/2} K M^{-1/2}$:

$$V = \begin{bmatrix} \mathbf{v}_1 & \mathbf{v}_2 & \cdots & \mathbf{v}_p \end{bmatrix} \tag{3.245}$$

Define a normalized matrix of eigenvectors, given by $S = M^{-1/2}V$. The decoupled system is then given by

$$S^T M S = I \tag{3.246a}$$

$$S^T K S = \text{diag}[\omega_i^2] \tag{3.246b}$$

$$S^T C S = \text{diag}[2\zeta_i \omega_i] \tag{3.246c}$$

where *modal frequencies* ω_i^2 are the eigenvalues of the matrix K and ζ_i are the *modal damping ratios*. The decoupled equations are given by

$$\ddot{y}_i + 2\zeta_i \omega_i \dot{y}_i + \omega_i^2 y_i = 0, \quad i = 1, 2, \ldots, p \tag{3.247}$$

The solution of eqn. (3.237) with $\mathbf{F} = \mathbf{0}$ can be found from $\mathbf{x}(t) = S\mathbf{y}(t)$.

This section presented the basic equations and concepts of vibration. The treatise shown here is not complete by any means. Other subjects such as distributed parameter systems, Hamilton's principle, Lagrange's equations, finite element methods, etc., can be found in the references provided in this section.

3.11 Summary

The essence-oriented discussion of differential equations and dynamical systems, while adequate background for following the discussion of Chapters 4 and 7, will likely prove incomplete in many applications. In particular, conspicuous by its lack of coverage here is perturbation theory; Refs. [13] and [14] document perturbation methods which are exceptionally valuable tools for solving weakly nonlinear differential equations. The results of the present chapter do provide an adequate basis for solving differential equations encountered in a substantial fraction of practical applications, and provide a foundation for further study.

A particularly useful tool for the practicing engineer is the state space approach to represent a system of ODEs. This tool will prove invaluable in representing high order systems, commonly found in many applications (e.g., vibration models of tall buildings). Equally valuable is the concept of observability introduced in §3.4. In many applications some states will be able to be "monitored" better than others. By examining the properties of the observability matrix in eqn. (3.105) one can deduce the relative degree of observability of each state. This provides a powerful and useful tool for making tradeoffs between sensor placement requirements and monitoring of states through state estimation techniques.

The terse review of dynamical systems covering spacecraft dynamics, orbital mechanics, aircraft flight dynamics, and vibration is adequate to provide the basic concepts required to demonstrate practical applications of estimation theory. This review serves as a springboard for the various branches in all areas of dynamics. The many fascinating recent discoveries, such as chaotic behavior, since the classical developments by Newton, Lagrange, and Hamilton (to name a few) provide an ongoing research venue in the foreseeable future. Indeed, it is our hope that the interested reader will be motivated to pursue these developments in the open literature.

A summary of the key formulas presented in this chapter is given below.

- State Space Approach

$$\begin{aligned}\dot{\mathbf{x}} &= F\,\mathbf{x} + B\,\mathbf{u}\\ \mathbf{y} &= H\,\mathbf{x} + D\,\mathbf{u}\end{aligned}$$

- Homogeneous Linear Systems

$$\begin{aligned}\dot{\mathbf{x}}(t) &= F(t)\,\mathbf{x}(t), \quad \mathbf{x}(t_0) \text{ known}\\ \mathbf{x}(t) &= \Phi(t,t_0)\,\mathbf{x}(t_0)\\ \Phi(t,t_0) &= I + \int_{t_0}^{t} F(\tau_1)\,\Phi(\tau_1,t_0)\,d\tau_1\\ \Phi(t,t_0) &= e^{F(t-t_0)}, \quad \text{for } F = \text{constant}\end{aligned}$$

- Forced Linear Systems

$$\begin{aligned}\dot{\mathbf{x}}(t) &= F(t)\,\mathbf{x}(t) + B(t)\,\mathbf{u}(t)\\ \mathbf{x}(t) &= \Phi(t,t_0)\,\mathbf{x}(t_0) + \int_{t_0}^{t} \Phi(t,\tau)\,B(\tau)\,\mathbf{u}(\tau)\,d\tau\end{aligned}$$

- Nonlinear Systems

$$\begin{aligned}\dot{\mathbf{x}} &= \mathbf{f}(t,\mathbf{x},\mathbf{u})\\ \mathbf{y} &= \mathbf{h}(t,\mathbf{x},\mathbf{u})\end{aligned}$$

$$\begin{aligned}\delta\dot{\mathbf{x}}(t) &= F(t)\,\delta\mathbf{x}(t) + B(t)\,\delta\mathbf{u}(t)\\ \delta\mathbf{y}(t) &= H(t)\,\delta\mathbf{x}(t) + D(t)\,\delta\mathbf{u}(t)\end{aligned}$$

$$\begin{aligned}F(t) &= \left.\frac{\partial\mathbf{f}}{\partial\mathbf{x}}\right|_{\mathbf{x}_N,\mathbf{u}_N}, \quad B(t) = \left.\frac{\partial\mathbf{f}}{\partial\mathbf{u}}\right|_{\mathbf{x}_N,\mathbf{u}_N}\\ H(t) &= \left.\frac{\partial\mathbf{h}}{\partial\mathbf{x}}\right|_{\mathbf{x}_N,\mathbf{u}_N}, \quad D(t) = \left.\frac{\partial\mathbf{h}}{\partial\mathbf{u}}\right|_{\mathbf{x}_N,\mathbf{u}_N}\end{aligned}$$

- Observability

$$\begin{aligned}\dot{\mathbf{x}} &= F\,\mathbf{x} + B\,\mathbf{u}\\ \mathbf{y} &= H\,\mathbf{x} + D\,\mathbf{u}\end{aligned}$$

$$\mathcal{O} = \begin{bmatrix} H \\ HF \\ HF^2 \\ \vdots \\ HF^{n-1} \end{bmatrix}$$

- Discrete-Time Systems

$$\mathbf{x}_{k+1} = \Phi \mathbf{x}_k + \Gamma \mathbf{u}_k$$
$$\mathbf{y}_k = H \mathbf{x}_k + D \mathbf{u}_k$$

$$\Phi = I + F\Delta t + \frac{1}{2!}F^2 \Delta t^2 + \frac{1}{3!}F^3 \Delta t^3 + \cdots$$
$$\Gamma = \left[I \Delta t + \frac{1}{2!}F \Delta t^2 + \frac{1}{3!}F^2 \Delta t^3 + \cdots \right] B$$

$$\mathcal{O}_d = \begin{bmatrix} H \\ H\Phi \\ H\Phi^2 \\ \vdots \\ H\Phi^{n-1} \end{bmatrix}$$

- Lyapunov Stability

$$F^T P + PF = -Q$$
$$\Phi^T P \Phi - P = -Q$$

- Spacecraft Dynamics

$$\dot{\mathbf{q}} = \frac{1}{2}\Omega(\omega)\mathbf{q}$$
$$J\dot{\omega} = -[\omega \times] J\omega + \mathbf{L}$$

- Orbital Mechanics

$$\ddot{\mathbf{r}} = -\frac{\mu}{||\mathbf{r}||^3}\mathbf{r}$$
$$M = E - e \sin E$$

- Aircraft Flight Dynamics

$$\theta = \alpha + \gamma$$

$$\alpha = \tan^{-1} \frac{v_3}{v_1}$$
$$\beta = \sin^{-1} \frac{v_2}{||\mathbf{v}||}$$

$$||\mathbf{v}|| = (v_1^2 + v_2^2 + v_3^2)^{1/2}$$

$$\begin{aligned}
T_1 - D\cos\alpha + L\sin\alpha - mg\sin\theta &= m(\dot{v}_1 + v_3\omega_2 - v_2\omega_3) \\
Y + mg\cos\theta\sin\phi &= m(\dot{v}_2 + v_1\omega_3 - v_3\omega_1) \\
T_3 - D\sin\alpha - L\cos\alpha + mg\cos\theta\cos\phi &= m(\dot{v}_3 + v_2\omega_1 - v_1\omega_2)
\end{aligned}$$

$$\begin{aligned}
D &= C_D\,\bar{q}\,S \\
Y &= C_Y\,\bar{q}\,S \\
L &= C_L\,\bar{q}\,S
\end{aligned}$$

$$\bar{q} = \frac{1}{2}\rho||\mathbf{v}||^2$$

$$\begin{aligned}
C_D &= C_{D_0} + C_{D_\alpha}\alpha + C_{D_{\delta_E}}\delta_E \\
C_Y &= C_{Y_0} + C_{Y_\beta}\beta + C_{Y_{\delta_R}}\delta_R + C_{Y_{\delta_A}}\delta_A \\
C_L &= C_{L_0} + C_{L_\alpha}\alpha + C_{L_{\delta_E}}\delta_E
\end{aligned}$$

$$\begin{aligned}
J_{11}\dot{\omega}_1 - J_{13}\dot{\omega}_3 - J_{13}\omega_1\omega_2 + (J_{33} - J_{22})\omega_2\omega_3 &= L_{A_1} + L_{T_1} \\
J_{22}\dot{\omega}_2 + (J_{11} - J_{33})\omega_1\omega_3 + J_{13}(\omega_1^2 - \omega_3^2) &= L_{A_2} + L_{T_2} \\
J_{33}\dot{\omega}_3 - J_{13}\dot{\omega}_1 + J_{13}\omega_2\omega_3 + (J_{22} - J_{11})\omega_1\omega_2 &= L_{A_3} + L_{T_3}
\end{aligned}$$

$$\begin{aligned}
L_{A_1} &= C_l\,\bar{q}\,S\,b \\
L_{A_2} &= C_m\,\bar{q}\,S\,\bar{c} \\
L_{A_3} &= C_n\,\bar{q}\,S\,b
\end{aligned}$$

$$\begin{aligned}
C_l &= C_{l_0} + C_{l_\beta}\beta + C_{l_{\delta_R}}\delta_R + C_{l_{\delta_A}}\delta_A + C_{l_p}\frac{\omega_1 b}{2||\mathbf{v}||^2} + C_{l_r}\frac{\omega_3 b}{2||\mathbf{v}||^2} \\
C_m &= C_{m_0} + C_{m_\alpha}\alpha + C_{m_{\delta_E}}\delta_E + C_{m_q}\frac{\omega_2\bar{c}}{2||\mathbf{v}||^2} \\
C_n &= C_{n_0} + C_{n_\beta}\beta + C_{n_{\delta_R}}\delta_R + C_{n_{\delta_A}}\delta_A + C_{n_p}\frac{\omega_1 b}{2||\mathbf{v}||^2} + C_{n_r}\frac{\omega_3 b}{2||\mathbf{v}||^2}
\end{aligned}$$

$$\begin{bmatrix} \dot{x} \\ \dot{y} \\ \dot{z} \end{bmatrix} = \begin{bmatrix} c\theta\,c\psi & s\phi\,s\theta\,c\psi - c\phi\,s\psi & c\phi\,s\theta\,c\psi + s\phi\,s\psi \\ c\theta\,s\psi & s\phi\,s\theta\,s\psi + c\phi\,c\psi & c\phi\,s\theta\,s\psi - s\phi\,c\psi \\ -s\theta & s\phi\,c\theta & c\phi\,c\theta \end{bmatrix} \begin{bmatrix} v_1 \\ v_2 \\ v_3 \end{bmatrix}$$

$$\begin{bmatrix} \dot{\phi} \\ \dot{\theta} \\ \dot{\psi} \end{bmatrix} = \begin{bmatrix} 1 & \sin\phi\tan\theta & \cos\phi\tan\theta \\ 0 & \cos\phi & -\sin\phi \\ 0 & \sin\phi\sec\theta & \cos\phi\sec\theta \end{bmatrix} \begin{bmatrix} \omega_1 \\ \omega_2 \\ \omega_3 \end{bmatrix}$$

- Vibration

$$s^2 + 2\zeta\omega_n s + w_n^2 = 0$$
$$s_{1,2} = -\zeta\omega_n \pm \omega_n\sqrt{\zeta^2 - 1}$$

$$\begin{aligned} \text{real and unequal} \quad & x = A_1 e^{s_1 t} + A_2 e^{s_2 t} \\ \text{real and repeated} \quad & x = A_1 e^{s_1 t} + t\, A_2 e^{s_1 t} \\ \text{complex conjugates} \quad & x = B e^{-at}\sin(bt + \phi) \end{aligned}$$

$$M\ddot{\mathbf{x}} + C\dot{\mathbf{x}} + K\mathbf{x} = \mathbf{F}$$

Exercises

3.1 Consider the following linear time-varying system: $\dot{\mathbf{x}}(t) = F(t)\mathbf{x}$. Denote the state transition matrix of $F(t)$ by $\Phi(t, t_0)$. The differential equation for $\Phi(t, t_0)$ obeys eqn. (3.19). Show that the differential equation for $\Phi(t_0, t)$ obeys

$$\dot{\Phi}(t_0, t) = -\Phi(t_0, t)\, F(t)$$

with $\Phi(t_0, t_0) = I$.

3.2 Consider the following system of equations:

$$\ddot{z} + 3\dot{z} - 2z = 0$$
$$\dot{y} - 3z - 3y = 0$$

Determine the state space matrices (F, B, H, D) with $\mathbf{x} = \left[z \; \dot{z} \; y\right]$ for an output y. Is this system observable? Is the system observable for an output z?

3.3 Consider the following system: $\dot{\mathbf{x}} = F\mathbf{x}$, with

$$F = \begin{bmatrix} a & 0 \\ 1 & 1 \end{bmatrix}$$

and the transformation $\mathbf{x} = T\mathbf{z}$, with

$$T = \begin{bmatrix} 1 & b \\ 0 & 1 \end{bmatrix}$$

Find a nonzero a and b such that the transformed equation $\dot{\mathbf{z}} = \Upsilon\mathbf{z}$ has the form given by

$$\Upsilon = \begin{bmatrix} 3 & -4 \\ 1 & -1 \end{bmatrix}$$

3.4 Consider the following state equations for a simple circuit:

$$\begin{bmatrix} \dot{x}_1 \\ \dot{x}_2 \end{bmatrix} = \begin{bmatrix} -1/(R_1 C) & 0 \\ 0 & -R_2/L \end{bmatrix} \begin{bmatrix} x_1 \\ x_2 \end{bmatrix} + \begin{bmatrix} 1/(R_1 C) \\ 1/L \end{bmatrix} u$$

$$y = \begin{bmatrix} -1/R_1 & 1 \end{bmatrix} \begin{bmatrix} x_1 \\ x_2 \end{bmatrix} + (1/R_1) u$$

For what value of L in terms of R_1, R_2, and C is the system unobservable?

3.5 Consider the following system matrices, which represent the linearized equations of motion for a spacecraft:

$$F = \begin{bmatrix} 0 & 1 & 0 & 0 \\ 3\omega_n^2 & 0 & 0 & 2\omega_n \\ 0 & 0 & 0 & 1 \\ 0 & -2\omega_n & 0 & 0 \end{bmatrix}, \quad H = \begin{bmatrix} 1 & 0 & 0 & 0 \\ 0 & 0 & 1 & 0 \end{bmatrix}$$

where ω_n is the angular frequency of the reference circular orbit. Also, the states x_1 and x_3 are radial and angular deviations for the reference circular orbit. Prove that this system is observable using both observations (i.e., using the full H matrix). Also, is the system observable using only one observation (try each one separately)?

3.6 Given the coupled nonlinear second-order system

$$\ddot{x} = -x + axy$$
$$\ddot{y} = -y + bxy$$

where a and b are constants. Rearrange these equations to the form of eqn. (3.73a). Also, determine the associated linear differential equations whose solutions yield the derivative matrices:

$$\Phi(t, t_0) = \begin{bmatrix} \frac{\partial x(t)}{\partial x(t_0)} & \frac{\partial x(t)}{\partial y(t_0)} & \frac{\partial x(t)}{\partial \dot{x}(t_0)} & \frac{\partial x(t)}{\partial \dot{y}(t_0)} \\ \frac{\partial y(t)}{\partial x(t_0)} & \frac{\partial y(t)}{\partial y(t_0)} & \frac{\partial y(t)}{\partial \dot{x}(t_0)} & \frac{\partial y(t)}{\partial \dot{y}(t_0)} \\ \frac{\partial \dot{x}(t)}{\partial x(t_0)} & \frac{\partial \dot{x}(t)}{\partial y(t_0)} & \frac{\partial \dot{x}(t)}{\partial \dot{x}(t_0)} & \frac{\partial \dot{x}(t)}{\partial \dot{y}(t_0)} \\ \frac{\partial \dot{y}(t)}{\partial x(t_0)} & \frac{\partial \dot{y}(t)}{\partial y(t_0)} & \frac{\partial \dot{y}(t)}{\partial \dot{x}(t_0)} & \frac{\partial \dot{y}(t)}{\partial \dot{y}(t_0)} \end{bmatrix}$$

and

$$\Psi(t,t_0) = \begin{bmatrix} \frac{\partial x(t)}{\partial a} & \frac{\partial x(t)}{\partial b} \\ \frac{\partial y(t)}{\partial a} & \frac{\partial y(t)}{\partial b} \\ \frac{\partial \dot{x}(t)}{\partial a} & \frac{\partial \dot{x}(t)}{\partial b} \\ \frac{\partial \dot{y}(t)}{\partial a} & \frac{\partial \dot{y}(t)}{\partial b} \end{bmatrix}$$

3.7 Consider the following continuous-time system:

$$\dot{\mathbf{x}} = \begin{bmatrix} 0 & 1 \\ -1 & 0 \end{bmatrix} \mathbf{x} + \begin{bmatrix} 0 \\ 1 \end{bmatrix} u$$
$$y = \begin{bmatrix} 1 & 0 \end{bmatrix} \mathbf{x}$$

Is the continuous system observable? Next, convert this system into the discrete-time representation shown in eqn. (3.111) for a sampling interval Δt. Check the discrete-time observability for various sampling intervals. Is the system observable for $\Delta t = 2\pi$ seconds? Explain your results by checking the discrete-time eigenvalues of the matrix Φ in eqn. (3.111a).

3.8 Prove that the observability of a system is invariant under a similarity transformation for both continuous-time and discrete-time systems.

3.9 Find the equilibrium points for the following systems and determine their stability by Lyapunov's linearization method:

(A) $\ddot{x} + \dot{x} = 0$

(B) $\dot{x} + 4x - x^3 = 0$

(C) $\ddot{x} + \dot{x} + \sin x = 0$

Can you show global stability for any of these systems using Lyapunov's direct method?

3.10 ♣ For the discrete matrix Lyapunov equation in eqn. (3.135) prove that if P is positive definite, then Q is positive definite if and only if all the eigenvalues of Φ are within the unit circle.

3.11 Show that the cross product matrix $[\mathbf{a}\times]$ is always singular. Also, show that the nonzero eigenvalues are given by $\pm||\mathbf{a}||j$.

3.12 Show that the matrix $(I \pm [\mathbf{a}\times])$ is always non-singular.

3.13 Prove the following identities:

(A) $[\mathbf{a}\times]\mathbf{a} = \mathbf{0}$

(B) $[\mathbf{a}\times][\mathbf{b}\times] = \mathbf{b}\mathbf{a}^T - (\mathbf{b}^T\mathbf{a})I$

(C) $[\mathbf{a}\times][\mathbf{b}\times] - [\mathbf{b}\times][\mathbf{a}\times] = \mathbf{b}\mathbf{a}^T - \mathbf{a}\mathbf{b}^T$

3.14 ♣ Prove that the following matrix: $-[\mathbf{a}\times]^2$, with $\mathbf{a}^T\mathbf{a}=1$ is a projection matrix (see §1.6.4).

3.15 Prove the identities in eqn. (3.163).

3.16 Show that the determinant of an orthogonal matrix is given by ± 1.

3.17 Show that the magnitude of any row or column of an orthogonal matrix is 1.

3.18 Derive the attitude matrix for a 3-1-3 rotation sequence. If the small angle approximation is used, what is the linear approximation for this attitude matrix? How does this matrix differ from eqn. (3.148)?

3.19 ♣ Show that $(I-[\mathbf{a}\times])(I+[\mathbf{a}\times])^{-1}=\frac{1}{1+\mathbf{a}^T\mathbf{a}}\left\{(1-\mathbf{a}^T\mathbf{a})I+2\mathbf{a}\mathbf{a}^T-2[\mathbf{a}\times]\right\}$, and show that this is an orthogonal matrix.

3.20 Show that the kinematics equation $\dot{A}=-[\boldsymbol{\omega}\times]A$ holds true for any orthogonal matrix A.

3.21 From the definitions of $\Xi(\mathbf{q})$, $\Psi(\mathbf{q})$, $\Omega(\boldsymbol{\omega})$, $\Gamma(\boldsymbol{\omega})$, and $A(\mathbf{q})$ in §3.7.1, prove the following identities:

$$\Omega(\boldsymbol{\omega})\Xi(\mathbf{q})=-\Xi(\mathbf{q})[\boldsymbol{\omega}\times]-\mathbf{q}\boldsymbol{\omega}^T$$

$$\Gamma(\boldsymbol{\omega})\Psi(\mathbf{q})=\Psi(\mathbf{q})[\boldsymbol{\omega}\times]-\mathbf{q}\boldsymbol{\omega}^T$$

$$\Omega(\boldsymbol{\omega})\Psi(\mathbf{q})=-\left\{\Xi(\mathbf{q})[\boldsymbol{\omega}\times]+\mathbf{q}\boldsymbol{\omega}^T\right\}A(\mathbf{q})$$

$$\Omega(\boldsymbol{\omega})\Psi(\mathbf{q})=\left[-q_4 I_{4\times 4}+\Omega(\boldsymbol{\varrho})\right]\begin{bmatrix}[\boldsymbol{\omega}\times]\\ \boldsymbol{\omega}^T\end{bmatrix}-\begin{bmatrix}2(\boldsymbol{\varrho}^T\boldsymbol{\omega})I_{3\times 3}\\ \mathbf{0}_{3\times 1}^T\end{bmatrix}$$

$$\Xi^T(\mathbf{q})\Omega(\boldsymbol{\omega})\Xi(\mathbf{q})=-[\boldsymbol{\omega}\times]$$

$$\Xi^T(\mathbf{q})\Gamma(\boldsymbol{\omega})\Xi(\mathbf{q})=[A(\mathbf{q})\boldsymbol{\omega}\times]$$

$$\Gamma(\boldsymbol{\omega})\Xi(\mathbf{q})=\Xi(\boldsymbol{\varpi})$$

$$\Omega(\boldsymbol{\omega})\Psi(\mathbf{q})=\Psi(\boldsymbol{\chi})$$

where $\boldsymbol{\varpi}\equiv\Psi(\mathbf{q})\boldsymbol{\omega}$ and $\boldsymbol{\chi}\equiv\Xi(\mathbf{q})\boldsymbol{\omega}$. Note that $\mathbf{q}^T\mathbf{q}=1$. Also, show that the matrices $\Omega(\boldsymbol{\omega})$ and $\Gamma(\boldsymbol{\lambda})$ commute, i.e., $\Omega(\boldsymbol{\omega})\Gamma(\boldsymbol{\lambda})=\Gamma(\boldsymbol{\lambda})\Omega(\boldsymbol{\omega})$ for any $\boldsymbol{\omega}$ and $\boldsymbol{\lambda}$.

3.22 A *symplectic matrix* A is a $2n\times 2n$ matrix with the defining property

$$A^T J A = J$$

where J is the matrix analogy of the scalar complex number $j^2=-1$; J is defined as the $2n\times 2n$ matrix

$$J=\begin{bmatrix}0 & I_{n\times n}\\ -I_{n\times n} & 0\end{bmatrix},\qquad J\,J=-I_{2n\times 2n}$$

An important consequence of the symplectic property is that the inverse can be obtained by the simple rearrangement of A's elements as

$$A^{-1} = -J\,A^T\,J = \begin{bmatrix} A_{22}^T & -A_{12}^T \\ -A_{21} & A_{11}^T \end{bmatrix}$$

where A is partitioned into $n \times n$ sub-matrices

$$A = \begin{bmatrix} A_{11} & A_{12} \\ A_{21} & A_{22} \end{bmatrix}$$

This non-numerical inversion is a most important computational advantage that symplectic matrices have in common with orthogonal matrices. The 6×6 state transition matrix $\Phi(t, t_0)$ for the orbit model in eqn. (3.198) satisfies

$$\dot{\Phi}(t, t_0) = \begin{bmatrix} 0 & I \\ G & 0 \end{bmatrix} \Phi(t, t_0)$$

Show that G is given by

$$G = \frac{3\mu}{||\mathbf{r}||^5} \begin{bmatrix} (r_1^2 - ||\mathbf{r}||^2/3) & r_1 r_2 & r_1 r_3 \\ r_1 r_2 & (r_2^2 - ||\mathbf{r}||^2/3) & r_2 r_3 \\ r_1 r_3 & r_2 r_3 & (r_3^2 - ||\mathbf{r}||^2/3) \end{bmatrix}$$

Next show that $\Phi(t, t))$ is symplectic.

3.23 In the torque-free response of spacecraft motion the "energy ellipsoid" is given by

$$1 = \frac{J_1^2 \omega_1^2}{2J_1 T} + \frac{J_2^2 \omega_2^2}{2J_2 T} + \frac{J_3^2 \omega_3^2}{2J_3 T}$$

where the kinetic energy T is given by

$$T = \frac{1}{2} J_1 \omega_1^2 + \frac{1}{2} J_2 \omega_2^2 + \frac{1}{2} J_3 \omega_3^2$$

The "momentum ellipsoid" is given by

$$||\mathbf{H}||^2 = J_1^2 \omega_1^2 + J_2^2 \omega_2^2 + J_3^2 \omega_3^2$$

In order for the angular velocity $\boldsymbol{\omega}$ to be feasible, the solution must satisfy both the energy and momentum ellipsoid equations. Show that eqn. (3.191) is a feasible solution.

3.24 Write a computer program to simulate the attitude dynamics of a spacecraft modelled by eqn. (3.184). Consider the following diagonal inertia matrix:

$$J = \begin{bmatrix} 100 & 0 & 0 \\ 0 & 100 & 0 \\ 0 & 0 & 50 \end{bmatrix} \text{N m s}$$

Integrate eqn. (3.184) for an 8-hour simulation. Use the identity quaternion for the initial attitude condition and set $\mathbf{L} = \mathbf{0}$. Use the following initial condition for the angular velocity: $\boldsymbol{\omega}(t_0) = \begin{bmatrix} 1 \times 10^{-3} & 1 \times 10^{-3} & 1 \times 10^{-3} \end{bmatrix}^T$ rad/sec.

Check your results with eqn. (3.191). Next, consider the following inertia matrix:

$$J = \begin{bmatrix} 150 & 0 & 0 \\ 0 & 100 & 0 \\ 0 & 0 & 50 \end{bmatrix} \text{ N m s}$$

Use the same initial attitude from before, but now try the following initial conditions for the initial angular velocity vector:

(A) $\omega(t_0) = \begin{bmatrix} 0 & 1\times10^{-3} & 0 \end{bmatrix}^T$.

(B) $\omega(t_0) = \begin{bmatrix} 1\times10^{-5} & 1\times10^{-3} & 1\times10^{-5} \end{bmatrix}^T$.

The first case is an intermediate axis spin with no perturbations in the other axes. The second case has slight perturbations in the other axes. Can you explain the vastly different results between these cases?

3.25 Program the analytical solution for the elliptic two-body given by eqns. (3.199) to (3.202). Compute the state histories at an interval of 10 seconds for 5000 seconds. The initial conditions are given by

$$\mathbf{r}(t_0) = \begin{bmatrix} 7000 & 10 & 20 \end{bmatrix}^T \text{ km}$$

$$\dot{\mathbf{r}}(t_0) = \begin{bmatrix} 4 & 7 & 2 \end{bmatrix}^T \text{ km/sec}$$

Compare the analytical solution with a numerical solution by integrating the nonlinear orbit model in eqn. (3.198).

3.26 Prove for an orbiting body that the angular momentum vector $\mathbf{h} = \mathbf{r}\times\dot{\mathbf{r}}$ is constant. This proves that a spacecraft's motion must be confined to a plane which is fixed in space since $\mathbf{r}$ and $\dot{\mathbf{r}}$ always remain in the same plane.

3.27 ♣ Prove Kepler's first law using eqn. (3.198).

3.28 ♣ Derive the coordinate transformations shown in eqns. (3.211) and (3.212).

3.29 In an aircraft, a trimmed condition exists if the forces and moments acting on the aircraft are in equilibrium. This is given when the pitching moment in eqn. (3.222b) is zero and when the lift force in eqn. (3.217c) is equal to mg. For this case determine expressions for the trimmed angle of attack α and elevator δ_E angles in terms of the dynamic pressure ($\bar{q}$), known reference area (S), mass (m), gravity (g), and aerodynamic coefficients.

3.30 Write a program to simulate the motion of a 747 aircraft using the equations of motion in §3.9. The aerodynamic coefficients, assuming a low cruise, for the 747 are given by

$$C_{D_0} = 0.0164 \quad C_{D_\alpha} = 0.20 \quad C_{D_{\delta_E}} = 0$$

$$C_{Y_0} = 0 \quad C_{Y_\beta} = -0.90 \quad C_{Y_{\delta_R}} = 0.120 \quad C_{Y_{\delta_A}} = 0$$

$$C_{L_0} = 0.21 \quad C_{L_\alpha} = 4.4 \quad C_{L_{\delta_E}} = 0.32$$

$$C_{l_0} = 0 \quad C_{l_\beta} = -0.160 \quad C_{l_{\delta_R}} = 0.008 \quad C_{l_{\delta_A}} = 0.013$$

$$C_{l_p} = -0.340 \quad C_{l_r} = 0.130$$

$$C_{m_0} = 0 \quad C_{m_\alpha} = -1.00 \quad C_{m_{\delta_E}} = -1.30 \quad C_{m_q} = -20.5$$

$$C_{n_0} = 0 \quad C_{n_\beta} = 0.160 \quad C_{n_{\delta_R}} = -0.100 \quad C_{n_{\delta_A}} = 0.0018$$

$$C_{n_p} = -0.026 \quad C_{n_r} = -0.280$$

The reference geometry quantities and density are given by

$$S = 510.97\,\text{m}^2 \quad \bar{c} = 8.321\,\text{m} \quad b = 59.74\,\text{m} \quad \rho = 0.6536033\,\text{kg/m}^3$$

The mass data and inertia quantities are given by

$$m = 288,674.58\,\text{kg} \quad J_{13} = 1,315,143\,\text{kg m}^2$$

$$J_{11} = 24,675,882\,\text{kg m}^2 \quad J_{22} = 44,877,565\,\text{kg m}^2 \quad J_{33} = 67,384,138\,\text{kg m}^2$$

The flight conditions for low cruise at an altitude of $6,096$ m are given by

$$||\mathbf{v}|| = 205.13\,\text{m/s} \quad \bar{q} = 13,751.2\,\text{N/m}^2$$

Using these flight conditions compute the trim values for the angle of attack and elevator (see exercise 3.29). Using these trim values compute the drag using eqn. (3.217a). Let the thrust equal the computed drag (assume that the thrust torque quantities in eqn. (3.220) are zero), and set the aileron and rudder angles to 0 degrees in your simulation. Integrate the equations of motion for a 200-second simulation for some initial linear velocities (let $\omega_0 = \mathbf{0}$, $x_0 = 0$, $y_0 = 0$, $z_0 = 6,096$, and $\phi_0 = 0$, $\theta_0 = 0$, and $\psi_0 = 0$). Next, perform a simple maneuver starting at 10 seconds in the simulation by setting the elevator angle equal to its trim value minus 1 degree, and set the aileron angle equal to 1 degree, holding each control surface for a 10-second interval (returning the elevator back to its trimmed condition and setting the aileron angle equal to 0 degrees after the interval). Show plots of aircraft position, velocity, orientation, etc. Perform other maneuvers by changing the thrust, elevator, etc.

3.31 Pick the correct form using eqn. (3.231) for the solution of the following second-order differential equations:

(A) $\ddot{x} + 2\dot{x} + x = 0$

(B) $\ddot{x} + 2\dot{x} + 2x = 0$

(C) $\ddot{x} + 3\dot{x} + 2x = 0$

(D) $\ddot{x} + 4x = 0$

3.32 Consider the following mass, damping, and stiffness matrices:

$$M = \begin{bmatrix} 9 & 0 \\ 0 & 1 \end{bmatrix}, \quad C = \begin{bmatrix} 9 & -1 \\ -1 & 1 \end{bmatrix}, \quad K = \begin{bmatrix} 27 & -3 \\ -3 & 3 \end{bmatrix}$$

Prove that this system is a normal mode system. Convert this system into state space form and numerically determine state trajectories for some given initial conditions. Compare the solutions with the decoupled solutions using eqn. (3.247).

3.33 ♣ Consider the following mass, damping, and stiffness matrices:

$$M = \begin{bmatrix} m_1 & 0 \\ 0 & m_2 \end{bmatrix}, \quad C = \begin{bmatrix} c_1 + c_2 & -c_2 \\ -c_2 & c_2 \end{bmatrix}, \quad K = \begin{bmatrix} k_1 + k_2 & -k_2 \\ -k_2 & k_2 \end{bmatrix}$$

Can you find values for m_1, m_2, c_1, c_2, k_1, and k_2 such that the system does not oscillate?

References

[1] Dorf, R.C. and Bishop, R.H., *Modern Control Systems*, Addison Wesley Longman, Menlo Park, CA, 1998.

[2] Palm, W.J., *Modeling, Analysis, and Control of Dynamic Systems*, John Wiley & Sons, New York, NY, 2nd ed., 1999.

[3] Franklin, G.F., Powell, J.D., and Workman, M., *Digital Control of Dynamic Systems*, Addison Wesley Longman, Menlo Park, CA, 3rd ed., 1998.

[4] Bélanger, P.R., *Control Engineering*, Saunders College Publishing, Fort Worth, TX, 1995.

[5] Shinners, S.M., *Modern Control System Theory and Design*, John Wiley & Sons, New York, NY, 2nd ed., 1999.

[6] Phillips, C.L. and Harbor, R.D., *Feedback Control Systems*, Prentice Hall, Englewood Cliffs, NJ, 1996.

[7] Kuo, B.C., *Automatic Control Systems*, Prentice Hall, Englewood Cliffs, NJ, 6th ed., 1991.

[8] Nise, N.S., *Control Systems Engineering*, Addison-Wesley Publishing, Menlo Park, CA, 2nd ed., 1995.

[9] Ogata, K., *Modern Control Engineering*, Prentice Hall, Upper Saddle River, NJ, 1997.

[10] LePage, W.R., *Complex Variables and the Laplace Transform for Engineers*, Dover Publications, New York, NY, 1980.

[11] Ince, E.L., *Ordinary Differential Equations*, Longmans, London, England, 1926.

[12] Chen, C.T., *Linear System Theory and Design*, Holt, Rinehart and Winston, New York, NY, 1984.

[13] Meirovitch, L., *Methods of Analytical Dynamics*, McGraw-Hill, New York, NY, 1970.

[14] Neyfeh, A.H., *Introduction to Perturbation Techniques*, John Wiley Interscience, New York, NY, 1981.

[15] Slotine, J.J.E. and Li, W., *Applied Nonlinear Control*, Prentice Hall, Englewood Cliffs, NJ, 1991.

[16] Garrard, W.L. and Jordan, J.M., "Design of Nonlinear Automatic Flight Control Systems," *Automatica*, Vol. 13, No. 5, Sept. 1977, pp. 497–505.

[17] Hermann, R. and Krener, A.J., "Nonlinear Controllability and Observability," *IEEE Transactions on Automatic Control*, Vol. AC-22, No. 5, Oct. 1977, pp. 728–740.

[18] Isidori, A., *Nonlinear Control System*, Springer-Verlag, Berlin, 3rd ed., 1990.

[19] Moler, C. and van Loan, C., "Nineteen Dubious Ways to Compute the Exponential of a Matrix," *SIAM Review*, Vol. 20, No. 4, 1978, pp. 801–836.

[20] Phillips, C.L. and Nagle, H.T., *Digital Control System Analysis and Design*, Prentice Hall, Englewood Cliffs, NJ, 2nd ed., 1990.

[21] Åström, K.J. and Wittenmark, B., *Computer-Controlled Systems*, Prentice Hall, Upper Saddle River, NJ, 3rd ed., 1997.

[22] Kalman, R.E. and Bertram, J., "Control System Analysis and Design via the Second Method of Lyapunov: II. Discrete-Time Systems," *Journal of Basic Engineering*, Vol. 82, No. 3, 1960, pp. 394–400.

[23] Schaub, H. and Junkins, J.L., *Analytical Mechanics of Aerospace Systems*, American Institute of Aeronautics and Astronautics, Inc., New York, NY, 2003.

[24] Goldstein, H., *Classical Mechanics*, Addison-Wesley Publishing Company, Reading, MA, 2nd ed., 1980.

[25] Shuster, M.D., "A Survey of Attitude Representations," *Journal of the Astronautical Sciences*, Vol. 41, No. 4, Oct.-Dec. 1993, pp. 439–517.

[26] Hamilton, W.R., *Elements of Quaternions*, Longmans, Green and Co., London, England, 1866.

[27] Lefferts, E.J., Markley, F.L., and Shuster, M.D., "Kalman Filtering for Spacecraft Attitude Estimation," *Journal of Guidance, Control, and Dynamics*, Vol. 5, No. 5, Sept.-Oct. 1982, pp. 417–429.

[28] Shepperd, S.W., "Quaternion from Rotation Matrix," *Journal of Guidance and Control*, Vol. 1, No. 3, May-June 1978, pp. 223–224.

[29] Kuipers, J.B., *Quaternions and Rotation Sequences: A Primer with Applications to Orbits, Aerospace, and Virtual Reality*, Princeton University Press, Princeton, NJ, 1999.

[30] Kibble, T.W.B. and Berkshire, F.H., *Classical Mechanics*, Addison Wesley Longman, Essex, England, 4th ed., 1996.

[31] Markley, F.L., "Attitude Dynamics," *Spacecraft Attitude Determination and Control*, edited by J.R. Wertz, chap. 16, Kluwer Academic Publishers, The Netherlands, 1978.

[32] Greenwood, D.T., *Principles of Dynamics*, Prentice Hall, Englewood Cliffs, NJ, 2nd ed., 1988.

[33] Thomson, W.T., *Introduction to Space Dynamics*, Dover Publications, New York, NY, 1986.

[34] Kane, T.R., Likens, P.W., and Levinson, D.A., *Spacecraft Dynamics*, McGraw-Hill, New York, NY, 1983.

[35] Hughes, P.C., *Spacecraft Attitude Dynamics*, Wiley, New York, NY, 1986.

[36] Kaplan, M.H., *Modern Spacecraft Dynamics and Control*, Wiley, New York, NY, 1976.

[37] Wiesel, W.E., *Spaceflight Dynamics*, McGraw-Hill, New York, NY, 2nd ed., 1997.

[38] Junkins, J.L. and Turner, J.D., *Optimal Spacecraft Rotational Maneuvers*, Elsevier, New York, NY, 1986.

[39] Battin, R.H., *An Introduction to the Mathematics and Methods of Astrodynamics*, American Institute of Aeronautics and Astronautics, Inc., New York, NY, 1987.

[40] Bate, R.R., Mueller, D.D., and White, J.E., *Fundamentals of Astrodynamics*, Dover Publications, New York, NY, 1971.

[41] Herrick, S., *Astrodynamics*, Vol. 1, Van Nostrand Reinhold, London, England, 1971.

[42] Nelson, R.C., *Flight Stability and Automatic Control*, McGraw-Hill, New York, NY, 1989.

[43] Roskam, J., *Airplane Flight Dynamics and Automatic Flight Controls*, Design, Analysis and Research Corporation, Lawrence, KS, 1994.

[44] Rao, S.S., *Mechanical Vibrations*, Addison-Wesley Publishing Company, Reading, MA, 2nd ed., 1990.

[45] Dimarogonas, A., *Vibration for Engineers*, Prentice Hall, Upper Saddle River, NJ, 2nd ed., 1996.

[46] Junkins, J.L. and Kim, Y., *Introduction to Dynamics and Control of Flexible Structures*, American Institute of Aeronautics and Astronautics, Inc., Washington, DC, 1993.

[47] Meirovitch, L., *Principles and Techniques of Vibrations*, Prentice Hall, Upper Saddle River, NJ, 1997.

[48] Inman, D.J., *Vibration with Control, Measurement, and Stability*, Prentice Hall, Englewood Cliffs, NJ, 1989.

[49] Weaver, W., Timoshenko, S.P., and Young, D.H., *Vibration Problems in Engineering*, John Wiley & Sons, New York, NY, 5th ed., 1990.

4

Parameter Estimation: Applications

Errors using inadequate data are much less than those using no data at all. Babbage, Charles

THE previous chapters laid down the foundation for the application of parameter estimation methods to dynamical systems. In this chapter several example applications are presented in which the methods of the first two chapters can be used to advantage with the class of dynamical systems discussed in the previous chapter. The problems and solutions are idealizations of "real-world" applications that are well-documented in the literature cited. First, the position of a vehicle is determined using Global Positioning System (GPS) signals transmitted from orbiting spacecraft. Then, spacecraft attitude determination is introduced using photographs of stars made from one or more spacecraft-fixed cameras. Next, spacecraft orbit determination from ground radar observations using a Gaussian Least Squares Differential Correction (GLSDC) is presented. Then, parameter estimation of an aircraft using various sensors is introduced. Finally, flexible structure modal realization using the Eigensystem Realization Algorithm (ERA) is studied. This chapter shows only the fundamental concepts of these applications; the emphasis here is upon the utility of the estimation methodology. However, the examples are presented in sufficient detail to serve as a foundation for each of the subject areas shown. The interested reader is encouraged to pursue these subjects in more depth by studying the many references cited in this chapter.

4.1 Global Positioning System Navigation

The Global Positioning System (GPS) constellation was originally developed to permit a wide variety of user vehicles an accurate means of determining position for autonomous navigation. The constellation includes 24 space vehicles (SVs) in known semi-synchronous (12-hour) orbits, providing a minimum of six SVs in view for ground-based navigation. The underlying principle involves geometric triangulation with the GPS SVs as known reference points to determine the user's position to a high degree of accuracy. The GPS was originally intended for ground-based and aviation applications, and is gaining much attention in the commercial community (e.g., automobile navigation, aircraft landing, etc.). However, in recent years there has been a growing interest in other applications, such as spacecraft navigation, attitude determination, and even as a vibration sensor. Since the GPS SVs are in approximately 20,000 km circular orbits, the position of any potential user below the constellation may be easily determined.

A minimum of four SVs are required so that, in addition to the three-dimensional position of the user, the time of the solution can be determined and in turn employed to correct the user's clock. Since its original inception, there have been many innovative improvements to the accuracy of the GPS determined position. These include using local area as well as wide area differential GPS and carrier-phase differential GPS. In particular, carrier-phase differential GPS measures the phase of the GPS carrier relative to the phase at a reference site, which dramatically improves the position accuracy. These innovative techniques allow for more accurate GPS determined positions.

The fundamental signal in GPS is the pseudo-random code (PRC) which is a complicated binary sequence of pulses. Each SV has its own complex PRC, which guarantees that the receiver won't be confused with another SV's signal. The GPS satellites transmit signals on two carrier frequencies: L1 at 1575.42 MHz and L2 at 1227.60 MHz. The modulated PRC at the L1 carrier is called the Coarse Acquisition (C/A) code, which repeats every 1023 bits and modulates at a 1MHz rate. The C/A code is the basis for civilian GPS use. Another PRC is called the Precise (P) code, which repeats on a seven-day cycle and modulates both the L1 and L2 carriers at a 10 MHz rate. This code is intended for military users and can be encrypted. Position location is made possible by comparing how late in time the SV's PRC appears to the receiver's code. Multiplying the travel time by the speed of light, one obtains the distance to the SV. This requires very accurate timing in the receiver, which is provided by using a fourth SV to correct a "clock bias" in the internal clock receiver.

There are many error sources that affect the GPS accuracy using the PRC. First, the GPS signal slows down slightly as it passes through the charged particles of the ionosphere and then through the water vapor in the troposphere. Second, the signal may bounce off various local obstructions before it arrives at the receiver (known as *multipath* errors). Third, SV ephemeris (i.e., known satellite position) errors can contribute to GPS location inaccuracy. Finally, the basic geometry on the available SVs can magnify errors, which is known as the Geometric Dilution of Precision (GDOP). A poor GDOP usually means that the SV sightlines to the receiver are close to being collinear, resulting in degraded accuracy. Many of the aforementioned errors can be minimized or even eliminated by using differential GPS.

Differential GPS (DGPS) involves the cooperation of two receivers, one that is stationary and another that is moving to make the position measurements. The basic principle incorporates the notion that two receivers will have virtually the same errors if they are fairly close to one another (within a few hundred kilometers). The stationary receiver uses its known (calibrated) position to calculate a timing difference (error correction) from the GPS determined position. This receiver then transmits this error information to the moving receiver, so that an updated position correction can be made. DGPS minimizes ionospheric and tropospheric errors, while virtually

Table 4.1: Levels of GPS Accuracy

Technique	Method	Accuracy
PRC	measure signal time-of-flight from each SV	10 to 100 m (absolute)
DGPS	difference of the time-of-flight between two receivers	1 to 5 m (relative)
CDGPS	reconstruct carrier and measure relative phase difference between two antennae	≤ 5 cm for kinematic (relative) ≤ 1 cm for static (relative)

eliminating SV clock errors, and ephemeris errors. Accuracies of 1 to 5 meters can be obtained using DGPS.

Carrier-Phase Differential GPS (CDGPS) can be used to further enhance the position determination performance. The PRC has a bit rate of about 1 MHz but its carrier frequency has a cycle rate of over 1 GHz. At the speed of light the 1.57 GHz GPS carrier signal has a wavelength of about 20 cm. Therefore, by obtaining 1% perfect phase, as is done in PRC receivers, accuracies in the mm region are possible. CDGPS measures the phase of the GPS carrier relative to the carrier phase at a reference site. If the GPS antennae are fixed, then the system is called static, and mm accuracies are typically possible since long averaging times can be used to filter any noise present. If the antennae are moving, then the system is kinematic, and cm accuracies are possible since shorter time constants are used in the averaging. Since phase differences are used, the correct number of integer wavelengths between a given pair of antennae must first be found (known as "integer ambiguity resolution"). CDGPS can also be used for attitude determination of static or moving vehicles. A chart summarizing the various levels of GPS accuracy is shown in Table 4.1.

The equations needed to be solved to determine a user's position (x, y, z) and clock bias τ (in equivalent distance) from GPS pseudorange measurements are given by

$$\boxed{\tilde{\rho}_i = [(e_{1i} - x)^2 + (e_{2i} - y)^2 + (e_{3i} - z)^2]^{1/2} + \tau + v_i, \quad i = 1, 2, \ldots, n} \tag{4.1}$$

where (e_{1i}, e_{2i}, e_{3i}) are the known i^{th} GPS satellite coordinates, n is the total number of observed GPS satellites, and v_i are the measurement errors which are assumed to be the same for each satellite and represented by a zero-mean Gaussian noise process with variance σ^2. Because the number of unknowns is four with $\mathbf{x} = \begin{bmatrix} x & y & z & \tau \end{bmatrix}^T$, at least four non-parallel SVs are required to solve eqn. (4.1).

Since eqn. (4.1) represents a nonlinear function of the unknowns, then nonlinear least squares must be utilized. The estimated pseudorange $\hat{\rho}$ is determined by using the current position estimates $(\hat{x}, \hat{y}, \hat{z})$ and clock bias $\hat{\tau}$ estimate, given by

$$\hat{\rho}_i = [(e_{1i} - \hat{x})^2 + (e_{2i} - \hat{y})^2 + (e_{3i} - \hat{z})^2]^{1/2} + \hat{\tau} \tag{4.2}$$

The ith row of H is formed by taking the partials of eqn. (4.1) with respect to the unknown variables, so that

$$H = \begin{bmatrix} \frac{\partial \hat{\rho}_1}{\partial \hat{x}} & \frac{\partial \hat{\rho}_1}{\partial \hat{y}} & \frac{\partial \hat{\rho}_1}{\partial \hat{z}} & 1 \\ \frac{\partial \hat{\rho}_2}{\partial \hat{x}} & \frac{\partial \hat{\rho}_2}{\partial \hat{y}} & \frac{\partial \hat{\rho}_2}{\partial \hat{z}} & 1 \\ \vdots & \vdots & \vdots & \vdots \\ \frac{\partial \hat{\rho}_n}{\partial \hat{x}} & \frac{\partial \hat{\rho}_n}{\partial \hat{y}} & \frac{\partial \hat{\rho}_n}{\partial \hat{z}} & 1 \end{bmatrix} \tag{4.3}$$

The partials are straightforward, with

$$\frac{\partial \hat{\rho}_i}{\partial \hat{x}} = -\frac{(e_{1i} - \hat{x})}{[(e_{1i} - \hat{x})^2 + (e_{2i} - \hat{y})^2 + (e_{3i} - \hat{z})^2]^{1/2}} \tag{4.4a}$$

$$\frac{\partial \hat{\rho}_i}{\partial \hat{y}} = -\frac{(e_{2i} - \hat{y})}{[(e_{1i} - \hat{x})^2 + (e_{2i} - \hat{y})^2 + (e_{3i} - \hat{z})^2]^{1/2}} \tag{4.4b}$$

$$\frac{\partial \hat{\rho}_i}{\partial \hat{z}} = -\frac{(e_{3i} - \hat{z})}{[(e_{1i} - \hat{x})^2 + (e_{2i} - \hat{y})^2 + (e_{3i} - \hat{z})^2]^{1/2}} \tag{4.4c}$$

Equations (4.2) to (4.4) are used in nonlinear least squares of §1.4 to determine the position of the user and clock bias. The covariance of the estimate errors is simply given by

$$\boxed{P = \sigma^2 (H^T H)^{-1}} \tag{4.5}$$

The matrix $A \equiv (H^T H)^{-1}$ can be used to define several DOP quantities,[1] including; geometrical DOP (GDOP), position DOP (PDOP), horizontal DOP (HDOP), vertical DOP (VDOP), and time DOP (TDOP), each given by

$$\text{GDOP} \equiv \sqrt{A_{11} + A_{22} + A_{33} + A_{44}} \tag{4.6a}$$

$$\text{PDOP} \equiv \sqrt{A_{11} + A_{22} + A_{33}} \tag{4.6b}$$

$$\text{HDOP} \equiv \sqrt{A_{11} + A_{22}} \tag{4.6c}$$

$$\text{VDOP} \equiv \sqrt{A_{33}} \tag{4.6d}$$

$$\text{TDOP} \equiv \sqrt{A_{44}} \tag{4.6e}$$

The quantity GDOP is most widely used since it gives an indication of the basic geometry of the available SVs and the effect of clock bias errors. The best possible value for GDOP with four available satellites is obtained when one satellite is directly overhead and the remaining are spaced equally at the minimum elevation angles around the horizon.[2] We note in passing that other observability measures are possible. For example, we could use the condition number of A, which is the ratio of the largest singular value to the least singular value of A. The smallest condition number is unity (for perfectly conditioned orthogonal matrices) and the largest is infinity (for singular matrices).

Example 4.1: In this example nonlinear least squares is employed to determine the position of a vehicle on the Earth from GPS pseudorange measurements. The vehicle is assumed to have coordinates of 38°N and 77°W (i.e., in Washington, DC). Converting this latitude and longitude into the Earth-Centered-Earth-Fixed (ECEF) frame[3] (see §7.1.1 for more details), and assuming a clock bias of 85,000 m gives the true vector as

$$\mathbf{x} = \begin{bmatrix} 1,132,049 & -4,903,445 & 3,905,453 & 85,000 \end{bmatrix}^T \text{ m}$$

At epoch the following GPS satellites and position vector in ECEF coordinates are available:

SV	e_1 (meters)	e_2 (meters)	e_3 (meters)
5	15,764,733	−1,592,675	21,244,655
13	6,057,534	−17,186,958	19,396,689
18	4,436,748	−25,771,174	1,546,041
22	−9,701,586	−19,687,467	15,359,118
26	23,617,496	−11,899,369	1,492,340
27	14,540,070	−12,201,965	18,352,632

The SV label is the specific GPS satellite number. Simulated pseudorange measurements are computed using eqn. (4.1) with a standard deviation on the measurement error of 5 meters. The nonlinear least squares routine is then initiated with starting conditions of 0 for all elements of $\hat{\mathbf{x}}$. The algorithm converges in five iterations. Results of the iterations are given below.

Iteration	$\hat{x}$ (meters)	$\hat{y}$ (meters)	$\hat{z}$ (meters)	Clock (meters)
0	0	0	0	0
1	1,417,486	−5,955,318	4,745,294	1,502,703
2	1,146,483	−4,944,222	3,938,182	143,265
3	1,132,071	−4,903,503	3,905,503	85,085
4	1,132,042	−4,903,436	3,905,448	85,000
5	1,132,042	−4,903,436	3,905,448	85,000

The 3σ estimate-error bounds are given by

$$3\sigma = \begin{bmatrix} 21.3 & 32.1 & 21.1 & 28.3 \end{bmatrix}^T \text{ m}$$

The estimate errors are clearly within the 3σ bounds. In general, the accuracy can be improved if more satellites are used in the solution.

4.2 Attitude Determination

Attitude determination refers to the identification of a proper orthogonal rotation matrix so that the measured observations in the sensor frame equal the reference frame observations mapped by that matrix into the sensor frame. If all the measured and reference vectors are error free, then the rotation (attitude) matrix is the same for all sets of observations. However, if measurement errors exist, then a least-squares type approach must be used to determine the attitude. Several attitude sensors exist, including: three-axis magnetometers, sun sensors, Earth-horizon sensors, global positioning system (GPS) sensors, and star cameras. In this next section we focus on vector measurement models for star cameras (which can also be applied to sun sensors, three-axis magnetometers and Earth-horizon sensors as well).

4.2.1 Vector Measurement Models

With reference to Figure 4.1, we consider the problem of determining the angular orientation of a space vehicle from photographs of the stars made from one or more spacecraft-fixed cameras. The stars are assumed to be inertially fixed neglecting the effects of proper motion and velocity abberation. The brightest 250,000 stars' spherical coordinate angles (α is the right ascension and δ is the declination, see Figure 4.2) are available in a computer accessible catalog.[4] Referring to Figures 4.2, 4.3, and 3.5, given the camera orientation angles (ϕ, θ, ψ), it is established in Ref. [5] that the photograph image plane coordinates of the jth star are determined by the stellar *collinearity equations*:

$$x_j = -f\left(\frac{A_{11}r_{x_j} + A_{12}r_{y_j} + A_{13}r_{z_j}}{A_{31}r_{x_j} + A_{32}r_{y_j} + A_{33}r_{z_j}}\right) \tag{4.7a}$$

$$y_j = -f\left(\frac{A_{21}r_{x_j} + A_{22}r_{y_j} + A_{23}r_{z_j}}{A_{31}r_{x_j} + A_{32}r_{y_j} + A_{33}r_{z_j}}\right) \tag{4.7b}$$

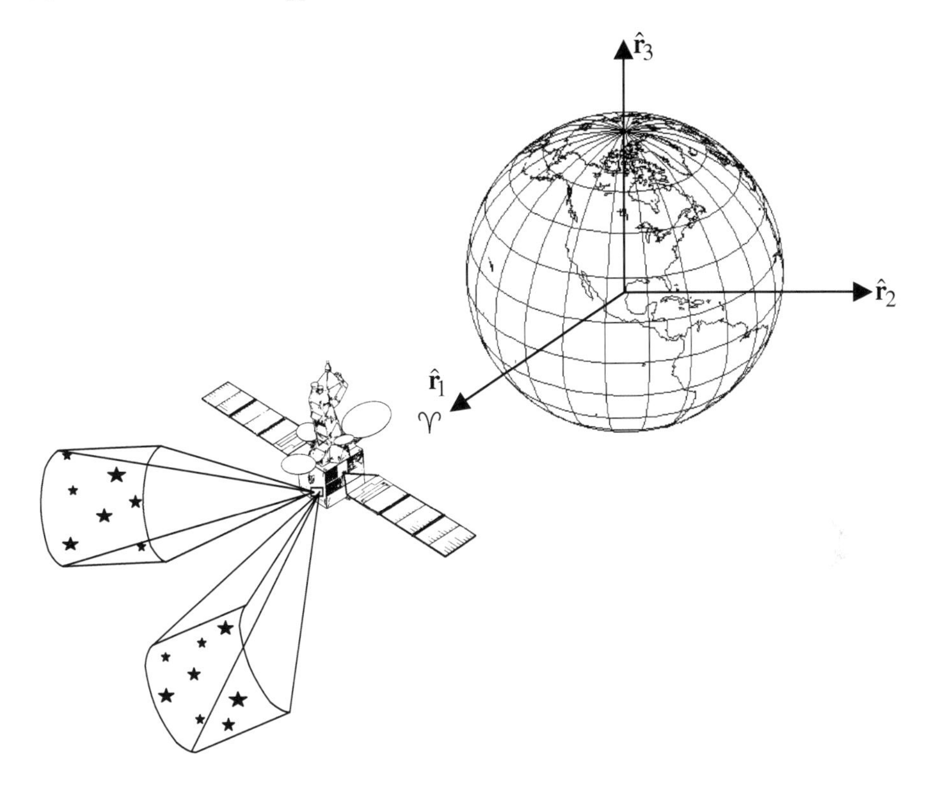

Figure 4.1: Spacecraft Attitude Estimation from Star Photography

where A_{ij} are elements of the attitude matrix A, and the inertial components of the vector toward the jth star are

$$
\begin{aligned}
r_{x_j} &= \cos\delta_j \cos\alpha_j \\
r_{y_j} &= \cos\delta_j \sin\alpha_j \\
r_{z_j} &= \sin\delta_j
\end{aligned}
\tag{4.8}
$$

and the camera focal length f is known from *a priori* calibration. Note that in this section the vector **r** denotes the reference frame, which may be any general frame (e.g., the ECEF frame). When using stars for attitude determination the reference frame coincides with the inertial frame shown in Figures 3.8 and 3.9.

Unfortunately, (ϕ, θ, ψ) are usually not known or poorly known, but if the measured stars can be identified* as specific cataloged stars, then the attitude matrix (and associated camera orientation angles) can be determined from the measured stars in image coordinates and identified stars in inertial coordinates. Clearly, this

*See Ref. [6] for a pattern recognition technique that can be employed to automate the association of the measured images with the cataloged stars.

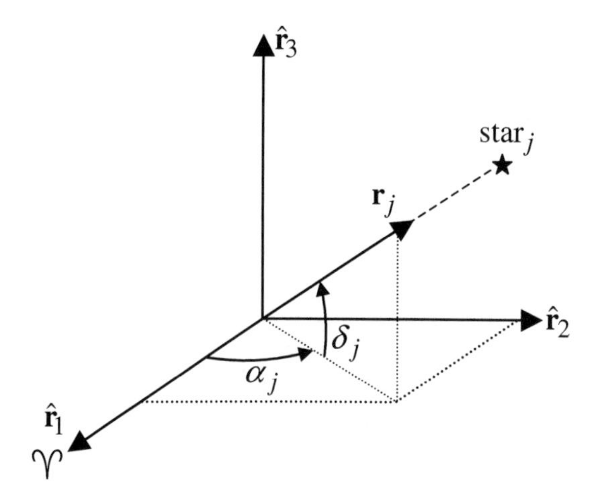

Figure 4.2: Spherical Coordinates Orienting the Line of Sight Vector to a Star

can be accomplished using the nonlinear least squares approach of §1.4. However, through judicious change of variables, a linear form of eqns. (4.7) can be constructed. Choosing the z-axis of the image coordinate system, consistent with Figure 4.3, to be directed outward along the boresight, then the star observation can be reconstructed in unit vector form as

$$\boxed{\mathbf{b}_j = A\mathbf{r}_j, \quad j = 1, 2, \ldots, N} \tag{4.9}$$

where

$$\mathbf{b}_j \equiv \frac{1}{\sqrt{f^2 + x_j^2 + y_j^2}} \begin{bmatrix} -x_j \\ -y_j \\ f \end{bmatrix} \tag{4.10a}$$

$$\mathbf{r}_j \equiv \left[r_{x_j} \; r_{y_j} \; r_{z_j} \right]^T \tag{4.10b}$$

and N is the total number of star observations. The components of $\mathbf{b}$ can be written using eqn. (3.142a). When measurement noise is present, Shuster[7] has shown that nearly all the probability of the errors is concentrated on a very small area about the direction of $A\mathbf{r}_j$, so the sphere containing that point can be approximated by a tangent plane, characterized by

$$\tilde{\mathbf{b}}_j = A\mathbf{r}_j + \boldsymbol{\upsilon}_j, \quad \boldsymbol{\upsilon}_j^T A\mathbf{r}_j = 0 \tag{4.11}$$

where $\tilde{\mathbf{b}}_j$ denotes the jth star measurement, and the sensor error $\boldsymbol{\upsilon}_j$ is approximately Gaussian which satisfies

$$E\left\{\boldsymbol{\upsilon}_j\right\} = \mathbf{0} \tag{4.12a}$$

$$E\left\{\boldsymbol{\upsilon}_j \boldsymbol{\upsilon}_j^T\right\} = \sigma_j^2 \left[I_{3\times 3} - (A\mathbf{r}_j)(A\mathbf{r}_j)^T \right] \tag{4.12b}$$

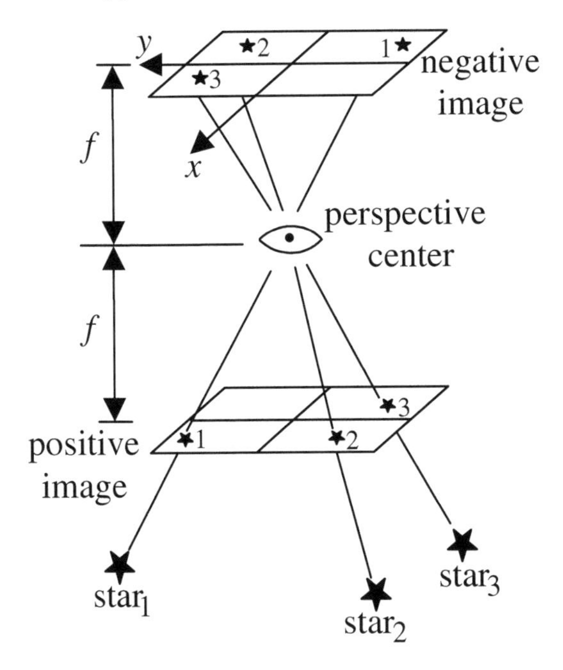

Figure 4.3: Collinearity of Perspective Center, Image, and Object

The measurement model in eqn. (4.11) is also valid for three-axis magnetometers and Earth-horizon sensors.

4.2.2 Maximum Likelihood Estimation

The maximum-likelihood approach for attitude estimation minimizes the following loss function:

$$\boxed{J(\hat{A}) = \frac{1}{2}\sum_{j=1}^{N} \sigma_j^{-2} \|\tilde{\mathbf{b}}_j - \hat{A}\mathbf{r}_j\|^2} \tag{4.13}$$

subject to the constraint

$$\hat{A}\hat{A}^T = I_{3\times 3} \tag{4.14}$$

This problem was first posed by Grace Wahba[8] in 1965. Although the least squares minimization in eqn. (4.13) seems to be straightforward, the equality constraint in eqn. (4.14) complicates the solution, which has lead to a wide area of linear algebra research for the computationally optimal solution since Wahba's original paper. Before proceeding with the solution to this problem, we first derive an estimate error covariance expression. This is accomplished by using results from maximum likelihood estimation of §2.4. Recall that the Fisher information matrix for a parameter vector $\mathbf{x}$ is given by

$$F = E\left\{\frac{\partial}{\partial \mathbf{x}\,\partial \mathbf{x}^T} J(\mathbf{x})\right\} \tag{4.15}$$

where $J(\mathbf{x})$ is the negative log-likelihood function, which is the loss function in this case (neglecting terms independent of A). Asymptotically, the Fisher information matrix tends to the inverse of the estimate error covariance so that

$$\lim_{N\to\infty} F = P^{-1} \tag{4.16}$$

The Fisher information for the attitude is expressed in terms of incremental error angles, $\boldsymbol{\delta\alpha}$, defined according to

$$\hat{A} = e^{-[\boldsymbol{\delta\alpha}\times]}A \approx (I_{3\times3} - [\boldsymbol{\delta\alpha}\times])\,A \tag{4.17}$$

where the 3×3 matrix $[\boldsymbol{\delta\alpha}\times]$ is a cross product matrix, see eqn. (3.149). Higher-order terms in the Taylor series expansion of the exponential function are not required since they do not contribute to the Fisher information matrix. The parameter vector is now given by $\mathbf{x} = \boldsymbol{\delta\alpha}$, and the covariance is defined by $P = E\left\{\mathbf{x}\mathbf{x}^T\right\} - E\{\mathbf{x}\}\,E^T\{\mathbf{x}\}$. Substituting eqn. (4.17) into eqn. (4.13), and after taking the appropriate partials the following optimal error covariance can be derived:

$$\boxed{P = \left(-\sum_{j=1}^{N}\sigma_j^{-2}[A\,\mathbf{r}_j\times]^2\right)^{-1}} \tag{4.18}$$

The attitude A is evaluated at its respective *true* value. In practice, though, $A\,\mathbf{r}_j$ is often replaced with the measurement $\tilde{\mathbf{b}}_j$, which allows a calculation of the covariance without computing an attitude! Equation (4.18) gives the Cramér-Rao lower bound (any estimator whose error covariance is equivalent to eqn. (4.18) is an *efficient*, i.e., optimal estimator). The Fisher information matrix is nonsingular only if at least two non-collinear observation vectors exist. This is due to the fact that one vector observation gives only two pieces of attitude information. To see this fact we first use the following identity:

$$-[A\,\mathbf{r}\times]^2 = ||\mathbf{r}||^2 I_{3\times3} - (A\,\mathbf{r})(A\,\mathbf{r})^T \tag{4.19}$$

This matrix has rank 2 and is the projection operator (see §1.6.4) onto the space perpendicular to $A\,\mathbf{r}$, which reflects the fact that an observation of a vector contains no information about rotations around an axis specified by that vector.

4.2.3 Optimal Quaternion Solution

One approach to determine the attitude involves using the Euler angle parameterization of the attitude matrix, shown in §3.7.1. Nonlinear least squares may be employed to determine the Euler angles; however, this is a highly iterative approach due to the nonlinear parameterization of the attitude matrix, which involve transcendental functions. A more elegant algorithm is given by Davenport, known as the *q-method*.[9] The loss function in eqn. (4.13) may be rewritten as

$$J(\hat{A}) = -\sum_{j=1}^{N} \sigma_j^{-2} \tilde{\mathbf{b}}_j^T \hat{A} \mathbf{r}_j + \text{constant terms} \tag{4.20}$$

This loss function is clearly a minimum when

$$J(\hat{A}) = \sum_{j=1}^{N} \sigma_j^{-2} \tilde{\mathbf{b}}_j^T \hat{A} \mathbf{r}_j \tag{4.21}$$

is a maximum (dropping the constant terms which are not needed). To determine the attitude we parameterize $\hat{A}$ in term of the quaternion using eqn. (3.154), so that eqn. (4.21) is rewritten as

$$J(\hat{\mathbf{q}}) = \sum_{j=1}^{N} \sigma_j^{-2} \tilde{\mathbf{b}}_j^T \Xi^T(\hat{\mathbf{q}}) \Psi(\hat{\mathbf{q}}) \mathbf{r}_j \tag{4.22}$$

Also, the orthogonality constraint in eqn. (4.14) reduces to $\hat{\mathbf{q}}^T \hat{\mathbf{q}} = 1$ for the quaternion. Using the identities in eqns. (3.161) and (3.164) leads to

$$J(\hat{\mathbf{q}}) = \hat{\mathbf{q}}^T K \hat{\mathbf{q}} \tag{4.23}$$

with

$$K \equiv -\sum_{j=1}^{N} \sigma_j^{-2} \Omega(\tilde{\mathbf{b}}_j) \Gamma(\mathbf{r}_j) \tag{4.24}$$

where $\Omega(\tilde{\mathbf{b}})$ and $\Gamma(\mathbf{r})$ are defined in eqns. (3.162) and (3.165), respectively. Note that these matrices commute so that $\Omega(\tilde{\mathbf{b}})\Gamma(\mathbf{r}) = \Gamma(\mathbf{r})\Omega(\tilde{\mathbf{b}})$. The extrema of $J(\hat{\mathbf{q}})$, subject to the normalization constraint $\hat{\mathbf{q}}^T \hat{\mathbf{q}} = 1$, is found by using the method of Lagrange multipliers (see Appendix C). The necessary conditions can be found by maximizing the following augmented function:

$$J(\hat{\mathbf{q}}) = \hat{\mathbf{q}}^T K \hat{\mathbf{q}} + \lambda(1 - \hat{\mathbf{q}}^T \hat{\mathbf{q}}) \tag{4.25}$$

where λ is a Lagrange multiplier. Therefore, as necessary conditions for constrained minimization of J, we have the following requirement:

$$\boxed{K \hat{\mathbf{q}} = \lambda \hat{\mathbf{q}}} \tag{4.26}$$

Equation (4.26) represents an eigenvalue decomposition of the matrix K, where the quaternion is an eigenvector of K and λ is an eigenvalue. Substituting eqn. (4.26) into eqn. (4.23) gives

$$J(\hat{\mathbf{q}}) = \lambda \tag{4.27}$$

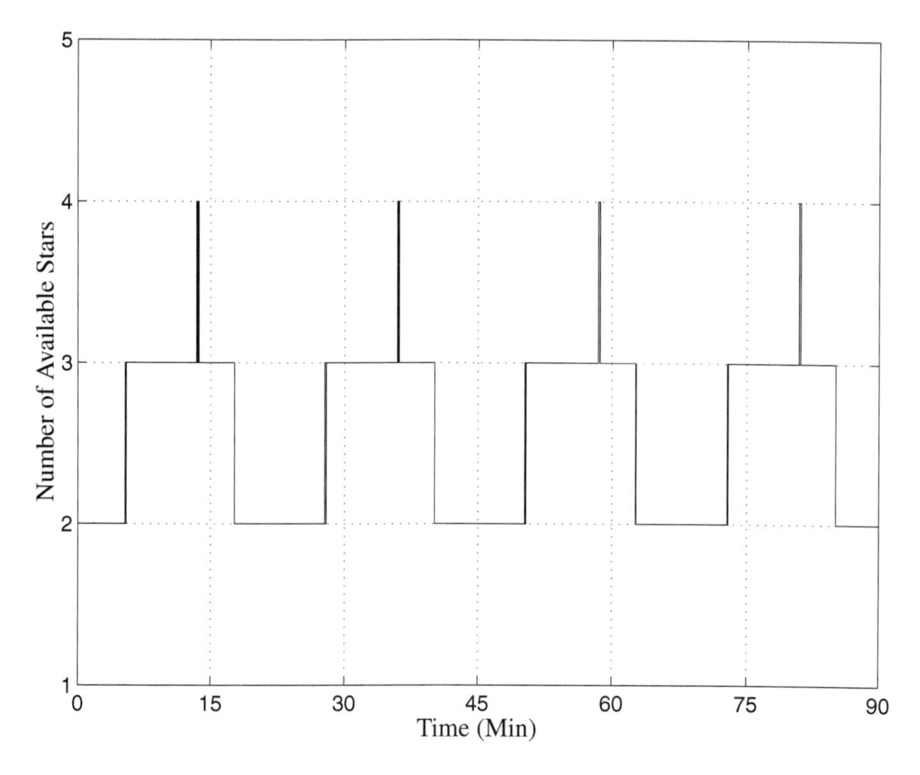

Figure 4.4: Availability of Stars

Thus, in order to maximize J the optimal quaternion $\hat{\mathbf{q}}$ is given by the eigenvector corresponding to the largest eigenvalue of K. It can be shown that if at least two non-collinear observation vectors exist, then the eigenvalues of K are distinct, which yields an unambiguous quaternion. Shuster[10] developed an algorithm, called QUEST (QUaternion ESTimator), that computes that quaternion without the necessity of performing an eigenvalue decomposition, which gives a very computationally efficient algorithm. This algorithm is widely used for many on-board spacecraft applications. Yet another efficient algorithm, developed by Mortari, called Estimator of Optimal Quaternion (ESOQ) is given in Ref. [11]. Also, Markley[12] develops an algorithm, using a singular value decomposition (SVD) approach, that determines the attitude matrix A directly.

Example 4.2: In this example a simulation using a typical star camera is used to determine the attitude of a rotating spacecraft. The star camera can sense up to 10 stars in a $6° \times 6°$ field-of-view. The catalog contains stars that can be sensed up to a magnitude of 5.0 (larger magnitudes indicate dimmer stars). The star camera's boresight is assumed to be along the z-axis pointed in the anti-nadir direction, and is initially aligned with the $\hat{\mathbf{r}}_1$ vector of the inertial reference frame shown in Figure

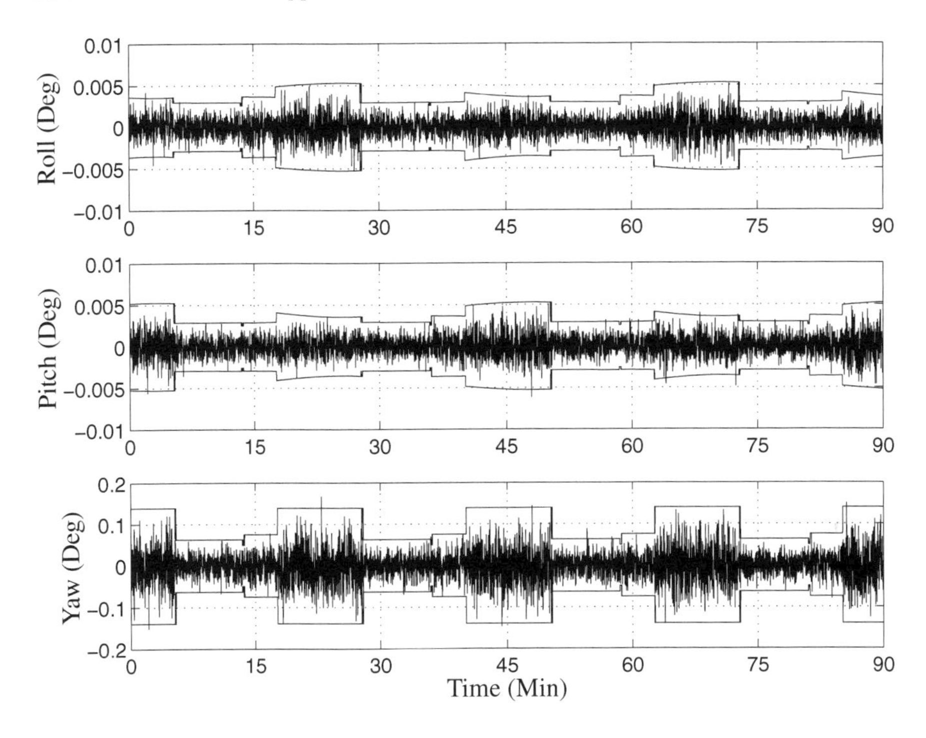

Figure 4.5: Attitude Errors and Boundaries

4.2. A rotation about the $\hat{\mathbf{r}}_3$ vector only is assumed and the spacecraft is in a 90-minute orbit (i.e., low Earth orbit). Star images are taken at 1-second intervals. A plot of the number of available stars over the full 360 degree rotation of the orbit is shown in Figure 4.4. The minimum number of available stars is two, which is also the minimum number required for attitude determination. In general, as the number of available stars decreases, the attitude accuracy degrades (although this is also dependent on the angle separation between stars). Generally, three or four stars are required for the first image, in order to reliably identify star patterns, associating each measured vector with the corresponding cataloged vector.

The star camera body observations are obtained by using eqn. (4.9), with an assumed focal length of 42.98 mm. Simulated measurements are derived using a zero-mean Gaussian noise process, which are added to the true values of x_j and y_j in eqn. (4.7):

$$\tilde{x}_j = x_j + v_{x_j}$$
$$\tilde{y}_j = y_j + v_{y_j}$$

where (v_{x_j}, v_{y_j}) are uncorrelated zero-mean Gaussian random variables each with a 3σ value of 0.005 degrees. We also assume that no sun obtrusions are present (although this is not truly realistic). At each time instant all available inertial star vectors and body measurements are used to form the K matrix in eqn. (4.24). Then, the quaternion estimate is found using eqn. (4.26). Furthermore, the attitude error-covariance is computed using eqn. (4.18), and the diagonal elements of this matrix are used to form 3σ boundaries on the attitude errors. A plot of the attitude errors and associated 3σ boundaries is shown in Figure 4.5. Clearly, the computed 3σ boundaries do indeed bound the attitude errors. Note that the yaw errors are much larger than the roll and pitch errors. This is due to the fact that the boresight of the star camera is along this yaw rotation axis. Also, as expected, the accuracy degrades as the number of available stars decreases, which is also illustrated in the covariance matrix. This covariance analysis provides valuable information to assess the expected performance of the attitude determination process (which can be calculated without any attitude knowledge!). In Chapter 7, we shall see how the accuracy can be significantly improved using rate gyroscope measurements in a Kalman filter.

4.2.4 Information Matrix Analysis

In this section an analysis of the observable attitude axes using the information matrix is shown. This analysis is shown for one and two vector observations. For one-vector observation the information matrix, which is the inverse of eqn. (4.18), is given by

$$F = -\sigma^{-2}[\mathbf{b}\times]^2 \tag{4.28}$$

where $\mathbf{b} \equiv A\,\mathbf{r}$. An eigenvalue/eigenvector decomposition can be useful to assess the observability of this system. Since F is a symmetric positive semi-definite matrix, then all of its eigenvalues are greater than or equal to zero (see Appendix A). Furthermore, the matrix of eigenvectors is orthogonal, which can be used to define a coordinate system. The eigenvalues of this matrix are given by $\lambda_1 = 0$ and $\lambda_{2,3} = \sigma^{-2}\mathbf{b}^T\mathbf{b}$. This indicates that rotations about one of the eigenvectors is not observable. The eigenvector associated with the zero eigenvalue is along $\mathbf{b}/||\mathbf{b}||$. Therefore, rotations about the boresight of the body vector are unknown, which intuitively makes sense. The other observable axes are perpendicular to this unobservable axis, which also intuitively makes sense.

A more interesting case involves two vector observations. The information matrix for this case is given by

$$F = -\sigma_1^{-2}[\mathbf{b}_1\times]^2 - \sigma_2^{-2}[\mathbf{b}_2\times]^2 \tag{4.29}$$

where $\mathbf{b}_1 \equiv A\,\mathbf{r}_1$ and $\mathbf{b}_2 \equiv A\,\mathbf{r}_2$. For any vector, $\mathbf{a}$, the following identity is true: $-[\mathbf{a}\times]^2 = (\mathbf{a}^T\mathbf{a})I_{3\times 3} - \mathbf{a}\,\mathbf{a}^T$. Using this identity simplifies eqn. (4.29) to

$$F = \sigma_1^{-2}\left[(\mathbf{b}_1^T\mathbf{b}_1)I_{3\times 3} - \mathbf{b}_1\mathbf{b}_1^T\right] + \sigma_2^{-2}\left[(\mathbf{b}_2^T\mathbf{b}_2)I_{3\times 3} - \mathbf{b}_2\mathbf{b}_2^T\right] \tag{4.30}$$

If two non-collinear vector observations exist, then the system is fully observable and no zero eigenvalues of F will exist. The maximum eigenvalue of F can be shown to

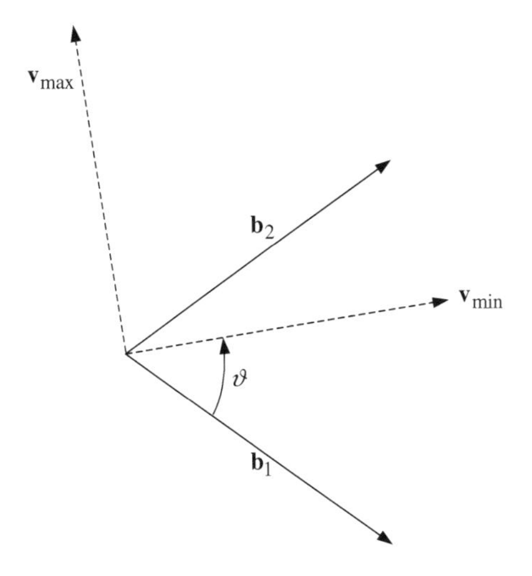

Figure 4.6: Observable Axes with Two Vector Observations

be given by

$$\lambda_{max} = \sigma_1^{-2}\mathbf{b}_1^T\mathbf{b}_1 + \sigma_2^{-2}\mathbf{b}_2^T\mathbf{b}_2 \tag{4.31}$$

Factoring this eigenvalue out of the characteristic equation, $|\lambda I_{3\times 3} - F|$, yields the following form for the remaining eigenvalues:

$$\lambda^2 - \lambda_{max}\lambda + \sigma_1^{-2}\sigma_2^{-2}||\mathbf{b}_1 \times \mathbf{b}_2||^2 = 0 \tag{4.32}$$

Therefore, the intermediate and minimum eigenvalues are given by

$$\lambda_{int} = \frac{\lambda_{max}(1+\chi)}{2} \tag{4.33a}$$

$$\lambda_{min} = \frac{\lambda_{max}(1-\chi)}{2} \tag{4.33b}$$

where

$$\chi = \left[\frac{\lambda_{max}^2 - 4\sigma_1^{-2}\sigma_2^{-2}||\mathbf{b}_1 \times \mathbf{b}_2||^2}{\lambda_{max}^2}\right]^{1/2} \tag{4.34}$$

Note that $\lambda_{max} = \lambda_{min} + \lambda_{int}$.

The eigenvectors of F are computed by solving $\lambda\mathbf{v} = F\mathbf{v}$ for each eigenvalue. The eigenvector associated with the maximum eigenvalue can be shown to be given by

$$\mathbf{v}_{max} = \pm\frac{\mathbf{b}_1 \times \mathbf{b}_2}{||\mathbf{b}_1 \times \mathbf{b}_2||} \tag{4.35}$$

The sign of this vector is not of consequence since we are only interested in rotations about this vector. This indicates that the most observable axis is perpendicular to the plane formed by $\mathbf{b}_1$ and $\mathbf{b}_2$, which intuitively makes sense. The remaining eigenvectors must surely lie in the $\mathbf{b}_1$-$\mathbf{b}_2$ plane. To determine the eigenvector associated with the minimum eigenvalue, we will perform a rotation about the $\mathbf{v}_{max}$ axis and determine the angle from $\mathbf{b}_1$. Using the Euler axis and angle parameterization in eqn. (3.151) gives

$$\mathbf{v}_{min} = \pm\left\{(\cos\vartheta)I_{3\times 3} + (1-\cos\vartheta)\mathbf{v}_{max}\mathbf{v}_{max}^T - \sin\vartheta[\mathbf{v}_{max}\times]\right\}\frac{\mathbf{b}_1}{||\mathbf{b}_1||} \tag{4.36}$$

where ϑ is the angle used to rotate $\mathbf{b}_1/||\mathbf{b}_1||$ to $\mathbf{v}_{min}$. Using the fact that $\mathbf{v}_{max}$ is perpendicular to $\mathbf{b}_1$ gives $\mathbf{v}_{max}^T\mathbf{b}_1 = 0$. Therefore, eqn. (4.36) reduces down to

$$\mathbf{v}_{min} = \pm\{(\cos\vartheta)I_{3\times 3} - \sin\vartheta[\mathbf{v}_{max}\times]\}\frac{\mathbf{b}_1}{||\mathbf{b}_1||} \tag{4.37}$$

Substituting eqn. (4.37) into $\lambda_{min}\mathbf{v}_{min} = F\mathbf{v}_{min}$ and using the property of the cross product matrix leads to the following equation for ϑ:

$$\tan\vartheta = \frac{a+b}{c} \tag{4.38}$$

where

$$a \equiv \lambda_{min}\sigma_1^{-2}\mathbf{b}_1^T\mathbf{b}_1 \tag{4.39a}$$

$$b \equiv \sigma_1^{-2}\sigma_2^{-2}\mathbf{b}_1^T[\mathbf{b}_2\times]^2\mathbf{b}_1 \tag{4.39b}$$

$$c \equiv -\frac{\sigma_1^{-2}\sigma_2^{-2}\mathbf{b}_1^T[\mathbf{b}_2\times]^2[\mathbf{b}_1\times]^2\mathbf{b}_2}{||\mathbf{b}_1\times\mathbf{b}_2||} \tag{4.39c}$$

Equation (4.38) can now be solved for ϑ, which can be used to determine $\mathbf{v}_{min}$ from eqns. (4.35) and (4.37). The intermediate axis is simply given by the cross product of $\mathbf{v}_{max}$ and $\mathbf{v}_{min}$:

$$\mathbf{v}_{int} = \pm\mathbf{v}_{max}\times\mathbf{v}_{min} \tag{4.40}$$

A plot of the minimum and intermediate axes is shown in Figure 4.6 for the case when the angle between $\mathbf{b}_1$ and $\mathbf{b}_2$ is less than 90 degrees. Intuitively, this analysis makes sense since we expect that the least determined axis, $\mathbf{v}_{min}$, is somewhere *between* $\mathbf{b}_1$ and $\mathbf{b}_2$ if these vector observations are less than 90 degrees apart.

The previous analysis greatly simplifies if the reference vectors are unit vectors and the variances of each observation are equal, so that $\sigma_1^2 = \sigma_2^2 \equiv \sigma^2$. These assumptions are valid for a single field-of-view star camera. The eigenvalues are now given by

$$\lambda_{max} = 2\sigma^{-2} \tag{4.41a}$$

$$\lambda_{int} = \sigma^{-2}(1+|\mathbf{b}_1^T\mathbf{b}_2|) \tag{4.41b}$$

$$\lambda_{min} = \sigma^{-2}(1-|\mathbf{b}_1^T\mathbf{b}_2|) \tag{4.41c}$$

The eigenvectors are now given by

$$\mathbf{v}_{max} = \pm \frac{\mathbf{b}_1 \times \mathbf{b}_2}{||\mathbf{b}_1 \times \mathbf{b}_2||} \tag{4.42a}$$

$$\mathbf{v}_{int} = \pm \frac{\mathbf{b}_1 - \text{sign}(\mathbf{b}_1^T \mathbf{b}_2)\mathbf{b}_2}{||\mathbf{b}_1 - \text{sign}(\mathbf{b}_1^T \mathbf{b}_2)\mathbf{b}_2||} \tag{4.42b}$$

$$\mathbf{v}_{min} = \pm \frac{\mathbf{b}_1 + \text{sign}(\mathbf{b}_1^T \mathbf{b}_2)\mathbf{b}_2}{||\mathbf{b}_1 + \text{sign}(\mathbf{b}_1^T \mathbf{b}_2)\mathbf{b}_2||} \tag{4.42c}$$

where $\text{sign}(\mathbf{b}_1^T \mathbf{b}_2)$ is used to ensure that the proper direction of the eigenvectors is determined when the angle between $\mathbf{b}_1$ and $\mathbf{b}_2$ is greater than 90 degrees. If this angle is less than 90 degrees then $\mathbf{v}_{min}$ is the *bisector* of $\mathbf{b}_1$ and $\mathbf{b}_2$. Intuitively this makes sense since we expect rotations perpendicular to the bisector of the two vector observations to be more observable than rotations about the bisector (again assuming that the vector observations are within 90 degrees of each other).

The analysis presented in this section is extremely useful for the visualization of the observability of the determined attitude. Closed-form solutions for special cases have been presented here. Still, in general, the eigenvalues and eigenvectors of the information matrix can be used to analyze the observability for cases involving multiple observations. An analytical observability analysis for a more complicated system is shown in Ref. [13].

4.3 Orbit Determination

In this section nonlinear least squares is used to determine the orbit of a spacecraft from range and line-of-sight (angle) observations. It is interesting to note that the original estimation problem motivating Gauss (i.e., determination of the planetary orbits from telescope/sextant observations) was nonlinear, and his methods (essentially §1.2) have survived as a standard operating procedure to this day.

Consider an observer (i.e., a radar site) that measures a range, azimuth, and elevation to a spacecraft in orbit. The geometry and common terminology associated with this observation is shown in Figure 4.7, where: ρ is the slant range, $\mathbf{r}$ is the radius vector locating the spacecraft, $\mathbf{R}$ is the radius vector locating the observer, α and δ is the right ascension and declination of the spacecraft, respectively, θ is the sidereal time of the observer, λ is the latitude of the observer, and ϕ is the east longitude from the observer to the spacecraft. The fundamental observation is given by

$$\boldsymbol{\rho} = \mathbf{r} - \mathbf{R} \tag{4.43}$$

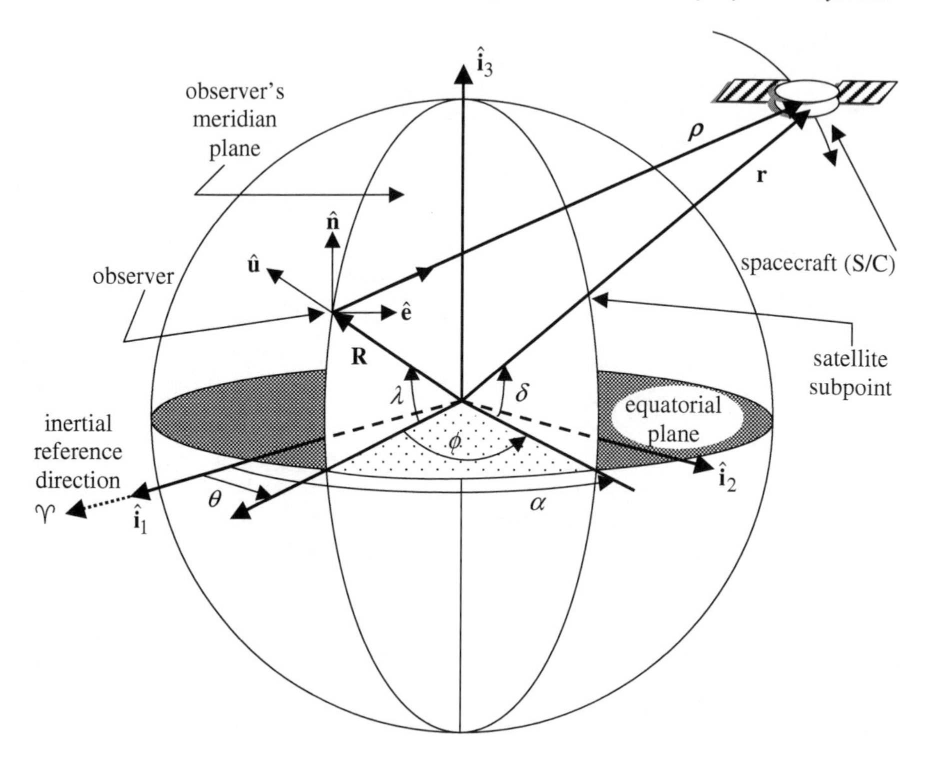

Figure 4.7: Geometry of Earth Observations of Spacecraft Motion

In non-rotating equatorial (inertial) components the vector $\boldsymbol{\rho}$ is given by

$$\boldsymbol{\rho} = \begin{bmatrix} x - ||\mathbf{R}|| \cos\lambda \cos\theta \\ y - ||\mathbf{R}|| \cos\lambda \sin\theta \\ z - ||\mathbf{R}|| \sin\lambda \end{bmatrix} \tag{4.44}$$

where x, y, and z are the components of the vector $\mathbf{r}$. The conversion from the inertial to the observer coordinate system ("up, east and north") is given by

$$\begin{bmatrix} \rho_u \\ \rho_e \\ \rho_n \end{bmatrix} = \begin{bmatrix} \cos\lambda & 0 & \sin\lambda \\ 0 & 1 & 0 \\ -\sin\lambda & 0 & \cos\lambda \end{bmatrix} \begin{bmatrix} \cos\theta & \sin\theta & 0 \\ -\sin\theta & \cos\theta & 0 \\ 0 & 0 & 1 \end{bmatrix} \boldsymbol{\rho} \tag{4.45}$$

Next, consider a radar site that measures the azimuth, az, elevation, el, and range, ρ. The observation equations are given by

$$||\boldsymbol{\rho}|| = (\rho_u^2 + \rho_e^2 + \rho_n^2)^{1/2} \tag{4.46a}$$

$$\text{az} = \tan^{-1}\left(\frac{\rho_e}{\rho_n}\right) \tag{4.46b}$$

$$\text{el} = \sin^{-1}\left(\frac{\rho_u}{||\boldsymbol{\rho}||}\right) \tag{4.46c}$$

The basic two-body orbital equation of motion is given by (see §3.8.2)

$$\ddot{\mathbf{r}} = -\frac{\mu}{||\mathbf{r}||^3}\mathbf{r} \tag{4.47}$$

The goal of orbit determination is to determine initial conditions for the position and velocity of $\mathbf{x}_0 = \begin{bmatrix}\mathbf{r}_0^T & \dot{\mathbf{r}}_0^T\end{bmatrix}^T$ from the observations. The nonlinear least square differential correction algorithm for orbit determination is shown in Figure 4.8. The model equation is given by eqn. (4.47) with $\mathbf{x} = \begin{bmatrix}\mathbf{r}^T & \dot{\mathbf{r}}^T\end{bmatrix}^T$, and also includes other parameters if desired, given by $\mathbf{p}$ (e.g., the parameter μ can also be determined if desired). The measurement equation is given by eqn. (4.46) with $\mathbf{y} = \begin{bmatrix}||\boldsymbol{\rho}|| & \text{az} & \text{el}\end{bmatrix}^T$. Other quantities, such as measurement biases or force model parameters, can be appended to the measurement observation equation through the vector $\mathbf{b}$. The matrices $\Phi(t,t_0)$, $\Psi(t,t_0)$, F, and G are defined as

$$\Phi(t,t_0) \equiv \frac{\partial \mathbf{x}(t)}{\partial \mathbf{x}_0}, \quad \Psi(t,t_0) \equiv \frac{\partial \mathbf{x}(t)}{\partial \mathbf{p}} \tag{4.48a}$$

$$F \equiv \frac{\partial \mathbf{f}}{\partial \mathbf{x}}, \quad G \equiv \frac{\partial \mathbf{f}}{\partial \mathbf{p}} \tag{4.48b}$$

which are evaluated at the current estimates. The matrix H is computed using

$$H = \begin{bmatrix} \frac{\partial \mathbf{h}}{\partial \mathbf{x}}\Phi(t,t_0) & \frac{\partial \mathbf{h}}{\partial \mathbf{x}}\Psi(t,t_0) & \frac{\partial \mathbf{h}}{\partial \mathbf{b}} \end{bmatrix} \tag{4.49}$$

which are again evaluated at the current estimates. Analytical expressions for $\Psi(t,t_0)$, F, and G are straightforward. The matrix F is given by

$$F = \begin{bmatrix} 0_{3\times3} & I_{3\times3} \\ F_{21} & 0_{3\times3} \end{bmatrix} \tag{4.50}$$

where

$$F_{21} = \begin{bmatrix} \frac{3\mu x^2}{||\mathbf{r}||^5} - \frac{\mu}{||\mathbf{r}||^3} & \frac{3\mu xy}{||\mathbf{r}||^5} & \frac{3\mu xz}{||\mathbf{r}||^5} \\ \frac{3\mu xy}{||\mathbf{r}||^5} & \frac{3\mu y^2}{||\mathbf{r}||^5} - \frac{\mu}{||\mathbf{r}||^3} & \frac{3\mu yz}{||\mathbf{r}||^5} \\ \frac{3\mu xz}{||\mathbf{r}||^5} & \frac{3\mu yz}{||\mathbf{r}||^5} & \frac{3\mu z^2}{||\mathbf{r}||^5} - \frac{\mu}{||\mathbf{r}||^3} \end{bmatrix} \tag{4.51}$$

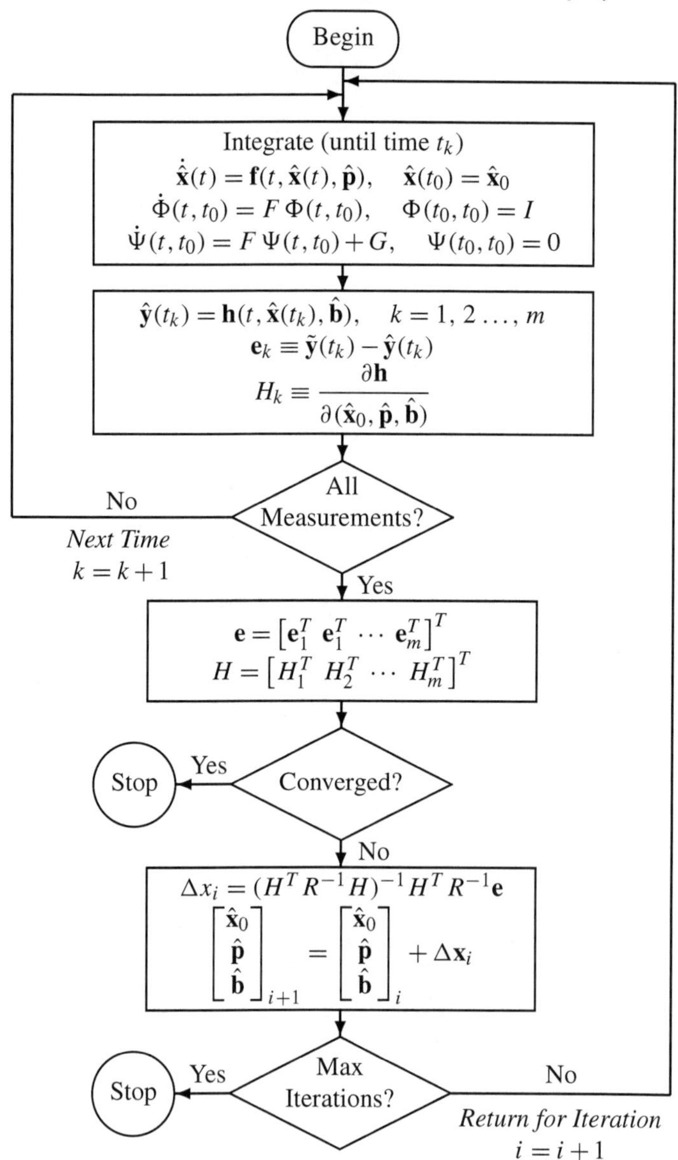

Figure 4.8: Least Squares Orbit Determination

For the general case of velocity dependent forces (such as drag), the lower right partition of eqn. (4.50) is nonzero. Analytical expressions for $\Phi(t, t_0)$ can be found in Refs. [14] and [15]. The "brute force" approach to determination of $\Phi(t, t_0)$ would be to attempt formal analytical or numerical solutions of the differential equation (3.88). However, we can make efficient use of the fact that the analytical solution is available for $\mathbf{x}(t)$, for Keplerian motion, (see §3.8.2) to determine the desired solution for $\Phi(t,t_0)$ by partial differentiation of the equations. The appropriate equations for the partials are given by[14]

$$\Phi(t,t_0) = \begin{bmatrix} \Phi_{11} & \Phi_{12} \\ \Phi_{21} & \Phi_{22} \end{bmatrix} \tag{4.52}$$

where

$$\Phi_{11} = \frac{||\mathbf{r}||}{\mu}(\dot{\mathbf{r}}-\dot{\mathbf{r}}_0)(\dot{\mathbf{r}}-\dot{\mathbf{r}}_0)^T + ||\mathbf{r}_0||^{-3}[||\mathbf{r}_0||(1-f)\mathbf{r}\mathbf{r}_0^T + c\dot{\mathbf{r}}\mathbf{r}_0^T] + f I_{3\times 3} \tag{4.53a}$$

$$\Phi_{12} = \frac{||\mathbf{r}_0||}{\mu}(1-f)[(\mathbf{r}-\mathbf{r}_0)\dot{\mathbf{r}}_0^T - (\dot{\mathbf{r}}-\dot{\mathbf{r}}_0)\mathbf{r}_0^T] + \frac{c}{\mu}\dot{\mathbf{r}}\dot{\mathbf{r}}_0^T + g I_{3\times 3} \tag{4.53b}$$

$$\begin{aligned}\Phi_{21} = &-||\mathbf{r}_0||^{-2}(\dot{\mathbf{r}}-\dot{\mathbf{r}}_0)\mathbf{r}_0^T - ||\mathbf{r}||^{-2}\mathbf{r}(\dot{\mathbf{r}}-\dot{\mathbf{r}}_0)^T - \frac{\mu c}{||\mathbf{r}||^3||\mathbf{r}_0||^3}\mathbf{r}\mathbf{r}_0^T \\ &+ \dot{f}\left[I_{3\times 3} - ||\mathbf{r}_0||^{-2}\mathbf{r}\mathbf{r}^T + \frac{1}{\mu||\mathbf{r}||}(\mathbf{r}\dot{\mathbf{r}}^T - \dot{\mathbf{r}}\mathbf{r}^T)\mathbf{r}(\dot{\mathbf{r}}-\dot{\mathbf{r}}_0)^T\right]\end{aligned} \tag{4.53c}$$

$$\Phi_{22} = \frac{||\mathbf{r}_0||}{\mu}(\dot{\mathbf{r}}-\dot{\mathbf{r}}_0)(\dot{\mathbf{r}}-\dot{\mathbf{r}}_0)^T + ||\mathbf{r}_0||^{-3}[||\mathbf{r}_0||(1-f)\mathbf{r}\mathbf{r}_0^T - c\mathbf{r}\dot{\mathbf{r}}_0^T] + \dot{g} I_{3\times 3} \tag{4.53d}$$

The variables f, g, $\dot{f}$, and $\dot{g}$ are given in eqn. (3.201). The symbol c is defined by

$$c = (3u_5 - \chi u_4 - \sqrt{\mu}(t-t_0)u_2)/\sqrt{\mu} \tag{4.54}$$

where χ is a *generalized anomaly* given by

$$\chi = \alpha\sqrt{\mu}(t-t_0) + \frac{\mathbf{r}^T\dot{\mathbf{r}}}{\sqrt{\mu}} - \frac{\mathbf{r}_0^T\dot{\mathbf{r}}_0}{\sqrt{\mu}} \tag{4.55}$$

where $\alpha = 1/a$, which is given by eqn. (3.199), and the *universal functions* for elliptic orbits are given by

$$u_2 = \frac{1-\cos(\sqrt{\alpha}\chi)}{\alpha} \tag{4.56a}$$

$$u_3 = \frac{\sqrt{\alpha}\chi - \sin(\sqrt{\alpha}\chi)}{\alpha\sqrt{\alpha}} \tag{4.56b}$$

$$u_4 = \frac{\chi^2}{2\alpha} - \frac{u_2}{\alpha} \tag{4.56c}$$

$$u_5 = \frac{\chi^3}{6\alpha} - \frac{u_3}{\alpha} \tag{4.56d}$$

Several interesting properties of the universal variables and functions $u_i(\alpha,\chi)$ can be found in Ref. [14], including universal algorithms to compute these functions for all species of two-body orbits. The partials for the observation, which are used to form $\partial \mathbf{h}/\partial \mathbf{x}$, are given by

$$\frac{\partial ||\boldsymbol{\rho}||}{\partial x} = (\rho_u \cos\lambda \cos\theta - \rho_e \sin\theta - \rho_n \sin\lambda \cos\theta)/||\boldsymbol{\rho}|| \tag{4.57a}$$

$$\frac{\partial ||\boldsymbol{\rho}||}{\partial y} = (\rho_u \cos\lambda \sin\theta + \rho_e \cos\theta - \rho_n \sin\lambda \sin\theta)/||\boldsymbol{\rho}|| \tag{4.57b}$$

$$\frac{\partial ||\boldsymbol{\rho}||}{\partial z} = (\rho_u \sin\lambda + \rho_n \cos\lambda)/||\boldsymbol{\rho}|| \tag{4.57c}$$

$$\frac{\partial \text{az}}{\partial x} = \frac{1}{(\rho_n^2+\rho_e^2)}(\rho_e \sin\lambda \cos\theta - \rho_n \sin\theta) \tag{4.58a}$$

$$\frac{\partial \text{az}}{\partial y} = \frac{1}{(\rho_n^2+\rho_e^2)}(\rho_e \sin\lambda \sin\theta + \rho_n \cos\theta) \tag{4.58b}$$

$$\frac{\partial \text{az}}{\partial z} = -\frac{1}{(\rho_n^2+\rho_e^2)}\rho_e \cos\lambda \tag{4.58c}$$

$$\frac{\partial \text{el}}{\partial x} = \frac{1}{||\boldsymbol{\rho}||(||\boldsymbol{\rho}||^2-\rho_u^2)^{1/2}}\left(||\boldsymbol{\rho}|| \cos\lambda \cos\theta - \rho_u \frac{\partial ||\boldsymbol{\rho}||}{\partial x}\right) \tag{4.59a}$$

$$\frac{\partial \text{el}}{\partial y} = \frac{1}{||\boldsymbol{\rho}||(||\boldsymbol{\rho}||^2-\rho_u^2)^{1/2}}\left(||\boldsymbol{\rho}|| \cos\lambda \sin\theta - \rho_u \frac{\partial ||\boldsymbol{\rho}||}{\partial y}\right) \tag{4.59b}$$

$$\frac{\partial \text{el}}{\partial z} = \frac{1}{||\boldsymbol{\rho}||(||\boldsymbol{\rho}||^2-\rho_u^2)^{1/2}}\left(||\boldsymbol{\rho}|| \sin\lambda - \rho_u \frac{\partial ||\boldsymbol{\rho}||}{\partial z}\right) \tag{4.59c}$$

The matrix $\partial \mathbf{h}/\partial \mathbf{x}$ is given by

$$\frac{\partial \mathbf{h}}{\partial \mathbf{x}} = \left[H_{11} \; 0_{3\times 3}\right] \tag{4.60}$$

where

$$H_{11} = \begin{bmatrix} \frac{\partial ||\boldsymbol{\rho}||}{\partial x} & \frac{\partial ||\boldsymbol{\rho}||}{\partial y} & \frac{\partial ||\boldsymbol{\rho}||}{\partial z} \\ \frac{\partial \text{az}}{\partial x} & \frac{\partial \text{az}}{\partial y} & \frac{\partial \text{az}}{\partial z} \\ \frac{\partial \text{el}}{\partial x} & \frac{\partial \text{el}}{\partial y} & \frac{\partial \text{el}}{\partial z} \end{bmatrix} \tag{4.61}$$

The least square differential correction process for orbit determination is as follows: integrate the equations of motion and partial derivatives until the observation time (t_k); next, compute the measurement residual $\mathbf{e}_k$ and observation partial equation; if all measurements are processed then proceed, otherwise continue to the next observation time; then, check convergence and stop if the convergence criterion is satisfied; otherwise, compute an updated correction and stop if the maximum number of iterations is given; continue the iteration process until a solution for the desired parameters is found.

Determining an initial estimate for the position and velocity is important to help achieve convergence (especially in the least squares approach). Several approaches exist for state determination from various sensor measurements (e.g., see Refs. [15] and [16]). We will show a popular approximate approach to determine the orbit given three observations of the range, azimuth, and elevation ($||\boldsymbol{\rho}||_k$, az_k, el_k, $k = 1, 2, 3$). Since $||\mathbf{R}||$, λ, and θ_k are known, then $\mathbf{R}_k$ can easily be computed by

$$\mathbf{R}_k = ||\mathbf{R}|| \begin{bmatrix} \cos\lambda\cos\theta_k \\ \cos\lambda\sin\theta_k \\ \sin\lambda \end{bmatrix} \quad k = 1, 2, 3 \tag{4.62}$$

Next compute

$$\boldsymbol{\rho}_k = \begin{bmatrix} \rho_u \\ \rho_e \\ \rho_n \end{bmatrix} = ||\boldsymbol{\rho}||_k \begin{bmatrix} \sin \text{el}_k \\ \cos \text{el}_k \sin \text{az}_k \\ \cos \text{el}_k \cos \text{az}_k \end{bmatrix} \quad k = 1, 2, 3 \tag{4.63}$$

The position is simply given by

$$\mathbf{r}_k = \begin{bmatrix} \cos\theta_k & -\sin\theta_k & 0 \\ \sin\theta_k & \cos\theta_k & 0 \\ 0 & 0 & 1 \end{bmatrix} \begin{bmatrix} \cos\lambda & 0 & -\sin\lambda \\ 0 & 1 & 0 \\ \sin\lambda & 0 & \cos\lambda \end{bmatrix} \boldsymbol{\rho}_k + \mathbf{R}_k \quad k = 1, 2, 3 \tag{4.64}$$

The velocity at second observation ($\dot{\mathbf{r}}_2$) can be determined from the three position vectors determined from eqn. (4.64). This is accomplished using a Taylor series expansion for the derivative. First, the following variables are computed:

$$\tau_{ij} = c\,(t_j - t_i) \tag{4.65a}$$

$$g_1 = \frac{\tau_{23}}{\tau_{12}\tau_{13}}, \quad g_3 = \frac{\tau_{12}}{\tau_{23}\tau_{13}}, \quad g_2 = g_1 - g_3 \tag{4.65b}$$

$$h_1 = \frac{\mu\tau_{23}}{12}, \quad h_3 = \frac{\mu\tau_{12}}{12}, \quad h_2 = h_1 - h_2 \tag{4.65c}$$

$$d_k = g_k + \frac{h_k}{||\mathbf{r}_k||^3}, \quad k = 1, 2, 3 \tag{4.65d}$$

where t_i and t_j are epoch times for $\mathbf{r}_i$ and $\mathbf{r}_j$, respectively, and $c = 1$, typically. The velocity is then given by[15]

$$\boxed{\dot{\mathbf{r}}_2 = -d_1\,\mathbf{r}_1 + d_2\,\mathbf{r}_2 + d_3\,\mathbf{r}_3} \tag{4.66}$$

This is known as the "Herrick-Gibbs" technique. The velocity is determined to within the order of $[(d^5||\mathbf{r}||/dt^5)/5!]\tau_{ij}^5$, which gives good results over short observation intervals. Typically, errors of a few kilometers in position and a few kilometers per second in velocity, for near Earth orbits, result in reliable convergence.

Example 4.3: In this example the least squares differential correction algorithm is used to determine the orbit of a spacecraft from range, azimuth, and elevation measurements. The true spacecraft position and velocity at epoch are given by

$$\mathbf{r}_0 = \begin{bmatrix} 7,000 & 1,000 & 200 \end{bmatrix}^T \text{ km}$$
$$\dot{\mathbf{r}}_0 = \begin{bmatrix} 4 & 7 & 2 \end{bmatrix}^T \text{ km/sec}$$

The latitude of the observer is given by $\lambda = 5°$, and the initial sidereal time is given by $\theta_0 = 10°$. Measurements are given at 10-second intervals over a 100-second simulation. The measurement errors are zero-mean Gaussian with a standard deviation of the range measurement error given by $\sigma_\rho = 1$ km, and a standard deviation of the angle measurements given by $\sigma_{\text{az}} = \sigma_{\text{el}} = 0.01°$. An initial estimate of the orbit parameters at the second time step is given by Herrick-Gibbs approach. The approximate results for position and velocity are given by

$$\hat{\mathbf{r}} = \begin{bmatrix} 7,038 & 1,070 & 221 \end{bmatrix}^T \text{ km}$$
$$\dot{\mathbf{r}} = \begin{bmatrix} 3.92 & 7.00 & 2.00 \end{bmatrix}^T \text{ km/sec}$$

The true position and velocity at the second time step are given by

$$\mathbf{r} = \begin{bmatrix} 7,040 & 1,070 & 220 \end{bmatrix}^T \text{ km}$$
$$\dot{\mathbf{r}} = \begin{bmatrix} 3.92 & 7.00 & 2.00 \end{bmatrix}^T \text{ km/sec}$$

which are in close agreement with the initial estimates. In order to assess the performance of the least squares differential correction algorithm the initial guesses for the position and velocity are given by $\hat{\mathbf{r}}_0 = \begin{bmatrix} 6,990 & 1 & 1 \end{bmatrix}^T$ km, and $\hat{\dot{\mathbf{r}}}_0 = \begin{bmatrix} 1 & 1 & 1 \end{bmatrix}^T$ km/sec. Results for the least square iterations are given in Table 4.2. The algorithm converges after seven iterations, and does well for large initial condition errors (the Levenberg-Marquardt method of §1.6.3 may also be employed if needed). The 3σ bounds (determined using the diagonal elements of the estimate error-covariance) for position are $3\boldsymbol{\sigma}_{\hat{\mathbf{r}}} = \begin{bmatrix} 1.26 & 0.25 & 0.51 \end{bmatrix}^T$ km, and for velocity are $3\boldsymbol{\sigma}_{\hat{\dot{\mathbf{r}}}} = \begin{bmatrix} 0.020 & 0.008 & 0.006 \end{bmatrix}^T$ km/sec. The bounds are useful to predict the performance of the algorithms.

A powerful technology for precise orbit determination is GPS. Differential GPS provides extremely accurate orbit estimates. The accuracy of GPS derived estimates ultimately depends on the orbit of the spacecraft and the geometry of the available GPS satellite in view of the spacecraft. More details on orbit determination using GPS can be found in Ref. [17].

Table 4.2: Least Squares Iterations for Orbit Determination

Iteration	Position (km)			Velocity (km/sec)		
0	6,990	1	1	1	1	1
1	7,496	1,329	-178	5.30	6.20	-18.42
2	7,183	609	27	12.66	22.63	12.69
3	6,842	905	490	6.65	13.73	-8.15
4	6,795	963	255	9.33	7.38	1.36
5	6,985	989	199	4.24	7.20	1.89
6	7,000	1,000	200	4.00	7.00	2.00
7	7,000	1,000	200	4.00	7.00	2.00

4.4 Aircraft Parameter Identification

For aircraft dynamics, parameter identification of unknown aerodynamic coefficients or stability and control derivatives is useful to quantify the performance of a particular aircraft using dynamic models introduced in §3.9. These models are often used to design control systems to provide increased maneuverability and for use in the design of automated unpiloted vehicles. In general, these coefficients are usually first determined using wind tunnel applications, and, as a newer approach, using computational fluid dynamics. Parameter identification using flight measurement data is useful to provide a final verification of these coefficients, and also update models for other applications such as adaptive control algorithms. This section introduces the basic concepts which incorporate estimation principles for aircraft parameter identification from flight data. For the interested reader, a more detailed discussion is given in Ref. [18].

Application of identification methods for aircraft coefficients dates back to the early 1920s, which involved basic detection of damping ratios and frequencies. In the 1940s and early 1950s these coefficients were fitted to frequency response data (magnitude and phase). Around the same time, linear least squares was applied using flight data, but gave poor results in the presence of measurement noise and gave biased estimates. Other methods, such as time vector techniques and analog matching methods, are described in Ref. [18]. The most popular approaches today for aircraft coefficient identification are based on maximum likelihood techniques as introduced in §2.3. The desirable attributes of these techniques, such as asymptotically unbiased and consistent estimates, are especially useful for the estimation of aircraft coefficients in the presence of measurement errors associated with flight data.

The aircraft equations of motion, derived in §3.9, can be written in continuous-discrete form as

$$\dot{\mathbf{x}} = \mathbf{f}(t, \mathbf{x}, \mathbf{p}) \tag{4.67a}$$

$$\tilde{\mathbf{y}}_k = \mathbf{h}(t_k, \mathbf{x}_k) + \mathbf{v}_k \tag{4.67b}$$

where $\mathbf{x}$ is the $n \times 1$ state vector (e.g., angle of attack, pitch angle, body rates, etc.), $\mathbf{p}$ is the $q \times 1$ vector of aircraft coefficients to be determined, $\mathbf{y}$ is the $m \times 1$ measurement vector, and $\mathbf{v}$ is the $m \times 1$ measurement-error vector which is assumed to be represented by a zero-mean Gaussian noise process with covariance R. Note that there is no noise associated with the state vector model. This will be addressed later in the Kalman filter of §5.3. Modelling errors may also be present, which lead to several obvious complications. However, the most common approach is to ignore it; any modelling error is most often treated as state or measurement noise, or both, in spite of the fact that the modelling error may be predominately deterministic rather than random.[18]

The maximum likelihood estimation approach minimizes the following loss function:

$$\boxed{J(\hat{\mathbf{p}}) = \frac{1}{2}\sum_{k=1}^{N}(\tilde{\mathbf{y}}_k - \hat{\mathbf{y}}_k)^T R^{-1}(\tilde{\mathbf{y}}_k - \hat{\mathbf{y}}_k)} \tag{4.68}$$

where $\hat{\mathbf{y}}_k$ is the estimated response of $\mathbf{y}$ at time t_k for a given value of the unknown parameter vector $\mathbf{p}$, and N is the total number of measurements. A common approach to minimize eqn. (4.68) for aircraft parameter identification involves using the Newton-Raphson algorithm. If i is the iteration number, then the $i+1$ estimate of $\mathbf{p}$, denoted by $\hat{\mathbf{p}}$, is obtained from the ith estimate by[18]

$$\boxed{\hat{\mathbf{p}}_{i+1} = \hat{\mathbf{p}}_i - [\nabla^2_{\hat{\mathbf{p}}} J(\hat{\mathbf{p}})]^{-1}[\nabla_{\hat{\mathbf{p}}} J(\hat{\mathbf{p}})]} \tag{4.69}$$

where the first and second gradients are defined as

$$[\nabla_{\hat{\mathbf{p}}} J(\hat{\mathbf{p}})] = -\sum_{k=1}^{N}[\nabla_{\hat{\mathbf{p}}}\hat{\mathbf{y}}_k]^T R^{-1}(\tilde{\mathbf{y}}_k - \hat{\mathbf{y}}_k) \tag{4.70a}$$

$$[\nabla^2_{\hat{\mathbf{p}}} J(\hat{\mathbf{p}})] = \sum_{k=1}^{N}[\nabla_{\hat{\mathbf{p}}}\hat{\mathbf{y}}_k]^T R^{-1}[\nabla_{\hat{\mathbf{p}}}\hat{\mathbf{y}}_k] - \sum_{k=1}^{N}[\nabla^2_{\hat{\mathbf{p}}}\hat{\mathbf{y}}_k]R^{-1}(\tilde{\mathbf{y}}_k - \hat{\mathbf{y}}_k) \tag{4.70b}$$

The Gauss-Newton approximation to the second gradient is given by

$$[\nabla^2_{\hat{\mathbf{p}}} J(\hat{\mathbf{p}})] \approx \sum_{k=1}^{N}[\nabla_{\hat{\mathbf{p}}}\hat{\mathbf{y}}_k]^T R^{-1}[\nabla_{\hat{\mathbf{p}}}\hat{\mathbf{y}}_k] \tag{4.71}$$

This approximation is easier to compute than eqn. (4.70b), and has the advantage of possible decreased convergence time.

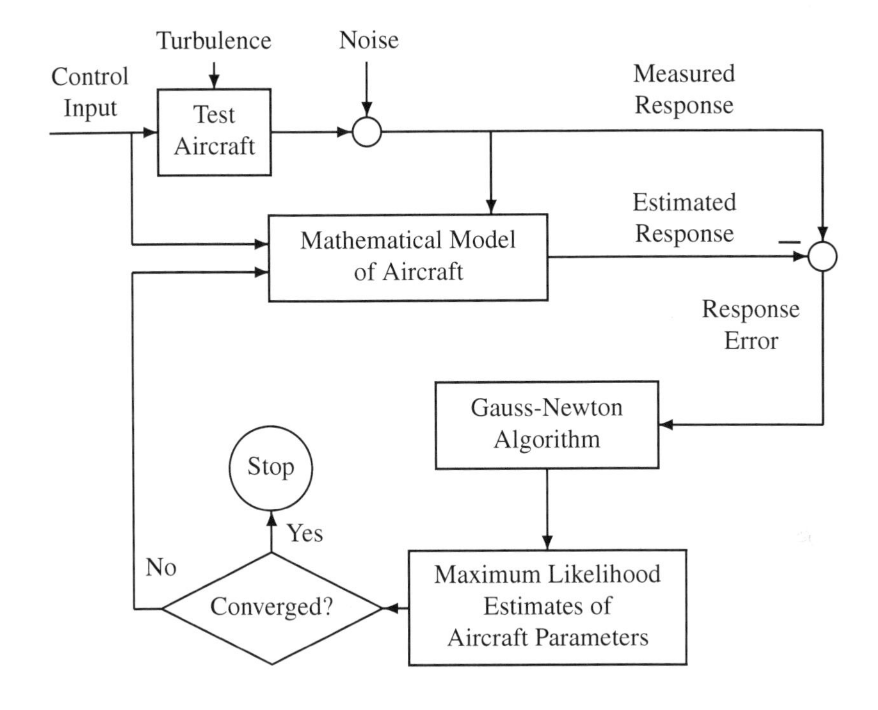

Figure 4.9: Aircraft Parameter Identification

The aircraft parameter identification process using maximum-likelihood is depicted in Figure 4.9.[18] First a control input is introduced to excite the motion. This input should be "rich" enough so that the test aircraft undergoes a general motion to allow sufficient observability of the to-be-identified parameters. For most applications, it is assumed that the control system inputs sufficiently dominate the motion in comparison to the effects of the turbulence and other unknown disturbances. An estimated response from the mathematical model is computed first using some initial guess of the aircraft parameters, which are usually obtained from ground-based wind tunnel data or by other means. A response error is computed from the estimated response and measured response. Then eqns. (4.69), (4.70a), and (4.71) are used to provide a Gauss-Newton update of the aircraft parameters. Next, the convergence is checked using some stopping criterion, e.g., eqn. (1.98). If the procedure has not converged then the previous aircraft parameters are replaced with the newly computed ones. These newly obtained aircraft parameters are used to compute a new estimated response from the mathematical model. The process continues until convergence is achieved. The error-covariance of the estimated parameters is given by the inverse of eqn. (4.71), which is also equivalent to within first-order terms to the Cramér-Rao lower bound.[18] Experiments are frequently repeated to confirm consistency. If the results are found to be consistent, then the measurements can be combined to obtain improved estimates.

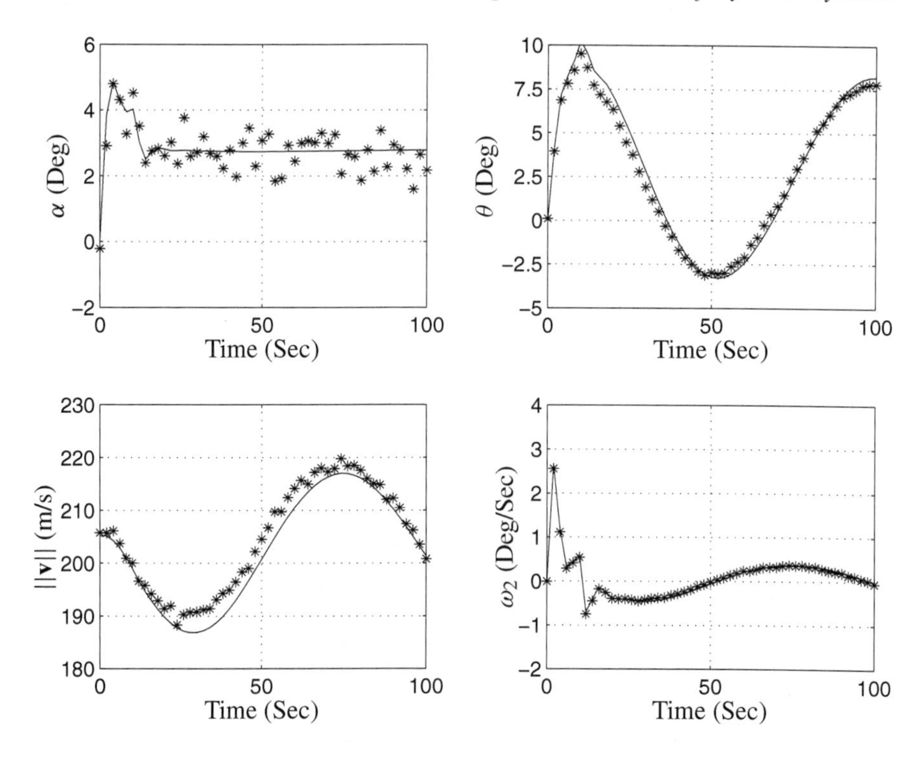

Figure 4.10: Simulated Aircraft Measurements and Estimated Trajectories

---

Example 4.4: To illustrate the power of maximum likelihood estimation, we show an example of identifying the longitudinal parameters of a simulated 747 aircraft. Decoupling the longitudinal motion equations from the lateral motion equations gives

$$\alpha = \tan^{-1}\frac{v_3}{v_1}$$
$$||\mathbf{v}|| = (v_1^2 + v_3^2)^{1/2}$$

$$T_1 - D\cos\alpha + L\sin\alpha - mg\sin\theta = m(\dot{v}_1 + v_3\omega_2)$$
$$T_3 - D\sin\alpha - L\cos\alpha + mg\cos\theta = m(\dot{v}_3 - v_1\omega_2)$$

$$D = C_D\,\bar{q}\,S$$
$$L = C_L\,\bar{q}\,S$$

$$\bar{q} = \frac{1}{2}\rho||\mathbf{v}||^2$$

$$C_D = C_{D_0} + C_{D_\alpha}\alpha + C_{D_{\delta_E}}\delta_E$$
$$C_L = C_{L_0} + C_{L_\alpha}\alpha + C_{L_{\delta_E}}\delta_E$$
$$J_{22}\dot{\omega}_2 = L_{A_2} + L_{T_2}$$
$$L_{A_2} = C_m\,\bar{q}\,S\,\bar{c}$$
$$C_m = C_{m_0} + C_{m_\alpha}\alpha + C_{m_{\delta_E}}\delta_E + C_{m_q}\frac{\omega_2\bar{c}}{2||\mathbf{v}||^2}$$
$$\begin{bmatrix}\dot{x}\\ \dot{z}\end{bmatrix} = \begin{bmatrix}\cos\theta & \sin\theta\\ -\sin\theta & \cos\theta\end{bmatrix}\begin{bmatrix}v_1\\ v_3\end{bmatrix}$$
$$\dot{\theta} = \omega_2$$

The longitudinal aerodynamic coefficients, assuming a low cruise, for the 747 are given by

$$C_{D_0} = 0.0164 \quad C_{D_\alpha} = 0.20 \quad C_{D_{\delta_E}} = 0$$
$$C_{L_0} = 0.21 \quad C_{L_\alpha} = 4.4 \quad C_{L_{\delta_E}} = 0.32$$
$$C_{m_0} = 0 \quad C_{m_\alpha} = -1.00 \quad C_{m_{\delta_E}} = -1.30 \quad C_{m_q} = -20.5$$

The reference geometry quantities and density are given by

$$S = 510.97\,\text{m}^2 \quad \bar{c} = 8.321\,\text{m} \quad b = 59.74\,\text{m} \quad \rho = 0.6536033\,\text{kg/m}^3$$

The mass data and inertia quantities are given by

$$m = 288,674.58\,\text{kg} \qquad J_{22} = 44,877,565\,\text{kg m}^2$$

The flight conditions for low cruise at an altitude of 6,096 m are given by

$$||\mathbf{v}|| = 205.13\,\text{m/s} \quad \bar{q} = 13,751.2\,\text{N/m}^2$$

Using these flight conditions the equations of motion are integrated for a 100-second simulation. The thrust is set equal to the computed drag, and the elevator is set to 1 degree down from the trim value for the first 10 seconds and then returned to the trimmed value thereafter. Measurements of angle of attack, α, pitch angle, θ, velocity, $||\mathbf{v}||$, and angular velocity, ω_2, are assumed with standard deviations of the measurement errors given by $\sigma_\alpha = 0.5$ degrees, $\sigma_\theta = 0.1$ degrees, $\sigma_{||\mathbf{v}||} = 1$ m/s, and $\sigma_{\omega_2} = 0.01$ deg/sec, respectively. A plot of the simulated measurements is shown in Figure 4.10. Clearly, the angle of attack measurements are very noisy due to the inaccuracy of the sensor. The quantities to be estimated are given by

$$\mathbf{p} = [C_{D_0}\ C_{L_0}\ C_{m_0}\ C_{D_\alpha}\ C_{L_\alpha}\ C_{m_\alpha}]^T$$

The initial guesses for these parameters are given by

$$C_{D_0} = 0.01 \quad C_{L_0} = 0.1 \quad C_{m_0} = 0.01$$
$$C_{D_\alpha} = 0.30 \quad C_{L_\alpha} = 3 \quad C_{m_\alpha} = -0.5$$

which represent a significant departure from the actual values. The partial derivatives used in the Gauss-Newton algorithms are computed using a simple first-order numerical derivative, for example:

$$\frac{\partial \alpha}{\partial C_{D_0}} \approx \frac{\alpha|_{C_{D_0}+\delta C_{D_0}} - \alpha|_{C_{D_0}}}{\delta C_{D_0}}$$

Results of the convergence history are summarized below.

Iteration	Aircraft Parameter					
	C_{D_0}	C_{L_0}	C_{m_0}	C_{D_α}	C_{L_α}	C_{m_α}
0	0.0100	0.1000	0.0100	0.3000	3.0000	−0.5000
1	−0.0191	0.4185	−0.0432	0.5215	2.7383	−0.4932
2	0.0113	0.3755	−0.0404	0.0125	2.9932	−0.5603
3	0.0117	0.3528	−0.0342	0.2809	3.4661	−0.6835
4	0.0104	0.2954	−0.0221	0.3029	4.1408	−0.8554
5	0.0146	0.2167	−0.0033	0.1965	4.5201	−1.0213
6	0.0167	0.2057	0.0012	0.1938	4.3779	−1.0035
7	0.0163	0.2070	0.0007	0.2026	4.4064	−1.0025
8	0.0164	0.2069	0.0007	0.2004	4.4038	−1.0027
9	0.0164	0.2069	0.0007	0.2006	4.4041	−1.0026
10	0.0164	0.2069	0.0007	0.2006	4.4041	−1.0026

The 3σ error bounds, derived from the inverse of eqn. (4.71), are given by

	Aircraft Parameter					
	C_{D_0}	C_{L_0}	C_{m_0}	C_{D_α}	C_{L_α}	C_{m_α}
3σ	0.0025	0.0070	0.0021	0.0515	0.0545	0.0104

The estimate errors are clearly within the 3σ values. A plot of the estimated trajectories using the converged values are also shown in Figure 4.10. The velocity estimated trajectory seems to be biased slightly. This is due to the fact that the long period motion (known as the *phugoid* mode) seen in pitch and linear velocity is not well excited by elevator inputs. A speed brake is commonly used to fully excite the phugoid mode. Also, some parameters can be estimated more accurately than others (see Ref. [19] for details).

This section introduced the basic concepts of aircraft parameter identification. As demonstrated here, the maximum likelihood technique is extremely useful to extract aircraft parameters from flight data. This approach has been used successfully for many years for a wide variety of aircraft ranging from transport vehicles to highly maneuverable aircraft. Although the example shown in this section is highly simplified it does capture the essence of all aircraft parameter identification approaches. The reader is highly encouraged to pursue actual applications in the references cited here and in the open literature.

4.5 Eigensystem Realization Algorithm

Experimental modelling of systems is required for both the design of control laws and the quantification of actual system performance. Modelling of linear systems can be divided into two categories: 1) realization of system model and order, and 2) identification of actual system parameters. Either approach can be used to develop mathematical models that reconstruct the input/output behavior of the actual system. However, identification is inherently more complex since actual model parameters are sought (e.g., stability derivatives of an aircraft as demonstrated in §4.4), while realization generates non-physical representations of a particular system.

The realization of system models can be achieved in either the time domain or frequency domain. Frequency domain methods are inherently robust with respect to noise sensitivity, but typically require extensive computation. Also, these methods generally require insight on model form. Time domain methods generally do not require *a priori* knowledge of system form, but may be sensitive to measurement noise. A few time-domain algorithms of particular interest include: AutoRegressive Moving Average (ARMA) models,[20] Least Squares algorithms,[21] the Impulse Response technique,[22] and Ibrahim's Time Domain technique.[23] The Eigensystem Realization Algorithm[24] (ERA) expands upon these algorithms by utilizing singular value decompositions in the least squares process. The advantages of the ERA over other algorithms include: 1) the realizations have matrices that are internally balanced (i.e., equivalent controllability and observability Grammians), 2) repeated eigenvalues are identifiable, and 3) the order of the system can be estimated from the singular values computed in the ERA.

The majority of time domain methods are based on discrete difference equations. These equations are used since general input/output histories can be represented as a linear function of the sampling interval and system matrices. Discrete realizations from input/output data can be found if the input persistently excites the dynamics of the system. The realization of system models can be performed from a number of time input histories, including: free response data, impulse response data, and random response data. A majority of the time domain techniques rely on impulse response data, which leads to the *Markov parameters*. These parameters can be obtained by applying a Fast Fourier Transform (FFT) and an inverse FFT of a random input and output response data set, or by time-domain techniques.[25]

The ERA is derived by using the discrete-time dynamic model in eqn. (4.72):

$$\mathbf{x}_{k+1} = \Phi\,\mathbf{x}_k + \Gamma\,\mathbf{u}_k \tag{4.72a}$$

$$\mathbf{y}_k = H\,\mathbf{x}_k + D\,\mathbf{u}_k \tag{4.72b}$$

where $\mathbf{x}$ is an $n \times 1$ state vector, $\mathbf{u}$ is a $p \times 1$ input vector, and $\mathbf{y}$ is an $m \times 1$ output vector. Consider the SISO system with an impulse input for u_k (i.e., $u_0 = 1$ and $u_k = 0$ for $k \geq 1$) and zero initial state conditions. The evolution of the output proceeds as

$$y_0 = D \tag{4.73}$$

$$y_1 = H\Gamma \tag{4.74}$$

$$y_2 = H\Phi\Gamma \tag{4.75}$$

$$y_3 = H\Phi^2\Gamma \tag{4.76}$$

$$\vdots \tag{4.77}$$

$$y_k = H\Phi^{k-1}\Gamma \tag{4.78}$$

Clearly a pattern has been established. For the MIMO system the pattern is identical, which leads to the following discrete Markov parameters:

$$Y_0 = D \tag{4.79a}$$

$$Y_k = H\Phi^{k-1}\Gamma, \quad k \geq 1 \tag{4.79b}$$

The first step in the ERA is to form a $(r \times s)$ block *Hankel matrix* composed of impulse response data:

$$\mathcal{H}_{k-1} = \begin{bmatrix} Y_k & Y_{k+m_1} & \cdots & Y_{k+m_{s-1}} \\ Y_{k+l_1} & Y_{k+l_1+m_1} & \cdots & Y_{k+l_1+m_{s-1}} \\ \vdots & \vdots & \ddots & \vdots \\ Y_{k+l_{r-1}} & Y_{k+l_{r-1}+m_1} & \cdots & Y_{k+l_{r-1}+m_{s-1}} \end{bmatrix} \tag{4.80}$$

where r and s are arbitrary integers satisfying the inequalities $rm \geq n$ and $sp \geq n$, and l_i $(i = 1, 2, \ldots, r-1)$ and m_j $(j = 1, 2, \ldots, s-1)$ are arbitrary integers. The kth order Hankel matrix can be shown to be given by

$$\mathcal{H}_k = V_r \Phi^k W_s \tag{4.81}$$

where

$$V_r = \begin{bmatrix} H \\ H\Phi^{l_1} \\ \vdots \\ H\Phi^{l_{r-1}} \end{bmatrix} \tag{4.82a}$$

$$W_s = \begin{bmatrix} \Gamma & \Phi^{m_1}\Gamma & \cdots & \Phi^{m_{s-1}}\Gamma \end{bmatrix} \tag{4.82b}$$

The matrices V_r and W_s are generalized observability and controllability matrices, respectively. The ERA is derived by using a singular value decomposition of $\mathcal{H}_0$, expressed as

$$\mathcal{H}_0 = P\,S\,Q^T \tag{4.83}$$

where P and Q are isometric matrices (i.e., all columns are orthonormal), with dimensions $rm \times n$ and $ps \times n$, respectively. Next, let $V_r = P\,S^{1/2}$ and $W_s = S^{1/2}Q^T$. For the equality $\mathcal{H}_1 = V_r \Phi W_s$ we now have

$$\mathcal{H}_1 = P\,S^{1/2}\Phi S^{1/2} Q^T \tag{4.84}$$

Next, we multiply the left-hand side of eqn. (4.84) by P^T and the right-hand side by Q. Therefore, since $P^T P = I$ and $Q^T Q = I$, and from the definitions of V_r and W_s we obtain the following system realization:

$$\Phi = S^{-1/2} P^T \mathcal{H}_1 Q\,S^{-1/2} \tag{4.85a}$$

$$\Gamma = S^{1/2} Q^T E_p \tag{4.85b}$$

$$H = E_m^T P\,S^{1/2} \tag{4.85c}$$

$$D = Y_0 \tag{4.85d}$$

where $E_m^T = [I_{m\times m},\, 0_{m\times m}, \ldots, 0_{m\times m}]$ and $E_p^T = [I_{p\times p},\, 0_{p\times p}, \ldots, 0_{p\times p}]$. The ERA is in fact a least squares minimization (see Ref. [24] for details).

The order of the system can be estimated by examining the magnitude of the singular values of the Hankel matrix. These singular values, with diagonal elements s_i, are arranged as

$$s_1 \ge s_2 \ge \cdots \ge s_n \ge s_{n+1} \ge \cdots \ge s_N \tag{4.86}$$

where N is the total number of singular values. However, the presence of noise often produces an indeterministic value for n. Subsequently, a cutoff magnitude is chosen below which the singular values are assumed to be in the bandwidth of the noise. Juang and Pappa[26] studied effects of noise on the ERA for the case of zero-mean Gaussian measurement errors. A suitable region for the rank of the Hankel matrix can be determined by $s_i^2 > 2N\sigma^2$ for $i = 1, 2, \ldots, n$, where σ is the standard deviation of the measurement error. Hence, a realization of order n is possible using this rank test scheme.

The natural frequencies and damping ratios of the continuous-time system are determined by first calculating the eigenvalue matrix Λ_d and eigenvector matrix Ψ_d of the realized discrete-time state matrix Φ, with

$$\Psi_d^{-1}[S^{-1/2} P^T \mathcal{H}_1 Q\,S^{-1/2}]\Psi_d = \Lambda_d \tag{4.87}$$

The modal damping ratios and damped natural frequencies are then calculated by observing the real and imaginary parts of the eigenvalues, after a transformation from the z-plane to the s-plane is completed:

$$s_i = \frac{[\ln(\lambda_i) + 2\pi j]}{\Delta t} \tag{4.88}$$

where λ_i corresponds to the ith eigenvalue of the matrix Λ_d, j corresponds to the imaginary component $\sqrt{-1}$, and Δt is the sampling interval. Although the eigenvalues and eigenvectors of the discrete-time system are usually complex, the transformation to the continuous-time domain can be performed by using a real algorithm since the realized state matrix has independent eigenvectors.[24]

The presence of random noise on the output measurements leads to a Hankel matrix that has a rank larger than the order of the system. The Modal Amplitude Coherence[24] (MAC) is used to estimate the degree of modal excitation (controllability) of each identified mode. Therefore, the MAC can be used to help distinguish the system modes from modes identified due to adverse noise effects or nonlinearities in the system. The MAC is defined as the coherence between the modal amplitude history and an ideal history formed by extrapolating the initial value of the history using the identified eigenvalue. The derivation begins by expressing the control input matrix and modal time history as

$$\Psi_d^{-1} S^{1/2} Q^T E_p = [\mathbf{b}_1, \mathbf{b}_2, \ldots, \mathbf{b}_n]^* \tag{4.89a}$$

$$\Psi_d^{-1} S^{1/2} Q^T = [\mathbf{q}_1, \mathbf{q}_2, \ldots, \mathbf{q}_n]^* \tag{4.89b}$$

where the asterisk is defined as the transpose complex conjugate, $\mathbf{b}_j$ is a column vector corresponding to the system eigenvalue s_j $(j = 1, 2, \ldots, n)$, and $\mathbf{q}_j$ represents the modal time history from the real measurement data obtained by the decomposition of the Hankel matrix. Equation (4.89) is used to form a sequence of idealized modal amplitudes in the complex domain, represented by

$$\bar{\mathbf{q}}_j^* = [\mathbf{b}_j^*, \exp(t\,\Delta t\, s_j)\mathbf{b}_j^*, \ldots, \exp(t_{s-1}\,\Delta t\, s_j)\mathbf{b}_j^*] \tag{4.90}$$

where t_j is the jth time shift defined in the Hankel matrix, and Δt is the sampling interval. The MAC coherence factor for the jth mode can be determined from

$$\boxed{\gamma_j = \frac{|\bar{\mathbf{q}}_j^* \mathbf{q}_j|}{\left(|\bar{\mathbf{q}}_j^* \bar{\mathbf{q}}_j||\mathbf{q}_j^* \mathbf{q}_j|\right)^{1/2}}} \tag{4.91}$$

The MAC factor must have a range between 0 and 1. As this factor approaches 1, the initial modal amplitude and realized eigenvalues approach the true values for the jth mode of the system. Conversely, a lower MAC factor indicates that the mode is not excited well during the testing procedure or is probably due to noise effects. Another factor, known as the Modal Phase Collinearity (MPC) can be used to indicate if the behavior of the identified modes exhibit normal mode characteristics (see Ref. [24] for details).

For vibratory systems, described in §3.10, determining the mass (M), stiffness (K), and damping (C) matrices is of interest. These matrices can be extracted from the realized system model given by the ERA. The MIMO state-space model considered for this process is assumed to be given by

$$\dot{\mathbf{x}} = \begin{bmatrix} 0 & I \\ -M^{-1}K & -M^{-1}C \end{bmatrix} \mathbf{x} + \begin{bmatrix} 0 \\ M^{-1} \end{bmatrix} \mathbf{u} \equiv F\mathbf{x} + B\mathbf{u} \tag{4.92a}$$

$$\mathbf{y} = \begin{bmatrix} I & 0 \end{bmatrix} \mathbf{x} \equiv H\mathbf{x} \tag{4.92b}$$

with obvious definitions for F, B, and H. The corresponding transfer function matrix from $\mathbf{u}$ to $\mathbf{y}$ is given by

$$H[sI - F]^{-1}B = [Ms^2 + Cs + K]^{-1} \equiv \Phi(s) \tag{4.93}$$

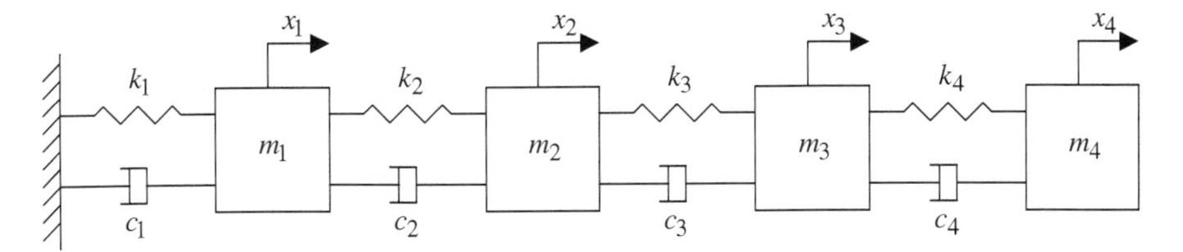

Figure 4.11: Mass-Stiffness-Damping System

Expanding the transfer function matrix in eqn. (4.93) as a power series yields

$$H[sI - F]^{-1}B = \frac{\phi_1}{s} + \frac{\phi_2}{s^2} + \frac{\phi_3}{s^3} + \cdots \tag{4.94}$$

where the continuous-time Markov parameters ϕ_i are given by

$$\phi_i = HF^{i-1}B \tag{4.95}$$

The continuous-time Markov parameters can be determined directly from the ERA. This is accomplished by first converting the discrete-time realization in eqn. (4.85) to a continuous-time realization using the methods described in §3.5. This continuous-time realization, denoted as $(\bar{F}, \bar{B}, \bar{H})$ may not necessarily be identical to the form in eqn. (4.92). However, both systems are similar, with

$$H[sI - F]^{-1}B = \bar{H}[sI - \bar{F}]^{-1}\bar{B} = \Phi(s) \tag{4.96a}$$

$$HF^{i-1}B = \bar{H}\bar{F}^{i-1}\bar{B} = \phi_i \tag{4.96b}$$

Therefore, there exists a similarity transformation T between the systems $(\bar{F}, \bar{B}, \bar{H})$ and (F, B, H). This similarity transformation can be used to determine the mass, stiffness, and damping matrices. Yeh and Yang[27] showed that the similarity transformation is determined by

$$F = T\bar{F}T^{-1} \tag{4.97a}$$

$$B = T\bar{B} \tag{4.97b}$$

$$H = \bar{H}T^{-1} \tag{4.97c}$$

where

$$T = \begin{bmatrix} \bar{H} \\ \bar{H}\bar{F} \end{bmatrix} \tag{4.98}$$

The mass, stiffness, and damping matrices are obtained by

$$M = [\bar{H}\bar{F}\bar{B}]^{-1} \tag{4.99a}$$

$$\begin{bmatrix} K & C \end{bmatrix} = -M\bar{H}\bar{F}^2T^{-1} \tag{4.99b}$$

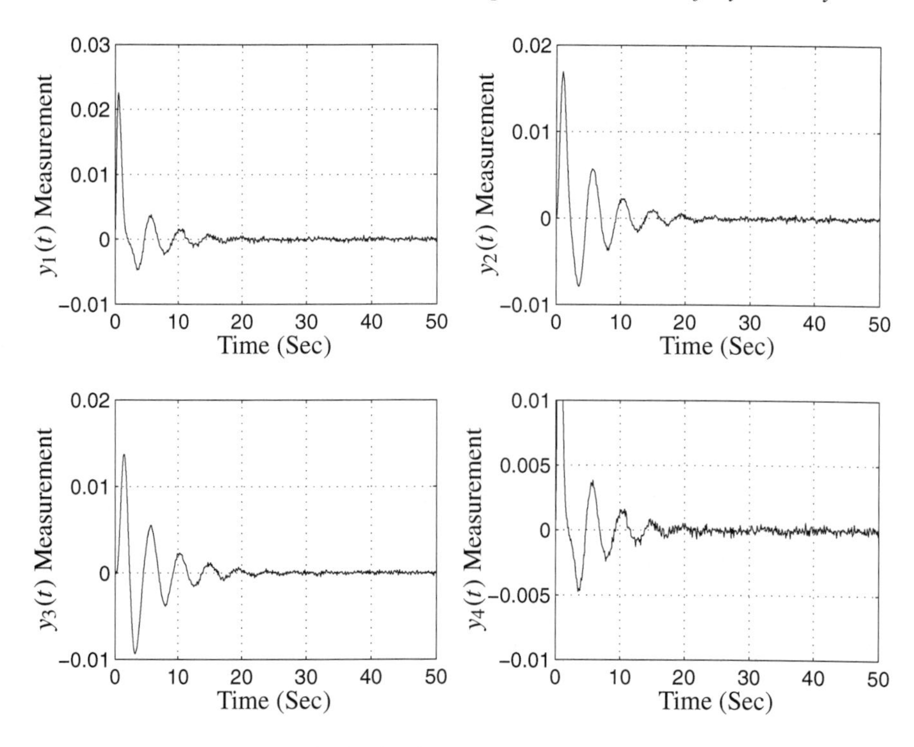

Figure 4.12: Simulated Position Measurements

Therefore, once a conversion of the ERA realized matrices from discrete-time to continuous-time is made, the modal properties and second-order matrix representations can be determined from eqn. (4.99). The ERA has been effectively used to determine system models for a wide variety of systems. More details on the ERA can be found in Ref. [28].

Example 4.5: In this example we will use the ERA to identify the mass, stiffness, and damping matrices of a 4 mode system from simulated mass-position measurements. This system is shown in Figure 4.11. The equations of motion can be found by using the techniques shown in §3.10. In this example the following mass-stiffness-damping matrices are used:

$$M = \begin{bmatrix} 1 & 0 & 0 & 0 \\ 0 & 1 & 0 & 0 \\ 0 & 0 & 1 & 0 \\ 0 & 0 & 0 & 1 \end{bmatrix}, \quad K = \begin{bmatrix} 10 & -5 & 0 & 0 \\ -5 & 10 & -5 & 0 \\ 0 & -5 & 10 & -5 \\ 0 & 0 & -5 & 10 \end{bmatrix}$$

$$C = \begin{bmatrix} 2 & -1 & 0 & 0 \\ -1 & 2 & -1 & 0 \\ 0 & -1 & 2 & -1 \\ 0 & 0 & -1 & 2 \end{bmatrix}$$

Note that proportional damping is given since $C = 1/5K$. In order to identify the system matrices using the ERA an impulse input is required at each mass, and the position of each mass must be measured. Therefore a total of 16 output measurements is required (4 position measurements for each impulse input). With the exact solution known, Gaussian white-noise of approximately 1% the size of the signal amplitude is added to simulate the output measurements. A 50-second simulation is performed, with measurements sampled every 0.1 seconds. A plot of the simulated position output measurements for an impulse input to the first mass is shown in Figure 4.12. Using all available measurements, the Hankel matrix in the ERA was chosen to be a 400×1600 dimension matrix. After computing the discrete-time state matrices using eqn. (4.85), a conversion to continuous-time state matrices is performed, and the mass, stiffness, and damping matrices are computed using eqn. (4.99). The results of this computation are

$$M = \begin{bmatrix} 1.0336 & -0.0144 & 0.0153 & -0.0071 \\ -0.0104 & 0.9857 & 0.0009 & -0.0013 \\ -0.0019 & 0.0208 & 0.9841 & 0.0060 \\ -0.0045 & 0.0067 & -0.0121 & 1.0166 \end{bmatrix}$$

$$K = \begin{bmatrix} 10.1728 & -5.1059 & 0.0709 & -0.0548 \\ -5.0897 & 9.9608 & -4.9498 & -0.0016 \\ 0.0281 & -4.9408 & 9.9469 & -5.0120 \\ -0.0656 & 0.0538 & -5.0408 & 10.0503 \end{bmatrix}$$

$$C = \begin{bmatrix} 1.9885 & -0.9877 & -0.0079 & 0.0004 \\ -0.9944 & 1.9855 & -0.9726 & -0.0222 \\ -0.0097 & -0.9461 & 1.9255 & -0.9612 \\ 0.0020 & -0.0073 & -1.0060 & 2.0195 \end{bmatrix}$$

These realized matrices are in close agreement to the true matrices. One drawback of the mass, stiffness, and damping identification method is that it does not produce matrices that are symmetric. A discussion on this issue is given in Ref. [29]. Obviously, the realized matrices are not physically consistent with the connectivity of Figure 4.11, and are simply a second-order representation of the system consistent with the measurements. Also, the true and identified natural frequencies and damping ratios are given below, which shows close agreement.

True		Identified	
ω_n	ζ	ω_n	ζ
1.3820	0.1382	1.3818	0.1381
2.6287	0.2629	2.6248	0.2622
3.6180	0.3618	3.5988	0.3686
4.2533	0.4253	4.2599	0.4129

4.6 Summary

In this chapter several applications of least squares methods have been presented for Global Positioning System navigation, spacecraft attitude determination from various sensor devices, orbit determination from ground-based sensors, aircraft parameter identification using on-board measurements, and modal identification of vibratory systems. These practical examples make extensive use of the tools derived in the previous chapters, and form the basis for "real-world" applications in dynamic systems. We anticipate that most readers, having gained computational and analytical experience from the examples of the first two chapters and elsewhere, will profit greatly by a careful study of these applications. The constraints imposed by the length of this text did not, however, permit an entirely self-contained and satisfactory development of the concepts introduced in the applications of this chapter. It will likely prove useful for the interested reader to pursue these important subjects in the cited literature.

A summary of the key formulas presented in this chapter is given below.

- GPS Pseudorange

$$\tilde{\rho}_i = [(s_{i1} - x)^2 + (s_{i2} - y)^2 + (s_{i3} - z)^2]^{1/2} + \tau + v_i, \; i = 1, 2, \ldots, n$$

- Vector Measurement Attitude Determination and Covariance

$$\mathbf{b} = A\mathbf{r}$$

$$J(\hat{A}) = \frac{1}{2}\sum_{j=1}^{N} \sigma_j^{-2} \|\tilde{\mathbf{b}}_j - \hat{A}\mathbf{r}_j\|^2, \quad \hat{A}\hat{A}^T = I_{3\times 3}$$

$$P = \left(-\sum_{j=1}^{N} \sigma_j^{-2} [A\mathbf{r}_j \times]^2 \right)^{-1}$$

- Davenport's Attitude Determination Algorithm

$$K \equiv -\sum_{j=1}^{N} \sigma_j^{-2} \Omega(\tilde{\mathbf{b}}_j)\Gamma(\mathbf{r}_j)$$

$$K\hat{\mathbf{q}} = \lambda\hat{\mathbf{q}}$$

- Orbit Determination

$$\ddot{\mathbf{r}} = -\frac{\mu}{||\mathbf{r}||^3}\mathbf{r}$$

$$\rho = \mathbf{r} - \mathbf{R} = \begin{bmatrix} x - ||\mathbf{R}|| \cos\lambda \cos\theta \\ y - ||\mathbf{R}|| \cos\lambda \sin\theta \\ z - ||\mathbf{R}|| \sin\lambda \end{bmatrix}$$

$$\begin{bmatrix} \rho_u \\ \rho_e \\ \rho_n \end{bmatrix} = \begin{bmatrix} \cos\lambda & 0 & \sin\lambda \\ 0 & 1 & 0 \\ -\sin\lambda & 0 & \cos\lambda \end{bmatrix} \begin{bmatrix} \cos\theta & \sin\theta & 0 \\ -\sin\theta & \cos\theta & 0 \\ 0 & 0 & 1 \end{bmatrix} \rho$$

$$||\rho|| = (\rho_u^2 + \rho_e^2 + \rho_n^2)^{1/2}$$
$$\text{az} = \tan^{-1}\left(\frac{\rho_e}{\rho_n}\right)$$
$$\text{el} = \sin^{-1}\left(\frac{\rho_u}{||\rho||}\right)$$

- Aircraft Parameter Identification

$$\dot{\mathbf{x}} = \mathbf{f}(t, \mathbf{x}, \mathbf{p})$$
$$\tilde{\mathbf{y}}_k = \mathbf{h}(t_k, \mathbf{x}_k) + \mathbf{v}_k$$

$$J(\hat{\mathbf{p}}) = \frac{1}{2} \sum_{k=1}^{N} (\tilde{\mathbf{y}}_k - \hat{\mathbf{y}}_k)^T R^{-1} (\tilde{\mathbf{y}}_k - \hat{\mathbf{y}}_k)$$

$$\hat{\mathbf{p}}_{i+1} = \hat{\mathbf{p}}_i - [\nabla_{\hat{\mathbf{p}}}^2 J(\hat{\mathbf{p}})]^{-1} [\nabla_{\hat{\mathbf{p}}} J(\hat{\mathbf{p}})]$$
$$[\nabla_{\hat{\mathbf{p}}} J(\hat{\mathbf{p}})] = -\sum_{k=1}^{N} [\nabla_{\hat{\mathbf{p}}} \hat{\mathbf{y}}_k]^T R^{-1} (\tilde{\mathbf{y}}_k - \hat{\mathbf{y}}_k)$$
$$[\nabla_{\hat{\mathbf{p}}}^2 J(\hat{\mathbf{p}})] \approx \sum_{k=1}^{N} [\nabla_{\hat{\mathbf{p}}} \hat{\mathbf{y}}_k]^T R^{-1} [\nabla_{\hat{\mathbf{p}}} \hat{\mathbf{y}}_k]$$

- Eigensystem Realization Algorithm

$$\mathbf{x}_{k+1} = \Phi \mathbf{x}_k + \Gamma \mathbf{u}_k$$
$$\mathbf{y}_k = H \mathbf{x}_k + D \mathbf{u}_k$$

$$Y_0 = D$$
$$Y_k = H \Phi^{k-1} \Gamma, \quad k > 1$$

$$\mathcal{H}_{k-1} = \begin{bmatrix} Y_k & Y_{k+m_1} & \cdots & Y_{k+m_{s-1}} \\ Y_{k+l_1} & Y_{k+l_1+m_1} & \cdots & Y_{k+l_1+m_{s-1}} \\ \vdots & \vdots & \ddots & \vdots \\ Y_{k+l_{r-1}} & Y_{k+l_{r-1}+m_1} & \cdots & Y_{k+l_{r-1}+m_{s-1}} \end{bmatrix}$$
$$\mathcal{H}_0 = P\, S\, Q^T$$

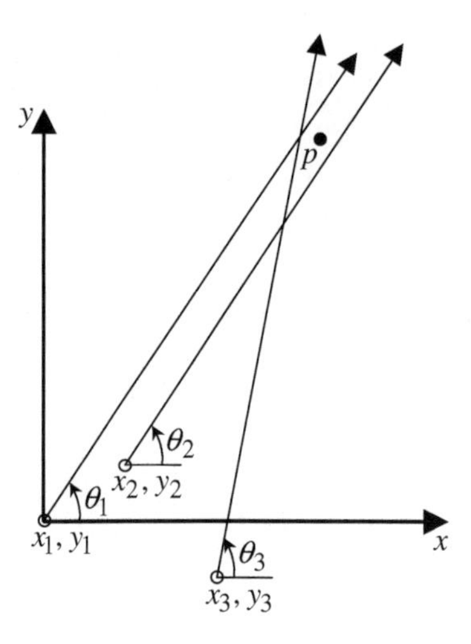

Figure 4.13: Planar Triangulation from Uncertain Base Points

$$\Phi = S^{-1/2} P^T \mathcal{H}_1 Q\, S^{-1/2}$$
$$\Gamma = S^{1/2} Q^T E_p$$
$$H = E_m^T P\, S^{1/2}$$
$$D = Y_0$$

Exercises

4.1 A problem closely related to the GPS position determination problem is planar triangulation. With reference to Figure 4.13, suppose a surveyor has collected data to estimate the location (x, y) of a point p. The point p is assumed, for simplicity, to lie in the $x-y$ plane. Suppose that the measurements consist of the azimuth θ of p from several imperfectly known points along a baseline (the x-axis). The first measurement base point is adopted as the origin $(x_1 = y_1 = 0)$ and the relative coordinates (x_2, y_2), (x_3, y_3) are admitted as four additional unknowns. The observations are modelled (refer to Figure 4.13) as

$$\tilde{\theta}_j = \tan^{-1}\left(\frac{y - y_j}{x - x_j}\right) + v_{\theta_j}, \quad j = 1, 2, 3$$
$$\tilde{x}_j = x_j + v_{x_j}, \quad j = 2, 3$$
$$\tilde{y}_j = y_j + v_{y_j}, \quad j = 2, 3$$

Thus, there are seven observed parameters $(\tilde{\theta}_1, \tilde{\theta}_2, \tilde{\theta}_3, \tilde{x}_2, \tilde{y}_2, \tilde{x}_3, \tilde{y}_3)$ and six unknown (to be estimated) parameters $(x, y, x_2, y_2, x_3, y_3)$. The dual role of (x_2, y_2, x_3, y_3) as observed and to-be-estimated parameters should present no particular conceptual difficulty if one recognizes that the measurement equations for these parameters are the simplest possible dependence of the observed parameters upon the unknown variables. The measurements and variances are given in the following table:

j	$\tilde{x}_j$	$\sigma^2_{x_j}$	$\tilde{y}_j$	$\sigma^2_{y_j}$	$\tilde{\theta}_j$	$\sigma^2_{\theta_j}$
1	0	0	0	0	30.1	0.01
2	500	100	50	144	45.0	0.01
3	1000	25	−100	100	73.6	0.01

Given the following starting estimates:

$$\begin{aligned}\mathbf{x}_c &= \left[x_c \ y_c \ x_{2_c} \ y_{2_c} \ x_{3_c} \ y_{3_c}\right]^T \\ &= \left[1210 \ 700 \ 500 \ 50 \ 1000 \ -100\right]^T\end{aligned}$$

and the measurements in the previous table, find estimates of the point p and base points using nonlinear least squares, and determine the associated covariance matrix. Also, program the Levenberg-Marquardt method of §1.6.3 and use this algorithm for improved convergence for various initial conditions.

4.2 Write a numerical algorithm based on the Levenberg-Marquardt method of §1.6.3 for the GPS navigation simulation in example 4.1. Can you achieve better convergence than nonlinear least squares for various starting conditions?

4.3 ♣ Consider the problem of determining the position and orientation of a vehicle using line-of-sight measurements from a vision-based beacon system based on Position Sensing Diode (PSD) technology,[30] depicted in Figure 4.14. If we choose the z-axis of the sensor coordinate system to be directed outward along the boresight of the PSD, then given object space (X, Y, Z) and image space (x, y, z) coordinate frames (see Figure 4.14), the ideal object to image space projective transformation (noiseless) can be written as follows:

$$x_i = -f\frac{A_{11}(X_i - X_c) + A_{12}(Y_i - Y_c) + A_{13}(Z_i - Z_c)}{A_{31}(X_i - X_c) + A_{32}(Y_i - Y_c) + A_{33}(Z_i - Z_c)}, \quad i = 1, 2, \ldots, N$$

$$y_i = -f\frac{A_{21}(X_i - X_c) + A_{22}(Y_i - Y_c) + A_{23}(Z_i - Z_c)}{A_{31}(X_i - X_c) + A_{32}(Y_i - Y_c) + A_{33}(Z_i - Z_c)}, \quad i = 1, 2, \ldots, N$$

where N is the total number of observations, (x_i, y_i) are the image space observations for the ith line-of-sight, (X_i, Y_i, Z_i) are the known object space locations of the ith beacon, (X_c, Y_c, Z_c) is the unknown object space location of the sensor, f is the known focal length, and A_{jk} are the unknown coefficients of the attitude matrix (A) associated to the orientation from the object plane to the image plane. The observation can be reconstructed in unit vector form as

$$\mathbf{b}_i = A\mathbf{r}_i, \quad i = 1, 2, \ldots, N$$

where

$$\mathbf{b}_i \equiv \frac{1}{\sqrt{f^2+x_i^2+y_i^2}}\begin{bmatrix}-x_i\\-y_i\\f\end{bmatrix}$$

$$\mathbf{r}_i \equiv \frac{1}{\sqrt{(X_i-X_c)^2+(Y_i-Y_c)^2+(Z_i-Z_c)^2}}\begin{bmatrix}X_i-X_c\\Y_i-Y_c\\Z_i-Z_c\end{bmatrix}$$

Write a nonlinear least squares program to determine the position and orientation from line-of-sight measurements. Assume the following six beacon locations:

$$\begin{aligned}
X_1 = 0.5\text{m},\quad & Y_1 = 0.5\text{m},\quad Z_1 = 0.0\text{m}\\
X_2 = -0.5\text{m},\quad & Y_2 = -0.5\text{m},\quad Z_2 = 0.0\text{m}\\
X_3 = -0.5\text{m},\quad & Y_3 = 0.5\text{m},\quad Z_3 = 0.0\text{m}\\
X_4 = 0.5\text{m},\quad & Y_4 = -0.5\text{m},\quad Z_4 = 0.0\text{m}\\
X_5 = 0.2\text{m},\quad & Y_5 = 0.0\text{m},\quad Z_5 = 0.1\text{m}\\
X_6 = 0.0\text{m},\quad & Y_6 = 0.2\text{m},\quad Z_6 = -0.1\text{m}
\end{aligned}$$

Any parameterization of the attitude matrix can be used, such as the Euler angles shown in §3.7.1; however, we suggest that the vector of modified Rodrigues parameters, $\mathbf{p}$, be used.[31] These parameters are closely related to the quaternions, with

$$\mathbf{p} = \frac{\boldsymbol{\varrho}}{1+q_4}$$

where the attitude matrix is given by

$$A(\mathbf{p}) = I_{3\times3} - \frac{4(1-\mathbf{p}^T\mathbf{p})}{(1+\mathbf{p}^T\mathbf{p})^2}[\mathbf{p}\times] + \frac{8}{(1+\mathbf{p}^T\mathbf{p})^2}[\mathbf{p}\times]^2$$

To help you along it can be shown that the partial of $A(\mathbf{p})\mathbf{r}$ with respect to $\mathbf{p}$ is given by[32]

$$\frac{\partial A(\mathbf{p})\mathbf{r}}{\partial \mathbf{p}} = \frac{4}{(1+\mathbf{p}^T\mathbf{p})^2}[A(\mathbf{p})\mathbf{r}\times]\left\{(1-\mathbf{p}^T\mathbf{p})I_{3\times3} - 2[\mathbf{p}\times] + 2\mathbf{p}\mathbf{p}^T\right\}$$

Consider a $1,800$-second simulation (i.e., $t_f = 1800$), and a focal length of $f = 1$. The true vehicle linear motion is given by $X_c = 30\exp[-(1/300)t]$ m, $Y_c = 30-(30/1800)t$ m, and $Z_c = 10-(10/1800)t$ m. The true angular motion is given by $\omega_1 = 0$ rad/sec, $\omega_2 = -0.0011$ rad/sec, and $\omega_3 = 0$ rad/sec, with zero initial conditions for the orientation angles. The measurement error is assumed to be zero-mean Gaussian with a standard deviation of $1/5000$ of the focal plane dimension, which for a 90 degree field-of-view corresponds to an angular resolution of $90/5000 \simeq 0.02$ degrees. For simplicity assume a measurement model given by $\tilde{\mathbf{b}} = A\mathbf{r} + \mathbf{v}$, where the covariance of $\mathbf{v}$ is assumed to be a diagonal matrix with elements given by $0.02\pi/180$. Find position and orientation estimates for this maneuver at 0.01-second intervals using the nonlinear least squares program, and determine the associated error-covariance matrix.

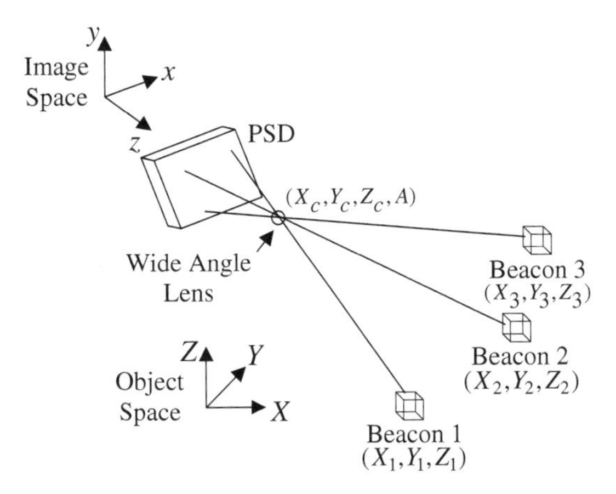

Figure 4.14: Vision Navigation System

4.4 Instead of determining the position of the PSD sensor shown in exercise 4.3, suppose we wish to determine a fixed attitude matrix, A, and focal length, f, given known positions X_c, Y_c, and Z_c over time. Develop a nonlinear least squares program to perform this calibration task using the true position location trajectories (X_c, Y_c, Z_c) shown in exercise 4.3. First, try determining the focal length only using some known fixed attitude. Then, try estimating both the fixed attitude matrix and focal length. How sensitive is your algorithm to initial guesses? Try various other known position motions to test the convergence properties of your algorithm. Also, try implementing the Levenberg-Marquardt algorithm of §1.6.3 to provide a more robust algorithm.

4.5 Given two non-parallel reference unit vectors $\mathbf{r}_1$ and $\mathbf{r}_2$ and the corresponding observation unit vectors $\mathbf{b}_1$ and $\mathbf{b}_2$, the TRIAD algorithm finds an orthogonal attitude matrix A that satisfies (in the noiseless case)

$$\mathbf{b}_1 = A\mathbf{r}_1, \quad \mathbf{b}_2 = A\mathbf{r}_2$$

This algorithm is given by first constructing two triads of manifestly orthonormal reference and observation vectors:

$$\mathbf{u}_1 = \mathbf{r}_1, \quad \mathbf{u}_2 = (\mathbf{r}_1 \times \mathbf{r}_2)/||(\mathbf{r}_1 \times \mathbf{r}_2)||$$
$$\mathbf{u}_3 = [\mathbf{r}_1 \times (\mathbf{r}_1 \times \mathbf{r}_2)]/||(\mathbf{r}_1 \times \mathbf{r}_2)||$$

$$\mathbf{v}_1 = \mathbf{b}_1, \quad \mathbf{v}_2 = (\mathbf{b}_1 \times \mathbf{b}_2)/||(\mathbf{b}_1 \times \mathbf{b}_2)||$$
$$\mathbf{v}_3 = [\mathbf{b}_1 \times (\mathbf{b}_1 \times \mathbf{b}_2)]/||(\mathbf{b}_1 \times \mathbf{b}_2)||$$

and then forming the following orthogonal matrices:

$$U = \begin{bmatrix}\mathbf{u}_1 & \mathbf{u}_2 & \mathbf{u}_3\end{bmatrix}, \quad V = \begin{bmatrix}\mathbf{v}_1 & \mathbf{v}_2 & \mathbf{v}_3\end{bmatrix}$$

Prove that U and V are orthogonal. Next, prove that the attitude matrix A is given by $A = V\,U^T$.

4.6 Using eqns. (4.15) to (4.17), prove that the attitude error covariance is given by the expression in eqn. (4.18).

4.7 ♣ Prove that the matrix K in eqn. (4.24) is also given by

$$K = \begin{bmatrix} S - \alpha I & \mathbf{z} \\ \mathbf{z}^T & \alpha \end{bmatrix}$$

where

$$B = \sum_{j=1}^{N} \sigma_j^{-2} \tilde{\mathbf{b}}_j \mathbf{r}_j^T$$

$$\alpha = \text{Tr}B = \sum_{j=1}^{N} \sigma_j^{-2} \tilde{\mathbf{b}}_j^T \mathbf{r}_j$$

$$S = B + B^T = \sum_{j=1}^{N} \sigma_j^{-2} (\tilde{\mathbf{b}}_j \mathbf{r}_j^T + \mathbf{r}_j \tilde{\mathbf{b}}_j^T)$$

$$\mathbf{z} = \sum_{j=1}^{N} \sigma_j^{-2} (\tilde{\mathbf{b}}_j \times \mathbf{r}_j)$$

4.8 Write a computer program to determine the optimal attitude from vector observations given by algorithms from Davenport in eqn. (4.26). Assuming a Gaussian distribution of stars, create a random sample of stars on a uniform sphere (note: the actual star distribution more closely follows a Poisson distribution[33]). Randomly pick 2 to 6 stars within an 8 degree field-of-view to simulate a star camera. Then, create synthetic body measurements with the measurement error for the camera given in example 4.2. Assuming a true attitude motion given by a constant angular velocity about the y-axis with $\omega = \begin{bmatrix} 0 & -0.0011 & 0 \end{bmatrix}^T$ rad /sec. Compute an attitude solution every second using both methods. Using the covariance expression in eqn. (4.18), numerically show that the 3σ bounds do indeed bound the attitude errors.

4.9 ♣ A problem that is closely related to the attitude determination problem involves determining ellipse parameters from measured data. Figure 4.15 depicts a general ellipse rotated by an angle θ. The basic equation of an ellipse is given by

$$\frac{(x' - x_0')^2}{a^2} + \frac{(y' - y_0')^2}{b^2} = 1$$

where (x_0', y_0') denotes the origin of the ellipse and (a, b) are positive values. The coordinate transformation follows

$$x' = x\cos\theta + y\sin\theta$$

$$y' = -x\sin\theta + y\cos\theta$$

Show that the ellipse equation can be rewritten as

$$Ax^2 + Bxy + Cy^2 + Dx + Ey + F = 0$$

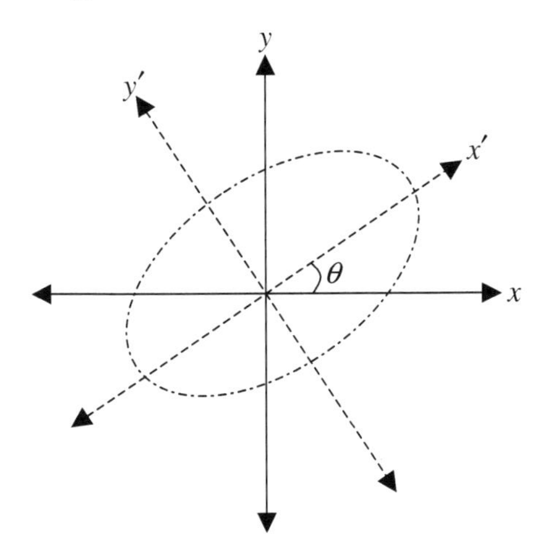

Figure 4.15: Ellipse with Rotation

Next, determine a form for the set of the coefficients so that the following constraint is always satisfied: $A^2 + 0.5B^2 + C^2 = 1$.[34]

Given a set of coefficients A, B, C, D, E, and F, show that the formulas for θ, a, b, x'_0, and y'_0 are given by

$$\cot(2\theta) = \frac{A-C}{B}$$
$$a = \sqrt{\frac{Q'}{A'}}, \quad b = \sqrt{\frac{Q'}{C'}}$$
$$x'_0 = -\frac{D'}{2A'}, \quad y'_0 = -\frac{E'}{2C'}$$

where

$$A' = A\cos^2\theta + B\sin\theta\cos\theta + C\sin^2\theta$$
$$B' = B(\cos^2\theta - \sin^2\theta) + 2(C-A)\sin\theta\cos\theta = 0$$
$$C' = A\sin^2\theta - B\sin\theta\cos\theta + C\cos^2\theta$$
$$D' = D\cos\theta + E\sin\theta$$
$$E' = -D\sin\theta + E\cos\theta$$
$$F' = F$$
$$Q' \equiv A'\left(\frac{D'}{2A'}\right)^2 + C'\left(\frac{E'}{2C'}\right)^2 - F'$$

(hint: show that the new variables follow the rotated ellipse equation: $A'x'^2 + B'x'y' + C'y'^2 + D'x' + E'y' + F' = 0$).

Suppose that a set of measurements for x and y exist, and we form the following vector of unknown parameters: $\mathbf{x} \equiv \begin{bmatrix} A & B & C & D & E & F \end{bmatrix}^T$. Our goal is to determine an estimate of $\mathbf{x}$ from this measured data set. Show that the minimum norm loss function can be written as

$$J(\hat{\mathbf{x}}) = \hat{\mathbf{x}}^T H^T H \hat{\mathbf{x}}$$

subject to

$$\hat{\mathbf{x}}^T Z \hat{\mathbf{x}} = 1$$

where the ith row of H is given by

$$H_i = \begin{bmatrix} \tilde{x}_i^2 & \tilde{x}_i \tilde{y}_i & \tilde{y}_i^2 & \tilde{x}_i & \tilde{y}_i & 1 \end{bmatrix}$$

Determine the matrix Z that satisfies the constraint. Using the eigenvalue method of §4.2 find the form for the optimal solution for $\hat{\mathbf{x}}$. Write a computer program for your derived solution and perform a simulation to test your algorithm. Note, a more robust approach involves using a reduced eigenvalue decomposition[35] or a singular value decomposition approach.[36]

4.10 A simple solution to the ellipse parameter identification system shown in exercise 4.9 involves using least squares. The ellipse parameter formulas shown in this problem are invariant under scalar multiplication (i.e., if we multiply A, B, C, etc., by a scalar then the formulas to determine θ, a, b, x_0', and y_0' remain unchanged). Therefore, we can assume that $F = 1$ without loss in generality. Derive an unconstrained least squares solution that estimates A, B, C, D, and E with the "measurement" given by $F = 1$. Test your algorithm using different simulation scenarios.

4.11 ♣ Consider the ellipse identification system shown in exercise 4.9. Using any estimation algorithm a set of reconstructed variables for x and y can be given by using the estimates of the coefficients A, B, C, D, E, and F. Suppose that $\hat{x}$ and $\hat{y}$ denote these estimated values, and $\tilde{x}$ and $\tilde{y}$ denote the measurement values. The current problem involves a method to check the consistency of the residuals between the measured and estimated x and y values. First, show that the measured data must satisfy the following inequalities in order for the data to conform to the ellipse model:

$$(B\tilde{x} + E)^2 - 4C(A\tilde{x}^2 + D\tilde{x} + F) > 0$$
$$(B\tilde{y} + D)^2 - 4A(C\tilde{y}^2 + E\tilde{y} + F) > 0$$

Suppose that the residual is defined as

$$f(\tilde{x}, \tilde{y}) \equiv A\tilde{x}^2 + B\tilde{x}\tilde{y} + C\tilde{y}^2 + D\tilde{x} + E\tilde{y} + F$$

Ideally $f(\tilde{x}, \tilde{y})$ should be zero, but this does not occur in practice due to measurement noise. Show that linearizing $f(\tilde{x}, \tilde{y})$ about $\hat{x}$ and $\hat{y}$ leads to

$$f(\tilde{x}, \tilde{y}) - f(\hat{x}, \hat{y}) = (2A\hat{x} + B\hat{y} + D)(\tilde{x} - \hat{x}) + (2C\hat{y} + B\hat{x} + E)(\tilde{y} - \hat{y})$$

Using this equation derive an expression for the variance of residual. Finally, using this expression derive a consistency test to remove extraneous measurement points (i.e., points outside some defined σ bound). Test your algorithm using simulated data points.

4.12 From the analysis of §4.2.4, show that the expressions for each of the eigenvalues in eqns. (4.31) and (4.33), and eigenvectors in eqns. (4.35), (4.37), and (4.40), do indeed satisfy $\lambda \mathbf{v} = F\mathbf{v}$.

4.13 Show that the expressions for the eigenvalues in eqn. (4.41), and eigenvectors in eqn. (4.42), reduce down from the eigenvalues in eqns. (4.31) and (4.33), and eigenvectors, in eqns. (4.35), (4.37), and (4.40), under the assumptions that $\mathbf{b}_1$ and $\mathbf{b}_2$ are unit vectors and $\sigma_1^2 = \sigma_2^2 \equiv \sigma^2$. Furthermore, prove that the vectors in eqn. (4.42) form an orthonormal set.

4.14 An alternative to using vector measurements to determine the attitude of a vehicle involves using GPS phase difference measurements.[37] The measurement model using GPS measurements is given by

$$\Delta\tilde{\phi}_{ij} = \mathbf{b}_i^T A \mathbf{s}_j + v_{ij}$$

where $\mathbf{s}_j$ is the known line-of-sight to the GPS spacecraft in reference-frame coordinates, $\mathbf{b}_i$ is the baseline vector between two antennae in body-frame coordinates, $\Delta\tilde{\phi}_{ij}$ denotes the phase difference measurement for the ith baseline and j^{th} sightline, and v_{ij} represents a zero-mean Gaussian measurement error with standard deviation σ_{ij} which is $0.5\text{cm}/\lambda = 0.026$ wavelengths for typical phase noise.[37] At each epoch it is assumed that m baselines and n sightlines exist.

Attitude determination using GPS signals involves finding the proper orthogonal matrix $\hat{A}$ that minimizes the following generalized loss function:

$$J(\hat{A}) = \frac{1}{2}\sum_{i=1}^{m}\sum_{j=1}^{n}\sigma_{ij}^{-2}(\Delta\tilde{\phi}_{ij} - \mathbf{b}_i^T \hat{A}\mathbf{s}_j)^2$$

Substitute eqn. (4.17) into this loss function, and after taking the appropriate partials show that the following optimal error covariance can be derived:

$$P = \left(\sum_{i=1}^{m}\sum_{j=1}^{n}\sigma_{ij}^{-2}[A\mathbf{s}_j\times]\mathbf{b}_i\,\mathbf{b}_i^T[A\mathbf{s}_j\times]^T\right)^{-1}$$

Note that the optimal covariance requires knowledge of the attitude matrix.

4.15 Consider the problem of converting the GPS attitude determination problem into a form given by Wahba's problem.[38] This is accomplished by converting the sightline vectors into the body frame, denoted by $\bar{\mathbf{s}}_j$. Assuming that at least three non-coplanar baselines exist, this conversion is given by

$$\bar{\mathbf{s}}_j = M_j^{-1}\mathbf{y}_j$$

where

$$M_j = \sum_{i=1}^{m}\sigma_{ij}^{-2}\mathbf{b}_i\,\mathbf{b}_i^T \quad \text{for } j = 1, 2, \ldots, n$$

$$\mathbf{y}_j = \sum_{i=1}^{m}\sigma_{ij}^{-2}\Delta\tilde{\phi}_{ij}\mathbf{b}_i \quad \text{for } j = 1, 2, \ldots, n$$

Then, given multiple (converted) body and known reference sightline vectors, Davenport's method of §4.2.3 can be employed to determine the attitude. It can be shown that this approach is suboptimal though. The covariance of this suboptimal approach is given by

$$P_s = \left(\sum_{j=1}^{n} a_j [\bar{\mathbf{s}}_j \times]^2\right)^{-1} \left(\sum_{j=1}^{n} a_j^2 [\bar{\mathbf{s}}_j \times] P_j [\bar{\mathbf{s}}_j \times]^T\right) \left(\sum_{j=1}^{n} a_j [\bar{\mathbf{s}}_j \times]^2\right)^{-1} \tag{4.100}$$

From the Cramér-Rao inequality we know that $P_s \geq P$, where P is given in exercise 4.14. Under what conditions does $P_s = P$? Prove your answer.

4.16 In this exercise you will simulate the performance of the conversion of the GPS attitude determination problem into a form given by Wahba's problem, discussed in exercise 4.15. Simulate the motion of a spacecraft as given in exercise 4.8. Assume that the spacecraft is always in the view of two GPS satellites with constant sightlines given by

$$\mathbf{s}_1 = (1/\sqrt{3})\left[1\ 1\ 1\right]^T, \quad \mathbf{s}_2 = (1/\sqrt{2})\left[0\ 1\ 1\right]^T$$

The three normalized baseline cases are given by the following:

Case 1:

$$\mathbf{b}_1 = (1/\sqrt{1.09})\left[1\ 0.3\ 0\right]^T, \quad \mathbf{b}_2 = \left[0\ 1\ 0\right]^T$$
$$\mathbf{b}_3 = \left[0\ 0\ 1\right]^T$$

Case 2:

$$\mathbf{b}_1 = (1/\sqrt{2})\left[1\ 1\ 0\right]^T, \quad \mathbf{b}_2 = \left[0\ 1\ 0\right]^T$$
$$\mathbf{b}_3 = \left[0\ 0\ 1\right]^T$$

Case 3:

$$\mathbf{b}_1 = (1/\sqrt{1.02})\left[0.1\ 1\ 0.1\right]^T, \quad \mathbf{b}_2 = \left[0\ 1\ 0\right]^T$$
$$\mathbf{b}_3 = \left[0\ 0\ 1\right]^T$$

The noise for each phase difference measurement is assumed to have a normalized standard deviation of $\sigma = 0.001$. To quantify the error introduced by the conversion to Wahba's form, use the following error factor:

$$f = \frac{1}{m_{\text{tot}}} \sum_{k=1}^{m_{\text{tot}}} \frac{\text{Tr}\left\{\text{diag}\left[P_s(t_k)^{1/2}\right]\right\}}{\text{Tr}\left\{\text{diag}\left[P(t_k)^{1/2}\right]\right\}}$$

where m_{tot} is the total number of measurements, P is given in exercise 4.14, and P_s is given in exercise 4.15. Compute the error factor f for each case. Also, show the 3σ bounds from P and P_s for each case. Which case produces the greatest errors?

4.17 Consider the problem of determining the state (position, $\mathbf{r}$, and velocity, $\dot{\mathbf{r}}$) and drag parameter of a vehicle at launch. The drag vector on the vehicle, which is modelled as a particle, is defined by

$$\mathbf{D} = -\left(\frac{1}{2}\rho V^2\right) C_D A \left(\frac{\dot{\mathbf{r}}}{V}\right)$$

where ρ is the density, $V \equiv ||\dot{\mathbf{r}}||$, C_D is the drag coefficient, and A is the projected area. This equation can be rewritten as

$$\mathbf{D} = -p\, m\, V\, \dot{\mathbf{r}}$$

where m is the mass of the vehicle and p is the drag parameter, given by

$$p \equiv \left(\frac{1}{2}\rho V^2\right) C_D A$$

Range and angle observations are assumed:

$$r = \sqrt{x^2 + y^2 + z^2}$$
$$\phi = \tan^{-1}\left(\frac{y}{x}\right)$$
$$\theta = \sin^{-1}\left(\frac{z}{r}\right)$$

with $\mathbf{r} = \begin{bmatrix} x & y & z \end{bmatrix}^T$. The equations of motion are given by

$$\ddot{x} = -p\,\dot{x}\,V$$
$$\ddot{y} = -p\,\dot{y}\,V$$
$$\ddot{z} = -g - p\,\dot{z}\,V$$

where $g = 9.81$ m/s^2. Create synthetic measurements sampled at 0.1-second intervals over a 20-second simulation by numerically integrating the equations of motion. Use a standard deviation of 10 m for the range measurement errors, and 0.01 rad for both angle measurement errors. Assume initial conditions of $\{x_0,\, y_0,\, z_0\} = \{-1000,\, -2000,\, 500\}$ m and $\{\dot{x}_0,\, \dot{y}_0,\, \dot{z}_0\} = \{100,\, 150,\, 50\}$ m/s. Also, set the drag parameter to

$$p = \frac{0.01}{\sqrt{\dot{x}_0^2 + \dot{y}_0^2 + \dot{z}_0^2}}$$

Using the nonlinear least-square differential correction algorithm depicted in Figure 4.8, estimate the initial conditions for position and velocity as well as the drag parameter (derive an analytical solution for the state transition matrix).

4.18 From eqns. (4.55) and (4.56) prove the following identity:

$$u_3^2 = \frac{1}{6}\chi^3 u_3 + u_5(u_1 - \chi)$$

4.19 ♣ Derive the Herrick-Gibbs formula in eqn. (4.66) by using the following Taylor series expansion:

$$\mathbf{r}_1 - \mathbf{r}_2 \approx -\tau_{12}\frac{d\mathbf{r}_2}{dt} + \frac{1}{2}\tau_{12}^2\frac{d^2\mathbf{r}_2}{dt^2} + \frac{1}{6}\tau_{12}^3\frac{d^3\mathbf{r}_2}{dt^3} + \frac{1}{24}\tau_{12}^4\frac{d^4\mathbf{r}_2}{dt^4}$$
$$\mathbf{r}_3 - \mathbf{r}_2 \approx -\tau_{23}\frac{d\mathbf{r}_2}{dt} + \frac{1}{2}\tau_{23}^2\frac{d^2\mathbf{r}_2}{dt^2} + \frac{1}{6}\tau_{23}^3\frac{d^3\mathbf{r}_2}{dt^3} + \frac{1}{24}\tau_{23}^4\frac{d^4\mathbf{r}_2}{dt^4}$$

Note, expressions for $\ddot{\mathbf{r}}_1$, $\ddot{\mathbf{r}}_2$, and $\ddot{\mathbf{r}}_3$ can be eliminated by using the inverse square law in eqn. (3.198).

4.20 Given the weakly coupled nonlinear oscillators

$$\ddot{x} = -\omega_1^2 x + \epsilon x z + A\cos\Omega_1 t$$
$$\ddot{z} = -\omega_1^2 z + \epsilon x z + B\cos\Omega_2 t$$

and the measurement model equation

$$\tilde{y}(t) = Cx + Dz + v \tag{4.101}$$

where ω_1^2, ω_2^2, Ω_1, Ω_2, A, B, C, D, and ϵ are constants, and $E\{v\} = 0$, $E\{v^2(t_j)\} = r$, and $E\{v(t_i)v(t_j)\} = 0$. Consider the following estimation problems:

(A) The model parameters $(\omega_1^2, \omega_2^2, \Omega_1, \Omega_2, A, B, C, D, \epsilon)$ are given constants, $\tilde{y}$ can be measured at m discrete instants; it is desired to estimate the initial state vector $\mathbf{x}(t_0) = \left[x(t_0)\ z(t_0)\ \dot{x}(t_0)\ \dot{z}(t_0)\right]^T$, given an initial estimate $\hat{\mathbf{x}}_a(t_0)$ and associated covariance matrix $P(t_0)$.

(B) The nine model parameters are uncertain, $\tilde{y}$ can be measured at m discrete instants, it is desired to estimate the initial state vector $\mathbf{x}(t_0)$ and the nine model parameters $(\omega_1^2, \omega_2^2, \Omega_1, \Omega_2, A, B, C, D, \epsilon)$, given *a priori* estimates and an associated covariance matrix.

Using the methods of the previous chapters, formulate minimal variance estimation algorithms for the aforementioned problems. Implement these algorithms as computer programs and study the performance of the algorithms (use synthetic measured data generated by adding zero-mean Gaussian distributed random numbers to perfect calculated y-values, see how well the true initial state and model parameter values are recovered).

4.21 Write a computer program to reproduce the orbit determination results in example 4.3. Also, write a numerical algorithm that replaces the nonlinear least squares iterations with the Levenberg-Marquardt method of §1.6.3. Can you achieve better results using this method over nonlinear least squares for poor initial guesses?

4.22 Consider the following nonlinear equations of motion for a highly maneuverable aircraft:

$$\dot{\alpha} = \dot{\theta} - \alpha^2\dot{\theta} - 0.09\alpha\dot{\theta} - 0.88\alpha + 0.47\alpha^2 + 3.85\alpha^3$$
$$- 0.22\delta_E + 0.28\delta_E\alpha^2 + 0.47\delta_E^2\alpha + 0.63\delta_E^3 - 0.02\theta^2$$

$$\ddot{\theta} = -0.396\dot{\theta} - 4.208\alpha - 0.470\alpha^2 - 3.564\alpha^3$$
$$-20.967\delta_E + 6.265\delta_E\alpha^2 + 46.00\delta_E^2 + 61.40\delta_E^3$$

Using a known "rich" input for δ_E create synthetic measurements of the angle of attack α and pitch angle θ with zero initial conditions. Assume standard deviations of the measurement errors to be the same as the ones given in exercise 4.4. Then use the results of §4.4 to identify various parameters of the above model. Which parameters can be most accurately identified?

4.23 Write a computer program to reproduce the aircraft parameter identification results in example 4.4. Compare the performance of the algorithm using the second gradient in eqn. (4.70b) and its approximation in eqn. (4.71). Also, expand upon the computer program for parameter identification of the lateral parameters of the simulated 747 aircraft (described in exercise 3.30). Finally, write a program that couples the longitudinal and lateral identification process.

4.24 Prove the similarity transformation for the identification of the mass, stiffness, and damping matrices in eqn. (4.99).

4.25 Write a general computer program for the Eigensystem Realization Algorithm, and the mass, stiffness, and damping matrix identification approach using eqn. (4.99). Use the computer program to reproduce the results in example 4.5.

References

[1] Axelrad, P. and Brown, R.G., "GPS Navigation Algorithms," *Global Positioning System: Theory and Applications*, edited by B. Parkinson and J. Spilker, Vol. 64 of *Progress in Astronautics and Aeronautics*, chap. 9, American Institute of Aeronautics and Astronautics, Washington, DC, 1996.

[2] Parkinson, B.W., "GPS Error Analysis," *Global Positioning System: Theory and Applications*, edited by B. Parkinson and J. Spilker, Vol. 64 of *Progress in Astronautics and Aeronautics*, chap. 11, American Institute of Aeronautics and Astronautics, Washington, DC, 1996.

[3] Bate, R.R., Mueller, D.D., and White, J.E., *Fundamentals of Astrodynamics*, Dover Publications, New York, NY, 1971.

[4] Slater, M.A., Miller, A.C., Warren, W.H., and Tracewell, D.A., "The New SKYMAP Master Catalog (Version 4.0)," *Advances in the Astronautical Sciences*, Vol. 90, Aug. 1995, pp. 67–81.

[5] Light, D.L., "Satellite Photogrammetry," *Manual of Photogrammetry*, edited by C.C. Slama, chap. 17, American Society of Photogrammetry, Falls Church, VA, 4th ed., 1980.

[6] Mortari, D., "Search-Less Algorithm for Star Pattern Recognition," *Journal of the Astronautical Sciences*, Vol. 45, No. 2, April-June 1997, pp. 179–194.

[7] Shuster, M.D., "Maximum Likelihood Estimation of Spacecraft Attitude," *The Journal of the Astronautical Sciences*, Vol. 37, No. 1, Jan.-March 1989, pp. 79–88.

[8] Wahba, G., "A Least-Squares Estimate of Satellite Attitude," *SIAM Review*, Vol. 7, No. 3, July 1965, pp. 409.

[9] Lerner, G.M., "Three-Axis Attitude Determination," *Spacecraft Attitude Determination and Control*, edited by J.R. Wertz, chap. 12, Kluwer Academic Publishers, The Netherlands, 1978.

[10] Shuster, M.D., "Attitude Determination from Vector Observations," *Journal of Guidance and Control*, Vol. 4, No. 1, Jan.-Feb. 1981, pp. 70–77.

[11] Mortari, D., "ESOQ: A Closed-Form Solution of the Wahba Problem," *Journal of the Astronautical Sciences*, Vol. 45, No. 2, April-June 1997, pp. 195–204.

[12] Markley, F.L., "Attitude Determination Using Vector Observations and the Singular Value Decomposition," *The Journal of the Astronautical Sciences*, Vol. 36, No. 3, July-Sept. 1988, pp. 245–258.

[13] Sun, D. and Crassidis, J.L., "Observability Analysis of Six-Degree-of-Freedom Configuration Determination Using Vector Observations," *Journal of Guidance, Control, and Dynamics*, Vol. 25, No. 6, Nov.-Dec. 2002, pp. 1149–1157.

[14] Battin, R.H., *An Introduction to the Mathematics and Methods of Astrodynamics*, American Institute of Aeronautics and Astronautics, Inc., New York, NY, 1987.

[15] Escobal, P.E., *Methods of Orbit Determination*, Krieger Publishing Company, Malabar, FL, 1965.

[16] Vallado, D.A. and McClain, W.D., *Fundamentals of Astrodynamics and Applications*, McGraw-Hill, New York, NY, 1997.

[17] Yunck, T.P., "Orbit Determination," *Global Positioning System: Theory and Applications*, edited by B. Parkinson and J. Spilker, Vol. 164 of *Progress in Astronautics and Aeronautics*, chap. 21, American Institute of Aeronautics and Astronautics, Washington, DC, 1996.

[18] Iliff, K.W., "Parameter Estimation of Flight Vehicles," *Journal of Guidance, Control, and Dynamics*, Vol. 12, No. 5, Sept.-Oct. 1989, pp. 261–280.

[19] Roskam, J., *Airplane Flight Dynamics and Automatic Flight Controls*, Design, Analysis and Research Corporation, Lawrence, KS, 1994.

[20] Aström, K.J. and Eykhoff, P., "System Identification-A Survey," *Automatica*, Vol. 7, No. 2, March 1971, pp. 123–162.

[21] Franklin, G.F., Powell, J.D., and Workman, M., *Digital Control of Dynamic Systems*, Addison Wesley Longman, Menlo Park, CA, 3rd ed., 1998.

[22] Yeh, F.B. and Yang, C.D., "New Time-Domain Identification Technique," *Journal of Guidance, Control, and Dynamics*, Vol. 10, No. 3, May-June 1987, pp. 313–316.

[23] Ibrahim, S.R. and Mikulcik, E.C., "A New Method for the Direct Identification of Vibration Parameters from the Free Response," *Shock and Vibration Bulletin*, Vol. 47, No. 4, Sept. 1977, pp. 183–198.

[24] Juang, J.N. and Pappa, R.S., "An Eigensystem Realization Algorithm for Modal Parameter Identification and Model Reduction," *Journal of Guidance, Control, and Dynamics*, Vol. 8, No. 5, Sept.-Oct. 1985, pp. 620–627.

[25] Juang, J.N., Phan, M., Horta, L.G., and Longman, R.W., "Identification of Observer/Kalman Filer Markov Parameters: Theory and Experiments," *Journal of Guidance, Control, and Dynamics*, Vol. 16, No. 2, March-April 1993, pp. 320–329.

[26] Juang, J.N. and Pappa, R.S., "Effects of Noise on Modal Parameters Identified by the Eigensystem Realization Algorithm," *Journal of Guidance, Control, and Dynamics*, Vol. 9, No. 3, May-June 1986, pp. 294–303.

[27] Yang, C.D. and Yeh, F.B., "Identification, Reduction, and Refinement of Model Parameters by the Eigensystem Realization Algorithm," *Journal of Guidance, Control, and Dynamics*, Vol. 13, No. 6, Nov.-Dec. 1990, pp. 1051–1059.

[28] Juang, J.N., *Applied System Identification*, Prentice Hall, Englewood Cliffs, NJ, 1994.

[29] Rajaram, S. and Junkins, J.L., "Identification of Vibrating Flexible Structures," *Journal of Guidance, Control, and Dynamics*, Vol. 8, No. 4, July-Aug. 1985, pp. 463–470.

[30] Junkins, J.L., Hughes, D.C., Wazni, K.P., and Pariyapong, V., "Vision-Based Navigation for Rendezvous, Docking and Proximity Operations," *22nd Annual AAS Guidance and Control Conference*, Breckenridge, CO, Feb. 1999, AAS 99-021.

[31] Shuster, M.D., "A Survey of Attitude Representations," *Journal of the Astronautical Sciences*, Vol. 41, No. 4, Oct.-Dec. 1993, pp. 439–517.

[32] Crassidis, J.L. and Markley, F.L., "Attitude Estimation Using Modified Rodrigues Parameters," *Proceedings of the Flight Mechanics/Estimation Theory*

Symposium, NASA-Goddard Space Flight Center, Greenbelt, MD, May 1996, pp. 71–83.

[33] Markley, F.L., Bauer, F.H., Deily, J.J., and Femiano, M.D., "Attitude Control System Conceptual Design for Geostationary Operational Environmental Satellite Spacecraft Series," *Journal of Guidance, Control, and Dynamics*, Vol. 18, No. 2, March-April 1995, pp. 247–255.

[34] Bookstein, F.L., "Fitting Conic Sections to Scattered Data," *Computer Graphics and Image Processing*, Vol. 9, 1979, pp. 56–71.

[35] Halíř, R. and Flusser, J., "Numerically Stable Direct Least Squares Fitting of Ellipses," *6th International Conference in Central Europe on Computer Graphics and Visualization, WSCG '98*, University of West Bohemia, Campus Bory, Plzen - Bory, Czech Republic, Feb. 1998, pp. 125–132.

[36] Gander, W., Golub, G.H., and Strebel, R., "Least-Squares Fitting of Circles and Ellipses," *Numerical analysis (in honour of Jean Meinguet)*, edited by editorial board Bulletin Belgian Mathematical Society, 1996, pp. 63–84.

[37] Cohen, C.E., "Attitude Determination," *Global Positioning System: Theory and Applications*, edited by B. Parkinson and J. Spilker, Vol. 64 of *Progress in Astronautics and Aeronautics*, chap. 19, American Institute of Aeronautics and Astronautics, Washington, DC, 1996.

[38] Crassidis, J.L. and Markley, F.L., "New Algorithm for Attitude Determination Using Global Positioning System Signals," *Journal of Guidance, Control, and Dynamics*, Vol. 20, No. 5, Sept.-Oct. 1997, pp. 891–896.

5

Sequential State Estimation

The advancement and perfection of mathematics are intimately connected with the prosperity of the State. Napoleon

IN the developments of the previous chapters, estimation concepts are formulated and applied to systems whose measured variables are related to the estimated parameters by *algebraic* equations. The present chapter extends these results to allow estimation of parameters embedded in the model of a *dynamical system*, where the model usually includes both *algebraic* and *differential* equations. We will find that the sequential estimation results of §1.3 and the probability concepts introduced in Chapter 2, developed for estimation of *algebraic* systems, remain valid for estimation of *dynamical* systems upon making the appropriate new interpretations of the matrices involved in the estimation algorithms. In the event that the differential equations have explicitly algebraic solutions, of course, the entire model becomes algebraic equations and the methods of the previous chapters apply immediately (see example 1.7 for instance). On the other hand, we'll find that the sequential estimation results of §1.3 must be extended to properly account for "motion" of the dynamical system between measurement and estimation epochs. We should now note that the words "sequential state estimation" and "filtering" are used synonymously throughout the remainder of the text. The concept of filtering is regularly stated when the time at which an estimate is desired coincides with the last measurement point.[1] In the examples presented in this chapter and in later chapters, sequential state estimation is often used to not only reconstruct state variables but also "filter" noisy measurement processes. Thus,"sequential state estimation" and "filtering" are often interchanged in the literature.

The formulations of the present chapter are developed as natural extensions of the estimation methods of the first two chapters using the differential equation models and notations of Chapter 4. We begin our discussion of sequential state estimation by showing a simple first-order sequential filtering process. Then we will introduce the concept of reconstructing all of the state variables in a dynamical system using Ackermann's formula. Next, the *Kalman filter* is derived for linear systems. We shall see that the filter structure remains unchanged from Ackermann's basic developments; however, the associated gain for the estimator in the Kalman filter is rigourously derived using the probability concepts introduced in Chapter 2. Then, the Kalman filter is expanded to include nonlinear dynamical models, which leads to the development of the *extended Kalman filter*. Formulations are presented for continuous-time measurements and models, discrete-time measurements and models, and discrete-time measurements with continuous-time models. Finally, several advanced topics are shown including: factorization methods, colored-noise Kalman filtering, adaptive filtering, error analysis, Unscented filtering, and robust filtering.

5.1 A Simple First-Order Filter Example

In the estimation formulations developed in the first two chapters, it has been assumed that a specific set of parameters are being estimated; additional data have been allowed, but the parameters being estimated remained unchanged. A more complicated situation arises whenever the set of parameters being estimated is allowed to change during the estimation process. To motivate the discussion, consider real-time estimation of the state of a maneuvering spacecraft. As each subset of observations becomes available, it is desired to obtain an optimal estimate of the state *at that instant* in order to, for example, provide the best current information to base control decisions upon.

In this section we introduce the concept of sequential state estimation by considering a simple first-order example that will be used to motivate the theoretical developments of this chapter. Suppose that a "truth" model is generated using the following first-order differential equation:

$$\dot{x}(t) = F\,x(t), \quad x(t_0) = 1 \tag{5.1a}$$

$$\tilde{y}(t) = H\,x(t) + v(t) \tag{5.1b}$$

Synthetic measurements are created for a 10-second time interval with $F = -1$ and $H = 1$, assuming that $v(t)$ is a zero-mean Gaussian noise process with the standard deviation given by 0.05. The measurements are shown in Figure 5.1.

Suppose now that we wish to estimate $x(t)$ using the available measurements and some dynamic model. In practice the actual "truth" model is unknown (if it were known exactly then we wouldn't need an estimator!). For this example, we will assume that the initial condition is known exactly, but the "modelled" value for F is given by $\bar{F} = -1.5$. Clearly, if we replace F with $\bar{F}$ in eqn. (5.1) and integrate this equation to find an estimate for $x(t)$, we would find that the estimated $x(t)$ is far from the truth. In order to produce better results, we shall use the age-old adage commonly spoken in control of dynamic systems: "when in doubt, use feedback!" Consider the following linear feedback system for the state and output estimates:

$$\dot{\hat{x}}(t) = \bar{F}\,\hat{x}(t) + K[\tilde{y}(t) - \bar{H}\,\hat{x}(t)], \quad \hat{x}(t_0) = 1 \tag{5.2a}$$

$$\hat{y}(t) = \bar{H}\,\hat{x}(t) \tag{5.2b}$$

where $\hat{x}(t)$ denotes the estimate of $x(t)$, K is a constant gain, and $\bar{H} = H = 1$. At this point we do not consider how to determine the value of K, but instead (since

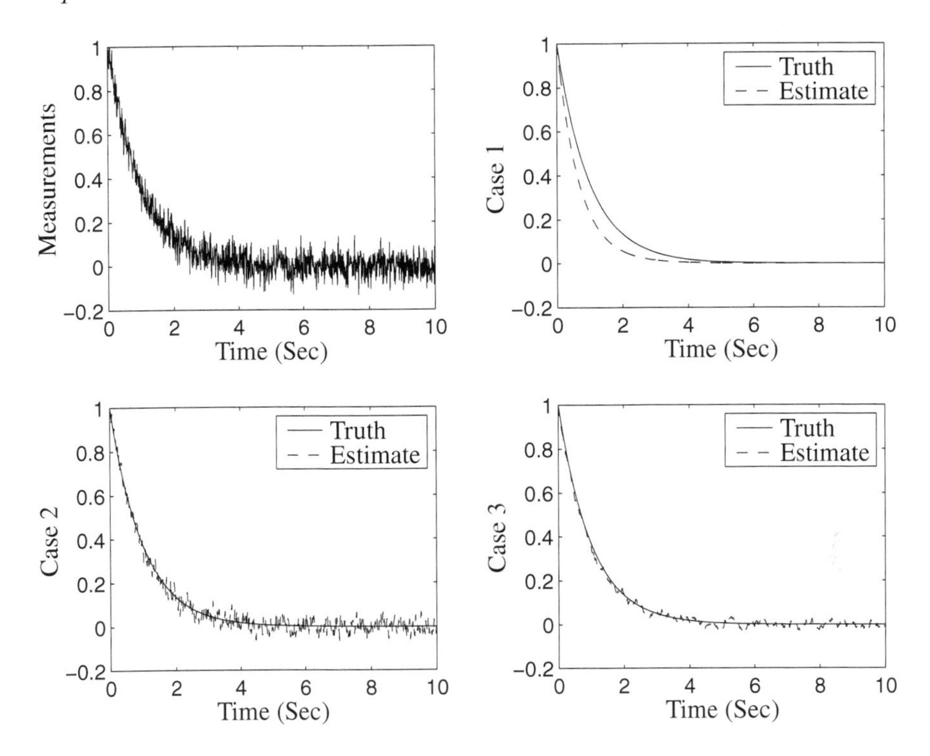

Figure 5.1: First-Order Filter Results

we know the truth) we will pick various values and compare the resulting $\hat{x}(t)$ with the true $x(t)$. Three cases are evaluated: Case 1 ($K = 0.1$), Case 2 ($K = 100$), and Case 3 ($K = 15$). The resulting estimates from each of these cases are shown in Figure 5.1. Clearly for small gains (such as Case 1) the estimates are far from the truth. Also, for large gains (such as Case 2) the estimates are very noisy. Case 3 depicts a gain that closely follows the truth, while at the same time providing filtered estimates.

This simple example illustrates the basic concepts used in state estimation and filtering. We can see from eqn. (5.2) that as the gain (K) decreases, measurements tend to be ignored and the system relies more heavily on the model (which in this case is incorrect leading to erroneous estimates). As the gain increases the estimates rely more on the measurements; however, if the gain is too large then the model tends to be ignored all together, as shown by Case 2. This concept can also be demonstrated using a frequency domain approach. The "filter dynamics" are given by $E = \bar{F} - K\bar{H}$ (here we assume that K is chosen so that the filter dynamics are stable), which is the inverse of the time constant of the system. In the frequency domain, the corner frequency (bandwidth) of the filter is given by $|E|$. As the gain K increases the corner frequency becomes larger, which yields a higher bandwidth in the system, thus allowing more high-frequency noise to enter into the estimate.

Conversely, as the gain K decreases the bandwidth decreases, which allows less noise through the filtered system. An "optimal" gain is one that both closely follows the model while at the same time provides filtered estimates.

5.2 Full-Order Estimators

In the previous section we showed a simple first-order filter. In the present section we expand the previous results to full-order (i.e., n^{th}-order) systems. For the first step we will assume that the plant dynamics (F, B, H), with $D = 0$, in eqn. (3.11) are known exactly; however, the initial condition $\mathbf{x}(t_0)$ is not known precisely. Expanding eqn. (5.2) for MIMO systems gives (assuming no errors in the plant dynamics)

$$\dot{\hat{\mathbf{x}}} = F\,\hat{\mathbf{x}} + B\,\mathbf{u} + K[\tilde{\mathbf{y}} - H\,\hat{\mathbf{x}}] \tag{5.3a}$$

$$\hat{\mathbf{y}} = H\,\hat{\mathbf{x}} \tag{5.3b}$$

Note that $\mathbf{u}$ is a deterministic quantity (such as a control input). The truth model is given by

$$\dot{\mathbf{x}} = F\,\mathbf{x} + B\,\mathbf{u} \tag{5.4a}$$

$$\mathbf{y} = H\,\mathbf{x} \tag{5.4b}$$

The measurement model follows

$$\tilde{\mathbf{y}} = H\,\mathbf{x} + \mathbf{v} \tag{5.5}$$

where $\mathbf{v}$ is a vector of measurement noise. In order to analyze the estimator's performance we can compute an error representing the difference between the estimated state and the true state:

$$\tilde{\mathbf{x}} \equiv \hat{\mathbf{x}} - \mathbf{x} \tag{5.6}$$

Taking the time derivative of eqn. (5.6) and substituting eqns. (5.3a) and (5.4a) into the resulting expression leads to

$$\dot{\tilde{\mathbf{x}}} = (F - K\,H)\tilde{\mathbf{x}} + K\,\mathbf{v} \tag{5.7}$$

Note that eqn. (5.7) is no longer a function of $\mathbf{u}$. Obviously, we must choose K so that $F - K\,H$ is stable. If the filter dynamics are stable and the measurements errors are negligibly small, then the error will decay to zero and remain there for any initial condition error. It is evident from the $K\,\mathbf{v}$ forcing term in eqn. (5.7) that if the gain K is large then the filter eigenvalues (poles) will be fast, but high-frequency noise can dominate the errors due to the measurements. If the gain K is too small then the errors may take too long to decay toward zero. We must choose K so that $F - K\,H$

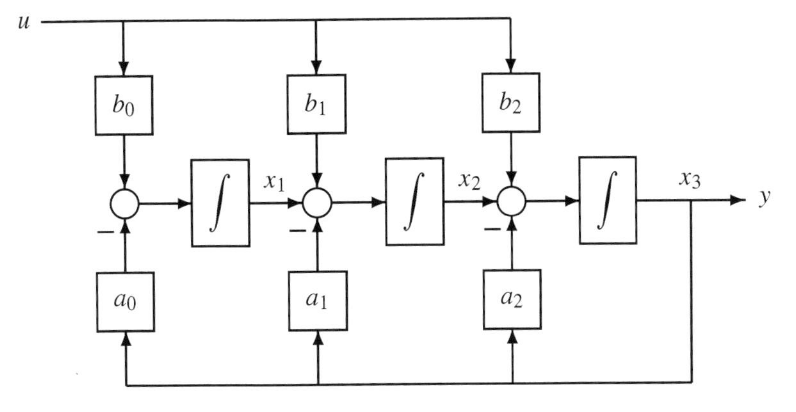

Figure 5.2: Third-Order Observer Canonical Form

is stable with reasonably fast eigenvalues, while at the same time providing filtered state estimates in the estimator.

One method to select K is to define a set of known estimator error-eigenvalue locations, and choose K so that these desired locations are achieved. This "pole-placement" concept is readily applied in the control of dynamic systems. We begin this concept by using the observer canonical form for SISO systems given by eqn. (3.98), which allows for a simple approach to place the estimator eigenvalues:

$$F_o = \begin{bmatrix} 0 & 0 & \cdots & 0 & -a_0 \\ 1 & 0 & \cdots & 0 & -a_1 \\ 0 & 1 & \cdots & 0 & -a_2 \\ \vdots & \vdots & \ddots & \vdots & \vdots \\ 0 & 0 & \cdots & 1 & -a_{n-1} \end{bmatrix} \tag{5.8a}$$

$$B_o = \begin{bmatrix} b_0 & b_1 & \cdots & b_{n-1} \end{bmatrix}^T \tag{5.8b}$$

$$H_o = \begin{bmatrix} 0 & 0 & \cdots & 1 \end{bmatrix} \tag{5.8c}$$

The coefficients of the characteristic equation are given by the last column of F_o.

Consider the third-order case, where the state matrix in eqn. (5.8a) reduces to

$$F_o = \begin{bmatrix} 0 & 0 & -a_0 \\ 1 & 0 & -a_1 \\ 0 & 1 & -a_2 \end{bmatrix} \tag{5.9}$$

Since we have assumed only a single measurement, then K reduces to a 3×1 vector. The estimator closed-loop state matrix $(F_o - K\,H_o)$ for this case is given by

$$F_o - K\,H_o = \begin{bmatrix} 0 & 0 & -(a_0 + k_1) \\ 1 & 0 & -(a_1 + k_2) \\ 0 & 1 & -(a_2 + k_3) \end{bmatrix} \tag{5.10}$$

where $K \equiv [k_1\ k_2\ k_3]^T$. A block diagram of this system is shown in Figure 5.2. This shows the advantage of this observer canonical form, since all of the feedback loops come from the output. The characteristic equation associated with the state matrix in eqn. (5.10) is given by

$$s^3 + (a_2 + k_3)\,s^2 + (a_1 + k_2)\,s + (a_0 + k_1) = 0 \tag{5.11}$$

Suppose that we have a desired characteristic equation formed from a set of desired eigenvalues in the estimator, given by

$$d(s) = s^3 + \delta_2\, s^2 + \delta_1\, s + \delta_0 = 0 \tag{5.12}$$

Then the gain matrix K can be obtained by comparing the corresponding coefficients in eqns. (5.11) and (5.12):

$$\begin{aligned} k_1 &= \delta_0 - a_0 \\ k_2 &= \delta_1 - a_1 \\ k_3 &= \delta_2 - a_2 \end{aligned} \tag{5.13}$$

This approach can easily be expanded to higher-order systems; however, this can become quite tedious and numerically inefficient. It would be useful if the gain K can be derived using the matrix F directly, without having to convert F into observer canonical form. Applying the Cayley-Hamilton theorem from eqn. (A.56), which states that every $n \times n$ matrix satisfies its own characteristic equation, to the matrix $E = F - KH$ in eqn. (5.12) leads to

$$d(E) = E^3 + \delta_2\, E^2 + \delta_1\, E + \delta_0\, I = 0 \tag{5.14}$$

Performing the multiplications for E^3 and E^2, and collecting terms gives

$$E^2 = F^2 - KHF - EKH \tag{5.15a}$$

$$E^3 = F^3 - KHF^2 - EKHF - E^2KH \tag{5.15b}$$

Substituting eqn. (5.15) into eqn. (5.14), and again collecting terms gives

$$\begin{aligned} &F^3 + \delta_2\, F^2 + \delta_1\, F + \delta_0\, I \\ &- \delta_1\, KH - \delta_2\, KHF - \delta_2\, EKH - KHF^2 - EKHF - E^2KH = 0 \end{aligned} \tag{5.16}$$

Since the first four terms are defined as $d(F)$, we can rewrite eqn. (5.16) as

$$d(F) = \left[(\delta_1\, K + \delta_2\, EK + E^2K)\ (\delta_2\, K + EK)\ K \right] \begin{bmatrix} H \\ HF \\ HF^2 \end{bmatrix} \tag{5.17}$$

Therefore, the gain K can be found from

$$K = d(F) \begin{bmatrix} H \\ HF \\ HF^2 \end{bmatrix}^{-1} \begin{bmatrix} 0 \\ 0 \\ 1 \end{bmatrix} \tag{5.18}$$

This can easily be extended for n^{th}-order systems to give *Ackermann's formula*:

$$K = d(F)\begin{bmatrix} H \\ HF \\ HF^2 \\ \vdots \\ HF^{n-1} \end{bmatrix}^{-1} \begin{bmatrix} 0 \\ 0 \\ 0 \\ \vdots \\ 1 \end{bmatrix} \equiv d(F)\mathcal{O}^{-1}\begin{bmatrix} 0 \\ 0 \\ 0 \\ \vdots \\ 1 \end{bmatrix} \tag{5.19}$$

where $\mathcal{O}$ is clearly the observability matrix derived in §3.4. Therefore, in order to place the eigenvalues of the estimator state matrix, the original system (F, H) must be observable.

Example 5.1: In this example we will demonstrate the usefulness of eqn. (5.19) to determine the required gain in the estimator for a simple second-order system. Consider the following general system matrices:

$$F = \begin{bmatrix} f_{11} & f_{12} \\ f_{21} & f_{22} \end{bmatrix}, \quad H = \begin{bmatrix} h_1 & h_2 \end{bmatrix}$$

where $f_{11}, f_{12}, f_{21}, f_{22}, h_1$, and h_2 are any real-valued numbers. The gain K is given by $K = \begin{bmatrix} k_1 & k_2 \end{bmatrix}^T$ for this case. The desired characteristic equation of the estimator is given by

$$d(s) = s^2 + \delta_1 s + \delta_0 = 0$$

Computing $\det(sI - F + KH) = 0$ allows us to solve for the gain K by comparing coefficients to the desired characteristic equation. Performing this operation gives

$$\delta_0 = (k_1 h_1 - f_{11})(k_2 h_2 - f_{22}) - (k_1 h_2 - f_{12})(k_2 h_1 - f_{21})$$
$$\delta_1 = k_1 h_1 + k_2 h_2 - f_{11} - f_{22}$$

Solving these two equations for k_1 and k_2 is not trivial (this is left as an exercise for the reader); however, using eqn. (5.19) the solution is straightforward leading to

$$k_1 = \frac{1}{b h_1 - a h_2}[d h_1 - c h_2 + \delta_1 (h_1 f_{12} - h_2 f_{11}) - \delta_0 h_2]$$
$$k_2 = \frac{1}{b h_1 - a h_2}[g h_1 - e h_2 + \delta_1 (h_1 f_{22} - h_2 f_{21}) + \delta_0 h_1]$$

where

$$a = h_1 f_{11} + h_2 f_{21}$$
$$b = h_1 f_{12} + h_2 f_{22}$$
$$c = f_{11}^2 + f_{12} f_{21}$$
$$d = f_{11} f_{12} + f_{12} f_{22}$$
$$e = f_{11} f_{21} + f_{21} f_{22}$$
$$g = f_{22}^2 + f_{12} f_{21}$$

Also, as $(bh_1 - ah_2) \to 0$ the gains k_1 and k_2 approach infinity. This is due to the fact that $(bh_1 - ah_2)$ is the determinant of the observability matrix. Therefore, as observability slips away the gains must increase in order to "see" the states. This can have a negative effect for noisy systems, as shown in §5.1.

If the system is in observer canonical form, then $h_1 = 0$, $h_2 = 1$, $f_{11} = 0$, and $f_{21} = 1$, and the gain expressions simplify significantly with $a = 1$, $b = f_{22}$, $c = f_{12}$, $d = f_{12}\, f_{22}$, $e = f_{22}$, and $g = f_{22}^2 + f_{12}$. Then the gains are given by

$$k_1 = f_{12} + \delta_0$$
$$k_2 = f_{22} + \delta_1$$

which is analogous to the expression shown in eqn. (5.11). This example clearly demonstrates the power of using Ackermann's to determine a gain K to match the desired characteristic equation in an estimator design.

5.2.1 Discrete-Time Estimators

We now will show Ackermann's formula for discrete-time system representations, given by eqn. (3.111). We can simply add a feedback term involving the difference between the measured and estimated output analogous to the continuous-time case; however, this gives an estimate at the current time based on the *previous* measurement (since $\hat{\mathbf{x}}_{k+1}$ will be used in the estimator). In order to provide a current estimate using the current measurement the discrete-time estimator is given by two coupled equations, given by

$$\boxed{\begin{aligned} \hat{\mathbf{x}}_{k+1}^- &= \Phi\,\hat{\mathbf{x}}_k^+ + \Gamma\,\mathbf{u}_k \\ \hat{\mathbf{x}}_k^+ &= \hat{\mathbf{x}}_k^- + K[\tilde{\mathbf{y}}_k - H\,\hat{\mathbf{x}}_k^-] \end{aligned}} \qquad \begin{aligned} &(5.20\text{a}) \\ &(5.20\text{b}) \end{aligned}$$

Equation (5.20a) is known as the *prediction* or *propagation* equation, and eqn. (5.20b) is known as the *update* equation. The truth model is given by

$$\mathbf{x}_{k+1} = \Phi\,\mathbf{x}_k + \Gamma\,\mathbf{u}_k \tag{5.21a}$$
$$\mathbf{y}_k = H\,\mathbf{x}_k \tag{5.21b}$$

A single estimator equation can be derived by simply substituting eqn. (5.20b) into eqn. (5.20a) giving

$$\hat{\mathbf{x}}_{k+1}^- = \Phi\,\hat{\mathbf{x}}_k^- + \Gamma\,\mathbf{u}_k + \Phi K[\tilde{\mathbf{y}}_k - H\,\hat{\mathbf{x}}_k^-] \tag{5.22}$$

The error states for the prediction and for the update are defined by

$$\tilde{\mathbf{x}}_k^- \equiv \hat{\mathbf{x}}_k^- - \mathbf{x}_k \tag{5.23a}$$
$$\tilde{\mathbf{x}}_k^+ \equiv \hat{\mathbf{x}}_k^+ - \mathbf{x}_k \tag{5.23b}$$

Taking one time-step ahead of eqn. (5.23) and substituting eqns. (5.20a) and (5.20b) into the resulting expressions leads to

$$\tilde{\mathbf{x}}_{k+1}^- = \Phi[I - KH]\tilde{\mathbf{x}}_k^- \quad (5.24a)$$

$$\tilde{\mathbf{x}}_{k+1}^+ = [I - KH]\Phi\tilde{\mathbf{x}}_k^+ \quad (5.24b)$$

Note that $\Phi[I - KH]$ and $[I - KH]\Phi$ have the same eigenvalues.

The discrete-time desired characteristic equation for the estimator is given by

$$d(z) = z^n + \delta_{n-1} z^{n-1} + \cdots + \delta_1 z + \delta_0 = 0 \quad (5.25)$$

The form for the estimator error in eqn. (5.24b) is similar to the continuous-time case in eqn. (5.7) with H replaced with $H\Phi$. Therefore, Ackermann's formula for the discrete-time case is given by

$$K = d(\Phi)\begin{bmatrix} H\Phi \\ H\Phi^2 \\ H\Phi^3 \\ \vdots \\ H\Phi^n \end{bmatrix}^{-1}\begin{bmatrix} 0 \\ 0 \\ 0 \\ \vdots \\ 1 \end{bmatrix} \equiv d(\Phi)\Phi^{-1}\mathcal{O}_d^{-1}\begin{bmatrix} 0 \\ 0 \\ 0 \\ \vdots \\ 1 \end{bmatrix} \quad (5.26)$$

where $\mathcal{O}_d$ is the discrete-time observability matrix given in eqn. (3.117). As in the continuous-time case, the discrete-time system must be observable for the inverse in eqn. (5.26) to exist.

The estimator design approach introduced in this section can be tedious and somewhat heuristic for higher-order systems since it is not commonly known where to properly place all the estimator eigenvalues. To overcome this difficulty, we can choose 2 of the n eigenvalues so that a dominant second-order system is produced. The remaining eigenvalues can be chosen to have real parts corresponding to a sufficiently damped response in the estimator.[2] Thus the higher-order estimator will mimic (and can be subsequently analyzed as) a second-order system. Thankfully, there is a better way, as will next be seen in the derivation of the Kalman filter.

5.3 The Discrete-Time Kalman Filter

The estimators derived in §5.2 require a desired characteristic equation in the filter dynamics. The answer to the obvious question "How do we choose the poles of the estimator?" is not trivial. In practice, this usually entails an ad hoc approach until a specified performance level is achieved. The *Kalman filter*[3] provides a rigorous theoretical approach to "place" the poles of the estimator, based upon stochastic processes for the measurement error and model error. As is shown in Chapter 2, we do not know the exact values for these errors; however, we do make some assumptions on the nature of the errors (e.g., a zero-mean Gaussian noise process). Three formulations will be given. The first, described in this section, assumes both discrete-time dynamic models and measurements; the second, described in the next section, assumes both continuous-time dynamic models and measurements; and the third assumes continuous-time dynamic models with discrete-time measurements.

5.3.1 Kalman Filter Derivation

We begin the derivation of the discrete-time Kalman filter assuming that both the model and measurements are available in discrete-time form. Suppose that the initial condition of a state $\mathbf{x}_0$ is unknown (as in §5.2); in addition suppose that the discrete-time model and measurements are corrupted by noise. The "truth" model for this case is given by

$$\boxed{\begin{aligned}\mathbf{x}_{k+1} &= \Phi_k \mathbf{x}_k + \Gamma_k \mathbf{u}_k + \Upsilon_k \mathbf{w}_k \qquad (5.27\text{a})\\ \tilde{\mathbf{y}}_k &= H_k \mathbf{x}_k + \mathbf{v}_k \qquad (5.27\text{b})\end{aligned}}$$

where $\mathbf{v}_k$ and $\mathbf{w}_k$ are assumed to be zero-mean Gaussian white-noise processes, which means that the errors are not correlated forward or backward in time so that

$$E\left\{\mathbf{v}_k \mathbf{v}_j^T\right\} = \begin{cases} 0 & k \neq j \\ R_k & k = j \end{cases} \tag{5.28}$$

and

$$E\left\{\mathbf{w}_k \mathbf{w}_j^T\right\} = \begin{cases} 0 & k \neq j \\ Q_k & k = j \end{cases} \tag{5.29}$$

This requirement preserves the block diagonal structure of the covariance and weight matrices introduced in §1.3. We further assume that $\mathbf{v}_k$ and $\mathbf{w}_k$ are *uncorrelated* so that $E\left\{\mathbf{v}_k \mathbf{w}_k^T\right\} = 0$ for all k. The quantity $\mathbf{w}_k$ is a forcing ("process") noise on the system of differential equations.

It is desired to update the current estimate of the state ($\hat{\mathbf{x}}_k$) to obtain ($\hat{\mathbf{x}}_{k+1}$) based upon all $k+1$ measurement subsets. We will still assume that the estimator form given by eqn. (5.20) is valid; however, the gain K can vary in time, so that

$$\boxed{\begin{aligned}\hat{\mathbf{x}}_{k+1}^- &= \Phi_k \hat{\mathbf{x}}_k^+ + \Gamma_k \mathbf{u}_k \qquad (5.30\text{a})\\ \hat{\mathbf{x}}_k^+ &= \hat{\mathbf{x}}_k^- + K_k[\tilde{\mathbf{y}}_k - H_k \hat{\mathbf{x}}_k^-] \qquad (5.30\text{b})\end{aligned}}$$

Proceeding from the developments of Chapter 2, we define the following error covariances:

$$P_k^- \equiv E\left\{\tilde{\mathbf{x}}_k^- \tilde{\mathbf{x}}_k^{-T}\right\}, \quad P_{k+1}^- \equiv E\left\{\tilde{\mathbf{x}}_{k+1}^- \tilde{\mathbf{x}}_{k+1}^{-T}\right\} \tag{5.31a}$$

$$P_k^+ \equiv E\left\{\tilde{\mathbf{x}}_k^+ \tilde{\mathbf{x}}_k^{+T}\right\}, \quad P_{k+1}^+ \equiv E\left\{\tilde{\mathbf{x}}_{k+1}^+ \tilde{\mathbf{x}}_{k+1}^{+T}\right\} \tag{5.31b}$$

where

$$\tilde{\mathbf{x}}_k^- \equiv \hat{\mathbf{x}}_k^- - \mathbf{x}_k, \quad \tilde{\mathbf{x}}_{k+1}^- \equiv \hat{\mathbf{x}}_{k+1}^- - \mathbf{x}_{k+1} \tag{5.32a}$$

$$\tilde{\mathbf{x}}_k^+ \equiv \hat{\mathbf{x}}_k^+ - \mathbf{x}_k, \quad \tilde{\mathbf{x}}_{k+1}^+ \equiv \hat{\mathbf{x}}_{k+1}^+ - \mathbf{x}_{k+1} \tag{5.32b}$$

are the state errors in the prediction and update, respectively. Our goal is to derive expressions for both P_{k+1}^- and P_{k+1}^+, and also derive an optimal expression for the gain K_k in eqn. (5.30b). Since eqn. (5.30a) is not a direct function of the gain K_k, the expression for P_{k+1}^- is fairly straightforward to derive. Substituting eqns. (5.27a) and (5.30a) into eqn. (5.32a), and using the definition of $\tilde{\mathbf{x}}_k^+$ in eqn. (5.32b) leads to

$$\tilde{\mathbf{x}}_{k+1}^- = \Phi_k \tilde{\mathbf{x}}_k^+ - \Upsilon_k \mathbf{w}_k \tag{5.33}$$

Note that eqn. (5.33) is not a function $\mathbf{u}_k$, since this term represents a known (deterministic) forcing input. Then P_{k+1}^- is given by

$$\begin{aligned} P_{k+1}^- &\equiv E\left\{\tilde{\mathbf{x}}_{k+1}^- \tilde{\mathbf{x}}_{k+1}^{-T}\right\} \\ &= E\left\{\Phi_k \tilde{\mathbf{x}}_k^+ \tilde{\mathbf{x}}_k^{+T} \Phi_k^T\right\} - E\left\{\Phi_k \tilde{\mathbf{x}}_k^+ \mathbf{w}_k^T \Upsilon_k^T\right\} \\ &- E\left\{\Upsilon_k \mathbf{w}_k \tilde{\mathbf{x}}_k^{+T} \Phi_k^T\right\} + E\left\{\Upsilon_k \mathbf{w}_k \mathbf{w}_k^T \Upsilon_k^T\right\} \end{aligned} \tag{5.34}$$

From eqn. (5.27a) we see that $\mathbf{w}_k$ and $\tilde{\mathbf{x}}_k^+$ are uncorrelated since $\tilde{\mathbf{x}}_{k+1}^+$ (not $\tilde{\mathbf{x}}_k^+$) directly depends on $\mathbf{w}_k$. Therefore $E\left\{\tilde{\mathbf{x}}_k^+ \mathbf{w}_k^T\right\} = E\left\{\mathbf{w}_k \tilde{\mathbf{x}}_k^{+T}\right\} = 0$. Using the definitions in eqns. (5.29) and (5.31b), eqn. (5.34) reduces to

$$\boxed{P_{k+1}^- = \Phi_k P_k^+ \Phi_k^T + \Upsilon_k Q_k \Upsilon_k^T} \tag{5.35}$$

with initial condition given by $P_0^- = E\left\{\tilde{\mathbf{x}}_0^- \tilde{\mathbf{x}}_0^{-T}\right\}$.

Our next step is to develop an optimal expression for P_k^+. Substituting eqn. (5.27b) into eqn. (5.30b), and then substituting the resulting expression into eqn. (5.32b) leads to

$$\tilde{\mathbf{x}}_k^+ = (I - K_k H_k)\hat{\mathbf{x}}_k^- + K_k H_k \mathbf{x}_k + K_k \mathbf{v}_k - \mathbf{x}_k \tag{5.36}$$

From the definition in eqn. (5.32a), eqn. (5.36) reduces to

$$\tilde{\mathbf{x}}_k^+ = (I - K_k H_k)\tilde{\mathbf{x}}_k^- + K_k \mathbf{v}_k \tag{5.37}$$

Then P_k^+ is given by

$$\begin{aligned} P_k^+ &\equiv E\left\{\tilde{\mathbf{x}}_k^+ \tilde{\mathbf{x}}_k^{+T}\right\} \\ &= E\left\{(I - K_k H_k)\tilde{\mathbf{x}}_k^- \tilde{\mathbf{x}}_k^{-T} (I - K_k H_k)^T\right\} \\ &+ E\left\{(I - K_k H_k)\tilde{\mathbf{x}}_k^- \mathbf{v}_k^T K_k^T\right\} \\ &+ E\left\{K_k \mathbf{v}_k \tilde{\mathbf{x}}_k^{-T} (I - K_k H_k)^T\right\} + E\left\{K_k \mathbf{v}_k \mathbf{v}_k^T K_k^T\right\} \end{aligned} \tag{5.38}$$

From eqn. (5.30b) we see that $\mathbf{v}_k$ and $\tilde{\mathbf{x}}_k^-$ are uncorrelated since $\tilde{\mathbf{x}}_k^+$ (not $\tilde{\mathbf{x}}_k^-$) directly depends on $\mathbf{v}_k$. Therefore $E\left\{\tilde{\mathbf{x}}_k^- \mathbf{v}_k^T\right\} = E\left\{\mathbf{v}_k \tilde{\mathbf{x}}_k^{-T}\right\} = 0$. Using the definition in eqns. (5.28) and (5.31a), then eqn. (5.38) reduces to

$$\boxed{P_k^+ = [I - K_k H_k] P_k^- [I - K_k H_k]^T + K_k R_k K_k^T} \tag{5.39}$$

In order to determine the gain K_k we minimize the trace of P_k^+, which is equivalent to minimizing the length of the estimation error vector:

$$\text{minimize} \quad J(K_k) = \text{Tr}(P_k^+) \tag{5.40}$$

Using the helpful trace identities in eqn. (2.37) with symmetric P_k^- and R_k leads to

$$\frac{\partial J}{\partial K_k} = 0 = -2(I - K_k H_k) P_k^- H_k^T + 2 K_k R_k \tag{5.41}$$

Solving eqn. (5.41) for K_k gives

$$\boxed{K_k = P_k^- H_k^T [H_k P_k^- H_k^T + R_k]^{-1}} \tag{5.42}$$

Substituting eqn. (5.42) into eqn. (5.39) yields

$$\begin{aligned} P_k^+ &= P_k^- - K_k H_k P_k^- - P_k^- H_k^T K_k^T + K_k [H_k P_k^- H_k^T + R_k] K_k^T \\ &= P_k^- - K_k H_k P_k^- \end{aligned} \tag{5.43}$$

Therefore

$$\boxed{P_k^+ = [I - K_k H_k] P_k^-} \tag{5.44}$$

Substituting eqn. (5.42) into eqn. (5.44) gives

$$P_k^+ = P_k^- - P_k^- H_k^T [H_k P_k^- H_k^T + R_k]^{-1} H_k P_k^- \tag{5.45}$$

An alternative form for the update P_k^+ is given by using the matrix inversion lemma in eqn. (1.69), which yields

$$P_k^+ = [(P_k^-)^{-1} + H_k^T R_k^{-1} H_k]^{-1} \tag{5.46}$$

Equation (5.45) implies that the update stage of the discrete-time Kalman filter *decreases* the covariance (while the propagation stage in eqn. (5.35) *increases* the covariance).[4] This observation is intuitively consistent since in general more measurements improve the state estimate.

The gain K_k in eqn. (5.42) can also be written as

$$\boxed{K_k = P_k^+ H_k^T R_k^{-1}} \tag{5.47}$$

To prove the identity we manipulate eqn. (5.42) as follows:

$$\begin{aligned} K_k &= P_k^- H_k^T [H_k P_k^- H_k^T + R_k]^{-1} \\ &= P_k^- H_k^T R_k^{-1} R_k [H_k P_k^- H_k^T + R_k]^{-1} \\ &= P_k^- H_k^T R_k^{-1} [I + H_k P_k^- H_k^T R_k^{-1}]^{-1} \end{aligned} \tag{5.48}$$

Equation (5.48) can now be rewritten as

$$K_k [I + H_k P_k^- H_k^T R_k^{-1}] = P_k^- H_k^T R_k^{-1} \tag{5.49}$$

Collecting terms now gives

$$\begin{aligned} K_k &= P_k^- H_k^T R_k^{-1} - K_k H_k P_k^- H_k^T R_k^{-1} \\ &= [I - K_k H_k] P_k^- H_k^T R_k^{-1} \end{aligned} \tag{5.50}$$

Substituting (5.44) into eqn. (5.50) proves the identity in eqn. (5.47).

A further expression can be derived for the state update in eqn. (5.30b). Equation (5.44) can be rearranged as

$$[I - K_k H_k] = P_k^+ \left(P_k^-\right)^{-1} \tag{5.51}$$

Also, the state update in eqn. (5.30b) can be rearranged as

$$\hat{\mathbf{x}}_k^+ = [I - K_k H_k] \hat{\mathbf{x}}_k^- + K_k \tilde{\mathbf{y}}_k \tag{5.52}$$

Substituting eqns. (5.47) and (5.51) into eqn. (5.52) gives

$$\boxed{\hat{\mathbf{x}}_k^+ = P_k^+ \left[\left(P_k^-\right)^{-1} \hat{\mathbf{x}}_k^- + H_k^T R_k^{-1} \tilde{\mathbf{y}}_k \right]} \tag{5.53}$$

Equation (5.53) is not particularly useful since the inverse of P_k^- is required, but its helpfulness will be shown in the derivation of the discrete-time fixed-interval smoother in Chapter 6.

The discrete-time Kalman filter is summarized in Table 5.1. First, initial conditions for the state and error covariance are given. If a measurement is given at the initial time then the state and covariance are updated using eqns. (5.42), (5.30b), and (5.44) with $\hat{\mathbf{x}}_0^- = \hat{\mathbf{x}}_0$ and $P_0^- = P_0$. Then, the state estimate and covariance are propagated to the next time step using eqns. (5.30a) and (5.35). If a measurement isn't given at the initial time then the estimate and covariance are propagated first to the next available measurement point with $\hat{\mathbf{x}}_0^+ = \hat{\mathbf{x}}_0$ and $P_0^+ = P_0$. The process is then repeated sequentially until all measurement times have been used in the filter.

We note that the structure of the discrete-time Kalman filter has the same form as the discrete estimator shown in §5.2.1, but the gain in the Kalman filter has been derived from an optimal probabilistic approach using methods from Chapter 2, namely a minimum variance approach. The propagation stage of the Kalman filter gives a

Table 5.1: Discrete-Time Linear Kalman Filter

Model	$\mathbf{x}_{k+1} = \Phi_k \mathbf{x}_k + \Gamma_k \mathbf{u}_k + \Upsilon_k \mathbf{w}_k, \quad \mathbf{w}_k \sim N(\mathbf{0}, Q_k)$ $\tilde{\mathbf{y}}_k = H_k \mathbf{x}_k + \mathbf{v}_k, \quad \mathbf{v}_k \sim N(\mathbf{0}, R_k)$
Initialize	$\hat{\mathbf{x}}(t_0) = \hat{\mathbf{x}}_0$ $P_0 = E\left\{\tilde{\mathbf{x}}(t_0)\,\tilde{\mathbf{x}}^T(t_0)\right\}$
Gain	$K_k = P_k^- H_k^T [H_k P_k^- H_k^T + R_k]^{-1}$
Update	$\hat{\mathbf{x}}_k^+ = \hat{\mathbf{x}}_k^- + K_k[\tilde{\mathbf{y}}_k - H_k \hat{\mathbf{x}}_k^-]$ $P_k^+ = [I - K_k H_k] P_k^-$
Propagation	$\hat{\mathbf{x}}_{k+1}^- = \Phi_k \hat{\mathbf{x}}_k^+ + \Gamma_k \mathbf{u}_k$ $P_{k+1}^- = \Phi_k P_k^+ \Phi_k^T + \Upsilon_k Q_k \Upsilon_k^T$

time update through a *prediction* of $\hat{\mathbf{x}}^-$ and covariance P^-. The measurement update stage of the Kalman filter gives a *correction* based on the measurement to yield a new *a posteriori* estimate $\hat{\mathbf{x}}^+$ and covariance P^+.[5] Together these equations form the *predictor-corrector* form of the Kalman filter.

The propagation and measurement update equations can be combined to form the *a priori* recursive form of the Kalman filter. This is accomplished by substituting eqn. (5.30b) into eqn. (5.30a), and substituting eqn. (5.44) into eqn. (5.35), giving

$$\hat{\mathbf{x}}_{k+1} = \Phi_k \hat{\mathbf{x}}_k + \Gamma_k \mathbf{u}_k + \Phi_k K_k [\tilde{\mathbf{y}}_k - H_k \hat{\mathbf{x}}_k] \tag{5.54a}$$

$$K_k = P_k H_k^T [H_k P_k H_k^T + R_k]^{-1} \tag{5.54b}$$

$$P_{k+1} = \Phi_k P_k \Phi_k^T - \Phi_k K_k H_k P_k \Phi_k^T + \Upsilon_k Q_k \Upsilon_k^T \tag{5.54c}$$

Equation (5.54c) is known as the *discrete Riccati equation.*

5.3.2 Stability and Joseph's Form

The filter stability can be proved by using Lyapunov's direct method, which is discussed for discrete-time systems in §3.6. We wish to show that the estimation error dynamics, $\tilde{\mathbf{x}}_k \equiv \hat{\mathbf{x}}_k - \mathbf{x}_k$, are stable. For the discrete-time Kalman filter we consider the following candidate Lyapunov function:

$$V(\tilde{\mathbf{x}}) = \tilde{\mathbf{x}}_k^T P_k^{-1} \tilde{\mathbf{x}}_k \tag{5.55}$$

Since P_k is required to be positive definite, then clearly its inverse exists and $V(\tilde{\mathbf{x}}) > 0$ for all $\tilde{\mathbf{x}}_k \neq \mathbf{0}$. The increment of $V(\tilde{\mathbf{x}})$ is given by

$$\Delta V(\tilde{\mathbf{x}}) = \tilde{\mathbf{x}}_{k+1}^T P_{k+1}^{-1} \tilde{\mathbf{x}}_{k+1} - \tilde{\mathbf{x}}_k^T P_k^{-1} \tilde{\mathbf{x}}_k \tag{5.56}$$

Stability is proven if we can show that $\Delta V(\tilde{\mathbf{x}}) < 0$. Substituting eqns. (5.27a) and (5.54a) into $\tilde{\mathbf{x}}_{k+1} = \hat{\mathbf{x}}_{k+1} - \mathbf{x}_{k+1}$, and collecting terms leads to

$$\tilde{\mathbf{x}}_{k+1} = \Phi_k [I - K_k H_k] \tilde{\mathbf{x}}_k + \Phi_k K_k \mathbf{v}_k - \Upsilon_k \mathbf{w}_k \tag{5.57}$$

We only need to consider the homogeneous part of eqn. (5.57) since the matrix $\Phi_k [I - K_k H_k]$ defines the stability of the filter. Substituting $\tilde{\mathbf{x}}_{k+1} = \Phi_k [I - K_k H_k] \tilde{\mathbf{x}}_k$ into eqn. (5.56) gives the following necessary condition for stability:

$$\tilde{\mathbf{x}}_k^T \left\{ [I - K_k H_k]^T \Phi_k^T P_{k+1}^{-1} \Phi_k [I - K_k H_k] - P_k^{-1} \right\} \tilde{\mathbf{x}}_k < 0 \tag{5.58}$$

Therefore, stability is achieved if the matrix within the brackets in eqn. (5.58) can be shown to be negative definite, i.e.,

$$[I - K_k H_k]^T \Phi_k^T P_{k+1}^{-1} \Phi_k [I - K_k H_k] - P_k^{-1} < 0 \tag{5.59}$$

Equation (5.59) can be rewritten as

$$I - P_{k+1} \Phi_k^{-T} [I - K_k H_k]^{-T} P_k^{-1} [I - K_k H_k]^{-1} \Phi_k^{-1} < 0 \tag{5.60}$$

Substituting eqn. (5.39) into eqn. (5.35) gives the following form for P_{k+1}:

$$P_{k+1} = \Phi_k [I - K_k H_k] P_k [I - K_k H_k]^T \Phi_k^T + \Phi_k K_k R K_k^T \Phi_k^T + \Upsilon_k Q_k \Upsilon_k^T \tag{5.61}$$

Substituting eqn. (5.61) into eqn. (5.60) gives

$$\begin{aligned} &- [\Phi_k K_k R_k K_k^T \Phi_k^T + \Upsilon_k Q_k \Upsilon_k^T] \\ &\times \Phi_k^{-T} [I - K_k H_k]^{-T} P_k^{-1} [I - K_k H_k]^{-1} \Phi_k^{-1} < 0 \end{aligned} \tag{5.62}$$

Since $\Phi_k^{-T} [I - K_k H_k]^{-T} P_k^{-1} [I - K_k H_k]^{-1} \Phi_k^{-1}$ is positive definite, eqn. (5.62) reduces down to

$$-[\Phi_k K_k R_k K_k^T \Phi_k^T + \Upsilon_k Q_k \Upsilon_k^T] < 0 \tag{5.63}$$

Clearly if R_k is positive definite and Q_k is at least positive semi-definite then the Lyapunov condition is satisfied and the discrete-time Kalman filter is stable.

In the previous derivations of the discrete-time Kalman filter the covariance matrix P_k must remain positive definite. We now show that if P_k is positive definite then P_{k+1} is also positive definite. Assuming that $Q_k = 0$ without loss in generality, from the recursive Riccati equation in eqn. (5.54c), P_{k+1} will remain positive definite if the following condition is true:

$$P_k > P_k H_k^T [H_k P_k H_k^T + R_k]^{-1} H_k P_k \tag{5.64}$$

Multiplying the left side and right side of eqn. (5.64) by H_k and H_k^T, respectively, gives

$$H_k P_k H_k^T > H_k P_k H_k^T [H_k P_k H_k^T + R_k]^{-1} H_k P_k H_k^T \tag{5.65}$$

Next, we assume that the inverse of $H_k P_k H_k^T$ exists (i.e., the number of measured observations is less than the number of states), which gives the following condition:

$$H_k P_k H_k^T + R_k > H_k P_k H_k^T \tag{5.66}$$

Clearly, if R_k is positive definite, then eqn. (5.66) is satisfied and P_{k+1} will be positive definite. Although this condition is theoretically true, numerical roundoff errors can still make P_{k+1} become negative definite. There are a number of numerical solutions to this problem, which will be further discussed in §5.7.1. One method involves using eqn. (5.39) instead of eqn. (5.44), which is referred to as the *Joseph stabilized version.*[6] This can be shown by substituting $K_k \rightarrow K_k + \delta K_k$ and $P_k^+ \rightarrow P_k^+ + \delta P_k^+$. Using these definitions eqn. (5.44) can be written as

$$P_k^+ + \delta P_k^+ = [I - K_k H_k - \delta K_k H_k] P_k^- \tag{5.67}$$

Therefore, from the definition of P_k^+ in eqn. (5.44) the perturbation δP_k^+ is given by

$$\boxed{\delta P_k^+ = -\delta K_k H_k P_k^-} \tag{5.68}$$

Equation (5.68) shows a first-order perturbation (i.e., δP_k^+ is a direct function of δK_k), which may produce roundoff errors in a computational algorithm. Substituting $K_k \rightarrow K_k + \delta K_k$ into eqn. (5.39) yields

$$\begin{aligned} \delta P^+ &= \delta K_k [H_k P_k^- H_k^T + R_k] \delta K_k^T \\ &+ \delta K_k [R_k K_k^T - H_k P_k^- (I - K_k H_k)^T] \\ &+ [K_k R_k - (I - K_k H_k) P_k^- H_k^T] \delta K_k^T \end{aligned} \tag{5.69}$$

We now will prove that $K_k R_k - (I - K_k H_k) P_k^- H_k^T = 0$. From the definition of P_k^+ in eqn. (5.44) we have

$$K_k R_k - (I - K_k H_k) P_k^- H_k^T = K_k R_k - P_k^+ H_k^T \tag{5.70}$$

Substituting the other definition of the gain K_k from eqn. (5.47) into eqn. (5.70) gives

$$K_k R_k - (I - K_k H_k) P_k^- H_k^T = P_k^+ H_k^T - P_k^+ H_k^T = 0 \tag{5.71}$$

Therefore, eqn. (5.69) reduces to

$$\boxed{\delta P_k^+ = \delta K_k [H_k P_k^- H_k^T + R_k] \delta K_k^T} \tag{5.72}$$

Equation (5.72) shows a second-order perturbation in δK_k, which provides a more robust approach in terms of numerical stability. However, Joseph's stabilized version has more computations than the form given by eqn. (5.44). Hence, a filter designer must trade off computational workload versus potential roundoff errors.

5.3.3 Information Filter and Sequential Processing

The gain K_k in eqn. (5.42) requires an inverse of order R_k, which may cause computational and numerical difficulties for large measurement sets. In order to circumvent these difficulties the *information* form of the Kalman filter can be used. The information matrix (denoted as $\mathcal{P}$) is simply the inverse of the covariance matrix P (i.e., $\mathcal{P} \equiv P^{-1}$). From eqn. (5.46) the update equation for $\mathcal{P}$ is given by

$$\boxed{\mathcal{P}_k^+ = \mathcal{P}_k^- + H_k^T R_k^{-1} H_k} \tag{5.73}$$

The information propagation is given from eqn. (5.35) by using the matrix inversion lemma in eqn. (1.69), which yields

$$\boxed{\mathcal{P}_{k+1}^- = \left[I - \Psi_k \Upsilon_k \left(\Upsilon_k^T \Psi_k \Upsilon_k + Q_k^{-1} \right)^{-1} \Upsilon_k^T \right] \Psi_k} \tag{5.74}$$

where

$$\Psi_k \equiv \Phi_k^{-T} \mathcal{P}_k^+ \Phi_k^{-1} \tag{5.75}$$

The gain can be computed from eqn. (5.47) directly as

$$\boxed{K_k = (\mathcal{P}_k^+)^{-1} H_k^T R_k^{-1}} \tag{5.76}$$

The information form clearly requires inverses of Φ_k and Q_k, which must exist. The inverse of Φ_k exists in most cases, unless a deadbeat response (i.e., a discrete pole at zero) is given in the model. However, Q_k may be zero in some cases, and the information filter cannot be used in this case. Also, if the initial state is known precisely then $P(t_0) = 0$, and the information filter cannot be initialized. Furthermore, the inverse of $\mathcal{P}_k^+$ is required in the gain calculation. The advantage of the information filter is that the largest dimension matrix inverse required is equivalent to the size of the state. Even though more inverses are needed, the information filter may be more computationally efficient than the traditional Kalman filter when the size of the measurement vector is much larger than the size of the state vector.

Another more commonly used approach to handle large measurement vectors in the Kalman filter is to use sequential processing.[4] This procedure involves processing one measurement at a time, repeated in sequence at each sampling instant. The gain and covariance are updated until all measurements at each sampling instant have been processed. The result produces estimates that are equivalent to processing all measurements together at one time instant. The underlying principle of this approach is rooted in the linearity of the Kalman filter update equation, where the rules of superposition in §3.1 apply unequivocally. This approach assumes that the measurements are uncorrelated at each time instant (i.e., R_k is a diagonal matrix). If this is not true then a linear transformation using the methods outlined in §3.1.4 can be used. We perform a linear transformation of the measurement $\tilde{\mathbf{y}}_k$ in eqn. (5.27b), giving a new measurement $\tilde{\mathbf{z}}_k$:

$$\tilde{\mathbf{z}}_k \equiv T_k \tilde{\mathbf{y}}_k = T_k H_k \mathbf{x}_k + T_k \mathbf{v}_k \tag{5.77a}$$

$$\equiv \mathcal{H}_k \mathbf{x}_k + \boldsymbol{\upsilon}_k \tag{5.77b}$$

where

$$\mathcal{H}_k \equiv T_k H_k \tag{5.78a}$$

$$\upsilon_k \equiv T_k \mathbf{v}_k \tag{5.78b}$$

Clearly, υ_k has zero mean and its covariance is given by $\mathcal{R}_k \equiv E\left\{\upsilon_k \upsilon_k^T\right\} = T_k^T R_k T_k$. Reference [7] shows that the eigenvectors of a real symmetric matrix are orthogonal. Therefore, using the results of §3.1.4, if T_k is chosen to be the matrix whose columns are the eigenvectors of R_k, then $\mathcal{R}_k$ is a diagonal matrix with elements given by the eigenvalues of R_k. Note that this decomposition has to be applied at each time instant; however, for many systems the measurement error process is *stationary* so that R_k is constant for all times, denoted simply by R. Therefore, in this case, the decomposition needs to be only performed once, which can significantly reduce the computational load. The Kalman gain and covariance update can now be performed using a sequential procedure, given by

$$\boxed{K_{i_k} = \frac{P_{i-1_k}^- \mathcal{H}_{i_k}^T}{\mathcal{H}_{i_k} P_{i-1_k}^- \mathcal{H}_{i_k}^T + \mathcal{R}_{i_k}}} \tag{5.79a}$$

$$\boxed{P_{i_k}^+ = [I - K_{i_k}\mathcal{H}_{i_k}]P_{i-1_k}^+, \quad P_{0_k}^+ = P_k^-} \tag{5.79b}$$

where i represents the ith measurement, $\mathcal{R}_i$ is the ith diagonal element of $\mathcal{R}$, and $\mathcal{H}_i$ is the ith row of $\mathcal{H}$. The process continues until all m measurements are processed (i.e., $i = 1, 2, \ldots, m$), with $P_k^+ = P_{m_k}^+$. The state update can now be computed using eqn. (5.30b):

$$\boxed{\hat{\mathbf{x}}_k^+ = \hat{\mathbf{x}}_k^- + P_k^+ \mathcal{H}_k^T \mathcal{R}_k^{-1}[\tilde{\mathbf{z}}_k - \mathcal{H}_k \hat{\mathbf{x}}_k^-]} \tag{5.80}$$

Note that the transformed measurement $\tilde{\mathbf{z}}_k$ is now used in the state update equation.

5.3.4 Steady-State Kalman Filter

The discrete Riccati equation in eqn. (5.54c) requires the propagation of an $n \times n$ matrix. Fortunately for time-invariant systems the error covariance P reaches a steady-state value very quickly. Therefore, a *constant* gain (K) in the filter can be pre-computed using the steady-state covariance, which can significantly reduce the computational burden. Although this approach is suboptimal in the strictest sense, the savings in computations compared to any loss in the estimated state quality makes the fixed-gain Kalman filter attractive in the design of many dynamical systems. The steady-state (autonomous) discrete-time Kalman filter is summarized in Table 5.2.

To determine the steady-state value for P we must solve the *discrete-time algebraic Riccati equation* in Table 5.2. The solution can be derived using the duality between estimation and optimal control theory (discussed in Chapter 8). The nonlinear Riccati equation can be processed using two sets of $n \times n$ matrices, given by

$$\boxed{P_k = S_k Z_k^{-1}} \tag{5.81}$$

Table 5.2: Discrete and Autonomous Linear Kalman Filter

Model	$\mathbf{x}_{k+1} = \Phi\,\mathbf{x}_k + \Gamma\,\mathbf{u}_k + \Upsilon\,\mathbf{w}_k, \quad \mathbf{w}_k \sim N(\mathbf{0}, Q)$ $\tilde{\mathbf{y}}_k = H\,\mathbf{x}_k + \mathbf{v}_k, \quad \mathbf{v}_k \sim N(\mathbf{0}, R)$
Initialize	$\hat{\mathbf{x}}(t_0) = \hat{\mathbf{x}}_0$
Gain	$K = P\,H^T[H\,P\,H^T + R]^{-1}$
Covariance	$P = \Phi\,P\,\Phi^T - \Phi\,P\,H^T[H\,P\,H^T + R]^{-1}H\,P\,\Phi^T + \Upsilon\,Q\,\Upsilon^T$
Estimate	$\hat{\mathbf{x}}_{k+1} = \Phi\,\hat{\mathbf{x}}_k + \Gamma\,\mathbf{u}_k + \Phi\,K\,[\tilde{\mathbf{y}}_k - H\,\hat{\mathbf{x}}_k]$

To determine linear equations for S_{k+1} and Z_{k+1} we first rewrite the discrete-time Riccati equation in eqn. (5.54c) using the matrix inversion lemma in eqn. (1.69), which yields

$$P_{k+1} = \Phi\,[\bar{H} + P_k^{-1}]^{-1}\Phi^T + \bar{Q} \tag{5.82}$$

where $\bar{H} \equiv H^T R^{-1} H$ and $\bar{Q} \equiv \Upsilon\,Q\,\Upsilon^T$. Factoring P_k and multiplying $\bar{Q}$ by an identity gives

$$P_{k+1} = \Phi\,P_k[\bar{H}\,P_k + I]^{-1}\Phi^T + \bar{Q}\,\Phi^{-T}\Phi^T \tag{5.83}$$

Rewriting eqn. (5.83) by factoring $[\bar{H}\,P_k + I]$ gives

$$P_{k+1} = \left\{\Phi\,P_k + \bar{Q}\,\Phi^{-T}[\bar{H}\,P_k + I]\right\}[\bar{H}\,P_k + I]^{-1}\Phi^T \tag{5.84}$$

Next collecting P_k terms gives

$$P_{k+1} = \left\{[\Phi + \bar{Q}\,\Phi^{-T}\bar{H}]P_k + \bar{Q}\,\Phi^{-T}]\right\}[\bar{H}\,P_k + I]^{-1}\Phi^T \tag{5.85}$$

Substituting eqn. (5.81) into eqn. (5.85) and factoring Z_k yields

$$P_{k+1} = \left\{[\Phi + \bar{Q}\,\Phi^{-T}\bar{H}]S_k + \bar{Q}\,\Phi^{-T}Z_k]\right\}Z_k^{-1}[\bar{H}\,S_k Z_k^{-1} + I]^{-1}\Phi^T \tag{5.86}$$

Finally, factoring Z_k^{-1} and Φ^T into the last inverse of eqn. (5.86) gives

$$P_{k+1} = \left\{[\Phi + \bar{Q}\,\Phi^{-T}\bar{H}]S_k + \bar{Q}\,\Phi^{-T}Z_k]\right\}[\Phi^{-T}Z_k + \Phi^{-T}\bar{H}\,S_k]^{-1} \tag{5.87}$$

Using a one-time step ahead of eqn. (5.81) yields the following relationship:

$$\begin{bmatrix} Z_{k+1} \\ S_{k+1} \end{bmatrix} = \mathcal{H}\begin{bmatrix} Z_k \\ S_k \end{bmatrix} \tag{5.88}$$

where the *Hamiltonian matrix* is defined as

$$\boxed{\mathcal{H} \equiv \begin{bmatrix} \Phi^{-T} & \Phi^{-T} H^T R^{-1} H \\ \Upsilon Q \Upsilon^T \Phi^{-T} & \Phi + \Upsilon Q \Upsilon^T \Phi^{-T} H^T R^{-1} H \end{bmatrix}} \tag{5.89}$$

We will now show that if λ is an eigenvalue of $\mathcal{H}$, then λ^{-1} is also an eigenvalue of $\mathcal{H}$ (i.e., $\mathcal{H}$ is a *symplectic* matrix[8]). The eigenvalues of $\mathcal{H}$ are determined by taking the determinant of the following equation and setting the resultant to zero:

$$\lambda I - \mathcal{H} = \begin{bmatrix} \lambda I - \Phi^{-T} & -\Phi^{-T} \bar{H} \\ -\bar{Q}\Phi^{-T} & \lambda I - \Phi - \bar{Q}\Phi^{-T} \bar{H} \end{bmatrix} \tag{5.90}$$

Next we multiply the right side of eqn. (5.90) by the following matrix:

$$\bar{H}_I \equiv \begin{bmatrix} I & -\bar{H} \\ 0 & I \end{bmatrix} \tag{5.91}$$

Since $\det(\bar{H}_I) = 1$ (see Appendix A), then the determinant of eqn. (5.90) is given by

$$\det(\lambda I - \mathcal{H}) = \det \begin{bmatrix} \lambda I - \Phi^{-T} & -\lambda \bar{H} \\ -\bar{Q}\Phi^{-T} & \lambda I - \Phi \end{bmatrix} = 0 \tag{5.92}$$

Next we use the following identity for square matrices A, B, C, and D:

$$\det \begin{bmatrix} A & B \\ C & D \end{bmatrix} = \det(D)\det(A - B\,D^{-1} C) \tag{5.93}$$

assuming that D^{-1} exists. This leads to

$$\det(\lambda I - \Phi)\det\left[(\lambda \Phi^T - I) - \bar{H}(I - \lambda^{-1}\Phi)^{-1}\bar{Q}\right] = 0 \tag{5.94}$$

where $\det(A\,B) = \det(A)\det(B)$ was used to factor out the term Φ^{-T}. Next, we factor the term $(\lambda\Phi^T - I)$ from the second term and multiply both sides of the resultant equation by λ^{-n}, where n is the order of Φ, to find

$$\alpha(\lambda)\alpha(\lambda^{-1})\det\left[I + (\lambda\Phi^T - I)^{-1}\bar{H}(\lambda^{-1}\Phi - I)^{-1}\bar{Q}\right] = 0 \tag{5.95}$$

where $\alpha(\lambda) \equiv \det(\lambda I - \Phi)$. Since both $\bar{H}$ and $\bar{Q}$ are symmetric matrices, they can be factored into $\bar{H} = \Xi^T \Xi$ and $\bar{Q} = \Theta^T \Theta$. Then using the identity $\det(I + A\,B) = \det(I + B\,A)$, with $A = (\lambda\Phi^T - I)^{-1}\Xi^T$, gives

$$\alpha(\lambda)\alpha(\lambda^{-1})\det\left[I + \Xi(\lambda^{-1}\Phi - I)^{-1}\Theta^T\Theta(\lambda\Phi^T - I)^{-1}\Xi^T\right] = 0 \tag{5.96}$$

Therefore, if λ is replaced by λ^{-1}, the result in eqn. (5.96) remains unchanged since the determinant of a matrix is equal to the determinant of its transpose. Thus the eigenvalues can be arranged in a diagonal matrix given by

$$\mathcal{H}_\Lambda = \begin{bmatrix} \Lambda & 0 \\ 0 & \Lambda^{-1} \end{bmatrix} \tag{5.97}$$

where Λ is a diagonal matrix of the n eigenvalues outside of the unit circle. Assuming that the eigenvalues are distinct, we can perform a linear state transformation, as shown in §3.1.4, such that

$$\mathcal{H}_\Lambda = W^{-1}\mathcal{H}W \tag{5.98}$$

where W is the matrix of eigenvectors, which can be represented in block form as

$$W = \begin{bmatrix} W_{11} & W_{12} \\ W_{21} & W_{22} \end{bmatrix} \tag{5.99}$$

At steady-state the unstable eigenvalues (Λ) will dominate the response of P_k. Using only the unstable eigenvalues we can partition eqn. (5.98) as

$$\begin{bmatrix} W_{11} \\ W_{21} \end{bmatrix} \Lambda = \mathcal{H} \begin{bmatrix} W_{11} \\ W_{21} \end{bmatrix} \tag{5.100}$$

If we make the analogy that $Z \to W_{11}$ and $S \to W_{21}$ from eqn. (5.88), then the steady-state solution for P with $k \to k+1$ is given by

$$\boxed{P = [W_{21}\Lambda][W_{11}\Lambda]^{-1} = W_{21}W_{11}^{-1}} \tag{5.101}$$

Therefore, the gain K in Table 5.2 can be computed off-line and remains constant. This can significantly reduce the on-board computational load on a computer.

Vaughan[9] has shown that a nonrecursive solution for P_k is given by

$$P_k = [W_{21} + W_{22}Y_k][W_{11} + W_{12}Y_k]^{-1} \tag{5.102}$$

where

$$Y_k = \Lambda^{-k} X \Lambda^{-k} \tag{5.103a}$$

$$X = -[W_{22} - P_0 W_{12}]^{-1}[W_{21} - P_0 W_{11}] \tag{5.103b}$$

The steady-state solution for P can be found by letting $k \to \infty$, which leads directly to eqn. (5.101).

5.3.5 Correlated Measurement and Process Noise

The derivations thus far have assumed that the measurement error is uncorrelated with the process noise (state error). In this section the correlated Kalman filter is derived. This correlation can be written mathematically by

$$\boxed{E\left\{\mathbf{w}_{k-1}\mathbf{v}_k^T\right\} = S_k} \tag{5.104}$$

Before proceeding, we must first explain why we wish to investigate the correlation between $\mathbf{w}_{k-1}$ and $\mathbf{v}_k$, not between $\mathbf{w}_k$ and $\mathbf{v}_k$. This is mainly due to the fact that the measurement at time t_k will be dependent on the state, deterministic input, and process noise at time t_{k-1}, as shown by eqn. (5.27). This is extremely useful for the correspondence between a sampled continuous-time system, since it represents correlation between the process noise over a sample period and the measurement at the end of the period.[5] Note that S_k is not a symmetric matrix in this case.

Equations (5.33) and (5.37) will be used to derive the filter equations. Clearly, when eqn. (5.33) is substituted into eqn. (5.37) at time t_k, the covariance update P_k^- in eqn. (5.35) remains unchanged since $E\left\{\mathbf{w}_k\mathbf{v}_k^T\right\} = E\left\{\mathbf{v}_k\mathbf{w}_k^T\right\} = 0$ from the assumptions in this section. However, the terms $E\left\{\tilde{\mathbf{x}}_k^-\mathbf{v}_k^T\right\}$ and $E\left\{\mathbf{v}_k\tilde{\mathbf{x}}_k^{-T}\right\}$ in eqn. (5.38) are no longer zero in this case. Performing the expectation for the previous expression gives

$$\begin{aligned} E\left\{\tilde{\mathbf{x}}_k^-\mathbf{v}_k^T\right\} &= E\left\{(\Phi_{k-1}\tilde{\mathbf{x}}_{k-1}^+ - \Upsilon_{k-1}\mathbf{w}_{k-1})\,\mathbf{v}_k^T\right\} \\ &= -\Upsilon_{k-1}S_k \end{aligned} \tag{5.105}$$

This is due to the fact that $\tilde{\mathbf{x}}_{k-1}^+$ is uncorrelated with $\mathbf{v}_k$. Therefore eqn. (5.38) becomes

$$\begin{aligned} P_k^+ &= [I - K_kH_k]P_k^-[I - K_kH_k]^T + K_kR_kK_k^T \\ &\quad - [I - K_kH_k]\Upsilon_{k-1}S_kK_k^T - K_kS_k^T\Upsilon_{k-1}^T[I - K_kH_k]^T \end{aligned} \tag{5.106}$$

This expression is valid for any gain K_k. To determine this gain we again minimize the trace of P_k^+, which leads to

$$\boxed{K_k = [P_k^-H_k^T + \Upsilon_{k-1}S_k][H_kP_k^-H_k^T + R_k + H_k\Upsilon_{k-1}S_k + S_k^T\Upsilon_{k-1}^TH_k^T]^{-1}} \tag{5.107}$$

Note that if $S_k = 0$ then the gain reduces to the standard form given in eqn. (5.42). Substituting eqn. (5.107) into eqn. (5.106), after some algebraic manipulations, yields

$$\boxed{P_k^+ = [I - K_kH_k]P_k^- - K_kS_k^T\Upsilon_{k-1}^T} \tag{5.108}$$

This again reduces to the standard form of the covariance update in eqn. (5.44) if $S_k = 0$. A summary of the correlated discrete-time Kalman filter is given in Table 5.3.

An excellent example of the usefulness of the correlated Kalman filter is an aircraft flying through a field of random turbulence.[4] The effect of turbulence in the aircraft's acceleration are complex, but can easily be modelled as process noise on $\mathbf{w}_{k-1}$. Since any sensor mounted on an aircraft is also corrupted by turbulence, the measurement error $\mathbf{v}_k$ is correlated with the process noise $\mathbf{w}_{k-1}$. Hence, the filter formulation presented in this section can be used directly to estimate aircraft state quantities in the face of turbulence disturbances.

Table 5.3: Correlated Discrete-Time Linear Kalman Filter

Model	$\mathbf{x}_{k+1} = \Phi_k \mathbf{x}_k + \Gamma_k \mathbf{u}_k + \Upsilon_k \mathbf{w}_k, \quad \mathbf{w}_k \sim N(\mathbf{0}, Q_k)$ $\tilde{\mathbf{y}}_k = H_k \mathbf{x}_k + \mathbf{v}_k, \quad \mathbf{v}_k \sim N(\mathbf{0}, R_k)$ $E\left\{\mathbf{w}_{k-1}\mathbf{v}_k^T\right\} = S_k$
Initialize	$\hat{\mathbf{x}}(t_0) = \hat{\mathbf{x}}_0$ $P_0 = E\left\{\tilde{\mathbf{x}}(t_0)\,\tilde{\mathbf{x}}^T(t_0)\right\}$
Gain	$K_k = [P_k^- H_k^T + \Upsilon_{k-1} S_k]$ $\times [H_k P_k^- H_k^T + R_k + H_k \Upsilon_{k-1} S_k + S_k^T \Upsilon_{k-1}^T H_k^T]^{-1}$
Update	$\hat{\mathbf{x}}_k^+ = \hat{\mathbf{x}}_k^- + K_k[\tilde{\mathbf{y}}_k - H_k \hat{\mathbf{x}}_k^-]$ $P_k^+ = [I - K_k H_k] P_k^- - K_k S_k^T \Upsilon_{k-1}^T$
Propagation	$\hat{\mathbf{x}}_{k+1}^- = \Phi_k \hat{\mathbf{x}}_k^+ + \Gamma_k \mathbf{u}_k$ $P_{k+1}^- = \Phi_k P_k^+ \Phi_k^T + \Upsilon_k Q_k \Upsilon_k^T$

5.3.6 Orthogonality Principle

One of the interesting aspects of the Kalman filter is the orthogonality of the estimate and its error,[1] which is stated mathematically as

$$\boxed{E\left\{\hat{\mathbf{x}}_k^+ \tilde{\mathbf{x}}_k^{+T}\right\} = 0} \tag{5.109}$$

This states that the estimate is uncorrelated from its error. To prove eqn. (5.109) set the time step to $k = 1$, and substitute eqn. (5.33) into eqn. (5.37), which gives

$$\tilde{\mathbf{x}}_1^+ = (\Phi_0 - K_1 H_1 \Phi_0)\,\tilde{\mathbf{x}}_0^+ + (K_1 H_1 - I)\,\Upsilon_0 \mathbf{w}_0 + K_1 \mathbf{v}_1 \tag{5.110}$$

Next, substituting eqn. (5.27a) into eqn. (5.27b), and then substituting the resultant into eqn. (5.30b) leads to the following state estimate update:

$$\hat{\mathbf{x}}_1^+ = \Phi_0 \hat{\mathbf{x}}_0^+ + \Gamma_0 \mathbf{u}_0 + K_1\left(H_1 \Upsilon_0 \mathbf{w}_0 + \mathbf{v}_1 - H_1 \Phi_0 \tilde{\mathbf{x}}_0^+\right) \tag{5.111}$$

Since the initial conditions are uncorrelated, then $E\left\{\hat{\mathbf{x}}_0^+ \tilde{\mathbf{x}}_0^{+T}\right\} = 0$, and we have

$$\begin{aligned} E\left\{\hat{\mathbf{x}}_1^+ \tilde{\mathbf{x}}_1^{+T}\right\} &= K_1 H_1 \Upsilon_0 Q_0 \Upsilon_0^T \left(H_1^T K_1^T - I\right) \\ &\quad + K_1 H_1 \Phi_0 P_0^+ \left(\Phi_0 H_1^T K_1^T - \Phi_0^T\right) + K_1 R_1 K_1^T \end{aligned} \tag{5.112}$$

Collecting terms yields

$$E\left\{\hat{\mathbf{x}}_1^+\tilde{\mathbf{x}}_1^{+T}\right\} = -K_1 H_1\left(\Phi_0 P_0^+ \Phi_0^T + \Upsilon_0 Q_0 \Upsilon_0^T\right) \\ + K_1 H_1\left(\Phi_0 P_0^+ \Phi_0^T + \Upsilon_0 Q_0 \Upsilon_0^T\right) H_1^T K_1^T + K_1 R_1 K_1^T \tag{5.113}$$

Using eqn. (5.35) in eqn. (5.113) gives

$$E\left\{\hat{\mathbf{x}}_1^+\tilde{\mathbf{x}}_1^{+T}\right\} = K_1 H_1 P_1^-\left(H_1^T K_1^T - I\right) + K_1 R_1 K_1^T \tag{5.114}$$

Next, using the definition of P_1^+ from eqn. (5.44) in eqn. (5.114) gives

$$E\left\{\hat{\mathbf{x}}_1^+\tilde{\mathbf{x}}_1^{+T}\right\} = -K_1 H_1 P_1^+ + K_1 R_1 K_1^T \tag{5.115}$$

Then substituting the gain K_1 from eqn. (5.47) into eqn. (5.115) yields

$$E\left\{\hat{\mathbf{x}}_1^+\tilde{\mathbf{x}}_1^{+T}\right\} = -P_1^+ H_1^T R^{-1} H_1 P_1^+ + P_1^+ H_1^T R^{-1} H_1 P_1^+ = 0 \tag{5.116}$$

The process is then repeated for the $k = 2$ case, and by induction the identity in eqn. (5.109) is proven. At first glance the Orthogonality Principle does not seem to have any practical value, but as we shall see it is extremely important in the derivation of the linear quadratic-Gaussian controller of §8.6.

Example 5.2: In this simple example the discrete-time Kalman filter is used to estimate a scalar state for a time-invariant system, whose truth model follows

$$x_{k+1} = \phi\, x_k + \gamma\, u_k + w_k$$
$$\tilde{y}_k = h\, x_k + v_k$$

where the random errors are assumed to be stationary noise processes with $w_k \sim N(0,q)$ and $v_k \sim N(0,r)$. Since the filter dynamics converge rapidly in this case we will use the steady-state Kalman filter, given in Table 5.2. The steady-state covariance equation gives the following second-order polynomial equation:

$$h^2 p^2 + (r - \phi^2 r - h^2 q)\, p - q\, r = 0$$

The closed-form solution for even this simple system is difficult to intuitively visualize; however, some simple forms can be given for two special cases. Consider the perfect-measurement case where $r = 0$, which simply yields $p = q$. Then the gain K in Table 5.2 is simply given by $1/h$, and the state estimate is given by

$$\hat{x}_{k+1} = \frac{\phi}{h}\,\tilde{y}_k + \gamma\, u_k$$

Note that the current state estimate $\hat{x}_{k+1}$ does not depend on the previous state estimate $\hat{x}_k$ in this case. This is due to the fact that with $r = 0$, the measurements are assumed perfect and the dynamics model can be ignored, which intuitively makes sense. Next, we consider the perfect-model case when $q = 0$, which simply yields $p = 0$. The gain is zero in this case and the state estimate is given by

$$\hat{x}_{k+1} = \phi\,\hat{x}_k + \gamma\,u_k$$

In this case the measurement is completely ignored, which again intuitively makes sense since the model is perfect with no errors.

Example 5.3: In this example the single axis attitude estimation problem using angle-attitude measurements and rate information from gyros is shown. We will demonstrate the power of the Kalman filter to update both the attitude-angle estimates and gyro drift rate. Angle measurements are corrupted with noise, which can be filtered by using rate information. However, all gyros inherently drift over time, which degrades the rate information over time. Two error sources are generally present in gyros.[10] The first is a short-term component of instability referred to as *random drift*, and the second is a random walk component referred to as *drift rate ramp*. The effects of both of these noise sources on the uncertainty of the gyro outputs can be compensated using a Kalman filter with attitude measurements. The attitude rate $\dot{\theta}$ is assumed to be related to the gyro output $\tilde{\omega}$ by

$$\dot{\theta} = \tilde{\omega} - \beta - \eta_v$$

where β is the gyro drift rate, and η_v is a zero-mean Gaussian white-noise process with variance given by σ_v^2. The drift rate is modelled by a random walk process, given by

$$\dot{\beta} = \eta_u$$

where η_u is a zero-mean Gaussian white-noise process with variance given by σ_u^2. The parameters σ_v^2 and σ_u^2 can be experimentally obtained using frequency response data from the gyro outputs. The estimated states clearly follow

$$\dot{\hat{\theta}} = \tilde{\omega} - \hat{\beta}$$
$$\dot{\hat{\beta}} = 0$$

Assuming a constant sampling interval in the gyro output, the discrete-time error propagation is given by[11]

$$\begin{bmatrix} \theta_{k+1} - \hat{\theta}_{k+1} \\ \beta_{k+1} - \hat{\beta}_{k+1} \end{bmatrix} = \Phi \begin{bmatrix} \theta_k - \hat{\theta}_k \\ \beta_k - \hat{\beta}_k \end{bmatrix} + \begin{bmatrix} p_k \\ q_k \end{bmatrix}$$

where the state transition matrix is given by

$$\Phi = \begin{bmatrix} 1 & -\Delta t \\ 0 & 1 \end{bmatrix}$$

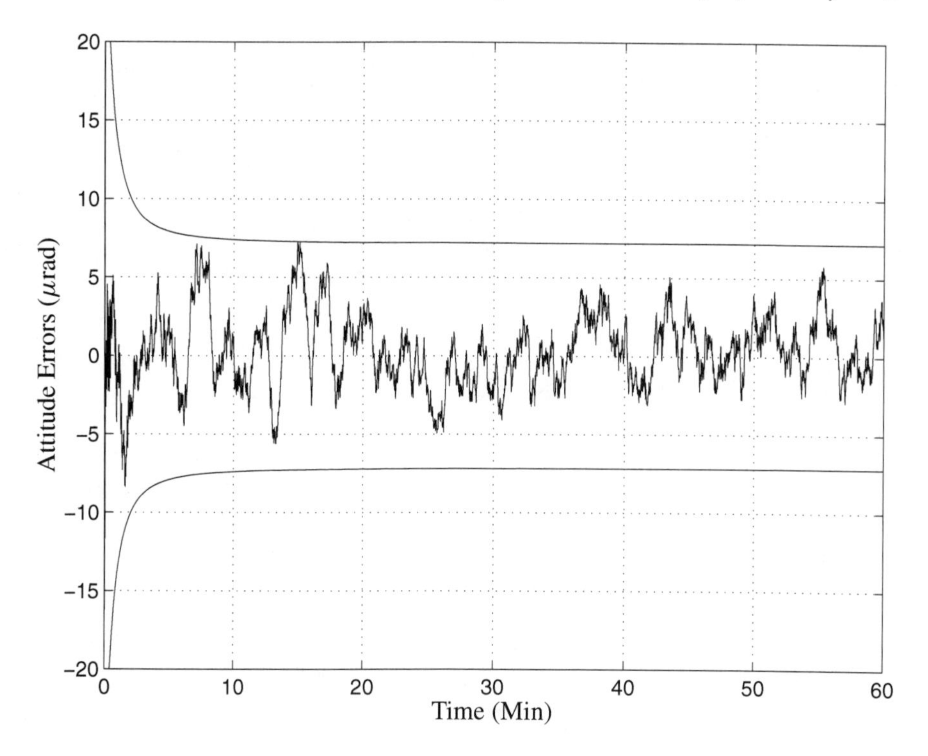

Figure 5.3: Kalman Filter Attitude Error and Bounds

where $\Delta t = t_{k+1} - t_k$ is the sampling interval, and

$$p_k = \int_{t_k}^{t_{k+1}} \left[-\eta_v(\tau) - (t_{k+1} - \tau)\eta_u(\tau) \right] d\tau$$

$$q_k = \int_{t_k}^{t_{k+1}} \eta_u(\tau)\, d\tau$$

The process noise covariance matrix Q can be computed as

$$Q = \begin{bmatrix} E\left\{p_k^2\right\} & E\left\{p_k q_k\right\} \\ E\left\{q_k p_k\right\} & E\left\{q_k^2\right\} \end{bmatrix}$$

$$= \begin{bmatrix} \sigma_v^2 \Delta t + \frac{1}{3}\sigma_u^2 \Delta t^3 & -\frac{1}{2}\sigma_u^2 \Delta t^2 \\ -\frac{1}{2}\sigma_u^2 \Delta t^2 & \sigma_u^2 \Delta t \end{bmatrix}$$

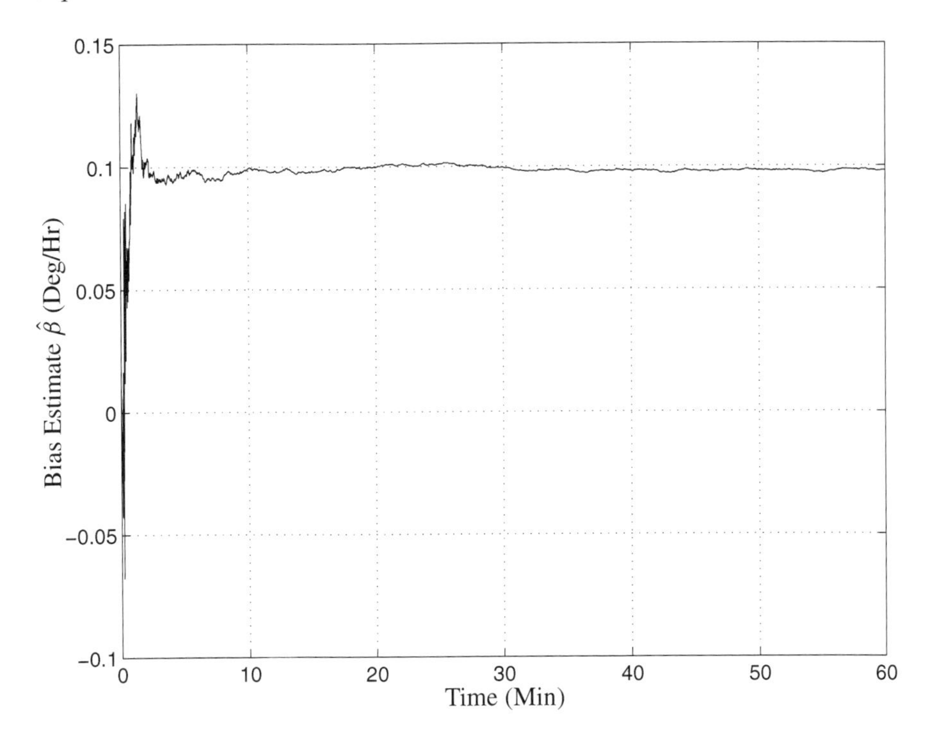

Figure 5.4: Kalman Filter Gyro Bias Estimate

which is independent of k since the sampling interval is assumed to be constant. The attitude-angle measurement is modelled by

$$\tilde{y}_k = \theta_k + v_k$$

where v_k is a zero-mean Gaussian white-noise process with variance given by $R = \sigma_n^2$. The discrete-time system used in the Kalman filter can now be written as

$$\begin{aligned} \mathbf{x}_{k+1} &= \Phi\, \mathbf{x}_k + \Gamma\, \tilde{\omega} + \mathbf{w}_k \\ \tilde{y}_k &= H\, \mathbf{x}_k + v_k \end{aligned}$$

where $\mathbf{x} = \begin{bmatrix} \theta \ \beta \end{bmatrix}^T$, $\Gamma = \begin{bmatrix} \Delta t \ 0 \end{bmatrix}^T$, $H = \begin{bmatrix} 1 \ 0 \end{bmatrix}$, and $E\left\{\mathbf{w}_k \mathbf{w}_k^T\right\} = Q$. We should note that the input to this system involves a measurement ($\tilde{\omega}$), which is counterintuitive but valid in the Kalman filter form and poses no problems in the estimation process. The discrete-time Kalman filter shown in Table 5.1 can now be applied to this system. Synthetic measurements are created using a true constant angle-rate given by $\dot{\theta} = 0.0011$ rad/sec and a sampling rate of 1 second. The noise parameters are given by $\sigma_n = 17 \times 10^{-6}$ rad, $\sigma_u = \sqrt{10} \times 10^{-10}$ rad/sec$^{3/2}$, and $\sigma_v = \sqrt{10} \times 10^{-7}$ rad/sec$^{1/2}$. The initial bias β_0 is given as 0.1 deg/hr, and the initial covariance matrix is set to $P_0 = \text{diag}\left[1 \times 10^{-4} \ 1 \times 10^{-12}\right]$. A plot of the attitude-angle error and 3σ bounds is shown in Figure 5.3. Clearly, the Kalman filter provides filtered estimates and the theoretical 3σ bounds do indeed bound the errors. A steady-state Kalman filter using the algebraic Riccati equation in Table 5.2 can also be used, which yields nearly identical results as the time-varying case. At steady-state the theoretical 3σ bound is given by 7.18 μrad. A plot of the estimated bias is shown in Figure 5.4. Clearly, the Kalman filter estimates the bias well. This example demonstrates the usefulness of the Kalman filter by fusing two sensors to produce estimates that are better than each sensor alone.

5.4 The Continuous-Time Kalman Filter

In this section the Kalman filter is derived using continuous-time models and measurements. The continuous-time Kalman filter is not widely used in practice due to the extensive use of digital computers in today's time; however, the derivation does provide some unique perspectives that are especially useful for small sampling intervals (i.e., well below Nyquist's limit). Two approaches are shown, which yield the same Kalman filter structure. The first uses the continuous-time structure directly, while the second uses the discrete-time formulation described in §5.4.1 to derive the corresponding continuous-time form.

5.4.1 Kalman Filter Derivation in Continuous Time

In this section the Kalman filter is derived directly from continuous-time models and measurements. Consider the following truth model:

$$\dot{\mathbf{x}}(t) = F(t)\,\mathbf{x}(t) + B(t)\,\mathbf{u}(t) + G(t)\,\mathbf{w}(t) \tag{5.117a}$$

$$\tilde{\mathbf{y}}(t) = H(t)\,\mathbf{x}(t) + \mathbf{v}(t) \tag{5.117b}$$

where $\mathbf{w}(t)$ and $\mathbf{v}(t)$ are zero-mean Gaussian noise processes with covariances given by

$$E\left\{\mathbf{w}(t)\,\mathbf{w}^T(\tau)\right\} = Q(t)\,\delta(t-\tau) \tag{5.118a}$$

$$E\left\{\mathbf{v}(t)\,\mathbf{v}^T(\tau)\right\} = R(t)\,\delta(t-\tau) \tag{5.118b}$$

$$E\left\{\mathbf{v}(t)\,\mathbf{w}^T(\tau)\right\} = 0 \tag{5.118c}$$

Equation (5.118c) implies that $\mathbf{v}(t)$ and $\mathbf{w}(t)$ are uncorrelated. Also, the control input $\mathbf{u}(t)$ is a deterministic quantity. The Kalman filter structure for the state and output estimate is given by

$$\dot{\hat{\mathbf{x}}}(t) = F(t)\,\hat{\mathbf{x}}(t) + B(t)\,\mathbf{u}(t) + K(t)[\tilde{\mathbf{y}}(t) - H(t)\,\hat{\mathbf{x}}(t)] \tag{5.119a}$$

$$\hat{\mathbf{y}}(t) = H(t)\,\hat{\mathbf{x}}(t) \tag{5.119b}$$

Defining the state error $\tilde{\mathbf{x}}(t) = \hat{\mathbf{x}}(t) - \mathbf{x}(t)$ and using eqns. (5.117) and (5.119) leads to

$$\dot{\tilde{\mathbf{x}}}(t) = E(t)\,\tilde{\mathbf{x}}(t) + \mathbf{z}(t) \tag{5.120}$$

where

$$E(t) = F(t) - K(t)\,H(t) \tag{5.121}$$

$$\mathbf{z}(t) = -G(t)\,\mathbf{w}(t) + K(t)\,\mathbf{v}(t) \tag{5.122}$$

Note that $\mathbf{u}(t)$ cancels in the error state. Since $\mathbf{v}(t)$ and $\mathbf{w}(t)$ are uncorrelated, we have

$$E\left\{\mathbf{z}(t)\,\mathbf{z}^T(\tau)\right\} = \left[G(t)\,Q(t)\,G^T(t) + K(t)\,R(t)\,K^T(t)\right]\delta(t-\tau) \tag{5.123}$$

Using the matrix exponential solution in eqn. (3.53) gives

$$\tilde{\mathbf{x}}(t) = \Phi(t,t_0)\,\tilde{\mathbf{x}}(t_0) + \int_{t_0}^{t} \Phi(t,\tau)\,\mathbf{z}(\tau)\,d\tau \tag{5.124}$$

The state error-covariance is defined by

$$P(t) \equiv E\left\{\tilde{\mathbf{x}}(t)\,\tilde{\mathbf{x}}^T(t)\right\} \tag{5.125}$$

Substituting eqn. (5.124) into eqn. (5.125), assuming that $\mathbf{z}(t)$ and $\tilde{\mathbf{x}}(t_0)$ are uncorrelated, leads to

$$\begin{aligned} P(t) &= \Phi(t,t_0)\,P(t_0)\,\Phi^T(t,t_0) \\ &\quad + \int_{t_0}^{t} \Phi(t,\tau)\left[G(\tau)\,Q(\tau)\,G^T(\tau) + K(\tau)\,R(\tau)\,K^T(\tau)\right]\Phi^T(t,\tau)\,d\tau \end{aligned} \tag{5.126}$$

Taking the time derivative of eqn. (5.126) gives

$$\begin{aligned} \dot{P}(t) &= \frac{\partial\Phi(t,t_0)}{\partial t}P(t_0)\Phi^T(t,t_0) + \Phi(t,t_0)P(t_0)\frac{\partial\Phi^T(t,t_0)}{\partial t} \\ &\quad + \int_{t_0}^{t} \frac{\partial\Phi(t,\tau)}{\partial t}\left[G(\tau)\,Q(\tau)\,G^T(\tau) + K(\tau)\,R(\tau)\,K^T(\tau)\right]\Phi^T(t,\tau)\,d\tau \\ &\quad + \int_{t_0}^{t} \Phi(t,\tau)\left[G(\tau)\,Q(\tau)\,G^T(\tau) + K(\tau)\,R(\tau)\,K^T(\tau)\right]\frac{\partial\Phi^T(t,\tau)}{\partial t}\,d\tau \\ &\quad + \Phi(t,t)\left[G(t)\,Q(t)\,G^T(t) + K(t)\,R(t)\,K^T(t)\right]\Phi^T(t,t) \end{aligned} \tag{5.127}$$

Using the properties of the matrix exponential in eqns. (3.17a) and (3.19) leads to

$$\begin{aligned}\dot{P}(t) &= E(t)\,\Phi(t,t_0)\,P(t_0)\Phi^T(t,t_0) + \Phi(t,t_0)P(t_0)\,\Phi^T(t,t_0)\,E^T(t)\\ &\quad + E(t)\int_{t_0}^{t}\Phi(t,\tau)\left[G(\tau)\,Q(\tau)\,G^T(\tau) + K(\tau)\,R(\tau)\,K^T(\tau)\right]\Phi^T(t,\tau)\,d\tau\\ &\quad + \int_{t_0}^{t}\Phi(t,\tau)\left[G(\tau)\,Q(\tau)\,G^T(\tau) + K(\tau)\,R(\tau)\,K^T(\tau)\right]\Phi^T(t,\tau)\,d\tau\,E^T(t)\\ &\quad + G(t)\,Q(t)\,G^T(t) + K(t)\,R(t)\,K^T(t)\end{aligned} \tag{5.128}$$

Using eqns. (5.121) and (5.126) in eqn. (5.128) simplifies the expression for $\dot{P}(t)$ significantly to

$$\begin{aligned}\dot{P}(t) &= [F(t) - K(t)\,H(t)]\,P(t) + P(t)\,[F(t) - K(t)\,H(t)]^T\\ &\quad + G(t)\,Q(t)\,G^T(t) + K(t)\,R(t)\,K^T(t)\end{aligned} \tag{5.129}$$

In order to determine the gain $K(t)$ we minimize the trace of $\dot{P}(t)$:

$$\text{minimize} \quad J[K(t)] = \text{Tr}[\dot{P}(t)] \tag{5.130}$$

The necessary conditions lead to

$$\frac{\partial J}{\partial K(t)} = 0 = 2K(t)\,R(t) - 2P(t)\,H^T(t) \tag{5.131}$$

Solving eqn. (5.131) for $K(t)$ gives

$$\boxed{K(t) = P(t)\,H^T(t)\,R^{-1}(t)} \tag{5.132}$$

Note the similarity of the gain $K(t)$ to the discrete-time case given in eqn. (5.47). Substituting eqn. (5.132) into eqn. (5.129) gives

$$\boxed{\begin{aligned}\dot{P}(t) &= F(t)\,P(t) + P(t)\,F^T(t)\\ &\quad - P(t)\,H^T(t)\,R^{-1}(t)H(t)\,P(t) + G(t)\,Q(t)\,G^T(t)\end{aligned}} \tag{5.133}$$

Equation (5.133) is known as the *continuous Riccati equation.*

A summary of the continuous-time Kalman filter is given in Table 5.4. First, initial conditions for the state and error covariances are given. Then, the gain $K(t)$ is computed using eqn. (5.132) with the initial covariance value. Next, the covariance in eqn. (5.133) and state estimate in eqn. (5.119a) are numerically integrated forward in time using the continuous-time measurement $\tilde{\mathbf{y}}(t)$ and known input $\mathbf{u}(t)$. The integration of the state estimate and covariance continues until the final measurement time is reached.

Table 5.4: Continuous-Time Linear Kalman Filter

Model	$\dot{\mathbf{x}}(t) = F(t)\mathbf{x}(t) + B(t)\mathbf{u}(t) + G(t)\mathbf{w}(t),\ \mathbf{w}(t) \sim N(\mathbf{0}, Q(t))$ $\tilde{\mathbf{y}}(t) = H(t)\mathbf{x}(t) + \mathbf{v}(t),\ \mathbf{v}(t) \sim N(\mathbf{0}, R(t))$
Initialize	$\hat{\mathbf{x}}(t_0) = \hat{\mathbf{x}}_0$ $P_0 = E\left\{\tilde{\mathbf{x}}(t_0)\,\tilde{\mathbf{x}}^T(t_0)\right\}$
Gain	$K(t) = P(t)\,H^T(t)\,R^{-1}(t)$
Covariance	$\dot{P}(t) = F(t)\,P(t) + P(t)\,F^T(t)$ $-P(t)\,H^T(t)\,R^{-1}(t)H(t)\,P(t) + G(t)\,Q(t)\,G^T(t)$
Estimate	$\dot{\hat{\mathbf{x}}}(t) = F(t)\,\hat{\mathbf{x}}(t) + B(t)\,\mathbf{u}(t)$ $+K(t)[\tilde{\mathbf{y}}(t) - H(t)\,\hat{\mathbf{x}}(t)]$

5.4.2 Kalman Filter Derivation from Discrete Time

The continuous-time Kalman filter can also be derived from the discrete-time version of §5.4.1. We must first find relationships between the discrete-time covariance matrices, Q_k and R_k, and continuous-time covariance matrices, $Q(t)$ and $R(t)$. From eqn. (5.118a) and from the theory of discrete-time systems in §3.5 we can write

$$\begin{aligned}
&\Upsilon_k E\left\{\mathbf{w}_k \mathbf{w}_k^T\right\}\Upsilon_k^T = \Upsilon_k Q_k \Upsilon_k^T \\
&= E\left\{\left[\int_{t_k}^{t_{k+1}} \Phi(t_{k+1},\tau)\,G(\tau)\,\mathbf{w}(\tau)\,d\tau\right]\left[\int_{t_k}^{t_{k+1}} \Phi(t_{k+1},\varsigma)\,G(\varsigma)\,\mathbf{w}(\varsigma)\,d\varsigma\right]^T\right\} \\
&= \int_{t_k}^{t_{k+1}}\int_{t_k}^{t_{k+1}} \Phi(t_{k+1},\tau)\,G(\tau)\,E\left\{\mathbf{w}(\tau)\,\mathbf{w}^T(\varsigma)\right\}G^T(\varsigma)\Phi^T(t_{k+1},\varsigma)\,d\tau\,d\varsigma
\end{aligned} \tag{5.134}$$

Substituting eqn. (5.118a) into eqn. (5.134) and using the property of the Dirac delta function leads to

$$\Upsilon_k Q_k \Upsilon_k^T = \int_{t_k}^{t_{k+1}} \Phi(t_{k+1},\tau)\,G(\tau)\,Q(\tau)G^T(\tau)\Phi^T(t_{k+1},\tau)\,d\tau \tag{5.135}$$

The integral in eqn. (5.135) is difficult to evaluate even for simple systems. However, we are only interested in the first-order terms, since in the limit as $\Delta t \to 0$ higher-order terms vanish. Therefore, for small Δt we have $\Phi \approx (I + \Delta t\,F)$, and integrating over the small Δt simply yields

$$\boxed{\Upsilon_k Q_k \Upsilon_k^T = \Delta t\,G(t)\,Q(t)\,G^T(t)} \tag{5.136}$$

where eqn. (5.118a) has been used, and terms of order Δt^2 and higher have been dropped. We should note here that the matrix Q_k is a covariance matrix; however, the matrix $Q(t)$ is a *spectral density matrix*.[1, 12] Multiplying $Q(t)$ by the delta function converts it into a covariance matrix.

The integral in eqn. (5.135) may be difficult to evaluate for complex systems. Fortunately, a numerical solution is given by van Loan[13, 14] for fixed-parameter systems, which includes a constant sampling interval and time invariant state and covariance matrices. First, the following $2n \times 2n$ matrix is formed:

$$\mathcal{A} = \begin{bmatrix} -F & G\,Q\,G^T \\ 0 & F^T \end{bmatrix} \Delta t \tag{5.137}$$

where Δt is the constant sampling interval, F is the constant continuous-time state matrix, and Q is the constant continuous-time process noise covariance. Then, the matrix exponential of eqn. (5.137) is computed:

$$\mathcal{B} = e^{\mathcal{A}} \equiv \begin{bmatrix} \mathcal{B}_{11} & \mathcal{B}_{12} \\ 0 & \mathcal{B}_{22} \end{bmatrix} = \begin{bmatrix} \mathcal{B}_{11} & \Phi^{-1}\mathcal{Q} \\ 0 & \Phi^T \end{bmatrix} \tag{5.138}$$

where Φ is the state transition matrix of F and $\mathcal{Q} = \Upsilon\, Q_k \Upsilon^T$ (note, this matrix is constant, but we maintain the subscript k in Q_k to distinguish Q_k from the continuous-time equivalent). An efficient numerical solution of eqn. (5.138) is given by using the series approach in eqn. (3.25). The state transition matrix is then given by

$$\boxed{\Phi = \mathcal{B}_{22}^T} \tag{5.139}$$

Also, the discrete-time process noise covariance is given by

$$\boxed{\mathcal{Q} = \Phi\,\mathcal{B}_{12}} \tag{5.140}$$

If the sampling interval is "small" enough, then eqn. (5.136) is a good approximation for the solution given by eqn. (5.140).

The relationship between the discrete measurement covariance and continuous measurement covariance is not as obvious as the process noise covariance case. Consider the following linear model:

$$\tilde{y}_k = x + v_k \tag{5.141}$$

where an estimate of x is desired. Suppose that the time interval Δt is broken into equal samples, denoted by δ. Using the principles of Chapter 1, the estimate of x, denoted by $\hat{x}$ for m measurement samples over the interval Δt, is given by

$$\hat{x} = \frac{1}{m}\sum_{j=1}^{m} \tilde{y}_j \tag{5.142}$$

The relationship between the discrete-time process v_k and the continuous-time process must surely involve the sampling interval. We consider the following relationship:

$$E\left\{v_k v_j^T\right\} = \begin{cases} 0 & k \neq j \\ \delta^d R & k = j \end{cases} \tag{5.143}$$

for some value of d. Then the estimate error-variance is given by

$$E\left\{(x-\hat{x})^2\right\} = \frac{\delta^d R}{m} \tag{5.144}$$

The limit $m \to \infty$, $\delta \to 0$, and $m\delta \to \Delta t$ gives

$$E\left\{(x-\hat{x})^2\right\} = \begin{cases} 0 & d < -1 \\ \infty & d > -1 \\ \dfrac{R}{\Delta t} & d = -1 \end{cases} \tag{5.145}$$

Therefore, if the continuous model $\tilde{y}(t) = x + v(t)$ is to be meaningful in the sense that the error-variance is nonzero but finite, we must choose $d = -1$.[15] Toward this end in the sampling process, the continuous-time measurement process must be averaged over the sampling interval Δt in order to determine the equivalent discrete sample (where $\mathbf{x}$ is approximated as a constant over the interval).[14] Then we have

$$\begin{aligned} \tilde{\mathbf{y}}_k &= \frac{1}{\Delta t}\int_{t_k}^{t_{k+1}} \tilde{\mathbf{y}}(t)\, dt = \frac{1}{\Delta t}\int_{t_k}^{t_{k+1}} [H(t)\mathbf{x}(t) + \mathbf{v}(t)]\, dt \\ &\approx H_k \mathbf{x}_k + \frac{1}{\Delta t}\int_{t_k}^{t_{k+1}} \mathbf{v}(t)\, dt \end{aligned} \tag{5.146}$$

Therefore, the discrete-to-continuous equivalence can be found by solving the following equation:

$$E\left\{\mathbf{v}_k \mathbf{v}_k^T\right\} \equiv R_k = \frac{1}{\Delta t^2}\int_{t_k}^{t_{k+1}}\int_{t_k}^{t_{k+1}} E\left\{\mathbf{v}(\tau)\mathbf{v}^T(\varsigma)\right\} d\tau\, d\varsigma \tag{5.147}$$

Substituting eqn. (5.118b) into eqn. (5.147) and using the property of the Dirac delta function leads to

$$\boxed{R_k = \frac{R(t)}{\Delta t}} \tag{5.148}$$

The implication of this relationship is that the discrete-time covariance approaches infinity in the continuous representation. This may be counterintuitive at first, but as shown in eqn. (5.145) the inverse time dependence of the discrete-time covariance and the continuous-time equivalent is the *only* relationship that yields a well-behaved process.

To derive the continuous-time Kalman filter we start with the discrete-time version summarized in eqn. (5.54):

$$\hat{\mathbf{x}}_{k+1} = \Phi_k \hat{\mathbf{x}}_k + \Gamma_k \mathbf{u}_k + \Phi K_k [\tilde{\mathbf{y}}_k - H_k \hat{\mathbf{x}}_k] \tag{5.149a}$$

$$K_k = P_k H_k^T [H_k P_k H_k^T + R_k]^{-1} \tag{5.149b}$$

$$P_{k+1} = \Phi_k P_k \Phi_k^T - \Phi_k K_k H_k P_k \Phi_k^T + \Upsilon_k Q_k \Upsilon_k^T \tag{5.149c}$$

Then, using the first-order approximation $\Phi = (I + \Delta t\, F)$ and the relationship in eqn. (5.136) gives the following discrete-time covariance update:

$$\begin{aligned} P_{k+1} &= [I + \Delta t\, F(t)] P_k [I + \Delta t\, F(t)]^T + \Delta t\, G(t)\, Q(t)\, G^T(t) \\ &\quad - [I + \Delta t\, F(t)] K_k H_k P_k [I + \Delta t\, F(t)]^T \end{aligned} \tag{5.150}$$

Dividing eqn. (5.150) by Δt and collecting terms yields

$$\begin{aligned} \frac{P_{k+1} - P_k}{\Delta t} &= F(t)\, P_k + P_k F^T(t) + \Delta t\, F(t)\, P_k F^T(t) \\ &\quad - F(t)\, K_k H_k P_k - K_k H_k P_k F^T(t) - \frac{1}{\Delta t} K_k H_k P_k \\ &\quad - \Delta t\, F(t)\, K_k H_k P_k F^T(t) + G(t)\, Q(t)\, G^T(t) \end{aligned} \tag{5.151}$$

From the definition of the gain K_k in eqn. (5.149b) and using the relationship in eqn. (5.148) we have

$$\begin{aligned} K_k &= P_k H_k^T \left[H_k P_k H_k^T + \frac{R(t)}{\Delta t} \right]^{-1} \\ &= \Delta t\, P_k H_k^T [\Delta t\, H_k P_k H_k^T + R(t)]^{-1} \end{aligned} \tag{5.152}$$

Therefore the limiting condition on K_k gives

$$\lim_{\Delta t \to 0} K_k = 0 \tag{5.153}$$

However when K_k is divided by Δt we have

$$\lim_{\Delta t \to 0} \frac{K_k}{\Delta t} = P(t)\, H^T(t)\, R^{-1}(t) \tag{5.154}$$

Hence in the limit as $\Delta t \to 0$ eqn. (5.151) reduces exactly to the continuous-time covariance propagation in Table 5.4.

Using the first-order approximations of $\Gamma = \Delta t\, B$ and $\Phi = (I + \Delta t\, F)$, the state estimate in eqn. (5.149a) becomes

$$\hat{\mathbf{x}}_{k+1} = [I + \Delta t\, F(t)]\hat{\mathbf{x}}_k + \Delta t\, B(t)\, \mathbf{u}_k + [I + \Delta t\, F(t)] K_k [\tilde{\mathbf{y}}_k - H_k \hat{\mathbf{x}}_k] \tag{5.155}$$

Dividing both sides of eqn. (5.155) by Δt and collecting terms leads to

$$\frac{\hat{\mathbf{x}}_{k+1} - \hat{\mathbf{x}}_k}{\Delta t} = F(t)\, \hat{\mathbf{x}}_k + B(t)\, \mathbf{u}_k + \left[\frac{K_k}{\Delta t} + F(t)\, K_k \right] [\tilde{\mathbf{y}}_k - H_k \hat{\mathbf{x}}_k] \tag{5.156}$$

Hence, using eqns. (5.153) and (5.154), in the limit as $\Delta t \to 0$ eqn. (5.156) reduces exactly to the continuous-time estimate propagation in Table 5.4.

5.4.3 Stability

The filter stability can be proved by using Lyapunov's direct method, which is discussed for continuous-time systems in §3.6. We wish to show that the estimation error dynamics, $\tilde{\mathbf{x}}(t) \equiv \hat{\mathbf{x}}(t) - \mathbf{x}(t)$, are stable. For the continuous-time Kalman filter we consider the following candidate Lyapunov function:

$$V[\tilde{\mathbf{x}}(t)] = \tilde{\mathbf{x}}^T(t)\, P^{-1}(t)\, \tilde{\mathbf{x}}(t) \tag{5.157}$$

Since $P(t)$ is required to be positive definite, then clearly its inverse exists and $V[\tilde{\mathbf{x}}(t)] > 0$ for all $\tilde{\mathbf{x}}(t) \neq \mathbf{0}$. We now need to determine an expression for $\dot{P}^{-1}(t)$ to evaluate the time derivative of eqn. (5.157). This is accomplished by taking the time derivative of $P(t)\, P^{-1}(t) = I$, which gives

$$\frac{d}{dt}\left[P(t)\, P^{-1}(t)\right] = \dot{P}(t)\, P^{-1}(t) + P(t)\, \dot{P}^{-1}(t) = 0 \tag{5.158}$$

Solving eqn. (5.158) for $\dot{P}^{-1}(t)$ gives

$$\dot{P}^{-1}(t) = -P^{-1}(t)\, \dot{P}(t)\, P^{-1}(t) \tag{5.159}$$

Substituting eqn. (5.133) into eqn. (5.159) gives

$$\begin{aligned}\dot{P}^{-1}(t) = &-P^{-1}(t)\, F(t) - F^T(t)\, P^{-1}(t) + H^T(t)\, R^{-1}(t)\, H(t) \\ &- P^{-1}(t)\, G(t)\, Q(t)\, G^T(t)\, P^{-1}(t)\end{aligned} \tag{5.160}$$

Taking the time derivative of eqn. (5.157) yields

$$\dot{V}[\tilde{\mathbf{x}}(t)] = \dot{\tilde{\mathbf{x}}}^T(t)\, P^{-1}(t)\, \tilde{\mathbf{x}}(t) + \tilde{\mathbf{x}}^T(t)\, P^{-1}(t)\, \dot{\tilde{\mathbf{x}}}(t) + \tilde{\mathbf{x}}^T(t)\, \dot{P}^{-1}(t)\, \tilde{\mathbf{x}}(t) \tag{5.161}$$

The continuous-time error dynamics are given by eqn. (5.120). Analogous to the discrete-time case the matrix $F(t) - K(t)\; H(t)$ defines the stability of the filter for the continuous-time case. Substituting $\dot{\tilde{\mathbf{x}}}(t) = [F(t) - K(t)\; H(t)]\tilde{\mathbf{x}}(t)$ and the inverse covariance propagation of eqn. (5.160) into eqn. (5.161), and simplifying leads to

$$\dot{V}[\tilde{\mathbf{x}}(t)] = -\tilde{\mathbf{x}}^T(t)\left[H^T(t)\, R^{-1}(t)\, H(t) + P^{-1}(t)\, G(t)\, Q(t)\, G^T(t)\, P^{-1}(t)\right]\tilde{\mathbf{x}}(t) \tag{5.162}$$

Clearly if $R(t)$ is positive definite and $Q(t)$ is at least positive semi-definite then the Lyapunov condition is satisfied and the continuous-time Kalman filter is stable.

5.4.4 Steady-State Kalman Filter

The continuous Riccati equation in eqn. (5.133) requires $n(n+1)/2$ nonlinear equations to be integrated numerically (normally an $n \times n$ matrix equation requires n^2 integrations, but we use the fact that $P(t)$ is symmetric to significantly reduce this number). Fortunately, analogous to the discrete-time case, for time-invariant systems

Table 5.5: Continuous and Autonomous Linear Kalman Filter

Model	$\dot{\mathbf{x}}(t) = F\,\mathbf{x}(t) + B\,\mathbf{u}(t) + G\,\mathbf{w}(t),\ \mathbf{w}(t) \sim N(\mathbf{0}, Q)$ $\tilde{\mathbf{y}}(t) = H\,\mathbf{x}(t) + \mathbf{v}(t),\ \mathbf{v}(t) \sim N(\mathbf{0}, R)$
Initialize	$\hat{\mathbf{x}}(t_0) = \hat{\mathbf{x}}_0$
Gain	$K = P\,H^T R^{-1}$
Covariance	$F\,P + P\,F^T - P\,H^T R^{-1} H\,P + G\,Q\,G^T = 0$
Estimate	$\dot{\hat{\mathbf{x}}}(t) = F\,\hat{\mathbf{x}}(t) + B\,\mathbf{u}(t) + K[\tilde{\mathbf{y}}(t) - H\,\hat{\mathbf{x}}(t)]$

the error covariance P reaches a steady-state value very quickly. The steady-state continuous-time Kalman filter is summarized in Table 5.5.

To determine the steady-state value for P we must solve the *continuous-time algebraic Riccati equation* in Table 5.5. A sufficient condition for the existence of a steady-state solution is complete observability.[3] Also, the solution is unique if complete controllability exists.[1] These conditions also hold true for the discrete-time Riccati equation in §5.3.4. The continuous-time Riccati equation is a nonlinear differential equation, but it can be transformed into two coupled linear differential equations. This is accomplished by writing P as a product of two matrices:[16]

$$\boxed{P(t) = S(t)\,Z^{-1}(t)} \tag{5.163}$$

or $P(t)\,Z(t) = S(t)$. Differentiating this equation leads to

$$\dot{P}(t)\,Z(t) + P(t)\dot{Z}(t) = \dot{S}(t) \tag{5.164}$$

Substituting eqn. (5.133) into eqn. (5.164) and collecting terms gives

$$\begin{aligned} &P(t)[F^T Z(t) - H^T R^{-1} H\,S(t) + \dot{Z}(t)] \\ &+ [G\,Q\,G^T Z(t) + F\,S(t) - \dot{S}(t)] = 0 \end{aligned} \tag{5.165}$$

Therefore, the following two matrix differential equations must be true to satisfy eqn. (5.165):

$$\dot{Z}(t) = -F^T Z(t) + H^T R^{-1} H\,S(t) \tag{5.166a}$$

$$\dot{S}(t) = G\,Q\,G^T Z(t) + F\,S(t) \tag{5.166b}$$

In order to satisfy eqn. (5.163), initial conditions of $Z(t_0) = I$ and $S(t_0) = P(t_0)$ can be used. Separating the columns of the $Z(t)$ and $S(t)$ gives

$$\begin{bmatrix} \dot{\mathbf{z}}_i(t) \\ \dot{\mathbf{s}}_i(t) \end{bmatrix} = \mathcal{H} \begin{bmatrix} \mathbf{z}_i(t) \\ \mathbf{s}_i(t) \end{bmatrix} \tag{5.167}$$

where $\mathbf{z}_i(t)$ and $\mathbf{s}_i(t)$ are the ith columns of $Z(t)$ and $S(t)$, respectively, and $\mathcal{H}$ is the *Hamiltonian matrix* defined by

$$\mathcal{H} \equiv \begin{bmatrix} -F^T & H^T R^{-1} H \\ G\,Q\,G^T & F \end{bmatrix} \tag{5.168}$$

It can be shown that if λ is an eigenvalue of $\mathcal{H}$, then $-\lambda$ is also an eigenvalue of $\mathcal{H}$, which is left as an exercise for the reader. Thus the eigenvalues can be arranged in a diagonal matrix given by

$$\mathcal{H}_\Lambda = \begin{bmatrix} \Lambda & 0 \\ 0 & -\Lambda \end{bmatrix} \tag{5.169}$$

where Λ is a diagonal matrix of the n eigenvalues in the right half-plane. Assuming that the eigenvalues are distinct, we can perform a linear state transformation, as shown in §3.1.4, such that

$$\mathcal{H}_\Lambda = W^{-1} \mathcal{H}\, W \tag{5.170}$$

where W is the matrix of eigenvectors, which can be represented in block form as

$$W = \begin{bmatrix} W_{11} & W_{12} \\ W_{21} & W_{22} \end{bmatrix} \tag{5.171}$$

The solutions for $\mathbf{z}_i(t)$ and $\mathbf{s}_i(t)$ can be found in terms of their eigensystems:

$$\mathbf{z}_i(t) = \mathbf{w}_1 e^{\lambda t} \tag{5.172a}$$

$$\mathbf{s}_i(t) = \mathbf{w}_2 e^{\lambda t} \tag{5.172b}$$

where $\mathbf{w}_1$ and $\mathbf{w}_2$ are eigenvectors that satisfy

$$(\lambda I - \mathcal{H}) \begin{bmatrix} \mathbf{w}_1 \\ \mathbf{w}_2 \end{bmatrix} = \mathbf{0} \tag{5.173}$$

Going forward in time the unstable eigenvalues dominate, so that

$$\mathbf{z}_i(t) \to W_{11}\, e^{\Lambda t}\, \mathbf{c}_i \tag{5.174a}$$

$$\mathbf{s}_i(t) \to W_{21}\, e^{\Lambda t}\, \mathbf{c}_i \tag{5.174b}$$

where $\mathbf{c}_i$ is an arbitrary constant, and W_{11} and W_{21} are the eigenvectors associated with the unstable eigenvalues. Then from eqn. (5.163) it follows that at steady-state, we have

$$\boxed{P = W_{21} W_{11}^{-1}} \tag{5.175}$$

This requires an inverse of an $n \times n$ matrix.

Vaughan[17] has also shown that a solution for $P(t)$ is given by

$$P(t) = [W_{21} + W_{22} Y(t)][W_{11} + W_{12} Y(t)]^{-1} \tag{5.176}$$

where

$$Y(t) = e^{-\Lambda t} X e^{-\Lambda t} \tag{5.177a}$$

$$X = -[W_{22} - F\, W_{12}]^{-1}[W_{21} - F\, W_{11}] \tag{5.177b}$$

The steady-state solution for P can be found from

$$P = \lim_{t \to \infty} P(t) = W_{21} W_{11}^{-1} \tag{5.178}$$

This result is identical to the steady-state solution derived independently by MacFarlane[18] and Potter,[19] which has been shown previously. Therefore, the gain K in eqn. (5.132) can be computed off-line and remains constant. As in the discrete-time case, this can significantly reduce the on-board computational load on a computer. As a final note, the steady-state solution for the Riccati equation can also be found using a *Schur decomposition*,[20, 21] which is more computationally efficient and more stable than the eigenvector approach. The interested reader is encouraged to pursue this approach, which is more widely used today.

Example 5.4: In this example a simple first-order system is analyzed. The truth model is given by

$$\dot{x}(t) = f\, x(t) + w(t)$$
$$y(t) = x(t) + v(t)$$

where f is a constant, and the variances of $w(t)$ and $v(t)$ are given by q and r, respectively. The first step involves solving the scalar version of the Riccati equation given in eqn. (5.133):

$$\dot{p}(t) = 2\, f\, p(t) - r^{-1} p(t)^2 + q, \quad p(t_0) = p_0 \tag{5.179}$$

To accomplish this task we use the approach given by eqns. (5.163) and (5.166). The Hamiltonian system is given by

$$\begin{bmatrix} \dot{z}(t) \\ \dot{s}(t) \end{bmatrix} = \begin{bmatrix} -f & r^{-1} \\ q & f \end{bmatrix} \begin{bmatrix} z(t) \\ s(t) \end{bmatrix}, \quad \begin{bmatrix} z(t_0) \\ s(t_0) \end{bmatrix} = \begin{bmatrix} 1 \\ p_0 \end{bmatrix}$$

The characteristic equation of this system is given by $s^2 - (f^2 + r^{-1} q) = 0$, which means the solutions for $z(t)$ and $s(t)$ involve hyperbolic functions. We assume that the solutions are given by

$$z(t) = \cosh(a\,t) + c_1 \sinh(a\,t)$$
$$s(t) = p_0 \cosh(a\,t) + c_2 \sinh(a\,t)$$

where $a = \sqrt{f^2 + r^{-1} q}$, and c_1 and c_2 are constants. The assumed solutions obviously satisfy the initial condition requirements. To determine the other constants we take time derivatives of $z(t)$ and $s(t)$ and compare them to the Hamiltonian system, which gives

$$c_1 = \frac{p_0 r^{-1} - f}{a}, \quad c_2 = \frac{p_0 f + q}{a}$$

Hence, using eqn. (5.163) the solution for $p(t)$ is given by

$$p(t) = \frac{p_0 a + (p_0 f + q)\tanh(a\,t)}{a + (p_0 r^{-1} - f)\tanh(a\,t)}$$

Clearly, even for this simple first-order system the solution to the Riccati equation involves complicated functions. Analytical solutions are extremely difficult (if not impossible!) to determine for higher-order systems, so numerical procedures are typically required to integrate the Riccati differential equation. The steady-state value for $p(t)$ is given by noting that as $t \to \infty$ the hyperbolic tangent function approaches one, so that

$$\lim_{t\to\infty} p(t) \equiv p = \frac{(a+f)p_0 + q}{r^{-1}p_0 + a - f} = r\,(a+f)$$

The steady-state value is independent of p_0, which is intuitively correct. This result is verified by solving the algebraic Riccati equation in Table 5.5. Hence, the continuous-time Kalman filter equations are given by

$$\begin{aligned}\dot{\hat{x}}(t) &= -a\,\hat{x}(t) + (a+f)\tilde{y}(t)\\ \hat{y}(t) &= \hat{x}(t)\end{aligned}$$

Note that the filter dynamics are always stable. Also, when $q = 0$ the solution for the steady-state gain is given by zero, and the measurements are completely ignored in the state estimate. Furthermore, the individual values for r and q are irrelevant; only their ratio is important in the filter design. In fact, one of the most arduous tasks in the Kalman filter design is the proper selection of q, which is often not well known. For some systems the filter designer may choose to select the gain K directly (often by trial and error), if the process noise covariance is not well known.

In the preceding example the final form of the steady-state estimator for the state takes the form of a first-order low-pass filter. In the Laplace domain the transfer function from the measured input to the state estimate output is given by

$$\frac{\hat{X}(s)}{\tilde{Y}(s)} = \frac{a+f}{s+a} \tag{5.180}$$

The time constant of this system is given by $1/a$. When q is large or r is small the time constant for the filter approaches zero, so that more high-frequency information is allowed into the state estimate by the filter (i.e., the bandwidth increases). The converse to this statement is also true. When q is small or r is large the time constant for the filter approaches a large value, so that less high-frequency information is allowed into the state estimate by the filter (i.e., the bandwidth decreases). This clearly demonstrates the relationship between the Kalman filter and frequency domain.

The design of the optimal gain using frequency domain methods is known as *Wiener* filtering*.[22] The Wiener filter obtains the best estimates by analyzing time series in the frequency domain using the Fourier transform. The Wiener and Kalman approach can be shown to be identical for the optimal steady-state filter.[5] Unfortunately, Wiener filters are difficult to derive for systems that involve time-varying models or MIMO models, which the Kalman filter handles with ease. Therefore, although a brief introduction of the Wiener filter is given here, we choose not to fully derive the appropriate Wiener (more commonly known as the Wiener-Hopf[5, 14]) filter equation. Still, Wiener filtering is widely used today for many applications in signal processing (e.g., digital image processing). The interested reader is encouraged to pursue Wiener filtering in the open literature.

5.4.5 Correlated Measurement and Process Noise

In this section the correlated Kalman filter for continuous-time models and measurements is derived. The procedure to derive the results of §5.3.5 can also be applied to the continuous-time case. However, an easier approach can be used.[1, 5] We consider the following correlation between the process and measurement noise:

$$\boxed{E\left\{\mathbf{w}(t)\,\mathbf{v}^T(t)\right\} = S(t)\,\delta(t-\tau)} \tag{5.181}$$

Next consider adding zero to the right-hand side of equation eqn. (5.117a), so that

$$\begin{aligned}\dot{\mathbf{x}}(t) &= F(t)\,\mathbf{x}(t) + B(t)\,\mathbf{u}(t) + G(t)\,\mathbf{w}(t)\\ &\quad + \mathcal{D}(t)[\tilde{\mathbf{y}}(t) - H(t)\,\mathbf{x}(t) - \mathbf{v}(t)]\end{aligned} \tag{5.182a}$$

$$\begin{aligned}&= [F(t) - \mathcal{D}(t)\,H(t)]\mathbf{x}(t) + B(t)\,\mathbf{u}(t)\\ &\quad + \mathcal{D}(t)\,\tilde{\mathbf{y}}(t) + [G(t)\,\mathbf{w}(t) - \mathcal{D}(t)\,\mathbf{v}(t)]\end{aligned} \tag{5.182b}$$

where $\mathcal{D}(t)$ is a nonzero matrix. The new process noise for this system is given by $G(t)\,\mathbf{w}(t) - \mathcal{D}(t)\,\mathbf{v}(t) \equiv \boldsymbol{\upsilon}(t)$, which has zero-mean and covariance, so

$$\begin{aligned}E\left\{\boldsymbol{\upsilon}(t)\,\boldsymbol{\upsilon}^T(\tau)\right\} &= \Big[G(t)\,Q(t)\,G^T(t) + \mathcal{D}(t)\,R(t)\,\mathcal{D}^T(t)\\ &\quad - \mathcal{D}(t)\,S(t)\,G^T(t) - G(t)\,S^T(t)\,\mathcal{D}^T(t)\Big]\delta(t-\tau)\end{aligned} \tag{5.183}$$

Any $\mathcal{D}(t)$ can be chosen since eqn. (5.182) will always be true. We choose $\mathcal{D}(t)$ so that $\boldsymbol{\upsilon}(t)$ and $\mathbf{v}(t)$ are uncorrelated. Specifically, if we choose

$$\mathcal{D}(t) = G(t)\,S^T(t)\,R^{-1}(t) \tag{5.184}$$

*Norbert Wiener developed this approach in response to some of the very practical technological problems to improve radar communication that arose during World War II.

Table 5.6: Correlated Continuous-Time Linear Kalman Filter

Model	$\dot{\mathbf{x}}(t) = F(t)\,\mathbf{x}(t) + B(t)\,\mathbf{u}(t) + G(t)\,\mathbf{w}(t),\ \mathbf{w}(t) \sim N(\mathbf{0}, Q)$ $\tilde{\mathbf{y}}(t) = H(t)\,\mathbf{x}(t) + \mathbf{v}(t),\ \mathbf{v}(t) \sim N(\mathbf{0}, R)$ $E\left\{\mathbf{w}(t)\,\mathbf{v}^T(t)\right\} = S(t)\,\delta(t-\tau)$
Initialize	$\hat{\mathbf{x}}(t_0) = \hat{\mathbf{x}}_0$ $P_0 = E\left\{\tilde{\mathbf{x}}(t_0)\,\tilde{\mathbf{x}}^T(t_0)\right\}$
Gain	$K(t) = \left[P(t)\,H^T(t) + G(t)\,S^T(t)\right]R^{-1}(t)$
Covariance	$\dot{P}(t) = F(t)\,P(t) + P(t)\,F^T(t)$ $-K(t)\,R(t)\,K^T(t) + G(t)\,Q(t)\,G^T(t)$
Estimate	$\dot{\hat{\mathbf{x}}}(t) = F(t)\,\hat{\mathbf{x}}(t) + B(t)\,\mathbf{u}(t)$ $+K(t)[\tilde{\mathbf{y}}(t) - H(t)\,\hat{\mathbf{x}}(t)]$

then

$$E\left\{\upsilon(t)\,\mathbf{v}^T(\tau)\right\} = [G(t)S^T(t) - \mathcal{D}(t)\,R(t)]\delta(t-\tau) = 0 \tag{5.185}$$

Hence the covariance of the new process noise $\upsilon(t)$ is given by

$$E\left\{\upsilon(t)\,\upsilon^T(\tau)\right\} = G(t)\left[Q(t) - S^T(t)\,R^{-1}(t)\,S(t)\right]G^T(t)\,\delta(t-\tau) \tag{5.186}$$

The derivation procedure of §5.4.1 can now be applied to eqn. (5.182b). The results are summarized in Table 5.6. Note that a nonzero $S(t)$ produces a smaller covariance than the uncorrelated case, which is due to the additional information provided by the cross-correlation between $\mathbf{w}(t)$ and $\mathbf{v}(t)$. Also, when $S(t) = 0$, i.e., $\mathbf{w}(t)$ and $\mathbf{v}(t)$ are uncorrelated, the correlated Kalman filter reduces exactly to the standard Kalman filter given in Table 5.4.

5.5 The Continuous-Discrete Kalman Filter

Most physical dynamical systems involve continuous-time models and discrete-time measurements taken from a digital signal processor. Therefore, the system model and measurement model are given by

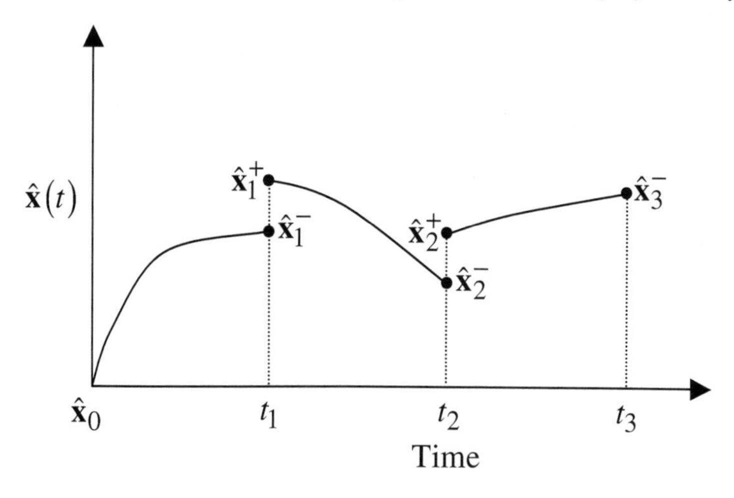

Figure 5.5: Mechanism for the Continuous-Discrete Kalman Filter

$$\boxed{\begin{aligned}\dot{\mathbf{x}}(t) &= F(t)\,\mathbf{x}(t) + B(t)\,\mathbf{u}(t) + G(t)\,\mathbf{w}(t) \qquad (5.187\text{a})\\ \tilde{\mathbf{y}}_k &= H_k\mathbf{x}_k + \mathbf{v}_k \qquad (5.187\text{b})\end{aligned}}$$

where the continuous-time covariance of $\mathbf{w}(t)$ is given by eqn. (5.118a) and the discrete-time covariance of $\mathbf{v}_k$ is given by eqn. (5.28).

The extension of the Kalman filter for this case is very straightforward. The mechanism of the filter approach for this case is illustrated in Figure 5.5. The state estimate model is propagated forward in time until a measurement occurs, given at time t_1. Then a discrete-time state update occurs, which updates the final value of the propagated state $\hat{\mathbf{x}}_1^-$ to the new state $\hat{\mathbf{x}}_1^+$. Finally this state is then used as the initial condition to propagate the state estimate model to time t_2. The scheme continues forward in time, updating the state when a measurement occurs.

A summary of the continuous-discrete Kalman filter is given in Table 5.7. Note that the continuous-time propagation model equation does not involve the measurement directly. Hence, the covariance propagation follows a continuous-time Lyapunov differential equation, which is a linear equation. When a measurement occurs both the state and the covariance are updated using the standard discrete-time updates. Also, if the state and measurement models are autonomous, and the measurements sampling interval is constant and well below Nyquist's limit, then a steady-state covariance expression can be found (this is left as an exercise for the reader).

We should note that the sample times of the measurements need not occur in regular intervals. In fact different measurement sets can be spread out over various time intervals. Whenever a measurement occurs then an update is invoked. The measurement set at that time may involve only one measurement or multiple measurements. The real beauty of the continuous-discrete Kalman filter is that it can handle different scattered measurement sets quite easily.

Table 5.7: Continuous-Discrete Kalman Filter

Model	$\dot{\mathbf{x}}(t) = F(t)\mathbf{x}(t) + B(t)\mathbf{u}(t) + G(t)\mathbf{w}(t),\ \mathbf{w}(t) \sim N(\mathbf{0}, Q(t))$ $\tilde{\mathbf{y}}_k = H_k\mathbf{x}_k + \mathbf{v}_k,\quad \mathbf{v}_k \sim N(\mathbf{0}, R_k)$
Initialize	$\hat{\mathbf{x}}(t_0) = \hat{\mathbf{x}}_0$ $P_0 = E\left\{\tilde{\mathbf{x}}(t_0)\tilde{\mathbf{x}}^T(t_0)\right\}$
Gain	$K_k = P_k^- H_k^T [H_k P_k^- H_k^T + R_k]^{-1}$
Update	$\hat{\mathbf{x}}_k^+ = \hat{\mathbf{x}}_k^- + K_k[\tilde{\mathbf{y}}_k - H_k\hat{\mathbf{x}}_k^-]$ $P_k^+ = [I - K_k H_k]P_k^-$
Propagation	$\dot{\hat{\mathbf{x}}}(t) = F(t)\hat{\mathbf{x}}(t) + B(t)\mathbf{u}(t)$ $\dot{P}(t) = F(t)P(t) + P(t)F^T(t) + G(t)Q(t)G^T(t)$

5.6 Extended Kalman Filter

A large class of estimation problems involve nonlinear models. For several reasons, state estimation for nonlinear systems is considerably more difficult and admits a wider variety of solutions than the linear problem.[1] A vast majority of nonlinear models are given in continuous-time. Therefore, we first consider the following common nonlinear truth model with continuous-time measurements:

$$\dot{\mathbf{x}}(t) = \mathbf{f}(\mathbf{x}(t), \mathbf{u}(t), t) + G(t)\mathbf{w}(t) \tag{5.188a}$$

$$\tilde{\mathbf{y}}(t) = \mathbf{h}(\mathbf{x}(t), t) + \mathbf{v}(t) \tag{5.188b}$$

where $\mathbf{f}(\mathbf{x}(t), \mathbf{u}(t), t)$ and $\mathbf{h}(\mathbf{x}(t), t)$ are assumed to be continuously differentiable, and $\mathbf{w}(t)$ and $\mathbf{v}(t)$ follow exactly from §5.4.1. The problem with this nonlinear model is that a Gaussian input does not necessarily produce a Gaussian output (unlike the linear case). Some of these problems are seen by considering the simple nonlinear and stochastic function

$$y(t) = \sin(t) + \upsilon(t) \tag{5.189}$$

The top plot of Figure 5.6 shows $y(t)$ with a Gaussian input ($\sigma = 1$), as a function of normalized time in degrees (360 degrees is equivalent to 2π seconds). Clearly, the probability density function of $\upsilon(t)$ is altered as it is transmitted through the nonlinear element. The exact probability density function can be determined using a transformation of variables[14, 23] (see Appendix B). But for small angles the output

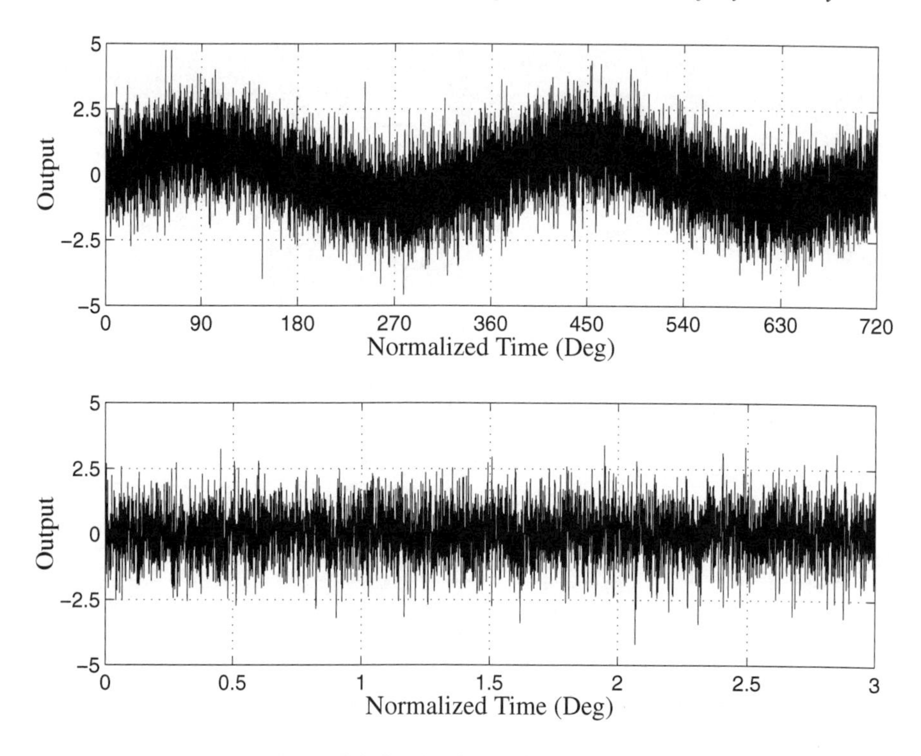

Figure 5.6: Stochastic Nonlinear Example

is *approximately* Gaussian, as shown by the bottom plot of Figure 5.6, where $\sin(t)$ can be approximated by t for small t. Also, $E\left\{y^2(t)\right\} \approx 1$ since terms in t^2 are second-order in nature, which can be ignored. This approach can be used to derive a Kalman filter using nonlinear models.

There are many possible ways to produce a linearized version of the Kalman filter.[1, 24] We will consider the most common approach, which is the *extended Kalman filter*. The extended Kalman filter, though not precisely "optimum," has been successfully applied to many nonlinear systems over the past many years. The fundamental concept of this filter involves the notion that the true state is sufficiently close to the estimated state. Therefore, the error dynamics can be represented fairly accurately by a linearized first-order Taylor series expansion. Consider the first-order expansion of $\mathbf{f}(\mathbf{x}(t), \mathbf{u}(t), t)$ about some nominal state $\bar{\mathbf{x}}(t)$:

$$\mathbf{f}(\mathbf{x}(t), \mathbf{u}(t), t) \cong \mathbf{f}(\bar{\mathbf{x}}(t), \mathbf{u}(t), t) + \left.\frac{\partial \mathbf{f}}{\partial \mathbf{x}}\right|_{\bar{\mathbf{x}}(t)} [\mathbf{x}(t) - \bar{\mathbf{x}}(t)] \tag{5.190}$$

where $\bar{\mathbf{x}}(t)$ is close to $\mathbf{x}(t)$. Also, the output in eqn. (5.188b) can also be expanded using

$$\mathbf{h}(\mathbf{x}(t), t) \cong \mathbf{h}(\bar{\mathbf{x}}(t), t) + \left.\frac{\partial \mathbf{h}}{\partial \mathbf{x}}\right|_{\bar{\mathbf{x}}(t)} [\mathbf{x}(t) - \bar{\mathbf{x}}(t)] \tag{5.191}$$

Table 5.8: Continuous-Time Extended Kalman Filter

Model	$\dot{\mathbf{x}}(t) = \mathbf{f}(\mathbf{x}(t), \mathbf{u}(t), t) + G(t)\,\mathbf{w}(t), \ \ \mathbf{w}(t) \sim N(\mathbf{0}, Q(t))$ $\tilde{\mathbf{y}}(t) = \mathbf{h}(\mathbf{x}(t), t) + \mathbf{v}(t), \ \ \mathbf{v}(t) \sim N(\mathbf{0}, R(t))$
Initialize	$\hat{\mathbf{x}}(t_0) = \hat{\mathbf{x}}_0$ $P_0 = E\left\{\tilde{\mathbf{x}}(t_0)\,\tilde{\mathbf{x}}^T(t_0)\right\}$
Gain	$K(t) = P(t)\,H^T(\hat{\mathbf{x}}(t), t)\,R^{-1}(t)$
Covariance	$\dot{P}(t) = F(\hat{\mathbf{x}}(t), t)\,P(t) + P(t)\,F^T(\hat{\mathbf{x}}(t), t)$ $-P(t)\,H^T(\hat{\mathbf{x}}(t), t)\,R^{-1}(t) H(\hat{\mathbf{x}}(t), t)\,P(t) + G(t)\,Q(t)\,G^T(t)$ $F(\hat{\mathbf{x}}(t), t) \equiv \left.\dfrac{\partial \mathbf{f}}{\partial \mathbf{x}}\right\vert_{\hat{\mathbf{x}}(t)}, \quad H(\hat{\mathbf{x}}(t), t) \equiv \left.\dfrac{\partial \mathbf{h}}{\partial \mathbf{x}}\right\vert_{\hat{\mathbf{x}}(t)}$
Estimate	$\dot{\hat{\mathbf{x}}}(t) = \mathbf{f}(\hat{\mathbf{x}}(t), \mathbf{u}(t), t) + K(t)[\tilde{\mathbf{y}}(t) - \mathbf{h}(\hat{\mathbf{x}}(t), t)]$

In the extended Kalman filter, the current estimate (i.e., conditional mean) is used for the nominal state estimate, so that $\bar{\mathbf{x}}(t) = \hat{\mathbf{x}}(t)$. Taking the expectation of both sides of eqns. (5.190) and (5.191), with $\bar{\mathbf{x}}(t) = \hat{\mathbf{x}}(t)$, gives

$$E\{\mathbf{f}(\mathbf{x}(t), \mathbf{u}(t), t)\} = \mathbf{f}(\hat{\mathbf{x}}(t), \mathbf{u}(t), t) \tag{5.192a}$$

$$E\{\mathbf{h}(\mathbf{x}(t), t)\} = \mathbf{h}(\hat{\mathbf{x}}(t), t) \tag{5.192b}$$

Therefore, the extended Kalman filter structure for the state and output estimate is given by

$$\boxed{\begin{aligned} \dot{\hat{\mathbf{x}}}(t) &= \mathbf{f}(\hat{\mathbf{x}}(t), \mathbf{u}(t), t) + K(t)[\tilde{\mathbf{y}}(t) - \mathbf{h}(\hat{\mathbf{x}}(t), t)] \quad (5.193a)\\ \hat{\mathbf{y}}(t) &= \mathbf{h}(\hat{\mathbf{x}}(t), t) \quad (5.193b) \end{aligned}}$$

Substituting eqns. (5.190) and (5.191), with $\bar{\mathbf{x}}(t) = \hat{\mathbf{x}}(t)$, into eqn. (5.193a), and using eqn. (5.188) leads to

$$\dot{\tilde{\mathbf{x}}}(t) = \left[F(\hat{\mathbf{x}}(t), t) - K(t)\,H(\hat{\mathbf{x}}(t), t)\right]\tilde{\mathbf{x}}(t) - G(t)\,\mathbf{w}(t) + K(t)\,\mathbf{v}(t) \tag{5.194}$$

where $\tilde{\mathbf{x}}(t) = \hat{\mathbf{x}}(t) - \mathbf{x}(t)$ and

$$\boxed{F(\hat{\mathbf{x}}(t), t) \equiv \left.\frac{\partial \mathbf{f}}{\partial \mathbf{x}}\right\vert_{\hat{\mathbf{x}}(t)}, \quad H(\hat{\mathbf{x}}(t), t) \equiv \left.\frac{\partial \mathbf{h}}{\partial \mathbf{x}}\right\vert_{\hat{\mathbf{x}}(t)}} \tag{5.195}$$

Equation (5.194) has the same structure as eqn. (5.120). Hence the covariance expression given by eqn. (5.133) can be used with $F(t)$ replaced by $F(\hat{\mathbf{x}}(t), t)$ and $H(t)$

replaced by $H(\hat{\mathbf{x}}(t), t)$. A summary of the continuous-time extended Kalman filter is given in Table 5.8. The matrices $F(\hat{\mathbf{x}}(t), t)$ and $H(\hat{\mathbf{x}}(t), t)$ will not be constant in general. Therefore, a steady-state gain cannot be found, which may significantly increase the computational burden since $n(n+1)/2$ nonlinear equations need to be integrated to determine $P(t)$.

Another approach involves linearizing about the nominal (*a priori*) state vector $\bar{\mathbf{x}}(t)$ instead of the current estimate $\hat{\mathbf{x}}(t)$. In this case taking the expectation of both sides of eqns. (5.190) and (5.191) gives

$$E\{\mathbf{f}(\mathbf{x}(t), \mathbf{u}(t), t)\} = \mathbf{f}(\bar{\mathbf{x}}(t), \mathbf{u}(t), t) + F(\bar{\mathbf{x}}(t), t)[\hat{\mathbf{x}}(t) - \bar{\mathbf{x}}(t)] \tag{5.196a}$$

$$E\{\mathbf{h}(\mathbf{x}(t), t)\} = \mathbf{h}(\bar{\mathbf{x}}(t), t) + H(\bar{\mathbf{x}}(t), t)[\hat{\mathbf{x}}(t) - \bar{\mathbf{x}}(t)] \tag{5.196b}$$

Therefore, the Kalman filter structure for the state and output estimate is given by

$$\boxed{\begin{aligned}\dot{\hat{\mathbf{x}}}(t) = {} & \mathbf{f}(\bar{\mathbf{x}}(t), \mathbf{u}(t), t) + F(\bar{\mathbf{x}}(t), t)[\hat{\mathbf{x}}(t) - \bar{\mathbf{x}}(t)] \\ & + K(t)\left\{\tilde{\mathbf{y}}(t) - \mathbf{h}(\bar{\mathbf{x}}(t), t) - H(\bar{\mathbf{x}}(t), t)[\hat{\mathbf{x}}(t) - \bar{\mathbf{x}}(t)]\right\}\end{aligned}} \tag{5.197a}$$

$$\boxed{\hat{\mathbf{y}}(t) = \mathbf{h}(\bar{\mathbf{x}}(t), t) + H(\bar{\mathbf{x}}(t), t)[\hat{\mathbf{x}}(t) - \bar{\mathbf{x}}(t)]} \tag{5.197b}$$

The covariance equation follows the form given in Table 5.8, with the partials evaluated at the nominal state instead of the current estimate. These equations form the *linearized Kalman filter*. In general, the linearized Kalman filter is less accurate than the extended Kalman filter since $\bar{\mathbf{x}}(t)$ is usually not as close to the truth as is $\hat{\mathbf{x}}(t)$.[1] However since the nominal state is known *a priori* the gain $K(t)$ can be pre-computed and stored, which reduces the on-line computational burden.

A summary of the continuous-discrete extended Kalman filter is given in Table 5.9. The approach used in the extended Kalman filter assumes that the true state is "close" to the estimated state. This restriction can prove to be especially damaging for highly nonlinear applications with large initial condition errors. Proving convergence in the extended Kalman filter is difficult (if not impossible!) even for simple systems where the initial condition is not well known. Even so, the extended Kalman filter is widely used in practice, and is often robust to initial condition errors, which can be often verified through simulation.

The current estimate in the extended Kalman filter can be improved by applying local iterations to repeatedly calculate $\hat{\mathbf{x}}_k^+$, P_k^+, and K_k, each time linearizing about the most recent estimate.[1, 23] This approach is known as the *iterated extended Kalman filter*. The iterations are given by

$$\hat{\mathbf{x}}_{k_i}^+ = \hat{\mathbf{x}}_k^- + K_{k_i}\left[\tilde{\mathbf{y}}_k - \mathbf{h}(\hat{\mathbf{x}}_{k_i}^+) - H_k(\hat{\mathbf{x}}_{k_i}^+)\left(\hat{\mathbf{x}}_k^- - \hat{\mathbf{x}}_{k_i}^+\right)\right] \tag{5.198a}$$

$$K_{k_i} = P_k^- H_k^T(\hat{\mathbf{x}}_{k_i}^+)\left[H_k(\hat{\mathbf{x}}_{k_i}^+) P_k^- H_k^T(\hat{\mathbf{x}}_{k_i}^+) + R_k\right]^{-1} \tag{5.198b}$$

$$P_{k_i}^+ = \left[I - K_{k_i} H_k(\hat{\mathbf{x}}_{k_i}^+)\right] P_k^- \tag{5.198c}$$

with $\hat{\mathbf{x}}_{k_0}^+ = \hat{\mathbf{x}}_k^-$. The iterations are continued until the estimate is no longer improved. The reference trajectory over $[t_{k-1}, t_k)$ can also be improved once the measurement

Table 5.9: Continuous-Discrete Extended Kalman Filter

Model	$\dot{\mathbf{x}}(t) = \mathbf{f}(\mathbf{x}(t), \mathbf{u}(t), t) + G(t)\,\mathbf{w}(t),\ \mathbf{w}(t) \sim N(\mathbf{0}, Q(t))$ $\tilde{\mathbf{y}}_k = \mathbf{h}(\mathbf{x}_k) + \mathbf{v}_k,\ \mathbf{v}_k \sim N(\mathbf{0}, R_k)$
Initialize	$\hat{\mathbf{x}}(t_0) = \hat{\mathbf{x}}_0$ $P_0 = E\left\{\tilde{\mathbf{x}}(t_0)\,\tilde{\mathbf{x}}^T(t_0)\right\}$
Gain	$K_k = P_k^- H_k^T(\hat{\mathbf{x}}_k^-)[H_k(\hat{\mathbf{x}}_k^-)P_k^- H_k^T(\hat{\mathbf{x}}_k^-) + R_k]^{-1}$ $H_k(\hat{\mathbf{x}}_k^-) \equiv \left.\frac{\partial \mathbf{h}}{\partial \mathbf{x}}\right\vert_{\hat{\mathbf{x}}_k^-}$
Update	$\hat{\mathbf{x}}_k^+ = \hat{\mathbf{x}}_k^- + K_k[\tilde{\mathbf{y}}_k - \mathbf{h}(\hat{\mathbf{x}}_k^-)]$ $P_k^+ = [I - K_k H_k(\hat{\mathbf{x}}_k^-)]P_k^-$
Propagation	$\dot{\hat{\mathbf{x}}}(t) = \mathbf{f}(\hat{\mathbf{x}}(t), \mathbf{u}(t), t)$ $\dot{P}(t) = F(\hat{\mathbf{x}}(t), t)\,P(t) + P(t)\,F^T(\hat{\mathbf{x}}(t), t) + G(t)\,Q(t)\,G^T(t)$ $F(\hat{\mathbf{x}}(t), t) \equiv \left.\frac{\partial \mathbf{f}}{\partial \mathbf{x}}\right\vert_{\hat{\mathbf{x}}(t)}$

$\tilde{\mathbf{y}}_k$ is taken. This is accomplished by applying a nonlinear smoother (see §6.1.3) backward to time t_{k-1}. This approach is known as an *iterated linearized filter-smoother*.[23, 24] The algorithm can also be iterated globally, having processed all measurements, by applying a smoother back to time t_0.[24]

Example 5.5: In this example we will demonstrate the usefulness of the extended Kalman filter to estimate the states of Van der Pol's equation, given by

$$m\,\ddot{x} + 2\,c\,(x^2 - 1)\,\dot{x} + k\,x = 0$$

where m, c, and k have positive values. This equation induces a limit cycle that is sustained by periodically releasing energy into and absorbing energy from the environment, through the damping term.[25] The system can be represented in first-order form by defining the following state vector $\mathbf{x} = \begin{bmatrix} x & \dot{x} \end{bmatrix}^T$:

$$\begin{aligned}\dot{x}_1 &= x_2 \\ \dot{x}_2 &= -2\,(c/m)(x_1^2 - 1)\,x_2 - (k/m)\,x_1\end{aligned}$$

The measurement output is position, so that $H = \begin{bmatrix} 1 & 0 \end{bmatrix}$. Synthetic states are generated using $m = c = k = 1$, with an initial condition of $\mathbf{x}_0 = \begin{bmatrix} 1 & 0 \end{bmatrix}^T$. The measure-

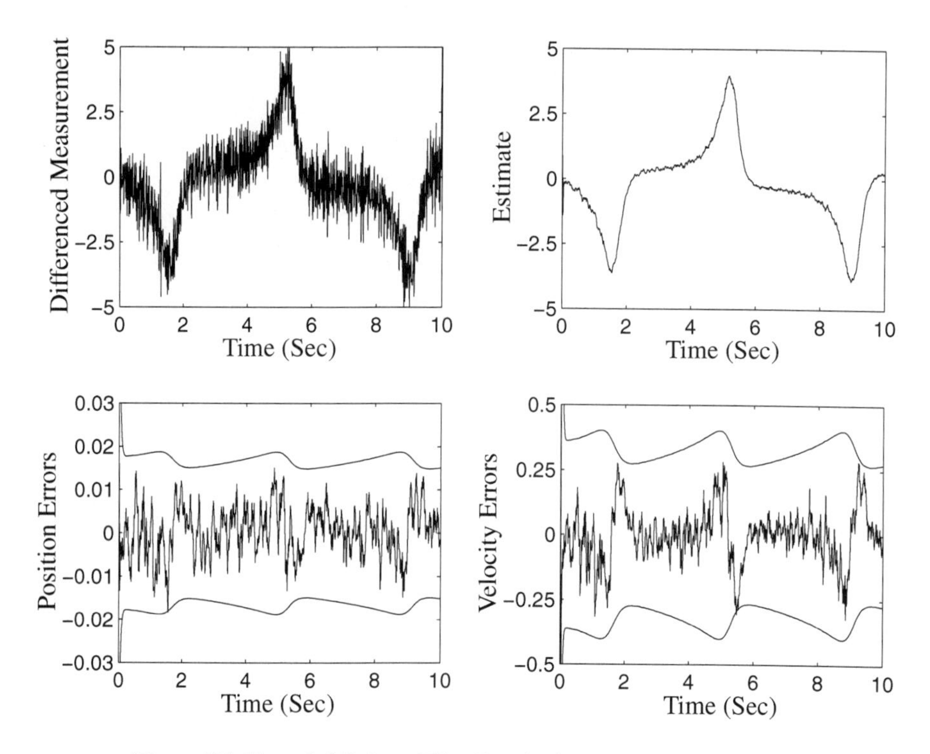

Figure 5.7: Extended Kalman Filter Results for Van der Pol's Equation

ments are sampled at $\Delta t = 0.01$-second intervals with a measurement-error standard deviation of $\sigma = 0.01$. The linearized model and G matrix used in the extended Kalman filter are given by

$$F = \begin{bmatrix} 0 & 1 \\ -4\,(c/m)\,\hat{x}_1\hat{x}_2 - (k/m) & -2\,(c/m)\,(\hat{x}_1^2 - 1) \end{bmatrix}, \quad G = \begin{bmatrix} 0 \\ 1 \end{bmatrix}$$

Note that no process noise (i.e., no error) is introduced into the first state. This is due to the fact that the first state is a kinematical relationship that is correct in theory and in practice (i.e., velocity is always the derivative of position). In the extended Kalman filter the model parameters are assumed to be given by $m = 1$, $c = 1.5$, and $k = 1.2$, which introduces errors in the assumed system, compared to the true system. The initial covariance is chosen to be $P_0 = 1000\,I$. The scalar $q \equiv Q(t)$ in the extended Kalman filter is then tuned until reasonable state estimates are achieved (this tuning process is often required in the design of a Kalman filter). The answer to the question "what are reasonable estimates?" is often left to the design engineer. Since for this simulation the truth is known, we can compare our estimates with the truth to tune q. It was found that $q = 0.2$ results in good estimates. The adaptive methods of §5.7.4 can also be employed to help determine q using measurement residuals.

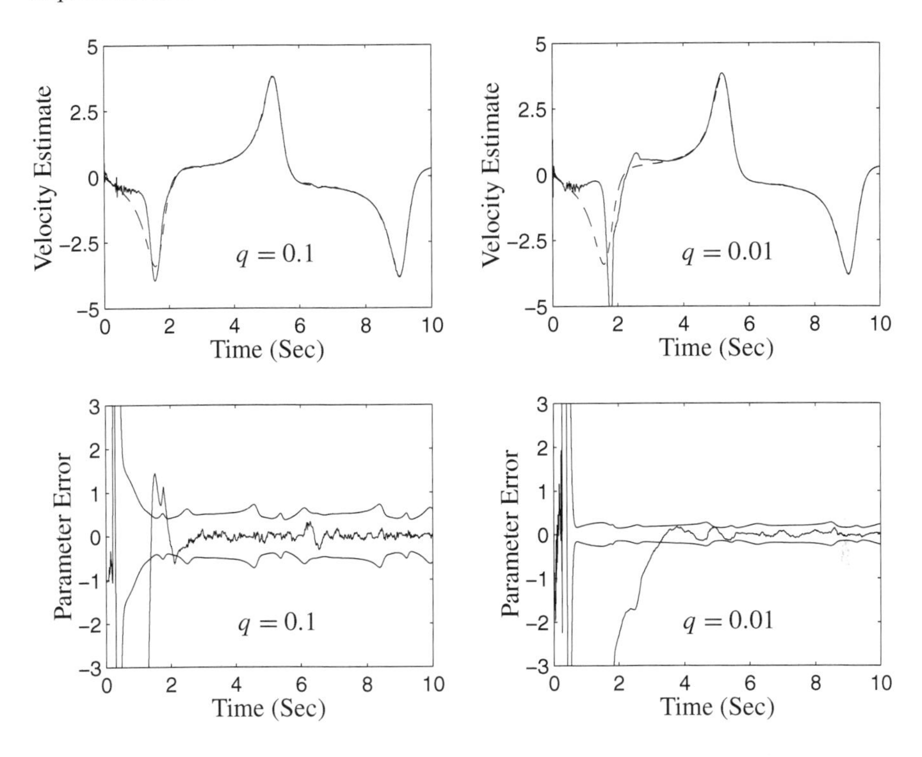

Figure 5.8: Extended Kalman Filter Parameter Identification Results

When first confronted with the position measurements, one may naturally choose to take a numerical finite-difference to derive a velocity estimate. The top left plot of Figure 5.7 shows the result of this approach (with the truth overlapped in the plot). Clearly the result is very noisy. The top right plot of Figure 5.7 shows the velocity estimate using the tuned extended Kalman filter. Clearly the state estimate is closer to the truth than using a numerical finite-difference approach. The bottom plots of Figure 5.7 show the state errors (estimate minus truth) with 3σ boundaries. The boundaries do provide a bound for the estimate errors. We should note that the estimate error does not look Gaussian. This is due to the fact that the process noise is in fact modelling errors in this example. However, the extended Kalman filter still works well even for this case. This example shows the power of the extended Kalman filter to provide accurate estimates for a highly nonlinear system.

Example 5.6: We next show the power of using the Kalman filter to estimate model parameters online. We will now assume that the damping coefficient c is unknown. This parameter can be estimated by appending the state vector of the assumed model in the extended Kalman filter. A common approach assumes a random-walk process, so that $\dot{\hat{c}} \equiv \dot{\hat{x}}_3 = 0$. The linearized model is now given by

$$F = \begin{bmatrix} 0 & 1 & 0 \\ -4(\hat{x}_3/m)\hat{x}_1\hat{x}_2 - (k/m) & -2(\hat{x}_3/m)(\hat{x}_1^2 - 1) & -(2/m)(\hat{x}_1^2 - 1)\hat{x}_2 \\ 0 & 0 & 0 \end{bmatrix}$$

In this case we assume that the model structure with $m = 1$ and $k = 1$ are known perfectly. Our objective is to find the parameter c, where the true value is $c = 1$. Therefore, the matrix G is assumed to be given by $G = \begin{bmatrix} 0 & 0 & 1 \end{bmatrix}^T$. The same measurements as before are used in this simulation. Also, the initial condition for the parameter estimate is set to zero ($\hat{c}(t_0) = 0$). Results using two different values for q are shown in Figure 5.8. The top plots show the estimated velocity states, while the bottom plots show the parameter error-states. When $q = 0.1$ the filter converges fairly rapidly as opposed to the case when $q = 0.01$. However, the estimate for c is more accurate using $q = 0.01$, since the covariance is smaller than the $q = 0.1$ case. Intuitively this makes sense since a smaller q relies more on the model, which implies better knowledge that leads to more accurate estimates. However, a price is paid in convergence, which may be a cause for concern if the model estimate is needed in an online control algorithm. This shows the classic tradeoff between convergence and accuracy when using the Kalman filter to identify model parameters.

5.7 Advanced Topics

In this section we will show some advanced topics used in the Kalman filter. As in previous chapters we encourage the interested reader to pursue these topics further in the references provided. These topics include: factorization methods, colored-noise Kalman filtering, consistency of the Kalman filter, adaptive filtering, error analysis, Unscented filtering, and robust filtering.

5.7.1 Factorization Methods

The linear and autonomous Kalman filter has been shown to be theoretically stable using Lyapunov's direct method (i.e., the estimates will not diverge from the true values), and provides accurate estimates under properly defined conditions. However, the numerical stability of the extended Kalman filter must be properly addressed before on-board implementation. Many factors affect filter stability for this case. One common problem is in the error covariance update and propagation, which may become semi-definite or even negative definite, chiefly due to computational instabilities. A measure of the potential for difficulty in an ill-conditioned matrix can be found by using the *condition number* (see Appendix A). This problem may be overcome by using the Joseph form shown in §5.3.2. Other methods described here decompose the covariance matrix P into better conditioned matrices, which attempt to overcome finite-word length computation errors. We should note that the methods described here do not increase the performance of the Kalman filter in theory. These methods are strictly used to provide a better conditioned Kalman filter in practice (i.e., in a computational sense).

Square Root Information Filter

The first method is based upon a square root factorization of P, given by

$$P = S\,S^T \tag{5.199}$$

One nice property of this factorization is that P is always positive semi-definite even if S is not. Unfortunately, the matrix S is not unique. The original idea for the square root filter is attributed to James E. Potter, and was developed only one year after Kalman's original paper.[26] So, the problem of computational stability was known from the onset. An estimator based on this approach was used extensively in the Apollo navigation system. The square root formulation requires half the significant digits of the standard covariance formulation.[4] Instead of the factorization shown in eqn. (5.199), we show a more robust approach by decomposing the inverse of P. This algorithm is known as the *Square Root Information Filter* (SRIF). The equations are described without derivation. We refer the readers to Refs. [23] and [27], which provide a thorough treatise on square root filtering. The SRIF uses the inverse of eqn. (5.199):

$$\boxed{\begin{aligned} \mathcal{P}_k^+ &\equiv (P_k^+)^{-1} = \mathcal{S}_k^{+T}\,\mathcal{S}_k^+ \qquad (5.200\text{a}) \\ \mathcal{P}_k^- &\equiv (P_k^-)^{-1} = \mathcal{S}_k^{-T}\,\mathcal{S}_k^- \qquad (5.200\text{b}) \end{aligned}}$$

where $\mathcal{S} \equiv S^{-1}$. A square root decomposition of the inverse measurement covariance and an eigenvalue decomposition of the process noise covariance is also used in the SRIF:

$$\boxed{\begin{aligned} R_k^{-1} &= \mathcal{V}_k^T\,\mathcal{V}_k \qquad (5.201\text{a}) \\ Q_k &= Z_k E_k Z_k^T \qquad (5.201\text{b}) \end{aligned}}$$

where $\mathcal{V}_k$ is the inverse of the matrix V_k in $R = V_k V_k^T$. The matrix Z_k is an $s \times s$ (where s is the dimension of the matrix Q_k) orthogonal matrix, and E_k is an $s \times s$ diagonal matrix of the eigenvalues of Q_k. Next, the following $(n+m) \times n$ matrix is formed:

$$\tilde{\mathcal{S}}_k^+ \equiv \begin{bmatrix} \mathcal{S}_k^- \\ \mathcal{V}_k H_k \end{bmatrix} \tag{5.202}$$

It can be shown that when a QR decomposition (see §1.6.1) of $\tilde{\mathcal{S}}_k^+$ is taken, then the updated matrix $\mathcal{S}_k^+$ can be extracted from

$$\boxed{\mathcal{Q}_k^T \tilde{\mathcal{S}}_k^+ = \begin{bmatrix} \mathcal{S}_k^+ \\ 0_{m\times n} \end{bmatrix}} \tag{5.203}$$

where $\mathcal{Q}_k$ is the orthogonal matrix from the QR decomposition of $\tilde{\mathcal{S}}_k^+$. In the SRIF the state is not explicitly estimated. Instead the following quantities are used:

$$\hat{\boldsymbol{\alpha}}_k^+ \equiv \mathcal{S}_k^+ \hat{\mathbf{x}}_k^+ \tag{5.204a}$$

$$\hat{\boldsymbol{\alpha}}_k^- \equiv \mathcal{S}_k^- \hat{\mathbf{x}}_k^- \tag{5.204b}$$

Note the updated and propagated state can easily be found by taking the inverse of eqn. (5.204). The update equation is given by

$$\boxed{\begin{bmatrix} \hat{\boldsymbol{\alpha}}_k^+ \\ \boldsymbol{\beta}_k \end{bmatrix} = \mathcal{Q}_k^T \begin{bmatrix} \hat{\boldsymbol{\alpha}}_k^- \\ \tilde{\mathbf{y}}_k \end{bmatrix}} \tag{5.205}$$

where $\boldsymbol{\beta}_k$ is an $m \times 1$ vector, which is the residual after processing the measurement, that is not required in the SRIF calculations. The following $n \times s$ matrix is now defined:

$$\Xi_k \equiv \Upsilon_k Z_k \tag{5.206}$$

where Υ_k is defined in the discrete-time Kalman filter (see Table 5.1). Let $\Xi_k(i)$ denote the ith column of Ξ_k and $E_k(i,i)$ denote the ith diagonal value of the matrix E_k. The propagated values are given by a set of s iterations:
for $i = 1$

$$\mathbf{a} = \mathcal{S}_k^+ \Phi_k^{-1} \Xi_k(1) \tag{5.207a}$$

$$b = \left[\mathbf{a}^T \mathbf{a} + 1/E_k(1,1)\right]^{-1} \tag{5.207b}$$

$$c = \left[1 + \sqrt{b/E_k(1,1)}\right]^{-1} \tag{5.207c}$$

$$\mathbf{d}^T = b\,\mathbf{a}^T \mathcal{S}_k^+ \Phi_k^{-1} \tag{5.207d}$$

$$\hat{\boldsymbol{\alpha}}_{k+1}^- = \hat{\boldsymbol{\alpha}}_k^+ - b\,c\,\mathbf{a}\mathbf{a}^T \hat{\boldsymbol{\alpha}}_k^+ \tag{5.207e}$$

$$\mathcal{S}_{k+1}^- = \mathcal{S}_k^+ \Phi_k^{-1} - c\,\mathbf{a}\mathbf{d}^T \tag{5.207f}$$

for $i > 1$

$$\mathbf{a} = \mathcal{S}_{k+1}^{-}\,\Xi_k(i) \tag{5.208a}$$

$$b = \left[\mathbf{a}^T\mathbf{a} + 1/E_k(i,i)\right]^{-1} \tag{5.208b}$$

$$c = \left[1 + \sqrt{b/E_k(i,i)}\right]^{-1} \tag{5.208c}$$

$$\mathbf{d}^T = b\,\mathbf{a}^T\,\mathcal{S}_{k+1}^{-} \tag{5.208d}$$

$$\hat{\alpha}_{k+1}^{-} \leftarrow \hat{\alpha}_{k+1}^{-} - b\,c\,\mathbf{a}\mathbf{a}^T\,\hat{\alpha}_{k+1}^{-} \tag{5.208e}$$

$$\mathcal{S}_{k+1}^{-} \leftarrow \mathcal{S}_{k+1}^{-} - c\,\mathbf{a}\mathbf{d}^T \tag{5.208f}$$

where Φ_k is the state matrix defined in the Kalman filter, and $\leftarrow$ denotes replacement. If a control input is present, then this can be added to $\hat{\alpha}_{k+1}^{-}$ after the final iteration, with $\hat{\alpha}_{k+1}^{-} \leftarrow \hat{\alpha}_{k+1}^{-} + \mathcal{S}_{k+1}^{-}\Gamma_k\mathbf{u}_k$.

U-D Filter

A typically more computationally efficient algorithm than the square root approach is given by the *U-D* filter.[28] The derivation is based on the sequential processing approach presented in §5.3.3. The *U-D* filter factors the covariance matrix using

$$P_{i_k}^{-} = U_{i_k}^{-}D_{i_k}^{-}U_{i_k}^{-T} = \left[U_{i_k}^{-}\left(D_{i_k}^{-}\right)^{1/2}\right]\left[U_{i_k}^{-}\left(D_{i_k}^{-}\right)^{1/2}\right]^T \equiv S_{i_k}^{-}S_{i_k}^{-T} \tag{5.209}$$

where $U_{i_k}^{-}$ is a unitary (with ones along the diagonal) upper triangular matrix and $D_{i_k}^{-}$ is a diagonal matrix. The main advantage of this approach is that the factorization is accomplished without taking square roots.[7] This leads to a formulation that approaches the standard Kalman filter in computational effort. The gain matrix, covariance propagation, and update are given in terms of these matrices. Using the factorization in eqn. (5.209) on the covariance update in eqn. (5.79b) leads to

$$P_{i_k}^{+} = U_{i_k}^{+}D_{i_k}^{+}U_{i_k}^{+T} = U_{i_k}^{-}\left[D_{i_k}^{-} - \frac{1}{\alpha_{i_k}}\mathbf{e}_{i_k}\mathbf{e}_{i_k}^T\right]U_{i_k}^{-T} \tag{5.210}$$

where

$$\alpha_{i_k} \equiv \mathcal{H}_{i_k}P_{i_k}^{-}\mathcal{H}_{i_k}^T + \mathcal{R}_{i_k} \tag{5.211a}$$

$$\mathbf{e}_{i_k} \equiv D_{i_k}^{-}U_{i_k}^{-T}\mathcal{H}_{i_k}^T \tag{5.211b}$$

Since the bracketed term in eqn. (5.210) is also symmetric, it can be factored into

$$\left[D_{i_k}^{-} - \frac{1}{\alpha_{i_k}}\mathbf{e}_{i_k}\mathbf{e}_{i_k}^T\right] = L_{i_k}^{-}\mathcal{E}_{i_k}^{-}L_{i_k}^{-T} \tag{5.212}$$

where $L_{i_k}^-$ is a unitary upper triangular matrix and $\mathcal{E}_{i_k}^-$ is a diagonal matrix. Therefore, eqn. (5.210) is given by

$$U_{i_k}^+ D_{i_k}^+ U_{i_k}^{+T} = \left[U_{i_k}^- L_{i_k}^- \right] \mathcal{E}_{i_k}^- \left[U_{i_k}^- L_{i_k}^- \right]^T \tag{5.213}$$

Since the matrix $\left[U_{i_k}^- L_{i_k}^- \right]$ is upper triangular and $\mathcal{E}_{i_k}^-$ is diagonal then the update matrices are simply given by

$$U_{i_k}^+ = U_{i-1_k}^+ L_{i_k}^-, \quad U_{0_k}^+ = U_k^- \tag{5.214a}$$

$$D_{i_k}^+ = \mathcal{E}_{i_k}^-, \quad D_{0_k}^+ = D_k^- \tag{5.214b}$$

The covariance update is given in terms of the factorized matrices U_k^+ and D_k^+ instead of using P_k^- directly, which leads to a more computationally stable algorithm. Also, filter gain is given by

$$K_{i_k} = \frac{1}{\alpha_{i_k}} U_{i_k}^- \mathbf{e}_{i_k} \tag{5.215}$$

The propagated values for U_{k+1}^- and D_{k+1}^- are computed by first defining the following variables:

$$W_{k+1}^- \equiv \left[\Phi_k U_k^+ \ \Xi_k \right] \tag{5.216a}$$

$$\tilde{D}_{k+1}^- \equiv \begin{bmatrix} D_k^+ & 0_{n\times s} \\ 0_{s\times n} & E_k \end{bmatrix} \tag{5.216b}$$

where Ξ_k is given by eqn. (5.206) and E_k is given by eqn. (5.201b). The matrix W_{k+1}^{-T} is partitioned into $(n+s)$ column vectors as

$$\left[\mathbf{w}(1) \ \mathbf{w}(2) \ \cdots \ \mathbf{w}(n) \right] = W_{k+1}^{-T} \tag{5.217}$$

First, the matrix U_{k+1}^- is initialized to be an $n \times n$ identity matrix and the matrix D_{k+1}^- is initialized to be an $n \times n$ matrix of zeros. Then, the following iterations are performed for $i = n, n-1, \ldots, 1$ to determine the upper triangular elements of U_{k+1}^- and the diagonal elements of D_{k+1}^-:

$$\mathbf{c}(i) = \tilde{D}_{k+1}^- \mathbf{w}(i) \tag{5.218a}$$

$$D_{k+1}^-(i,i) = \mathbf{w}^T(i)\,\mathbf{c}(i) \tag{5.218b}$$

$$\mathbf{d}(i) = \mathbf{c}(i)/D_{k+1}^-(i,i) \tag{5.218c}$$

$$U_{k+1}^-(j,i) = \mathbf{w}^T(j)\,\mathbf{d}(i), \quad j = 1, 2, \ldots, i-1 \tag{5.218d}$$

$$\mathbf{w}(j) \leftarrow \mathbf{w}(j) - U_{k+1}^-(j,i)\,\mathbf{w}(i), \quad j = 1, 2, \ldots, i-1 \tag{5.218e}$$

On the last iteration, for $i = 1$, only the first two equations in eqn. (5.218) need to be processed. The state propagation still follows eqn. (5.30a).

Finding a tractable solution for any numerical issues in the propagation equation is often problem dependent, and often relies on other factors such as computational load. In general, the factorization algorithms presented in this section should always be employed if the computational load is not burdensome. The SRIF algorithm is less computationally efficient than the *U-D* filter, but the SRIF is computationally competitive if the number of measurements is large (see Ref. [23] for more details). With the rapid progress in computer technology today the methods shown in this section have nearly become obsolete. Still, they should be employed as a first step to investigate any anomalous behaviors in the Kalman filter, especially in nonlinear systems.

5.7.2 Colored-Noise Kalman Filtering

A critical assumption required in the derivation of the Kalman filter in §5.3 is that both the process and measurement noise are represented by zero-mean Gaussian white-noise processes. If this assumption is invalid then the filter may be suboptimal and even produce biased estimates. An example of this scenario involves spacecraft attitude determination using three-axis magnetometers (TAMs). The TAM sensor measurement error itself can adequately be represented by a white-noise process, but the errors in the actual Earth's magnetic field cannot be modelled by a white-noise process. These errors appear in the actual measurement equation.[29] For many spacecraft missions the state errors introduced from the colored measurement process may not cause any concerns; however, other systems may require the need to provide increased accuracy for colored (non-white) errors. Fortunately an exact Kalman filter can still be designed by using *shaping filters* that are driven by zero-mean white-noise processes. However, this is generally at the expense of increased complexity in the filter. Still for many systems a suboptimal filter should have its performance compared with that of the optimal filter.[30]

In this section colored-noise filters are designed for both the process noise and measurement noise. Only discrete-time systems are discussed here since the extension to continuous-time models is fairly straightforward. We first consider the case of a colored process noise. Consider the discrete-time autonomous system given in Table 5.2. Next, we assume that the process noise vector $\mathbf{w}_k$ is not white, but is uncorrelated with the initial condition and measurement noise. A shaping filter for $\mathbf{w}_k$ is given by[5, 30]

$$\chi_{k+1} = \Psi \chi_k + \mathcal{V} \omega_k \tag{5.219a}$$

$$\mathbf{w}_k = \mathcal{H} \chi_k + \mathcal{D} \omega_k \tag{5.219b}$$

where χ_k is the shaping filter state, and ω_k is a zero-mean Gaussian white-noise process with covariance given by $\mathcal{Q}$ (in general we can assume that $\mathcal{Q} = I$ and use $\mathcal{D}$ in the filter design to yield identical results for any general covariance matrix).

The system matrices Ψ, $\mathcal{V}$, $\mathcal{H}$, and $\mathcal{D}$ are used to "shape" the process noise into a colored-noise process. The augmented system that includes the state $\mathbf{x}_k$ is given by

$$\begin{bmatrix} \mathbf{x}_{k+1} \\ \chi_{k+1} \end{bmatrix} = \begin{bmatrix} \Phi & \Upsilon\mathcal{H} \\ 0 & \Psi \end{bmatrix} \begin{bmatrix} \mathbf{x}_k \\ \chi_k \end{bmatrix} + \begin{bmatrix} \Gamma \\ 0 \end{bmatrix} \mathbf{u}_k + \begin{bmatrix} \Upsilon\mathcal{D} \\ \mathcal{V} \end{bmatrix} \omega_k \tag{5.220a}$$

$$\tilde{\mathbf{y}}_k = \begin{bmatrix} H & 0 \end{bmatrix} \begin{bmatrix} \mathbf{x}_k \\ \chi_k \end{bmatrix} + \mathbf{v}_k \tag{5.220b}$$

The discrete-time Kalman filter in Table 5.2 can now be employed on the augmented system given in eqn. (5.220). Clearly the new system order is equal to the order of the original system plus the order of the shaping filter, which increases the complexity of the filter design. However, better performance may be possible if the shaping filter can adequately "model" the colored-noise process.

We now consider the case of colored measurement noise, where the measurement noise $\mathbf{v}_k$ is modelled by the following shaping filter:

$$\chi_{k+1} = \Psi\chi_k + \mathcal{V}\omega_k \tag{5.221a}$$

$$\mathbf{v}_k = \mathcal{H}\chi_k + \mathcal{D}\omega_k + \nu_k \tag{5.221b}$$

where ω_k and ν_k are both zero-mean Gaussian white-noise processes with covariances given by $\mathcal{Q}$ and $\mathcal{R}$, respectively. Assuming that ω_k and ν_k are uncorrelated, the new measurement noise covariance is given by

$$\begin{aligned} R &\equiv E\left\{(\mathcal{D}\omega_k + \nu_k)(\mathcal{D}\omega_k + \nu_k)^T\right\} \\ &= \mathcal{D}\mathcal{Q}\mathcal{D}^T + \mathcal{R} \end{aligned} \tag{5.222}$$

The augmented system that includes the state $\mathbf{x}_k$ is given by

$$\begin{bmatrix} \mathbf{x}_{k+1} \\ \chi_{k+1} \end{bmatrix} = \begin{bmatrix} \Phi & 0 \\ 0 & \Psi \end{bmatrix} \begin{bmatrix} \mathbf{x}_k \\ \chi_k \end{bmatrix} + \begin{bmatrix} \Gamma \\ 0 \end{bmatrix} \mathbf{u}_k + \begin{bmatrix} \Upsilon & 0 \\ 0 & \mathcal{V} \end{bmatrix} \begin{bmatrix} \mathbf{w}_k \\ \omega_k \end{bmatrix} \tag{5.223a}$$

$$\tilde{\mathbf{y}}_k = \begin{bmatrix} H & \mathcal{H} \end{bmatrix} \begin{bmatrix} \mathbf{x}_k \\ \chi_k \end{bmatrix} + \mathcal{D}\omega_k + \nu_k \tag{5.223b}$$

Assuming that $\mathbf{w}_k$ and ω_k are uncorrelated, the new process noise covariance matrix is given by

$$E\left\{\begin{bmatrix} \mathbf{w}_k \\ \omega_k \end{bmatrix} \begin{bmatrix} \mathbf{w}_k^T & \omega_k^T \end{bmatrix}\right\} = \begin{bmatrix} Q & 0 \\ 0 & \mathcal{Q} \end{bmatrix} \tag{5.224}$$

However for the augmented system in eqn. (5.223), the new process noise and measurement noise are now correlated. This correlation is given by

$$S \equiv E\left\{(\mathcal{D}\omega_k + \nu_k)\begin{bmatrix} \mathbf{w}_k^T & \omega_k^T \end{bmatrix}\right\} = \begin{bmatrix} 0 & \mathcal{D}\mathcal{Q} \end{bmatrix} \tag{5.225}$$

Therefore, the correlated Kalman filter in Table 5.3 should be employed in this case. However, for many practical systems $\mathcal{D} = 0$ so that the standard Kalman filter in Table 5.2 can be used.

As in the colored process noise case the state vector for the colored measurement noise case can also be augmented by the shaping filter state. However, an alternative to this augmentation is possible if the shaping filter can be generated by the following expression:[5]

$$\boxed{\begin{aligned}\chi_{k+1} &= \Psi\,\chi_k + \mathcal{V}\,\omega_k \\ \mathbf{v}_k &= \chi_k\end{aligned}} \qquad \begin{aligned}&(5.226a)\\&(5.226b)\end{aligned}$$

Note that the order of the shaping filter is the same as the dimension of the measurement noise vector. Next we define the following derived measurement:

$$\boxed{\tilde{\gamma}_{k+1} \equiv \tilde{\mathbf{y}}_{k+1} - \Psi\,\tilde{\mathbf{y}}_k - H\,\Gamma\,\mathbf{u}_k} \tag{5.227}$$

Substituting eqn. (5.27b) into eqn. (5.227) gives

$$\tilde{\gamma}_{k+1} = H\,\mathbf{x}_{k+1} + \mathbf{v}_{k+1} - \Psi\,H\,\mathbf{x}_k - \Psi\,\mathbf{v}_k - H\,\Gamma\,\mathbf{u}_k \tag{5.228}$$

Finally substituting eqns. (5.27a) and (5.226) into eqn. (5.228), and collecting terms yields

$$\tilde{\gamma}_{k+1} = \mathcal{H}\,\mathbf{x}_k + \mathcal{V}\,\omega_k + H\,\Upsilon\,\mathbf{w}_k \tag{5.229}$$

where

$$\mathcal{H} \equiv H\,\Phi - \Psi\,H \tag{5.230}$$

Assuming that ω_k and $\mathbf{w}_k$ are uncorrelated, the new measurement noise covariance is given by

$$\boxed{\begin{aligned}R &\equiv E\left\{(\mathcal{V}\,\omega_k + H\,\Upsilon\,\mathbf{w}_k)\,(\mathcal{V}\,\omega_k + H\,\Upsilon\,\mathbf{w}_k)^T\right\} \\ &= \mathcal{V}\,\mathcal{Q}\,\mathcal{V}^T + H\,\Upsilon\,Q\,\Upsilon^T\,H^T\end{aligned}} \tag{5.231}$$

However, the new process noise and measurement noise are correlated with

$$\boxed{S \equiv E\left\{(\mathcal{V}\,\omega_k + H\,\Upsilon\,\mathbf{w}_k)\,\mathbf{w}_k^T\right\} = H\,\Upsilon\,Q} \tag{5.232}$$

For this case the correlation S is rarely zero, so the correlated Kalman filter in Table 5.3 needs to be employed. However, the order of the system does not increase, which leads to a computational efficient routine, assuming that the colored measurement noise can be adequately modelled by eqn. (5.226).

Example 5.7: In this example a colored-noise filter will be designed using the longitudinal short-period dynamics of an aircraft. The approximate dynamical equations are given by a harmonic oscillator model:[5, 31]

$$\begin{bmatrix}\dot{\theta}(t)\\ \ddot{\theta}(t)\end{bmatrix} = \begin{bmatrix}0 & 1\\ -\omega_n^2 & -2\,\zeta\,\omega_n\end{bmatrix}\begin{bmatrix}\theta(t)\\ \dot{\theta}(t)\end{bmatrix} + \begin{bmatrix}0\\ 1\end{bmatrix} w(t)$$

$$\tilde{y}(t) = \theta(t) + v(t)$$

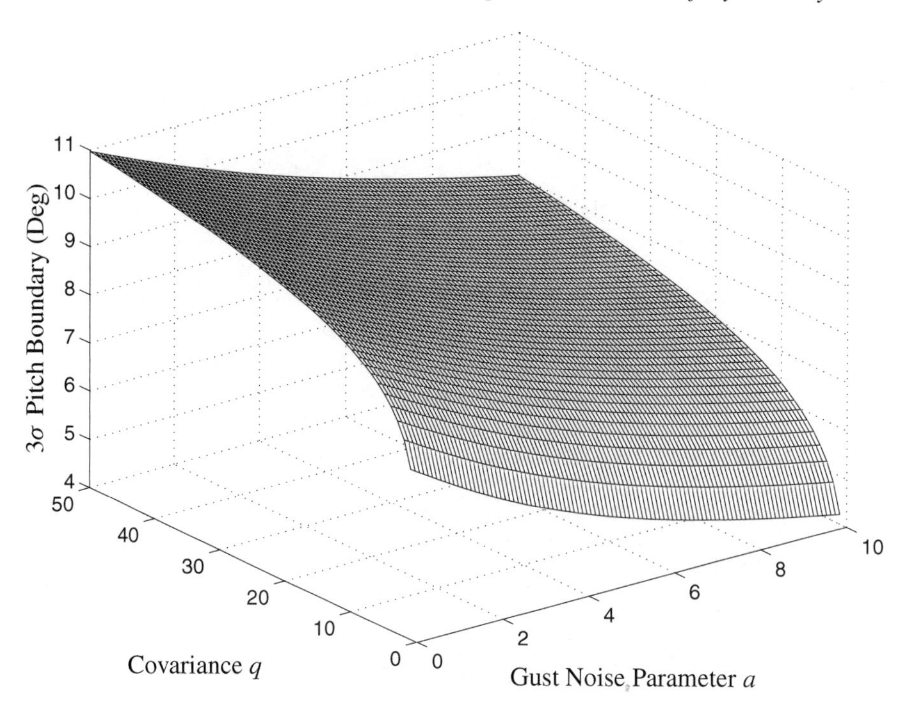

Figure 5.9: Colored-Noise Covariance Analysis

where $\theta(t)$ is the pitch angle, and ω_n and ζ are the short-period natural frequency and damping ratio, respectively. The process noise $w(t)$ now represents a wind gust input that is not white. This gust noise can be approximated by a first-order shaping filter, given by

$$\dot{\chi}(t) = -a\,\chi(t) + \omega(t)$$
$$w(t) = \chi(t)$$

where $\omega(t)$ is a zero-mean Gaussian white-noise process with variance q, and a dictates the "edge" of the gust profile. A larger value of a produces a shaper-edged gust. Also, the takeoff and landing performance of an aircraft can be shown to be a function of the wing loading. Aircraft designed for minimum runway requirements such as short-takeoff-and-landing aircraft will have low wing loadings compared with conventional transport aircraft and, therefore, should be more responsive to wind gusts.[31]

Augmenting the aircraft model by the shaping filter gives the following Kalman filter model-form:

$$\begin{bmatrix} \dot{\theta}(t) \\ \ddot{\theta}(t) \\ \chi(t) \end{bmatrix} = \begin{bmatrix} 0 & 1 & 0 \\ -\omega_n^2 & -2\zeta\,\omega_n & 1 \\ 0 & 0 & -a \end{bmatrix} \begin{bmatrix} \theta(t) \\ \dot{\theta}(t) \\ \chi(t) \end{bmatrix} + \begin{bmatrix} 0 \\ 0 \\ 1 \end{bmatrix} \omega(t)$$

$$\tilde{y}(t) = \begin{bmatrix} 1 & 0 & 0 \end{bmatrix} \begin{bmatrix} \theta(t) \\ \dot{\theta}(t) \\ \chi(t) \end{bmatrix} + v(t)$$

A discrete-time version of this model can easily be derived with a known sampling rate. As an example of the performance tradeoffs in the colored-noise Kalman filter we will consider the case where $\omega_n = 1$ rad/sec and $\zeta = \sqrt{2}/2$, and the standard deviation of the measurement noise process is given by 1 degree. A plot of the 3σ bound for $\theta(t)$, derived using the steady-state covariance equation in Table 5.2, with various values of a and p is shown in Figure 5.9. Clearly as q decreases more accurate pitch estimates are provided by the Kalman filter, which intuitively makes sense since the magnitude of the wind gust is smaller. As a increases better estimates are also provided. This is due to the effect of the gust edge on the observability of the pitch motion. As a increases more pitch motion from the wind gust is prevalent.

5.7.3 Consistency of the Kalman Filter

As discussed in example 5.5, a tuning process is usually required in the Kalman filter to achieve reasonable state estimates. In this section we show methods that can help answer the question: "what are reasonable estimates?" In practice the truth is never known, but there are still checks available to the design engineer that can (at the very least) provide mechanisms to show that a Kalman filter is not performing in an optimal fashion. For example, several tests can be applied to check the *consistency* of the Kalman filter from the desired characteristics of the measurement residuals. These include: the normalized error square (NES) test, the autocorrelation test, and the normalized mean error (NME) test.[32]

Suppose that some discrete error process $\mathbf{e}_k$ with dimension $m \times 1$ is known to be a zero-mean Gaussian white-noise process with covariance given by E_k. This process may be the state error or the measurement residual in the Kalman filter. Define the following NES:

$$\epsilon_k \equiv \mathbf{e}_k^T E_k^{-1} \mathbf{e}_k \tag{5.233}$$

The NES can be shown to have a chi-square distribution with n degrees of freedom (see Appendix B). A suitable check for the NES is to numerically show that the following condition is met with some level of confidence:

$$E\{\epsilon_k\} = m \tag{5.234}$$

This can be accomplished by using *Hypothesis testing*, which incorporates a degree of plausibility specified by a confidence interval.[33] A 95% confidence interval is most commonly used in practice, which is specified using $100(1-\alpha)$, where $\alpha = 0.05$ in this case. In practice a *two-sided probability region* is used (cutting off both 2.5% tails). Suppose that M Monte Carlo runs are taken, and the following average NES is computed:

$$\bar{\epsilon}_k = \frac{1}{M}\sum_{i=1}^{M}\epsilon_k(i) = \frac{1}{M}\sum_{i=1}^{M}\mathbf{e}_k^T(i)\,E_k^{-1}(i)\,\mathbf{e}_k(i) \tag{5.235}$$

where $\epsilon_k(i)$ denotes the ith run at time t_k. Then $M\bar{\epsilon}_k$ will have a chi-square density with Mm degrees of freedom.[32] This condition can be checked using a *chi-square test.* The hypothesis is accepted if the following condition is satisfied:

$$\bar{\epsilon}_k \in [\zeta_1, \zeta_2] \tag{5.236}$$

where ζ_1 and ζ_2 are derived from the tail probabilities of the chi-square density. For example, for $m = 2$ and $M = 100$, using eqn. (B.51), we have $\chi^2_{Mm}(0.025) = 162$ and $\chi^2_{Mm}(0.975) = 241$. This gives $\zeta_1 = \chi^2_{Mm}(0.025)/M = 1.62$ and $\zeta_2 = \chi^2_{Mm}(0.975)/M = 2.41$.

Another test for consistency is given by a *test for whiteness.* This is accomplished by using the following sample autocorrelation:[32]

$$\bar{\rho}_{k,j} = \frac{1}{\sqrt{m}}\sum_{i=1}^{M}\mathbf{e}_k^T(i)\left[\sum_{i=1}^{M}\mathbf{e}_k(i)\mathbf{e}_k^T(i)\sum_{i=1}^{M}\mathbf{e}_k(j)\mathbf{e}_k^T(j)\right]^{-1/2}\mathbf{e}_k(j) \tag{5.237}$$

For M large enough, $\bar{\rho}_{k,j}$ for $k \neq j$ is zero mean with variance given by $1/M$. A normal approximation can now be used with the central limit theorem.[33] With a 95% acceptance interval we have

$$\bar{\rho}_{k,j} \in \left[-\frac{1.96}{\sqrt{M}}, \frac{1.96}{\sqrt{M}}\right] \tag{5.238}$$

The hypothesis is accepted if eqn. (5.238) is satisfied.

The final consistency test is given by the NME for the jth element of $\mathbf{e}_k$:

$$[\bar{\boldsymbol{\mu}}_k]_j = \frac{1}{M}\sum_{i=1}^{M}\frac{[\mathbf{e}_k]_j}{\sqrt{[E_k]_{jj}}}, \quad j = 1, 2, \ldots, m \tag{5.239}$$

Then, since the variance of $[\bar{\boldsymbol{\mu}}_k]_j$ is $1/M$, for a 95% acceptance interval we have

$$[\bar{\boldsymbol{\mu}}_k]_j \in \left[-\frac{1.96}{\sqrt{M}}, \frac{1.96}{\sqrt{M}}\right] \tag{5.240}$$

The hypothesis is accepted if eqn. (5.240) is satisfied.

The NES, autocorrelation, and NME tests can all be performed with a single run using N data points, which is useful when a set of data cannot be collected more than once. From our example of $m = 2$ with $M = 1$, the two-sided 95% confidence interval is [0.05, 7.38], which is much wider than the $M = 100$ case. This illustrates the variability reduction with multiple runs. A low variability test statistic, which can be executed in real time, can be developed using a time-average approach. The time-average NES is given by

$$\bar{\epsilon} = \frac{1}{N}\sum_{k=1}^{N} \mathbf{e}_k^T E_k^{-1} \mathbf{e}_k \tag{5.241}$$

If $\mathbf{e}_k$ is a zero-mean, white noise process, then $N\bar{\epsilon}$ has a chi-square density distribution with Nm degrees of freedom. The whiteness test for $\mathbf{e}_k$ that are j steps apart, from a single run is derived by computing the time-average autocorrelation:

$$\bar{\rho}_j = \frac{1}{\sqrt{n}}\sum_{k=1}^{N} \mathbf{e}_k^T \mathbf{e}_{k+j} \left[\sum_{k=1}^{N} \mathbf{e}_k^T \mathbf{e}_k \sum_{k=1}^{N} \mathbf{e}_{k+j}^T \mathbf{e}_{k+j}\right]^{-1/2} \tag{5.242}$$

For N large enough, $\bar{\rho}_j$ is zero mean with variance given by $1/N$. With a 95% acceptance interval we have

$$\bar{\rho}_j \in \left[-\frac{1.96}{\sqrt{N}}, \frac{1.96}{\sqrt{N}}\right] \tag{5.243}$$

The hypothesis is accepted if eqn. (5.243) is satisfied. These tests can be applied to the Kalman filter residuals or the state errors through simulated runs to check the necessary consistency for filter optimality. If these tests are not satisfied then the Kalman filter is not running optimally, and the design needs to be investigated to identify the source of the problem for the particular system.

Example 5.8: In this example single run consistency tests will be performed on the residual between a scalar measurement and the estimated output of a Kalman filter. The discrete-time system for this example is given by

$$\mathbf{x}_{k+1} = \begin{bmatrix} 0.9999 & 0.0099 \\ -0.0296 & 0.9703 \end{bmatrix} \mathbf{x}_k + \begin{bmatrix} 0 \\ 0.01 \end{bmatrix} w_k$$

$$\tilde{y}_k = \begin{bmatrix} 1 & 0 \end{bmatrix} \mathbf{x}_k + v_k$$

where the true covariances of w_k and r_k are given by $q = 10$ and $r = 0.01$, respectively. The initial condition is given by $\mathbf{x}_0 = \begin{bmatrix} 1 & 1 \end{bmatrix}^T$. A steady-state Kalman filter shown in Table 5.2 is executed for various values of assumed q with 1001 synthetic measurements. The single run consistency checks involving the time-average NES and autocorrelation tests are performed on the last 500 points, which is well after the filter has converged. With $N = 500$ the two-sided 95% region for the NES test is [0.88, 1.125], and the 95% upper limit for the autocorrelation test is $1.96/\sqrt{500} = 0.0877$.

The true state is always generated using $q = 10$, and the same measurement set is used for the consistency tests. Various values of assumed q in the Kalman filter,

Table 5.10: Results of the Kalman Filter Consistency Tests

q	$\bar{\epsilon}$	$\lvert\bar{\rho}_1\rvert$
0.1	1.9334	0.4752
0.5	1.3501	0.2408
1	1.2065	0.1463
10	1.0367	0.0015
20	1.0231	0.0100
100	1.0006	0.0224
1000	0.9817	0.0424
1×10^4	0.9372	0.0888
1×10^5	0.8607	0.1739

ranging from 0.1 to 1×10^5, are tested. For the consistency tests involving the measurement residual we use $\mathbf{e}_k = \tilde{\mathbf{y}}_k - H_k \mathbf{x}_k$, where for our case $\mathbf{e}_k$, $\tilde{\mathbf{y}}_k$ are scalars and $H_k = \begin{bmatrix} 1 & 0 \end{bmatrix}$. The covariance of $\mathbf{e}_k$, denoted by E_k, can be shown to be given by

$$E_k = H_k^T P_k H_k + R_k$$

which is used in the NES test. Table 5.10 gives numerical values for the computed NES and autocorrelation values. The NES values are outside the region when q is larger than about 1×10^5 or smaller than about 1. The autocorrelation is computed using a one time-step ahead sample. Table 5.10 shows that the autocorrelation test gives about the same level of confidence as the NES test. From both a theoretical and practical point of view the best results are obtained with an autocorrelation near zero and an NES close to one. Table 5.10 indicates that these conditions are met with q values of 10 or 20. Therefore, since the true value of q is 10, we can conclude the consistency tests provide a good means to find q.

5.7.4 Adaptive Filtering

The results of §5.7.3 can be used to manually tune the Kalman filter. In this section a common approach used to automatically identify the process noise and measurement-error noise covariances is shown. The theoretical aspects of the Kalman filter for linear systems are very sound, derived from a rigorous analysis. In practice "tuning" a Kalman filter can be arduous and very time-consuming. Usually, the measurement-error covariance is fairly well known, derived from statistical inferences of the hardware sensing device. However, the process noise covariance is usually not well known and is often derived from experiences gained by the design engineer based on intimate knowledge of the particular system. The approach is based on "residual whitening."[34, 35] The approach presented in this section is applicable to time-invariant systems with stationary noise processes only. Consider the following discrete-time residual equation:

$$\boxed{\begin{aligned}\mathbf{e}_k &\equiv \tilde{\mathbf{y}}_k - H\,\hat{\mathbf{x}}_k^- \\ &= -H\,\tilde{\mathbf{x}}_k^- + \mathbf{v}_k\end{aligned}} \tag{5.244}$$

where eqn. (5.32a) and (5.27b) have been used in eqn. (5.244). The following autocorrelation function matrix can be computed:

$$C_i = \begin{cases} H\,E\left\{\tilde{\mathbf{x}}_k^-\tilde{\mathbf{x}}_{k-i}^{-T}\right\}H^T - H\,E\left\{\tilde{\mathbf{x}}_k^-\mathbf{v}_{k-i}^{-T}\right\} & i>0 \\ H\,P\,H^T + R & i=0 \end{cases} \tag{5.245}$$

where $C_i \equiv E\left\{\mathbf{e}_k\mathbf{e}_{k-i}^T\right\}$, and P is the steady-state covariance obtained from

$$P = \Phi[(I-K\,H)\,P\,(I-K\,H) + K\,R\,K^T]\,\Phi^T + \Upsilon\,Q\,\Upsilon^T \tag{5.246}$$

Note the use of a suboptimal gain K in eqn. (5.246), but an optimal Q and R.[35] Substituting eqn. (5.37) into eqn. (5.33) leads to

$$\tilde{\mathbf{x}}_k^- = \Phi\,(I-K\,H)\,\tilde{\mathbf{x}}_{k-1}^- + \Phi\,K\,\mathbf{v}_{k-1} - \Upsilon\,\mathbf{w}_{k-1} \tag{5.247}$$

Carrying eqn. (5.247) i steps back yields

$$\begin{aligned}\tilde{\mathbf{x}}_k^- &= [\Phi\,(I-K\,H)]^i\,\tilde{\mathbf{x}}_{k-i}^- + \sum_{j=1}^{i}[\Phi\,(I-K\,H)]^{j-1}\Phi\,K\,\mathbf{v}_{k-j} \\ &\quad - \sum_{j=1}^{i}[\Phi\,(I-K\,H)]^{j-1}\Upsilon\,\mathbf{w}_{k-j}\end{aligned} \tag{5.248}$$

Then, the following expectations are easily given:

$$E\left\{\tilde{\mathbf{x}}_k^-\tilde{\mathbf{x}}_{k-i}^{-T}\right\} = [\Phi\,(I-K\,H)]^i\,P \tag{5.249a}$$

$$E\left\{\tilde{\mathbf{x}}_k^-\mathbf{v}_{k-i}^{-T}\right\} = [\Phi\,(I-K\,H)]^{i-1}\Phi\,K\,R \tag{5.249b}$$

Hence, substituting eqn. (5.249) into eqn. (5.245), the autocorrelation is now given by

$$\boxed{C_i = \begin{cases} H\,[\Phi\,(I-K\,H)]^{i-1}\Phi[P\,H^T - K\,C_0] & i>0 \\ H\,P\,H^T + R & i=0 \end{cases}} \tag{5.250}$$

where the definition of C_0 is used to simplify the resulting substitution process leading to eqn. (5.250). Note that if the optimal gain K is used, given by eqn. (5.50), then $C_i = 0$ for $i \neq 0$.

A test for whiteness can now be computed based on the autocorrelation matrix. Note that if $\mathbf{e}_k$ is a white-noise process, then $C_i = 0$ for $i \neq 0$, which means that the filter is performing in an optimal fashion. An estimate of C_i is given by

$$\boxed{\hat{C}_i = \frac{1}{N}\sum_{j=i}^{N} \mathbf{e}_j \mathbf{e}_{j-i}^T} \tag{5.251}$$

where N is sufficiently large. The estimate for C_i is biased, which can be removed by dividing by $N-i$ instead of N, but the original form may be preferable to an unbiased estimate since less mean-square error is given. The diagonal elements of C_i are of particular interest. These can be normalized by their zero-lag elements leading to the following autocorrelation coefficients:

$$[\rho_i]_{jj} \equiv \frac{[\hat{C}_i]_{jj}}{[\hat{C}_0]_{jj}} \tag{5.252}$$

where the subscript jj denotes a diagonal element of $\hat{C}$. The numbered values for $[\rho_i]_{jj}$ range between 0 and 1. A 95% confidence interval on $[\rho_i]_{jj}$ for $i \neq 0$ is given by

$$\boxed{|[\rho_i]_{jj}| \leq 1.96/N^{1/2}} \tag{5.253}$$

Therefore, if less than 5% of the values of $[\rho_i]_{jj}$ exceed the threshold given by eqn. (5.253), then the jth residual is a white-noise process.

Our first goal is to determine an estimate for $Z \equiv P\,H^T$. Writing out the autocorrelation matrix in eqn. (5.250) for $i > 0$ gives

$$\begin{aligned}
C_1 &= H\,\Phi\,P\,H^T - H\,\Phi\,K\,C_0 \\
C_2 &= H\,\Phi^2 P\,H^T - H\,\Phi\,K\,C_1 - H\,\Phi^2 K\,C_0 \\
&\vdots \\
C_n &= H\,\Phi^n P\,H^T - H\,\Phi\,K\,C_{n-1} - \cdots - H\,\Phi^n K\,C_0
\end{aligned} \tag{5.254}$$

Using the methods of Chapter 1 the following least-squares estimate for Z is obtained:

$$\hat{Z} = (M^T M)^{-1} M^T \begin{bmatrix} \hat{C}_1 + H\,\Phi\,K\,\hat{C}_0 \\ \hat{C}_2 + H\,\Phi\,K\,\hat{C}_1 + H\,\Phi^2 K\,\hat{C}_0 \\ \vdots \\ \hat{C}_n + H\,\Phi\,K\,\hat{C}_{n-1} + \cdots + H\,\Phi^n K\,\hat{C}_0 \end{bmatrix} \tag{5.255}$$

where M is the product of the observability matrix in eqn. (3.117) and the transition matrix Φ, i.e., $M \equiv \mathcal{O}_d\,\Phi$. Note that the dynamical system must be observable in order for the inverse in eqn. (5.255) to exist. Therefore, using eqn. (5.250) an estimate for R is given by

$$\boxed{\hat{R} = \hat{C}_0 - H\,\hat{Z}} \tag{5.256}$$

Determining an estimate for Q is not as straightforward as the R case. If the number of unknown elements of Q is $n \times m$ or less, then a unique solution is possible. We first rewrite eqn. (5.246) as

$$P = \Phi P \Phi^T + \Omega + \Upsilon Q \Upsilon^T \tag{5.257}$$

where

$$\Omega \equiv \Phi [K C_0 K^T - P H^T K^T - K H P] \Phi^T \tag{5.258}$$

Substituting back for P n times on the right-hand side of eqn. (5.257) yields

$$\sum_{j=0}^{i-1} \Phi^j \Upsilon Q \Upsilon^T (\Phi^j)^T = P - \Phi^i P (\Phi^i)^T - \sum_{j=0}^{i-1} \Phi^j \Omega (\Phi^j)^T, \quad i = 1, 2, \ldots, n \tag{5.259}$$

Multiplying the left-hand side of eqn. (5.259) by H and multiplying the right-hand side by $(\Phi^{-i})^T H^T$, and using estimated quantities leads to

$$\boxed{\begin{aligned} \sum_{j=0}^{i-1} H \Phi^j \Upsilon \hat{Q} \Upsilon^T (\Phi^{j-i})^T H^T &= \hat{Z}^T (\Phi^{-i})^T H^T - H \Phi^i \hat{Z} \\ &\quad - \sum_{j=0}^{i-1} H \Phi^j \hat{\Omega} (\Phi^{j-i})^T H^T, \quad i = 1, 2, \ldots, n \end{aligned}} \tag{5.260}$$

where

$$\hat{\Omega} \equiv \Phi [K \hat{C}_0 K^T - \hat{Z} K^T - K \hat{Z}^T] \Phi^T \tag{5.261}$$

Once the right-hand side of eqn. (5.260) has been evaluated then $\hat{Q}$ can be extracted. Note that the equations for the elements of $\hat{Q}$ are not linearly independent, and one has to choose a linearly independent subset of these equations.[34]

If the number of unknown elements of Q is greater than $n \times m$, then a unique solution is not possible. To overcome this case, the gain K can be estimated directly, which is denoted by K^*. Then the optimal covariance P^* follows

$$P^* = \Phi (P^* - K^* H P^*) \Phi^T + \Upsilon Q \Upsilon^T \tag{5.262}$$

Defining $\delta P = P^* - P$ and using eqns. (5.246) and (5.262) yields[35]

$$\begin{aligned} \delta P = \Phi \Big[&\delta P - (P H^T + \delta P H^T)(C_0 + H \delta P H^T)^{-1} (H P + H \delta P) \\ &+ K H P + P H^T K^T - K C_0 K^T \Big] \Phi^T \end{aligned} \tag{5.263}$$

where $C_0 = H P H^T + R$ is used to eliminate R. An optimal estimate for δP, denoted by $\delta \hat{P}$ is obtained by using $\hat{C}_0$ from eqn. (5.251) and $\hat{Z}$ from eqn. (5.255), so that

$$\begin{aligned} \delta \hat{P} = \Phi \Big[&\delta P - (\hat{Z} + \delta P H^T)(\hat{C}_0 + H \delta P H^T)^{-1} (\hat{Z}^T + H \delta P) \\ &+ K \hat{Z}^T + \hat{Z} K^T - K \hat{C}_0 K^T \Big] \Phi^T \end{aligned} \tag{5.264}$$

which can now be solved for $\delta \hat{P}$. The optimal gain is given by

$$\begin{aligned} K^* &= P^* H^T [H P^* H^T + R]^{-1} \\ &= \left[(P+\delta P) H^T\right]\left[H P H^T + H \delta P H^T + R\right]^{-1} \\ &= \left[P H^T + \delta P H^T\right]\left[C_0 + H \delta P H^T\right]^{-1} \end{aligned} \tag{5.265}$$

Therefore the estimate of the optimal gain is given by

$$\boxed{\hat{K}^* = \left[\hat{Z} + \delta \hat{P} H^T\right]\left[\hat{C}_0 + H \delta \hat{P} H^T\right]^{-1}} \tag{5.266}$$

For batch-type applications, local iterations on the estimates $\hat{C}_0$, $\hat{Z}$, $\delta \hat{P}$, and $\hat{K}^*$ are possible on the same set of N measurements, which could improve these estimates, where the residual sequence becomes increasingly more white.[35] Also, care must be given when estimating for the gain directly since no guarantees can be made about the stability of the resulting filter. Reference [34] provides an example involving an inertial navigation problem to estimate components of the matrices Q and R. Asymptotic convergence of the estimates toward their true values has been shown in this example. Other adaptive methods, such as covariance matching, can be found in Refs. [4] and [35].

5.7.5 Error Analysis

The optimality of the Kalman filter hinges on many factors. First, although precise knowledge of the process noise and measurements inputs is not required, we must have accurate knowledge of their respective covariance values. When these covariances are not well known then the methods in §5.7.4 can be applied to estimate them online. Also, errors in the assumed model may be present. Determining these errors is a formidable (and nearly impossible!) task. This section shows an analysis on how the error-covariance of the nominal system is changed with the aforementioned errors. This new covariance can be used to assess the performance of the nominal Kalman filter given bounds on the model and noises quantities, which may provide insight to filter performance and sensitivity to various errors. The development in this section is based on continuous-time models and measurements. Also, in this section we eliminate the explicit dependence on time for notational brevity. Consider the following nominal system, which will be used to derive the Kalman filter:

$$\dot{\bar{\mathbf{x}}} = \bar{F}\,\bar{\mathbf{x}} + B\,\mathbf{u} + \bar{G}\,\bar{\mathbf{w}} \tag{5.267a}$$

$$\tilde{\mathbf{y}} = \bar{H}\,\bar{\mathbf{x}} + \bar{\mathbf{v}} \tag{5.267b}$$

where $\bar{F}$, $\bar{G}$, and $\bar{H}$ are the nominal model matrices (note we assume that the control input and its associated input matrix are known exactly). The Kalman filter for this system is given by

$$\boxed{\dot{\hat{\bar{\mathbf{x}}}} = \bar{F}\,\hat{\bar{\mathbf{x}}} + B\,\mathbf{u} + \bar{K}[\tilde{\mathbf{y}} - \bar{H}\,\hat{\bar{\mathbf{x}}}]} \tag{5.268}$$

with

$$\bar{K} = \bar{P}\,\bar{H}^T\,\bar{R}^{-1} \tag{5.269a}$$

$$\dot{\bar{P}} = \bar{F}\,\bar{P} + \bar{P}\,\bar{F}^T - \bar{P}\,\bar{H}^T\,\bar{R}^{-1}\,\bar{H}\,\bar{P} + \bar{G}\,\bar{Q}\,\bar{G}^T \tag{5.269b}$$

where $\bar{Q}$ and $\bar{R}$ are the nominal process noise and measurement noise covariances, respectively.

The actual system is given by

$$\dot{\mathbf{x}} = F\,\mathbf{x} + B\,\mathbf{u} + G\,\mathbf{w} \tag{5.270a}$$

$$\tilde{\mathbf{y}} = H\,\mathbf{x} + \mathbf{v} \tag{5.270b}$$

We now define the following variables: $\bar{\tilde{\mathbf{x}}} \equiv \mathbf{x} - \bar{\hat{\mathbf{x}}}$, $\Delta F \equiv F - \bar{F}$, and $\Delta H \equiv H - \bar{H}$, where $\bar{\tilde{\mathbf{x}}}$ is the error between the truth and the estimate using the assumed nominal model. Taking the time derivative of $\bar{\tilde{\mathbf{x}}}$ yields

$$\dot{\bar{\tilde{\mathbf{x}}}} = (\bar{F} - \bar{K}\,\bar{H})\,\bar{\tilde{\mathbf{x}}} + (\Delta F - \bar{K}\,\Delta H)\mathbf{x} + G\,\mathbf{w} - \bar{K}\,\mathbf{v} \tag{5.271}$$

The covariance of $\bar{\tilde{\mathbf{x}}}$ can be shown to be given by[36, 37]

$$\boxed{P_{\bar{\tilde{\mathbf{x}}}} = V_{\bar{\tilde{\mathbf{x}}}} + \mu_{\bar{\tilde{\mathbf{x}}}}\mu_{\bar{\tilde{\mathbf{x}}}}^T} \tag{5.272}$$

where

$$\boxed{\begin{aligned} \dot{V}_{\bar{\tilde{\mathbf{x}}}} = {} & (\bar{F} - \bar{K}\,\bar{H})\,V_{\bar{\tilde{\mathbf{x}}}} + V_{\bar{\tilde{\mathbf{x}}}}(\bar{F} - \bar{K}\,\bar{H})^T + V^T(\Delta F - \bar{K}\,\Delta H)^T \\ & + (\Delta F - \bar{K}\,\Delta H)\,V + G\,Q\,G^T + \bar{K}\,R\,\bar{K}^T \end{aligned}} \tag{5.273}$$

The matrix V is determined from

$$\dot{V} = F\,V + V\,(\bar{F} - \bar{K}\,\bar{H})^T + V_{\mathbf{x}}(\Delta F - \bar{K}\,\Delta H)^T + G\,Q\,G^T \tag{5.274a}$$

$$\dot{V}_{\mathbf{x}} = F\,V_{\mathbf{x}} + V_{\mathbf{x}}F^T + G\,Q\,G^T \tag{5.274b}$$

The mean of the estimation error $\mu_{\bar{\tilde{\mathbf{x}}}}$ is determined from

$$\dot{\mu}_{\bar{\tilde{\mathbf{x}}}} = (\bar{F} - \bar{K}\,\bar{H})\,\mu_{\bar{\tilde{\mathbf{x}}}} + (\Delta F - \bar{K}\,\Delta H)\,\mu_{\mathbf{x}} \tag{5.275a}$$

$$\dot{\mu}_{\mathbf{x}} = F\,\mu_{\mathbf{x}} \tag{5.275b}$$

where $\mu_{\mathbf{x}}$ is the system mean. The initial conditions for the differential equations are left to the discretion of the filter designer.

The procedure to determine $P_{\bar{\tilde{\mathbf{x}}}}$ is as follows. First, compute $\mu_{\mathbf{x}}$ and $V_{\mathbf{x}}$ using eqns. (5.275b) and (5.274b), respectively. Note these variables require knowledge of the true system matrices. Then compute $\mu_{\bar{\tilde{\mathbf{x}}}}$ and V using eqns. (5.275a) and (5.274a), respectively. Next, compute $V_{\bar{\tilde{\mathbf{x}}}}$ using eqn. (5.273), and finally compute $P_{\bar{\tilde{\mathbf{x}}}}$ using eqn. (5.272). Note that if $\Delta F - \bar{K}\,\Delta H = 0$, then both V and $V_{\mathbf{x}}$ do not need to be computed. A more useful quantity involves rewriting eqn. (5.272) as

$$\boxed{P_{\tilde{\mathbf{x}}} = \bar{P} + \Delta V_{\tilde{\mathbf{x}}} + \mu_{\tilde{\mathbf{x}}}\mu_{\tilde{\mathbf{x}}}^T} \tag{5.276}$$

where $(\Delta V_{\tilde{\mathbf{x}}} + \mu_{\tilde{\mathbf{x}}}\mu_{\tilde{\mathbf{x}}}^T)$ is now the covariance difference between total error-covariance and the nominal error-covariance. The quantity $\Delta V_{\tilde{\mathbf{x}}}$ can be found from

$$\boxed{\begin{aligned}\Delta \dot{V}_{\tilde{\mathbf{x}}} = (\bar{F} - \bar{K}\,\bar{H})\,\Delta V_{\tilde{\mathbf{x}}} + V_{\tilde{\mathbf{x}}}(\bar{F} - \bar{K}\,\bar{H})^T + V^T(\Delta F - \bar{K}\,\Delta H)^T \\ + (\Delta F - \bar{K}\,\Delta H)\,V + (G\,Q\,G^T - \bar{G}\,\bar{Q}\,\bar{G}^T) + \bar{K}\,(R - \bar{R})\,\bar{K}^T\end{aligned}} \tag{5.277}$$

Under steady-state conditions, for time-invariant stable systems, both $\mu_{\mathbf{x}}$ and $\mu_{\tilde{\mathbf{x}}}$ are zero. Also, eqns. (5.274b) and (5.272) can be found using an algebraic Lyapunov equation, which has the same form as given by eqn. (3.125). Other forms can be given in which the system estimation error is separated into optimum and non-optimum error components.[36, 37]

5.7.6 Unscented Filtering

The problem of filtering using nonlinear dynamic and/or measurement models is inherently more difficult than for the case of linear models. The extended Kalman filter in §5.6 typically works well only in the region where the first-order Taylor-series linearization adequately approximates the nonlinear probability distribution. The primary area of concern for this application is during the initialization stage, where the estimated initial state may be far from the true state. This may lead to instabilities in the extended Kalman filter. To overcome these instabilities a Kalman filter can be used based upon including second-order terms in the Taylor-series.[1, 35] Improved performance can be achieved in many cases, but at the expense of an increased computational burden. Maybeck[35] also suggests that a first-order filter with bias correction terms, without altering the covariance and gain expressions, may be generated to obtain the essential benefits of second-order filtering with the computational penalty of additional second-moment calculations. An exact nonlinear filter has been developed by Daum,[38] which reduces to the standard Kalman filter in linear systems. However, Daum's theory may be difficult to implement on practical systems due to the nature of the requirement to solve a partial differential equation (known as the Fokker-Planck equation). Therefore, the standard form of the extended Kalman filter has remained the most popular method for nonlinear estimation to this day, and other designs are investigated only when the performance of the standard form is not sufficient.

In this section a new approach that has been developed by Julier, Uhlmann, and Durrant-Whyte[39, 40] is shown as an alternative to the extended Kalman filter. This approach, which they called the *Unscented filter* (UF), typically involves more computations than the extended Kalman filter, but has several advantages, including: 1) the expected error is lower than the extended Kalman filter, 2) the new filter can be applied to non-differentiable functions, 3) the new filter avoids the derivation of Jacobian matrices, and 4) the new filter is valid to higher-order expansions than the standard extended Kalman filter. The Unscented filter works on the premise that with a fixed number of parameters it should be easier to approximate a Gaussian distribution than to approximate an arbitrary nonlinear function. The filter presented in Ref. [39] is derived for discrete-time nonlinear equations, where the system model is given by

$$\boxed{\begin{aligned}\mathbf{x}_{k+1} &= \mathbf{f}(\mathbf{x}_k, \mathbf{w}_k, \mathbf{u}_k, k) \\ \tilde{\mathbf{y}}_k &= \mathbf{h}(\mathbf{x}_k, \mathbf{u}_k, \mathbf{v}_k, k)\end{aligned}} \tag{5.278a, 5.278b}$$

Note that a continuous-time model can always be written using eqn. (5.278a) through an appropriate numerical integration scheme. It is again assumed that $\mathbf{w}_k$ and $\mathbf{v}_k$ are zero-mean Gaussian noise processes with covariances given by Q_k and R_k, respectively. We first rewrite the Kalman filter update equations in Table 5.9 as[41]

$$\boxed{\hat{\mathbf{x}}_k^+ = \hat{\mathbf{x}}_k^- + K_k \upsilon_k} \tag{5.279a}$$

$$\boxed{P_k^+ = P_k^- - K_k P_k^{\upsilon\upsilon} K_k^T} \tag{5.279b}$$

where υ_k is the *innovations process*, given by

$$\begin{aligned}\upsilon_k &\equiv \tilde{\mathbf{y}}_k - \hat{\mathbf{y}}_k^- \\ &= \tilde{\mathbf{y}}_k - \mathbf{h}(\hat{\mathbf{x}}_k^-, \mathbf{u}_k, k)\end{aligned} \tag{5.280}$$

The covariance of υ_k is defined by $P_k^{\upsilon\upsilon}$. The gain K_k is computed by

$$\boxed{K_k = P_k^{xy}(P_k^{\upsilon\upsilon})^{-1}} \tag{5.281}$$

where P_k^{xy} is the cross-correlation matrix between $\hat{\mathbf{x}}_k^-$ and $\hat{\mathbf{y}}_k^-$.

The Unscented filter uses a different propagation than the form given by the standard extended Kalman filter. Given an $n \times n$ covariance matrix P, a set of order n points can be generated from the columns (or rows) of the matrices $\pm\sqrt{nP}$. The set of points is zero-mean, but if the distribution has mean μ, then simply adding μ to each of the points yields a symmetric set of $2n$ points having the desired mean and covariance.[39] Due to the symmetric nature of this set, its odd central moments are zero, so its first three moments are the same as the original Gaussian distribution. This is the foundation for the Unscented filter. A complete derivation of this filter is beyond the scope of the present text, so only the final results are presented here. Various methods can be used to handle the process noise and measurement noise in the Unscented filter. One approach involves augmenting the covariance matrix with

$$P_k^a = \begin{bmatrix} P_k^+ & P_k^{xw} & P_k^{xv} \\ (P_k^{xw})^T & Q_k & P_k^{wv} \\ (P_k^{xv})^T & (P_k^{wv})^T & R_k \end{bmatrix} \tag{5.282}$$

where P_k^{xw} is the correlation between the state error and process noise, P_k^{xv} is the correlation between the state error and measurement noise, and P_k^{wv} is the correlation between the process noise and measurement noise, which are all zero for most systems. Augmenting the covariance requires the computation of $2(q+l)$ additional sigma points (where q is the dimension of $\mathbf{w}_k$ and l is the dimension of $\mathbf{v}_k$, which does not necessarily have to be the same dimension, m, as the output in this case), but the effects of the process and measurement noise in terms of the impact on the mean and covariance are introduced with the same order of accuracy as the uncertainty in the state.

The general formulation for the propagation equations are given as follows. First, the following set of *sigma points* are computed:

$$\boldsymbol{\sigma}_k \leftarrow 2L \text{ columns from } \pm\gamma\sqrt{P_k^a} \tag{5.283a}$$

$$\boldsymbol{\chi}_k^a(0) = \hat{\mathbf{x}}_k^a \tag{5.283b}$$

$$\boldsymbol{\chi}_k^a(i) = \boldsymbol{\sigma}_k(i) + \hat{\mathbf{x}}_k^a \tag{5.283c}$$

where $\hat{\mathbf{x}}_k^a$ is an augmented state defined by

$$\mathbf{x}_k^a = \begin{bmatrix} \mathbf{x}_k \\ \mathbf{w}_k \\ \mathbf{v}_k \end{bmatrix}, \quad \hat{\mathbf{x}}_k^a = \begin{bmatrix} \hat{\mathbf{x}}_k \\ \mathbf{0}_{q\times 1} \\ \mathbf{0}_{m\times 1} \end{bmatrix} \tag{5.284}$$

and L is the size of the vector $\hat{\mathbf{x}}_k^a$. The parameter γ is given by

$$\gamma = \sqrt{L+\lambda} \tag{5.285}$$

where the composite scaling parameter, λ, is given by

$$\lambda = \alpha^2(L+\kappa) - L \tag{5.286}$$

The constant α determines the spread of the sigma points and is usually set to a small positive value (e.g., $1\times 10^{-4} \le \alpha \le 1$).[42] Also, the significance of the parameter κ will be discussed shortly. Efficient methods to compute the matrix square root can be found by using the Cholesky decomposition (see Appendix A) or using eqn. (5.209). If an orthogonal matrix square root is used, then the sigma points lie along the eigenvectors of the covariance matrix. Note that there are a total of $2L$ values for $\boldsymbol{\sigma}_k$ (the positive and negative square roots). The transformed set of sigma points are evaluated for each of the points by

$$\boldsymbol{\chi}_{k+1}(i) = \mathbf{f}(\boldsymbol{\chi}_k^x(i), \boldsymbol{\chi}_k^w(i), \mathbf{u}_k, k) \tag{5.287}$$

where $\boldsymbol{\chi}_k^x(i)$ is a vector of the first n elements of $\boldsymbol{\chi}_k^a(i)$, and $\boldsymbol{\chi}_k^w(i)$ is a vector of the next q elements of $\boldsymbol{\chi}_k^a(i)$, with

$$\boldsymbol{\chi}_k^a(i) = \begin{bmatrix} \boldsymbol{\chi}_k^x(i) \\ \boldsymbol{\chi}_k^w(i) \\ \boldsymbol{\chi}_k^v(i) \end{bmatrix} \tag{5.288}$$

where $\chi_k^v(i)$ is a vector of the last l elements of $\chi_k^a(i)$, which will be used to compute the output covariance. We now define the following weights:

$$W_0^{\text{mean}} = \frac{\lambda}{L+\lambda} \tag{5.289a}$$

$$W_0^{\text{cov}} = \frac{\lambda}{L+\lambda} + (1-\alpha^2+\beta) \tag{5.289b}$$

$$W_i^{\text{mean}} = W_i^{\text{cov}} = \frac{1}{2(L+\lambda)}, \quad i = 1, 2, \ldots, 2L \tag{5.289c}$$

where β is used to incorporate prior knowledge of the distribution (a good starting guess is $\beta = 2$).

The predicted mean for the state estimate is calculated using a weighted sum of the points $\chi_{k+1}^x(i)$, which is given by

$$\boxed{\hat{\mathbf{x}}_{k+1}^- = \sum_{i=0}^{2L} W_i^{\text{mean}} \chi_{k+1}^x(i)} \tag{5.290}$$

The predicted covariance is given by

$$\boxed{P_{k+1}^- = \sum_{i=0}^{2L} W_i^{\text{cov}} [\chi_{k+1}^x(i) - \hat{\mathbf{x}}_{k+1}^-][\chi_{k+1}^x(i) - \hat{\mathbf{x}}_{k+1}^-]^T} \tag{5.291}$$

The mean observation is given by

$$\boxed{\hat{\mathbf{y}}_{k+1}^- = \sum_{i=0}^{2L} W_i^{\text{mean}} \gamma_{k+1}(i)} \tag{5.292}$$

where

$$\gamma_{k+1}(i) = \mathbf{h}(\chi_{k+1}^x(i), \mathbf{u}_{k+1}, \chi_{k+1}^v(i), k+1) \tag{5.293}$$

The output covariance is given by

$$\boxed{P_{k+1}^{yy} = \sum_{i=0}^{2L} W_i^{\text{cov}} [\gamma_{k+1}(i) - \hat{\mathbf{y}}_{k+1}^-][\gamma_{k+1}(i) - \hat{\mathbf{y}}_{k+1}^-]^T} \tag{5.294}$$

Then the innovations covariance is simply given by

$$\boxed{P_{k+1}^{vv} = P_{k+1}^{yy}} \tag{5.295}$$

Finally the cross correlation matrix is determined using

$$\boxed{P_{k+1}^{xy} = \sum_{i=0}^{2L} W_i^{\text{cov}} [\chi_{k+1}^x(i) - \hat{\mathbf{x}}_{k+1}^-][\gamma_{k+1}(i) - \hat{\mathbf{y}}_{k+1}^-]^T} \tag{5.296}$$

The filter gain is then computed using eqn. (5.281), and the state vector can now be updated using eqn. (5.279). Even though propagations on the order of $2n$ are required for the Unscented filter, the computations may be comparable to the extended Kalman filter (especially if the continuous-time covariance equation needs to be integrated and a numerical Jacobian matrix is evaluated). Also, if the measurement noise, $\mathbf{v}_k$, appears linearly in the output (with $l = m$), then the augmented state can be reduced because the system state does not need to augmented with the measurement noise. In this case the covariance of the measurement error is simply added to the innovations covariance, with $P^{\upsilon\upsilon}_{k+1} = P^{yy}_{k+1} + R_{k+1}$. This can greatly reduce the computational requirements in the Unscented filter.

The scalar κ in the previous set of equations is a convenient parameter for exploiting knowledge (if available) about the higher moments of the given distribution.[41] In scalar systems (i.e., for $L = 1$), a value of $\kappa = 2$ leads to errors in the mean and variance that are sixth order. For higher-dimensional systems choosing $\kappa = 3 - L$ minimizes the mean-squared-error up to the fourth order.[39] However, caution should be exercised when κ is negative since a possibility exists that the predicted covariance can become non-positive semi-definite. A modified form has been suggested for this case (see Ref. [39]). Also, a square root version of the Unscented filter is presented in Ref. [42] that avoids the need to re-factorize at each step. Furthermore, Ref. [42] presents an Unscented Particle filter, which makes no assumptions on the form of the probability densities, i.e., full nonlinear, non-Gaussian estimation.

Example 5.9: In this example a comparison is made between the extended Kalman filter and the Unscented filter to estimate the altitude, velocity, and ballistic coefficient of a vertically falling body.[43] The geometry of the problem is shown in Figure 5.10, where $x_1(t)$ is the altitude, $x_2(t)$ is the downward velocity, $r(t)$ is the range (measured by a radar), M is the horizontal distance, and Z is the radar altitude. The truth model is given by

$$\begin{aligned}
\dot{x}_1(t) &= -x_2(t) \\
\dot{x}_2(t) &= -e^{-\alpha x_1(t)} x_2^2(t) x_3(t) \\
\dot{x}_3(t) &= 0
\end{aligned}$$

where $x_3(t)$ is the (constant) ballistic coefficient and α is a constant (5×10^{-5}) that relates the air density with altitude. The discrete-time range measurement at time t_k is given by

$$\tilde{y}_k = \sqrt{M^2 + (x_{1_k} - Z)^2} + \upsilon_k$$

where the variance of υ_k is given by 1×10^4, and $M = Z = 1 \times 10^5$. Note that the dynamic model contains no process noise so that $Q_k = 0$.

The extended Kalman filter requires various partials to be computed. The matrix F from Table 5.9 is given by

$$F = e^{-\alpha \hat{x}_1} \begin{bmatrix} 0 & -e^{\alpha \hat{x}_1} & 0 \\ \alpha \hat{x}_2^2 \hat{x}_3 & -2\hat{x}_2 \hat{x}_3 & -\hat{x}_2^2 \\ 0 & 0 & 0 \end{bmatrix}$$

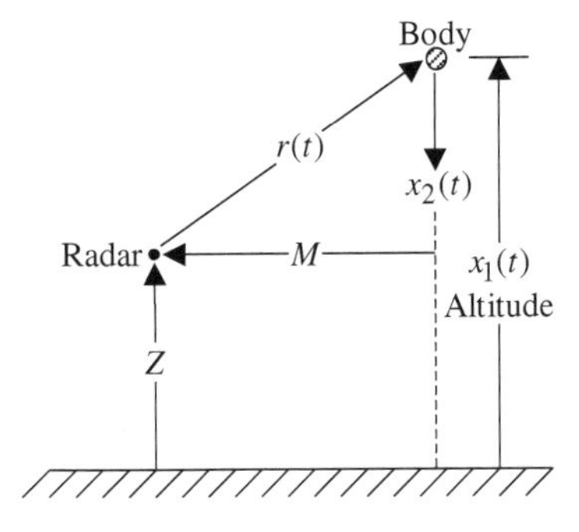

Figure 5.10: Vertically Falling Body Example

The matrix H is given by

$$H = \left[\frac{\hat{x}_1 - Z}{\sqrt{M^2 + (\hat{x}_1 - Z)^2}} \; 0 \; 0 \right]$$

The Kalman filter covariance propagation is carried out by converting F into discrete-time form with the known sampling interval, using eqn. (5.35) to propagate to P_{k+1}^-. For the Unscented filter, since $n = 3$ then $\kappa = 0$, which minimizes the maximum error up to fourth order. The true state and initial estimates are given by

$$\begin{aligned} x_1(0) &= 3 \times 10^5 & \hat{x}_1(0) &= 3 \times 10^5 \\ x_2(0) &= 2 \times 10^4 & \hat{x}_2(0) &= 2 \times 10^4 \\ x_3(0) &= 1 \times 10^{-3} & \hat{x}_3(0) &= 3 \times 10^{-5} \end{aligned}$$

Clearly, an error is present in the ballistic coefficient value. Physically this corresponds to assuming that the body is "heavy" whereas in reality the body is "light." The initial covariance for both filters is given by

$$P(0) = \begin{bmatrix} 1 \times 10^6 & 0 & 0 \\ 0 & 4 \times 10^6 & 0 \\ 0 & 0 & 1 \times 10^{-4} \end{bmatrix}$$

Measurements are sampled at 1-second intervals. In the original test[43] all differential equations were integrated using a fourth-order Runge-Kutta method with a step size of 1/64 second. In our simulations only the truth trajectory has been generated in this manner. The integration step size in both filters has been set to the measurement sample interval (1 second), which further stresses both filters.

Figure 5.11 depicts the average magnitude of the position error by each filter using a Monte Carlo simulation consisting of 100 runs. At the beginning stage where the altitude is high there is little difference between both filters. We should note that correct estimation of x_3 cannot take place at high altitudes due to the small air density.[43] The most severe nonlinearities start taking effect at about 9 seconds,

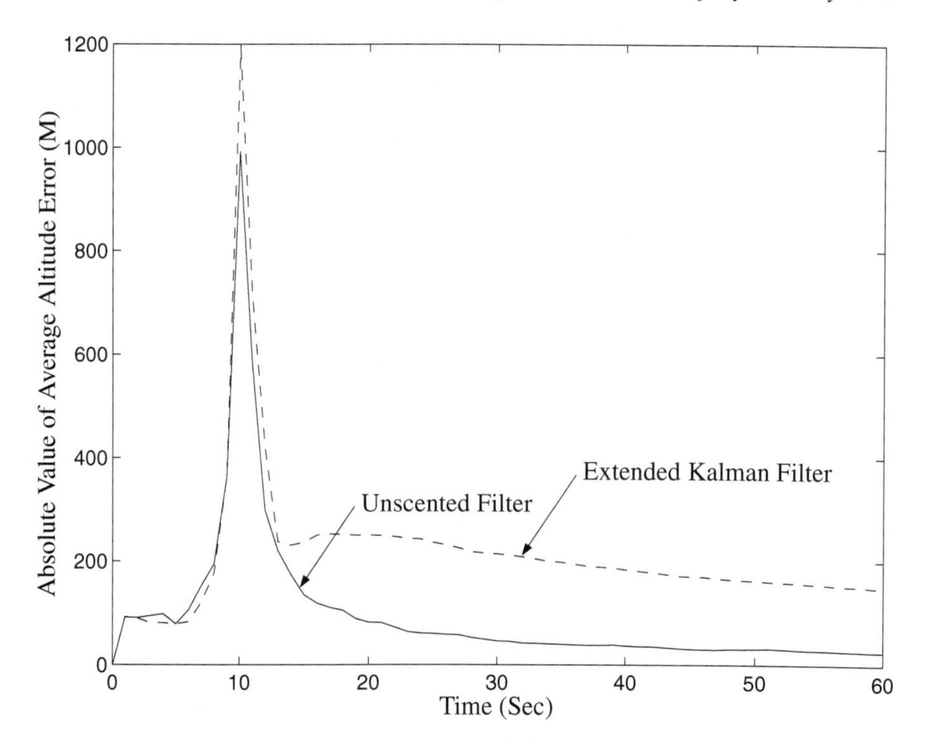

Figure 5.11: Absolute Mean Position Error

where the effects of drag become significant. Large errors are present in both filters, which corresponds to the time when the altitude of the body is the same as the radar (this occurs at 10 seconds where the system is nearly unobservable). However, the Unscented filter has a smaller error-spike than the extended Kalman filter. Finally, the extended Kalman filter converges much slower than the Unscented filter, which is due to the highly nonlinear nature of the model. Similar results are also obtained for the other states. For the x_3 state the extended Kalman filter converges to an order of magnitude larger than the Unscented filter, which attests to the power of using the Unscented filter for highly nonlinear systems.

5.7.7 Robust Filtering

The design of robust filters attempts to maintain filter responses and error signals to within some tolerances despite the effects of uncertainty on the system. Uncertainty may take many forms, but among the most common involve noise (structural) uncertainty and system model uncertainties. The basic idea of one of these designs, called H_∞ filtering, minimizes a "worst-case" loss function, which can be shown to be a *minimax problem* where the maximum "energy" in the error is minimized over all noise trajectories that lead to the same problem.[44] Unfortunately the mathematics behind this theory is intense, involving Hilbert spaces, and well beyond the present text. Therefore, only a brief introduction is presented here.

A good introduction to the H_∞ theory is provided in Refs. [45] and [46]. Before we present the main results of robust filtering we first give an introduction to the operator norms $||G(s)||_2$ and $||G(s)||_\infty$, where $G(s)$ is a proper rational transfer function. The 2-norm of $G(s)$ is defined by

$$||G(s)||_2 = \left\{ \frac{1}{2\pi} \int_{-\infty}^{\infty} \mathrm{Tr}[G(j\omega)\, G^T(-j\omega)]\, d\omega \right\}^{1/2} \tag{5.297}$$

The ∞-norm of $G(s)$ is defined by

$$||G(s)||_\infty = \sup_{\omega} \bar{\sigma}[G(j\omega)] \tag{5.298}$$

where sup denotes the supremum and $\bar{\sigma}$ is the largest singular value. One way to compute $||G(s)||_\infty$ is to take the supremum of the largest singular value of $[G(j\omega)]$ over all frequencies ω. Also, from $y(s) = G(s)\,u(s)$, if $||u(s)||_2 < \infty$ and $G(s)$ is proper with no poles on the imaginary axis, then[45]

$$||G(s)||_\infty = \sup_{u} \frac{||y||_2}{||u||_2} \tag{5.299}$$

A closed-form solution for computing $||G(s)||_2$ is possible, derived using either the controllability or observability Gramians, but a closed-form solution of $||G(s)||_\infty$ is not possible in general. Consider the state-space representation of $G(s)$, given by eqn. (3.11). The procedure to compute $||G(s)||_\infty$ involves searching for the scalar $\gamma > 0$ that yields $||G(s)||_\infty < \gamma$, if and only if $\bar{\sigma}(D) < \gamma$ and the following matrix has no eigenvalues on the imaginary axis:

$$\mathcal{H} \equiv \begin{bmatrix} F + B\,W^{-1} D^T H & B\,W^{-1} B^T \\ -H^T (I + D\,W^{-1} D^T)\,H & -(A + B\,W^{-1} D^T H)^T \end{bmatrix} \tag{5.300}$$

where $W = \gamma^2 I - D^T D$. A proof of this result can be found in Ref. [46]. An iterative solution for γ can be found using a bisection algorithm.[46]

The filtering results presented in this section involve continuous-time models and measurements. Discrete-time systems are discussed in Ref. [8]. We first rewrite the system in eqn. (5.117) as

$$\dot{\mathbf{x}}(t) = F(t)\,\mathbf{x}(t) + B(t)\,\mathbf{u}(t) + G(t)\,\mathbf{w}(t) \tag{5.301a}$$

$$\tilde{\mathbf{y}}(t) = H(t)\,\mathbf{x}(t) + D(t)\,\mathbf{w}(t) \tag{5.301b}$$

Note that the same noise term $\mathbf{w}(t)$ is added in the dynamic model and measurement equations. But the covariance of the measurement noise can be derived directly using $D(t)$, i.e., $R(t) = D(t)\,Q(t)\,D^T(t)$. Also, without loss in generality we can assume that measurement noise can be normalized so that $D(t)\,D^T(t) = I$. Finally it is assumed that the process and measurement noise are uncorrelated so that $D(t)\,G^T(t) = 0$.

The following worst-case loss function is now defined with known initial conditions:

$$J = \sup_{\mathbf{0}\neq\mathbf{w}} \frac{||\mathbf{x}-\hat{\mathbf{x}}||_2^2}{||\mathbf{w}||_2^2} \tag{5.302}$$

with $\mathbf{x}(t_0) = \mathbf{0}$. Note that if the initial condition is not zero, since the system is linear by subtracting the contribution from the nonzero initial condition, then the assumption is valid without loss in generality. Our goal is to determine a filter, given $\gamma > 0$, such that $J < \gamma^2$. Reference [44] has shown that the following filter achieves this condition:

$$\begin{aligned}\dot{\hat{\mathbf{x}}}(t) &= F(t)\,\hat{\mathbf{x}}(t) + B(t)\,\mathbf{u}(t)\\ &\quad + P(t)\,H^T(t)[\tilde{\mathbf{y}}(t) - H(t)\,\hat{\mathbf{x}}(t)], \quad \hat{\mathbf{x}}(t_0) = \mathbf{0}\end{aligned} \tag{5.303}$$

where

$$\begin{aligned}\dot{P}(t) &= F(t)\,P(t) + P(t)\,F^T(t) - P(t)\,[H^T(t)\,H(t) - \gamma^{-2}I]\,P(t)\\ &\quad + G(t)\,G^T(t), \quad P(t_0) = 0\end{aligned} \tag{5.304}$$

Notice that the H_∞ filter bears a striking resemblance to the classical Kalman filter in §5.4. As $\gamma \to \infty$ eqn. (5.304) becomes the corresponding Kalman filter Riccati equation with known initial conditions. For time-invariant systems a steady-state approach can be used. In this case γ can be chosen to be as small as possible such that the Hamiltonian matrix corresponding to the algebraic version of eqn. (5.304), with $\dot{P}(t) = 0$, does not have any eigenvalues on the imaginary axis. Therefore a bisection approach discussed previously can be used to determine γ. If the initial condition is not known then the following loss function is used:

$$J = \sup_{\mathbf{0}\neq\mathbf{w}} \frac{||\mathbf{x}-\hat{\mathbf{x}}||_2^2}{||\mathbf{w}||_2^2 + \mathbf{x}_0^T\,S\,\mathbf{x}_0} \tag{5.305}$$

with $\mathbf{x}(t_0) = \mathbf{x}_0$ and where S is a positive definite symmetric matrix. The solution to this problem is equivalent to eqn. (5.304) but with $P(t_0) = S^{-1}$. Also, the correlated case can be constructed by using the following modifications:

$$F(t) \leftarrow F(t) - G(t)\,D^T(t)\,H(t) \tag{5.306a}$$

$$G(t) \leftarrow G(t)\,[I - D^T(t)\,D(t)] \tag{5.306b}$$

and the filters are obtained simply by superposition, treating $G(t)\,D^T(t)\tilde{\mathbf{y}}(t)$ as a known quantity.[44]

Example 5.10: In this simple example the performance characteristics of the H_∞ filter approach are investigated for a simple scalar and autonomous system, given by

$$\dot{x}(t) = f\,x(t) + g\,w(t)$$
$$\tilde{y}(t) = h\,x(t) + v(t)$$

Note that $w(t)$ and $v(t)$ are not correlated. The steady-state value for $p \equiv P(t)$ in eqn. (5.304) can be found by solving the following algebraic Riccati equation:

$$2fp - (h^2 - \gamma^{-2})\,p^2 + g^2 = 0$$

which gives

$$p = \frac{f \pm \sqrt{f^2 + g^2(h^2 - \gamma^{-2})}}{h^2 - \gamma^{-2}}$$

Consider the case where p has non-complex values, given by the following condition:

$$f^2 + g^2(h^2 - \gamma^{-2}) \geq 0$$

which yields

$$\gamma^2 \geq \frac{g^2}{f^2 + g^2 h^2} \tag{5.307}$$

If we choose the limiting case where γ^2 is equal to the previous expression, then $p = -g^2/f$. Note that p is positive only when f is negative. Therefore, the original system must be stable, which is an undesired consequence of the H_∞ filter approach.

In order to maintain non-complex values for p we can choose γ^2 to be given by

$$\gamma^2 = \frac{g^2}{f^2 + g^2 h^2 - \alpha^2}$$

where α^2 is a scalar that must satisfy $0 \leq \alpha^2 < (f^2 + g^2 h^2)$. When α^2 approaches its upper bound then γ^{-2} approaches 0, which yields the standard Kalman filter Riccati equation. When $\alpha = 0$ then $p = -g^2/f$. Substituting γ^2 into the Riccati equation yields

$$p = \frac{g^2(f \pm \alpha)}{(\alpha + f)(\alpha - f)}$$

Since f is required to be negative then

$$p = \frac{g^2}{\alpha - f}$$

For the range of valid α the following inequality is true (which is left as an exercise for the reader):

$$\frac{g^2}{\alpha - f} > \frac{f + \sqrt{f^2 + g^2 h^2}}{h^2}$$

Note that the right-hand side of the previous equation is the solution of p for the standard Kalman filter. This inequality shows that the gain in the H_∞ filter will always be larger than the gain in the Kalman filter, which means that the bandwidth of the H_∞ filter is larger than the Kalman filter. Therefore, the H_∞ filter relies more on the measurements than the *a priori* state to obtain the state estimate, which is more robust to modelling errors, but allows more high-frequency noise in the estimate.

This section has introduced the basic concepts of robust filtering. This subject area (as well as robust control) is currently an evolving theory for which the benefits are yet unknown. Still the relationship between the H_∞ filter and Kalman filter is interesting, and in some multi-dimensional cases the H_∞ filter may provide some significant advantages over the Kalman filter. Other areas such as H_∞ adaptive filtering and nonlinear H_∞ filtering may be found in the references provided in this section, as well as the open literature. The reader is encouraged to pursue these references in order to evaluate the performance of robust filtering approaches for the reader's particular dynamical system studies.

5.8 Summary

The results of §5.2 provide the basis for all state estimation algorithms. One of the most fascinating aspects of the estimators developed in §5.2 is the similarity to the sequential estimation results in §1.3. This is truly remarkable since the results of Chapter 1 are applied to constant parameter estimation, while the results of this chapter are applied to parameters that are allowed to change during the estimation process. Another important aspect of state estimation is the similarity to feedback control, where the measurement is the quantity to be "tracked" by the feedback system. This similarity between control and estimation will be further expanded upon in Chapter 6.

The discrete-time Kalman filter developments of §5.3 are based upon the discrete-time sequential estimator of §5.2.1. The only difference between them is in how the gain matrix is derived. The driving force of any estimator is the location of the estimator poles. If these poles are well-known then Ackermann's formula should be employed to determine the gain matrix. However, in practice this is hardly ever the case. The Kalman filter also is a "pole-placement" method, but these poles are selected through rigorous use of known statistical properties of the process noise and measurement noise.

Several theoretical aspects of the Kalman filter are given in this chapter. One of the most important is the stability of the closed-loop Kalman filter state matrix, which is rigorously proved using Lyapunov's theorem. This stability is especially appeasing, since even if the model state matrix is unstable the Kalman filter will always be stable. Several other important aspects of the Kalman filter are shown in this chapter, including: the information filter form, sequential processing, the steady-state Kalman filter, correlated measurement and process noise cases, and the orthogonality principle. The derivation of the continuous-time Kalman filter is shown from two different approaches. The first approach is based upon a continuous-time covariance derivation, and the second approach is shown by applying a limiting argument to the discrete-time formulas. We believe that both approaches are important in understanding the intricacies of the linear Kalman filter.

The Kalman filter is probably one of the most studied algorithms to date. This fact is attested to by the plethora of publications in journals and books. Its popularity will continue for many years to come. An excellent overview of the history behind general filtering theory is given in Ref. [47]. A few of the several notable research results are presented in §5.7, which includes factorization methods, colored-noise Kalman filtering, adaptive filtering, error analysis, Unscented filtering, and robust filtering. This section has merely "scratched the surface" of the flood of research results obtained by studying the Kalman filter. Our own experiences have shown that every time we implement the Kalman filter or study its theoretical foundation, new insights are brought to the surface. The reader is also encouraged to further study the Kalman filter in the open literature (decentralized Kalman filtering,[14] approximate Kalman filtering,[48] and Particle filtering[42, 49, 50] are also interesting topics).

A summary of the key formulas presented in this chapter is given below.

- Ackermann's formula (Continuous-Time)

$$\dot{\hat{\mathbf{x}}} = F\hat{\mathbf{x}} + B\mathbf{u} + K[\tilde{\mathbf{y}} - H\hat{\mathbf{x}}]$$
$$\hat{\mathbf{y}} = H\hat{\mathbf{x}}$$
$$K = d(F)\begin{bmatrix} H \\ HF \\ HF^2 \\ \vdots \\ HF^{n-1} \end{bmatrix}^{-1}\begin{bmatrix} 0 \\ 0 \\ 0 \\ \vdots \\ 1 \end{bmatrix} \equiv d(F)\mathcal{O}^{-1}\begin{bmatrix} 0 \\ 0 \\ 0 \\ \vdots \\ 1 \end{bmatrix}$$

- Ackermann's formula (Discrete-Time)

$$\hat{\mathbf{x}}_{k+1}^- = \Phi\hat{\mathbf{x}}_k^+ + \Gamma\mathbf{u}_k$$
$$\hat{\mathbf{x}}_k^+ = \hat{\mathbf{x}}_k^- + K[\tilde{\mathbf{y}}_k - H\hat{\mathbf{x}}_k^-]$$
$$K = d(\Phi)\begin{bmatrix} H\Phi \\ H\Phi^2 \\ H\Phi^3 \\ \vdots \\ H\Phi^n \end{bmatrix}^{-1}\begin{bmatrix} 0 \\ 0 \\ 0 \\ \vdots \\ 1 \end{bmatrix} \equiv d(\Phi)\Phi^{-1}\mathcal{O}_d^{-1}\begin{bmatrix} 0 \\ 0 \\ 0 \\ \vdots \\ 1 \end{bmatrix}$$

- Kalman Filter (Discrete-Time)

$$\hat{\mathbf{x}}_{k+1}^- = \Phi_k \hat{\mathbf{x}}_k^+ + \Gamma_k \mathbf{u}_k$$
$$P_{k+1}^- = \Phi_k P_k^+ \Phi_k^T + \Upsilon_k Q_k \Upsilon_k^T$$

$$\hat{\mathbf{x}}_k^+ = \hat{\mathbf{x}}_k^- + K_k[\tilde{\mathbf{y}}_k - H_k \hat{\mathbf{x}}_k^-]$$
$$P_k^+ = [I - K_k H_k] P_k^-$$

$$K_k = P_k^- H_k^T [H_k P_k^- H_k^T + R_k]^{-1}$$

- Alternative Gain and Update Forms

$$K_k = P_k^+ H_k^T R_k^{-1}$$

$$\hat{\mathbf{x}}_k^+ = P_k^+ \left[\left(P_k^-\right)^{-1} \hat{\mathbf{x}}_k^- + H_k^T R_k^{-1} \tilde{\mathbf{y}}_k \right]$$

- Joseph's Form

$$P_k^+ = [I - K_k H_k] P_k^- [I - K_k H_k]^T + K_k R_k K_k^T$$

- Information Filter

$$\mathcal{P}_k^+ = \mathcal{P}_k^- + H_k^T R_k^{-1} H_k$$
$$\mathcal{P}_{k+1}^- = \left[I - \Psi_k \Upsilon_k \left(\Upsilon_k^T \Psi_k \Upsilon_k + Q_k^{-1} \right)^{-1} \Upsilon_k^T \right] \Psi_k$$
$$\Psi_k \equiv \Phi_k^{-T} \mathcal{P}_k^+ \Phi_k^{-1}$$
$$K_k = (\mathcal{P}_k^+)^{-1} H_k^T R_k^{-1}$$

- Sequential Processing

$$\tilde{\mathbf{z}}_k \equiv T_k \tilde{\mathbf{y}}_k = T_k H_k \mathbf{x}_k + T_k \mathbf{v}_k$$
$$\equiv \mathcal{H}_k \mathbf{x}_k + \boldsymbol{\upsilon}_k$$
$$\hat{\mathbf{x}}_k^+ = \hat{\mathbf{x}}_k^- + P_k^+ \mathcal{H}_k^T \mathcal{R}_k^{-1} [\tilde{\mathbf{z}}_k - \mathcal{H}_k \hat{\mathbf{x}}_k^-]$$

$$K_{i_k} = \frac{P_{i-1_k}^- \mathcal{H}_{i_k}^T}{\mathcal{H}_{i_k} P_{i-1_k}^- \mathcal{H}_{i_k}^T + \mathcal{R}_{i_k}}$$
$$P_{i_k}^+ = [I - K_{i_k} \mathcal{H}_{i_k}] P_{i-1_k}^T, \quad P_{0_k}^+ = P_k^-$$

- Autonomous Kalman Filter (Discrete-Time)

$$\hat{\mathbf{x}}_{k+1} = \Phi \hat{\mathbf{x}}_k + \Gamma \mathbf{u}_k + \Phi K [\tilde{\mathbf{y}}_k - H \hat{\mathbf{x}}_k]$$
$$P = \Phi P \Phi^T - \Phi P H^T [H P H^T + R]^{-1} H P \Phi^T + \Upsilon Q \Upsilon^T$$
$$K = P H^T [H P H^T + R]^{-1}$$

- Correlated Kalman Filter (Discrete-Time)

$$\hat{\mathbf{x}}_{k+1}^- = \Phi_k \hat{\mathbf{x}}_k^+ + \Gamma_k \mathbf{u}_k$$
$$P_{k+1}^- = \Phi_k P_k^+ \Phi_k^T + \Upsilon_k Q_k \Upsilon_k^T$$

$$\hat{\mathbf{x}}_k^+ = \hat{\mathbf{x}}_k^- + K_k[\tilde{\mathbf{y}}_k - H_k \hat{\mathbf{x}}_k^-]$$
$$P_k^+ = [I - K_k H_k] P_k^- - K_k S_k^T \Upsilon_{k-1}^T$$

$$\begin{aligned} K_k &= [P_k^- H_k^T + \Upsilon_{k-1} S_k] \\ &\quad \times [H_k P_k^- H_k^T + R_k + H_k \Upsilon_{k-1} S_k + S_k^T \Upsilon_{k-1}^T H_k^T]^{-1} \end{aligned}$$

- Continuous-Time to Discrete-Time Covariance Calculation

$$\mathcal{A} = \begin{bmatrix} -F & G\,Q\,G^T \\ 0 & F^T \end{bmatrix} \Delta t$$

$$\mathcal{B} = e^{\mathcal{A}} \equiv \begin{bmatrix} \mathcal{B}_{11} & \mathcal{B}_{12} \\ 0 & \mathcal{B}_{22} \end{bmatrix} = \begin{bmatrix} \mathcal{B}_{11} & \Phi^{-1}\mathcal{Q} \\ 0 & \Phi^T \end{bmatrix}$$

$$\Phi = \mathcal{B}_{22}^T$$
$$\mathcal{Q} = \Phi\,\mathcal{B}_{12}$$

- Kalman Filter (Continuous-Time)

$$\dot{\hat{\mathbf{x}}}(t) = F(t)\,\hat{\mathbf{x}}(t) + B(t)\,\mathbf{u}(t) + K(t)[\tilde{\mathbf{y}}(t) - H(t)\,\hat{\mathbf{x}}(t)]$$
$$\begin{aligned} \dot{P}(t) &= F(t)\,P(t) + P(t)\,F^T(t) \\ &\quad - P(t)\,H^T(t)\,R^{-1}(t) H(t)\,P(t) + G(t)\,Q(t)\,G^T(t) \end{aligned}$$
$$K(t) = P(t)\,H^T(t)\,R^{-1}(t)$$

- Autonomous Kalman Filter (Continuous-Time)

$$\dot{\hat{\mathbf{x}}}(t) = F\,\hat{\mathbf{x}}(t) + B\,\mathbf{u}(t) + K[\tilde{\mathbf{y}}(t) - H\,\hat{\mathbf{x}}(t)]$$
$$F\,P + P\,F^T - P\,H^T\,R^{-1} H\,P + G\,Q\,G^T = 0$$
$$K = P\,H^T R^{-1}$$

- Correlated Kalman Filter (Continuous-Time)

$$\dot{\hat{\mathbf{x}}}(t) = F(t)\,\hat{\mathbf{x}}(t) + B(t)\,\mathbf{u}(t) + K(t)[\tilde{\mathbf{y}}(t) - H(t)\,\hat{\mathbf{x}}(t)]$$
$$\begin{aligned} \dot{P}(t) &= F(t)\,P(t) + P(t)\,F^T(t) \\ &\quad - K(t)\,R(t)\,K^T(t) + G(t)\,Q(t)\,G^T(t) \end{aligned}$$
$$K(t) = \left[P(t)\,H^T(t) + G(t)\,S^T(t)\right] R^{-1}(t)$$

- Continuous-Discrete Kalman Filter

$$\dot{\hat{\mathbf{x}}}(t) = F(t)\,\hat{\mathbf{x}}(t) + B(t)\,\mathbf{u}(t)$$
$$\dot{P}(t) = F(t)\,P(t) + P(t)\,F^T(t) + G(t)\,Q(t)\,G^T(t)$$

$$\hat{\mathbf{x}}_k^+ = \hat{\mathbf{x}}_k^- + K_k[\tilde{\mathbf{y}}_k - H_k\hat{\mathbf{x}}_k^-]$$
$$P_k^+ = [I - K_k H_k]P_k^-$$
$$K_k = P_k^- H_k^T[H_k P_k^- H_k^T + R_k]^{-1}$$

- Extended Kalman Filter (Continuous-Time)

$$\dot{\hat{\mathbf{x}}}(t) = \mathbf{f}(\hat{\mathbf{x}}(t),\,\mathbf{u}(t),\,t) + K(t)[\tilde{\mathbf{y}}(t) - \mathbf{h}(\hat{\mathbf{x}}(t),\,t)]$$

$$\begin{aligned}\dot{P}(t) = {} & F(\hat{\mathbf{x}}(t),\,t)\,P(t) + P(t)\,F^T(\hat{\mathbf{x}}(t),\,t) \\ & - P(t)\,H^T(\hat{\mathbf{x}}(t),\,t)\,R^{-1}(t)H(\hat{\mathbf{x}}(t),\,t)\,P(t) + G(t)\,Q(t)\,G^T(t)\end{aligned}$$
$$F(\hat{\mathbf{x}}(t),\,t) \equiv \left.\frac{\partial \mathbf{f}}{\partial \mathbf{x}}\right|_{\hat{\mathbf{x}}(t)}, \quad H(\hat{\mathbf{x}}(t),\,t) \equiv \left.\frac{\partial \mathbf{h}}{\partial \mathbf{x}}\right|_{\hat{\mathbf{x}}(t)}$$
$$K(t) = P(t)\,H^T(\hat{\mathbf{x}}(t),\,t)\,R^{-1}(t)$$

- Continuous-Discrete Extended Kalman Filter

$$\dot{\hat{\mathbf{x}}}(t) = \mathbf{f}(\hat{\mathbf{x}}(t),\,\mathbf{u}(t),\,t)$$
$$\dot{P}(t) = F(\hat{\mathbf{x}}(t),\,t)\,P(t) + P(t)\,F^T(\hat{\mathbf{x}}(t),\,t) + G(t)\,Q(t)\,G^T(t)$$
$$F(\hat{\mathbf{x}}(t),\,t) \equiv \left.\frac{\partial \mathbf{f}}{\partial \mathbf{x}}\right|_{\hat{\mathbf{x}}(t)}$$

$$\hat{\mathbf{x}}_k^+ = \hat{\mathbf{x}}_k^- + K_k[\tilde{\mathbf{y}}_k - \mathbf{h}(\hat{\mathbf{x}}_k^-)]$$
$$P_k^+ = [I - K_k H_k(\hat{\mathbf{x}}_k^-)]P_k^-$$
$$H_k(\hat{\mathbf{x}}_k^-) \equiv \left.\frac{\partial \mathbf{h}}{\partial \mathbf{x}}\right|_{\hat{\mathbf{x}}_k^-}$$
$$K_k = P_k^- H_k^T(\hat{\mathbf{x}}_k^-)[H_k(\hat{\mathbf{x}}_k^-)P_k^- H_k^T(\hat{\mathbf{x}}_k^-) + R_k]^{-1}$$

- Iterated Extended Kalman Filter

$$\hat{\mathbf{x}}_{k_i}^+ = \hat{\mathbf{x}}_k^- + K_{k_i}\left[\tilde{\mathbf{y}}_k - \mathbf{h}(\hat{\mathbf{x}}_{k_i}^+) - H_k(\hat{\mathbf{x}}_{k_i}^+)\left(\hat{\mathbf{x}}_k^- - \hat{\mathbf{x}}_{k_i}^+\right)\right]$$
$$K_{k_i} = P_k^- H_k^T(\hat{\mathbf{x}}_{k_i}^+)\left[H_k(\hat{\mathbf{x}}_{k_i}^+)P_k^- H_k^T(\hat{\mathbf{x}}_{k_i}^+) + R_k\right]^{-1}$$
$$P_{k_i}^+ = \left[I - K_{k_i} H_k(\hat{\mathbf{x}}_{k_i}^+)\right]P_k^-$$
$$\hat{\mathbf{x}}_{k_0}^+ = \hat{\mathbf{x}}_k^-$$

- Square Root Information Filter

$$\mathcal{P}_k^+ \equiv (P_k^+)^{-1} = \mathcal{S}_k^{+T}\mathcal{S}_k^+$$
$$\mathcal{P}_k^- \equiv (P_k^-)^{-1} = \mathcal{S}_k^{-T}\mathcal{S}_k^-$$
$$R_k^{-1} = \mathcal{V}_k^T\mathcal{V}_k$$
$$Q_k = Z_k E_k Z_k^T$$
$$\Xi_k \equiv \Upsilon_k Z_k$$
$$\hat{\boldsymbol{\alpha}}_k^+ \equiv \mathcal{S}_k^+\hat{\mathbf{x}}_k^+$$
$$\hat{\boldsymbol{\alpha}}_k^- \equiv \mathcal{S}_k^-\hat{\mathbf{x}}_k^-$$
$$\mathcal{Q}_k^T \begin{bmatrix} \mathcal{S}_k^- \\ \mathcal{V}_k H_k \end{bmatrix} = \begin{bmatrix} \mathcal{S}_k^+ \\ 0_{m\times n} \end{bmatrix}$$

for $i = 1$

$$\mathbf{a} = \mathcal{S}_k^+\Phi_k^{-1}\Xi_k(1)$$
$$b = \left[\mathbf{a}^T\mathbf{a} + 1/E_k(1,1)\right]^{-1}$$
$$c = \left[1 + \sqrt{b/E_k(1,1)}\right]^{-1}$$
$$\mathbf{d}^T = b\,\mathbf{a}^T\mathcal{S}_k^+\Phi_k^{-1}$$
$$\hat{\boldsymbol{\alpha}}_{k+1}^- = \hat{\boldsymbol{\alpha}}_k^+ - b\,c\,\mathbf{a}\mathbf{a}^T\hat{\boldsymbol{\alpha}}_k^+$$
$$\mathcal{S}_{k+1}^- = \mathcal{S}_k^+\Phi_k^{-1} - c\,\mathbf{a}\mathbf{d}^T$$

for $i > 1$

$$\mathbf{a} = \mathcal{S}_{k+1}^-\Xi_k(i)$$
$$b = \left[\mathbf{a}^T\mathbf{a} + 1/E_k(i,i)\right]^{-1}$$
$$c = \left[1 + \sqrt{b/E_k(i,i)}\right]^{-1}$$
$$\mathbf{d}^T = b\,\mathbf{a}^T\mathcal{S}_{k+1}^-$$
$$\hat{\boldsymbol{\alpha}}_{k+1}^- \leftarrow \hat{\boldsymbol{\alpha}}_{k+1}^- - b\,c\,\mathbf{a}\mathbf{a}^T\hat{\boldsymbol{\alpha}}_{k+1}^-$$
$$\mathcal{S}_{k+1}^- \leftarrow \mathcal{S}_{k+1}^- - c\,\mathbf{a}\mathbf{d}^T$$

- *U-D* Filter

$$P_{i_k}^- = U_{i_k}^- D_{i_k}^- U_{i_k}^{-T}$$
$$P_{i_k}^+ = U_{i_k}^+ D_{i_k}^+ U_{i_k}^{+T} = U_{i_k}^-\left[D_{i_k}^- - \frac{1}{\alpha_{i_k}}\mathbf{e}_{i_k}\mathbf{e}_{i_k}^T\right]U_{i_k}^{-T}$$
$$\alpha_{i_k} \equiv \mathcal{H}_{i_k}P_{i_k}^-\mathcal{H}_{i_k}^T + \mathcal{R}_{i_k}$$
$$\mathbf{e}_{i_k} \equiv D_{i_k}^- U_{i_k}^{-T}\mathcal{H}_{i_k}^T$$

$$\left[D_{i_k}^- - \frac{1}{\alpha_{i_k}}\mathbf{e}_{i_k}\mathbf{e}_{i_k}^T\right] = L_{i_k}^- \mathcal{E}_{i_k}^- L_{i_k}^{-T}$$
$$U_{i_k}^+ = U_{i-1_k}^+ L_{i_k}^-, \quad U_{0_k}^+ = U_k^-$$
$$D_{i_k}^+ = \mathcal{E}_{i_k}^-, \quad D_{0_k}^+ = D_k^-$$

$$K_{i_k} = \frac{1}{\alpha_{i_k}} U_{i_k}^- \mathbf{e}_{i_k}$$

$$W_{k+1}^- \equiv \left[\Phi_k U_k^+ \ \Xi_k\right]$$
$$\tilde{D}_{k+1}^- \equiv \begin{bmatrix} D_k^+ & 0_{n\times s} \\ 0_{s\times n} & E_k \end{bmatrix}$$
$$\left[\mathbf{w}(1)\ \mathbf{w}(2)\ \cdots\ \mathbf{w}(n)\right] = W_{k+1}^{-T}$$

$$\mathbf{c}(i) = \tilde{D}_{k+1}^- \mathbf{w}(i)$$
$$D_{k+1}^-(i,i) = \mathbf{w}^T(i)\,\mathbf{c}(i)$$
$$\mathbf{d}(i) = \mathbf{c}(i)/D_{k+1}^-(i,i)$$
$$U_{k+1}^-(j,i) = \mathbf{w}^T(j)\,\mathbf{d}(i), \quad j = 1, 2, \ldots, i-1$$
$$\mathbf{w}(j) \leftarrow \mathbf{w}(j) - U_{k+1}^-(j,i)\,\mathbf{w}(i), \quad j = 1, 2, \ldots, i-1$$

- Process-Noise Colored-Filter

$$\begin{bmatrix} \mathbf{x}_{k+1} \\ \chi_{k+1} \end{bmatrix} = \begin{bmatrix} \Phi & \Upsilon\mathcal{H} \\ 0 & \Psi \end{bmatrix} \begin{bmatrix} \mathbf{x}_k \\ \chi_k \end{bmatrix} + \begin{bmatrix} \Gamma \\ 0 \end{bmatrix} \mathbf{u}_k + \begin{bmatrix} \Upsilon\mathcal{D} \\ \mathcal{V} \end{bmatrix} \omega_k$$
$$\tilde{\mathbf{y}}_k = \left[H\ 0\right] \begin{bmatrix} \mathbf{x}_k \\ \chi_k \end{bmatrix} + \mathbf{v}_k$$

- Measurement-Noise Colored-Filter

$$\begin{bmatrix} \mathbf{x}_{k+1} \\ \chi_{k+1} \end{bmatrix} = \begin{bmatrix} \Phi & 0 \\ 0 & \Psi \end{bmatrix} \begin{bmatrix} \mathbf{x}_k \\ \chi_k \end{bmatrix} + \begin{bmatrix} \Gamma \\ 0 \end{bmatrix} \mathbf{u}_k + \begin{bmatrix} \Upsilon & 0 \\ 0 & \mathcal{V} \end{bmatrix} \begin{bmatrix} \mathbf{w}_k \\ \omega_k \end{bmatrix}$$
$$\tilde{\mathbf{y}}_k = \left[H\ \mathcal{H}\right] \begin{bmatrix} \mathbf{x}_k \\ \chi_k \end{bmatrix} + \mathcal{D}\omega_k + \boldsymbol{\nu}_k$$

$$E\left\{\begin{bmatrix} \mathbf{w}_k \\ \omega_k \end{bmatrix} \left[\mathbf{w}_k^T\ \omega_k^T\right]\right\} = \begin{bmatrix} Q & 0 \\ 0 & \mathcal{Q} \end{bmatrix}$$
$$R = \mathcal{D}\mathcal{Q}\mathcal{D}^T + \mathcal{R}$$
$$S = \left[0\ \mathcal{D}\mathcal{Q}\right]$$

- Measurement-Noise Colored-Filter (Restricted Case)

$$\chi_{k+1} = \Psi\,\chi_k + \mathcal{V}\,\omega_k$$
$$\mathbf{v}_k = \chi_k$$

$$\tilde{\gamma}_{k+1} \equiv \tilde{\mathbf{y}}_{k+1} - \Psi\,\tilde{\mathbf{y}}_k - H\,\Gamma\,\mathbf{u}_k$$
$$= \mathcal{H}\mathbf{x}_k + \mathcal{V}\,\omega_k + H\,\Upsilon\,\mathbf{w}_k$$
$$\mathcal{H} \equiv H\,\Phi - \Psi\,H$$

$$R = \mathcal{V}\,\mathcal{Q}\,\mathcal{V}^T + H\,\Upsilon\,Q\,\Upsilon^T H^T$$
$$S = H\,\Upsilon\,Q$$

- Consistency of the Kalman Filter

$$\bar{\epsilon}_k = \frac{1}{M}\sum_{i=1}^{M}\epsilon_k(i) = \frac{1}{M}\sum_{i=1}^{M}\mathbf{e}_k^T(i)\,E_k^{-1}(i)\,\mathbf{e}_k(i)$$

$$\bar{\rho}_{k,j} = \frac{1}{\sqrt{m}}\sum_{i=1}^{M}\mathbf{e}_k^T(i)\left[\sum_{i=1}^{M}\mathbf{e}_k(i)\mathbf{e}_k^T(i)\sum_{i=1}^{M}\mathbf{e}_k(j)\mathbf{e}_k^T(j)\right]^{-1/2}\mathbf{e}_k(j)$$

$$[\bar{\mu}_k]_j = \frac{1}{M}\sum_{i=1}^{M}\frac{[\mathbf{e}_k]_j}{\sqrt{[E_k]_{jj}}}, \quad j = 1, 2, \ldots, m$$

$$\bar{\epsilon} = \frac{1}{N}\sum_{k=1}^{N}\mathbf{e}_k^T\,E_k^{-1}\,\mathbf{e}_k$$

$$\bar{\rho}_j = \frac{1}{\sqrt{n}}\sum_{k=1}^{N}\mathbf{e}_k^T\,\mathbf{e}_{k+j}\left[\sum_{k=1}^{N}\mathbf{e}_k^T\,\mathbf{e}_k\sum_{k=1}^{N}\mathbf{e}_{k+j}^T\mathbf{e}_{k+j}\right]^{-1/2}$$

- Adaptive Filtering

$$\hat{C}_i = \frac{1}{N}\sum_{j=i}^{N}\mathbf{e}_j\mathbf{e}_{j-i}^T$$
$$\mathbf{e}_k \equiv \tilde{\mathbf{y}}_k - H\,\hat{\mathbf{x}}_k^-$$

$$\hat{R} = \hat{C}_0 - H\,\hat{Z}$$

$$\hat{Z} = (M^T M)^{-1} M^T \begin{bmatrix} \hat{C}_1 + H\,\Phi\,K\,\hat{C}_0 \\ \hat{C}_2 + H\,\Phi\,K\,\hat{C}_1 + H\,\Phi^2 K\,\hat{C}_0 \\ \vdots \\ \hat{C}_n + H\,\Phi\,K\,\hat{C}_{n-1} + \cdots + H\,\Phi^n K\,\hat{C}_0 \end{bmatrix}$$

$$\delta\hat{P} = \Phi\Big[\delta P - (\hat{Z} + \delta P\, H^T)(\hat{C}_0 + H\,\delta P\, H^T)^{-1}(\hat{Z}^T + H\,\delta P)$$
$$+ K\,\hat{Z}^T + \hat{Z}\,K^T - K\,\hat{C}_0\,K^T\Big]\Phi^T$$
$$\hat{K}^* = \Big[\hat{Z} + \delta\hat{P}\,H^T\Big]\Big[\hat{C}_0 + H\,\delta\hat{P}\,H^T\Big]^{-1}$$

- Error Analysis

$$\dot{\hat{\bar{\mathbf{x}}}} = \bar{F}\,\hat{\bar{\mathbf{x}}} + B\,\mathbf{u} + \bar{K}[\tilde{\mathbf{y}} - \bar{H}\,\hat{\bar{\mathbf{x}}}]$$
$$\dot{\bar{P}} = \bar{F}\,\bar{P} + \bar{P}\,\bar{F}^T - \bar{P}\,\bar{H}^T\,\bar{R}^{-1}\,\bar{H}\,\bar{P} + \bar{G}\,\bar{Q}\,\bar{G}^T$$
$$\bar{K} = \bar{P}\,\bar{H}^T\,\bar{R}^{-1}$$

$$P_{\tilde{\mathbf{x}}} = \bar{P} + \Delta V_{\tilde{\mathbf{x}}} + \boldsymbol{\mu}_{\tilde{\mathbf{x}}}\boldsymbol{\mu}_{\tilde{\mathbf{x}}}^T$$

$$\dot{\boldsymbol{\mu}}_{\tilde{\mathbf{x}}} = (\bar{F} - \bar{K}\,\bar{H})\,\boldsymbol{\mu}_{\tilde{\mathbf{x}}} + (\Delta F - \bar{K}\,\Delta H)\,\boldsymbol{\mu}_{\mathbf{x}}$$
$$\dot{\boldsymbol{\mu}}_{\mathbf{x}} = F\,\boldsymbol{\mu}_{\mathbf{x}}$$
$$\Delta F \equiv F - \bar{F}, \quad \Delta H \equiv H - \bar{H}$$

$$\Delta V_{\tilde{\mathbf{x}}} = (\bar{F} - \bar{K}\,\bar{H})\,\Delta V_{\tilde{\mathbf{x}}} + V_{\tilde{\mathbf{x}}}(\bar{F} - \bar{K}\,\bar{H})^T + V^T(\Delta F - \bar{K}\,\Delta H)^T$$
$$+ (\Delta F - \bar{K}\,\Delta H)\,V + (G\,Q\,G^T - \bar{G}\,\bar{Q}\,\bar{G}^T) + \bar{K}\,(R - \bar{R})\,\bar{K}^T$$

- Unscented Filtering

$$\mathbf{x}_{k+1} = \mathbf{f}(\mathbf{x}_k, \mathbf{w}_k, \mathbf{u}_k, k)$$
$$\tilde{\mathbf{y}}_k = \mathbf{h}(\mathbf{x}_k, \mathbf{u}_k, \mathbf{v}_k, k)$$

$$\hat{\mathbf{x}}_k^+ = \hat{\mathbf{x}}_k^- + K_k \boldsymbol{\upsilon}_k$$
$$P_k^+ = P_k^- - K_k P_k^{\upsilon\upsilon} K_k^T$$
$$\boldsymbol{\upsilon}_k \equiv \tilde{\mathbf{y}}_k - \hat{\mathbf{y}}_k^-$$
$$= \tilde{\mathbf{y}}_k - \mathbf{h}(\hat{\mathbf{x}}_k^-, \mathbf{u}_k, k)$$

$$K_k = P_k^{xy}(P_k^{\upsilon\upsilon})^{-1}$$

$$P_k^a = \begin{bmatrix} P_k^+ & P_k^{xw} & P_k^{xv} \\ (P_k^{xw})^T & Q_k & P_k^{wv} \\ (P_k^{xv})^T & (P_k^{wv})^T & R_k \end{bmatrix}$$
$$\boldsymbol{\sigma}_k \leftarrow 2L \text{ columns from } \pm\gamma\sqrt{P_k^a}$$
$$\boldsymbol{\chi}_k^a(0) = \hat{\mathbf{x}}_k^a$$
$$\boldsymbol{\chi}_k^a(i) = \boldsymbol{\sigma}_k(i) + \hat{\mathbf{x}}_k^a$$

$$\mathbf{x}_k^a = \begin{bmatrix} \mathbf{x}_k \\ \mathbf{w}_k \\ \mathbf{v}_k \end{bmatrix}, \quad \hat{\mathbf{x}}_k^a = \begin{bmatrix} \hat{\mathbf{x}}_k \\ \mathbf{0}_{q\times 1} \\ \mathbf{0}_{m\times 1} \end{bmatrix}$$

$$W_0^{\text{mean}} = \frac{\lambda}{L+\lambda}$$

$$W_0^{\text{cov}} = \frac{\lambda}{L+\lambda} + (1-\alpha^2+\beta)$$

$$W_i^{\text{mean}} = W_i^{\text{cov}} = \frac{1}{2(L+\lambda)}, \quad i = 1, 2, \ldots, 2L$$

$$\chi_{k+1}(i) = \mathbf{f}(\chi_k^x(i), \chi_k^w(i), \mathbf{u}_k, k)$$

$$\hat{\mathbf{x}}_{k+1}^- = \sum_{i=0}^{2L} W_i^{\text{mean}} \chi_{k+1}^x(i)$$

$$P_{k+1}^- = \sum_{i=0}^{2L} W_i^{\text{cov}} [\chi_{k+1}^x(i) - \hat{\mathbf{x}}_{k+1}^-][\chi_{k+1}^x(i) - \hat{\mathbf{x}}_{k+1}^-]^T$$

$$\gamma_{k+1}(i) = \mathbf{h}(\chi_{k+1}^x(i), \mathbf{u}_{k+1}, \chi_{k+1}^v(i), k+1)$$

$$\hat{\mathbf{y}}_{k+1}^- = \sum_{i=0}^{2L} W_i^{\text{mean}} \gamma_{k+1}(i)$$

$$P_{k+1}^{yy} = \sum_{i=0}^{2L} W_i^{\text{cov}} [\gamma_{k+1}(i) - \hat{\mathbf{y}}_{k+1}^-][\gamma_{k+1}(i) - \hat{\mathbf{y}}_{k+1}^-]^T$$

$$P_{k+1}^{vv} = P_{k+1}^{yy}$$

$$P_{k+1}^{xy} = \sum_{i=0}^{2L} W_i^{\text{cov}} [\chi_{k+1}^x(i) - \hat{\mathbf{x}}_{k+1}^-][\gamma_{k+1}(i) - \hat{\mathbf{y}}_{k+1}^-]^T$$

- Robust Filtering

$$\begin{aligned}\dot{\hat{\mathbf{x}}}(t) &= F(t)\,\hat{\mathbf{x}}(t) + B(t)\,\mathbf{u}(t) \\ &\quad + P(t)\,H^T(t)[\tilde{\mathbf{y}}(t) - H(t)\,\hat{\mathbf{x}}(t)], \quad \hat{\mathbf{x}}(t_0) = \mathbf{0}\end{aligned}$$

$$\begin{aligned}\dot{P}(t) &= F(t)\,P(t) + P(t)\,F^T(t) - P(t)\,[H^T(t)\,H(t) - \gamma^{-2} I]\,P(t) \\ &\quad + G(t)\,G^T(t), \quad P(t_0) = S^{-1}\end{aligned}$$

Exercises

5.1 Write a general computer routine for Ackermann's formula in eqn. (5.19).

5.2 Design an estimator for a simple pendulum model, given by

$$\dot{\mathbf{x}}(t) = \begin{bmatrix} 0 & 1 \\ -\omega_n^2 & 0 \end{bmatrix} \mathbf{x}(t)$$
$$y(t) = \begin{bmatrix} 1 & 0 \end{bmatrix} \mathbf{x}(t)$$

where both estimator eigenvalues are at $-10\omega_n$. Convert your estimator into discrete-time. Pick any initial conditions and simulate the performance of the estimator using synthetic measurements ($\tilde{y}_k = y_k + v_k$), with various values for the measurement-error variance. How do your estimates change as more noise is introduced into the measurement? Also, try changing the pole locations of the estimator for various noise levels.

5.3 The stick-fixed lateral equations of motion for a general aviation aircraft are given by[31]

$$\begin{bmatrix} \Delta\dot{\beta}(t) \\ \Delta\dot{p}(t) \\ \Delta\dot{r}(t) \\ \Delta\dot{\phi}(t) \end{bmatrix} = \begin{bmatrix} -0.254 & 0 & -1.0 & 0.182 \\ -16.02 & -8.40 & -2.19 & 0 \\ 4.488 & -0.350 & -0.760 & 0 \\ 0 & 1 & 0 & 0 \end{bmatrix} \begin{bmatrix} \Delta\beta(t) \\ \Delta p(t) \\ \Delta r(t) \\ \Delta\phi(t) \end{bmatrix}$$
$$y(t) = \Delta\phi(t)$$

where $\Delta\beta(t)$, $\Delta p(t)$, $\Delta r(t)$, and $\Delta\phi(t)$ are perturbations in sideslip, lateral angular velocities quantities, and roll angle, respectively. Determine the open-loop eigenvalues and the observability of the system. Design an estimator that places the poles at $s_1 = -10$, $s_2 = -20$, and $s_{3,4} = -10 \pm 2j$. Check the performance of this estimator through simulated runs for various initial condition errors.

5.4 In example 5.1 prove that the solutions for k_1 and k_2 solve the desired characteristic equation.

5.5 Consider the following system to be controlled:

$$\dot{\mathbf{x}}(t) = F\,\mathbf{x}(t) + B\,u(t)$$

Let $u(t) = -K\,\mathbf{x}(t)$, where K is a $1 \times n$ matrix. The closed-loop system matrix is given by $F - B\,K$ {compare this to eqn. (5.7)}. Suppose that a desired closed-loop characteristic equation is sought, with $d(s) = 0$. Following the steps in §5.2 derive Ackermann's formula for this control system. Also, derive an equivalent formula for a discrete-time system. What condition is required for K to exist (note: this control problem is the dual of the estimator design)?

5.6 Equation (5.22) represents an estimator for the predicted state. Derive a similar equation for the updated state using eqn. (5.20). Compare your result to eqn. (5.22).

5.7 ♣ Prove that $\Phi[I - KH]$ and $[I - KH]\Phi$ have the same eigenvalues.

5.8 In order to design a discrete-time estimator in eqn. (5.26), the system must be observable and the inverse of Φ must exist. Discuss the physical connotations for the inverse of Φ to exist.

5.9 Consider the following second-order continuous-time system:

$$\dot{\mathbf{x}} = \begin{bmatrix} 0 & 1 \\ 0 & 0 \end{bmatrix} \mathbf{x} + \begin{bmatrix} 0 \\ 1 \end{bmatrix} w \equiv F\,\mathbf{x} + G\,w$$

where $\mathbf{x} \equiv \begin{bmatrix} \theta & \omega \end{bmatrix}^T$ and the variance of w is given by q. Suppose we have measurements of θ only, so that $H = \begin{bmatrix} 1 & 0 \end{bmatrix}$. A simple method to study the behavior of discrete-time measurements is to assume continuous-time measurements with variance given by $R(t) = \sigma_{\text{sensor}}^2 \Delta t$, where Δt is the sampling interval. Note the relation to eqn. (5.148) for this substitution. This will be a reasonable approximation if the sampling interval is much shorter than the time constants of interest. Using this approximation, solve for all the elements of the 2×2 continuous-time steady-state covariance matrix, P, shown in Table 5.5 in terms of q, σ_{sensor}^2 and Δt.

5.10 Consider the following first-order discrete-time system:

$$x_{k+1} = \phi\, x_k + w_k$$

where w_k is a zero-mean Gaussian noise process with variance q. Derive a closed-form expression for the variance of x_k, where $p_k \equiv E\left\{x_k^2\right\}$. What is the steady-state variance? Also, discuss the properties of the steady-state value in terms of the stability of the system (i.e., in terms of ϕ).

5.11 Consider the following discrete-time model:

$$x_{k+1} = x_k$$
$$\tilde{y}_k = x_k + v_k$$

where v_k is a zero-mean Gaussian noise process with variance r. Note that this system has no process noise, so $Q = 0$. Using the discrete-time Kalman filter equations in Table 5.1 derive a closed-form recursive solution for the gain K in terms of r, P_0 (the initial error-variance) and k (the time index). Discuss the properties of this simple Kalman filter as k increases.

5.12 Consider the following truth model for a simple second-order system:

$$\mathbf{x}_{k+1} = \begin{bmatrix} 9.9985 \times 10^{-1} & 9.8510 \times 10^{-3} \\ -2.9553 \times 10^{-2} & 9.7030 \times 10^{-1} \end{bmatrix} \mathbf{x}_k + \begin{bmatrix} 4.9502 \times 10^{-5} \\ 9.8510 \times 10^{-3} \end{bmatrix} w_k$$

$$\tilde{y}_k = \begin{bmatrix} 1 & 0 \end{bmatrix} \mathbf{x}_k + v_k$$

where the sampling interval is given by 0.01 seconds. Using initial conditions of $\mathbf{x}_0 = \begin{bmatrix} 1 & 1 \end{bmatrix}^T$, create a set of 1001 synthetic measurements with the following variances for the process noise and measurement noise: $Q = 1$ and $R = 0.01$. Run the Kalman filter in Table 5.1 with the given model and assumed values for Q and R. Test the convergence of the filter for various state and covariance initial condition errors. Also, compare the computed state errors with their respective 3σ bounds computed from the covariance matrix P_k.

5.13 Repeat the simulation in exercise 5.12 using the same state model but with the following measurement model:

$$\tilde{\mathbf{y}}_k = \begin{bmatrix} 1 & 0 \\ 0 & 1 \\ 1 & 1 \end{bmatrix} \mathbf{x}_k + \mathbf{v}_k$$

where $R = \text{diag}\begin{bmatrix} 0.01 & 0.01 & 0.01 \end{bmatrix}$. Do the added measurements yield better estimates (compare the values of P_k with the previous simulation)?

5.14 Repeat the simulation in exercise 5.13 using the information filter and sequential processing algorithm shown in §5.3.3. Compare the computational loads (in terms of Floating Point Operations) of the conventional Kalman filter with both the information filter and sequential processing algorithm.

5.15 Using the truth model in exercise 5.12, with initial conditions of $\mathbf{x}_0 = \begin{bmatrix} 1 & 1 \end{bmatrix}^T$, create a set of 1001 synthetic measurements with the following variances for the process noise and measurement noise: $Q = 0$ and $R = 0.01$. Run the Kalman filter in Table 5.1 with the following assumed model:

$$\Phi = \begin{bmatrix} 9.9990 \times 10^{-1} & 9.8512 \times 10^{-3} \\ -1.9702 \times 10^{-2} & 9.7035 \times 10^{-1} \end{bmatrix}, \quad \Upsilon = \begin{bmatrix} 4.9503 \times 10^{-5} \\ 9.8512 \times 10^{-3} \end{bmatrix}$$

$$H = \begin{bmatrix} 1 & 0 \end{bmatrix}$$

Can you pick a value for Q that yields accurate estimates with this incorrect model (try various values to "tune" Q)? Compare your estimate errors with the theoretical 3σ bounds.

5.16 In example 5.3 the discrete-time process-noise covariance is shown without derivation. Fully derive this expression. Also, reproduce the results of this example using your own simulation.

5.17 Write a general program that solves the discrete-time algebraic Riccati equation using the eigenvalue/eigenvector decomposition algorithm of the Hamiltonian matrix derived in §5.3.4. Compare the steady-state values computed from your program to the values computed by the Kalman filter covariance propagation and update in problems 5.12 and 5.13.

5.18 Consider the following delayed-state measurement problem:

$$\mathbf{x}_k = \Phi_{k-1}\mathbf{x}_{k-1} + \Gamma_{k-1}\mathbf{u}_{k-1} + \Upsilon_{k-1}\mathbf{w}_{k-1}$$
$$\tilde{\mathbf{y}}_k = H_k\mathbf{x}_k + J_k\mathbf{x}_{k-1} + \mathbf{v}_k$$

where $\mathbf{w}_{k-1}$ and $\mathbf{v}_k$ are uncorrelated. Show that the measurement model can be rewritten as

$$\tilde{\mathbf{y}}_k = (H_k + J_k\Phi_{k-1}^{-1})\mathbf{x}_k + (\mathbf{v}_k - J_k\Phi_{k-1}^{-1}\mathbf{w}_{k-1})$$

What is the covariance of the new measurement error? What is the correlation between the new measurement error and process noise? Derive a correlated Kalman filter for the delayed-state measurement problem that is independent of Φ_{k-1}^{-1} (hint: use the following equation: $\Upsilon_{k-1}Q_{k-1}\Upsilon_{k-1}^T = P_k^- - \Phi_{k-1}P_{k-1}^+\Phi_{k-1}^T$).

5.19 ♣ Prove that the covariance for the correlated discrete-time Kalman filter in §5.3.5 is lower when $S_k \neq 0$ than with $S_k = 0$. Why is this true?

5.20 Fully show that the first-order approximation of eqn. (5.135) is given by eqn. (5.136).

5.21 Use the numerical solution in eqn. (5.140) to prove the analytical solution of the discrete-time process noise covariance in example 5.3.

5.22 Prove that the continuous-time Kalman filter estimation error is orthogonal to the state estimate, i.e., $E\left\{\hat{\mathbf{x}}(t)\tilde{\mathbf{x}}^T(t)\right\} = 0$, where $\tilde{\mathbf{x}}(t) \equiv \hat{\mathbf{x}}(t) - \mathbf{x}(t)$.

5.23 Using the methods of §5.4.2 find the relationship between the discrete-time correlation matrix S_k in eqn. (5.104) and the continuous-time correlation matrix $S(t)$ in eqn. (5.181).

5.24 Consider the steady-state continuous-time Kalman filter in Table 5.5 for a second-order system with $Q \equiv \text{diag}\begin{bmatrix} q_1 & q_2 \end{bmatrix}$ and $R = I$. Using the dynamical model in exercise 5.2, find closed-form values for q_1 and q_2 in terms of ω_n that yield estimator eigenvalues at $-10\omega_n$. Discuss the aspects of using the Kalman filter over Ackermann's formula for pole-placement (which method do you think is easier)?

5.25 Prove that the eigenvalues of the Hamiltonian matrix in eqn. (5.168) are symmetric about the imaginary axis (i.e., if λ is an eigenvalue of $\mathcal{H}$, then $-\lambda$ is also an eigenvalue of $\mathcal{H}$).

5.26 Write a general program that solves the continuous-time algebraic Riccati equation using the eigenvalue/eigenvector decomposition algorithm of the Hamiltonian matrix derived in §5.4.4. Check your program for the solution you found in exercise 5.24 (use any value for ω_n).

5.27 The solution for the steady-state variance in example 5.4 is given by $p = r(a+f)$, where $a = \sqrt{f^2 + r^{-1}q}$. Show that another solution is given by $p = q/(a-f)$.

5.28 ♣ Prove that the covariance for the correlated continuous-time Kalman filter in §5.4.5 is lower when $S(t) \neq 0$ than with $S(t) = 0$.

5.29 Consider the following continuous-time model with discrete-time measurements (where the state quantities are explained in exercise 5.3):

$$\begin{bmatrix} \Delta\dot{\beta}(t) \\ \Delta\dot{p}(t) \\ \Delta\dot{r}(t) \\ \Delta\dot{\phi}(t) \end{bmatrix} = \begin{bmatrix} -0.254 & 0 & -1.0 & 0.182 \\ -16.02 & -8.40 & -2.19 & 0 \\ 4.488 & -0.350 & -0.760 & 0 \\ 0 & 1 & 0 & 0 \end{bmatrix} \begin{bmatrix} \Delta\beta(t) \\ \Delta p(t) \\ \Delta r(t) \\ \Delta\phi(t) \end{bmatrix} + \begin{bmatrix} 1 \\ 0 \\ 0 \\ 0 \end{bmatrix} w(t)$$

$$\tilde{y}_k = \Delta\phi_k + v_k$$

Assume that the measurements are sampled every 0.01 seconds. Using initial conditions of $\left[\pi/180 \;\; \pi/180 \;\; \pi/180 \;\; \pi/180\right]^T$ radians, create a set of 1001 synthetic measurements with the following variances for the process noise and measurement noise: $Q = 0.001$ and $R = (0.1\pi/180)^2$ (note: Q is the continuous-time variance and R is the discrete-time covariance). Run the Kalman filter in Table 5.7 with the given model and assumed values for Q and R. Test the convergence of the filter for various state and covariance initial condition errors. Also, compare the computed state errors with their respective 3σ bounds computed from the covariance matrix $P(t)$.

5.30 Consider a linear Kalman filter with no measurements. Discuss the stability of the propagated covariance matrix with no state updates for stable, unstable, and marginally stable system-state matrices.

5.31 ♣ Using the approximations shown in §5.4.2 derive an algebraic Riccati equation for the continuous-discrete Kalman filter in Table 5.7, assuming that the system matrices F, G, and H are constants and that the noise processes are stationary. Compare your result to the algebraic Riccati equation in Table 5.5. Write a program that solves the algebraic Riccati equation you derived. Compare the steady-state values computed from your program to the values computed by the Kalman filter covariance propagation and update in exercise 5.29.

5.32 Consider the following first-order system:

$$\dot{x}(t) = x^2(t) + w(t)$$

$$\tilde{y}_k = x_k^{-1} + v_k$$

where $w(t)$ and v_k are zero-mean Gaussian noise processes with variances q and r, respectively. Derive the continuous-discrete extended Kalman filter equations in Table 5.9 for this system. Create synthetic measurements of this system for various values of x_0, P_0, q, and r. Test the performance of the extended Kalman filter using simulated computer runs. Compare the computed state errors with their respective 3σ bounds computed from the covariance matrix $P(t)$. Also, try changing the sampling interval in your simulations. Discuss the effects of the sampling interval on the overall covariance $P(t)$.

5.33 Consider the following model that is used to simulate the demodulation of angle-modulated signals:[30]

$$\begin{bmatrix} \dot{\lambda}(t) \\ \dot{\theta}(t) \end{bmatrix} \begin{bmatrix} -1/\beta & 0 \\ 1 & 0 \end{bmatrix} \begin{bmatrix} \lambda(t) \\ \theta(t) \end{bmatrix} + \begin{bmatrix} 1 \\ 0 \end{bmatrix} w(t)$$

$$\tilde{y}_k = \sqrt{2}\sin(\omega_c t_k + \theta_k) + v_k$$

where the message $\lambda(t)$ has a first-order Butterworth spectrum, being modulated as the output of a first-order, time-invariant linear system with one real pole driven by a continuous zero-mean Gaussian noise process, $w(t)$, with variance q. This message is then passed through an integrator to give $\theta(t)$, which is then employed to phase modulate a carrier signal with frequency ω_c. The measurement noise process v_k is also zero-mean Gaussian noise with variance r.

Create 1001 synthetic measurements, sampled every 0.01 seconds, of the aforementioned system using the following parameters: $\omega_c = 5$ (rad/sec), $\beta = 1$, $q = 0.5$, $r = 1$, and initial conditions of $\lambda_0 = \pi$ (rad/sec) and $\theta_0 = \pi/6$ (rad). Run the extended Kalman filter in Table 5.9 with the given model and assumed values for Q and R. Test the convergence of the filter for various initial condition errors and values for P_0. Also, compare the computed state errors with their respective 3σ bounds computed from the covariance matrix $P(t)$. Finally, is it possible to use a fully discrete-time version of the extended Kalman filter on this system?

5.34 Consider the following second-order system:

$$\dot{\mathbf{x}}(t) = \begin{bmatrix} 0 & 1 \\ -a & -b \end{bmatrix} \mathbf{x}(t) + \begin{bmatrix} 0 \\ 1 \end{bmatrix} u(t)$$

$$\tilde{y}_k = \begin{bmatrix} 1 & 0 \end{bmatrix} \mathbf{x}_k + v_k$$

Create 1001 synthetic measurements, sampled every 0.01 seconds, of the aforementioned system using the following parameters: $a = b = 3$, $R = 0.0001$, $u(t) = 0$, and $\mathbf{x}_0 = \begin{bmatrix} 1 & 1 \end{bmatrix}^T$. Append the model to include states to estimate the parameters a and b, so that the Kalman filter propagation model is given by

$$\dot{\hat{\mathbf{x}}}(t) = \begin{bmatrix} \hat{x}_2(t) \\ -\hat{x}_1(t)\,\hat{x}_3(t) - \hat{x}_2(t)\,\hat{x}_4(t) \\ 0 \\ 0 \end{bmatrix} + \begin{bmatrix} 0 \\ 1 \\ 0 \\ 0 \end{bmatrix} u(t)$$

$$\hat{y}_k = \begin{bmatrix} 1 & 0 & 0 & 0 \end{bmatrix} \hat{\mathbf{x}}_k$$

where $\hat{x}_3$ and $\hat{x}_4$ are estimates of a and b, respectively. Run the extended Kalman filter given in Table 5.9 with the given model to estimate a and b.

Use the following matrices for G and Q:

$$G = \begin{bmatrix} 0 & 0 \\ 0 & 0 \\ 1 & 0 \\ 0 & 1 \end{bmatrix}, \quad Q = \begin{bmatrix} q & 0 \\ 0 & q \end{bmatrix}$$

Try various values for q to test the performance of the extended Kalman filter. Also, compare the computed state errors with their respective 3σ bounds computed from the covariance matrix $P(t)$. Try adding a nonzero control input into the system, e.g., let $u(t) = 10\sin(t) - 8\cos(t) + 5\sin(2t) + 3\cos(2t)$. Does this help the observability of the system? Finally, try increasing R by an order of magnitude (as well as other values) and repeat the entire procedure.

5.35 Reproduce the results using the extended Kalman filter with Van Der Pol's model in examples 5.5 and 5.6 using your own simulation. Check the sensitivity of the extended Kalman filter for various initial condition errors. Can you find initial conditions that cause the filter to become unstable? For the parameter identification simulation, pick various values of q and discuss the performance of the identification results.

5.36 Consider the following first-order nonlinear system:

$$\dot{x}(t) = 0$$
$$\tilde{y}_k = \sin(x_k t_k) + v_k$$

Create 201 synthetic measurements, sampled every 0.1 seconds, of the aforementioned system using the following parameters: $t_0 = 0$, $x_k = 1$ for all time and $R = 0.1$. Develop an extended Kalman filter to estimate the frequency x_k with the following starting conditions: $\hat{x}_0 = 10$ and $P_0 = 1$ (note: $\hat{x}_{k+1}^- = \hat{x}_k^+$ and $P_{k+1}^- = P_k^+$ for this system). How does your EKF perform for this problem? Next, try an iterated Kalman filter using eqns. (5.198). Compare the performance of the iterated Kalman filter to the standard extended Kalman filter.

5.37 Consider the following formulas to simulate the effects of roundoff errors in a Kalman filter:

$$1 + \epsilon \overset{r}{\neq} 1$$
$$1 + \epsilon^2 \overset{r}{=} 1$$

where $\overset{r}{=}$ means equal to rounding and $\epsilon << 1$. Consider a scalar measurement update of a two-state problem with the following characteristics:

$$P_k^- = \begin{bmatrix} 1 & 0 \\ 0 & 1 \end{bmatrix}, \quad H = \begin{bmatrix} 1 & 0 \end{bmatrix}, \quad R = \epsilon^2$$

The exact covariance update is given by

$$P_k^+ = \begin{bmatrix} \epsilon^2/(1+\epsilon^2) & 0 \\ 0 & 1 \end{bmatrix}$$

Using the roundoff errors introduced previously compute the update covariance using: 1) the conventional Kalman filter form in eqn. (5.44), 2) Joseph's form in eqn. (5.39), 3) the SRIF factorization using eqn. (5.203), and 4) the U-D factorization using eqn. (5.214). Discuss the performance characteristics of each approach. Also, redo the problem with $H = \begin{bmatrix} 1 & 1 \end{bmatrix}$.

5.38 An SRIF approach can also be implemented for the extended Kalman filter. Derive this filter by using the inverse linearized dynamics to eliminate the state in the extended SRIF. Note: in the linear discrete-time SRIF Φ_k^{-1} is used to eliminate the state (use the inverse linearized dynamics in the extended SRIF).

5.39 Derive continuous-time versions of the colored-noise filters shown in §5.7.2.

5.40 Create synthetic measurements using the dynamical model and shaping filter discussed in example 5.7. Pick various values for the gust noise parameter a and process noise q, and run the Kalman filter given by Table 5.7 for the full model (including the shaping filter). With the same synthetic measurements run the standard Kalman filter using only the dynamical model without the shaping filter, tuning q until reasonable estimates are achieved. Under what cases does the "reduced-order" Kalman filter provide good estimates (i.e., when colored-noise process noise exists, but when only the standard Kalman filter is used)?

5.41 In example 5.8 the covariance of the measurement residual is stated to be given by

$$E_k = H_k^T P_k H_k + R_k$$

Prove that this covariance is correct.

5.42 Reproduce the results of example 5.8 using a single run of measurements. Also, try multiple (Monte Carlo) runs and use eqns. (5.235), (5.237), and (5.239) to check the consistency of the Kalman filter for various values of q. Since the truth is known for the example, consistency tests can also be applied to the state error with $\mathbf{e}_k = \hat{\mathbf{x}}_k - \mathbf{x}_k$ and $E_k = P_k$. Check the consistency of the Kalman filter using the state error with a single run and with multiple runs.

5.43 Write a computer program that computes the autocorrelation coefficients using eqns. (5.251) and (5.252). Create Gaussian noise values for $\mathbf{e}_k$ using a random noise generator and numerically check the confidence limit given by eqn. (5.253). Also, try non-Gaussian values for $\mathbf{e}_k$.

5.44 Using the adaptive methods of §5.7.4 estimate the measurement and process noise variances from exercise 5.12. How well do your estimates compare with their respective true values? Also, use the adaptive approach to find an "optimal" value for q using the synthetic measurements created with the model in exercise 5.12, but with the model in exercise 5.15 in the Kalman filter.

5.45 ♣ Derive the error analysis results of §5.7.5.

5.46 Consider the following nominal model:

$$\bar{F} = \begin{bmatrix} 0 & 1 \\ -3 & -3 \end{bmatrix}, \quad \bar{G} = \begin{bmatrix} 0 \\ 1 \end{bmatrix}, \quad \bar{H} = \begin{bmatrix} 1 & 0 \end{bmatrix}$$

with $\bar{R} = \bar{Q} = 1$. Compute the steady-state continuous-time covariance using eqn. (5.269b). Next consider the following actual system model:

$$F = \begin{bmatrix} 0 & 1 \\ -a & -3 \end{bmatrix}$$

where $a > 0$. Compute the covariance of the error introduced by this modelling error using the methods shown in §5.7.5. Also, for this system evaluate the performance of the Kalman filter for the following error cases: 1) errors in Q alone, 2) errors in Q and a together, and 3) errors in R and a together. Which case seems to be the most sensitive in the Kalman filter design?

5.47 ♣ Consider the following one-dimensional random variable y that is related to x by the following nonlinear transformation:

$$y = x^2$$

where x is a Gaussian noise process with mean μ and variance σ_x^2. Prove that the true variance of y is given by

$$\sigma_y^2 = 2\sigma_x^4 + 4\mu\sigma_x^2$$

Compute an approximation of the true σ_y^2 by linearizing the nonlinear transformation. Next, compute an approximation of the true σ_y^2 by using the methods described in §5.7.6. Which approach yields better results?

5.48 Reproduce the results using the extended Kalman filter and Unscented filter of the vertically falling-body problem in example 5.9. Check the performance of both algorithms for various sampling intervals.

5.49 Implement the Unscented filter to estimate the damping coefficient c for Van der Pol's equation in examples 5.5 and 5.6. How does the performance of the Unscented filter compare to the extended Kalman filter for various initial condition errors?

5.50 Implement the Unscented filter to estimate the frequency of the model shown in exercise 5.36. Try various values of α in your Unscented filter (even outside the recommended upper bound of 1). Compare the performance of the Unscented filter to the iterated Kalman filter and standard extended Kalman filter.

5.51 ♣ Derive the H_∞ filtering results of §5.7.7.

5.52 Derive the last inequality shown in example 5.10.

5.53 Create synthetic measurements using the model described in example 5.10. Using known errors in f compare the performance of the standard Kalman filter to the performance of the H_∞ filter. Is the H_∞ filter more robust?

5.54 Using the synthetic measurements created in exercise 5.29, run the standard Kalman filter and H_∞ filter with various errors in the assumed model. Can you find a parameter change in the assumed model that yields better performance characteristics using the using H_∞ filter over the standard Kalman filter? Discuss the effect of the parameter γ on the performance of the H_∞ filter for this system.

References

[1] Gelb, A., editor, *Applied Optimal Estimation*, The MIT Press, Cambridge, MA, 1974.

[2] Franklin, G.F., Powell, J.D., and Workman, M., *Digital Control of Dynamic Systems*, Addison Wesley Longman, Menlo Park, CA, 3rd ed., 1998.

[3] Kalman, R.E. and Bucy, R.S., "New Results in Linear Filtering and Prediction Theory," *Journal of Basic Engineering*, March 1961, pp. 95–108.

[4] Stengle, R.F., *Optimal Control and Estimation*, Dover Publications, New York, NY, 1994.

[5] Lewis, F.L., *Optimal Estimation with an Introduction to Stochastic Control Theory*, John Wiley & Sons, New York, NY, 1986.

[6] Kalman, R.E. and Joseph, P.D., *Filtering for Stochastic Processes with Applications to Guidance*, Interscience Publishers, New York, NY, 1968.

[7] Golub, G.H. and Van Loan, C.F., *Matrix Computations*, The Johns Hopkins University Press, Baltimore, MD, 3rd ed., 1996.

[8] Kailath, T., Sayed, A.H., and Hassibi, B., *Linear Estimation*, Prentice Hall, Upper Saddle River, NJ, 2000.

[9] Vaughan, D.R., "A Nonrecursive Algebraic Solution for the Discrete Riccati Equation," *IEEE Transactions on Automatic Control*, Vol. AC-15, No. 5, Oct. 1970, pp. 597–599.

[10] Fallon, L., "Gyroscopes," *Spacecraft Attitude Determination and Control*, edited by J.R. Wertz, chap. 6.5, Kluwer Academic Publishers, The Netherlands, 1978.

[11] Farrenkopf, R.L., "Analytic Steady-State Accuracy Solutions for Two Common Spacecraft Attitude Estimators," *Journal of Guidance and Control*, Vol. 1, No. 4, July-Aug. 1978, pp. 282–284.

[12] Bendat, J.S. and Piersol, A.G., *Engineering Applications of Correlation and Spectral Analysis*, John Wiley & Sons, New York, NY, 1980.

[13] van Loan, C.F., "Computing Integrals Involving the Matrix Exponential," *IEEE Transactions on Automatic Control*, Vol. AC-23, No. 3, June 1978, pp. 396–404.

[14] Brown, R.G. and Hwang, P.Y.C., *Introduction to Random Signals and Applied Kalman Filtering*, John Wiley & Sons, New York, NY, 3rd ed., 1997.

[15] Schweppe, F.C., *Uncertain Dynamic Systems*, Prentice Hall, Englewood Cliffs, NJ, 1973.

[16] Reid, W.T., *Riccati Differential Equations*, Academic Press, New York, NY, 1972.

[17] Vaughan, D.R., "A Negative Exponential Solution for the Matrix Riccati Equation," *IEEE Transactions on Automatic Control*, Vol. AC-14, No. 1, Feb. 1969, pp. 72–75.

[18] MacFarlane, A.G.J., "An Eigenvector Solution of the Optimal Linear Regulator," *Journal of Electronics and Control*, Vol. 14, No. 6, June 1963, pp. 643–654.

[19] Potter, J.E., "Matrix Quadratic Solutions," *SIAM Journal of Applied Mathematics*, Vol. 14, No. 3, May 1966, pp. 496–501.

[20] Laub, A.J., "A Schur Method for Solving Algebraic Riccati Equations," *IEEE Transactions on Automatic Control*, Vol. AC-24, No. 6, Dec. 1979, pp. 913–921.

[21] Bittanti, S., Laub, A., and Willems, J., editors, *The Riccati Equation*, Communications and Control Engineering Series, Springer-Verlag, Berlin, 1991.

[22] Wiener, N., *Extrapolation, Interpolation, and Smoothing of Stationary Time Series*, John Wiley, New York, NY, 1949.

[23] Maybeck, P.S., *Stochastic Models, Estimation, and Control*, Vol. 1, Academic Press, New York, NY, 1979.

[24] Jazwinski, A.H., *Stochastic Processes and Filtering Theory*, Academic Press, San Diego, CA, 1970.

[25] Slotine, J.J.E. and Li, W., *Applied Nonlinear Control*, Prentice Hall, Englewood Cliffs, NJ, 1991.

[26] Battin, R.H., *Astronautical Guidance*, McGraw Hill, New York, NY, 1964.

[27] Kaminski, P.G., Bryson, A.E., and Schmidt, S.F., "Discrete Square Root Filtering: A Survey of Current Techniques," *IEEE Transactions on Automatic Control*, Vol. AC-16, No. 5, Dec. 1971, pp. 727–735.

[28] Bierman, G.J., *Factorization Methods for Discrete Sequential Estimation*, Academic Press, Orlando, FL, 1977.

[29] Crassidis, J.L., Andrews, S.F., Markley, F.L., and Ha, K., "Contingency Designs for Attitude Determination of TRMM," *Proceedings of the Flight Mechanics/Estimation Theory Symposium*, NASA-Goddard Space Flight Center, Greenbelt, MD, May 1995, pp. 419–433.

[30] Anderson, B.D.O. and Moore, J.B., *Optimal Filtering*, Prentice Hall, Englewood Cliffs, NJ, 1979.

[31] Nelson, R.C., *Flight Stability and Automatic Control*, McGraw-Hill, New York, NY, 1989.

[32] Bar-Shalom, Y., Li, X.R., and Kirubarajan, T., *Estimation with Applications to Tracking and Navigation*, John Wiley & Sons, New York, NY, 2001.

[33] Devore, J.L., *Probability and Statistics for Engineering and Sciences*, Duxbury Press, Pacific Grove, CA, 1995.

[34] Mehra, R.K., "On the Identification of Variances and Adaptive Kalman Filtering," *IEEE Transactions on Automatic Control*, Vol. AC-15, No. 2, April 1970, pp. 175–184.

[35] Maybeck, P.S., *Stochastic Models, Estimation, and Control*, Vol. 2, Academic Press, New York, NY, 1982.

[36] Brown, R.J. and Sage, A.P., "Error Analysis of Modeling and Bias Errors in Continuous Time State Estimation," *Automatica*, Vol. 7, No. 5, Sept. 1971, pp. 577–590.

[37] Sage, A.P. and White, C.C., *Optimum Systems Control*, Prentice Hall, Englewood Cliffs, NJ, 2nd ed., 1977.

[38] Daum, F.E., "Exact Finite-Dimensional Nonlinear Filters," *IEEE Transactions on Automatic Control*, Vol. AC-31, No. 7, July 1986, pp. 616–622.

[39] Julier, S.J., Uhlmann, J.K., and Durrant-Whyte, H.F., "A New Approach for Filtering Nonlinear Systems," *American Control Conference*, Seattle, WA, June 1995, pp. 1628–1632.

[40] Julier, S.J., Uhlmann, J.K., and Durrant-Whyte, H.F., "A New Method for the Nonlinear Transformation of Means and Covariances in Filters and Estimators," *IEEE Transactions on Automatic Control*, Vol. AC-45, No. 3, March 2000, pp. 477–482.

[41] Bar-Shalom, Y. and Fortmann, T.E., *Tracking and Data Association*, Academic Press, Boston, MA, 1988.

[42] Wan, E. and van der Merwe, R., "The Unscented Kalman Filter," *Kalman Filtering and Neural Networks*, edited by S. Haykin, chap. 7, Wiley, 2001.

[43] Athans, M., Wishner, R.P., and Bertolini, A., "Suboptimal State Estimation for Continuous-Time Nonlinear Systems from Discrete Noisy Measurements," *IEEE Transactions on Automatic Control*, Vol. AC-13, No. 5, Oct. 1968, pp. 504–514.

[44] Nagpal, K.M. and Khargonekar, P.P., "Filtering and Smoothing in an H_∞ Setting," *IEEE Transactions on Automatic Control*, Vol. AC-36, No. 2, Feb. 1991, pp. 152–166.

[45] Francis, B.A., *A Course in H_∞ Control Theory*, Springer-Verlag, Berlin, 1987.

[46] Zhou, K., Doyle, J.C., and Glover, K., *Robust and Optimal Control*, Prentice Hall, Upper Saddle River, NJ, 1996.

[47] Kailath, T., "A View of Three Decades of Linear Filtering Theory," *IEEE Transactions on Information Theory*, Vol. IT-20, No. 2, March 1974, pp. 146–181.

[48] Chen, G., editor, *Approximate Kalman Filtering*, World Scientific, Singpore, China, 1993.

[49] Maskell, S. and Gordon, N., "A Tutorial on Particle Filters for On-line Nonlinear/Non-Gaussian Bayesian Tracking," *IEEE Transactions on Signal Processing*, Vol. 50, No. 2, Feb. 2002, pp. 174–189.

[50] Doucet, A., de Freitas, N., and Gordan, N., editors, *Sequential Monte Carlo Methods in Practice*, Springer-Verlag, New York, NY, 2001.

6

Batch State Estimation

A state without the means of some change is without the means of its conservation. Burke, Edmund

THE previous chapter allows estimation of the states in the model of a dynamical system using *sequential measurements*. We found that the sequential estimation results of §1.3 and the probability concepts introduced in Chapter 2, developed for estimation of *algebraic* systems, remain valid for estimation of *dynamical* systems upon making the appropriate new interpretations of the matrices involved in the estimation algorithms. Specifically, taking a measurement at the current time and an estimate of the state at the previous time with knowledge of its error properties, the methods of Chapter 5 are used to produce a state estimate of the dynamic system at the current time. In this chapter the results of the previous chapter are extended to batch state estimation. The disadvantage of batch estimation methods is they cannot be implemented in real time; however, they have the advantage of providing state estimates with a lower error-covariance than sequential methods. This may be extremely helpful when accuracy is an issue, but real-time application is not required.

The batch methods shown in this chapter are also known as *smoothers*, since they typically are used to "smooth" out the effects of measurement noise. Basically, smoothers are used to estimate the state quantities using measurements made before and after a certain time t. To accomplish this task, two filters are usually used (see Figure 6.1): a forward-time filter and a backward-time filter.[1] Three types of smoothers are usually defined:

1. *Fixed-Interval Smoothing*. This smoother uses the entire batch of measurements over a fixed interval to estimate all the states in the interval. The times 0 and T are fixed and t varies from time 0 to T in this formulation. Since the entire batch of measurements are used to produce an estimate, this smoother provides the best possible estimate over the interval.

2. *Fixed-Point Smoothing*. This smoother estimates the state at a specific fixed point in time t, given a batch of measurements up to the current time T. This smoother is often used to estimate the state at only one time point in the interval.

3. *Fixed-Lag Smoothing*. This smoother estimates the state at a fixed time interval that lags the time of the current measurement at time T. This smoother is often used to refine the optimal forward filter estimate.

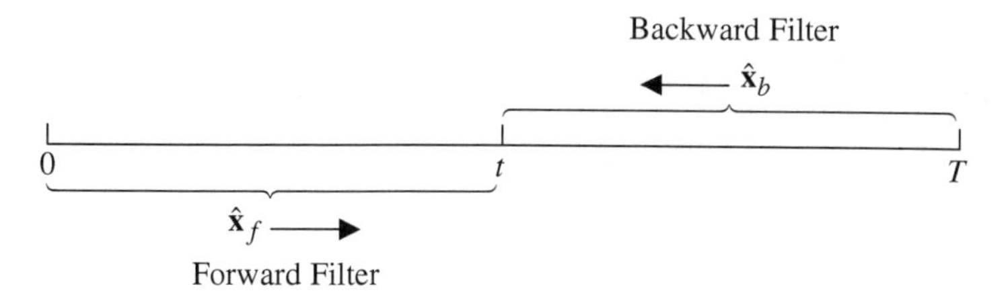

Figure 6.1: Forward-Time and Backward-Time Filtering

The fixed-point and fixed-lag smoothers are batch processes only in the sense that they require measurements up to the current time. The derivation of all of these smoothers can be given from the Kalman filter. In fact, all smoothers use the Kalman filter for forward-time filtering.

The history of smoothing actually predates the Kalman filter. Wiener[2] solved the original fixed-lag smoothing problem in the 1940s, but he only considered the stationary case where the smoother assumes that the entire past history of the input is available for weighting in its estimate.[3] The first practical smoothing algorithms are attributed to Bryson and Frazier,[4] as well as Rauch, Tung, and Striebel (RTS).[5] In particular, the RTS smoothing algorithm has maintained its popularity since the initial paper, and is most likely the most widely used algorithm for smoothing to date.

6.1 Fixed-Interval Smoothing

As mentioned previously, fixed-interval smoothing uses the entire batch of measurements over a fixed interval to estimate all the states in the interval. Fraser and Potter[6] have shown that this smoother can be derived from a combination of two Kalman filters, one of which works forward over the data and the other of which works backward over the fixed interval. Together these two filters use all the available information to provide optimal estimates. Earlier work[4, 5] gives the smoother estimate as a correction to the Kalman filter estimate for the same point, and others[7, 8] do not have the appearance of a correction to the Kalman filter estimate. All are mathematically equivalent, but the required computations are different for each approach.[9]

6.1.1 Discrete-Time Formulation

We begin our introduction of fixed-interval smoothing by considering discrete-time models and measurements, where the true system is modelled by eqn. (5.27):

$$\mathbf{x}_{k+1} = \Phi_k \mathbf{x}_k + \Gamma_k \mathbf{u}_k + \Upsilon_k \mathbf{w}_k \tag{6.1a}$$
$$\tilde{\mathbf{y}}_k = H_k \mathbf{x}_k + \mathbf{v}_k \tag{6.1b}$$

where $\mathbf{w}_k \sim N(\mathbf{0}, Q_k)$ and $\mathbf{v}_k \sim N(\mathbf{0}, R_k)$. The optimal smoother is given by a combination of the estimates of two filters: one, denoted by $\hat{\mathbf{x}}_{fk}$, is given from a filter that runs from the beginning of the data interval to time t, and the other, denoted by $\hat{\mathbf{x}}_{bk}$, that works backward from the end of the time interval. The first step of the optimal smoother involves using the forward Kalman filter summarized in Table 5.1:

forward filter

$$\hat{\mathbf{x}}^-_{fk+1} = \Phi_k \hat{\mathbf{x}}^+_{fk} + \Gamma_k \mathbf{u}_k \tag{6.2a}$$
$$P^-_{fk+1} = \Phi_k P^+_{fk} \Phi_k^T + \Upsilon_k Q_k \Upsilon_k^T \tag{6.2b}$$

$$\hat{\mathbf{x}}^+_{fk} = \hat{\mathbf{x}}^-_{fk} + K_{fk}[\tilde{\mathbf{y}}_k - H_k \hat{\mathbf{x}}^-_{fk}] \tag{6.2c}$$
$$P^+_{fk} = [I - K_{fk} H_k] P^-_{fk} \tag{6.2d}$$

$$K_{fk} = P^-_{fk} H_k^T [H_k P^-_{fk} H_k^T + R_k]^{-1} \tag{6.2e}$$

The basic Kalman filter structure incorporates a measurement update at time t_k to give $\hat{\mathbf{x}}^+_{fk}$. To derive the backward filter we solve eqn. (6.1a) for $\mathbf{x}_k$, which gives

$$\mathbf{x}_k = \Phi_k^{-1} \mathbf{x}_{k+1} - \Phi_k^{-1} \Gamma_k \mathbf{u}_k - \Phi_k^{-1} \Upsilon_k \mathbf{w}_k \tag{6.3}$$

Clearly, the inverse of Φ must exist, meaning that the state matrix has no zero eigenvalues, but we shall see that the final form of the backward filter does not depend on this condition. The backward estimate is provided by the backward-running filter just *before* the measurement at time t_k.[10] Hence, the backward-time state propagation, denoted by $\hat{\mathbf{x}}^-_{bk}$, is given by

$$\hat{\mathbf{x}}^-_{bk} = \Phi_k^{-1} \hat{\mathbf{x}}^+_{bk+1} - \Phi_k^{-1} \Gamma_k \mathbf{u}_k \tag{6.4}$$

Comparing eqn. (6.4) with eqn. (6.2a) indicates that the backward filter time update and propagation roles are reversed from the forward filter, which is due to the measurement at time t_k going backward in time.

We seek a smoothed estimate that is a function of $\hat{\mathbf{x}}^+_{fk}$ and $\hat{\mathbf{x}}^-_{bk}$. Specifically, using methods similar to the methods of §2.1.2, we seek an optimal estimate that is a linear combination of the forward and backward estimates, given by

$$\hat{\mathbf{x}}_k = M_k \hat{\mathbf{x}}^+_{fk} + N_k \hat{\mathbf{x}}^-_{bk} \tag{6.5}$$

Next, following the error state definitions in §5.3.1, with $\tilde{\mathbf{x}}_k = \hat{\mathbf{x}}_k - \mathbf{x}_k$, $\tilde{\mathbf{x}}^+_{fk} = \hat{\mathbf{x}}^+_{fk} - \mathbf{x}_k$, and $\tilde{\mathbf{x}}^-_{bk} = \hat{\mathbf{x}}^-_{bk} - \mathbf{x}_k$, leads to

$$\tilde{\mathbf{x}}_k = [M_k + N_k - I]\mathbf{x}_k + M_k \tilde{\mathbf{x}}^+_{fk} + N_k \tilde{\mathbf{x}}^-_{bk} \tag{6.6}$$

Clearly an unbiased state estimate (see §2.2) requires

$$N_k = I - M_k \tag{6.7}$$

Therefore, substituting eqn. (6.7) into eqn. (6.5) yields

$$\hat{\mathbf{x}}_k = M_k \hat{\mathbf{x}}_{fk}^{+} + [I - M_k]\hat{\mathbf{x}}_{bk}^{-} \tag{6.8}$$

We now define the following covariance expressions:

$$P_k \equiv E\left\{\tilde{\mathbf{x}}_k \tilde{\mathbf{x}}_k^T\right\} \tag{6.9a}$$

$$P_{fk}^{+} \equiv E\left\{\tilde{\mathbf{x}}_{fk}^{+} \tilde{\mathbf{x}}_{fk}^{+T}\right\} \tag{6.9b}$$

$$P_{bk}^{-} \equiv E\left\{\tilde{\mathbf{x}}_{bk}^{-} \tilde{\mathbf{x}}_{bk}^{-T}\right\} \tag{6.9c}$$

where P_k is the smoother error covariance, P_{fk}^{+} is the forward-filter error covariance, and P_{bk}^{-} is the backward-filter error covariance. Since the forward and backward processes are uncorrelated, then from eqns. (6.6) and (6.9), the smoother covariance can be written as

$$P_k = M_k P_{fk}^{+} M_k^T + [I - M_k] P_{bk}^{-} [I - M_k]^T \tag{6.10}$$

The optimal expression for M_k is given by minimizing the trace of P_k. The necessary conditions, i.e., differentiating with respect to M_k, lead to

$$0 = 2M_k P_{fk}^{+} - 2[I - M_k] P_{bk}^{-} \tag{6.11}$$

Solving eqn. (6.11) for M_k gives

$$M_k = P_{bk}^{-} [P_{fk}^{+} + P_{bk}^{-}]^{-1} \tag{6.12}$$

Also, $I - M_k$ is given by

$$\begin{aligned} I - M_k &= [P_{fk}^{+} + P_{bk}^{-}][P_{fk}^{+} + P_{bk}^{-}]^{-1} - P_{bk}^{-}[P_{fk}^{+} + P_{bk}^{-}]^{-1} \\ &= P_{fk}^{+}[P_{fk}^{+} + P_{bk}^{-}]^{-1} \end{aligned} \tag{6.13}$$

Substituting eqns. (6.12) and (6.13) into eqn. (6.10) and performing some algebraic manipulations (which are left as an exercise for the reader) yields

$$\boxed{P_k = \left[(P_{fk}^{+})^{-1} + (P_{bk}^{-})^{-1}\right]^{-1}} \tag{6.14}$$

Let us consider the physical connotation of eqn. (6.14). For scalar systems eqn. (6.14) reduces down to

$$p_k = \frac{p_{fk}^{+}\, p_{bk}^{-}}{p_{fk}^{+} + p_{bk}^{-}} \tag{6.15}$$

Equation (6.15) clearly shows that $p_k \le p_{fk}^+$ and $p_k \le p_{bk}^-$, which indicates that the smoother error covariance is always less than or equal to either the forward or backward covariance. Therefore, the smoother estimate is always better than either filter alone. This analysis can easily be expanded to higher-order systems.

Equation (6.14) involves matrix inverses of both P_{fk}^+ and P_{bk}^-. The inverse of P_{fk}^+ can be avoided though. We first define the following quantities: $\mathcal{P}_{bk}^- \equiv (P_{bk}^-)^{-1}$ and $\mathcal{P}_{fk}^+ \equiv (P_{fk}^+)^{-1}$. Then, using the matrix inversion lemma in eqn. (1.70) with $A = \mathcal{P}_{fk}^+$, $B = \mathcal{P}_{bk}^-$, and $C = D = I$ leads to

$$\boxed{P_k = P_{fk}^+ - P_{fk}^+ \mathcal{P}_{bk}^- [I + P_{fk}^+ \mathcal{P}_{bk}^-]^{-1} P_{fk}^+} \tag{6.16}$$

Note that eqn. (6.16) requires only one matrix inverse. Equation (6.16) can be further expanded into a symmetric form:

$$P_k = [I - W_k \mathcal{P}_{bk}^-] P_{fk}^+ [I - W_k \mathcal{P}_{bk}^-]^T + W_k \mathcal{P}_{bk}^- W_k^T \tag{6.17}$$

where

$$W_k = P_{fk}^+ [I + P_{fk}^+ \mathcal{P}_{bk}^-]^{-T} \tag{6.18}$$

Equation (6.17) is the sum of two positive definite matrices, which is equivalent to Joseph's stabilized version shown by eqn. (5.39), and provides a more robust approach in terms of numerical stability.

Substituting eqns. (6.12) and (6.13) into eqn. (6.8) and using eqn. (6.14) leads to

$$\hat{\mathbf{x}}_k = P_k \left[(P_{fk}^+)^{-1} \hat{\mathbf{x}}_{fk}^+ + (P_{bk}^-)^{-1} \hat{\mathbf{x}}_{bk}^- \right] \tag{6.19}$$

Equation (6.19) shows the optimal weighting of the forward and backward state estimates to produce the smoothed estimate. Equation (6.19) is also known as *Millman's theorem*,[11] which is also an exact analog to maximum likelihood of a scalar with independent measurements (see exercise 2.7 in Chapter 2 with $\rho = 0$). Equation (6.19) also involves matrix inverses of both P_{fk}^+ and P_{bk}^-. The inverse of P_{fk}^+ can be avoided by substituting eqn. (6.14) into eqn. (6.19) and factoring, which yields

$$\hat{\mathbf{x}}_k = [I + P_{fk}^+ \mathcal{P}_{bk}^-]^{-1} \hat{\mathbf{x}}_{fk}^+ + P_k \mathcal{P}_{bk}^- \hat{\mathbf{x}}_{bk}^- \tag{6.20}$$

Using the matrix inversion lemma in eqn. (1.70) with $A = I$, $B = P_{fk}^+ \mathcal{P}_{bk}^-$, and $C = D = I$ leads to

$$\boxed{\hat{\mathbf{x}}_k = [I - K_k] \hat{\mathbf{x}}_{fk}^+ + P_k \mathcal{P}_{bk}^- \hat{\mathbf{x}}_{bk}^-} \tag{6.21}$$

where the smoother gain is defined by

$$\boxed{K_k \equiv P_{fk}^+ \mathcal{P}_{bk}^- [I + P_{fk}^+ \mathcal{P}_{bk}^-]^{-1}} \tag{6.22}$$

Equation (6.21) gives the desired form for the smoothed state estimate using the combined forward and backward state estimates.

With the definitions of $\mathcal{P}_{bk}^{-}$ and $\mathcal{P}_{bk}^{+}$, the inverse of the backward update covariance follows directly from the information filter of §5.3.3, given by eqn. (5.73):

$$\mathcal{P}_{bk}^{+} = \mathcal{P}_{bk}^{-} + H_k^T R_k^{-1} H_k \tag{6.23}$$

To derive a backward recursion for $\mathcal{P}_{bk}^{-}$ we first subtract eqn. (6.3) from eqn. (6.4), and use the error definitions $\tilde{\mathbf{x}}_{bk}^{-} = \hat{\mathbf{x}}_{bk}^{-} - \mathbf{x}_k$ and $\tilde{\mathbf{x}}_{bk}^{+} = \hat{\mathbf{x}}_{bk}^{+} - \mathbf{x}_k$ to give

$$\tilde{\mathbf{x}}_{bk}^{-} = \Phi_k^{-1} \tilde{\mathbf{x}}_{bk+1}^{+} + \Phi_k^{-1} \Upsilon_k \mathbf{w}_k \tag{6.24}$$

Since $\tilde{\mathbf{x}}_{bk}^{+}$ and $\mathbf{w}_k$ are uncorrelated, then applying the definition in eqn. (6.9c) with eqn. (6.24) leads to the following backward covariance propagation:

$$P_{bk}^{-} = \Phi_k^{-1} [P_{bk+1}^{+} + \Upsilon_k Q_k \Upsilon_k^T] \Phi_k^{-T} \tag{6.25}$$

The inverse of eqn. (6.25) gives the desired result; however, straightforward implementation of this scheme requires computing P_{bk+1}^{+}, which is given by the inverse of eqn. (6.23). To overcome this undesired aspect of the smoother covariance the matrix inversion lemma in eqn. (1.70) is again used with $A = P_{bk+1}^{+}$, $B = \Upsilon_k$, $C = Q_k$ and $D = \Upsilon_k^T$, which leads to

$$\boxed{\mathcal{P}_{bk}^{-} = \Phi_k^T [I - K_{bk} \Upsilon_k^T] \mathcal{P}_{bk+1}^{+} \Phi_k} \tag{6.26}$$

where the gain K_{bk} is defined as

$$\boxed{K_{bk} = \mathcal{P}_{bk+1}^{+} \Upsilon_k [\Upsilon_k^T \mathcal{P}_{bk+1}^{+} \Upsilon_k + Q_k^{-1}]^{-1}} \tag{6.27}$$

Equation (6.27) involves the inverse of Q_k. However, Fraser[8] showed that only those states that are controllable by the process noise driving the system are smoothable (this will be clearly shown in §6.4.1 using the duality between control and estimation). Therefore, in practice Q_k must have an inverse, otherwise this controllable condition is violated. Another form of eqn. (6.27) is given by (which is left as an exercise for the reader):

$$K_{bk} = \Phi_k^{-T} \mathcal{P}_{bk}^{-} \Phi_k^{-1} \Upsilon_k Q_k \tag{6.28}$$

Equation (6.26) can be further expanded into a symmetric form (which is again left as an exercise for the reader):

$$\mathcal{P}_{bk}^{-} = \Phi_k^T [I - K_{bk} \Upsilon_k^T] \mathcal{P}_{bk+1}^{+} [I - K_{bk} \Upsilon_k^T]^T \Phi_k + \Phi_k^T K_{bk} Q_k^{-1} K_{bk}^T \Phi_k \tag{6.29}$$

Equation (6.29) is the sum of two positive definite matrices, which provides a more robust approach in terms of numerical stability.

Before we can continue with the backward filter update, we must first discuss boundary conditions. The forward filter is implemented using the same initial conditions as given in Table 5.1, with state and covariance initial conditions of $\hat{\mathbf{x}}_{f0}$ and

P_{f0}, respectively, which can be applied to either the updated or propagation state estimate (depending on whether or not a measurement occurs at the initial time). Let t_N denote the terminal time. Since at time $t_k = t_N$ the smoother estimate must be the same as the forward Kalman filter, this clearly requires that $\hat{\mathbf{x}}_N = \hat{\mathbf{x}}_{fN}^+$ and $P_N = P_{fN}^+$. From eqn. (6.14) the covariance condition at the terminal time can only be satisfied when $(P_{bN}^-)^{-1} \equiv \mathcal{P}_{bN}^- = 0$. However, the backward terminal state boundary condition, $\hat{\mathbf{x}}_{bN}$, is yet unknown for the following backward measurement update:

$$\hat{\mathbf{x}}_{bk}^+ = \hat{\mathbf{x}}_{bk}^- + K_{bk}[\tilde{\mathbf{y}}_k - H_k\hat{\mathbf{x}}_{bk}^-] \tag{6.30}$$

To overcome this difficulty consider the alternative state update form that is given by eqn. (5.53), rewritten as

$$\hat{\mathbf{x}}_{bk}^+ = P_{bk}^+[\mathcal{P}_{bk}^-\hat{\mathbf{x}}_{bk}^- + H_k^T R_k^{-1}\tilde{\mathbf{y}}_k] \tag{6.31}$$

where the definition of $\mathcal{P}_{bk}^-$ has been used. Left multiplying both sides of eqn. (6.31) by the inverse of P_{bk}^+, and using the definition of $\mathcal{P}_{bk}^+$, gives

$$\mathcal{P}_{bk}^+\hat{\mathbf{x}}_{bk}^+ = \mathcal{P}_{bk}^-\hat{\mathbf{x}}_{bk}^- + H_k^T R_k^{-1}\tilde{\mathbf{y}}_k \tag{6.32}$$

Define the following new variables:

$$\hat{\chi}_{bk}^+ \equiv \mathcal{P}_{bk}^+\hat{\mathbf{x}}_{bk}^+ \tag{6.33a}$$

$$\hat{\chi}_{bk}^- \equiv \mathcal{P}_{bk}^-\hat{\mathbf{x}}_{bk}^- \tag{6.33b}$$

Using the definitions in eqn. (6.33), then eqn. (6.32) can be rewritten as

$$\boxed{\hat{\chi}_{bk}^+ = \hat{\chi}_{bk}^- + H_k^T R_k^{-1}\tilde{\mathbf{y}}_k} \tag{6.34}$$

Since $\mathcal{P}_{bN}^- = 0$ then from eqn. (6.33b) we have $\hat{\chi}_{bN}^- = \mathbf{0}$, which is valid for any value of $\hat{\mathbf{x}}_{bN}^-$. The backward update is given by eqn. (6.34). A backward propagation must now be derived. Substituting eqn. (6.30) into eqn. (6.33b), and using the definition in eqn. (6.33a) yields

$$\hat{\chi}_{bk}^- = \mathcal{P}_{bk}^-\Phi_k^{-1}\left[(\mathcal{P}_{bk+1}^+)^{-1}\hat{\chi}_{bk+1}^+ - \Gamma_k\mathbf{u}_k\right] \tag{6.35}$$

Substituting eqn. (6.26) into eqn. (6.35) gives the desired form:

$$\boxed{\hat{\chi}_{bk}^- = \Phi_k^T[I - K_{bk}\Upsilon_k^T][\hat{\chi}_{bk+1}^+ - \mathcal{P}_{bk+1}^+\Gamma_k\mathbf{u}_k]} \tag{6.36}$$

Equations (6.23), (6.26), (6.27), (6.34), and (6.36) define the backward filter.

A summary of the discrete-time fixed-interval smoother is given in Table 6.1. First, the basic discrete-time Kalman filter is executed forward in time on the data set using eqn. (6.2). Then, the backward filter is run with the gain given by eqn. (6.27). In order to avoid undesirable matrix inversions, the backward updates are implemented

Table 6.1: Discrete-Time Fixed-Interval Smoother

Model	$\mathbf{x}_{k+1} = \Phi_k \mathbf{x}_k + \Gamma_k \mathbf{u}_k + \Upsilon_k \mathbf{w}_k, \quad \mathbf{w}_k \sim N(\mathbf{0}, Q_k)$ $\tilde{\mathbf{y}}_k = H_k \mathbf{x}_k + \mathbf{v}_k, \quad \mathbf{v}_k \sim N(\mathbf{0}, R_k)$
Forward Initialize	$\hat{\mathbf{x}}_f(t_0) = \hat{\mathbf{x}}_{f0}$ $P_f(t_0) = E\{\tilde{\mathbf{x}}_f(t_0)\tilde{\mathbf{x}}_f^T(t_0)\}$
Gain	$K_{fk} = P_{fk}^- H_k^T [H_k P_{fk}^- H_k^T + R_k]^{-1}$
Forward Update	$\hat{\mathbf{x}}_{fk}^+ = \hat{\mathbf{x}}_{fk}^- + K_{fk}[\tilde{\mathbf{y}}_k - H_k \hat{\mathbf{x}}_{fk}^-]$ $P_{fk}^+ = [I - K_{fk} H_k] P_{fk}^-$
Forward Propagation	$\hat{\mathbf{x}}_{fk+1}^- = \Phi_k \hat{\mathbf{x}}_{fk}^+ + \Gamma_k \mathbf{u}_k$ $P_{fk+1}^- = \Phi_k P_{fk}^+ \Phi_k^T + \Upsilon_k Q_k \Upsilon_k^T$
Backward Initialize	$\hat{\chi}_{bN}^- = \mathbf{0}$ $\mathcal{P}_{bN}^- = 0$
Gain	$K_{bk} = \mathcal{P}_{bk+1}^+ \Upsilon_k [\Upsilon_k^T \mathcal{P}_{bk+1}^+ \Upsilon_k + Q_k^{-1}]^{-1}$
Backward Update	$\hat{\chi}_{bk}^+ = \hat{\chi}_{bk}^- + H_k^T R_k^{-1} \tilde{\mathbf{y}}_k$ $\mathcal{P}_{bk}^+ = \mathcal{P}_{bk}^- + H_k^T R_k^{-1} H_k$
Backward Propagation	$\hat{\chi}_{bk}^- = \Phi_k^T [I - K_{bk} \Upsilon_k^T][\hat{\chi}_{bk+1}^+ - \mathcal{P}_{bk+1}^+ \Gamma_k \mathbf{u}_k]$ $\mathcal{P}_{bk}^- = \Phi_k^T [I - K_{bk} \Upsilon_k^T] \mathcal{P}_{bk+1}^+ \Phi_k$
Gain	$K_k = P_{fk}^+ \mathcal{P}_{bk}^- [I + P_{fk}^+ \mathcal{P}_{bk}^-]^{-1}$
Covariance	$P_k = [I - K_k] P_{fk}^+$
Estimate	$\hat{\mathbf{x}}_k = [I - K_k] \hat{\mathbf{x}}_{fk}^+ + P_k \hat{\chi}_{bk}^-$

using eqns. (6.23) and (6.34), and the backward propagations are implemented using eqns. (6.26) [or using eqn. (6.29) if numerical stability is of concern] and (6.36). The forward and backward covariances and estimates must be stored in order to evaluate the smoother covariance and estimate. The optimal smoother covariance is computed using eqn. (6.16) [or using eqn. (6.17) if numerical stability is of concern]. Finally, the optimal smoother estimate is computed using eqn. (6.21).

6.1.1.1 Steady-State Fixed-Interval Smoother

If the system matrices and covariance are time-invariant, then a steady-state (i.e., constant gain) smoother can be used, which significantly reduces the computational burden. The steady-state forward filter has been derived in §5.3.4. The only issue for the backward filter is the steady-state Riccati equation for $\mathcal{P}_b^-$. At steady-state from eqn. (6.26) we have

$$\mathcal{P}_b^- = \Phi^T \mathcal{P}_b^+ \Phi - \Phi^T \mathcal{P}_b^+ \Upsilon \left[\Upsilon^T \mathcal{P}_b^+ \Upsilon + Q^{-1} \right]^{-1} \Upsilon^T \mathcal{P}_b^+ \Phi \tag{6.37}$$

Using eqn. (6.23) in eqn. (6.37) yields

$$\mathcal{P}_b^+ = \Phi^T \mathcal{P}_b^+ \Phi - \Phi^T \mathcal{P}_b^+ \Upsilon \left[\Upsilon^T \mathcal{P}_b^+ \Upsilon + Q^{-1} \right]^{-1} \Upsilon^T \mathcal{P}_b^+ \Phi + H^T R^{-1} H \tag{6.38}$$

Comparing eqn. (6.38) to the Riccati (covariance) equation in Table 5.2 and using a similar transformation as eqn. (5.81) yields the following Hamiltonian matrix:

$$\boxed{\mathcal{H} \equiv \begin{bmatrix} \Phi^{-1} & \Phi^{-1} \Upsilon Q \Upsilon^T \\ H^T R^{-1} H \Phi^{-1} & \Phi^T + H^T R^{-1} H \Phi^{-1} \Upsilon Q \Upsilon^T \end{bmatrix}} \tag{6.39}$$

An eigenvalue/eigenvector decomposition of eqn. (6.39) gives

$$\mathcal{H} = \begin{bmatrix} W_{11} & W_{12} \\ W_{21} & W_{22} \end{bmatrix} \begin{bmatrix} \Lambda & 0 \\ 0 & \Lambda^{-1} \end{bmatrix} \begin{bmatrix} W_{11} & W_{12} \\ W_{21} & W_{22} \end{bmatrix}^{-1} \tag{6.40}$$

where Λ is a diagonal matrix of the n eigenvalues outside of the unit circle, and W_{11}, W_{21}, W_{12}, and W_{22} are block elements of the eigenvector matrix. From the derivations of §5.3.4 the steady-state value for $\mathcal{P}_b^+$ is given by

$$\boxed{\mathcal{P}_b^+ = W_{21} W_{11}^{-1}} \tag{6.41}$$

which requires an inverse of an $n \times n$ matrix. To determine the steady-state value for $\mathcal{P}_b^-$ we simply use eqn. (6.23), with

$$\boxed{\mathcal{P}_b^- = \mathcal{P}_b^+ - H^T R^{-1} H} \tag{6.42}$$

The smoother covariance and estimate can now be computed using the steady-state values for P_f^+ and $\mathcal{P}_b^-$. Note that the steady-state value for P in Table 5.2 gives P_f^-, but P_f^+ can be calculated by using eqn. (5.44).

6.1.1.2 RTS Fixed-Interval Smoother

Several other forms of the fixed-interval smoother exist. One of the most convenient forms is given by Rauch, Tung, and Striebel (RTS),[5] who combine the backward filter and smoother into one single backward recursion. Our first task is to determine a recursive expression for the smoother covariance that is independent of the backward covariance. To accomplish this task eqn. (6.16) is rewritten as

$$P_k = P^+_{fk} - P^+_{fk}[P^+_{fk} + P^-_{bk}]^{-1}P^+_{fk} \tag{6.43}$$

We now concentrate our attention on the matrix inverse expression in eqn. (6.43). Substituting eqn. (6.25) into this matrix inversion expression and factoring out Φ_k on both sides yields

$$[P^+_{fk} + P^-_{bk}]^{-1} = \Phi_k^T[\Phi_k P^+_{fk}\Phi_k^T + P^+_{bk+1} + \Upsilon_k Q_k \Upsilon_k^T]^{-1}\Phi_k \tag{6.44}$$

Using eqn. (6.2b) in eqn. (6.44) gives

$$[P^+_{fk} + P^-_{bk}]^{-1} = \Phi_k^T[P^-_{fk+1} + P^+_{bk+1}]^{-1}\Phi_k \tag{6.45}$$

A more convenient form for P^+_{bk+1} is required. Solving eqn. (5.73) for $H_k^T R_k^{-1} H_k$, and substituting the resultant into eqn. (6.23) yields

$$P^+_{bk} = [\mathcal{P}^-_{bk} + \mathcal{P}^+_{fk} - \mathcal{P}^-_{fk}]^{-1} \tag{6.46}$$

Using eqn. (6.14) in eqn. (6.46) yields

$$P^+_{bk} = [P_k^{-1} - \mathcal{P}^-_{fk}]^{-1} \tag{6.47}$$

Taking one time-step ahead of eqn. (6.47), and substituting the resulting expression into eqn. (6.45) gives

$$[P^+_{fk} + P^-_{bk}]^{-1} = \Phi_k^T \left\{P^-_{fk+1} + [P^{-1}_{k+1} - \mathcal{P}^-_{fk+1}]^{-1}\right\}^{-1}\Phi_k \tag{6.48}$$

Factoring $\mathcal{P}^-_{fk+1}$ yields

$$[P^+_{fk} + P^-_{bk}]^{-1} = \Phi_k^T \mathcal{P}^-_{fk+1}\left\{\mathcal{P}^-_{fk+1} + \mathcal{P}^-_{fk+1}[P^{-1}_{k+1} - \mathcal{P}^-_{fk+1}]^{-1}\mathcal{P}^-_{fk+1}\right\}^{-1}\mathcal{P}^-_{fk+1}\Phi_k \tag{6.49}$$

Then, using the matrix inversion lemma in eqn. (1.70) with $A = P^-_{fk+1}$, $B = D = I$ and $C = -P_{k+1}$ leads to

$$[P^+_{fk} + P^-_{bk}]^{-1} = \Phi_k^T \mathcal{P}^-_{fk+1}[P^-_{fk+1} - P_{k+1}]\mathcal{P}^-_{fk+1}\Phi_k \tag{6.50}$$

Substituting eqn. (6.50) into eqn. (6.43) yields

$$\boxed{P_k = P^+_{fk} - \mathcal{K}_k[P^-_{fk+1} - P_{k+1}]\mathcal{K}_k^T} \tag{6.51}$$

where the gain matrix $\mathcal{K}_k$ is defined as

$$\boxed{\mathcal{K}_k \equiv P^+_{fk}\Phi_k^T(P^-_{fk+1})^{-1}} \tag{6.52}$$

Note that eqn. (6.51) is no longer a function of the backward covariance P_{bk}^+ or P_{bk}^-. Therefore, the smoother covariance can be solved directly from knowledge of the forward covariance alone, which provides a very computationally efficient algorithm.

The RTS smoother state-estimate equation is given by

$$\boxed{\hat{\mathbf{x}}_k = \hat{\mathbf{x}}_{fk}^+ + \mathcal{K}_k[\hat{\mathbf{x}}_{k+1} - \hat{\mathbf{x}}_{fk+1}^-]} \tag{6.53}$$

The proof of this form begins by comparing this eqn. (6.53) to eqn. (6.21). From this comparison we need to prove that the following relationship is true:

$$-K_k\hat{\mathbf{x}}_{fk}^+ + P_k\hat{\chi}_{bk}^- = \mathcal{K}_k[\hat{\mathbf{x}}_{k+1} - \hat{\mathbf{x}}_{fk+1}^-] \tag{6.54}$$

Substituting eqns. (6.22), (6.51), and (6.52) into eqn. (6.54), and simplifying gives

$$\begin{aligned}&-\mathcal{P}_{bk}^-[I + P_{fk}^+\mathcal{P}_{bk}^-]^{-1}\hat{\mathbf{x}}_{fk}^+ + \hat{\chi}_{bk}^- - \Phi_k^T\mathcal{P}_{fk+1}^-[P_{fk+1}^- - P_{k+1}]\mathcal{P}_{fk+1}^-\Phi_k P_{fk}^+\hat{\chi}_{bk}^- \\ &= \Phi_k^T\mathcal{P}_{fk+1}^-[\hat{\mathbf{x}}_{k+1} - \hat{\mathbf{x}}_{fk+1}^-]\end{aligned} \tag{6.55}$$

We will return to eqn. (6.55), but for the time being let's concentrate on determining a more useful expression for $\hat{\mathbf{x}}_{k+1}$, which will be used to help simplify eqn. (6.55). Taking one time-step ahead of eqn. (6.19) gives

$$\hat{\mathbf{x}}_{k+1} = P_{k+1}\mathcal{P}_{fk+1}^+\hat{\mathbf{x}}_{fk+1}^+ + P_{k+1}\hat{\chi}_{bk+1}^- \tag{6.56}$$

Taking one time-step ahead of eqn. (6.34) and solving for $\hat{\chi}_{bk+1}^-$ gives

$$\hat{\chi}_{bk+1}^- = \hat{\chi}_{bk+1}^+ - H_{k+1}^T R_{k+1}^{-1}\tilde{\mathbf{y}}_{k+1} \tag{6.57}$$

Taking one time-step ahead of eqn. (5.30b), with the gain given by eqn. (5.47), and substituting the resultant and eqn. (6.57) into eqn. (6.56) yields

$$\hat{\mathbf{x}}_{k+1} = P_{k+1}\left[\mathcal{P}_{fk+1}^+ - H_{k+1}^T R_{k+1}^{-1} H_{k+1}\right]\hat{\mathbf{x}}_{fk+1}^- + P_{k+1}\hat{\chi}_{bk+1}^+ \tag{6.58}$$

Using one time-step ahead of eqn. (5.73) in eqn. (6.58) now gives a simpler form:

$$\hat{\mathbf{x}}_{k+1} = P_{k+1}\mathcal{P}_{fk+1}^-\hat{\mathbf{x}}_{fk+1}^- + P_{k+1}\hat{\chi}_{bk+1}^+ \tag{6.59}$$

Subtracting $\hat{\mathbf{x}}_{fk+1}^-$ from both sides of eqn. (6.59) and factoring out $\mathcal{P}_{fk+1}^-$ yields

$$\hat{\mathbf{x}}_{k+1} - \hat{\mathbf{x}}_{fk+1}^- = [P_{k+1} - P_{fk+1}^-]\mathcal{P}_{fk+1}^-\hat{\mathbf{x}}_{fk+1}^- + P_{k+1}\hat{\chi}_{bk+1}^+ \tag{6.60}$$

Next, rewrite the forward-time prediction, given by eqn. (5.30a), as

$$\hat{\mathbf{x}}_{fk}^+ = \Phi_k^{-1}\hat{\mathbf{x}}_{fk+1}^- - \Phi_k^{-1}\Gamma_k\mathbf{u}_k \tag{6.61}$$

Substituting eqns. (6.60) and (6.61) into eqn. (6.55), and multiplying in $\mathcal{P}_{bk}^-$ yields

$$\begin{aligned}
&-[P_{bk}^- + P_{fk}^+]^{-1}\Phi_k^{-1}\hat{\mathbf{x}}_{fk+1}^- + [P_{bk}^- + P_{fk}^+]^{-1}\Phi_k^{-1}\Gamma_k\mathbf{u}_k \\
&+\hat{\chi}_{bk}^- - \Phi_k^T\mathcal{P}_{fk+1}^-[P_{fk+1}^- - P_{k+1}]\mathcal{P}_{fk+1}^-\Phi_k P_{fk}^+\hat{\chi}_{bk}^- \\
&= \Phi_k^T\mathcal{P}_{fk+1}^-[P_{k+1} - P_{fk+1}^-]\mathcal{P}_{fk+1}^-\hat{\mathbf{x}}_{fk+1}^- + \Phi_k^T\mathcal{P}_{fk+1}^- P_{k+1}\hat{\chi}_{bk+1}^+
\end{aligned} \tag{6.62}$$

Using eqn. (6.50) in eqn. (6.62), and simplifying yields

$$[P_{bk}^- + P_{fk}^+]^{-1}\Phi_k^{-1}\Gamma_k\mathbf{u}_k + \hat{\chi}_{bk}^- - [P_{bk}^- + P_{fk}^+]^{-1}P_{fk}^+\hat{\chi}_{bk}^- = \Phi_k^T\mathcal{P}_{fk+1}^- P_{k+1}\hat{\chi}_{bk+1}^+ \tag{6.63}$$

Using eqn. (6.45) in eqn. (6.63), and left multiplying both sides of the resulting equation by $[P_{fk+1}^- + P_{bk+1}^+]\Phi_k^{-T}$ yields

$$\Gamma_k\mathbf{u}_k + \left([P_{fk+1}^- + P_{bk+1}^+]\Phi_k^{-T} - \Phi_k P_{fk}^+\right)\hat{\chi}_{bk}^- = [P_{fk+1}^- + P_{bk+1}^+]\mathcal{P}_{fk+1}^- P_{k+1}\hat{\chi}_{bk+1}^+ \tag{6.64}$$

Next, rewrite the forward-time covariance prediction, given by eqn. (5.35), as

$$P_{fk}^+ = \Phi_k^{-1}P_{fk+1}^-\Phi_k^{-T} - \Phi_k^{-1}\Upsilon_k Q_k \Upsilon_k^T\Phi_k^{-T} \tag{6.65}$$

Substituting eqn. (6.65) into eqn. (6.64), left multiplying both sides of the resulting equation by $\mathcal{P}_{bk+1}^+$, using eqn. (6.47) with one time-step ahead and solving for $\hat{\chi}_{bk}^-$ yields

$$\hat{\chi}_{bk}^- = \Phi_k^T[I + \mathcal{P}_{bk+1}^+\Upsilon_k Q_k \Upsilon_k^T]^{-1}[\hat{\chi}_{bk+1}^+ - \mathcal{P}_{bk+1}^+\Gamma_k\mathbf{u}_k] \tag{6.66}$$

Finally, using the matrix inversion lemma in eqn. (1.70) with $A = I$, $B = \mathcal{P}_{bk+1}^+\Upsilon_k$, $C = Q_k$, and $D = \Upsilon_k^T$ gives the same form as eqn. (6.36), which completes the proof.

A summary of the RTS smoother is given in Table 6.2. As before, the forward Kalman filter is executed using the measurements until time T. Storing the propagated and updated state estimates from the forward filter, the smoothed estimate is then determined by executing eqn. (6.53) backward in time. In order to determine the RTS smoothed estimate, the forward filter covariance update and propagation, as well as the state matrix, do not need to be stored. This is due to the fact that the gain in eqn. (6.52) can be computed during the forward filter process and stored to be used in the smoother estimate equation. One of the extraordinary results of the smoother state estimate is the fact that the smoother state in eqn. (6.53) does not involve the smoother covariance P_k! Therefore, eqn. (6.51) is only used to derive the smoother covariance, which may be required for analysis purposes, but is not used to find the optimal smoother state estimate. For all these reasons the RTS smoother is more widely used in practice over the formulation given in Table 6.1. Note, in §6.4.1 we will derive the RTS smoother from optimal control theory, which shows the duality between control and estimation.

6.1.1.3 Stability

The backward state matrix in the RTS smoother defines the stability of the system, which is given by $P_{fk}^+\Phi_k^T\mathcal{P}_{fk+1}^-$. Note that the smoother state estimate in eqn. (6.53)

Table 6.2: Discrete-Time RTS Smoother

Model	$\mathbf{x}_{k+1} = \Phi_k \mathbf{x}_k + \Gamma_k \mathbf{u}_k + \Upsilon_k \mathbf{w}_k, \quad \mathbf{w}_k \sim N(\mathbf{0}, Q_k)$ $\tilde{\mathbf{y}}_k = H_k \mathbf{x}_k + \mathbf{v}_k, \quad \mathbf{v}_k \sim N(\mathbf{0}, R_k)$
Forward Initialize	$\hat{\mathbf{x}}_f(t_0) = \hat{\mathbf{x}}_{f0}$ $P_f(t_0) = E\{\tilde{\mathbf{x}}_f(t_0)\tilde{\mathbf{x}}_f^T(t_0)\}$
Gain	$K_{fk} = P_{fk}^- H_k^T [H_k P_{fk}^- H_k^T + R_k]^{-1}$
Forward Update	$\hat{\mathbf{x}}_{fk}^+ = \hat{\mathbf{x}}_{fk}^- + K_{fk}[\tilde{\mathbf{y}}_k - H_k \hat{\mathbf{x}}_{fk}^-]$ $P_{fk}^+ = [I - K_{fk} H_k] P_{fk}^-$
Forward Propagation	$\hat{\mathbf{x}}_{fk+1}^- = \Phi_k \hat{\mathbf{x}}_{fk}^+ + \Gamma_k \mathbf{u}_k$ $P_{fk+1}^- = \Phi_k P_{fk}^+ \Phi_k^T + \Upsilon_k Q_k \Upsilon_k^T$
Smoother Initialize	$\hat{\mathbf{x}}_N = \hat{\mathbf{x}}_{fN}^+$ $P_N = P_{fN}^+$
Gain	$\mathcal{K}_k \equiv P_{fk}^+ \Phi_k^T (P_{fk+1}^-)^{-1}$
Covariance	$P_k = P_{fk}^+ - \mathcal{K}_k [P_{fk+1}^- - P_{k+1}] \mathcal{K}_k^T$
Estimate	$\hat{\mathbf{x}}_k = \hat{\mathbf{x}}_{fk}^+ + \mathcal{K}_k [\hat{\mathbf{x}}_{k+1} - \hat{\mathbf{x}}_{fk+1}^-]$

is a backward recursion, which is stable if and only if all the eigenvalues of the state matrix are within the unit circle. The reader should not be confused by the fact that eqn. (6.53) is executed backward in time. All discrete-time recursions, whether executed forward or backward in time, must have state matrix eigenvalues within the unit circle to be stable. Considering only the homogeneous part of eqn. (6.53), the RTS smoother is stable if the following recursion is stable:

$$\hat{\mathbf{x}}_k = P_{fk}^+ \Phi_k^T \mathcal{P}_{fk+1}^- \hat{\mathbf{x}}_{k+1} \tag{6.67}$$

The smoother stability can be proved by using Lyapunov's direct method, which is discussed for discrete-time systems in §3.6. For the discrete-time RTS smoother we consider the following candidate Lyapunov function:

$$V(\hat{\mathbf{x}}) = \hat{\mathbf{x}}_{k+1}^T \mathcal{P}_{fk+1}^+ \hat{\mathbf{x}}_{k+1} \tag{6.68}$$

The increment of $V(\hat{\mathbf{x}})$ is given by

$$\Delta V(\hat{\mathbf{x}}) = \hat{\mathbf{x}}_{k+1}^T \mathcal{P}_{fk+1}^+ \hat{\mathbf{x}}_{k+1} - \hat{\mathbf{x}}_k^T \mathcal{P}_{fk}^+ \hat{\mathbf{x}}_k \tag{6.69}$$

Substituting eqn. (6.67) into eqn. (6.69) gives

$$\Delta V(\hat{\mathbf{x}}) = \hat{\mathbf{x}}_{k+1}^T \left[\mathcal{P}_{fk+1}^- \Phi_k P_{fk}^+ \Phi_k^T \mathcal{P}_{fk+1}^- - \mathcal{P}_{fk+1}^+ \right] \hat{\mathbf{x}}_{k+1} \tag{6.70}$$

Substituting eqn. (6.65) into eqn. (6.70) gives

$$\Delta V(\hat{\mathbf{x}}) = \hat{\mathbf{x}}_{k+1}^T \left[\mathcal{P}_{fk+1}^- - \mathcal{P}_{fk+1}^- \Upsilon_k Q_k \Upsilon_k^T \mathcal{P}_{fk+1}^- - \mathcal{P}_{fk+1}^+ \right] \hat{\mathbf{x}}_{k+1} \tag{6.71}$$

Taking one time-step ahead of the expression in eqn. (5.73) and substituting the resultant into eqn. (6.71) leads to

$$\Delta V(\hat{\mathbf{x}}) = -\hat{\mathbf{x}}_{k+1}^T \left[H_{k+1}^T R_{k+1}^{-1} H_{k+1} + \mathcal{P}_{fk+1}^- \Upsilon_k Q_k \Upsilon_k^T \mathcal{P}_{fk+1}^- \right] \hat{\mathbf{x}}_{k+1} \tag{6.72}$$

Clearly if R_{k+1} is positive definite and Q_k is at least positive semi-definite, then the Lyapunov condition is satisfied and the discrete-time RTS smoother is stable.

Example 6.1: In this example the model used in example 5.3 is used to demonstrate the power of the fixed-point smoother. For this simulation we are interested in investigating the covariance of the smoother. Therefore, both the forward-time updated and propagated covariance must be stored. The smoothed state estimates and covariance are computed using the RTS formulation. A plot of the smoother attitude-angle error and 3σ bounds is shown in Figure 6.2. Comparing the smoother 3σ bounds with the ones shown in Figure 5.3 indicates that the smoother clearly provides better estimates than the Kalman filter alone. Note that the steady-state covariance can be used for this system with little loss in accuracy. Using the methods of §5.3.4 the steady-state value for the steady-state forward-time propagated covariance, P_f^-, can be computed by solving the algebraic Riccati equation in Table 5.2. Then, the steady-state forward-time updated covariance, P_f^+, can be computed from eqn. (5.44). Finally, the steady-state smoother covariance can be computed by solving the following Lyapunov equation:

$$P = K\,P\,K^T + \left(P_f^+ - K\,P_f^-\,K^T\right)$$

with

$$K = P_f^+ \Phi^T \mathcal{P}_f^-$$

Performing these calculations give a 3σ attitude bound of 4.9216 μrad, which is verified by Figure 6.2. A more dramatic result for the advantages of using the smoother is shown for the bias estimate, given by the bottom plot of Figure 6.3 (the top plot shows the Kalman filter estimate). Clearly, the smoother estimate is far superior than the Kalman filter estimate, which can be very useful for calibration purposes.

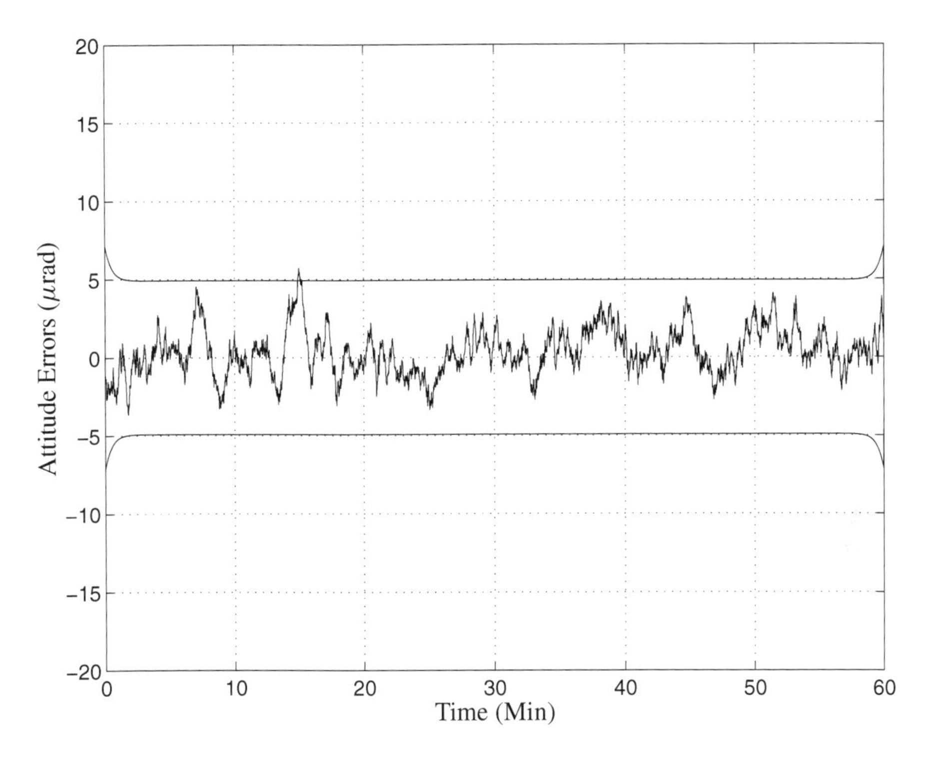

Figure 6.2: Smoother Attitude Error and Bounds

6.1.2 Continuous-Time Formulation

The true system for the continuous-time models and measurements is given by eqn. (5.117):

$$\frac{d}{dt}\mathbf{x}(t) = F(t)\,\mathbf{x}(t) + B(t)\,\mathbf{u}(t) + G(t)\,\mathbf{w}(t) \tag{6.73a}$$

$$\tilde{\mathbf{y}}(t) = H(t)\,\mathbf{x}(t) + \mathbf{v}(t) \tag{6.73b}$$

where $\mathbf{w}(t) \sim N(\mathbf{0}, Q(t))$ and $\mathbf{v}(t) \sim N(\mathbf{0}, R(t))$. The optimal smoother is again given by a combination of the estimates of two filters: one, denoted by $\hat{\mathbf{x}}_f(t)$, is given from a filter that runs from the beginning of the data interval to time t, and the other, denoted by $\hat{\mathbf{x}}_b(t)$, that works backward from the end of the time interval. These two filters follow the continuous-time form of the Kalman filter, given in §5.4.1:

forward filter

$$\frac{d}{dt}\hat{\mathbf{x}}_f(t) = F(t)\,\hat{\mathbf{x}}_f(t) + B(t)\,\mathbf{u}(t) + K_f(t)[\tilde{\mathbf{y}}(t) - H(t)\,\hat{\mathbf{x}}_f(t)] \tag{6.74a}$$

$$K_f(t) = P_f(t)\,H^T(t)\,R^{-1}(t) \tag{6.74b}$$

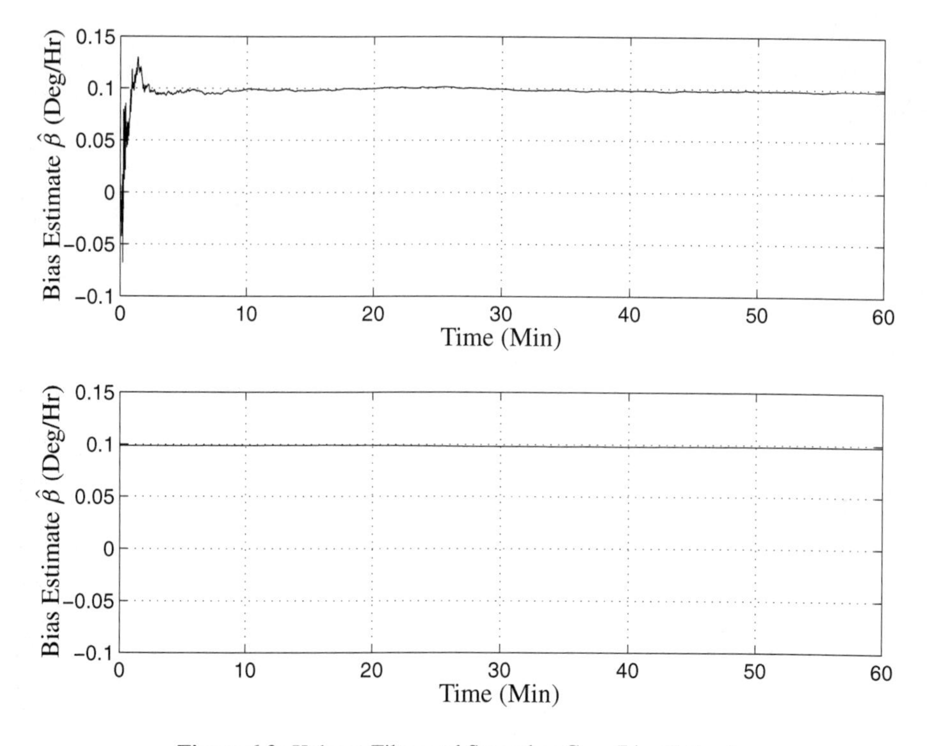

Figure 6.3: Kalman Filter and Smoother Gyro Bias Estimates

$$
\begin{aligned}
\frac{d}{dt}P_f(t) = F(t)\,P_f(t) + P_f(t)\,F^T(t) \\
- P_f(t)\,H^T(t)\,R^{-1}(t)\,H(t)\,P_f(t) + G(t)\,Q(t)\,G^T(t)
\end{aligned}
\tag{6.74c}
$$

backward filter

$$
\frac{d}{dt}\hat{\mathbf{x}}_b(t) = F(t)\,\hat{\mathbf{x}}_b(t) + B(t)\,\mathbf{u}(t) + K_b(t)[\tilde{\mathbf{y}}(t) - H(t)\,\hat{\mathbf{x}}_b(t)] \tag{6.75a}
$$

$$
K_b(t) = P_b(t)\,H^T(t)\,R^{-1}(t) \tag{6.75b}
$$

$$
\begin{aligned}
\frac{d}{dt}P_b(t) = F(t)\,P_b(t) + P_b(t)\,F^T(t) \\
- P_b(t)\,H^T(t)\,R^{-1}(t)\,H(t)\,P_b(t) + G(t)\,Q(t)\,G^T(t)
\end{aligned}
\tag{6.75c}
$$

Equation (6.75) must be integrated backward in time. In order to express this integration in a more convenient form, it is convenient to set $\tau = T - t$,[1] where T is the terminal time of the data interval. Since $d\mathbf{x}/dt = -d\mathbf{x}/d\tau$, writing eqn. (6.73a) in terms of τ gives

$$
\frac{d}{d\tau}\mathbf{x}(t) = -F(t)\,\mathbf{x}(t) - B(t)\,\mathbf{u}(t) - G(t)\,\mathbf{w}(t) \tag{6.76}
$$

Therefore the backward filter equations can be written in terms of τ by replacing $F(t)$ with $-F(t)$, $B(t)$ with $-B(t)$, and $G(t)$ with $-G(t)$, which leads to
backward filter

$$\frac{d}{d\tau}\hat{\mathbf{x}}_b(t) = -F(t)\,\hat{\mathbf{x}}_b(t) - B(t)\,\mathbf{u}(t) + K_b(t)[\tilde{\mathbf{y}}(t) - H(t)\,\hat{\mathbf{x}}_b(t)] \tag{6.77a}$$

$$K_b(t) = P_b(t)\,H^T(t)\,R^{-1}(t) \tag{6.77b}$$

$$\begin{aligned}\frac{d}{d\tau}P_b(t) = &-F(t)\,P_b(t) - P_b(t)\,F^T(t)\\ &- P_b(t)\,H^T(t)\,R^{-1}(t)\,H(t)\,P_b(t) + G(t)\,Q(t)\,G^T(t)\end{aligned} \tag{6.77c}$$

Therefore, from this point forward whenever $d/d\tau$ is used, this will denote a backward differentiation. We should note that if $F(t)$ is stable going forward in time, then $-F(t)$ is stable going backward in time.

The continuous-time smoother combination of the forward and backward state estimates follows exactly from the discrete-time equivalent of §6.1.1. The continuous-time equivalent of eqn. (6.14) is simply given by

$$\boxed{P(t) = \left[P_f^{-1}(t) + P_b^{-1}(t)\right]^{-1}} \tag{6.78}$$

Also, the continuous-time equivalent of eqn. (6.19) is simply given by

$$\boxed{\hat{\mathbf{x}}(t) = P(t)\left[P_f^{-1}(t)\,\hat{\mathbf{x}}_f(t) + P_b^{-1}(t)\,\hat{\mathbf{x}}_b(t)\right]} \tag{6.79}$$

Equations (6.74), (6.77), (6.78), and (6.79) summarize the basic equations for the smoother. We must now define the boundary conditions. Since at time $t = T$ the smoother estimate must be the same as the forward Kalman filter, this clearly requires that $\hat{\mathbf{x}}(T) = \hat{\mathbf{x}}_f(T)$ and $P(T) = P_f(T)$. From eqn. (6.78) the covariance condition at the terminal time can only be satisfied when $P_b^{-1}(T) = 0$. Therefore, $P_b(t)$ is not finite at the terminal time. To overcome this difficulty consider taking the time derivative of $P_b^{-1}(t)\,P_b(t) = I$, which gives

$$\left[\frac{d}{d\tau}P_b^{-1}(t)\right]P_b(t) + P_b^{-1}(t)\left[\frac{d}{d\tau}P_b(t)\right] = 0 \tag{6.80}$$

Rearranging eqn. (6.80) yields

$$\left[\frac{d}{d\tau}P_b^{-1}(t)\right] = -P_b^{-1}(t)\left[\frac{d}{d\tau}P_b(t)\right]P_b^{-1}(t) \tag{6.81}$$

Substituting (6.77c) into eqn. (6.81) yields

$$\begin{aligned}\frac{d}{d\tau}P_b^{-1}(t) = &\,P_b^{-1}(t)\,F(t) + F^T(t)\,P_b^{-1}(t)\\ &- P_b^{-1}(t)\,G(t)\,Q(t)\,G^T(t)\,P_b^{-1}(t) + H^T(t)\,R^{-1}(t)\,H(t)\end{aligned} \tag{6.82}$$

which can be integrated backward in time with the appropriate boundary condition of $P_b^{-1}(T) = 0$.

Even with the matrix inverse expression for $P_b^{-1}(t)$, eqn. (6.78) still requires the calculation of two matrix inverses, which is generally not desirable. To overcome this aspect of the smoother covariance, the matrix inversion lemma in eqn. (1.70) is used with $A = P_f^{-1}(t)$, $B = D = I$, and $C = P_b^{-1}(t)$, which leads to

$$\boxed{P(t) = P_f(t) - P_f(t)\,P_b^{-1}(t)[I + P_f(t)\,P_b^{-1}(t)]^{-1}P_f(t)} \tag{6.83}$$

Note that eqn. (6.83), in conjunction with eqn. (6.82), requires only one matrix inverse. Equation (6.83) can be further expanded into a symmetric form:

$$P(t) = \left[I - W(t)\,P_b^{-1}(t)\right] P_f(t)[I - W(t)\,P_b^{-1}(t)]^T + W(t)\,P_b^{-1}(t)W^T(t) \tag{6.84}$$

where

$$W(t) = P_f(t)\,[I + P_f(t)\,P_b^{-1}(t)]^{-T} \tag{6.85}$$

As with the discrete symmetric form, eqn. (6.84) is the sum of two positive definite matrices, which provides a more robust approach in terms of numerical stability.

As previously mentioned, the boundary condition for the smoother state is $\hat{\mathbf{x}}(T) = \hat{\mathbf{x}}_f(T)$, but the boundary condition for $\hat{\mathbf{x}}_b(T)$ is still unknown. This difficulty may be overcome by defining a new variable:

$$\hat{\chi}_b(t) \equiv P_b^{-1}(t)\,\hat{\mathbf{x}}_b(t) \tag{6.86}$$

where $\hat{\chi}_b(T) = \mathbf{0}$ since $P_b^{-1}(T) = 0$ and $\hat{\mathbf{x}}_b(T)$ is finite. Differentiating eqn. (6.86) with respect to time and substituting eqns. (6.77a) and (6.82) into the resulting expression yields

$$\boxed{\begin{aligned}\frac{d}{d\tau}\hat{\chi}_b(t) = \left[F(t) - G(t)\,Q(t)\,G^T(t)\,P_b^{-1}(t)\right]^T \hat{\chi}_b(t)\\ - P_b^{-1}(t)\,B(t)\,\mathbf{u}(t) + H^T(t)\,R^{-1}(t)\,\tilde{\mathbf{y}}(t)\end{aligned}} \tag{6.87}$$

The continuous-time equivalent of eqn. (6.21) is now given by

$$\boxed{\hat{\mathbf{x}}(t) = [I - K(t)]\hat{\mathbf{x}}_f(t) + P(t)\,\hat{\chi}_b(t)} \tag{6.88}$$

where the continuous smoother gain is defined by

$$K(t) \equiv P_f(t)\,P_b^{-1}(t)[I + P_f(t)\,P_b^{-1}(t)]^{-1} \tag{6.89}$$

Note that the definition of $\hat{\chi}_b(t)$ has been used in eqn. (6.88).

A summary of the continuous-time fixed-interval smoother is given in Table 6.3. First, the basic continuous-time Kalman filter is executed forward in time on the data set using eqn. (6.74). Then, the backward filter is run using eqns. (6.82) and (6.87), which avoids undesirable matrix inversions. The forward and backward covariances and estimates must be stored in order to evaluate the smoother covariance and estimate. The optimal smoother covariance is computed using eqn. (6.83), or using eqn. (6.84) if numerical stability is of concern. Finally, the optimal smoother estimate is computed using eqn. (6.88).

Table 6.3: Continuous-Time Fixed-Interval Smoother

Model	$\frac{d}{dt}\mathbf{x}(t) = F(t)\mathbf{x}(t) + B(t)\mathbf{u}(t) + G(t)\mathbf{w}(t),\ \mathbf{w}(t) \sim N(\mathbf{0}, Q(t))$ $\tilde{\mathbf{y}}(t) = H(t)\mathbf{x}(t) + \mathbf{v}(t),\ \mathbf{v}(t) \sim N(\mathbf{0}, R(t))$
Forward Covariance	$\frac{d}{dt}P_f(t) = F(t)\,P_f(t) + P_f(t)\,F^T(t)$ $-P_f(t)\,H^T(t)\,R^{-1}(t)H(t)\,P_f(t)$ $+G(t)\,Q(t)\,G^T(t),$ $P_f(t_0) = E\{\tilde{\mathbf{x}}_f(t_0)\,\tilde{\mathbf{x}}_f^T(t_0)\}$
Forward Filter	$\frac{d}{dt}\hat{\mathbf{x}}_f(t) = F(t)\,\hat{\mathbf{x}}_f(t) + B(t)\,\mathbf{u}(t)$ $+P_f(t)\,H^T(t)\,R^{-1}(t)[\tilde{\mathbf{y}}(t) - H(t)\,\hat{\mathbf{x}}_f(t)], \quad \hat{\mathbf{x}}_f(t_0) = \hat{\mathbf{x}}_{f0}$
Backward Covariance	$\frac{d}{d\tau}P_b^{-1}(t) = P_b^{-1}(t)\,F(t) + F^T(t)\,P_b^{-1}(t)$ $-P_b^{-1}(t)\,G(t)\,Q(t)\,G^T(t)\,P_b^{-1}(t)$ $+H^T(t)\,R^{-1}(t)\,H(t), \quad P_b^{-1}(T) = 0$
Backward Filter	$\frac{d}{d\tau}\hat{\chi}_b(t) = \left[F(t) - G(t)\,Q(t)\,G^T(t)\,P_b^{-1}(t)\right]^T\hat{\chi}_b(t)$ $-P_b^{-1}(t)\,B(t)\,\mathbf{u}(t) + H^T(t)\,R^{-1}(t)\,\tilde{\mathbf{y}}(t), \quad \hat{\chi}_b(T) = 0$
Gain	$K(t) = P_f(t)\,P_b^{-1}(t)\left[I + P_f(t)\,P_b^{-1}(t)\right]^{-1}$
Covariance	$P(t) = [I - K(t)]P_f(t)$
Estimate	$\hat{\mathbf{x}}(t) = [I - K(t)]\hat{\mathbf{x}}_f(t) + P(t)\,\hat{\chi}_b(t)$

6.1.2.1 Steady-State Fixed-Interval Smoother

If the system matrices and covariance are time-invariant, then a steady-state (i.e., constant gain) smoother can be used, which significantly reduces the computational burden. The steady-state forward filter has been derived in §5.4.4. The only issue for the backward filter is the steady-state Riccati equation, given by

$$P_b^{-1}F + F^T P_b^{-1} - P_b^{-1}G\,Q\,G^T P_b^{-1} + H^T R^{-1} H = 0 \tag{6.90}$$

Comparing eqn. (6.90) to the Riccati (covariance) equation in Table 5.5 and using a similar transformation as eqn. (5.163) yields the following Hamiltonian matrix:

$$\boxed{\mathcal{H} \equiv \begin{bmatrix} -F & G\,Q G^T \\ H^T R^{-1} H & F^T \end{bmatrix}} \tag{6.91}$$

An eigenvalue/eigenvector decomposition of eqn. (6.91) gives

$$\mathcal{H} = \begin{bmatrix} W_{11} & W_{12} \\ W_{21} & W_{22} \end{bmatrix} \begin{bmatrix} \Lambda & 0 \\ 0 & -\Lambda \end{bmatrix} \begin{bmatrix} W_{11} & W_{12} \\ W_{21} & W_{22} \end{bmatrix}^{-1} \tag{6.92}$$

where Λ is a diagonal matrix of the n eigenvalues in the right half-plane, and W_{11}, W_{21}, W_{12}, and W_{22} are block elements of the eigenvector matrix. From the derivations of §5.4.4 the steady-state value for P_b^{-1} is given by

$$\boxed{P_b^{-1} = W_{21} W_{11}^{-1}} \tag{6.93}$$

which requires an inverse of an $n \times n$ matrix. Also, the nonlinear (extended) version of the smoother is straightforward, replacing the space matrices with their equivalent Jacobian matrices evaluated at the current estimate. These equations will be summarized in §6.1.3.

6.1.2.2 RTS Fixed-Interval Smoother

As with the discrete-time smoother shown in §6.1.1 an RTS form can also be derived for the continuous-time smoother, which combines the backward filter and smoother into one single backward recursion. Taking the derivative of $P^{-1}(t) = P_f^{-1}(t) + P_b^{-1}(t)$, and using eqn. (6.81) for the derivative of $P_f^{-1}(t)$ leads to

$$\frac{d}{d\tau} P^{-1}(t) = -P_f^{-1}(t) \left[\frac{d}{d\tau} P_f(t) \right] P_f^{-1}(t) + \frac{d}{d\tau} P_b^{-1}(t) \tag{6.94}$$

Next, using $dP_f/dt = -dP_f/d\tau$ gives

$$\frac{d}{d\tau} P^{-1}(t) = P_f^{-1}(t) \left[\frac{d}{dt} P_f(t) \right] P_f^{-1}(t) + \frac{d}{d\tau} P_b^{-1}(t) \tag{6.95}$$

Substituting eqns. (6.74c) and (6.82) into eqn. (6.95) gives

$$\begin{aligned} \frac{d}{d\tau} P^{-1}(t) = {} & P_f^{-1}(t)\, F(t) + F^T(t)\, P_f^{-1}(t) + P_f^{-1}(t)\, G(t)\, Q(t)\, G^T(t)\, P_f^{-1}(t) \\ & + P_b^{-1}(t)\, F(t) + F^T(t)\, P_b^{-1}(t) - P_b^{-1}(t)\, G(t)\, Q(t)\, G^T(t)\, P_b^{-1}(t) \end{aligned} \tag{6.96}$$

Using $P^{-1}(t) = P_f^{-1}(t) + P_b^{-1}(t)$, then eqn. (6.96) can be rewritten as

$$\frac{d}{d\tau} P^{-1}(t) = P^{-1}(t)\,F(t) + F^T(t)\,P^{-1}(t) + P_f^{-1}(t)\,G(t)\,Q(t)\,G^T(t)\,P_f^{-1}(t) - \left[P^{-1}(t) - P_f^{-1}(t)\right] G(t)\,Q(t)\,G^T(t) \left[P^{-1}(t) - P_f^{-1}(t)\right] \tag{6.97}$$

Substituting the following relation into eqn. (6.97):

$$\left[\frac{d}{d\tau} P^{-1}(t)\right] = P^{-1}(t) \left[\frac{d}{dt} P(t)\right] P^{-1}(t) \tag{6.98}$$

and then multiplying both sides of the resulting expression by $P(t)$ yields

$$\boxed{\frac{d}{dt} P(t) = \left[F(t) + G(t)\,Q(t)\,G^T(t)\,P_f^{-1}(t)\right] P(t) + P(t)\left[F(t) + G(t)\,Q(t)\,G^T(t)\,P_f^{-1}(t)\right]^T - G(t)\,Q(t)\,G^T(t)} \tag{6.99}$$

Since $P_b^{-1}(T) = 0$, then eqn. (6.99) is integrated backward in time with the boundary condition $P(T) = P_f(T)$. This form clearly has significant computational advantages over integrating the backward filter covariance and using eqn. (6.83). Similar to eqn. (6.83) only one matrix inverse is required in eqn. (6.99); however, the smoother covariance is calculated directly without the need to first calculate the backward filter covariance. Also, at steady-state eqn. (6.99) reduces down to an algebraic Lyapunov equation, which is a linear equation.

To derive an expression for the smoother state estimate, we begin with eqn. (6.79), which can be rewritten as

$$P^{-1}(t)\hat{\mathbf{x}}(t) = P_f^{-1}(t)\hat{\mathbf{x}}_f(t) + \hat{\chi}_b(t) \tag{6.100}$$

Taking the time derivative of eqn. (6.100), and using eqn. (6.81) for the derivative of $P^{-1}(t)$ and $P_f^{-1}(t)$ leads to

$$P^{-1}(t)\left[\frac{d}{dt}\hat{\mathbf{x}}(t)\right] = P^{-1}(t)\left[\frac{d}{dt} P(t)\right] P^{-1}(t)\hat{\mathbf{x}}(t) + P_f^{-1}(t)\left[\frac{d}{dt}\hat{\mathbf{x}}_f(t)\right] - P_f^{-1}(t)\left[\frac{d}{dt} P_f(t)\right] P_f^{-1}(t)\hat{\mathbf{x}}_f(t) + \frac{d}{dt}\hat{\chi}_b(t) \tag{6.101}$$

Substituting the relations in eqns. (6.74) and (6.87) with $d\hat{\chi}_b/d\tau = -d\hat{\chi}_b/dt$, and (6.99) into eqn. (6.101), and after considerable algebra manipulations (which are left as an exercise for the reader), yields

$$\boxed{\frac{d}{dt}\hat{\mathbf{x}}(t) = F(t)\hat{\mathbf{x}}(t) + B(t)\mathbf{u}(t) + G(t)\,Q(t)\,G^T(t)\,P_f^{-1}(t)\left[\hat{\mathbf{x}}(t) - \hat{\mathbf{x}}_f(t)\right]} \tag{6.102}$$

Table 6.4: Continuous-Time RTS Smoother

Model	$\frac{d}{dt}\mathbf{x}(t) = F(t)\mathbf{x}(t) + B(t)\mathbf{u}(t) + G(t)\mathbf{w}(t),\ \mathbf{w}(t) \sim N(\mathbf{0}, Q(t))$ $\tilde{\mathbf{y}}(t) = H(t)\mathbf{x}(t) + \mathbf{v}(t),\ \mathbf{v}(t) \sim N(\mathbf{0}, R(t))$
Forward Covariance	$\frac{d}{dt}P_f(t) = F(t)\,P_f(t) + P_f(t)\,F^T(t)$ $-P_f(t)\,H^T(t)\,R^{-1}(t)H(t)\,P_f(t)$ $+G(t)\,Q(t)\,G^T(t),$ $P_f(t_0) = E\{\tilde{\mathbf{x}}_f(t_0)\,\tilde{\mathbf{x}}_f^T(t_0)\}$
Forward Filter	$\frac{d}{dt}\hat{\mathbf{x}}_f(t) = F(t)\,\hat{\mathbf{x}}_f(t) + B(t)\,\mathbf{u}(t)$ $+P_f(t)\,H^T(t)\,R^{-1}(t)[\tilde{\mathbf{y}}(t) - H(t)\,\hat{\mathbf{x}}_f(t)],\quad \hat{\mathbf{x}}_f(t_0) = \hat{\mathbf{x}}_{f0}$
Smoother Covariance	$\frac{d}{d\tau}P(t) = -[F(t) + G(t)\,Q(t)\,G^T(t)\,P_f^{-1}(t)]P(t)$ $-P(t)[F(t) + G(t)\,Q(t)\,G^T(t)\,P_f^{-1}(t)]^T$ $+G(t)\,Q(t)\,G^T(t),\quad P(T) = P_f(T)$
Smoother Estimate	$\frac{d}{d\tau}\hat{\mathbf{x}}(t) = -F(t)\,\hat{\mathbf{x}}(t) - B(t)\,\mathbf{u}(t)$ $-G(t)\,Q(t)\,G^T(t)\,P_f^{-1}(t)\left[\hat{\mathbf{x}}(t) - \hat{\mathbf{x}}_f(t)\right],\quad \hat{\mathbf{x}}(T) = \hat{\mathbf{x}}_f(T)$

Equation (6.102) is integrated backward in time with the boundary condition $\hat{\mathbf{x}}(T) = \hat{\mathbf{x}}_f(T)$.

A summary of the RTS smoother is given in Table 6.4. As before, the forward Kalman filter is executed using the measurements until time T. Storing the estimated states from the forward filter, the smoothed estimate is then determined by integrating eqn. (6.102) backward in time. Similar to the discrete-time RTS smoother, eqn. (6.99) is only used to derive the smoother covariance, which is not used to find the optimal smoother state estimate. Also, eqn. (6.102) does not involve the measurement directly, but still uses the forward filter state estimate. For all these reasons the RTS smoother is more widely used in practice over the formulation given in Table 6.3.

6.1.2.3 Stability

The backward state matrix in the RTS smoother defines the stability of the system, which is given by $[F(t) + G(t)\,Q(t)\,G^T(t)\,P_f^{-1}(t)]$. A backward integration is stable if all the eigenvalues lie in the right-hand plane. This can be re-evaluated using the negative of the RTS smoother state-matrix, so that its eigenvalues must lie in the left-hand plane for stability. Then the backward smoother stability can be evaluated by investigating the dynamics of the following system:

$$\frac{d}{d\tau}\hat{\mathbf{x}}(t) = -[F(t)+G(t)\,Q(t)\,G^T(t)\,P_f^{-1}(t)]\,\hat{\mathbf{x}}(t) \tag{6.103}$$

The smoother stability can be proved by using Lyapunov's direct method, which is discussed for continuous-time systems in §3.6. For the continuous-time RTS smoother we consider the following candidate Lyapunov function:

$$V[\hat{\mathbf{x}}(t)] = \hat{\mathbf{x}}^T(t)\,P_f^{-1}(t)\,\hat{\mathbf{x}}(t) \tag{6.104}$$

Taking a time derivative of eqn. (6.104) gives

$$\begin{aligned}\frac{d}{d\tau}V[\hat{\mathbf{x}}(t)] = &\left[\frac{d}{d\tau}\hat{\mathbf{x}}(t)\right]^T P_f^{-1}(t)\,\hat{\mathbf{x}}(t) + \hat{\mathbf{x}}^T(t)\left[\frac{d}{d\tau}P_f^{-1}(t)\right]\hat{\mathbf{x}}(t)\\ &+\hat{\mathbf{x}}^T(t)\,P_f^{-1}(t)\left[\frac{d}{d\tau}\hat{\mathbf{x}}(t)\right]\end{aligned} \tag{6.105}$$

Using eqn. (6.81) for $P_f^{-1}(t)$ with $dP_f^{-1}/dt = -dP_f^{-1}/d\tau$, and substituting the resulting expression and eqn. (6.103) into eqn. (6.105) leads to

$$\frac{d}{d\tau}V[\hat{\mathbf{x}}(t)] = -\hat{\mathbf{x}}^T(t)\left[H^T(t)\,R^{-1}(t)H(t) + P_f^{-1}(t)\,G(t)\,Q(t)\,G^T(t)\,P_f^{-1}(t)\right]\hat{\mathbf{x}}(t) \tag{6.106}$$

Clearly if $R(t)$ is positive definite and $Q(t)$ is at least positive semi-definite, then the Lyapunov condition is satisfied and the continuous-time RTS smoother is stable.

Example 6.2: We consider the simple first-order system shown in example 5.4, where the truth model is given by

$$\begin{aligned}\dot{x}(t) &= f\,x(t) + w(t)\\ y(t) &= x(t) + v(t)\end{aligned}$$

where f is a constant, and the variances of $w(t)$ and $v(t)$ are given by q and r, respectively. In the current example the steady-state smoother covariance is investigated. From eqn. (6.99) this value can be determined by solving the following linear differential equation:

$$\frac{d}{dt}p(t) = 2[f + q\,p_f^{-1}(t)]\,p(t) - q$$

where $p_f(t)$ is defined in example 5.4. Since q is a constant, then the steady-state value for $p(t)$ is simply given by

$$\lim_{t\to\infty} p(t) \equiv p = \frac{q}{2\left(f + q\,p_f^{-1}\right)}$$

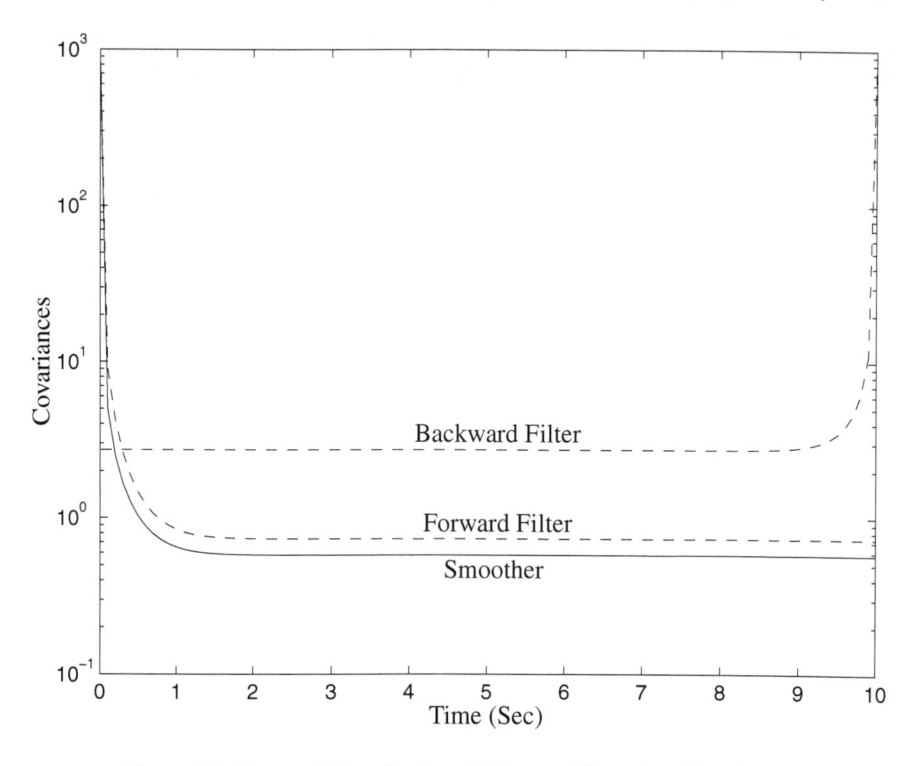

Figure 6.4: Forward Filter, Backward Filter, and Smoother Covariances

Substituting $p_f^{-1} = r^{-1}/(a+f)$, where $a \equiv \sqrt{f^2 + r^{-1}q}$, into the above expression, and after some algebraic manipulations yields

$$p = \frac{q}{2a}$$

From eqn. (6.82) the steady-state backward filter covariance (defined by p_b) can be determined by solving the following quadratic equation:

$$q\, p_b^{-2} - 2\, f\, p_b^{-1} - r^{-1} = 0$$

Taking the positive root yields

$$p_b = \frac{q}{a+f}$$

This can also be easily verified from $p^{-1} = p_f^{-1} + p_b^{-1}$. An interesting aspect of the backward filter covariance is that it is zero when $q = 0$, so that the smoother covariance is equivalent to the forward filter covariance. Hence, for this case the smoother offers no improvements over the forward filter, which is fully proved by Fraser.[8] For all other positive values of q it can be shown that $p \leq p_f$ and $p \leq p_b$, which is left as an exercise for the reader. Consider the following values: $f = -1$,

$q = 2$, and $r = 1$, with an initial condition of $p_f(t_0) = 1,000$. Plots of the forward filter, backward filter, and smoother covariances given by integrating eqns. (6.74c), (6.82), and (6.99), respectively, are shown in Figure 6.4. The analytical steady-state values are given by: $p_f = (\sqrt{3}-1)/1 = 0.7321$, $p_b = 2/(\sqrt{3}-1) = 2.7321$, and $p = 1/\sqrt{3} = 0.5774$, which all agree with the plots in Figure 6.4. An interesting case occurs when $f = 0$, which gives $p_f = p_b = \sqrt{r\,q}$ and $p = \sqrt{r\,q}/2$. From eqn. (6.79) the smoother state estimate for this case is given by

$$\hat{x}(t) = \frac{1}{2}\left[\hat{x}_f(t) + \hat{x}_b(t)\right]$$

Therefore, using the steady-state smoother the optimal estimate of $x(t)$ is the average of the forward and backward filter estimates. This simple example clearly shows the power of the fixed-interval smoother to provide better estimates (i.e., estimates with lower error-covariances) than the standard Kalman filter alone.

6.1.3 Nonlinear Smoothing

In this section the fixed-interval smoothing algorithms derived previously are extended for nonlinear systems. Most modern-day nonlinear applications involve systems with discrete-time measurements and continuous-time models. The first step in the nonlinear smoother involves applying the extended Kalman filter shown in Table 5.9. In order to perform the backward-time integration and measurement updates, straightforward application of the methods in §6.1.2 cannot be applied directly to nonlinear systems. This is due to the fact that we must linearize the backward-time filter about the forward-time filter estimate, not the backward-time filter estimate! Hence, the linearized Kalman filter form shown in §5.6 will be used to derive the backward-time smoother, where the nominal (*a priori*) estimate is given by the forward-time extended Kalman filter. A more formal treatment of nonlinear smoothing is given in Ref. [12].

The derivation of the nonlinear smoother can be shown by using the same procedure leading to the forward/backward filters shown previously. However, we will only show the RTS version of this smoother, since it has clear advantages over the two filter solution, which is given in Ref. [1]. A rigorous proof of the nonlinear RTS smoother is possible using similar methods shown to derive the Kalman filter in §5.5. A detailed derivation for the linear case is given by Bierman.[13] We will prove the nonlinear smoother using variational calculus in §6.4.1.3.

The actual implementation of the RTS nonlinear smoother state estimate is fairly simple. Note that the extended Kalman filter in Table 5.9 provides continuous-time estimates. Therefore, the nonlinear version of eqn. (6.102) can be used directly to determine the smoother state estimate. First, we linearize $\mathbf{f}(\hat{\mathbf{x}}(t), \mathbf{u}(t), t)$ about $\hat{\mathbf{x}}_f(t)$.

Table 6.5: Continuous-Discrete Nonlinear RTS Smoother

Model	$\frac{d}{d\tau}\mathbf{x}(t) = \mathbf{f}(\mathbf{x}(t),\, \mathbf{u}(t),\, t) + G(t)\,\mathbf{w}(t),\ \mathbf{w}(t) \sim N(\mathbf{0},\, Q(t))$ $\tilde{\mathbf{y}}_k = \mathbf{h}(\mathbf{x}_k) + \mathbf{v}_k,\ \mathbf{v}_k \sim N(\mathbf{0},\, R_k)$
Forward Initialize	$\hat{\mathbf{x}}_f(t_0) = \hat{\mathbf{x}}_{f0}$ $P_{f0} = E\left\{\tilde{\mathbf{x}}_f(t_0)\,\tilde{\mathbf{x}}_f^T(t_0)\right\}$
Forward Gain	$K_{fk} = P_{fk}^- H_k^T(\hat{\mathbf{x}}_{fk}^-)[H_k(\hat{\mathbf{x}}_{fk}^-)P_{fk}^- H_k^T(\hat{\mathbf{x}}_{fk}^-) + R_k]^{-1}$ $H_k(\hat{\mathbf{x}}_{fk}^-) \equiv \left.\frac{\partial \mathbf{h}}{\partial \mathbf{x}}\right\|_{\hat{\mathbf{x}}_{fk}^-}$
Forward Update	$\hat{\mathbf{x}}_{fk}^+ = \hat{\mathbf{x}}_{fk}^- + K_{fk}[\tilde{\mathbf{y}}_k - \mathbf{h}(\hat{\mathbf{x}}_{fk}^-)]$ $P_{fk}^+ = [I - K_{fk}H_k(\hat{\mathbf{x}}_{fk}^-)]P_{fk}^-$
Forward Propagation	$\frac{d}{dt}\hat{\mathbf{x}}_f(t) = \mathbf{f}(\hat{\mathbf{x}}_f(t),\, \mathbf{u}(t),\, t)$ $\frac{d}{dt}P_f(t) = F(\hat{\mathbf{x}}_f(t),\, t)\, P_f(t) + P_f(t)\, F^T(\hat{\mathbf{x}}_f(t),\, t)$ $+G(t)\, Q(t)\, G^T(t)$ $F(\hat{\mathbf{x}}_f(t),\, t) \equiv \left.\frac{\partial \mathbf{f}}{\partial \mathbf{x}}\right\|_{\hat{\mathbf{x}}_f(t)}$
Gain	$K(t) \equiv G(t)\, Q(t)\, G^T(t)\, P_f^{-1}(t)$
Smoother Covariance	$\frac{d}{d\tau}P(t) = -[F(\hat{\mathbf{x}}_f(t),\, t) + K(t)]P(t)$ $-P(t)[F(\hat{\mathbf{x}}_f(t),\, t) + K(t)]^T$ $+G(t)\, Q(t)\, G^T(t), \quad P(T) = P_f(T)$
Smoother Estimate	$\frac{d}{d\tau}\hat{\mathbf{x}}(t) = -\left[F(\hat{\mathbf{x}}_f(t),\, t) + K(t)\right]\left[\hat{\mathbf{x}}(t) - \hat{\mathbf{x}}_f(t)\right]$ $-\mathbf{f}(\hat{\mathbf{x}}_f(t),\, \mathbf{u}(t),\, t), \quad \hat{\mathbf{x}}(T) = \hat{\mathbf{x}}_f(T)$

Then, using $d\mathbf{x}/dt = -d\mathbf{x}/d\tau$ to denote the backward-time integration leads to

$$\boxed{\frac{d}{d\tau}\hat{\mathbf{x}}(t) = -\left[F(\hat{\mathbf{x}}_f(t),\, t) + K(t)\right]\left[\hat{\mathbf{x}}(t) - \hat{\mathbf{x}}_f(t)\right] - \mathbf{f}(\hat{\mathbf{x}}_f(t),\, \mathbf{u}(t),\, t)} \tag{6.107}$$

where

$$\boxed{K(t) \equiv G(t)\, Q(t)\, G^T(t)\, P_f^{-1}(t)} \tag{6.108}$$

and

$$F(\hat{\mathbf{x}}_f(t), t) \equiv \left. \frac{\partial \mathbf{f}}{\partial \mathbf{x}} \right|_{\hat{\mathbf{x}}_f(t)} \tag{6.109}$$

Equation (6.107) must be integrated backward in time with a boundary condition of $\hat{\mathbf{x}}(T) = \hat{\mathbf{x}}_f(T)$. Note that eqn. (6.107) is a linear equation in $\hat{\mathbf{x}}(t)$, which allows us to use linear integration methods. Also, the smoother covariance follows the following equation:

$$\boxed{\begin{aligned} \frac{d}{d\tau} P(t) = & -\left[F(\hat{\mathbf{x}}_f(t), t) + K(t)\right] P(t) - P(t) \left[F(\hat{\mathbf{x}}_f(t), t) + K(t)\right]^T \\ & + G(t)\, Q(t)\, G^T(t) \end{aligned}} \tag{6.110}$$

Equation (6.110) must also be integrated backward in time with a boundary condition of $P(T) = P_f(T)$.

A summary of continuous-discrete nonlinear RTS smoother is given in Table 6.5. First, the extended Kalman filter is executed forward in time on the data set. Then, eqn. (6.107) is integrated backward in time using the stored forward-filter state estimate and covariance. The smoother state and covariance are clearly a function of the inverse of the forward-time covariance. One method to overcome this inverse is to use the information matrix version of the Kalman filter, shown in §5.3.3. Another smoother form that does not require the inverse of the covariance matrix is presented by Bierman.[13] This form uses an "adjoint variable," $\boldsymbol{\lambda}(t)$, to derive the smoother equation. We will derive this form directly using variational calculus in §6.4.1.3. The propagation equations are given by

$$\boxed{\frac{d}{d\tau} \boldsymbol{\lambda}(t) = F^T(\hat{\mathbf{x}}(t), t)\, \boldsymbol{\lambda}(t)} \tag{6.111a}$$

$$\boxed{\frac{d}{d\tau} \Lambda(t) = F^T(\hat{\mathbf{x}}_f(t), t)\, \Lambda(t) + \Lambda(t)\, F(\hat{\mathbf{x}}_f(t), t)} \tag{6.111b}$$

where $\Lambda(t)$ is the covariance of $\boldsymbol{\lambda}(t)$. The backward updates are given by

$$\boxed{\boldsymbol{\lambda}_k^- = \left[I - H_k^T(\hat{\mathbf{x}}_{fk}^-)\, K_{fk}^T\right] \boldsymbol{\lambda}_k^+ - H_k^T(\hat{\mathbf{x}}_{fk}^-)\, D_{fk}^{-1} \left[\tilde{\mathbf{y}}_k - \mathbf{h}_k(\hat{\mathbf{x}}_{fk}^-)\right]} \tag{6.112a}$$

$$\boxed{\begin{aligned} \Lambda_k^- = & \left[I - K_{fk}\, H_k(\hat{\mathbf{x}}_{fk}^-)\right]^T \Lambda_k^+ \left[I - K_{fk}\, H_k(\hat{\mathbf{x}}_{fk}^-)\right] \\ & + H_k^T(\hat{\mathbf{x}}_{fk}^-)\, D_{fk}^{-1}\, H_k(\hat{\mathbf{x}}_{fk}^-) \end{aligned}} \tag{6.112b}$$

where

$$D_{fk} \equiv H_k(\hat{\mathbf{x}}_{fk}^-)\, P_{fk}^-\, H_k^T(\hat{\mathbf{x}}_{fk}^-) + R_k \tag{6.113}$$

Note that in this formulation $\boldsymbol{\lambda}_k^-$ is used to denote the backward update just before the measurement is processed. If $T \equiv t_N$ is an observation time, then the boundary conditions are given by

$$\boldsymbol{\lambda}_N^- = -H_N^T(\hat{\mathbf{x}}_{fN}^-)\, D_{fN}^{-1}\left[\tilde{\mathbf{y}}_N - \mathbf{h}_N(\hat{\mathbf{x}}_{fN}^-)\right] \tag{6.114a}$$

$$\Lambda_N^- = H_T N^T(\hat{\mathbf{x}}_{fN}^-)\, D_{fN}^{-1} H_N(\hat{\mathbf{x}}_{fN}^-) \tag{6.114b}$$

If T is not an observation time, then $\boldsymbol{\lambda}$ and Λ simply have boundary conditions of zero. Finally, the smoother state and covariance can be constructed via

$$\hat{\mathbf{x}}_k = \hat{\mathbf{x}}_{fk}^{\pm} - P_{fk}^{\pm}\boldsymbol{\lambda}_k^{\pm} \tag{6.115a}$$

$$P_k = P_{fk}^{\pm} - P_{fk}^{\pm}\Lambda_k^{\pm}P_{fk}^{\pm} \tag{6.115b}$$

where the propagated or updated variables yield the same result. The matrix D_{fk}^{-1} is used directly in the forward-time Kalman filter, which can be stored directly. Therefore, an extra inverse is not required by this alternative approach.

Example 6.3: In this example the model used in example 5.5 is used to demonstrate the power of the RTS nonlinear smoother using continuous-time models with discrete-time measurements. The smoother given in Table 6.5 is used to determine optimal state estimates. The parameters used for this simulation are identical to the parameters given in example 5.5. First, the forward-time extended Kalman filter is executed using the measured data. Then, the smoother state estimate and covariance are determined by integrating eqns. (6.107) and (6.110) backward in time.

A plot of the results is shown in Figure 6.5. Clearly, the smoother covariance and estimate errors are much smaller than the forward-time estimates. Although the smoother estimates cannot be given in real time, these estimates can often provide very useful information. For example, in Ref. [14] a nonlinear smoother algorithm has been used to show uncontrolled motions ("nutation") of a spacecraft, which are not visible in the forward-time estimates. This information may be used to redesign a controller or filter if these nutations lead to unacceptable pointing errors.

6.2 Fixed-Point Smoothing

In this section the fixed-point smoothing algorithm is shown for both discrete-time and continuous-time models. Meditch[15] provides an excellent example on the usefulness of fixed-point smoothing, which we will summarize here. Suppose that a spacecraft is tracked by a ground-based radar, and we implement an orbit determination algorithm using an extended Kalman filter to determine a state estimate of $\mathbf{x}_N$ at some time t_N. The estimate is derived from the measurements $\tilde{\mathbf{y}}_k$, where

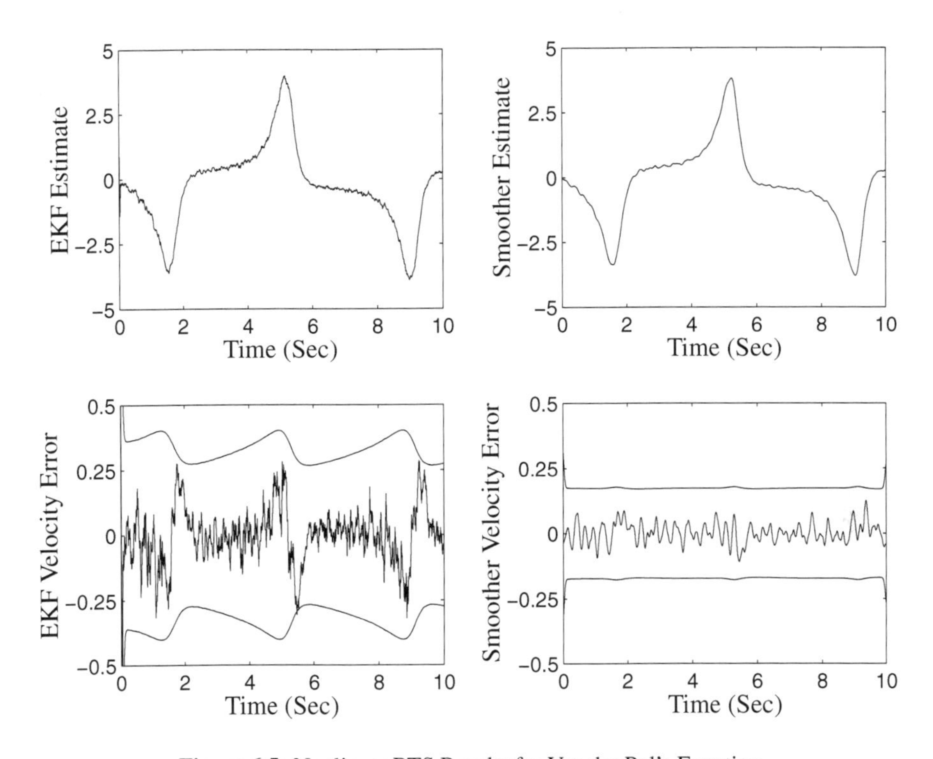

Figure 6.5: Nonlinear RTS Results for Van der Pol's Equation

$k = 1, 2, \ldots, N$, and is denoted by $\hat{\mathbf{x}}_{N|N}$. Suppose now that additional orbital data becomes available, say after an orbital burn, and that we wish to estimate the state at later times. Thus we seek to determine $\hat{\mathbf{x}}_{N|N+1}$, $\hat{\mathbf{x}}_{N|N+2}$, etc., taking the estimate at time t_N into account. Using notation from §2.6, for some fixed N we wish to determine the following quantity:

$$\hat{\mathbf{x}}_{k|N} \equiv E\left\{\hat{\mathbf{x}}_k|[\tilde{\mathbf{y}}_1, \tilde{\mathbf{y}}_2 \ldots, \tilde{\mathbf{y}}_N]\right\} \tag{6.116}$$

for $N > k$, where the notation $k|N$ denotes the smoothed estimate at time t_k, given measurements up to time t_N.

6.2.1 Discrete-Time Formulation

To derive the necessary relations for the discrete-time fixed-point smoother, we start with the measurement residual equation given by eqn. (5.280):

$$\begin{aligned} \boldsymbol{\upsilon}_{fk} &= \tilde{\mathbf{y}}_k - H_k \hat{\mathbf{x}}_{fk}^- \\ &= -H_k \tilde{\mathbf{x}}_{fk}^- + \mathbf{v}_k \end{aligned} \tag{6.117}$$

where $\tilde{\mathbf{x}}_{fk}^- \equiv \hat{\mathbf{x}}_{fk}^- - \mathbf{x}_k$. Meditch[15] shows that the single-stage optimal smoothing relation follows the following equation:

$$\hat{\mathbf{x}}_{k|k+1} = \hat{\mathbf{x}}_{k|k} + P_k^{xv}(P_{k+1}^{vv})^{-1}\boldsymbol{v}_{fk+1} \tag{6.118}$$

where

$$P_k^{xv} = E\left\{\mathbf{x}_k \boldsymbol{v}_{fk+1}^T\right\} \tag{6.119a}$$

$$P_{k+1}^{vv} = E\left\{\boldsymbol{v}_{fk+1} \boldsymbol{v}_{fk+1}^T\right\} \tag{6.119b}$$

Note the similarities between eqn. (6.118) and eqn. (5.279a). Substituting the one time-step ahead of eqn. (6.117) into eqn. (6.119a) and using the fact that $\mathbf{v}_{k+1}$ has zero mean leads to

$$P_k^{xv} = -E\left\{\mathbf{x}_k \tilde{\mathbf{x}}_{fk+1}^{-T}\right\} H_{k+1}^T \tag{6.120}$$

Substituting eqn. (5.33) into eqn. (6.120) and using the fact that $\mathbf{w}_k$ has zero mean leads to

$$P_k^{xv} = -E\left\{\mathbf{x}_k \tilde{\mathbf{x}}_{fk}^{+T}\right\} \Phi_k^T H_{k+1}^T \tag{6.121}$$

Substituting the relationship $\mathbf{x}_k = \hat{\mathbf{x}}_{fk}^+ - \tilde{\mathbf{x}}_{fk}^+$ into eqn. (6.121) and using the orthogonality principle given by eqn. (5.109) yields

$$P_k^{xv} = P_{fk}^+ \Phi_k^T H_{k+1}^T \tag{6.122}$$

The covariance of the innovations process can easily be derived using eqn. (6.117), which is given by

$$P_{k+1}^{vv} = H_{k+1} P_{fk+1}^- H_{k+1}^T + R_{k+1} \tag{6.123}$$

Substituting eqn. (5.30a) into the one time-step ahead of eqn. (6.117), and then substituting the resultant together with eqns. (6.122) and (6.123) into eqn. (6.118) yields

$$\hat{\mathbf{x}}_{k|k+1} = \hat{\mathbf{x}}_{k|k} + \mathcal{M}_{k|k+1}\left[\tilde{\mathbf{y}}_{k+1} - H_{k+1}\Phi_k \hat{\mathbf{x}}_{fk}^+ - H_{k+1}\Gamma_k \mathbf{u}_k\right] \tag{6.124}$$

where

$$\mathcal{M}_{k|k+1} \equiv P_k^{xv}(P_{k+1}^{vv})^{-1} = P_{fk}^+ \Phi_k^T H_{k+1}^T \left[H_{k+1} P_{fk+1}^- H_{k+1}^T + R_{k+1}\right]^{-1} \tag{6.125}$$

The expression in eqn. (6.124) can be rewritten by using the definition of the forward gain, K_{fk} in Table 6.2, which gives

$$H_{k+1}^T \left[H_{k+1} P_{fk+1}^- H_{k+1}^T + R_{k+1}\right]^{-1} = (P_{fk+1}^-)^{-1} K_{fk+1} \tag{6.126}$$

Also, from eqns. (5.30a) and (5.30b) we have

$$K_{fk+1}\left[\tilde{\mathbf{y}}_{k+1} - H_{k+1}\Phi_k \hat{\mathbf{x}}_{fk}^+ - H_{k+1}\Gamma_k \mathbf{u}_k\right] = \hat{\mathbf{x}}_{fk+1}^+ - \hat{\mathbf{x}}_{fk+1}^- \tag{6.127}$$

Therefore, eqn. (6.124) can be rewritten as

$$\hat{\mathbf{x}}_{k|k+1} = \hat{\mathbf{x}}_{k|k} + \mathcal{K}_k[\hat{\mathbf{x}}_{fk+1}^{+} - \hat{\mathbf{x}}_{fk+1}^{-}] \tag{6.128}$$

where

$$\mathcal{K}_k \equiv P_{fk}^{+}\Phi_k^T(P_{fk+1}^{-})^{-1} \tag{6.129}$$

Note that the gain in eqn. (6.129) is the same exact gain used in the discrete-time RTS smoother given in Table 6.2. In fact the RTS smoother can be derived directly from eqn. (6.128) (which is left as an exercise for the reader).

We now develop an expression for the double-stage optimal smoother relationship. This relationship can be derived from the double-stage version of eqn. (6.118):

$$\hat{\mathbf{x}}_{k|k+2} = \hat{\mathbf{x}}_{k|k+1} + P_{k+1}^{x\upsilon}(P_{k+2}^{\upsilon\upsilon})^{-1}\boldsymbol{\upsilon}_{fk+2} \tag{6.130}$$

where

$$P_{k+1}^{x\upsilon} = E\left\{\mathbf{x}_k\boldsymbol{\upsilon}_{fk+2}^T\right\} \tag{6.131a}$$

$$P_{k+2}^{\upsilon\upsilon} = E\left\{\boldsymbol{\upsilon}_{fk+2}\boldsymbol{\upsilon}_{fk+2}^T\right\} \tag{6.131b}$$

Implementing the same procedure that has been used to derive eqn. (6.121), then $P_{k+1}^{x\upsilon}$ is easily shown to be given by

$$P_{k+1}^{x\upsilon} = -E\left\{\mathbf{x}_k\tilde{\mathbf{x}}_{fk+1}^{T+}\right\}\Phi_{k+1}^T H_{k+2} \tag{6.132}$$

Substituting the one time-step ahead of eqn. (5.37) into eqn. (6.132) and using the fact that $\mathbf{v}_{k+1}$ has zero mean leads to

$$P_{k+1}^{x\upsilon} = -E\left\{\mathbf{x}_k\tilde{\mathbf{x}}_{fk+1}^{-T}[I - K_{fk+1}H_{k+1}]^T\right\}\Phi_{k+1}^T H_{k+2} \tag{6.133}$$

Using eqns. (6.120) and (6.122) in eqn. (6.133) yields

$$P_{k+1}^{x\upsilon} = P_{fk}^{+}\Phi_k^T[I - K_{fk+1}H_{k+1}]^T\Phi_{k+1}^T H_{k+2} \tag{6.134}$$

Using one time-step ahead of eqn. (6.123) with eqn. (6.134) yields

$$\begin{aligned}\mathcal{M}_{k|k+2} \equiv P_{k+1}^{x\upsilon}(P_{k+2}^{\upsilon\upsilon})^{-1} &= P_{fk}^{+}\Phi_k^T[I - K_{fk+1}H_{k+1}]^T\Phi_{k+1}^T H_{k+2} \\ &\times\left[H_{k+2}P_{fk+2}^{-}H_{k+2}^T + R_{k+2}\right]^{-1}\end{aligned} \tag{6.135}$$

Note that comparing eqns. (6.125) and (6.135) indicates $\mathcal{M}_{k|k+2}$ is not simply the one time-step ahead of $\mathcal{M}_{k|k+1}$. From eqn. (5.44) we have

$$[I - K_{fk+1}H_{k+1}] = P_{fk+1}^{+}(P_{fk+1}^{-})^{-1} \tag{6.136}$$

Substituting eqn. (6.136) and the one time-step ahead of eqn. (6.126) into eqn. (6.135) yields

$$\mathcal{M}_{k|k+2} = \mathcal{K}_k\mathcal{K}_{k+1}K_{fk+2} \tag{6.137}$$

where $\mathcal{K}_{k+1}$ is clearly the one time-step ahead of $\mathcal{K}_k$. Hence, the double-stage optimal smoother follows the following equation:

$$\hat{\mathbf{x}}_{k|k+2} = \hat{\mathbf{x}}_{k|k+1} + \mathcal{K}_k \mathcal{K}_{k+1}[\hat{\mathbf{x}}_{fk+2}^{+} - \hat{\mathbf{x}}_{fk+2}^{-}] \tag{6.138}$$

By induction the discrete-time fixed-point optimal smoother equation follows

$$\boxed{\hat{\mathbf{x}}_{k|N} = \hat{\mathbf{x}}_{k|N-1} + \mathcal{B}_N[\hat{\mathbf{x}}_{fN}^{+} - \hat{\mathbf{x}}_{fN}^{-}]} \tag{6.139}$$

where

$$\boxed{\mathcal{B}_N = \prod_{i=k}^{N-1} \mathcal{K}_i = \mathcal{B}_{N-1}\mathcal{K}_{N-1}} \tag{6.140}$$

and

$$\boxed{\mathcal{K}_i = P_{fi}^{+} \Phi_i^T (P_{fi+1}^{-})^{-1}} \tag{6.141}$$

with boundary condition given by $\hat{\mathbf{x}}_{k|k} = \hat{\mathbf{x}}_{fk}^{+}$.

The covariance of the discrete-time fixed-point smoother can be derived from

$$P_{k|N} \equiv E\left\{\hat{\mathbf{x}}_{k|N}\hat{\mathbf{x}}_{k|N}^T\right\} \tag{6.142}$$

First, the following error state is defined:

$$\tilde{\mathbf{x}}_{k|N} = \hat{\mathbf{x}}_{k|N} - \mathbf{x}_k \tag{6.143}$$

Substituting eqn. (6.139) into eqn. (6.143) yields

$$\tilde{\mathbf{x}}_{k|N} = \tilde{\mathbf{x}}_{k|N-1} - \mathcal{B}_N[\hat{\mathbf{x}}_{fN}^{+} - \hat{\mathbf{x}}_{fN}^{-}] \tag{6.144}$$

which is rearranged into

$$\tilde{\mathbf{x}}_{k|N} - \mathcal{B}_N\hat{\mathbf{x}}_{fN}^{+} = \tilde{\mathbf{x}}_{k|N-1} + \mathcal{B}_N\hat{\mathbf{x}}_{fN}^{-} \tag{6.145}$$

Since the terms $\tilde{\mathbf{x}}_{k|N-1}$, $\hat{\mathbf{x}}_{fN}^{+}$, and $\hat{\mathbf{x}}_{fN}^{-}$ are all uncorrelated, the covariance is simply given by

$$P_{k|N} = P_{k|N-1} + \mathcal{B}_N[P_{fN}^{\hat{x}\hat{x}-} - P_{fN}^{\hat{x}\hat{x}+}]\mathcal{B}_N^T \tag{6.146}$$

where

$$P_{fN}^{\hat{x}\hat{x}-} = E\left\{\hat{\mathbf{x}}_{fN}^{-}\hat{\mathbf{x}}_{fN}^{-T}\right\} \tag{6.147a}$$

$$P_{fN}^{\hat{x}\hat{x}+} = E\left\{\hat{\mathbf{x}}_{fN}^{+}\hat{\mathbf{x}}_{fN}^{+T}\right\} \tag{6.147b}$$

Next the following relationship is used (the proof is left as an exercise for the reader):

$$P_{fN}^{\hat{x}\hat{x}-} - P_{fN}^{\hat{x}\hat{x}+} = P_{fN}^{+} - P_{fN}^{-} \tag{6.148}$$

Table 6.6: Discrete-Time Fixed-Point Smoother

Model	$\mathbf{x}_{k+1} = \Phi_k \mathbf{x}_k + \Gamma_k \mathbf{u}_k + \Upsilon_k \mathbf{w}_k, \quad \mathbf{w}_k \sim N(\mathbf{0}, Q_k)$ $\tilde{\mathbf{y}}_k = H_k \mathbf{x}_k + \mathbf{v}_k, \quad \mathbf{v}_k \sim N(\mathbf{0}, R_k)$
Forward Initialize	$\hat{\mathbf{x}}_f(t_0) = \hat{\mathbf{x}}_{f0}$ $P_f(t_0) = E\{\tilde{\mathbf{x}}_f(t_0)\tilde{\mathbf{x}}_f^T(t_0)\}$
Gain	$K_{fk} = P_{fk}^- H_k^T [H_k P_{fk}^- H_k^T + R_k]^{-1}$
Forward Update	$\hat{\mathbf{x}}_{fk}^+ = \hat{\mathbf{x}}_{fk}^- + K_{fk}[\tilde{\mathbf{y}}_k - H_k \hat{\mathbf{x}}_{fk}^-]$ $P_{fk}^+ = [I - K_{fk} H_k] P_{fk}^-$
Forward Propagation	$\hat{\mathbf{x}}_{fk+1}^- = \Phi_k \hat{\mathbf{x}}_{fk}^+ + \Gamma_k \mathbf{u}_k$ $P_{fk+1}^- = \Phi_k P_{fk}^+ \Phi_k^T + \Upsilon_k Q_k \Upsilon_k^T$
Smoother Initialize	$\hat{\mathbf{x}}_{k\|k} = \hat{\mathbf{x}}_{fk}^+$ $P_{k\|k} = P_{fk}^+$
Gain	$\mathcal{B}_N = \prod_{i=k}^{N-1} \mathcal{K}_i, \quad \mathcal{K}_i = P_{fi}^+ \Phi_i^T (P_{fi+1}^-)^{-1}$
Covariance	$P_{k\|N} = P_{k\|N-1} + \mathcal{B}_N [P_{fN}^+ - P_{fN}^-] \mathcal{B}_N^T$
Estimate	$\hat{\mathbf{x}}_{k\|N} = \hat{\mathbf{x}}_{k\|N-1} + \mathcal{B}_N [\hat{\mathbf{x}}_{fN}^+ - \hat{\mathbf{x}}_{fN}^-]$

Substituting eqn. (6.148) into eqn. (6.146) gives

$$\boxed{P_{k|N} = P_{k|N-1} + \mathcal{B}_N [P_{fN}^+ - P_{fN}^-] \mathcal{B}_N^T} \tag{6.149}$$

with boundary condition of $P_{k|k} = P_{fk}^+$. A summary of the discrete-time fixed-point smoother is given in Table 6.6. As with the discrete-time RTS smoother the fixed-point smoother begins by implementing the standard forward-time Kalman filter. The smoother state at a fixed point is simply given by using eqn. (6.139). The smoother covariance at the desired point is computed using eqn. (6.149). Once again the smoother does not require the computation of the covariance to determine the estimate, analogous to the RTS smoother.

Table 6.7: Continuous-Time Fixed-Point Smoother

Model	$\frac{d}{dt}\mathbf{x}(t) = F(t)\,\mathbf{x}(t) + B(t)\,\mathbf{u}(t) + G(t)\,\mathbf{w}(t),\ \mathbf{w}(t) \sim N(\mathbf{0}, Q(t))$ $\tilde{\mathbf{y}}(t) = H(t)\,\mathbf{x}(t) + \mathbf{v}(t),\ \mathbf{v}(t) \sim N(\mathbf{0}, R(t))$
Forward Covariance	$\frac{d}{dt}P_f(t) = F(t)\,P_f(t) + P_f(t)\,F^T(t)$ $-P_f(t)\,H^T(t)\,R^{-1}(t)H(t)\,P_f(t)$ $+G(t)\,Q(t)\,G^T(t),$ $P_f(t_0) = E\{\tilde{\mathbf{x}}_f(t_0)\,\tilde{\mathbf{x}}_f^T(t_0)\}$
Forward Filter	$\frac{d}{dt}\hat{\mathbf{x}}_f(t) = F(t)\,\hat{\mathbf{x}}_f(t) + B(t)\,\mathbf{u}(t)$ $+P_f(t)\,H^T(t)\,R^{-1}(t)[\tilde{\mathbf{y}}(t) - H(t)\,\hat{\mathbf{x}}_f(t)],\quad \hat{\mathbf{x}}_f(t_0) = \hat{\mathbf{x}}_{f0}$
Transition Matrix	$\frac{d}{dT}\Phi(t,T) = -\Phi(t,T)[F(T) + G(T)\,Q(T)\,G^T(T)\,P_f^{-1}(T)],$ $\Phi(t,t) = I$
Smoother Covariance	$\frac{d}{dT}P(t\|T) = -\Phi(t,T)\,P_f(T)\,H^T(T)\,R^{-1}(T)\,P_f(T)\,\Phi^T(t,T),$ $P(t\|t) = P_f(t)$
Smoother Estimate	$\frac{d}{dT}\hat{\mathbf{x}}(t\|T) = \Phi(t,T)\,P_f(T)\,H^T(T)\,R^{-1}(T)[\tilde{\mathbf{y}}(T) - H(T)\,\hat{\mathbf{x}}_f(T)],$ $\hat{\mathbf{x}}(t\|t) = \hat{\mathbf{x}}_f(t)$

6.2.2 Continuous-Time Formulation

The continuous-time fixed-point smoother can be derived from the discrete-time version. A simpler way involves rewriting eqn. (6.102) in terms as the smoother estimate at time t given a state estimate at time T:[1]

$$\begin{aligned}\frac{d}{dt}\hat{\mathbf{x}}(t|T) &= [F(t) + G(t)\,Q(t)\,G^T(t)\,P_f^{-1}(t)]\,\hat{\mathbf{x}}(t|T) + B(t)\,\mathbf{u}(t)\\ &\quad - G(t)\,Q(t)\,G^T(t)\,P_f^{-1}(t)\,\hat{\mathbf{x}}_f(t)\end{aligned} \tag{6.150}$$

where $T \geq t$ and $\hat{\mathbf{x}}(t|t) = \hat{\mathbf{x}}_f(t)$. The solution of eqn. (6.150) is given by using the methods described in §3.1.3, which is given by

$$\begin{aligned}\hat{\mathbf{x}}(t|T) &= \Phi(t,T)\,\hat{\mathbf{x}}_f(T) + \int_T^t \Phi(t,\tau)\,B(\tau)\,\mathbf{u}(\tau)\,d\tau\\ &\quad - \int_T^t \Phi(t,\tau)\,G(\tau)\,Q(\tau)\,G^T(\tau)\,P_f^{-1}(\tau)\,\hat{\mathbf{x}}_f(\tau)\,d\tau\end{aligned} \tag{6.151}$$

where $\Phi(t,T)$ is the state transition matrix of $F(t)+G(t)\,Q(t)\,G^T(t)\,P_f^{-1}(t)$, which clearly must obey

$$\frac{d}{dt}\Phi(t,T) = [F(t)+G(t)\,Q(t)\,G^T(t)\,P_f^{-1}(t)]\,\Phi(t,T), \quad \Phi(t,t)=I \tag{6.152}$$

For the fixed-point smoother we consider the case where t is fixed and allow T to vary. Therefore, in order to derive an expression for the fixed-point smoother estimate, eqn. (6.151) must be differentiated with respect to T, which yields

$$\begin{aligned}\frac{d}{dT}\hat{\mathbf{x}}(t|T) = &\frac{d\Phi(t,T)}{dT}\hat{\mathbf{x}}_f(T)+\Phi(t,T)\frac{d\hat{\mathbf{x}}_f(T)}{dT} - \Phi(t,T)\,B(T)\,\mathbf{u}(T)\\ &+\Phi(t,T)\,G(T)\,Q(T)\,G^T(T)\,P_f^{-1}(T)\,\hat{\mathbf{x}}_f(T)\end{aligned} \tag{6.153}$$

The expression for the derivative of $\Phi(t,T)$ in eqn. (6.153) is given by differentiating $\Phi(t,T)\,\Phi(T,t)=I$ with respect to t, which yields

$$\begin{aligned}\frac{d\Phi(T,t)}{dt} &= -\Phi^{-1}(t,T)\frac{d\Phi(t,T)}{dt}\Phi(T,t)\\ &= -\Phi(T,t)\frac{d\Phi(t,T)}{dt}\Phi^{-1}(t,T)\end{aligned} \tag{6.154}$$

Substituting eqn. (6.152) into eqn. (6.154) yields (after some notational changes)

$$\boxed{\frac{d}{dT}\Phi(t,T) = -\Phi(t,T)[F(T)+G(T)\,Q(T)\,G^T(T)\,P_f^{-1}(T)], \quad \Phi(t,t)=I} \tag{6.155}$$

Hence, substituting the forward-time state filter equation from Table 6.3 and the expression in eqn. (6.155) into eqn. (6.153) leads to

$$\boxed{\frac{d}{dT}\hat{\mathbf{x}}(t|T) = \Phi(t,T)\,P_f(T)\,H^T(T)\,R^{-1}(T)[\tilde{\mathbf{y}}(T)-H(T)\,\hat{\mathbf{x}}_f(T)]} \tag{6.156}$$

Applying the same concepts leading toward eqn. (6.156) to the covariance yields (which is left as an exercise for the reader)

$$\boxed{\frac{d}{dT}P(t|T) = -\Phi(t,T)\,P_f(T)\,H^T(T)\,R^{-1}(T)\,P_f(T)\,\Phi^T(t,T)} \tag{6.157}$$

with $P(t|t)=P_f(t)$.

A summary of the continuous-time fixed-point smoother is given in Table 6.7. As with the discrete-time RTS smoother the fixed-point smoother begins by implementing the standard forward-time Kalman filter. The smoother state at a fixed point is simply given by using eqns. (6.155) and (6.156). The smoother covariance at the desired point is computed using eqn. (6.157), which is not required to determine the state estimate.

Example 6.4: We again consider the simple first-order system shown in example 6.2. Assuming that the forward-pass covariance has reached a steady-state value, given by p_f, the state transition matrix using eqn. (6.155) reduces down to

$$\frac{d\phi(t,T)}{dT} = -\beta\,\phi(t,T), \quad \phi(t,t) = 1$$

where $\beta \equiv (f + q/p_f)$. Note that t is fixed and $T \geq t$. The solution for $\phi(t,T)$ is given by

$$\phi(t,T) = e^{-\beta\,(T-t)}$$

Then, the smoother covariance using eqn. (6.157) reduces down to

$$\frac{dp(t|T)}{dT} = -\frac{p_f^2}{r} e^{-2\beta\,(T-t)}$$

The solution for $p(t|T)$ can be shown to be given by (which is left as an exercise for the reader)

$$p(t|T) = p_f\, e^{-2\beta\,(T-t)} + \frac{q}{2a}\left[1 - e^{-2\beta\,(T-t)}\right]$$

where $a \equiv \sqrt{f^2 + r^{-1}q}$. Consider when the point of interest is far enough in the past, so that $p(t|T)$ is at steady-state (e.g., after four times the time constant, i.e., when $T - t \geq 2/\beta$).[1] Then, the fixed-point smoother covariance at steady-state is given by $q/(2a)$, which is equivalent to the smoother steady-state covariance given in example 6.2. The smoother state estimate using eqn. (6.156) follows the following differential equation:

$$\frac{d\hat{x}(t|T)}{dT} = \frac{p_f}{r} e^{-\beta\,(T-t)}[\tilde{y}(T) - \hat{x}_f(T)], \quad \hat{x}(t|t) = \hat{x}_f(t)$$

This differential equation is integrated forward in time from time t until the present time T.

6.3 Fixed-Lag Smoothing

In this section the fixed-lag smoothing algorithm is shown for both discrete-time and continuous-time models. This smoother can be used for estimating the state where a lag is allowable between the current measurement and the estimate. Thus, the fixed-lag smoother is used to determine $\hat{\mathbf{x}}_{k|k+N}$ that is intuitively "better" than $\hat{\mathbf{x}}_{k|k}$, which is obtained through a Kalman filter. The fixed-lag estimate is defined by

$$\hat{\mathbf{x}}_{k|k+N} \equiv E\left\{\hat{\mathbf{x}}_k | [\tilde{\mathbf{y}}_1, \tilde{\mathbf{y}}_2 \ldots, \tilde{\mathbf{y}}_k, \tilde{\mathbf{y}}_{k+1}, \ldots, \tilde{\mathbf{y}}_{k+N}]\right\} \tag{6.158}$$

Thus, the point of time at which we seek the state estimate that lags the most recent measurement time by a fixed interval of time N, so that $t_{k+N} - t_k = \text{constant} > 0$.[15]

6.3.1 Discrete-Time Formulation

To derive the necessary relations for the discrete-time fixed-lag smoother, we start by rewriting the fixed-point smoother given by eqn. (6.53) as

$$\hat{\mathbf{x}}_{k|N} = \hat{\mathbf{x}}_{fk}^{+} + \mathcal{K}_k[\hat{\mathbf{x}}_{k+1|N} - \hat{\mathbf{x}}_{fk+1}^{-}] \tag{6.159}$$

where the notation in eqn. (6.158) has been used. Assuming that $\mathcal{K}_k^{-1}$ exists, then eqn. (6.159) can be solved for $\hat{\mathbf{x}}_{k+1|N}$, giving

$$\hat{\mathbf{x}}_{k+1|N} = \hat{\mathbf{x}}_{fk+1}^{-} + \mathcal{K}_k^{-1}[\hat{\mathbf{x}}_{k|N} - \hat{\mathbf{x}}_{fk}^{+}] \tag{6.160}$$

Substituting the relation for $\hat{\mathbf{x}}_{fk+1}^{-}$ in Table 6.2 into eqn. (6.160) gives

$$\hat{\mathbf{x}}_{k+1|N} = \Phi_k \hat{\mathbf{x}}_{fk}^{+} + \Gamma_k \mathbf{u}_k + \mathcal{K}_k^{-1}[\hat{\mathbf{x}}_{k|N} - \hat{\mathbf{x}}_{fk}^{+}] \tag{6.161}$$

Adding and subtracting $\Phi_k \hat{\mathbf{x}}_{k|N}$ from the right-hand side of eqn. (6.161) yields

$$\hat{\mathbf{x}}_{k+1|N} = \Phi_k \hat{\mathbf{x}}_{k|N} + \Gamma_k \mathbf{u}_k + [\mathcal{K}_k^{-1} - \Phi_k][\hat{\mathbf{x}}_{k|N} - \hat{\mathbf{x}}_{fk}^{+}] \tag{6.162}$$

Let us concentrate our attention on $\mathcal{K}_k^{-1} - \Phi_k$. From eqn. (6.52) we have

$$U_k \equiv \mathcal{K}_k^{-1} - \Phi_k = P_{fk+1}^{-} \Phi_k^{-T} (P_{fk}^{+})^{-1} - \Phi_k \tag{6.163}$$

Substituting the relation for P_{fk+1}^{-} in Table 6.2 into eqn. (6.163) gives

$$U_k = \Upsilon_k Q_k \Upsilon_k^T \Phi_k^{-T} (P_{fk}^{+})^{-1} \tag{6.164}$$

Therefore, substituting eqn. (6.164) into eqn. (6.162) gives

$$\hat{\mathbf{x}}_{k+1|N} = \Phi_k \hat{\mathbf{x}}_{k|N} + \Gamma_k \mathbf{u}_k + U_k[\hat{\mathbf{x}}_{k|N} - \hat{\mathbf{x}}_{fk}^{+}] \tag{6.165}$$

We now allow the right endpoint of the interval to be variable by replacing N by $k+N$, which gives

$$\hat{\mathbf{x}}_{k+1|k+N} = \Phi_k \hat{\mathbf{x}}_{k|k+N} + \Gamma_k \mathbf{u}_k + U_k[\hat{\mathbf{x}}_{k|k+N} - \hat{\mathbf{x}}_{fk}^{+}] \tag{6.166}$$

Equation (6.166) will be used to compute the fixed-lag state estimate.

From the results of §6.2.1, replacing k by $k+1$ and N by $k+1+N$ in eqn. (6.139) and using the measurement residual form from eqn. (6.118) yields

$$\hat{\mathbf{x}}_{k+1|k+1+N} = \hat{\mathbf{x}}_{k+1|k+N} + \mathcal{M}_{k+1|k+1+N} \boldsymbol{\upsilon}_{fk+1+N} \tag{6.167}$$

where

$$\mathcal{M}_{k+1|k+1+N} = \mathcal{B}_{k+1+N} K_{fk+1+N} \tag{6.168}$$

with

$$\mathcal{B}_{k+1+N} = \prod_{i=k+1}^{k+N} \mathcal{K}_i \tag{6.169}$$

Substituting eqn. (6.166) into eqn. (6.168) gives

$$\begin{aligned}\hat{\mathbf{x}}_{k+1|k+1+N} &= \Phi_k \hat{\mathbf{x}}_{k|k+N} + \Gamma_k \mathbf{u}_k + U_k[\hat{\mathbf{x}}_{k|k+N} - \hat{\mathbf{x}}_{fk}^+] \\ &\quad + \mathcal{M}_{k+1|k+1+N} \boldsymbol{\upsilon}_{fk+1+N}\end{aligned} \tag{6.170}$$

Using the definition of the residual $\boldsymbol{\upsilon}_{fk+1+N}$ from eqn. (6.117) and the forward-time state propagation in Table 6.2 in eqn. (6.170) leads to

$$\boxed{\begin{aligned}\hat{\mathbf{x}}_{k+1|k+1+N} &= \Phi_k \hat{\mathbf{x}}_{k|k+N} + \Gamma_k \mathbf{u}_k \\ &\quad + \Upsilon_k Q_k \Upsilon_k^T \Phi_k^{-T} (P_{fk}^+)^{-1} [\hat{\mathbf{x}}_{k|k+N} - \hat{\mathbf{x}}_{fk}^+] \\ &\quad + \mathcal{B}_{k+1+N} K_{fk+1+N} \{\tilde{\mathbf{y}}_{k+1+N} \\ &\quad - H_{k+1+N} \Phi_{k+N} [\hat{\mathbf{x}}_{fk+N}^+ + \Gamma_{k+N} \mathbf{u}_{k+N}]\}\end{aligned}} \tag{6.171}$$

where the initial condition for eqn. (6.171) is given by $\hat{\mathbf{x}}_{0|N}$. This initial condition is obtained from the optimal fixed-point smoother starting with $\hat{\mathbf{x}}_{0|0}$, processing measurements to obtain $\hat{\mathbf{x}}_{0|N}$. Then the Kalman filter is employed, where its gain, covariance, and state estimate are used in the fixed-lag smoother.

The fixed-lag smoother covariance is derived by first rewriting eqn. (6.51) as

$$P_{k+1|N} = P_{fk+1}^- - \mathcal{K}_k^{-1} [P_{fk}^+ - P_{k|N}] \mathcal{K}_k^{-T} \tag{6.172}$$

Replacing N by $k+N$ in eqn. (6.172) gives

$$P_{k+1|k+N} = P_{fk+1}^- - \mathcal{K}_k^{-1} [P_{fk}^+ - P_{k|k+N}] \mathcal{K}_k^{-T} \tag{6.173}$$

Next, using eqn. (5.44) we can write

$$P_{fN}^+ - P_{fN}^- = -K_{fN} H_N P_{fN}^- \tag{6.174}$$

Substituting eqn. (6.174) into eqn. (6.149), and replacing k by $k+1$ and N by $k+1+N$ leads to

$$P_{k+1|k+1+N} = P_{k+1|k+N} - \mathcal{B}_{k+1+N} K_{fk+1+N} H_{k+1+N} P_{fk+1+N}^- \mathcal{B}_{k+1+N}^T \tag{6.175}$$

Substituting eqn. (6.173) into eqn. (6.175) yields

$$\boxed{\begin{aligned}P_{k+1|k+1+N} &= P_{fk+1}^- - \mathcal{K}_k^{-1} [P_{fk}^+ - P_{k|k+N}] \mathcal{K}_k^{-T} \\ &\quad - \mathcal{B}_{k+1+N} K_{fk+1+N} H_{k+1+N} P_{fk+1+N}^- \mathcal{B}_{k+1+N}^T\end{aligned}} \tag{6.176}$$

where the initial condition for eqn. (6.176) is given by $P_{0|N}$, which is given by the optimal fixed-point covariance.

Table 6.8: Discrete-Time Fixed-Lag Smoother

Model	$\mathbf{x}_{k+1} = \Phi_k \mathbf{x}_k + \Gamma_k \mathbf{u}_k + \Upsilon_k \mathbf{w}_k, \quad \mathbf{w}_k \sim N(\mathbf{0}, Q_k)$ $\tilde{\mathbf{y}}_k = H_k \mathbf{x}_k + \mathbf{v}_k, \quad \mathbf{v}_k \sim N(\mathbf{0}, R_k)$
Forward Initialize	$\hat{\mathbf{x}}_f(t_0) = \hat{\mathbf{x}}_{f0}$ $P_f(t_0) = E\{\tilde{\mathbf{x}}_f(t_0)\,\tilde{\mathbf{x}}_f^T(t_0)\}$
Gain	$K_{fk} = P_{fk}^- H_k^T [H_k P_{fk}^- H_k^T + R_k]^{-1}$
Forward Update	$\hat{\mathbf{x}}_{fk}^+ = \hat{\mathbf{x}}_{fk}^- + K_{fk}[\tilde{\mathbf{y}}_k - H_k \hat{\mathbf{x}}_{fk}^-]$ $P_{fk}^+ = [I - K_{fk} H_k] P_{fk}^-$
Forward Propagation	$\hat{\mathbf{x}}_{fk+1}^- = \Phi_k \hat{\mathbf{x}}_{fk}^+ + \Gamma_k \mathbf{u}_k$ $P_{fk+1}^- = \Phi_k P_{fk}^+ \Phi_k^T + \Upsilon_k Q_k \Upsilon_k^T$
Smoother Initialize	$\hat{\mathbf{x}}_{0\|N}$ from fixed-point smoother $P_{0\|N}$ from fixed-point smoother
Gain	$\mathcal{B}_{k+1+N} = \prod_{i=k+1}^{k+N} \mathcal{K}_i, \quad \mathcal{K}_i = P_{fi}^+ \Phi_i^T (P_{fi+1}^-)^{-1}$
Covariance	$P_{k+1\|k+1+N} = P_{fk+1}^- - \mathcal{K}_k^{-1}[P_{fk}^+ - P_{k\|k+N}]\mathcal{K}_k^{-T}$ $-\mathcal{B}_{k+1+N} K_{fk+1+N} H_{k+1+N} P_{fk+1+N}^- \mathcal{B}_{k+1+N}^T$
Estimate	$\hat{\mathbf{x}}_{k+1\|k+1+N} = \Phi_k \hat{\mathbf{x}}_{k\|k+N} + \Gamma_k \mathbf{u}_k$ $+\Upsilon_k Q_k \Upsilon_k^T \Phi_k^{-T} (P_{fk}^+)^{-1}[\hat{\mathbf{x}}_{k\|k+N} - \hat{\mathbf{x}}_{fk}^+]$ $+\mathcal{B}_{k+1+N} K_{fk+1+N} \{\tilde{\mathbf{y}}_{k+1+N}$ $-H_{k+1+N} \Phi_{k+N} [\hat{\mathbf{x}}_{fk+N}^+ + \Gamma_{k+N} \mathbf{u}_{k+N}]\}$

A summary of the discrete-time fixed-lag smoother is given in Table 6.8. Equation (6.171) incorporates two correction terms. One is applied to the residual between the optimal fixed-lag smoother estimate, $\hat{\mathbf{x}}_{k|k+N}$, and the optimal Kalman filter estimate, $\hat{\mathbf{x}}_{fk}^+$, at time t_k. The other is applied to the measurement residual directly. The first correction reflects the residual back to the fixed-lag estimate. When no process noise is present (i.e., when $Q_k = 0$), this term has no effect on the fixed-lag smoother estimate, which intuitively makes sense. The second correction comes after a "waiting period"[15] where the fixed-lag smoother is dormant over the interval $[0, N]$. Then the fixed-lag smoother depends on the Kalman filter, which leads to the measurement residual in the fixed-lag smoother estimate. Finally, we should note that the fixed-lag smoother can be actually implemented in real time once it has been initialized. However, we still consider this "filter" to be a batch smoother since the sought estimate is not provided in real time, which is derived from future data points.

6.3.2 Continuous-Time Formulation

The continuous-time fixed-lag smoother can be derived using the same methods to derive the continuous-time fixed-point smoother in §6.2.2.[1] Suppose that we seek a smoother solution that lags the most recent measurement by a constant time delay Δ. Replacing t with $T-\Delta$ in eqn. (6.150) gives

$$\begin{aligned}\frac{d}{dt}\hat{\mathbf{x}}(T-\Delta|T) = {} & [F(T-\Delta)+G(T-\Delta)\,Q(T-\Delta)\,G^T(T-\Delta)\,P_f^{-1}(T-\Delta)] \\ & \times \hat{\mathbf{x}}(T-\Delta|T)+B(T-\Delta)\,\mathbf{u}(T-\Delta) \\ & -G(T-\Delta)\,Q(T-\Delta)\,G^T(T-\Delta)\,P_f^{-1}(T-\Delta)\,\hat{\mathbf{x}}_f(T-\Delta)\end{aligned} \tag{6.177}$$

The solution of eqn. (6.177) is given by

$$\begin{aligned}\hat{\mathbf{x}}(T-\Delta|T) = {} & \Psi(T-\Delta,T)\,\hat{\mathbf{x}}_f(T)+\int_T^{T-\Delta}\Psi(T-\Delta,\tau)\,B(\tau)\,\mathbf{u}(\tau)\,d\tau \\ & -\int_T^{T-\Delta}\Psi(T-\Delta,\tau)\,G(\tau)\,Q(\tau)\,G^T(\tau)\,P_f^{-1}(\tau)\,\hat{\mathbf{x}}_f(\tau)\,d\tau\end{aligned} \tag{6.178}$$

where $\Psi(T-\Delta,T)$ is the state transition matrix, which clearly must obey

$$\begin{aligned}\frac{d}{dt}\Psi(T-\Delta,T) = {} & [F(T-\Delta)+G(T-\Delta)\,Q(T-\Delta)\,G^T(T-\Delta)\,P_f^{-1}(T-\Delta)] \\ & \times\Psi(T-\Delta,T), \quad \Psi(t,t)=I\end{aligned} \tag{6.179}$$

Note, the matrix $\Phi(t,T)$ from §6.2.2 and the matrix $\Psi(T-\Delta,T)$ are related by

$$\Psi(T-\Delta,T)=\Phi(T-\Delta,t)\,\Phi(t,T) \tag{6.180}$$

with $\Phi(0,T)=\Psi(0,T)$. Differentiating eqn. (6.180) with respect to T and using eqn. (6.152) yields

$$\boxed{\begin{aligned}\frac{d}{dT}\Psi(T-\Delta,T) = {} & [F(T-\Delta)+G(T-\Delta)\,Q(T-\Delta) \\ & \times G^T(T-\Delta)\,P_f^{-1}(T-\Delta)]\Psi(T-\Delta,T) \\ & -\Psi(T-\Delta,T)[F(T)+G(T)\,Q(T)\,G^T(T)\,P_f^{-1}(T)]\end{aligned}} \tag{6.181}$$

Table 6.9: Continuous-Time Fixed-Lag Smoother

Model	$\frac{d}{dt}\mathbf{x}(t) = F(t)\,\mathbf{x}(t) + B(t)\,\mathbf{u}(t) + G(t)\,\mathbf{w}(t),\ \mathbf{w}(t) \sim N(\mathbf{0}, Q(t))$ $\tilde{\mathbf{y}}(t) = H(t)\,\mathbf{x}(t) + \mathbf{v}(t),\ \mathbf{v}(t) \sim N(\mathbf{0}, R(t))$
Forward Covariance	$\frac{d}{dt}P_f(t) = F(t)\,P_f(t) + P_f(t)\,F^T(t)$ $-P_f(t)\,H^T(t)\,R^{-1}(t)H(t)\,P_f(t)$ $+G(t)\,Q(t)\,G^T(t),$ $P_f(t_0) = E\{\tilde{\mathbf{x}}_f(t_0)\,\tilde{\mathbf{x}}_f^T(t_0)\}$
Forward Filter	$\frac{d}{dt}\hat{\mathbf{x}}_f(t) = F(t)\,\hat{\mathbf{x}}_f(t) + B(t)\,\mathbf{u}(t)$ $+P_f(t)\,H^T(t)\,R^{-1}(t)[\tilde{\mathbf{y}}(t) - H(t)\,\hat{\mathbf{x}}_f(t)], \quad \hat{\mathbf{x}}_f(t_0) = \hat{\mathbf{x}}_{f0}$
Smoother Initialize	$\hat{\mathbf{x}}(0\|\Delta)$ from fixed-point smoother $P(0\|\Delta)$ from fixed-point smoother
Transition Matrix	$\frac{d}{dT}\Psi(T-\Delta, T) = [F(T-\Delta) + G(T-\Delta)\,Q(T-\Delta)$ $\times G^T(T-\Delta)\,P_f^{-1}(T-\Delta)]\Psi(T-\Delta, T)$ $-\Psi(T-\Delta, T)[F(T) + G(T)\,Q(T)\,G^T(T)\,P_f^{-1}(T)],$ $\Psi(0, \Delta) = \Phi(0, \Delta)$
Smoother Covariance	$\frac{d}{dT}P(T-\Delta\|T) = [F(T-\Delta) + G(T-\Delta)\,Q(T-\Delta)$ $\times G^T(T-\Delta)\,P_f^{-1}(T-\Delta)]P(T-\Delta\|T) + P(T-\Delta\|T)$ $\times[F(T-\Delta) + G(T-\Delta)\,Q(T-\Delta)\,G^T(T-\Delta)\,P_f^{-1}(T-\Delta)]^T$ $-\Psi(T-\Delta, T)\,P_f(T)\,H^T(T)\,R^{-1}(T)\,P_f(T)\,\Psi(T-\Delta, T)$ $-G(T-\Delta)\,Q(T-\Delta)\,G^T(T-\Delta)$
Smoother Estimate	$\frac{d}{dT}\hat{\mathbf{x}}(T-\Delta\|T) = [F(T-\Delta) + G(T-\Delta)\,Q(T-\Delta)$ $\times G^T(T-\Delta)\,P_f^{-1}(T-\Delta)]\hat{\mathbf{x}}(T-\Delta\|T)$ $-G(T-\Delta)\,Q(T-\Delta)\,G^T(T-\Delta)\,P_f^{-1}(T-\Delta)\,\hat{\mathbf{x}}_f(T-\Delta)$ $+\Psi(T-\Delta, T)\,P_f(T)\,H^T(T)\,R^{-1}(T)[\tilde{\mathbf{y}}(T) - H(T)\,\hat{\mathbf{x}}_f(T)]$

Taking the derivative of eqn. (6.178) with respect to T, and substituting the forward-time state filter equation from Table 6.3 and eqn. (6.181) into the resulting equation leads to (the details are left as an exercise for the reader)

$$\begin{aligned}\frac{d}{dT}\hat{\mathbf{x}}(T-\Delta|T) = {} & [F(T-\Delta)+G(T-\Delta)\,Q(T-\Delta) \\ & \times G^T(T-\Delta)\,P_f^{-1}(T-\Delta)]\hat{\mathbf{x}}(T-\Delta|T) \\ & - G(T-\Delta)\,Q(T-\Delta)\,G^T(T-\Delta)\,P_f^{-1}(T-\Delta)\,\hat{\mathbf{x}}_f(T-\Delta) \\ & + \Psi(T-\Delta,T)\,P_f(T)\,H^T(T)\,R^{-1}(T)[\tilde{\mathbf{y}}(T)-H(T)\,\hat{\mathbf{x}}_f(T)]\end{aligned} \tag{6.182}$$

where the initial condition for eqn. (6.182) is given by $\hat{\mathbf{x}}(0|\Delta)$. This initial condition is obtained from the optimal fixed-point smoother starting with $\hat{\mathbf{x}}(0|0)$, processing measurements to obtain $\hat{\mathbf{x}}(0|\Delta)$. The covariance can be shown to be given by (the details are left as an exercise for the reader)

$$\begin{aligned}\frac{d}{dT}P(T-\Delta|T) = {} & [F(T-\Delta)+G(T-\Delta)\,Q(T-\Delta) \\ & \times G^T(T-\Delta)\,P_f^{-1}(T-\Delta)]P(T-\Delta|T)+P(T-\Delta|T) \\ & \times [F(T-\Delta)+G(T-\Delta)\,Q(T-\Delta)\,G^T(T-\Delta)\,P_f^{-1}(T-\Delta)]^T \\ & - \Psi(T-\Delta,T)\,P_f(T)\,H^T(T)\,R^{-1}(T)\,P_f(T)\,\Psi(T-\Delta,T) \\ & - G(T-\Delta)\,Q(T-\Delta)\,G^T(T-\Delta)\end{aligned} \tag{6.183}$$

with initial condition given by $P(0|\Delta)$ from the optimal fixed-point smoother covariance. A summary of the continuous-time fixed-lag smoother is given in Table 6.9. The initial conditions and smoother implementation follow exactly like the discrete-time fixed-lag smoother, but the continuous-time equations are integrated in order to provide the state estimate.

Example 6.5: We again consider the simple first-order system shown in example 6.2. Assuming that the forward-pass covariance has reached a steady-state value, given by p_f, since $p(T-\Delta) = p(T) = p_f$ the state transition matrix using eqn. (6.181) reduces down to

$$\frac{d\psi(T-\Delta,T)}{dT} = 0$$

Note that Δ is fixed and $T \geq \Delta$. The solution for $\psi(t,T)$ is given by using the initial condition from the fixed-point smoother state transition matrix in example 6.4, which gives

$$\psi(T-\Delta,T) = e^{-\beta\Delta}$$

where $\beta \equiv (f + q/p_f)$. Then, the smoother covariance using eqn. (6.183) reduces down to

$$\frac{dp(T-\Delta|T)}{dT} = 2\beta\, p(T-\Delta|T) - \left[r^{-1} p_f^2 e^{-2\beta\Delta} + q\right]$$

Using the initial condition $p(0|\Delta)$, the solution for $p(T-\Delta|T)$ is given by

$$p(T-\Delta|T) = p(0|\Delta)\,e^{2\beta(T-\Delta)} + \frac{r^{-1}p_f^2 e^{-2\beta\,\Delta}+q}{2\beta}\left[1-e^{2\beta(T-\Delta)}\right]$$

where $p(0|\Delta)$ is evaluated from example 6.4, which leads to

$$p(0|\Delta) = p_f\,e^{-2\beta\,\Delta} + \frac{q}{2a}\left[1-e^{-2\beta\,\Delta}\right]$$

where $a \equiv \sqrt{f^2+r^{-1}q}$. Then, the solution for $p(T-\Delta|T)$ can be shown to be given by $p(T-\Delta|T) = p(0|\Delta)$ (which is left as an exercise for the reader). Note, this only occurs since $p_f(t)$ is at steady-state. Consider when Δ is sufficiently large so that the exponential terms decay to near-zero (e.g., after four times the time constant, i.e., when $\Delta \geq 2/\beta$). Then, the fixed-lag smoother covariance at steady-state is given by $q/(2a)$, which is equivalent to the smoother steady-state covariance given in example 6.2. This intuitively makes sense since the accuracy of the fixed-lag smoother should be equivalent to the fixed-interval smoother at steady-state.

6.4 Advanced Topics

In this section we will show some advanced topics used in smoothers. As in previous chapters we encourage the interested reader to pursue these topics further in the references provided. These topics include the duality between estimation and control, and new derivations of the fixed-interval smoothers based on the innovations process.

6.4.1 Estimation/Control Duality

One of the most fascinating aspects of the fixed-interval RTS smoother is that it can be completely derived from optimal control theory. This mathematical *duality* between estimation and control arises from solving the two-point-boundary-value-problem (TPBVP) associated with optimal control theory.[9, 16] In this section we assume that the reader is familiar with the variational approach, which transforms the minimization problem into a TPBVP. More details on the variational approach can be found in Chapter 8. We will derive the discrete-time and continuous-time cases here, as well as a new derivation of the nonlinear RTS smoother with continuous-time models and discrete-time measurements.

6.4.1.1 Discrete-Time Formulation

Consider minimizing the following discrete-time loss function:

$$J(\mathbf{w}_k) = \frac{1}{2}\sum_{k=1}^{N}[\tilde{\mathbf{y}}_k - H_k\mathbf{x}_k]^T R_k^{-1}[\tilde{\mathbf{y}}_k - H_k\mathbf{x}_k] + \mathbf{w}_k^T Q_k^{-1}\mathbf{w}_k$$
$$+\frac{1}{2}[\hat{\mathbf{x}}_{f0} - \mathbf{x}_0]^T P_{f0}^{-1}[\hat{\mathbf{x}}_{f0} - \mathbf{x}_0] \tag{6.184}$$

subject to the dynamic constraint

$$\mathbf{x}_{k+1} = \Phi_k\mathbf{x}_k + \Gamma_k\mathbf{u}_k + \Upsilon_k\mathbf{w}_k \tag{6.185}$$

Note that $\hat{\mathbf{x}}_{f0}$ is the *a priori* estimate of $\mathbf{x}_0$, with error-covariance P_{f0}, and $J(\mathbf{w}_k)$ is the negative log-likelihood function.[9] Also, we treat $\mathbf{u}_k$ as a deterministic input. Finally, the inverse of Q_k must exist in order to achieve controllability in this minimization problem, which is also discussed in §6.1.1. Let us denote the best estimate of $\mathbf{x}$ as $\hat{\mathbf{x}}$. Then, the minimization of eqn. (6.184) yields the following TPBVP (see §8.4):[16, 17]

$$\hat{\mathbf{x}}_{k+1} = \Phi_k\hat{\mathbf{x}}_k + \Gamma_k\mathbf{u}_k + \Upsilon_k\mathbf{w}_k \tag{6.186a}$$
$$\boldsymbol{\lambda}_k = \Phi_k^T\boldsymbol{\lambda}_{k+1} + H_k^T R_k^{-1} H_k\hat{\mathbf{x}}_k - H_k^T R_k^{-1}\tilde{\mathbf{y}}_k \tag{6.186b}$$
$$\mathbf{w}_k = -Q_k\Upsilon_k^T\boldsymbol{\lambda}_{k+1} \tag{6.186c}$$

where $\boldsymbol{\lambda}_k$ is known as the *costate vector*, which arises from using a Lagrange multiplier for the equality constraint in eqn. (6.185). The boundary conditions are given by

$$\boldsymbol{\lambda}_N = \mathbf{0} \tag{6.187a}$$
$$\boldsymbol{\lambda}_0 = P_{f0}^{-1}[\hat{\mathbf{x}}_{f0} - \hat{\mathbf{x}}_0] \tag{6.187b}$$

Substituting eqn. (6.186c) into eqn. (6.186a) gives the following TPBVP:

$$\boxed{\begin{aligned}\hat{\mathbf{x}}_{k+1} &= \Phi_k\hat{\mathbf{x}}_k + \Gamma_k\mathbf{u}_k - \Upsilon_k Q_k\Upsilon_k^T\boldsymbol{\lambda}_{k+1} \\ \boldsymbol{\lambda}_k &= \Phi_k^T\boldsymbol{\lambda}_{k+1} + H_k^T R_k^{-1} H_k\hat{\mathbf{x}}_k - H_k^T R_k^{-1}\tilde{\mathbf{y}}_k\end{aligned}} \quad \begin{aligned}(6.188a)\\(6.188b)\end{aligned}$$

Equation (6.188) will be used to derive the discrete-time RTS smoother solution.

In order to decouple the state and costate vectors in eqn. (6.188) we use the following inhomogeneous Riccati transformation:

$$\boxed{\hat{\mathbf{x}}_k = \hat{\mathbf{x}}_{fk} - P_{fk}\boldsymbol{\lambda}_k} \tag{6.189}$$

where P_{fk} is an $n \times n$ matrix and $\hat{\mathbf{x}}_{fk}$ is the inhomogeneous vector. We will show in the subsequent derivation that $\hat{\mathbf{x}}_{fk}$ is indeed the forward-time Kalman filter state estimate and $\hat{\mathbf{x}}_k$ is the smoother state estimate. Comparing eqn. (6.189) with eqn. (6.187) indicates that in order for $\boldsymbol{\lambda}_N = \mathbf{0}$ to be satisfied, then $\hat{\mathbf{x}}_N = \hat{\mathbf{x}}_{fN}$. Substituting eqn. (6.189) into eqn. (6.188b), collecting terms and factoring out P_{fk}^{-1} yields

$$\boldsymbol{\lambda}_k = P_{fk}^{-1} Z_k [\Phi_k^T \boldsymbol{\lambda}_{k+1} + H_k^T R_k^{-1} H_k \hat{\mathbf{x}}_{fk} - H_k^T R_k^{-1} \tilde{\mathbf{y}}_k] \tag{6.190}$$

where

$$Z_k \equiv [P_{fk}^{-1} + H_k^T R_k^{-1} H_k]^{-1} \tag{6.191}$$

Taking one time-step ahead of eqn. (6.189) gives

$$\hat{\mathbf{x}}_{k+1} = \hat{\mathbf{x}}_{fk+1} - P_{fk+1} \boldsymbol{\lambda}_{k+1} \tag{6.192}$$

Substituting eqns. (6.189) and (6.192) into eqn. (6.188a) and rearranging gives

$$[P_{fk+1} - \Upsilon_k Q_k \Upsilon_k^T] \boldsymbol{\lambda}_{k+1} - \Phi_k P_{fk} \boldsymbol{\lambda}_k - \hat{\mathbf{x}}_{fk+1} + \Phi_k \hat{\mathbf{x}}_{fk} + \Gamma_k \mathbf{u}_k = \mathbf{0} \tag{6.193}$$

Substituting eqn. (6.190) into eqn. (6.193) and collecting terms yields

$$\begin{aligned} &[P_{fk+1} - \Phi_k Z_k \Phi_k^T - \Upsilon_k Q_k \Upsilon_k^T] \boldsymbol{\lambda}_{k+1} \\ &+ \Phi_k \hat{\mathbf{x}}_{fk} + \Gamma_k \mathbf{u}_k + \Phi_k Z_k H_k^T R_k^{-1} [\tilde{\mathbf{y}}_k - H_k \hat{\mathbf{x}}_{fk}] - \hat{\mathbf{x}}_{fk+1} = \mathbf{0} \end{aligned} \tag{6.194}$$

Avoiding the trivial solution of $\boldsymbol{\lambda}_{k+1} = \mathbf{0}$ gives the following two equations:

$$P_{fk+1} = \Phi_k Z_k \Phi_k^T + \Upsilon_k Q_k \Upsilon_k^T \tag{6.195a}$$

$$\hat{\mathbf{x}}_{fk+1} = \Phi_k \hat{\mathbf{x}}_{fk} + \Gamma_k \mathbf{u}_k + \Phi_k Z_k H_k^T R_k^{-1} [\tilde{\mathbf{y}}_k - H_k \hat{\mathbf{x}}_{fk}] \tag{6.195b}$$

We now prove the following identity:

$$Z_k H_k^T R_k^{-1} = K_{fk} \equiv P_{fk} H_k^T [H_k P_{fk} H_k^T + R_k]^{-1} \tag{6.196}$$

Using the matrix inversion lemma in eqn. (1.70) with $A = \mathcal{P}_{fk}^{-1}$, $B = H_k^T$, $C = R_k^{-1}$, and $D = H_k$ leads to the following form for eqn. (6.196):

$$\left\{ P_{fk} - P_{fk} H_k^T [H_k P_{fk} H_k^T + R_k]^{-1} H_k P_{fk} \right\} H_k^T R_k^{-1} = K_{fk} \tag{6.197}$$

Next, using the definition of the forward-time gain K_{fk} and right-multiplying both sides of eqn. (6.197) by R_k leads to

$$P_{fk} H_k^T - K_{fk} H_k P_{fk} H_k^T = K_{fk} R_k \tag{6.198}$$

Collecting terms reduces eqn. (6.198) to

$$P_{fk} H_k^T = K_{fk} [H_k P_{fk} H_k^T + R_k] \tag{6.199}$$

Finally, using the definition of the gain K_{fk} proves the identity. Therefore, we can write eqn. (6.195) as

$$P_{fk+1} = \Phi_k P_{fk} \Phi_k^T - \Phi_k P_{fk} H_k^T [H_k P_{fk} H_k^T + R_k]^{-1} H_k P_{fk} \Phi_k^T + \Upsilon_k Q_k \Upsilon_k^T \tag{6.200a}$$

$$\hat{\mathbf{x}}_{fk+1} = \Phi_k \hat{\mathbf{x}}_{fk} + \Gamma_k \mathbf{u}_k + \Phi_k K_{fk} [\tilde{\mathbf{y}}_k - H_k \hat{\mathbf{x}}_{fk}] \tag{6.200b}$$

Equation (6.200) constitutes the forward-time Kalman filter covariance and state estimate with $P_{fk} \equiv P_{fk}^-$ and $\hat{\mathbf{x}}_{fk} \equiv \hat{\mathbf{x}}_{fk}^-$.

We now need an expression for the state estimate $\hat{\mathbf{x}}_k$. Solving eqns. (6.189) and (6.190) for $\boldsymbol{\lambda}_k$ and $\boldsymbol{\lambda}_{k+1}$, respectively, and substituting the resulting expressions into eqn. (6.186b) gives

$$P_{fk}^{-1}[\hat{\mathbf{x}}_{fk} - \hat{\mathbf{x}}_k] = \Phi_k^T P_{fk+1}^{-1}[\hat{\mathbf{x}}_{fk+1} - \hat{\mathbf{x}}_{k+1}] + H_k^T R_k^{-1} H_k \hat{\mathbf{x}}_k - H_k^T R_k^{-1} \tilde{\mathbf{y}}_k \quad (6.201)$$

Solving eqn. (6.201) for $\hat{\mathbf{x}}_k$ yields

$$\hat{\mathbf{x}}_k = L_k \hat{\mathbf{x}}_{fk} + L_k P_{fk} H_k^T R_k^{-1} \tilde{\mathbf{y}}_k + L_k P_{fk} \Phi_k^T P_{fk+1}^{-1}[\hat{\mathbf{x}}_{k+1} - \hat{\mathbf{x}}_{fk+1}] \quad (6.202)$$

where

$$L_k \equiv [I + P_{fk} H_k^T R_k^{-1} H_k]^{-1} = [P_{fk}^{-1} + H_k^T R_k^{-1} H_k]^{-1} P_{fk}^{-1} \quad (6.203)$$

Now, consider the following identities (which are left as an exercise for the reader):

$$L_k P_{fk} = P_{fk}^+ \quad (6.204a)$$

$$L_k = I - K_{fk} H_k \quad (6.204b)$$

Substituting eqn. (6.204) into eqn. (6.202) yields

$$\hat{\mathbf{x}}_k = \hat{\mathbf{x}}_{fk} - K_{fk} H_k \hat{\mathbf{x}}_{fk} + P_{fk}^+ H_k^T R_k^{-1} \tilde{\mathbf{y}}_k + P_{fk}^+ \Phi_k^T P_{fk+1}^{-1}[\hat{\mathbf{x}}_{k+1} - \hat{\mathbf{x}}_{fk+1}] \quad (6.205)$$

Finally, using the definitions of the gain K_{fk} from eqn. (5.47), and $P_{fk+1} \equiv P_{fk+1}^-$ and $\hat{\mathbf{x}}_{fk} \equiv \hat{\mathbf{x}}_{fk}^-$ leads to

$$\hat{\mathbf{x}}_k = \hat{\mathbf{x}}_{fk}^+ + \mathcal{K}_k[\hat{\mathbf{x}}_{k+1} - \hat{\mathbf{x}}_{fk+1}^-] \quad (6.206)$$

where

$$\mathcal{K}_k \equiv P_{fk}^+ \Phi_k^T (P_{fk+1}^-)^{-1} \quad (6.207)$$

Equation (6.206) is exactly the discrete-time RTS smoother.

6.4.1.2 Continuous-Time Formulation

The continuous-time formulation is much easier to derive than the discrete-time system. Consider minimizing the following continuous-time loss function:

$$\begin{aligned} J[\mathbf{w}(t)] = \frac{1}{2}\int_{t_0}^{t_N} & \Big\{[\tilde{\mathbf{y}}(t) - H(t)\mathbf{x}(t)]^T R^{-1}(t)[\tilde{\mathbf{y}}(t) - H(t)\mathbf{x}(t)] \\ & + \mathbf{w}^T(t) Q^{-1}(t)\mathbf{w}(t)\Big\}\, dt \\ & + \frac{1}{2}[\hat{\mathbf{x}}_f(t_0) - \mathbf{x}(t_0)]^T P_f^{-1}(t_0)[\hat{\mathbf{x}}_f(t_0) - \mathbf{x}(t_0)] \end{aligned} \quad (6.208)$$

subject to the dynamic constraint

$$\frac{d}{dt}\mathbf{x}(t) = F(t)\,\mathbf{x}(t) + B(t)\,\mathbf{u}(t) + G(t)\,\mathbf{w}(t) \tag{6.209}$$

Note that for the continuous-time case the loss function in eqn. (6.208) becomes infinite if the measurement and process noises are represented by white noise. However, since white noise can be formulated as a limiting case of nonwhite noise, the derivation of the final results can be achieved by avoiding stochastic calculus.[9] Let us again denote the best estimate of $\mathbf{x}$ as $\hat{\mathbf{x}}$. Then, the minimization of eqn. (6.208) yields the following TPBVP (see §8.2):[16, 17]

$$\frac{d}{dt}\hat{\mathbf{x}}(t) = F(t)\,\hat{\mathbf{x}}(t) + B(t)\,\mathbf{u}(t) + G(t)\,\mathbf{w}(t) \tag{6.210a}$$

$$\frac{d}{dt}\boldsymbol{\lambda}(t) = -F^T(t)\,\boldsymbol{\lambda}(t) - H^T(t)\,R^{-1}(t)\,H(t)\,\hat{\mathbf{x}}(t) + H^T(t)\,R^{-1}(t)\,\tilde{\mathbf{y}}(t) \tag{6.210b}$$

$$\mathbf{w}(t) = -Q(t)\,G^T(t)\,\boldsymbol{\lambda}(t) \tag{6.210c}$$

The boundary conditions are given by

$$\boldsymbol{\lambda}(T) = \mathbf{0} \tag{6.211a}$$

$$\boldsymbol{\lambda}(t_0) = P_f^{-1}(t_0)\,[\hat{\mathbf{x}}_f(t_0) - \hat{\mathbf{x}}(t_0)] \tag{6.211b}$$

Substituting eqn. (6.210c) into eqn. (6.210a) gives the following TPBVP:

$$\boxed{\begin{aligned}\frac{d}{dt}\hat{\mathbf{x}}(t) &= F(t)\,\hat{\mathbf{x}}(t) + B(t)\,\mathbf{u}(t) - G(t)\,Q(t)\,G^T(t)\boldsymbol{\lambda}(t) && (6.212a)\\ \frac{d}{dt}\boldsymbol{\lambda}(t) &= -F^T(t)\,\boldsymbol{\lambda}(t) - H^T(t)\,R^{-1}(t)\,H(t)\,\hat{\mathbf{x}}(t) \\ &\quad + H^T(t)\,R^{-1}(t)\,\tilde{\mathbf{y}}(t) && (6.212b)\end{aligned}}$$

Equation (6.212) will be used to derive the continuous-time RTS smoother solution.

As with the discrete-time case we consider the following inhomogeneous Riccati transformation:

$$\hat{\mathbf{x}}(t) = \hat{\mathbf{x}}_f(t) - P_f(t)\,\boldsymbol{\lambda}(t) \tag{6.213}$$

Comparing eqn. (6.213) with eqn. (6.211) indicates that in order for $\boldsymbol{\lambda}(T) = \mathbf{0}$ to be satisfied, then $\hat{\mathbf{x}}(T) = \hat{\mathbf{x}}_f(T)$. Taking the time-derivative of eqn. (6.213) gives

$$\frac{d}{dt}\hat{\mathbf{x}}(t) = \frac{d}{dt}\hat{\mathbf{x}}_f(t) - \left[\frac{d}{dt}P_f(t)\right]\boldsymbol{\lambda}(t) - P_f(t)\left[\frac{d}{dt}\boldsymbol{\lambda}(t)\right] \tag{6.214}$$

Substituting eqn. (6.212) into eqn. (6.214) gives

$$\begin{aligned}&F(t)\,\hat{\mathbf{x}}(t) + B(t)\,\mathbf{u}(t) - G(t)\,Q(t)\,G^T(t)\boldsymbol{\lambda}(t)\\ &-\frac{d}{dt}\hat{\mathbf{x}}_f(t) + \left[\frac{d}{dt}P_f(t)\right]\boldsymbol{\lambda}(t) - P_f(t)\,F^T(t)\,\boldsymbol{\lambda}(t)\\ &-P_f(t)\,H^T(t)\,R^{-1}(t)\,H(t)\,\hat{\mathbf{x}}(t) + P_f(t)\,H^T(t)\,R^{-1}(t)\,\tilde{\mathbf{y}}(t) = \mathbf{0}\end{aligned} \tag{6.215}$$

Substituting eqn. (6.213) into eqn. (6.215), and collecting terms gives

$$\begin{aligned}&\Big[\frac{d}{dt}P_f(t) - F(t)\,P_f(t) - P_f(t)\,F^T(t) + P_f(t)\,H^T(t)\,R^{-1}(t)\,H(t)\,P_f(t)\\ &- G(t)\,Q(t)\,G^T(t)\Big]\boldsymbol{\lambda}(t) + F(t)\,\hat{\mathbf{x}}_f(t) + B(t)\,\mathbf{u}(t)\\ &+ P_f(t)\,H^T(t)\,R^{-1}(t)[\tilde{\mathbf{y}}(t) - H(t)\,\hat{\mathbf{x}}_f(t)] - \frac{d}{dt}\hat{\mathbf{x}}_f(t) = \mathbf{0}\end{aligned} \tag{6.216}$$

Avoiding the trivial solution of $\boldsymbol{\lambda}(t) = \mathbf{0}$ gives the following two equations:

$$\begin{aligned}\frac{d}{dt}P_f(t) = F(t)\,P_f(t) + P_f(t)\,F^T(t) - P_f(t)\,H^T(t)\,R^{-1}(t)\,H(t)\,P_f(t)\\ + G(t)\,Q(t)\,G^T(t)\end{aligned} \tag{6.217a}$$

$$\frac{d}{dt}\hat{\mathbf{x}}_f(t) = F(t)\,\hat{\mathbf{x}}_f(t) + B(t)\,\mathbf{u}(t) + K_f(t)[\tilde{\mathbf{y}}(t) - H(t)\,\hat{\mathbf{x}}_f(t)] \tag{6.217b}$$

where

$$K_f(t) \equiv P_f(t)\,H^T(t)\,R^{-1}(t) \tag{6.218}$$

Equation (6.217) constitutes the forward-time Kalman filter covariance and state estimate. The smoother equation is easily given by solving eqn. (6.213) for $\boldsymbol{\lambda}(t)$ and substituting the resulting expression into eqn. (6.212a), which yields

$$\frac{d}{dt}\hat{\mathbf{x}}(t) = F(t)\,\hat{\mathbf{x}}(t) + B(t)\,\mathbf{u}(t) + G(t)\,Q(t)\,G^T(t)\,P_f^{-1}(t)\left[\hat{\mathbf{x}}(t) - \hat{\mathbf{x}}_f(t)\right] \tag{6.219}$$

Equation (6.219) is exactly the continuous-time RTS smoother.

6.4.1.3 Nonlinear Formulation

In this section the results of §6.1.3 will fully be derived. The literature for nonlinear smoothing involving continuous-time models and discrete-time measurements is sparse though. An algorithm is presented in Ref. [1] without proof or reference. This algorithm relies upon the computation and use of the discrete-time model state-transition matrix as well as the discrete-time process noise covariance. From a practical point of view this approach may become unstable if the measurement frequency is not within Nyquist's limit. McReynolds[9] fills in many of the gaps in the derivations of early linear fixed-interval smoothers. In this current section McReynolds' results are extended for nonlinear continuous-discrete time systems. Consider minimizing the following mixed continuous-discrete loss function:

$$\begin{aligned}J = &\frac{1}{2}\sum_{k=1}^{N}\left[\tilde{\mathbf{y}}_k - \mathbf{h}_k(\mathbf{x}_k)\right]^T R_k^{-1}\left[\tilde{\mathbf{y}}_k - \mathbf{h}_k(\mathbf{x}_k)\right]\\ &+ \frac{1}{2}\int_{t_0}^{t_N} \mathbf{w}^T(t)\,Q^{-1}(t)\,\mathbf{w}(t)\,dt\\ &+ \frac{1}{2}\left[\hat{\mathbf{x}}_f(t_0) - \mathbf{x}(t_0)\right]^T P_f(t_0)^{-1}\left[\hat{\mathbf{x}}_f(t_0) - \mathbf{x}(t_0)\right]\end{aligned} \tag{6.220}$$

subject to the dynamic constraint

$$\dot{\mathbf{x}}(t) = \mathbf{f}(\mathbf{x}(t), \mathbf{u}(t), t) + G(t)\,\mathbf{w}(t) \tag{6.221a}$$

$$\tilde{\mathbf{y}}_k = \mathbf{h}_k(\mathbf{x}_k) + \mathbf{v}_k \tag{6.221b}$$

Let us again denote the best estimate of $\mathbf{x}$ as $\hat{\mathbf{x}}$. The optimal control theory for continuous-time loss functions including discrete state penalty terms has been studied by Geering.[18] Using this theory the minimization of eqn. (6.220) yields the following TPBVP:[19]

$$\frac{d}{dt}\hat{\mathbf{x}}(t) = \mathbf{f}(\hat{\mathbf{x}}(t), \mathbf{u}(t), t) - G(t)\,Q(t)\,G^T(t)\boldsymbol{\lambda}(t) \tag{6.222a}$$

$$\dot{\boldsymbol{\lambda}}(t) = -F^T(\hat{\mathbf{x}}(t), t)\,\boldsymbol{\lambda}(t) \tag{6.222b}$$

$$\boldsymbol{\lambda}_k^+ = \boldsymbol{\lambda}_k^- + H_k^T(\hat{\mathbf{x}}_k)\,R_k^{-1}\left[\tilde{\mathbf{y}}_k - \mathbf{h}_k(\hat{\mathbf{x}}_k)\right] \tag{6.222c}$$

where

$$F(\hat{\mathbf{x}}(t), t) \equiv \left.\frac{\partial \mathbf{f}}{\partial \mathbf{x}}\right|_{\hat{\mathbf{x}}(t)}, \quad H_k(\hat{\mathbf{x}}_k) \equiv \left.\frac{\partial \mathbf{h}}{\partial \mathbf{x}}\right|_{\hat{\mathbf{x}}_k} \tag{6.223}$$

The boundary conditions are given by

$$\boldsymbol{\lambda}(T) = \mathbf{0} \tag{6.224a}$$

$$\boldsymbol{\lambda}(t_0) = P_f^{-1}(t_0)[\hat{\mathbf{x}}_f(t_0) - \hat{\mathbf{x}}(t_0)] \tag{6.224b}$$

Note that discrete-joints are present in the costate vector at the measurement times, but these discontinuities do not directly appear in the state vector estimate equation.

In order to decouple the state and costate vectors in eqn. (6.222) we use the inhomogeneous Riccati transformation given by eqns. (6.213) and (6.214). Our first step in the derivation of the smoother equation is to linearize $\mathbf{f}(\hat{\mathbf{x}}(t), \mathbf{u}(t), t)$ about $\hat{\mathbf{x}}_f(t)$, which yields

$$\begin{aligned}\frac{d}{dt}\hat{\mathbf{x}}(t) = \mathbf{f}(\hat{\mathbf{x}}_f(t), \mathbf{u}(t), t) + F(\hat{\mathbf{x}}_f(t), t)\left[\hat{\mathbf{x}}(t) - \hat{\mathbf{x}}_f(t)\right] \\ - G(t)\,Q(t)\,G^T(t)\,\boldsymbol{\lambda}(t)\end{aligned} \tag{6.225}$$

Substituting eqn. (6.213) into eqn. (6.225), and then substituting the resulting expression and eqn. (6.222b) into eqn. (6.214) yields

$$\begin{aligned}&\left[\frac{d}{dt}P_f(t) - F(\hat{\mathbf{x}}_f(t), t)\,P_f(t) - P_f(t)\,F^T(\hat{\mathbf{x}}(t), t) - G(t)\,Q(t)\,G^T(t)\right]\boldsymbol{\lambda}(t) \\ &+ \left[\mathbf{f}(\hat{\mathbf{x}}_f(t), \mathbf{u}(t), t) - \frac{d}{dt}\hat{\mathbf{x}}_f(t)\right] = \mathbf{0}\end{aligned} \tag{6.226}$$

Avoiding the trivial solution of $\boldsymbol{\lambda}(t) = \mathbf{0}$ for all time leads to the following two equations:

$$\frac{d}{dt}\hat{\mathbf{x}}_f(t) = \mathbf{f}(\hat{\mathbf{x}}_f(t), \mathbf{u}(t), t) \tag{6.227a}$$

$$\frac{d}{dt}P_f(t) = F(\hat{\mathbf{x}}_f(t), t)\,P_f(t) + P_f(t)\,F^T(\hat{\mathbf{x}}_f(t), t) + G(t)\,Q(t)\,G^T(t) \tag{6.227b}$$

Equation (6.227) represents the Kalman filter propagation, where $\hat{\mathbf{x}}_f(t)$ denotes the forward-time estimate and $P_f(t)$ denotes the forward-time covariance. Note that $F^T(\hat{\mathbf{x}}(t), t)$ has been replaced by $F^T(\hat{\mathbf{x}}_f(t), t)$ in eqn. (6.227b). This substitution results in second-order error effects, which are neglected in the linearization assumption.

We now investigate the costate update equation. Solving eqn. (6.189) for $\boldsymbol{\lambda}_k$ and substituting the resulting expression into eqn. (6.222c) gives

$$\mathcal{P}_{fk}^{+}[\hat{\mathbf{x}}_{fk}^{+} - \hat{\mathbf{x}}_k] = \mathcal{P}_{fk}^{-}[\hat{\mathbf{x}}_{fk}^{-} - \hat{\mathbf{x}}_k] + H_k^T(\hat{\mathbf{x}}_k)\,R_k^{-1}[\tilde{\mathbf{y}}_k - \mathbf{h}_k(\hat{\mathbf{x}}_k)] \tag{6.228}$$

where $\mathcal{P}_{fk}^{+}$ is the matrix inverse of P_{fk}^{+}, and $\mathcal{P}_{fk}^{-}$ is the matrix inverse of P_{fk}^{-}. Note that the smoother state $\hat{\mathbf{x}}_k$ does not contain discontinuities at the measurement points, but its derivative is discontinuous due to the update in the costate. Linearizing $\mathbf{h}_k(\hat{\mathbf{x}}_k)$ about $\hat{\mathbf{x}}_{fk}^{-}$ gives

$$\mathbf{h}_k(\hat{\mathbf{x}}_k) = \mathbf{h}_k(\hat{\mathbf{x}}_{fk}^{-}) + H_k(\hat{\mathbf{x}}_{fk}^{-})[\hat{\mathbf{x}}_k - \hat{\mathbf{x}}_{fk}^{-}] \tag{6.229}$$

Substituting eqn. (6.229) into eqn. (6.228), and replacing $H_k^T(\hat{\mathbf{x}}_k)$ by $H_k^T(\hat{\mathbf{x}}_{fk}^{-})$, which again leads to second-order errors that are neglected, yields

$$\begin{aligned}&[\mathcal{P}_{fk}^{-} - \mathcal{P}_{fk}^{+} + H_k^T(\hat{\mathbf{x}}_{fk}^{-})\,R_k^{-1}\,H_k(\hat{\mathbf{x}}_{fk}^{-})]\hat{\mathbf{x}}_k \\ &+ \mathcal{P}_{fk}^{+}\hat{\mathbf{x}}_{fk}^{+} - \mathcal{P}_{fk}^{-}\hat{\mathbf{x}}_{fk}^{-} - H_k^T(\hat{\mathbf{x}}_{fk}^{-})R_k^{-1}[\tilde{\mathbf{y}}_k - \mathbf{h}_k(\hat{\mathbf{x}}_{fk}^{-}) + H_k(\hat{\mathbf{x}}_{fk}^{-})\hat{\mathbf{x}}_{fk}^{-}] = \mathbf{0}\end{aligned} \tag{6.230}$$

Avoiding the trivial solution of $\hat{\mathbf{x}}_k = \mathbf{0}$ for all time leads to the following two equations:

$$\mathcal{P}_{fk}^{-} = \mathcal{P}_{fk}^{+} - H_k^T(\hat{\mathbf{x}}_{fk}^{-})\,R_k^{-1}\,H_k(\hat{\mathbf{x}}_{fk}^{-}) \tag{6.231a}$$

$$\mathcal{P}_{fk}^{+}\hat{\mathbf{x}}_{fk}^{+} = \mathcal{P}_{fk}^{-}\hat{\mathbf{x}}_{fk}^{-} + H_k^T(\hat{\mathbf{x}}_{fk}^{-})R_k^{-1}[\tilde{\mathbf{y}}_k - \mathbf{h}_k(\hat{\mathbf{x}}_{fk}^{-}) + H_k(\hat{\mathbf{x}}_{fk}^{-})\hat{\mathbf{x}}_{fk}^{-}] \tag{6.231b}$$

Note that eqn. (6.231a) is the information form of the covariance update shown in §5.3.3. Substituting eqn. (6.231a) into eqn. (6.231b) leads to

$$\hat{\mathbf{x}}_{fk}^{+} = \hat{\mathbf{x}}_{fk}^{-} + K_{fk}[\tilde{\mathbf{y}}_k - \mathbf{h}_k(\hat{\mathbf{x}}_{fk}^{-})] \tag{6.232}$$

where the Kalman gain K_{fk} is given by

$$K_{fk} = P_{fk}^{+}H_k^T(\hat{\mathbf{x}}_{fk}^{-})\,R_k^{-1} \equiv V_{fk}D_{fk}^{-1} \tag{6.233}$$

with

$$V_{fk} \equiv P_{fk}^{-} H_k^T(\hat{\mathbf{x}}_{fk}^{-}) \tag{6.234a}$$

$$D_{fk} \equiv H_k(\hat{\mathbf{x}}_{fk}^{-})\, V_{fk} + R_k \tag{6.234b}$$

Equation (6.232) gives the forward-time Kalman filter update, which is used to update the filter propagation in eqn. (6.227a). The covariance update is easily derived using the matrix inversion lemma on eqn. (6.231a), which yields

$$P_{fk}^{+} = [I - K_{fk} H_k(\hat{\mathbf{x}}_{fk}^{-})] P_{fk}^{-} \tag{6.235}$$

Equations (6.227), (6.232), (6.233), and (6.235) constitute the standard extended Kalman filter equations.

Since no jump discontinuities exist in the smoother state estimate equation, the smoother estimate can simply be found by solving eqn. (6.213) for $\boldsymbol{\lambda}(t)$ and substituting the resulting expression into eqn. (6.225), which yields

$$\frac{d}{dt}\hat{\mathbf{x}}(t) = [F(\hat{\mathbf{x}}_f(t), t) + K(t)][\hat{\mathbf{x}}(t) - \hat{\mathbf{x}}_f(t)] + \mathbf{f}(\hat{\mathbf{x}}_f(t), \mathbf{u}(t), t) \tag{6.236}$$

where

$$K(t) \equiv G(t)\, Q(t)\, G^T(t)\, P_f^{-1}(t) \tag{6.237}$$

Equation (6.236) must be integrated backward in time with a boundary condition of $\hat{\mathbf{x}}(T) = \hat{\mathbf{x}}_f(T)$, which satisfies eqn. (6.224a). Note that eqn. (6.222a) can be used instead of eqn. (6.236), but the advantage of using eqn. (6.236) is that a linear integration can be implemented.

The smoother state estimate shown in eqn. (6.107) requires an inversion of the propagated forward-time Kalman filter covariance. This can be overcome by using the information matrix version of the Kalman filter of §5.3.3, which directly involves the inverse of the covariance matrix. Another approach that avoids this matrix inversion involves using the costate equation directly to derive the smoother state.[13] This approach can easily be extended for the nonlinear case. Substituting eqn. (6.229) into eqn. (6.222c) gives

$$\boldsymbol{\lambda}_k^{+} = \boldsymbol{\lambda}_k^{-} + H_k^T(\hat{\mathbf{x}}_k)\, R_k^{-1}\{\tilde{\mathbf{y}}_k - \mathbf{h}_k(\hat{\mathbf{x}}_{fk}^{-}) - H_k(\hat{\mathbf{x}}_{fk}^{-})[\hat{\mathbf{x}}_k - \hat{\mathbf{x}}_{fk}^{-}]\} \tag{6.238}$$

Solving eqn. (6.189) for $\boldsymbol{\lambda}_k$ and substituting the resulting expression into eqn. (6.238), and once again replacing $H_k^T(\hat{\mathbf{x}}_k)$ by $H_k^T(\hat{\mathbf{x}}_{fk}^{-})$ yields

$$\boldsymbol{\lambda}_k^{+} = [I + H_k^T(\hat{\mathbf{x}}_{fk}^{-})\, R_k^{-1} H_k(\hat{\mathbf{x}}_{fk}^{-})\, P_{fk}^{-}]\boldsymbol{\lambda}_k^{-} + H_k^T(\hat{\mathbf{x}}_{fk}^{-})\, R_k^{-1}[\tilde{\mathbf{y}}_k - \mathbf{h}_k(\hat{\mathbf{x}}_{fk}^{-})] \tag{6.239}$$

Solving eqn. (6.239) for $\boldsymbol{\lambda}_k^{-}$ and using the matrix inversion lemma leads to

$$\boldsymbol{\lambda}_k^{-} = [I - H_k^T(\hat{\mathbf{x}}_{fk}^{-})\, K_{fk}^T]\boldsymbol{\lambda}_k^{+} - H_k^T(\hat{\mathbf{x}}_{fk}^{-})\, D_{fk}^{-1}[\tilde{\mathbf{y}}_k - \mathbf{h}_k(\hat{\mathbf{x}}_{fk}^{-})] \tag{6.240}$$

which is used to update the backward integration of eqn. (6.222b). The covariance of the costate follows (which is left as an exercise for the reader)

$$\frac{d}{dt}\Lambda(t) = -F^T(\hat{\mathbf{x}}_f(t),t)\,\Lambda(t) - \Lambda(t)\,F(\hat{\mathbf{x}}_f(t),t) \tag{6.241a}$$

$$\Lambda_k^- = [I - K_{fk}\,H_k(\hat{\mathbf{x}}_{fk}^-)]^T\,\Lambda_k^+[I - K_{fk}\,H_k(\hat{\mathbf{x}}_{fk}^-)] + H_k^T(\hat{\mathbf{x}}_{fk}^-)\,D_{fk}^{-1}\,H_k(\hat{\mathbf{x}}_{fk}^-) \tag{6.241b}$$

The boundary conditions are given by

$$\boldsymbol{\lambda}_N^- = -H_N^T(\hat{\mathbf{x}}_{fN}^-)\,D_{fN}^{-1}[\tilde{\mathbf{y}}_N - \mathbf{h}_k(\hat{\mathbf{x}}_{fN}^-)]\delta_{t_n,N} \tag{6.242a}$$

$$\Lambda_N^- = H_N^T(\hat{\mathbf{x}}_{fN}^-)\,D_{fN}^{-1}\,H_N(\hat{\mathbf{x}}_{fN}^-)\,\delta_{t_n,N} \tag{6.242b}$$

where $\delta_{t_n,N}$ is the Kronecker symbol (if N is not an observation time, then $\boldsymbol{\lambda}$ and Λ have end boundary conditions of zero). Finally, the smoother state and covariance can be constructed via

$$\hat{\mathbf{x}}_k = \hat{\mathbf{x}}_{fk}^{\pm} - P_{fk}^{\pm}\boldsymbol{\lambda}_k^{\pm} \tag{6.243a}$$

$$P_k = P_{fk}^{\pm} - P_{fk}^{\pm}\Lambda_k^{\pm}P_{fk}^{\pm} \tag{6.243b}$$

where the propagated or updated variables yield the same result. The nonlinear algorithm derived in this section does not require the computation of the discrete-time model state-transition matrix nor the discrete-time process noise covariance, which has clear advantages over the algorithm presented in Ref. [1]. Also, when linear models are used the smoothing solution reduces to the classical smoothing algorithms shown in Refs. [9] and [13]. For example, in the linear case the costate vector integration in eqn. (6.222b) with jump discontinuities given by eqn. (6.240) is equivalent to the adjoint filter variable given by Bierman.[13]

6.4.2 Innovations Process

In §6.4.1 the RTS smoother has been derived from optimal control theory. From this theory the costate vector (adjoint variable) is seen to be directly related to the process noise vector, shown by eqns. (6.186c) and (6.210c). Although this provides a nice mathematical representation of the smoothing problem, the physical meaning of the adjoint variable is somewhat unclear from this framework. In this section a different derivation of the TPBVP is shown, which helps to provide some physical meaning to the adjoint variable. This derivation is based upon the innovations process, which can be used to derive the Kalman filter. Here we will use the innovations process to directly derive the TPBVP associated with the RTS smoother. More details on this approach can be found in Refs. [20] and [21].

6.4.2.1 Discrete-Time Formulation

For the discrete-time case, we begin the derivation of the smoother by considering the following innovations process:

$$\begin{aligned} \mathbf{e}_{fk} &\equiv \tilde{\mathbf{y}}_k - \hat{\mathbf{y}}_{fk} \\ &= -H_k \tilde{\mathbf{x}}_{fk} + \mathbf{v}_k \end{aligned} \tag{6.244}$$

where $\tilde{\mathbf{x}}_{fk} \equiv \hat{\mathbf{x}}_{fk} - \mathbf{x}_k$ and $\mathbf{v}_k$ is the measurement noise. The covariance of the error in eqn. (6.244) is given by eqn. (5.245), so that

$$\boxed{E\left\{\mathbf{e}_{fk}\mathbf{e}_{fk}^T\right\} = H_k P_{fk} H_k^T + R_k \equiv \mathcal{R}_{fk}} \tag{6.245}$$

To derive the smoother state estimate we use the following general formula for state estimation given the innovations process:[21]

$$\boxed{\hat{\mathbf{x}}_k = \sum_{i=0}^{N} E\left\{\mathbf{x}_k\, \mathbf{e}_{fi}^T\right\} \mathcal{R}_{fi}^{-1} \mathbf{e}_{fi}} \tag{6.246}$$

This relation can be directly derived from the orthogonality of the innovations, which is closely related to the projection property in least squares estimation (see §1.6.4). Setting $N = k-1$ in eqn. (6.246) gives the state estimate $\hat{\mathbf{x}}_{fk}$. This implies that the summation for $i \geq k$ can be broken up as

$$\hat{\mathbf{x}}_k = \hat{\mathbf{x}}_{fk} + \sum_{i=k}^{N} E\left\{\mathbf{x}_k\, \mathbf{e}_{fi}^T\right\} \mathcal{R}_{fi}^{-1} \mathbf{e}_{fi} \tag{6.247}$$

We now concentrate our attention to the expectation in eqn. (6.247). Substituting eqn. (6.244) into the expectation in eqn. (6.247) gives

$$E\left\{\mathbf{x}_{fk}\, \mathbf{e}_{fi}^T\right\} = -E\left\{\mathbf{x}_k\, \tilde{\mathbf{x}}_{fi}^T\right\} H_i^T + E\left\{\mathbf{x}_k \mathbf{v}_i^T\right\} \tag{6.248}$$

Substituting $\mathbf{x}_k = \hat{\mathbf{x}}_{fk} - \tilde{\mathbf{x}}_{fk}$ into eqn. (6.248) gives

$$E\left\{\mathbf{x}_{fk}\, \mathbf{e}_{fi}^T\right\} = E\left\{\tilde{\mathbf{x}}_{fk}\, \tilde{\mathbf{x}}_{fi}^T\right\} H_i^T - E\left\{\hat{\mathbf{x}}_{fk}\, \tilde{\mathbf{x}}_{fi}^T\right\} H_i^T + E\left\{\mathbf{x}_k \mathbf{v}_i^T\right\} \tag{6.249}$$

Since the state estimate is orthogonal to its error (see §5.3.6) and since the measurement noise is uncorrelated with the true state, then eqn. (6.249) reduces down to

$$E\left\{\mathbf{x}_{fk}\, \mathbf{e}_{fi}^T\right\} = E\left\{\tilde{\mathbf{x}}_{fk}\, \tilde{\mathbf{x}}_{fi}^T\right\} H_i^T \equiv P_{fk,i}\, H_i^T \tag{6.250}$$

where $P_{fk,i} \equiv E\left\{\tilde{\mathbf{x}}_{fk}\, \tilde{\mathbf{x}}_{fi}^T\right\}$. Therefore, substituting eqn. (6.250) into eqn. (6.247) gives

$$\boxed{\hat{\mathbf{x}}_k = \hat{\mathbf{x}}_{fk} + \sum_{i=k}^{N} P_{fk,i}\, H_i^T \mathcal{R}_{fi}^{-1} \mathbf{e}_{fi}} \tag{6.251}$$

Note that $P_{fk,i}$ gives the correlation between the error states at different times. If $i = k$ then $P_{fk,i}$ is exactly the forward-time Kalman filter error-covariance.

The smoother error covariance can be derived by first subtracting $\mathbf{x}_k$ from both sides of eqn. (6.251), which leads to

$$\tilde{\mathbf{x}}_k = \tilde{\mathbf{x}}_{fk} + \sum_{i=k}^{N} P_{fk,i}\, H_i^T \mathcal{R}_{fi}^{-1} \mathbf{e}_{fi} \tag{6.252}$$

Substituting eqn. (6.244) into eqn. (6.252) and performing the covariance operation $P_k \equiv E\left\{\tilde{\mathbf{x}}_k\, \tilde{\mathbf{x}}_k^T\right\}$ yields

$$\boxed{P_k = P_{fk} - \sum_{i=k}^{N} P_{fk,i}\, H_i^T \mathcal{R}_{fi}^{-1} H_i\, P_{fk,i}^T} \tag{6.253}$$

Note the sign difference between eqns. (6.252) and (6.253).

Our next step involves determining a relationship between $P_{fk,i}$ and P_{fk}. Substituting eqn. (5.37) into eqn. (5.33) gives the forward-time state error:

$$\tilde{\mathbf{x}}_{fk+1} = Z_{fk}\tilde{\mathbf{x}}_{fk} + \mathbf{b}_{fk} \tag{6.254}$$

where

$$Z_{fk} \equiv \Phi_k[I - K_{fk}H_k] \tag{6.255a}$$

$$\mathbf{b}_{fk} \equiv \Phi_k K_{fk}\mathbf{v}_k - \Upsilon_k \mathbf{w}_k \tag{6.255b}$$

Taking one time-step ahead of eqn. (6.254) gives

$$\begin{aligned}\tilde{\mathbf{x}}_{fk+2} &= Z_{fk+1}\tilde{\mathbf{x}}_{fk+1} + \mathbf{b}_{k+1} \\ &= Z_{fk+1}Z_{fk}\tilde{\mathbf{x}}_{fk} + Z_{fk+1}\mathbf{b}_{fk} + \mathbf{b}_{fk+1}\end{aligned} \tag{6.256}$$

where eqn. (6.254) has been used. Taking more time-steps ahead leads to the following relationship for $i \geq k$:

$$\tilde{\mathbf{x}}_{fi} = \mathcal{Z}_{fi,k}\tilde{\mathbf{x}}_{fk} + \sum_{j=k}^{i-1} \mathcal{Z}_{fi,j+1}\mathbf{b}_{fj} \tag{6.257}$$

where

$$\mathcal{Z}_{fi,k} = \begin{cases} Z_{fi-1}Z_{fi-2}\cdots Z_{fk} & \text{for } i > k \\ I & \text{for } i = k \end{cases} \tag{6.258}$$

Then the relationship between $P_{fk,i}$ and P_{fk} is simply given by

$$P_{fk,i} \equiv E\left\{\tilde{\mathbf{x}}_{fk}\,\tilde{\mathbf{x}}_{fi}^T\right\} = P_{fk}\mathcal{Z}_{fi,k}^T \tag{6.259}$$

where eqn. (6.259) is valid for $i \geq k$. Substituting eqn. (6.259) into eqn. (6.251) gives

$$\hat{\mathbf{x}}_k = \hat{\mathbf{x}}_{fk} - P_{fk}\boldsymbol{\lambda}_k \tag{6.260}$$

where

$$\boxed{\boldsymbol{\lambda}_k \equiv -\sum_{i=k}^{N} \mathcal{Z}_{fi,k}^T H_i^T \mathcal{R}_{fi}^{-1} \mathbf{e}_{fi}} \tag{6.261}$$

This result clearly shows the relationship between the adjoint variable $\boldsymbol{\lambda}_k$ and the forward-time residual. Comparing this result with eqn. (6.186c) shows an interesting relationship between the process noise and the innovations process in the adjoint variable.

Using the definition of $\mathcal{Z}_{fi,k}$ in eqn. (6.258) immediately implies that $\boldsymbol{\lambda}_k$ can be given by the following backward recursion:

$$\boldsymbol{\lambda}_k = Z_{fk}^T \boldsymbol{\lambda}_{k+1} - H_k^T \mathcal{R}_{fk}^{-1} \mathbf{e}_{fk}, \quad \boldsymbol{\lambda}_N = \mathbf{0} \tag{6.262}$$

Substituting eqn. (6.244) into eqn. (6.262), and using the definitions of Z_{fk} from eqn. (6.255a) and $\mathcal{R}_{fk}$ from eqn. (6.245) gives

$$\boxed{\boldsymbol{\lambda}_k = [I - K_{fk} H_k]^T \Phi_k^T \boldsymbol{\lambda}_{k+1} + H_k^T [H_k P_{fk} H_k^T + R_k]^{-1} [H_k \hat{\mathbf{x}}_{fk} - \tilde{\mathbf{y}}_k]} \tag{6.263}$$

Next, solving eqn. (6.260) for $\hat{\mathbf{x}}_{fk}$ and substituting the resulting expression into eqn. (6.263) yields

$$\begin{aligned} \boldsymbol{\lambda}_k &= [I - K_{fk} H_k]^T \Phi_k^T \boldsymbol{\lambda}_{k+1} \\ &\quad + H_k^T [H_k P_{fk} H_k^T + R_k]^{-1} [H_k \hat{\mathbf{x}}_k - \tilde{\mathbf{y}}_k + H_k P_{fk} \boldsymbol{\lambda}_k] \end{aligned} \tag{6.264}$$

Finally, collecting terms and solving eqn. (6.264) for $\boldsymbol{\lambda}_k$ leads to

$$\begin{aligned} \boldsymbol{\lambda}_k = [I - W_{fk}]^{-1} \Big\{ & [I - K_{fk} H_k]^T \Phi_k^T \boldsymbol{\lambda}_{k+1} \\ & + H_k^T [H_k P_{fk} H_k^T + R_k]^{-1} [H_k \hat{\mathbf{x}}_k - \tilde{\mathbf{y}}_k] \Big\} \end{aligned} \tag{6.265}$$

where

$$W_{fk} \equiv H_k^T [H_k P_{fk} H_k^T + R_k]^{-1} H_k P_{fk} \tag{6.266}$$

At first glance eqn. (6.188b) and eqn. (6.265) do not appear to be equivalent. However, upon further inspection the following identities can be proven (which are left as an exercise for the reader):

$$[I - W_{fk}]^{-1} [I - K_{fk} H_k]^T = I \tag{6.267a}$$

$$[I - W_{fk}]^{-1} H_k^T [H_k P_{fk} H_k^T + R_k]^{-1} = H_k^T R_k \tag{6.267b}$$

Hence, eqn. (6.188b) and eqn. (6.265) are indeed equivalent. The state equation can be derived by taking one time-step ahead of eqn. (6.260) and substituting the forward-time Kalman filter equations from eqn. (5.54), which leads to

$$\begin{aligned} \hat{\mathbf{x}}_{k+1} &= \Phi_k \hat{\mathbf{x}}_{fk} + \Gamma_k \mathbf{u}_k + \Phi_k K_{fk} [\tilde{\mathbf{y}}_k - H_k \hat{\mathbf{x}}_{fk}] \\ &\quad - [\Phi_k P_{fk} \Phi_k^T - \Phi_k K_k H_k P_{fk} \Phi_k^T + \Upsilon_k Q_k \Upsilon_k^T] \boldsymbol{\lambda}_{k+1} \end{aligned} \tag{6.268}$$

Now, solving eqn. (6.260) for $\hat{\mathbf{x}}_{fk}$ and substituting the resulting expression into eqn. (6.268) yields

$$\begin{aligned}\hat{\mathbf{x}}_{k+1} &= \Phi_k\hat{\mathbf{x}}_k + \Gamma_k\mathbf{u}_k - \Upsilon_k Q_k \Upsilon_k^T \boldsymbol{\lambda}_{k+1} + \Phi_k P_{fk}\boldsymbol{\lambda}_k \\ &\quad - \Phi_k K_{fk}[H_k\hat{\mathbf{x}}_{fk} - \tilde{\mathbf{y}}_k] - [\Phi_k P_{fk}\Phi_k^T - \Phi_k K_k H_k P_{fk}\Phi_k^T]\boldsymbol{\lambda}_{k+1}\end{aligned} \tag{6.269}$$

Substituting eqn. (6.263) into eqn. (6.269) simply produces the state equation in eqn. (6.188a).

The innovations process leads to some important conclusions. For example, as previously mentioned, the innovations process can be used to directly derive the Kalman filter, where the filter is used to *whiten* the innovations process (see Refs. [11] and [21] for more details). The derivations provided in this section give a very important result, since they show yet another approach to derive the RTS smoother. In fact, a form of the RTS smoother has been derived more directly than the lengthy algebraic process shown in §6.1.1. This form involves using eqn. (6.263) to solve for the adjoint variable directly from the forward-time Kalman filter quantities. Then, the smoothed estimate can be found directly from eqn. (6.260), which can be implemented in conjunction with the adjoint variable calculation.

6.4.2.2 Continuous-Time Formulation

For the continuous-time case, we begin the derivation of the smoother by considering the following innovations process:

$$\begin{aligned}\mathbf{e}_f(t) &\equiv \tilde{\mathbf{y}}(t) - \hat{\mathbf{y}}_f(t) \\ &= -H(t)\tilde{\mathbf{x}}_f(t) + \mathbf{v}(t)\end{aligned} \tag{6.270}$$

where $\tilde{\mathbf{x}}_f(t) = \hat{\mathbf{x}}_f(t) - \mathbf{x}(t)$ and $\mathbf{v}(t)$ is the measurement noise. One's first natural instinct is to assume that the innovations covariance is just the continuous-time version of eqn. (6.245). However, a rigorous derivation of the continuous-time covariance for the innovations process is far more complicated than the discrete-time case. It can be shown that this covariance obeys[21]

$$\boxed{E\left\{\mathbf{e}_f(t)\mathbf{e}_f^T(\tau)\right\} = R(t)\,\delta(t-\tau)} \tag{6.271}$$

Equation (6.271) can be proven in many ways, i.e., using the orthogonality conditions or by working with white noise replaced with a Wiener process and using martingale theory.[21] Also, since the innovations process is uncorrelated between different times, then the expression in eqn. (6.271) is valid for all time.

The continuous-time version of eqn. (6.246) is given by

$$\boxed{\hat{\mathbf{x}}(t) = \int_0^T E\left\{\mathbf{x}(t)\mathbf{e}_f^T(\tau)\right\} R^{-1}(\tau)\,\mathbf{e}_f(\tau)\,d\tau} \tag{6.272}$$

As with the discrete-time case, eqn. (6.272) can be broken up into two parts, given by

$$\hat{\mathbf{x}}(t) = \hat{\mathbf{x}}_f(t) + \int_t^T E\left\{\mathbf{x}(t)\mathbf{e}_f^T(\tau)\right\} R^{-1}(\tau)\,\mathbf{e}_f(\tau)\,d\tau \tag{6.273}$$

for $0 \le t \le T$. We now concentrate our attention on the expectation in eqn. (6.273). Substituting eqn. (6.270) into the expectation in eqn. (6.273) gives

$$E\left\{\mathbf{x}(t)\,\mathbf{e}_f^T(\tau)\right\} = -E\left\{\mathbf{x}(t)\,\tilde{\mathbf{x}}_f^T(\tau)\right\}H^T(\tau) + E\left\{\mathbf{x}(t)\,\mathbf{v}^T(\tau)\right\} \tag{6.274}$$

Substituting $\mathbf{x}(t) = \hat{\mathbf{x}}_f(t) - \tilde{\mathbf{x}}_f(t)$ into eqn. (6.274) gives

$$\begin{aligned} E\left\{\mathbf{x}(t)\,\mathbf{e}_f^T(\tau)\right\} &= E\left\{\tilde{\mathbf{x}}_f(t)\,\tilde{\mathbf{x}}_f^T(\tau)\right\}H^T(\tau) - E\left\{\hat{\mathbf{x}}_f(t)\,\tilde{\mathbf{x}}_f^T(\tau)\right\}H^T(\tau) \\ &\quad + E\left\{\mathbf{x}(t)\,\mathbf{v}^T(\tau)\right\} \end{aligned} \tag{6.275}$$

Since the state estimate is orthogonal to its error and since the measurement noise is uncorrelated with the true state, then eqn. (6.275) reduces down to

$$E\left\{\mathbf{x}(t)\,\mathbf{e}_f^T(\tau)\right\} = E\left\{\tilde{\mathbf{x}}_f(t)\,\tilde{\mathbf{x}}_f^T(\tau)\right\}H^T(\tau) \equiv P_f(t,\,\tau)\,H^T(\tau) \tag{6.276}$$

where $P_f(t,\,\tau) \equiv E\left\{\tilde{\mathbf{x}}_f(t)\,\tilde{\mathbf{x}}_f^T(\tau)\right\}$. Substituting eqn. (6.276) into eqn. (6.273) gives

$$\boxed{\hat{\mathbf{x}}(t) = \hat{\mathbf{x}}_f(t) + \int_t^T P_f(t,\,\tau)\,H^T(\tau)\,R^{-1}(\tau)\,\mathbf{e}_f(\tau)\,d\tau} \tag{6.277}$$

Note that $P_f(t,\,\tau)$ gives the correlation between the error states at different times. If $\tau = t$ then $P_f(t,\,\tau)$ is exactly the forward-time Kalman filter error-covariance. The smoother error covariance can be shown to be given by (which is left as an exercise for the reader)

$$\boxed{P(t) = P_f(t) - \int_t^T P_f(t,\,\tau)\,H^T(\tau)\,R^{-1}(\tau)\,H(\tau)\,P_f(t,\,\tau)\,d\tau} \tag{6.278}$$

As with the discrete-time case, note the sign difference between eqns. (6.277) and (6.278).

Our next step involves determining a relationship between $P_f(t,\,\tau)$ and $P_f(t)$. From eqn. (5.120) we can write

$$\frac{d}{d\tau}\tilde{\mathbf{x}}_f(\tau) = E_f(\tau)\,\tilde{\mathbf{x}}_f(\tau) + \mathbf{z}_f(\tau) \tag{6.279}$$

where

$$E_f(\tau) = F(\tau) - K_f(\tau)\,H(\tau) \tag{6.280}$$

$$\mathbf{z}_f(\tau) = -G(\tau)\,\mathbf{w}(\tau) + K_f(\tau)\,\mathbf{v}(\tau) \tag{6.281}$$

The solution for $\tilde{\mathbf{x}}_f(\tau)$ in eqn. (6.279) is given by

$$\tilde{\mathbf{x}}_f(\tau) = \Psi(\tau,\,t)\,\tilde{\mathbf{x}}_f(t) + \int_t^\tau \Phi(\tau,\,\eta)\,\mathbf{z}_f(\eta)\,d\eta \tag{6.282}$$

where $\Psi(\tau, t)$ is the state transition matrix of $E_f(\tau)$, which follows

$$\frac{d}{d\tau}\Psi(\tau, t) = [F(\tau) - K_f(\tau) H(\tau)]\Psi(\tau, t), \quad \Psi(\tau, \tau) = I \tag{6.283}$$

Substituting eqn. (6.283) into $P_f(t, \tau) = E\left\{\tilde{\mathbf{x}}_f(t)\tilde{\mathbf{x}}_f^T(\tau)\right\}$ leads to

$$P_f(t, \tau) = E\left\{\tilde{\mathbf{x}}_f(t)\tilde{\mathbf{x}}_f^T(\tau)\Psi^T(\tau, t)\right\} = P_f(t)\,\Psi^T(\tau, t) \tag{6.284}$$

where the fact that $\tilde{\mathbf{x}}_f(t)$ is uncorrelated to $\mathbf{z}_f(\tau)$ has been used to yield eqn. (6.284). Substituting eqn. (6.284) into eqn. (6.277) gives

$$\boxed{\hat{\mathbf{x}}(t) = \hat{\mathbf{x}}_f(t) - P_f(t)\,\boldsymbol{\lambda}(t)} \tag{6.285}$$

where

$$\boxed{\begin{aligned}\boldsymbol{\lambda}(t) &\equiv -\int_t^T \Psi^T(\tau, t)\,H^T(\tau)\,R^{-1}(\tau)\,\mathbf{e}_f(\tau)\,d\tau \\ &= \int_T^t \Psi^T(\tau, t)\,H^T(\tau)\,R^{-1}(\tau)\,\mathbf{e}_f(\tau)\,d\tau\end{aligned}} \tag{6.286}$$

The physical interpretation of the continuous-time adjoint variable is analogous to the discrete-time case, which is an intuitively pleasing result. Taking the time derivative of eqn. (6.286) leads to

$$\frac{d}{dt}\boldsymbol{\lambda}(t) = \int_T^t \left[\frac{d}{dt}\Psi^T(\tau, t)\right] H^T(\tau)\,R^{-1}(\tau)\,\mathbf{e}_f(\tau)\,d\tau + H^T(t)\,R^{-1}(t)\,\mathbf{e}_f(t) \tag{6.287}$$

Using the result shown in exercise 3.1 as well as the definition of $\boldsymbol{\lambda}$ in eqn. (6.286), with the definitions of $\mathbf{e}_f(t)$ in eqn. (6.270) and $E(t)$ in eqn. (6.280), leads to

$$\boxed{\begin{aligned}\frac{d}{dt}\boldsymbol{\lambda}(t) &= -[F(t) - K_f(t)\,H(t)]^T\boldsymbol{\lambda}(t) \\ &\quad - H^T(t)\,R^{-1}(t)H(t)\hat{\mathbf{x}}_f(t) + H^T(t)\,R^{-1}(t)\,\tilde{\mathbf{y}}(t)\end{aligned}} \tag{6.288}$$

Solving eqn. (6.285) for $\boldsymbol{\lambda}(t)$ and substituting the resulting expression into the differential equation of eqn. (6.288) gives

$$\begin{aligned}\frac{d}{dt}\boldsymbol{\lambda}(t) &= -F^T(t)\,\boldsymbol{\lambda}(t) + H^T(t)\,K_f^T(t)\,P_f^{-1}(t)[\hat{\mathbf{x}}_f(t) - \hat{\mathbf{x}}(t)] \\ &\quad - H^T(t)\,R^{-1}(t)H(t)\hat{\mathbf{x}}_f(t) + H^T(t)\,R^{-1}(t)\,\tilde{\mathbf{y}}(t)\end{aligned} \tag{6.289}$$

Substituting eqn. (6.218) into eqn. (6.289) leads exactly to eqn. (6.212b). Also, the differential equation for $\hat{\mathbf{x}}(t)$ follows directly from the steps leading to eqn. (6.219). Equation (6.288) can be used to solve for the adjoint variable directly from the forward-time Kalman filter quantities. The results in the section validate the associated TPBVP shown in eqn. (6.212) using the innovations process, which leads to the continuous-time RTS smoother.

6.5 Summary

In this chapter several smoothing algorithms have been presented that are based on using a batch set of measurement data. The advantage of using a smoother has been clearly shown by the fact that its associated error-covariance is always less than (or equal to) the Kalman filter error covariance. This indicates that better estimates can be achieved by using the optimal smoother; however, a significant disadvantage of a smoother is that a real-time estimate is not possible. The fixed-interval smoother of §6.1 is particularly useful though for many applications, such as sensor bias calculations and parameter estimation.

Intrinsic in all smoothing algorithms presented in this chapter is the Kalman filter. The fixed-interval smoother can conceptually be divided into two separate filters: a forward-time Kalman filter and a backward time recursion. For the fixed-interval smoother the backward-time recursion has been derived two different ways. One uses a backward-time Kalman filter-type implementation, where the smoother estimate is given by an optimally-derived combination of both filters. The other uses a direct computation of the smoother estimate without the need of combining the forward-time and backward-time estimates. Each approach is equivalent to one another from a theoretical point of view. However, depending on the particular situation, one approach may provide a computational advantage of another. A comparison of the computational requirements in the various smoother equation approaches is given by McReynolds.[9]

Several theoretical aspects of the optimal smoother are given in this chapter. For example, a formal proof of the stability of the RTS smoother has been provided using a Lyapunov stability analysis. Fairly complete derivations of the fixed-point and fixed-lag smoothers have also been provided so that the reader can better understand the intricacies of the properties of these smoothers. One of the most interesting aspects of smoothing is the dual relationship with optimal control, which has been presented in §6.4.1. At first glance one might not realize this relationship, but after closer examination we realize than *any* estimation problem can be rewritten as a control problem. The results of §6.4.2 further strengthen this statement. Several references have been provided in this chapter, and the reader is strongly encouraged to further study smoothing approaches in the open literature.

A summary of the key formulas presented in this chapter is given below. All variables with the subscript f denote the forward-time Kalman filter.

- Fixed-Interval Smoother (Discrete-Time)

$$K_{bk} = \mathcal{P}^{+}_{bk+1}\Upsilon_k[\Upsilon_k^T\mathcal{P}^{+}_{bk+1}\Upsilon_k + Q_k^{-1}]^{-1}$$

$$\hat{\chi}^{+}_{bk} = \hat{\chi}^{-}_{bk} + H_k^T R_k^{-1}\tilde{\mathbf{y}}_k$$
$$\mathcal{P}^{+}_{bk} = \mathcal{P}^{-}_{bk} + H_k^T R_k^{-1} H_k$$

$$\hat{\chi}_{bk}^- = \Phi_k^T [I - K_{bk} \Upsilon_k^T][\hat{\chi}_{bk+1}^+ - \mathcal{P}_{bk+1}^+ \Gamma_k \mathbf{u}_k]$$
$$\mathcal{P}_{bk}^- = \Phi_k^T [I - K_{bk} \Upsilon_k^T] \mathcal{P}_{bk+1}^+ \Phi_k$$

$$K_k = P_{fk}^+ \mathcal{P}_{bk}^- [I + P_{fk}^+]$$

$$\hat{\mathbf{x}}_k = [I - K_k] \hat{\mathbf{x}}_{fk}^+ + P_k \hat{\chi}_{bk}^-$$
$$P_k = [I - K_k] P_{fk}^+$$

- RTS Smoother (Discrete-Time)

$$\mathcal{K}_k \equiv P_{fk}^+ \Phi_k^T (P_{fk+1}^-)^{-1}$$

$$\hat{\mathbf{x}}_k = \hat{\mathbf{x}}_{fk}^+ + \mathcal{K}_k [\hat{\mathbf{x}}_{k+1} - \hat{\mathbf{x}}_{fk+1}^-]$$
$$P_k = P_{fk}^+ - \mathcal{K}_k [P_{fk+1}^- - P_{k+1}] \mathcal{K}_k^T$$

- Fixed-Interval Smoother (Continuous-Time)

$$\frac{d}{d\tau} P_b^{-1}(t) = P_b^{-1}(t) F(t) + F^T(t) P_b^{-1}(t) - P_b^{-1}(t) G(t) Q(t) G^T(t) P_b^{-1}(t) + H^T(t) R^{-1}(t) H(t)$$

$$\frac{d}{d\tau} \hat{\chi}_b(t) = \left[F(t) - G(t) Q(t) G^T(t) P_b^{-1}(t) \right]^T \hat{\chi}_b(t) - P_b^{-1}(t) B(t) \mathbf{u}(t) + H^T(t) R^{-1}(t) \tilde{\mathbf{y}}(t)$$

$$K(t) = P_f(t) P_b^{-1}(t) \left[I + P_f(t) P_b^{-1}(t) \right]^{-1}$$

$$\hat{\mathbf{x}}(t) = [I - K(t)] \hat{\mathbf{x}}_f(t) + P(t) \hat{\chi}_b(t)$$
$$P(t) = [I - K(t)] P_f(t)$$

- RTS Smoother (Continuous-Time)

$$\frac{d}{d\tau} \hat{\mathbf{x}}(t) = -F(t) \hat{\mathbf{x}}(t) - B(t) \mathbf{u}(t) - G(t) Q(t) G^T(t) P_f^{-1}(t) \left[\hat{\mathbf{x}}(t) - \hat{\mathbf{x}}_f(t) \right]$$

$$\frac{d}{d\tau} P(t) = -[F(t) + G(t) Q(t) G^T(t) P_f^{-1}(t)] P(t) - P(t) [F(t) + G(t) Q(t) G^T(t) P_f^{-1}(t)]^T + G(t) Q(t) G^T(t)$$

- Nonlinear RTS Smoother

$$K(t) \equiv G(t)\,Q(t)\,G^T(t)\,P_f^{-1}(t)$$

$$\frac{d}{d\tau}\hat{\mathbf{x}}(t) = -\left[F(\hat{\mathbf{x}}_f(t),\,t) + K(t)\right]\left[\hat{\mathbf{x}}(t) - \hat{\mathbf{x}}_f(t)\right] - \mathbf{f}(\hat{\mathbf{x}}_f(t),\,\mathbf{u}(t),\,t)$$

$$\frac{d}{d\tau}P(t) = -[F(\hat{\mathbf{x}}_f(t),t) + K(t)]P(t) - P(t)[F(\hat{\mathbf{x}}_f(t),t) + K(t)]^T + G(t)\,Q(t)\,G^T(t)$$

- Fixed-Point Smoother (Discrete-Time)

$$\mathcal{B}_N = \prod_{i=k}^{N-1} \mathcal{K}_i$$

$$\mathcal{K}_i = P_{fi}^+\Phi_i^T(P_{fi+1}^-)^{-1}$$

$$\hat{\mathbf{x}}_{k|N} = \hat{\mathbf{x}}_{k|N-1} + \mathcal{B}_N[\hat{\mathbf{x}}_{fN}^+ - \hat{\mathbf{x}}_{fN}^-]$$

$$P_{k|N} = P_{k|N-1} + \mathcal{B}_N[P_{fN}^+ - P_{fN}^-]\mathcal{B}_N^T$$

- Fixed-Point Smoother (Continuous-Time)

$$\frac{d}{dT}\Phi(t,T) = -\Phi(t,T)[F(T) + G(T)\,Q(T)\,G^T(T)\,P_f^{-1}(T)],$$

$$\frac{d}{dT}\hat{\mathbf{x}}(t|T) = \Phi(t,T)\,P_f(T)\,H^T(T)\,R^{-1}(T)[\tilde{\mathbf{y}}(T) - H(T)\,\hat{\mathbf{x}}_f(T)]$$

$$\frac{d}{dT}P(t|T) = -\Phi(t,T)\,P_f(T)\,H^T(T)\,R^{-1}(T)\,P_f(T)\,\Phi^T(t,T)$$

- Fixed-Lag Smoother (Discrete-Time)

$$\mathcal{B}_{k+1+N} = \prod_{i=k+1}^{k+N} \mathcal{K}_i$$

$$\mathcal{K}_i = P_{fi}^+\Phi_i^T(P_{fi+1}^-)^{-1}$$

$$\begin{aligned}\hat{\mathbf{x}}_{k+1|k+1+N} = {} & \Phi_k\hat{\mathbf{x}}_{k|k+N} + \Gamma_k\mathbf{u}_k \\ & + \Upsilon_k Q_k \Upsilon_k^T \Phi_k^{-T}(P_{fk}^+)^{-1}[\hat{\mathbf{x}}_{k|k+N} - \hat{\mathbf{x}}_{fk}^+] \\ & + \mathcal{B}_{k+1+N} K_{fk+1+N}\{\tilde{\mathbf{y}}_{k+1+N} \\ & - H_{k+1+N}\Phi_{k+N}[\hat{\mathbf{x}}_{fk+N}^+ + \Gamma_{k+N}\mathbf{u}_{k+N}]\}\end{aligned}$$

$$\begin{aligned}P_{k+1|k+1+N} = {} & P_{fk+1}^- - \mathcal{K}_k^{-1}[P_{fk}^+ - P_{k|k+N}]\mathcal{K}_k^{-T} \\ & - \mathcal{B}_{k+1+N} K_{fk+1+N} H_{k+1+N} P_{fk+1+N}^- \mathcal{B}_{k+1+N}^T\end{aligned}$$

- Fixed-Lag Smoother (Continuous-Time)

$$\begin{aligned}\frac{d}{dT}\Psi(T-\Delta,T) &= [F(T-\Delta)+G(T-\Delta)\,Q(T-\Delta)\\ &\quad\times G^T(T-\Delta)\,P_f^{-1}(T-\Delta)]\Psi(T-\Delta,T)\\ &\quad-\Psi(T-\Delta,T)[F(T)+G(T)\,Q(T)\,G^T(T)\,P_f^{-1}(T)]\end{aligned}$$

$$\begin{aligned}\frac{d}{dT}\hat{\mathbf{x}}(T-\Delta|T) &= [F(T-\Delta)+G(T-\Delta)\,Q(T-\Delta)\\ &\quad\times G^T(T-\Delta)\,P_f^{-1}(T-\Delta)]\hat{\mathbf{x}}(T-\Delta|T)\\ &\quad-G(T-\Delta)\,Q(T-\Delta)\,G^T(T-\Delta)\,P_f^{-1}(T-\Delta)\,\hat{\mathbf{x}}_f(T-\Delta)\\ &\quad+\Psi(T-\Delta,T)\,P_f(T)\,H^T(T)\,R^{-1}(T)[\tilde{\mathbf{y}}(T)-H(T)\,\hat{\mathbf{x}}_f(T)]\end{aligned}$$

$$\begin{aligned}\frac{d}{dT}P(T-\Delta|T) &= [F(T-\Delta)+G(T-\Delta)\,Q(T-\Delta)\\ &\quad\times G^T(T-\Delta)\,P_f^{-1}(T-\Delta)]P(T-\Delta|T)+P(T-\Delta|T)\\ &\quad\times[F(T-\Delta)+G(T-\Delta)\,Q(T-\Delta)\,G^T(T-\Delta)\,P_f^{-1}(T-\Delta)]^T\\ &\quad-\Psi(T-\Delta,T)\,P_f(T)\,H^T(T)\,R^{-1}(T)\,P_f(T)\,\Psi(T-\Delta,T)\\ &\quad-G(T-\Delta)\,Q(T-\Delta)\,G^T(T-\Delta)\end{aligned}$$

- Innovations Process (Discrete-Time)

$$\boldsymbol{\lambda}_k \equiv -\sum_{i=k}^{N} \mathcal{Z}_{fi,k}^T H_i^T \mathcal{R}_{fi}^{-1}\mathbf{e}_{fi}$$

$$\mathcal{Z}_{fi,k} = \begin{cases} Z_{fi-1}Z_{fi-2}\cdots Z_{fk} & \text{for } i>k\\ I & \text{for } i=k\end{cases}$$

$$Z_{fk} \equiv \Phi_k[I-K_{fk}H_k]$$

$$\mathcal{R}_{fk} \equiv H_k P_{fk} H_k^T + R_k$$

$$\mathbf{e}_{fk} \equiv \tilde{\mathbf{y}}_k - H_k\hat{\mathbf{x}}_{fk}$$

$$\boldsymbol{\lambda}_k = Z_{fk}^T\boldsymbol{\lambda}_{k+1} - H_k^T\mathcal{R}_{fk}^{-1}\mathbf{e}_{fk}, \quad \boldsymbol{\lambda}_N = \mathbf{0}$$

- Innovations Process (Continuous-Time)

$$\boldsymbol{\lambda}(t) \equiv -\int_t^T \Psi^T(\tau,t)\,H^T(\tau)\,R^{-1}(\tau)\,\mathbf{e}_f(\tau)\,d\tau$$

$$\frac{d}{d\tau}\Psi(\tau,t) = [F(\tau)-K_f(\tau)\,H(\tau)]\Psi(\tau,t), \quad \Psi(\tau,\tau)=I$$

$$\mathbf{e}_f(t) \equiv \tilde{\mathbf{y}}(t) - H(t)\,\hat{\mathbf{x}}_f(t)$$

$$\frac{d}{dt}\boldsymbol{\lambda}(t) = -[F(t)+K_f(t)\,H(t)]^T\,\boldsymbol{\lambda}(t) \\ -H^T(t)\,R^{-1}(t)H(t)\hat{\mathbf{x}}_f(t)+H^T(t)\,R^{-1}(t)\,\tilde{\mathbf{y}}(t)$$

Exercises

6.1 After substituting eqns. (6.12) and (6.13) into eqn. (6.10) prove that the expression in eqn. (6.14) is valid.

6.2 Prove that the backward gain expressions given in eqns. (6.27) and (6.28) are equivalent to each other. Also, prove that the backward inverse covariance expressions given in eqns. (6.26) and (6.29) are equivalent to each other.

6.3 Write a general program that solves the discrete-time algebraic Riccati equation using the eigenvalue/eigenvector decomposition algorithm of the Hamiltonian matrix given by eqn. (6.39). Compare the steady-state values computed from your program to the values computed by the backward propagation in eqn. (6.29). Pick any order system with various values for Φ, H, Q, Υ, and R to test your program.

6.4 Reproduce the results of example 6.1 using your own simulation. Also, instead of using the RTS smoother form, use the two-filter algorithm shown in Table 6.1. Do you obtain the same results as the RTS smoother? Compute the steady-state values for P_{fk}^+ and P_{bk}^- using the eigenvalue/eigenvector decompositions of eqns. (5.89) and (6.39). Next, from these values compute the steady-state value for the smoother covariance P_k. Compare the 3σ attitude bound from this approach with the solution given in example 6.1.

6.5 Use the discrete-time fixed-interval smoother to provide smoothed estimates for the system described in problems 5.12 and 5.13.

6.6 Show that the solution for the optimal smoother estimate given by eqn. (6.79) can be derived by minimizing the following loss function:

$$J[\hat{\mathbf{x}}(t)] = [\hat{\mathbf{x}}(t)-\hat{\mathbf{x}}_f(t)]^T\,P_f^{-1}(t)[\hat{\mathbf{x}}(t)-\hat{\mathbf{x}}_f(t)] \\ +[\hat{\mathbf{x}}(t)-\hat{\mathbf{x}}_b(t)]^T\,P_b^{-1}(t)[\hat{\mathbf{x}}(t)-\hat{\mathbf{x}}_b(t)]$$

What are the physical connotations of this result?

6.7 Write a general program that solves the continuous-time algebraic Riccati equation using the eigenvalue/eigenvector decomposition algorithm of the Hamiltonian matrix given by eqn. (6.91). Compare the steady-state values computed from your program to the values computed by the backward propagation in eqn. (6.29). Pick any order system with various values for F, H, Q, G, and R to test your program.

6.8 After substituting the relations given in eqns. (6.74) and (6.87) with $d\hat{\chi}_b/d\tau = -d\hat{\chi}_b/dt$, and (6.99) into eqn. (6.101), prove that the expression given in eqn. (6.102) is valid.

6.9 What changes need to be made (if any) to the RTS smoother equations if the process noise and measurement noise are correlated? Discuss both the discrete-time and continuous-time cases.

6.10 ♣ Using the approach outlined in §5.4.2, beginning with the discrete-time fixed-interval smoother shown in Table 6.1, derive the continuous-time version shown in Table 6.3. Also, perform the same derivation for the RTS version of the smoother.

6.11 In example 6.2 show that at steady-state the smoother variance p is always less than half the forward-time filter variance p_f. Also, show $p \leq p_b$.

6.12 The nonlinear RTS smoother shown in Table 6.5 is also valid for linear systems with continuous-time models and discrete-time measurements. Use the smoother to provide smoothed estimates for the system described in exercise 5.29.

6.13 Use the nonlinear RTS smoother to provide smoothed estimates for the system described in exercise 5.33.

6.14 Use the nonlinear RTS smoother to provide a smoothed estimate for the damping coefficient described in the parameter identification problem shown in example 5.6.

6.15 ♣ Fully derive the expression shown for the single-stage optimal smoother in eqn. (6.118).

6.16 Derive the discrete-time RTS smoother directly from eqn. (6.128).

6.17 Prove the expression shown in eqn. (6.148).

6.18 Use the fixed-point discrete-time shown in Table 6.6 to find a fixed-point smoother estimate at some time reference for the system described in example 5.3.

6.19 ♣ Using the approach outlined in §5.4.2, beginning with the discrete-time fixed-point smoother shown in Table 6.6, derive the continuous-time version shown in Table 6.7.

6.20 Prove the covariance expression shown in eqn. (6.157) using similar steps outlined to obtain eqn. (6.156).

6.21 Prove that the solution for $p(t|T)$ given in example 6.4 satisfies its differential equation.

6.22 Use the fixed-lag discrete-time shown in Table 6.8 to find a fixed-lag smoother estimate for the system described in example 5.3. Choose any constant lag in your simulation.

6.23 ♣ Using the approach outlined in §5.4.2, beginning with the discrete-time fixed-lag smoother shown in Table 6.8, derive the continuous-time version shown in Table 6.9.

6.24 After taking the derivative of eqn. (6.178) with respect to T, and substituting the forward-time state filter equation from Table 6.3 and eqn. (6.181) into the resulting equation, prove that the expression in eqn. (6.182) is valid.

6.25 Starting with the fixed-lag estimate in eqn. (6.182) derive the covariance expression given in eqn. (6.183).

6.26 In example 6.5 verify that the fixed-lag smoother variance solution is given $p(T-\Delta|T) = p(0|\Delta)$.

6.27 Prove the identities given in eqn. (6.204).

6.28 Starting with the costate differential equation shown in eqn. (6.222b) and update shown in eqn. (6.222c), prove that the covariance of the costate is given by eqn. (6.241).

6.29 ♣ Using the approach outlined in §5.4.2, beginning with the discrete-time TPBVP shown in eqn. (6.188), derive the continuous-time version shown eqn. (6.212).

6.30 ♣ In the nonlinear formulation of §6.4.1 the quantity $\hat{\mathbf{x}}(t)$ has been replaced by $\hat{\mathbf{x}}_f(t)$ in a number of cases, e.g., in eqn. (6.227b). Prove that this substitution leads to second-order errors that can be ignored in the linearization assumption.

6.31 Prove the identities shown in eqn. (6.267).

6.32 ♣ The general linear least-mean-square estimator for $\mathbf{x}$, given a set of N measurements, can be represented by

$$\hat{\mathbf{x}} = \sum_{k=0}^{N} E\left\{\mathbf{x}\,\mathbf{e}_k^T\right\} ||\mathbf{e}_k||^{-2}\mathbf{e}_k$$

where $\mathbf{e}_k \equiv \tilde{\mathbf{y}}_k - \hat{\mathbf{y}}_k$. Prove this relationship using the orthogonality of the innovations process.

6.33 ♣ Derive the forward-time discrete-time Kalman filter beginning with the following basic formula for state estimation:

$$\hat{\mathbf{x}}_{f\,k+1} = \sum_{i=0}^{N} E\left\{\mathbf{x}_{k+1}\,\mathbf{e}_{fi}^T\right\} \mathcal{R}_{fi}^{-1}\mathbf{e}_{fi} \tag{6.290}$$

where $\mathbf{e}_{fi} \equiv \tilde{\mathbf{y}}_i - H_i \hat{\mathbf{x}}_{fi}$, and $\mathcal{R}_{fi}$ is the covariance of $\mathbf{e}_{fi}$.

6.34 Starting with the state estimate given in eqn. (6.277), prove that the smoother error covariance is given by eqn. (6.278). How can eqn. (6.278) be used to verify that the smoother error covariance is always less than or equal to the forward-time error covariance?

6.35 ♣ Using the results of §5.7.5, derive error equations for the continuous-time fixed-interval, fixed-point, and fixed-lag smoothers.

6.36 Intrinsic in all smoothing algorithms derived in this chapter is the forward-time Kalman filter. However, a better approach may involve using the Unscented filter shown in §5.7.6 as the forward-time filter. Using the model of a vertically falling body in example 5.9, compare the performance of the RTS nonlinear smoother using the forward-time Kalman filter versus the Unscented filter.

References

[1] Gelb, A., editor, *Applied Optimal Estimation*, The MIT Press, Cambridge, MA, 1974.

[2] Wiener, N., *Extrapolation, Interpolation, and Smoothing of Stationary Time Series*, John Wiley, New York, NY, 1949.

[3] Brown, R.G. and Hwang, P.Y.C., *Introduction to Random Signals and Applied Kalman Filtering*, John Wiley & Sons, New York, NY, 3rd ed., 1997.

[4] Bryson, A.E. and Frazier, M., "Smoothing for Linear and Nonlinear Dynamic Systems," Tech. Rep. TDR-63-119, Aeronautical Systems Division, Wright-Patterson Air Force Base, Ohio, Sept. 1962.

[5] Rauch, H.E., Tung, F., and Striebel, C.T., "Maximum Likelihood Estimates of Linear Dynamic Systems," *AIAA Journal*, Vol. 3, No. 8, Aug. 1965, pp. 1445–1450.

[6] Fraser, D.C. and Potter, J.E., "The Optimum Smoother as a Combination of Two Opimum Linear Filters," *IEEE Transactions on Automatic Control*, Vol. AC-14, No. 4, Aug. 1969, pp. 387–390.

[7] Mayne, D.Q., "A Solution to the Smoothing Problem for Linear Dynamic Systems," *Automatica*, Vol. 4, No. 6, Dec. 1966, pp. 73–92.

[8] Fraser, D.C., *A New Technique for the Optimal Smoothing of Data*, Sc.D. thesis, Massachusetts Institute of Technology, Cambridge, Massachusetts, 1967.

[9] McReynolds, S.R., "Fixed Interval Smoothing: Revisited," *Journal of Guidance, Control, and Dynamics*, Vol. 13, No. 5, Sept.-Oct. 1990, pp. 913–921.

[10] Maybeck, P.S., *Stochastic Models, Estimation, and Control*, Vol. 2, Academic Press, New York, NY, 1982.

[11] Lewis, F.L., *Optimal Estimation with an Introduction to Stochastic Control Theory*, John Wiley & Sons, New York, NY, 1986.

[12] Leondes, C.T., Peller, J.B., and Stear, E.B., "Nonlinear Smoothing Theory," *IEEE Transactions on Systems Science and Cybernetics*, Vol. SSC-6, No. 1, Jan. 1970, pp. 63–71.

[13] Bierman, G.J., "Fixed Interval Smoothing with Discrete Measurements," *International Journal of Control*, Vol. 18, No. 1, July 1973, pp. 65–75.

[14] Crassidis, J.L. and Markley, F.L., "A Minimum Model Error Approach for Attitude Estimation," *Journal of Guidance, Control, and Dynamics*, Vol. 20, No. 6, Nov.-Dec. 1997, pp. 1241–1247.

[15] Meditch, J.S., *Stochastic Optimal Linear Estimation and Control*, McGraw-Hill, New York, NY, 1969.

[16] Bryson, A.E. and Ho, Y.C., *Applied Optimal Control*, Taylor & Francis, London, England, 1975.

[17] Sage, A.P. and White, C.C., *Optimum Systems Control*, Prentice Hall, Englewood Cliffs, NJ, 2nd ed., 1977.

[18] Geering, H.P., "Continuous-Time Optimal Control Theory for Cost Functionals Including Discrete State Penalty Terms," *IEEE Transactions on Automatic Control*, Vol. AC-21, No. 12, Dec. 1976, pp. 866–869.

[19] Mook, D.J. and Junkins, J.L., "Minimum Model Error Estimation for Poorly Modeled Dynamics Systems," *Journal of Guidance, Control, and Dynamics*, Vol. 11, No. 3, May-June 1988, pp. 256–261.

[20] Kailath, T., "An Innovations Approach to Least-Squares Estimation, Part 1: Linear Filtering in Additive White Noise," *IEEE Transactions on Automatic Control*, Vol. AC-13, No. 6, Dec. 1968, pp. 646–655.

[21] Kailath, T., Sayed, A.H., and Hassibi, B., *Linear Estimation*, Prentice Hall, Upper Saddle River, NJ, 2000.

7

Estimation of Dynamic Systems: Applications

In theory, there is no difference between theory and practice. But, in practice, there is. van de Snepscheut, Jan

THE previous chapters provided the basic concepts for state estimation of dynamic systems. The foundations of these chapters still rely on the estimation results of Chapter 1 and the probability concepts introduced in Chapter 2. Applications of the fundamental concepts have been shown for various systems in Chapter 4. In this chapter these applications are extended to demonstrate the power of the sequential Kalman filter and batch estimation algorithms. As with Chapter 4, this chapter shows only the fundamental concepts of these applications, where the emphasis is upon the utility of the estimation methodologies. The interested reader is encouraged to pursue these applications in more depth by studying the references cited in this chapter.

7.1 GPS Position Estimation

In §4.1 nonlinear least squares has been used to determine the position of a vehicle using Global Positioning System (GPS) pseudorange measurements. An application of this concept has been demonstrated in example 4.1 using simulated GPS satellite position locations. In the example the GPS locations are shown in Earth-Centered-Earth-Fixed (ECEF), which provides an easy approach to convert the position of a vehicle into longitude and latitude. However, example 4.1 shows only a point-by-point solution approach (i.e., only one specific solution in time). The results of this chapter give a trajectory of solutions over an entire time interval. Hence, the evolution of the positions of the GPS satellites over time must be discussed here. Using these GPS positions a Kalman filter application will then be shown to determine filtered position and velocity estimates.

7.1.1 GPS Coordinate Transformations

In this section we first review the basic concepts of the Earth-Centered-Inertial (ECI) to ECEF transformation, which can be used to propagate the GPS satellite

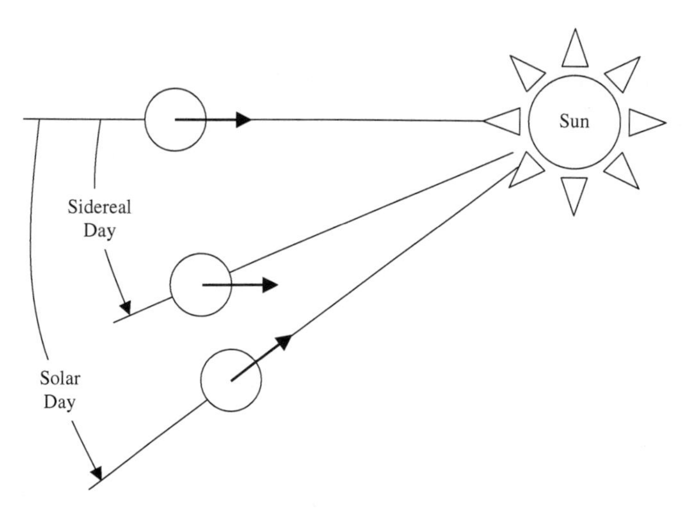

Figure 7.1: Solar and Sidereal Day

positions over time. Also, specific formulas for the conversion from ECEF coordinates to longitude and latitude are shown. The ECEF coordinate system, denoted by $\{\hat{\mathbf{e}}_1, \hat{\mathbf{e}}_2, \hat{\mathbf{e}}_3\}$, is similar to the ECI coordinate system shown in Figure 4.7. In fact the $\hat{\mathbf{i}}_3$ axis is identical, so that $\hat{\mathbf{e}}_3 = \hat{\mathbf{i}}_3$. The only difference between these two coordinates systems is that while the ECI $\hat{\mathbf{i}}_1$ axis is inertially fixed, the ECEF $\hat{\mathbf{e}}_1$ axis points at the Earth's prime meridian, which rotates with the Earth. Also, $\hat{\mathbf{e}}_2$ rotates through the same angle as $\hat{\mathbf{e}}_1$. This rotation angle is denoted by θ in Figure 4.7. The transformation from ECI to ECEF follows

$$\boxed{\begin{bmatrix} e_1 \\ e_2 \\ e_3 \end{bmatrix} = \begin{bmatrix} \cos\theta & \sin\theta & 0 \\ -\sin\theta & \cos\theta & 0 \\ 0 & 0 & 1 \end{bmatrix} \begin{bmatrix} i_1 \\ i_2 \\ i_3 \end{bmatrix}} \tag{7.1}$$

where i_1, i_2, and i_3 are the components of the ECI position vector, and e_1, e_2, and e_3 are the components of the ECEF position vector.

In order to determine the ECEF position vector we must first determine the angle θ, which is related to time. Suppose that we know our current time in Universal Time (UT), which is defined by the Greenwich hour angle augmented by 12 hours of a fictitious Sun uniformly orbiting in the equatorial plane.[1] At first glance one might believe that θ equals zero at midnight UT, which corresponds directly to the solar day time. To see why this thinking is erroneous consider the exaggerated angular movement of a solar day shown in Figure 7.1. A solar day is the length of time that elapses between the Sun reaching its highest point in the sky two consecutive times. However, the ECI coordinate system is fixed relative to the stars, not the Sun. A *sidereal day* is the length of time that passes between a given fixed star in the sky crossing a given projected meridian. From Figure 7.1 a sidereal day is clearly shorter than a solar day. This difference is about 4 minutes.[2]

The Greenwich Mean Sidereal Time (GMST) is the mean sidereal time at zero longitude, which can be given by the angle θ. Several formulas for the conversion from UT to GMST are given in the open literature (e.g., see Ref. [3]). One of the most widely-used formulas is presented by Meeus.[4] First, given UT year y, month m, day d, hour h, minute min, and second s, compute the days past or before the year 2000 using

$$\boxed{\begin{aligned} d_{2000} =& 367y - \text{INT}\left\{\frac{7\{y+\text{INT}[(m+9)/12]\}}{4}\right\} \\ &+\text{INT}\left\{\frac{275m}{9}\right\}+\frac{h+min/60+s/3600}{24}+d-730531.5 \end{aligned}} \tag{7.2}$$

where INT is the integer part of the fraction (e.g., INT(23.8) $=$ 23). The angle θ in degrees is given by

$$\boxed{\theta = 280.46061837 + 360.98564736628 \times d_{2000}} \tag{7.3}$$

Another formula that takes into account the precession and nutation of the Earth that occurs as a result of the Moon's motion is given in Ref. [4], which is accurate to within 0.03 seconds up to the year 2050. However, eqn. (7.3) is accurate enough for simulation purposes.

The GPS satellite information is usually given by a GPS almanac, which provides orbital element information, including: eccentricity, inclination, semimajor axis, right ascension, argument of perigee, and mean anomaly.* These parameters can be converted to an initial ECI position and velocity using the method described in §3.8.2. The ECI position and velocity at any time can be computed using eqn. (3.202). It should be noted that GPS time is based on the atomic standard time and is continuous without the leap seconds of UT, due to the non-smooth rotation of the Earth. GPS epoch is midnight of January 6, 1980, and GPS time is conventionally represented in weeks and seconds from this epoch. The GPS week is represented by an integer from 0 to 1023. A rollover occurred on August 22, 1999, so that 1024 needs to be added for references past this date. For simulation purposes counting the days past GPS epoch to determine UT is adequate (ignoring leap seconds, but not leap days). More details on the conversion from GPS week to UT can be found in Refs. [1] and [5]. With the known UT reference the ECEF position vector can be determined by first computing θ using eqn. (7.3). Then, the ECEF position vector is computed using eqn. (7.1). Another, more direct method to determine the ECEF position vector involves using the ephemeris parameters, which are broadcasted by the satellites and are available from the receiver.[6]

As previously mentioned the ECEF position vector is useful since this gives a simple approach to determine the longitude and latitude of a user. The Earth's geoid can be approximated by an ellipsoid of revolution about its minor axis. A common

*The U.S. Coast Guard Navigation Center maintains a website that contains GPS almanacs, and as of this writing this website is given by http://www.navcen.uscg.gov/.

ellipsoid model is given by the World Geodetic System 1984 model (WGS-84), with semimajor axis $a = 6,378,137.0$ m and semiminor axis $b = 6,356,752.3142$ m. The eccentricity of this ellipsoid is given by $e = 0.0818$. The geodetic coordinates are given by the longitude ϕ, latitude λ, and height h. To determine the ECEF position vector, the length of the normal to the ellipsoid is first computed, given by[6]

$$\boxed{N = \frac{a}{\sqrt{1 - e^2 \sin^2 \lambda}}} \tag{7.4}$$

Then, given the observer geodetic quantities ϕ, λ, and h, the observer ECEF position coordinates are computed using

$$x = (N + h)\cos\lambda\cos\phi \tag{7.5a}$$

$$y = (N + h)\cos\lambda\sin\phi \tag{7.5b}$$

$$z = [N(1 - e^2) + h]\sin\lambda \tag{7.5c}$$

The conversion from ECEF to geodetic coordinates is not that straightforward. A complicated closed-form solution is given in Ref. [6], but a good approximation up to low Earth orbit is given by[1]

$$p = \sqrt{x^2 + y^2} \tag{7.6a}$$

$$\psi = \text{atan}\left(\frac{z\,a}{p\,b}\right) \tag{7.6b}$$

$$\bar{e}^2 = \frac{a^2 - b^2}{b^2} \tag{7.6c}$$

$$\lambda = \text{atan}\left(\frac{z + \bar{e}^2 b \sin^3\psi}{p - e^2 a \cos^3\psi}\right) \tag{7.6d}$$

$$\phi = \text{atan2}(y, x) \tag{7.6e}$$

$$h = \frac{p}{\cos\lambda} - N \tag{7.6f}$$

where N is given by eqn. (7.4) and atan2 is a four quadrant inverse tangent function.

Another useful quantity is the availability of a particular GPS satellite at a given observer longitude and latitude. The solution to this problem involves computing the "up" vector shown in Figure 7.2, which is given by

$$\boxed{\mathbf{u} = \begin{bmatrix} \cos\lambda\cos\phi \\ \cos\lambda\sin\phi \\ \sin\lambda \end{bmatrix}} \tag{7.7}$$

The zenith angle, ξ, for the ith GPS satellite is given by

$$\boxed{\cos\xi_i = \boldsymbol{\rho}_i^T \mathbf{u}} \tag{7.8}$$

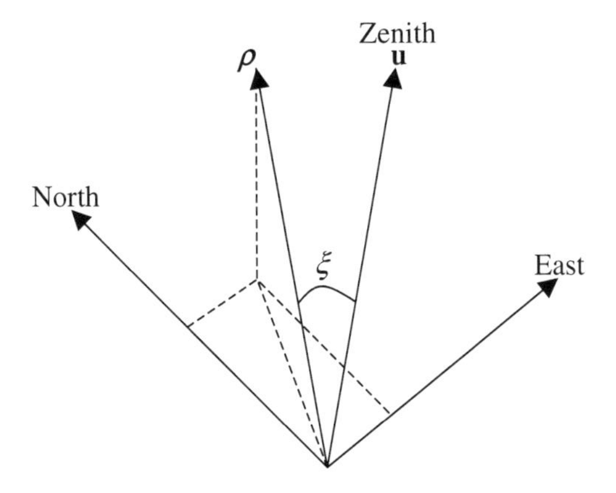

Figure 7.2: Definition of the Zenith Angle

From the assumed observer longitude and latitude of the user, its ECEF coordinates can be computed using eqn. (7.5), which is now defined by the vector $\mathbf{r} \equiv \begin{bmatrix} x & y & z \end{bmatrix}^T$. The ECEF coordinates of the ith GPS satellite, denoted by $\mathbf{e}_i$, is computed using the ECI to ECEF coordinate transformation shown by eqn. (7.1). Next, the following vector is computed:

$$\boxed{\rho_i = \frac{\mathbf{e}_i - \mathbf{r}}{||\mathbf{e}_i - \mathbf{r}||}} \tag{7.9}$$

Note that the pseudorange is given by $||\mathbf{e}_i - \mathbf{r}||$ plus the clock bias. Then, the following vertical (elevation) angle is computed:

$$\boxed{E_i = 90^\circ - \xi_i} \tag{7.10}$$

An adequate elevation cutoff for an observer on the Earth is given by 15 degrees.[1] Therefore, the ith GPS satellite is available if $E_i > 15^\circ$ is satisfied.

7.1.2 Extended Kalman Filter Application to GPS

In this section an application of the EKF is shown for GPS position and velocity estimation. In this formulation we assume that an Inertial Navigation System (INS), including gyroscopes and/or accelerometers, is not present. More details on a combined GPS/INS integrated architecture can be found in Ref. [7]. In our formulation we assume a random walk process for the acceleration of the vehicle and derivative of the clock bias, so that the truth model is given by

$$\ddot{\mathbf{r}}(t) = \mathbf{w}_r(t) \tag{7.11a}$$

$$\dot{\tau}(t) = w_\tau(t) \tag{7.11b}$$

where $\mathbf{r} \equiv \begin{bmatrix} x & y & z \end{bmatrix}^T$ is a 3×1 vector of the vehicle position in ECEF coordinates, and τ is the GPS clock bias defined in §4.1. The state-space model of eqn. (7.11) is given by

$$\dot{\mathbf{x}}(t) = \begin{bmatrix} 0_{3\times3} & I_{3\times3} & 0_{3\times1} \\ 0_{3\times3} & 0_{3\times3} & 0_{3\times1} \\ 0_{1\times3} & 0_{1\times3} & 0 \end{bmatrix} \mathbf{x}(t) + \begin{bmatrix} 0_{3\times4} \\ I_{4\times4} \end{bmatrix} \mathbf{w}(t) \tag{7.12}$$

where $\mathbf{x} \equiv \left[\mathbf{r}^T \; \dot{\mathbf{r}}^T \; \tau\right]^T$ and the process noise is defined by $\mathbf{w} \equiv \left[\mathbf{w}_r^T \; w_\tau\right]^T$. The discrete-time version of the system shown in eqn. (7.12) is given by

$$\mathbf{x}_{k+1} = \begin{bmatrix} I_{3\times3} & \Delta t I_{3\times3} & 0_{3\times1} \\ 0_{3\times3} & I_{3\times3} & 0_{3\times1} \\ 0_{1\times3} & 0_{1\times3} & 1 \end{bmatrix} \mathbf{x}_k + \begin{bmatrix} \Delta t^2/2 I_{3\times3} & 0_{3\times1} \\ \Delta t I_{3\times3} & 0_{3\times1} \\ 0_{1\times3} & \Delta t \end{bmatrix} \mathbf{w}_k \tag{7.13}$$

The ith row of the output vector is the pseudorange, given by eqn. (4.2). Therefore, matrix $H(\hat{\mathbf{x}})$ used in the EKF (see Table 5.9) is given by

$$H(\hat{\mathbf{x}}) = \begin{bmatrix} \frac{\partial \hat{\rho}_1}{\partial \hat{x}} & \frac{\partial \hat{\rho}_1}{\partial \hat{y}} & \frac{\partial \hat{\rho}_1}{\partial \hat{z}} & 0 & 0 & 0 & 1 \\ \frac{\partial \hat{\rho}_2}{\partial \hat{x}} & \frac{\partial \hat{\rho}_2}{\partial \hat{y}} & \frac{\partial \hat{\rho}_2}{\partial \hat{z}} & 0 & 0 & 0 & 1 \\ \vdots & \vdots & \vdots & \vdots & \vdots & \vdots & \vdots \\ \frac{\partial \hat{\rho}_n}{\partial \hat{x}} & \frac{\partial \hat{\rho}_n}{\partial \hat{y}} & \frac{\partial \hat{\rho}_n}{\partial \hat{z}} & 0 & 0 & 0 & 1 \end{bmatrix} \tag{7.14}$$

where the partials are given by eqn. (4.4).

The procedure to simulate the application of the EKF to GPS position determination is as follows. First, from the known orbital elements of the GPS satellites, determine the ECI position and velocity coordinates using eqns. (3.211) and (3.212). The GPS ECI vectors can then be propagated forward in time using eqn. (3.202). Then, using a chosen time reference, the GPS ECI position vectors are converted to ECEF coordinates using eqns. (7.1) and (7.3). Next, convert the chosen longitude, latitude, and height of the vehicle position on the Earth to ECEF coordinates using eqn. (7.5). Then, check the availability of each GPS satellite using the $15°$ cutoff in the elevation angles computed in eqn. (7.10). If the ith satellite is available, then compute a pseudorange measurement using eqn. (4.1). Next, compute pseudorange measurements at different time steps using the aforementioned procedure with a new vehicle location due to its movement on the Earth. Finally, pick a value for the process noise covariance and run the EKF to determine the vehicle's position and velocity. Typically, $\mathbf{w}_r$ and w_τ are uncorrelated, and the variance of w_τ is very well known, which is derived from the clock stability. As a first pass consider the covariance of $\mathbf{w}_r$ to be given by a scalar times the identity matrix, where the scalar is used as a tuning parameter.

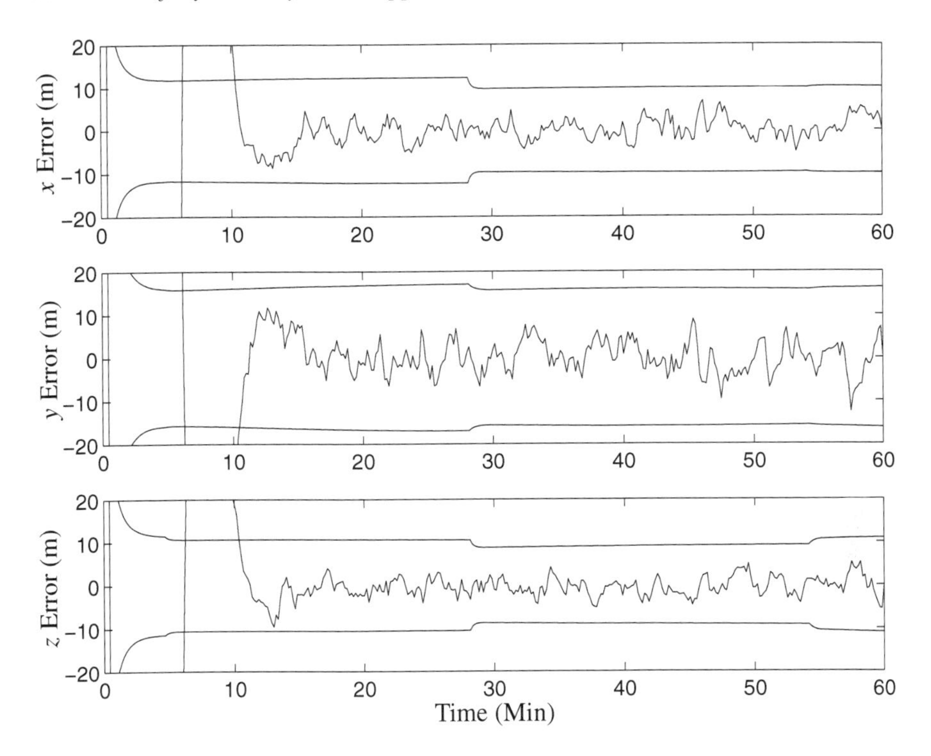

Figure 7.3: Vehicle Position Error Estimates

Example 7.1: In this example the extended Kalman filter is employed to determine the position and velocity of a vehicle on the Earth from GPS pseudorange measurements. The initial parameters of the ground vehicle are identical to example 4.1, where the vehicle is assumed to initially have coordinates of 38°N and 77°W (i.e., in Washington, DC). For simulation purposes the vehicle moves in the south direction at a rate of 100 km/hr. The variance of the continuous-time clock-bias process noise is assumed to be 200, which is used to generate the clock drift over time using eqn. (7.11b). The simulation runs over a 60-minute interval with a sampling interval of 10 seconds. The initial conditions for the EKF are given by $\hat{\mathbf{x}}_0 = \mathbf{0}$, which tests the convergence properties of the filter.

The continuous-time process noise covariance for the position error-process, $\mathbf{w}_r(t)$, shown in eqn. (7.11) is given by a scalar times a 3×3 identity matrix. The simulation uses only the true position movement of the vehicle (i.e., due south motion) to create the actual position. The scalar parameter in the process noise is only used in the EKF as a tuning factor to produce filtered estimates. After some trial and error a value of 1×10^{-5} for this tuning parameter is found to produce adequate results. Note that this value may be different for various vehicle motions. The adaptive methods of §5.7.4 may be employed to determine the optimal value if needed.

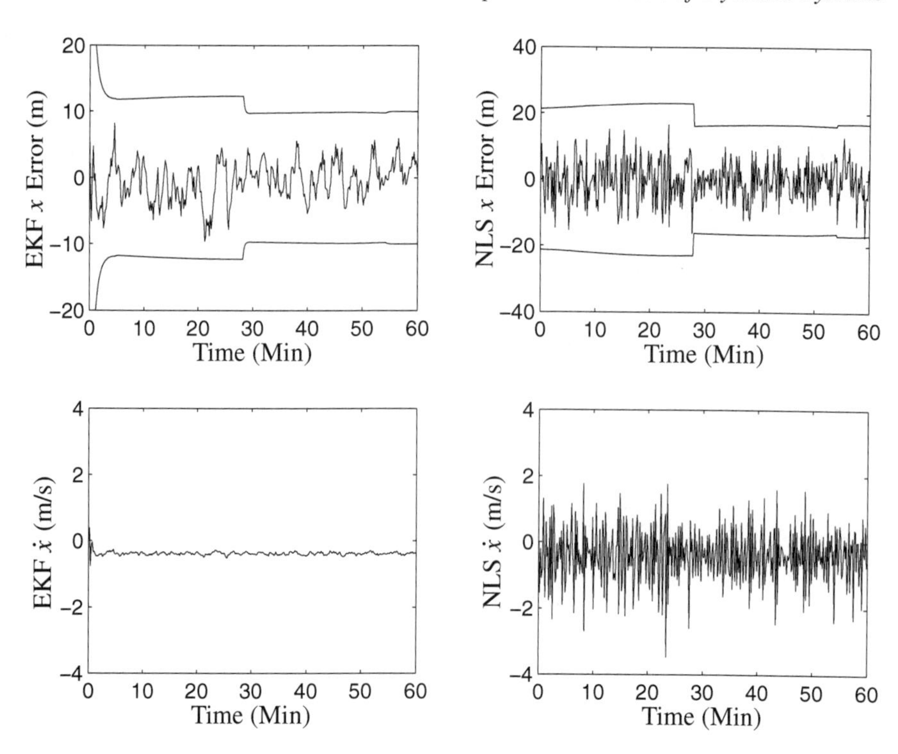

Figure 7.4: EKF and Nonlinear Least Squares Comparison

A plot of the EKF position error estimates with 3σ bounds is shown in Figure 7.3. The EKF converges in about 10 minutes, and the EKF accuracy in position is about 15 meters. Note that the 3σ bounds become smaller just before 30 minutes. This is due to the addition of another available GPS satellite in the vehicle's view (for this simulation the number of available satellites is 7 at the beginning of the simulation time). This result is intuitively correct, although the accuracy also depends on the GPS satellite geometry at the time of interest. A comparison of the EKF accuracy versus the nonlinear least squares deterministic solution of §4.1 is shown in Figure 7.4. For this simulation run the EKF estimates are initialized using the nonlinear least squares solution. The velocity estimates using the nonlinear least squares approach are determined by a finite difference of the determined position. The position estimates show a modest performance increase using the EKF; however, the true power of the EKF is shown in the velocity estimates, which are much smoother than the finite difference approach in nonlinear least squares. Similar results are obtained for the other axes. This example clearly shows how a Kalman filter can be employed to provide better performance capabilities using GPS pseudorange measurements to determine the position and velocity of a vehicle.

7.2 Attitude Estimation

In this section an extended Kalman filter is used to sequentially estimate the attitude and rate of a vehicle with attitude sensor measurements and three-axis strapdown gyroscopes. Several parameterizations can be used to represent the attitude, such as Euler angles,[8] quaternions,[9] modified Rodrigues parameters,[10] and even the rotation vector.[11] Quaternions are especially appealing since no singularities are present and the kinematics equation is bilinear. However, the quaternion must obey a normalization constraint, which can be violated by the linear measurement-updates associated with the standard EKF approach. The most common approach to overcome this shortfall involves using a multiplicative error quaternion, where after neglecting higher-order terms, the four-component quaternion can effectively be replaced by a three-component error vector.[9] Under ideal circumstances, such as small attitude errors, this approach works extremely well. Also, a useful variation to this filter is shown, which processes a single vector measurement at one time. This approach substantially reduces the computational burden.

7.2.1 Multiplicative Quaternion Formulation

The extended Kalman filter for attitude estimation begins with the quaternion kinematics model, shown previously in §3.7.1 as

$$\dot{\mathbf{q}} = \frac{1}{2}\Xi(\mathbf{q})\boldsymbol{\omega} = \frac{1}{2}\Omega(\boldsymbol{\omega})\mathbf{q} \tag{7.15}$$

The quaternion, $\mathbf{q} \equiv \begin{bmatrix}\boldsymbol{\varrho}^T & q_4\end{bmatrix}^T$, must obey a normalization constraint given by $\mathbf{q}^T\mathbf{q} = 1$. The most straightforward method for the filter design is to use eqn. (7.15) directly in the extended Kalman filter of Table 5.9; however, this "additive" approach can destroy normalization. This is clearly seen by example. Consider a true quaternion of $\mathbf{q} = [0 \;\; 0 \;\; \sqrt{0.001} \;\; \sqrt{0.999}]^T$, and assume that the estimated quaternion is given by $\hat{\mathbf{q}} = [0 \;\; 0 \;\; 0 \;\; 1]^T$. The additive error quaternion is given by the difference $\hat{\mathbf{q}} - \mathbf{q} = [0 \;\; 0 \;\; -\sqrt{0.001} \;\; 1-\sqrt{0.999}]^T$, which clearly is not close to being a unit vector. This can cause significant difficulties during the filtering process. A more physical (true to nature) approach involves using a multiplicative error quaternion in the body frame, given by

$$\delta\mathbf{q} = \mathbf{q} \otimes \hat{\mathbf{q}}^{-1} \tag{7.16}$$

with $\delta\mathbf{q} \equiv \begin{bmatrix}\delta\boldsymbol{\varrho}^T & \delta q_4\end{bmatrix}^T$. Also, the quaternion inverse is defined by eqn. (3.169). Taking the time derivative of eqn. (7.16) gives

$$\delta\dot{\mathbf{q}} = \dot{\mathbf{q}} \otimes \hat{\mathbf{q}}^{-1} + \mathbf{q} \otimes \dot{\hat{\mathbf{q}}}^{-1} \tag{7.17}$$

We now need to determine an expression for $\dot{\hat{\mathbf{q}}}^{-1}$. The estimated quaternion kinematics model follows

$$\boxed{\dot{\hat{\mathbf{q}}} = \frac{1}{2}\Xi(\hat{\mathbf{q}})\hat{\boldsymbol{\omega}} = \frac{1}{2}\Omega(\hat{\boldsymbol{\omega}})\hat{\mathbf{q}}} \tag{7.18}$$

Taking the time derivative of $\hat{\mathbf{q}} \otimes \hat{\mathbf{q}}^{-1} = \begin{bmatrix} 0 & 0 & 0 & 1 \end{bmatrix}^T$ gives

$$\dot{\hat{\mathbf{q}}} \otimes \hat{\mathbf{q}}^{-1} + \hat{\mathbf{q}} \otimes \dot{\hat{\mathbf{q}}}^{-1} = \mathbf{0} \tag{7.19}$$

Substituting eqn. (7.18) into eqn. (7.19) gives

$$\frac{1}{2}\Omega(\hat{\boldsymbol{\omega}})\hat{\mathbf{q}} \otimes \hat{\mathbf{q}}^{-1} + \hat{\mathbf{q}} \otimes \dot{\hat{\mathbf{q}}}^{-1} = \mathbf{0} \tag{7.20}$$

Since $\hat{\mathbf{q}} \otimes \hat{\mathbf{q}}^{-1} = \begin{bmatrix} 0 & 0 & 0 & 1 \end{bmatrix}^T$, and using the definition of $\Omega(\hat{\boldsymbol{\omega}})$ in eqn. (3.162), then eqn. (7.20) reduces down to

$$\frac{1}{2}\begin{bmatrix} \hat{\boldsymbol{\omega}} \\ 0 \end{bmatrix} + \hat{\mathbf{q}} \otimes \dot{\hat{\mathbf{q}}}^{-1} = \mathbf{0} \tag{7.21}$$

Solving eqn. (7.21) for $\dot{\hat{\mathbf{q}}}^{-1}$ yields

$$\dot{\hat{\mathbf{q}}}^{-1} = -\frac{1}{2}\hat{\mathbf{q}}^{-1} \otimes \begin{bmatrix} \hat{\boldsymbol{\omega}} \\ 0 \end{bmatrix} \tag{7.22}$$

Also, a useful identity is given by

$$\dot{\mathbf{q}} = \frac{1}{2}\Omega(\boldsymbol{\omega})\mathbf{q} = \frac{1}{2}\begin{bmatrix} \boldsymbol{\omega} \\ 0 \end{bmatrix} \otimes \mathbf{q} \tag{7.23}$$

This identity can easily be verified using the definitions of $\Omega(\boldsymbol{\omega})$ in eqn. (3.162) and quaternion multiplication in eqn. (3.168). Substituting eqns. (7.22) and (7.23) into eqn. (7.17), and using the definition of the error quaternion in eqn. (7.16) gives

$$\delta\dot{\mathbf{q}} = \frac{1}{2}\left\{\begin{bmatrix} \boldsymbol{\omega} \\ 0 \end{bmatrix} \otimes \delta\mathbf{q} - \delta\mathbf{q} \otimes \begin{bmatrix} \hat{\boldsymbol{\omega}} \\ 0 \end{bmatrix}\right\} \tag{7.24}$$

We now define the following error angular velocity: $\delta\boldsymbol{\omega} \equiv \boldsymbol{\omega} - \hat{\boldsymbol{\omega}}$. Substituting $\boldsymbol{\omega} = \hat{\boldsymbol{\omega}} + \delta\boldsymbol{\omega}$ into eqn. (7.24) leads to

$$\delta\dot{\mathbf{q}} = \frac{1}{2}\left\{\begin{bmatrix} \hat{\boldsymbol{\omega}} \\ 0 \end{bmatrix} \otimes \delta\mathbf{q} - \delta\mathbf{q} \otimes \begin{bmatrix} \hat{\boldsymbol{\omega}} \\ 0 \end{bmatrix}\right\} + \frac{1}{2}\begin{bmatrix} \delta\boldsymbol{\omega} \\ 0 \end{bmatrix} \otimes \delta\mathbf{q} \tag{7.25}$$

Next, consider the following helpful identities:

$$\begin{bmatrix} \hat{\boldsymbol{\omega}} \\ 0 \end{bmatrix} \otimes \delta\mathbf{q} = \Omega(\hat{\boldsymbol{\omega}})\delta\mathbf{q} \tag{7.26a}$$

$$\delta\mathbf{q} \otimes \begin{bmatrix} \hat{\boldsymbol{\omega}} \\ 0 \end{bmatrix} = \Gamma(\hat{\boldsymbol{\omega}})\delta\mathbf{q} \tag{7.26b}$$

where $\Gamma(\hat{\omega})$ is given by eqn. (3.165). Substituting eqn. (7.26) into eqn. (7.25), and after some algebraic manipulations (which are left as an exercise for the reader), leads to

$$\delta\dot{\mathbf{q}} = -\begin{bmatrix} [\hat{\omega}\times]\delta\boldsymbol{\varrho} \\ 0 \end{bmatrix} + \frac{1}{2}\begin{bmatrix} \delta\omega \\ 0 \end{bmatrix} \otimes \delta\mathbf{q} \tag{7.27}$$

where the cross product matrix $[\hat{\omega}\times]$ is defined by eqn. (3.149). Note that eqn. (7.27) is an exact relationship since no linearizations have been performed yet. The nonlinear term is present only in the last term on the right-hand side of eqn. (7.27). Its first-order approximation is given by[9]

$$\frac{1}{2}\begin{bmatrix} \delta\omega \\ 0 \end{bmatrix} \otimes \delta\mathbf{q} \approx \frac{1}{2}\begin{bmatrix} \delta\omega \\ 0 \end{bmatrix} \tag{7.28}$$

Substituting eqn. (7.28) into eqn. (7.27) leads to the following linearized model:

$$\delta\dot{\boldsymbol{\varrho}} = -[\hat{\omega}\times]\delta\boldsymbol{\varrho} + \frac{1}{2}\delta\omega \tag{7.29a}$$

$$\delta\dot{q}_4 = 0 \tag{7.29b}$$

Note that the fourth error-quaternion component is constant. The first-order approximation, which assumes that the true quaternion is "close" to the estimated quaternion, gives $\delta q_4 \approx 1$. This allows us to reduce the order of the system in the EKF by one state. The linearization using eqn. (7.16) maintains quaternion normalization to within first-order if the estimated quaternion is "close" to the true quaternion, which is within the first-order approximation in the EKF.

A common sensor that measures the angular rate is a rate-integrating gyro. For this sensor, a widely used model is given by[12]

$$\omega = \tilde{\omega} - \boldsymbol{\beta} - \boldsymbol{\eta}_v \tag{7.30a}$$

$$\dot{\boldsymbol{\beta}} = \boldsymbol{\eta}_u \tag{7.30b}$$

where $\boldsymbol{\eta}_v$ and $\boldsymbol{\eta}_u$ are zero-mean Gaussian white-noise processes with covariances usually given by $\sigma_v^2 I_{3\times3}$ and $\sigma_u^2 I_{3\times3}$, respectively, $\boldsymbol{\beta}$ is a bias vector, and $\tilde{\omega}$ is the measured observation. The estimated angular velocity is given by

$$\hat{\omega} = \tilde{\omega} - \hat{\boldsymbol{\beta}} \tag{7.31}$$

Also, the estimated bias differential equation follows

$$\dot{\hat{\boldsymbol{\beta}}} = \mathbf{0} \tag{7.32}$$

Substituting eqns. (7.30a) and (7.31) into $\delta\omega \equiv \omega - \hat{\omega}$ gives

$$\delta\omega = -(\Delta\boldsymbol{\beta} + \boldsymbol{\eta}_v) \tag{7.33}$$

where $\Delta\boldsymbol{\beta} \equiv \boldsymbol{\beta} - \hat{\boldsymbol{\beta}}$. Substituting eqn. (7.33) into eqn. (7.29a) gives

$$\delta\dot{\boldsymbol{\varrho}} = -[\hat{\omega}\times]\delta\boldsymbol{\varrho} - \frac{1}{2}(\Delta\boldsymbol{\beta} + \boldsymbol{\eta}_v) \tag{7.34}$$

A common simplification, which is discussed in §3.7.1, is given by the small angle approximation $\delta\boldsymbol{\varrho} \approx \delta\boldsymbol{\alpha}/2$, where $\delta\boldsymbol{\alpha}$ has components of roll, pitch, and yaw error-angles for any rotation sequence. Using this simplification in eqn. (7.34) gives

$$\delta\dot{\boldsymbol{\alpha}} = -[\hat{\boldsymbol{\omega}}\times]\delta\boldsymbol{\alpha} - (\Delta\boldsymbol{\beta} + \boldsymbol{\eta}_v) \tag{7.35}$$

This approach minimizes the use of factors of $1/2$ and 2 in the EKF, and also gives a direct physical meaning to the state error-covariance, which can be used to directly determine the 3σ bounds of the actual attitude errors. The EKF error model is now given by

$$\Delta\dot{\tilde{\mathbf{x}}}(t) = F(\hat{\mathbf{x}}(t), t)\,\Delta\tilde{\mathbf{x}}(t) + G(t)\,\mathbf{w}(t) \tag{7.36}$$

where $\Delta\tilde{\mathbf{x}}(t) \equiv \left[\delta\boldsymbol{\alpha}^T(t)\ \Delta\boldsymbol{\beta}^T(t)\right]^T$, $\mathbf{w}(t) \equiv \left[\boldsymbol{\eta}_v^T(t)\ \boldsymbol{\eta}_u^T(t)\right]^T$, and $F(\hat{\mathbf{x}}(t), t)$, $G(t)$, and $Q(t)$ are given by

$$F(\hat{\mathbf{x}}(t), t) = \begin{bmatrix} -[\hat{\boldsymbol{\omega}}(t)\times] & -I_{3\times3} \\ 0_{3\times3} & 0_{3\times3} \end{bmatrix} \tag{7.37a}$$

$$G(t) = \begin{bmatrix} -I_{3\times3} & 0_{3\times3} \\ 0_{3\times3} & I_{3\times3} \end{bmatrix} \tag{7.37b}$$

$$Q(t) = \begin{bmatrix} \sigma_v^2 I_{3\times3} & 0_{3\times3} \\ 0_{3\times3} & \sigma_u^2 I_{3\times3} \end{bmatrix} \tag{7.37c}$$

Note that these matrices are 6×6 matrices now, since the order of the system has been reduced by one state.

Our next step involves the determination of the sensitivity matrix $H_k(\hat{\mathbf{x}}_k^-)$ used in the EKF. Discrete-time attitude observations for a single sensor are given by eqn. (4.11). Multiple, n, vector measurements can be concatenated to form

$$\tilde{\mathbf{y}}_k = \left.\begin{bmatrix} A(\mathbf{q})\mathbf{r}_1 \\ A(\mathbf{q})\mathbf{r}_2 \\ \vdots \\ A(\mathbf{q})\mathbf{r}_n \end{bmatrix}\right|_{t_k} + \left.\begin{bmatrix} \boldsymbol{\nu}_1 \\ \boldsymbol{\nu}_2 \\ \vdots \\ \boldsymbol{\nu}_n \end{bmatrix}\right|_{t_k} \equiv \mathbf{h}_k(\hat{\mathbf{x}}_k) + \mathbf{v}_k \tag{7.38a}$$

$$R = \text{diag}\left[\sigma_1^2 I_{3\times3}\ \sigma_2^2 I_{3\times3}\ \ldots\ \sigma_n^2 I_{3\times3}\right] \tag{7.38b}$$

where diag denotes a diagonal matrix of appropriate dimension. The actual attitude matrix, $A(\mathbf{q})$, is related to the propagated attitude, $A(\delta\mathbf{q})$, through

$$A(\mathbf{q}) = A(\delta\mathbf{q})A(\hat{\mathbf{q}}^-) \tag{7.39}$$

The first-order approximation of the error-attitude matrix is given by (see §3.7.1)

$$A(\delta\mathbf{q}) \approx I_{3\times3} - [\delta\boldsymbol{\alpha}\times] \tag{7.40}$$

where $\boldsymbol{\delta\alpha}$ is again the small angle approximation. For a single sensor the true and estimated body vectors are given by

$$\mathbf{b} = A(\mathbf{q})\mathbf{r} \tag{7.41a}$$

$$\hat{\mathbf{b}}^{-} = A(\hat{\mathbf{q}}^{-})\mathbf{r} \tag{7.41b}$$

Substituting eqns. (7.39) and (7.40) into eqn. (7.41) yields

$$\Delta\mathbf{b} = [A(\hat{\mathbf{q}}^{-})\mathbf{r}\times]\boldsymbol{\delta\alpha} \tag{7.42}$$

where $\Delta\mathbf{b} \equiv \mathbf{b} - \hat{\mathbf{b}}^{-}$. The sensitivity matrix for all measurement sets is therefore given by

$$\boxed{H_k(\hat{\mathbf{x}}_k^{-}) = \left.\begin{bmatrix} \left[A(\hat{\mathbf{q}}^{-})\mathbf{r}_1\times\right] & 0_{3\times3} \\ \left[A(\hat{\mathbf{q}}^{-})\mathbf{r}_2\times\right] & 0_{3\times3} \\ \vdots & \vdots \\ \left[A(\hat{\mathbf{q}}^{-})\mathbf{r}_n\times\right] & 0_{3\times3} \end{bmatrix}\right|_{t_k}} \tag{7.43}$$

Note that the number of columns of $H_k(\hat{\mathbf{x}}_k^{-})$ is six, which is the dimension of the reduced-order state.

The final part in the EKF involves the quaternion and bias updates. The error-state update follows

$$\boxed{\Delta\hat{\tilde{\mathbf{x}}}_k^{+} = K_k[\tilde{\mathbf{y}}_k - \mathbf{h}_k(\hat{\mathbf{x}}_k^{-})]} \tag{7.44}$$

where $\Delta\hat{\tilde{\mathbf{x}}}_k^{+} \equiv \left[\boldsymbol{\delta\hat{\alpha}}_k^{+T} \;\; \Delta\hat{\boldsymbol{\beta}}_k^{+T}\right]^T$, $\tilde{\mathbf{y}}_k$ is the measurement output, and $\mathbf{h}_k(\hat{\mathbf{x}}_k^{-})$ is the estimate output, given by

$$\boxed{\mathbf{h}_k(\hat{\mathbf{x}}_k^{-}) = \left.\begin{bmatrix} A(\hat{\mathbf{q}}^{-})\mathbf{r}_1 \\ A(\hat{\mathbf{q}}^{-})\mathbf{r}_2 \\ \vdots \\ A(\hat{\mathbf{q}}^{-})\mathbf{r}_n \end{bmatrix}\right|_{t_k}} \tag{7.45}$$

The gyro bias update is simply given by

$$\boxed{\hat{\boldsymbol{\beta}}_k^{+} = \hat{\boldsymbol{\beta}}_k^{-} + \Delta\hat{\boldsymbol{\beta}}_k^{+}} \tag{7.46}$$

The quaternion update is more complicated. As previously mentioned the fourth component of $\boldsymbol{\delta}\mathbf{q}$ is nearly one. Therefore, to within first-order the quaternion update is given by

$$\hat{\mathbf{q}}_k^{+} = \begin{bmatrix} \frac{1}{2}\boldsymbol{\delta\hat{\alpha}}_k^{+} \\ 1 \end{bmatrix} \otimes \hat{\mathbf{q}}_k^{-} \tag{7.47}$$

Table 7.1: Extended Kalman Filter for Attitude Estimation

Initialize	$\hat{\mathbf{q}}(t_0) = \hat{\mathbf{q}}_0, \quad \beta(t_0) = \beta_0$ $P(t_0) = P_0$
Gain	$K_k = P_k^- H_k^T(\hat{\mathbf{x}}_k^-)[H_k(\hat{\mathbf{x}}_k^-)P_k^- H_k^T(\hat{\mathbf{x}}_k^-) + R]^{-1}$ $H_k(\hat{\mathbf{x}}_k^-) = \left.\begin{bmatrix} [A(\hat{\mathbf{q}}^-)\mathbf{r}_1\times] & 0_{3\times 3} \\ \vdots & \vdots \\ [A(\hat{\mathbf{q}}^-)\mathbf{r}_n\times] & 0_{3\times 3} \end{bmatrix}\right\|_{t_k}$
Update	$P_k^+ = [I - K_k H_k(\hat{\mathbf{x}}_k^-)]P_k^-$ $\Delta\hat{\tilde{\mathbf{x}}}_k^+ = K_k[\tilde{\mathbf{y}}_k - \mathbf{h}_k(\hat{\mathbf{x}}_k^-)]$ $\Delta\hat{\tilde{\mathbf{x}}}_k^+ \equiv \begin{bmatrix} \delta\hat{\alpha}_k^{+T} & \Delta\hat{\beta}_k^{+T} \end{bmatrix}^T$ $\mathbf{h}_k(\hat{\mathbf{x}}_k^-) = \left.\begin{bmatrix} A(\hat{\mathbf{q}}^-)\mathbf{r}_1 \\ A(\hat{\mathbf{q}}^-)\mathbf{r}_2 \\ \vdots \\ A(\hat{\mathbf{q}}^-)\mathbf{r}_n \end{bmatrix}\right\|_{t_k}$ $\hat{\mathbf{q}}_k^+ = \hat{\mathbf{q}}_k^- + \frac{1}{2}\Xi(\hat{\mathbf{q}}_k^-)\delta\hat{\alpha}_k^+$, re-normalize quaternion $\hat{\beta}_k^+ = \hat{\beta}_k^- + \Delta\hat{\beta}_k^+$
Propagation	$\hat{\omega}(t) = \tilde{\omega}(t) - \hat{\beta}(t)$ $\dot{\hat{\mathbf{q}}}(t) = \frac{1}{2}\Xi\left(\hat{\mathbf{q}}(t)\right)\hat{\omega}(t)$ $\dot{P}(t) = F(\hat{\mathbf{x}}(t), t)\,P(t) + P(t)\,F^T(\hat{\mathbf{x}}(t), t) + G(t)\,Q(t)\,G^T(t)$ $F(\hat{\mathbf{x}}(t), t) = \begin{bmatrix} -[\hat{\omega}(t)\times] & -I_{3\times 3} \\ 0_{3\times 3} & 0_{3\times 3} \end{bmatrix}, \quad G(t) = \begin{bmatrix} -I_{3\times 3} & 0_{3\times 3} \\ 0_{3\times 3} & I_{3\times 3} \end{bmatrix}$

Note that the small angle approximation has been used to define the vector part of the error-quaternion. Using the quaternion multiplication rule of eqn. (3.168) in eqn. (7.47) gives

$$\boxed{\hat{\mathbf{q}}_k^+ = \hat{\mathbf{q}}_k^- + \frac{1}{2}\Xi(\hat{\mathbf{q}}_k^-)\delta\hat{\alpha}_k^+} \tag{7.48}$$

This updated quaternion is a unit vector to within first-order; however, a brute-force normalization should be performed to insure $\hat{\mathbf{q}}_k^{+T}\hat{\mathbf{q}}_k^+ = 1$.

The attitude estimation algorithm is summarized in Table 7.1. The filter is first initialized with a known state (the bias initial condition is usually assumed zero) and error-covariance matrix. The first three diagonal elements of the error-covariance matrix correspond to attitude errors. Then, the Kalman gain is computed using the measurement-error covariance R and sensitivity matrix in eqn. (7.43). The state error-covariance follows the standard EKF update, while the error-state update is computed using eqn. (7.44). The bias and quaternion updates are now given by eqns. (7.46) and (7.48). Also, the updated quaternion is re-normalized by brute force. Finally, the estimated angular velocity is used to propagate the quaternion kinematics model in eqn. (7.18) and standard error-covariance in the EKF. Note that the gyro bias propagation is constant as shown by eqn. (7.32).

7.2.2 Discrete-Time Attitude Estimation

The propagation of the state and covariance can be accomplished by using numerical integration techniques. However, in general, the gyro observations are sampled at a high rate (usually higher than or at least at the same rate as the vector attitude observations). Therefore, a discrete propagation is usually sufficient. Discrete propagation of the quaternion model in eqn. (7.18) can be derived by using a power series approach:[13]

$$\begin{aligned}\exp\left[\frac{1}{2}\Omega(\hat{\omega})t\right] &= \sum_{j=0}^{\infty}\frac{\left[\frac{1}{2}\Omega(\hat{\omega})t\right]^{j}}{j!}\\ &= \sum_{k=0}^{\infty}\left\{\frac{\left[\frac{1}{2}\Omega(\hat{\omega})t\right]^{2k}}{(2k)!}+\frac{\left[\frac{1}{2}\Omega(\hat{\omega})t\right]^{2k+1}}{(2k+1)!}\right\}\end{aligned} \tag{7.49}$$

Next, consider the following identities:

$$\Omega^{2k}(\hat{\omega}) = (-1)^{k}||\hat{\omega}||^{2k}I_{4\times 4} \tag{7.50a}$$

$$\Omega^{2k+1}(\hat{\omega}) = (-1)^{k}||\hat{\omega}||^{2k}\Omega(\hat{\omega}) \tag{7.50b}$$

Substituting eqn. (7.50) into eqn. (7.49) gives

$$\begin{aligned}\exp\left[\frac{1}{2}\Omega(\hat{\omega})t\right] &= I_{4\times 4}\sum_{k=0}^{\infty}\frac{(-1)^{k}\left(\frac{1}{2}||\hat{\omega}||t\right)^{2k}}{(2k)!}\\ &\quad+||\hat{\omega}||^{-1}\Omega(\hat{\omega})\sum_{k=0}^{\infty}\frac{(-1)^{k}\left(\frac{1}{2}||\hat{\omega}||t\right)^{2k+1}}{(2k+1)!}\end{aligned} \tag{7.51}$$

Recognizing that the first series in eqn. (7.51) is the cosine function and that the second series in eqn. (7.51) is the sine function yields

$$\exp\left[\frac{1}{2}\Omega(\hat{\omega})\,t\right] = I_{4\times4}\cos\left(\frac{1}{2}||\hat{\omega}||\,t\right) + \Omega(\hat{\omega})\frac{\sin\left(\frac{1}{2}||\hat{\omega}||\,t\right)}{||\hat{\omega}||} \tag{7.52}$$

Hence, given post-update estimates $\hat{\omega}_k^+$ and $\hat{\mathbf{q}}_k^+$, the propagated quaternion is found using

$$\boxed{\hat{\mathbf{q}}_{k+1}^- = \bar{\Omega}(\hat{\omega}_k^+)\hat{\mathbf{q}}_k^+} \tag{7.53}$$

with

$$\boxed{\bar{\Omega}(\hat{\omega}_k^+) \equiv \begin{bmatrix} \cos\left(\frac{1}{2}||\hat{\omega}_k^+||\,\Delta t\right) I_{3\times3} - \left[\hat{\boldsymbol{\psi}}_k^+\times\right] & \hat{\boldsymbol{\psi}}_k^+ \\ -\hat{\boldsymbol{\psi}}_k^{+T} & \cos\left(\frac{1}{2}||\hat{\omega}_k^+||\,\Delta t\right) \end{bmatrix}} \tag{7.54}$$

where

$$\boxed{\hat{\boldsymbol{\psi}}_k^+ \equiv \frac{\sin\left(\frac{1}{2}||\hat{\omega}_k^+||\,\Delta t\right)\hat{\omega}_k^+}{||\hat{\omega}_k^+||}} \tag{7.55}$$

and Δt is the sampling interval in the gyro. In the standard EKF formulation, given a post-update estimate $\hat{\boldsymbol{\beta}}_k^+$, the post-update angular velocity and propagated gyro bias follow

$$\hat{\omega}_k^+ = \tilde{\omega}_k - \hat{\boldsymbol{\beta}}_k^+ \tag{7.56a}$$

$$\hat{\boldsymbol{\beta}}_{k+1}^- = \hat{\boldsymbol{\beta}}_k^+ \tag{7.56b}$$

Note that the propagated gyro-bias estimate is equal to the previous update, which is due to the propagation model in eqn. (7.32).

The discrete propagation of the covariance equation is given by

$$P_{k+1}^- = \Phi_k P_k^+ \Phi_k^T + G_k Q_k G_k^T \tag{7.57}$$

where G_k is given by

$$G_k = \begin{bmatrix} -I_{3\times3} & 0_{3\times3} \\ 0_{3\times3} & I_{3\times3} \end{bmatrix} \tag{7.58}$$

The discrete error-state transition matrix can also be derived using a power series approach (which is left as an exercise for the reader):

$$\Phi = \begin{bmatrix} \Phi_{11} & \Phi_{12} \\ \Phi_{21} & \Phi_{22} \end{bmatrix} \tag{7.59a}$$

$$\Phi_{11} = I_{3\times 3} - [\hat{\omega}\times]\frac{\sin(||\hat{\omega}||\,\Delta t)}{||\hat{\omega}||} + [\hat{\omega}\times]^2 \frac{\{1-\cos(||\hat{\omega}||\,\Delta t)\}}{||\hat{\omega}||^2} \tag{7.59b}$$

$$\begin{aligned} \Phi_{12} = [\hat{\omega}\times]\frac{\{1-\cos(||\hat{\omega}||\,\Delta t)\}}{||\hat{\omega}||^2} - I_{3\times 3}\Delta t \\ - [\hat{\omega}\times]^2 \frac{\{||\hat{\omega}||\,\Delta t - \sin(||\hat{\omega}||\,\Delta t)\}}{||\hat{\omega}||^3} \end{aligned} \tag{7.59c}$$

$$\Phi_{21} = 0_{3\times 3} \tag{7.59d}$$

$$\Phi_{22} = I_{3\times 3} \tag{7.59e}$$

The discrete process noise covariance has already been derived in example 5.3, which is given by

$$Q_k = \begin{bmatrix} \left(\sigma_v^2 \Delta t + \frac{1}{3}\sigma_u^2 \Delta t^3\right) I_{3\times 3} & -\left(\frac{1}{2}\sigma_u^2 \Delta t^2\right) I_{3\times 3} \\ -\left(\frac{1}{2}\sigma_u^2 \Delta t^2\right) I_{3\times 3} & \left(\sigma_u^2 \Delta t\right) I_{3\times 3} \end{bmatrix} \tag{7.60}$$

Therefore, the continuous-time propagations of eqns. (7.18), (7.32), and covariance propagation can be replaced by their discrete-time equivalents of eqns. (7.53), (7.56b), and (7.57), respectively. These discrete-time forms make the EKF especially suitable for on-board implementation.

7.2.3 Murrell's Version

The only problem for the filter shown in Table 7.1 occurs in the gain calculation, which requires an inverse of a $3n \times 3n$ matrix. In order to overcome this difficulty a variation to this filter can be used, based on an algorithm by Murrell.[14] Even though the extended Kalman filter involves nonlinear models, a linear update is still performed. Therefore, linear tools such as the principle of superposition (see §3.1) can still be used. Murrell's filter uses this principle to process one 3×1 vector observation at a time. A flow diagram of Murrell's approach is given in Figure 7.5. The first step involves propagating the quaternion, gyro bias, and error-covariance to the current observation time. Then, the attitude matrix is computed. The propagated state vector is now initialized to zero. Next, the error-covariance and state quantities are updated using a single vector observation. This procedure is continued (replacing the propagated error-covariance and state vector with the updated values) until all vector observations are processed. Finally, the updated values are used to propagate the error-covariance and state quantities to the next observation time. Therefore, this approach reduces taking an inverse of a $3n \times 3n$ matrix to taking an inverse of a 3×3 matrix n times, which can significantly decrease the computational load.

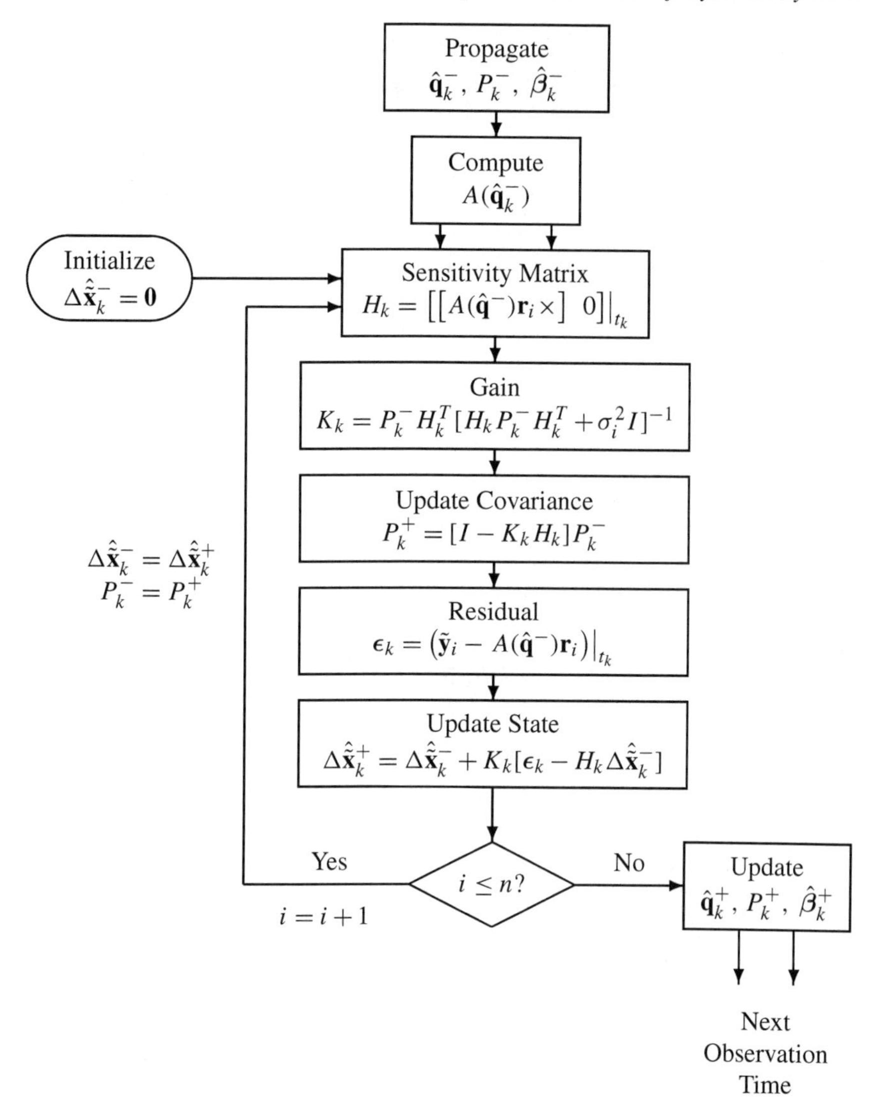

Figure 7.5: Computationally Efficient Attitude Estimation Algorithm

---

Example 7.2: In this example the extended Kalman filter algorithm shown in Table 7.1 is employed for attitude estimation using the simulation parameters shown by example 4.2. The attitude determination results of the deterministic approach (i.e., without using a filter) are shown in Figure 4.5. The goals of the EKF application involve the estimation of the gyro biases for all three axes and the filtering of the attitude star camera measurements. The standard deviation of the star camera

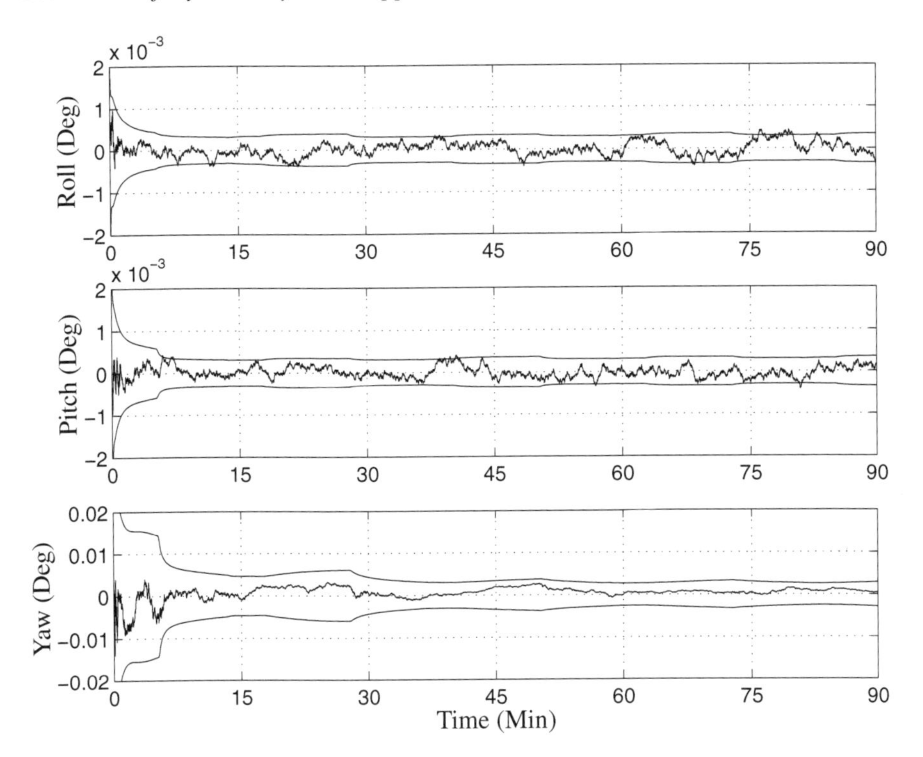

Figure 7.6: Attitude Errors and Boundaries

measurement error is the same as given in example 4.2. The noise parameters for the gyro measurements are given by $\sigma_u = \sqrt{10} \times 10^{-10}$ rad/sec$^{3/2}$ and $\sigma_v = \sqrt{10} \times 10^{-7}$ rad/sec$^{1/2}$. The initial bias for each axis is given by 0.1 deg/hr. Also, the gyro measurements are sampled at the same rate as the star camera measurements (i.e., at 1 Hz). We should note that in practice the gyros are sampled at a much higher frequency, which is usually required for jitter control. The initial covariance for the attitude error is set to 0.1^2 deg^2, and the initial covariance for the gyro drift is set to 0.2^2 (deg/hr)^2. Converting these quantities to radians gives the following initial attitude and gyro drift covariances for each axis: $P_0^a = 3.0462 \times 10^{-6}$ and $P_0^b = 9.4018 \times 10^{-13}$, so that the initial covariance is given by

$$P_0 = \text{diag}\left[P_0^a \; P_0^a \; P_0^a \; P_0^b \; P_0^b \; P_0^b\right]$$

The initial attitude condition for the EKF is given by the deterministic quaternion from example 4.2. The initial gyro bias conditions in the EKF are set to zero.

A plot of the attitude errors and associated 3σ boundaries is shown in Figure 7.6. Clearly, the computed 3σ boundaries do indeed bound the attitude errors. Comparing Figure 4.5 to Figure 7.6 shows a vast improvement (by an order of magnitude) in the attitude accuracy. This is due to the combination of the attitude measurements with

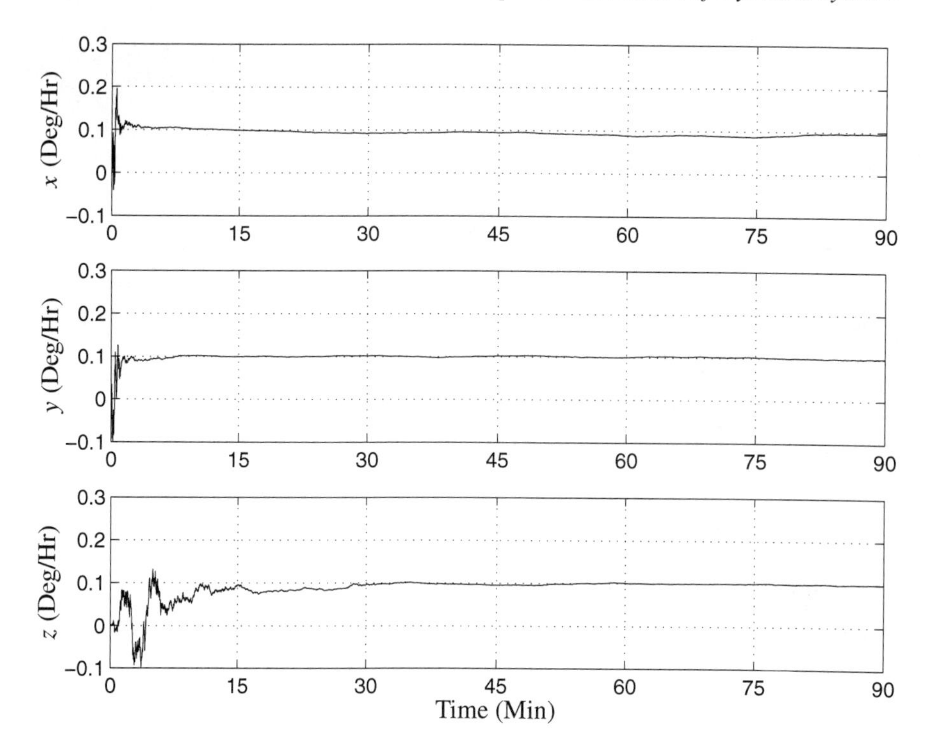

Figure 7.7: Gyro Drift Estimates

an accurate three-axis gyro. As with the deterministic solution the EKF results show that the yaw errors are much larger than the roll and pitch errors, which is intuitively correct. Also, the accuracy degrades as the number of available stars decreases, although this effect is not as pronounced with EKF results as with the deterministic results. This is due to the effect of filtering on the measurements. A plot of the gyro drift estimates is shown in Figure 7.7. The EKF is able to accurately estimate the initial bias errors. Also, the "drift" in this plot looks very steady, which is due to the fact that a high-grade three-axis gyro has been used in the simulation. A single axis analysis that can be used to access the performance of the EKF with various gyros will be shown in §7.2.4. This example clearly shows the power of the EKF for attitude estimation, which has been successfully applied to many spacecraft (e.g., see Ref. [15]). Another more robust approach to initial condition errors involves the application of the Unscented filter of §5.7.6, which may be found in Ref. [16].

7.2.4 Farrenkopf's Steady-State Analysis

The predicted performance of the attitude estimation can be found by checking the diagonal elements of the attitude error covariance. If a sensor is used to measure the integrated rates directly (i.e., assuming that the error angles can be decoupled) with standard deviation of the measurement error process given by σ_n, then a steady-state covariance given can be used. The model used for a single-axis analysis is shown in example 5.3, which is repeated here for completeness. The attitude rate $\dot{\theta}$ is assumed to be related to the gyro output $\tilde{\omega}$ by

$$\boxed{\dot{\theta} = \tilde{\omega} - \beta - \eta_v} \tag{7.61}$$

where β is the gyro drift, and η_v is a zero-mean Gaussian white-noise process with variance given by σ_v^2. The drift rate is modelled by a random walk process, given by

$$\boxed{\dot{\beta} = \eta_u} \tag{7.62}$$

where η_u is a zero-mean Gaussian white-noise process with variance given by σ_u^2. The state transition matrix and process noise covariance are shown in example 5.3. The discrete-time system used in the Kalman filter is given by

$$\mathbf{x}_{k+1} = \Phi\,\mathbf{x}_k + \Gamma\,\tilde{\omega}_k + \mathbf{w}_k \tag{7.63a}$$

$$\tilde{y}_k = H\,\mathbf{x}_k + v_k \tag{7.63b}$$

where $\mathbf{x} = \begin{bmatrix}\theta \; \beta\end{bmatrix}^T$, $\Gamma = \begin{bmatrix}\Delta t \; 0\end{bmatrix}^T$, $H = \begin{bmatrix}1 \; 0\end{bmatrix}$, and $E\left\{\mathbf{w}_k \mathbf{w}_k^T\right\} = Q$. The matrices Q and Φ are given in example 5.3:

$$Q = \begin{bmatrix} \sigma_v^2 \Delta t + \frac{1}{3}\sigma_u^2 \Delta t^3 & -\frac{1}{2}\sigma_u^2 \Delta t^2 \\ -\frac{1}{2}\sigma_u^2 \Delta t^2 & \sigma_u^2 \Delta t \end{bmatrix} \tag{7.64a}$$

$$\Phi = \begin{bmatrix} 1 & -\Delta t \\ 0 & 1 \end{bmatrix} \tag{7.64b}$$

Using the model in eqn. (7.63) a solution to the resulting steady-state algebraic Riccati equation shown in Table 5.2 can be determined for the attitude and gyro drift estimate variances. Farrenkopf[12] obtained analytic solutions to the resulting Riccati equation. First, define the following the propagated and updated covariances:

$$P^- \equiv \begin{bmatrix} p_{\theta\theta}^- & p_{\theta\beta}^- \\ p_{\theta\beta}^- & p_{\beta\beta}^- \end{bmatrix}, \quad P^+ \equiv \begin{bmatrix} p_{\theta\theta}^+ & p_{\theta\beta}^+ \\ p_{\theta\beta}^+ & p_{\beta\beta}^+ \end{bmatrix} \tag{7.65}$$

Next, define the following variables:

$$\xi \equiv p_{\theta\beta}^- \Delta t \Big/ \sigma_n^2 \tag{7.66a}$$

$$S_u \equiv \sigma_u \Delta t^{3/2} \Big/ \sigma_n \tag{7.66b}$$

$$S_v \equiv \sigma_v \Delta t^{1/2} \Big/ \sigma_n \tag{7.66c}$$

Using the defined matrices in this section for Φ, Q, H, and $R = \sigma_n^2$, from the steady-state Riccati equation in Table 5.2 the following equation can be derived for ξ in terms of S_u and S_v (note, the procedure to determine this equation is outlined in §7.4.1):

$$\xi^4 + S_u^2 \xi^3 + S_u^2 \left[(S_u^2/6) - S_v^2 - 2 \right] \xi^2 + S_u^4 \xi + S_u^4 = 0 \tag{7.67}$$

This a quartic equation, but it can be simplified significantly since it is actually the product of two quadratic equations:

$$\xi^2 + \left[(S_u^2/2) \pm \vartheta \right] \xi + S_u^2 = 0 \tag{7.68}$$

where

$$\boxed{\vartheta = \left[S_u^2 (4 + S_v^2) + S_u^4/12 \right]^{1/2}} \tag{7.69}$$

The root of physical significance is the maximally negative root, assuming $+\vartheta$ in eqn. (7.68), so that

$$\boxed{\xi = -\frac{1}{2} \left[\left(\frac{S_u^2}{2} + \vartheta \right) + \sqrt{\left(\frac{S_u^2}{2} + \vartheta \right)^2 - 4S_u^2} \right]} \tag{7.70}$$

Then the solution for $p_{\theta\beta}^-$ is given using eqn. (7.66a). Once $p_{\theta\beta}^-$ is determined then the solutions for $p_{\theta\theta}^-$ and $p_{\beta\beta}^-$ are fairly straightforward (which are left as an exercise for the reader):

$$p_{\theta\theta}^- = \sigma_n^2 \left[\left(\frac{\xi}{S_u} \right)^2 - 1 \right] \tag{7.71a}$$

$$p_{\beta\beta}^- = \left(\frac{\sigma_n}{\Delta t} \right)^2 \left[S_u^2 \left(\frac{1}{\xi} + \frac{1}{2} \right) - \xi \right] \tag{7.71b}$$

The updated variances can be determined using the steady-state version of eqn. (5.44), which yields

$$p_{\theta\theta}^+ = \sigma_n^2 \left[1 - \left(\frac{S_u}{\xi} \right)^2 \right] \tag{7.72a}$$

$$p_{\beta\beta}^+ = \left(\frac{\sigma_n}{\Delta t} \right)^2 \left[S_u^2 \left(\frac{1}{\xi} - \frac{1}{2} \right) - \xi \right] \tag{7.72b}$$

Equations (7.71) and (7.72) can be used to determine 3σ bounds on the expected attitude and bias errors.

In the limiting case of very frequent updates, the pre-update and post-update attitude error standard deviations both approach the continuous-update limit, given by

$$\sqrt{p_{\theta\theta}^-} = \sqrt{p_{\theta\theta}^+} \equiv \sigma_c = \Delta t^{1/4} \sigma_n^{1/2} \left(\sigma_v^2 + 2\sigma_u \sigma_v \Delta t^{1/2} \right)^{1/4} \tag{7.73}$$

The even simpler limiting form when the contribution of σ_u to the attitude error is negligible is given by

$$\sigma_c = \Delta t^{1/4} \sigma_n^{1/2} \sigma_v^{1/2} \tag{7.74}$$

which indicates a one-half power dependence on both σ_n and σ_v, and a one-fourth power dependence on the update time Δt. This shows why it is extremely difficult to improve the attitude performance by simply increasing the update frequency. Farrenkopf's equations are useful for an initial estimate on attitude performance. Using the noise parameters from example 5.3 in eqn. (7.74) gives an approximate 3σ bound of 6.96 μrad for the attitude error, which is very close the actual solution of 7.18 μrad. Even though the observation model is not realistic, it can provide relative accuracies for various gyro parameters and sampling intervals. Converting 6.96 μrad to degrees gives 4×10^{-4} deg, which closely matches the roll and pitch errors of the results shown in Figure 7.6.

7.3 Orbit Estimation

In §4.3 a nonlinear least squares approach is shown to determine the initial state of an orbiting vehicle from range and line-of-sight (angle) observations. Another approach for orbit determination incorporates an *iterated Kalman filter*. This procedure uses the extended Kalman filter shown in Table 5.9 with $Q = 0$ to process the data forward with some initial condition guess, and then process the data backward to epoch. Initial conditions for the state are then given by previous pass results (e.g., the backward pass uses the final state from the forward pass for its initial condition). Also, the covariance must be reset after each forward or backward pass (this is required since no "new" information is given with each pass). The algorithm for orbit determination is essentially equivalent to the nonlinear fixed-point smoother in §6.1.3 with a covariance reset. The truth model used in the EKF is given by (see §3.8.2)

$$\boxed{\ddot{\mathbf{r}}(t) = -\frac{\mu}{||\mathbf{r}(t)||^3}\mathbf{r}(t) + \mathbf{w}(t)} \tag{7.75}$$

where $\mathbf{r}(t)$ is the orbital position and $\mathbf{w}(t)$ is the process noise, which is assumed to be zero. The discrete-time measurements include the azimuth, elevation, and range. The observation equations are given by eqn. (4.46). The goal of orbit determination is to determine initial conditions for the position and velocity of $\mathbf{x}_0 = \left[\mathbf{r}_0^T \; \dot{\mathbf{r}}_0^T\right]^T$ from the observations. The model equation is given by (7.75) with $\mathbf{x} = \left[\mathbf{r}^T \; \dot{\mathbf{r}}^T\right]^T$. Unlike the Gaussian Least Squares Differential Correction (GLSDC) shown in §4.3, the only analytical computations for the orbital EKF are the evaluations for the partial derivatives of eqns. (4.47) and (4.46) with respect to the state vector $\mathbf{x}$. These Jacobian, F, and sensitivity, H, matrix expressions are given by eqns. (4.50) and (4.60), respectively, which are evaluated at the current estimated state. Therefore,

Table 7.2: Extended Kalman Filter Iterations for Orbit Determination

Iteration	Position (km)			Velocity (km/sec)		
0	6,990	1	1	1	1	1
1	7,121	1,046	192	-0.07	5.70	1.67
2	7,000	1,000	200	4.00	7.00	2.00
3	7,000	1,000	200	4.00	7.00	2.00

the implementation of the EKF algorithm for orbit estimation at epoch is much more straightforward than the GLSDC.

Example 7.3: In this example the EKF algorithm is used to determine the orbit of a spacecraft from range, azimuth, and elevation measurements. The parameters used for the simulation are equivalent to the ones shown in example 4.3, but are repeated here for completeness. The true spacecraft position and velocity at epoch are given by

$$\mathbf{r}_0 = \begin{bmatrix}7,000 \; 1,000 \; 200\end{bmatrix}^T \text{ km}$$
$$\dot{\mathbf{r}}_0 = \begin{bmatrix}4 \; 7 \; 2\end{bmatrix}^T \text{ km/sec}$$

The latitude of the observer is given by $\lambda = 5°$, and the initial sidereal time is given by $\theta_0 = 10°$. Measurements are given at 10-second intervals over a 100-second simulation. The measurement errors are zero-mean Gaussian with a standard deviation of the range measurement error given by $\sigma_\rho = 1$ km, and a standard deviation of the angle measurements given by $\sigma_{az} = \sigma_{el} = 0.01°$.

A plot of a typical EKF iteration for the first position and velocity states is shown in Figure 7.8 (an iteration is one forward and one backward pass). The discontinuous jumps are due to the discrete-time measurement updates in the EKF. Note how these measurement updates help to reduce the error due to the propagation. Results for the EKF iterations are given in Table 7.2. Clearly, the EKF converges much faster than the least-square approach. This is due to the fact that the EKF uses a sequential process to update the estimates with each new measurement, while the GLSDC approach considers the entire batch of data to make a correction. The 3σ boundaries (determined using the diagonal elements of the estimate error-covariance) for position are $3\boldsymbol{\sigma}_\mathbf{r} = \begin{bmatrix}1.26 \; 0.25 \; 0.51\end{bmatrix}^T$ km, and for velocity are $3\boldsymbol{\sigma}_{\dot{\mathbf{r}}} = \begin{bmatrix}0.020 \; 0.008 \; 0.006\end{bmatrix}^T$ km/sec. The covariance results for the GLSDC in example 4.3 and EKF approaches are nearly identical, within the assumed applicability of linear error theory. The boundaries are useful to predict the performance of the algorithms.

The algorithm presented in this section uses a batch of data to determine the initial state of an orbit. The advantage of the Kalman filter approach is that the matrix $\Phi(t,t_0)$ used in the GLSDC is not required. The disadvantage of using a Kalman filter is that other quantities, such as biases, need to be appended into an augmented state vector. Another use of the Kalman filter involves the *navigation* problem that implements only a forward pass in the filter to determine the states in real time (typically with a nonzero value for Q), which can be used for control purposes. Modern-day navigation approaches predominately use GPS data to determine an orbit estimate, while differential GPS uses the on-board data with data collected from multiple ground stations. More details on orbit determination using GPS can be found in Ref. [17].

7.4 Target Tracking of Aircraft

One of the most useful early-day applications of the Kalman filter involves target tracking of aircraft from radar observations. Kalman filtering for target tracking has two main purposes. The first involves actual filtering of the radar measurements to obtain accurate range estimates. The second involves the estimation of velocity (and possibly acceleration). Velocity information is extremely important for air traffic control radar, which is used to avoid aircraft collisions when tracking multiple targets. Accurate velocity information can be used to predict ahead of time where multiple targets are expected in future radar scans in order to make a correct association of each target. A 3σ bound from the error covariance can be used to access the validity of the radar scan at future times.[18] This is used to ensure that the same target is actually tracked, thus avoiding incorrect target associations of multiple vehicles. In this section several tracking filters are introduced. The first two, called the α-β and α-β-γ filters, use kinematical models to derive the state estimate, which usually involves the aircraft's position and its derivatives. The third incorporates a dynamics-based model, which will be used to estimate the dynamical parameters of an aircraft from various observations, but can also be used to provide enhanced aircraft tracking capabilities.

7.4.1 The α-β Filter

One of the simplest target trackers is known as the α-β filter, which is used to estimate the position and velocity (usually range and range rate) of a vehicle. To derive this filter we begin with the following simple truth model in continuous-time:

$$\dot{\mathbf{x}}(t) = \begin{bmatrix} 0 & 1 \\ 0 & 0 \end{bmatrix} \mathbf{x}(t) + \begin{bmatrix} 0 \\ 1 \end{bmatrix} w(t) \tag{7.76}$$

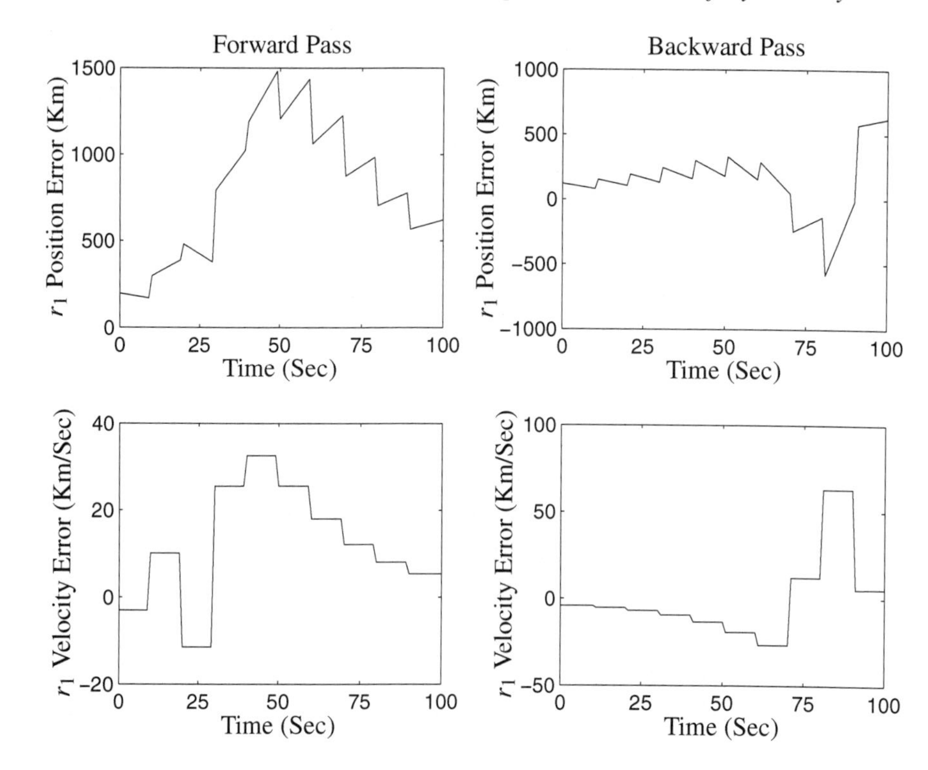

Figure 7.8: Extended Kalman Filter Iteration

where $w(t)$ is the process noise with variance q, and the states $\mathbf{x} \equiv \begin{bmatrix} x_1 & x_2 \end{bmatrix}^T$ are position and velocity, denoted by r and $\dot{r}$, respectively. Note that the first state does not contain any process noise in this formulation. This is due to the fact that this state represents a kinematical relationship that is valid in theory and in the real-world, since velocity is always the derivative of position. Discrete-time measurements of position are assumed, so that

$$\tilde{y}_k = \begin{bmatrix} 1 & 0 \end{bmatrix} \mathbf{x}_k + v_k \equiv H\mathbf{x}_k + v_k \tag{7.77}$$

where v_k is the measurement noise, which is assumed to be modelled by a zero-mean Gaussian white-noise process with variance σ_n^2. The α-β filter uses a discrete-time model, which is easy to derive for the model in eqn. (7.76). The state transition matrix can be computed using eqn. (3.25). Since $F^2 = 0$ for the model in eqn. (7.76), then the discrete-time state matrix is given by

$$\Phi = I + \Delta t\, F = \begin{bmatrix} 1 & \Delta t \\ 0 & 1 \end{bmatrix} \tag{7.78}$$

where Δt is the sampling interval.

Our next step in the derivation of the α-β filter involves the determination of the discrete-time process noise covariance. This can be accomplished using eqn. (5.135).

Performing a change of variables gives an equivalent integral for constant sampling with constant G and Q matrices:

$$\Upsilon\, Q\, \Upsilon^T = \int_0^{\Delta t} \Phi(\tau)\, G\, Q G^T\, \Phi^T(\tau)\, d\tau \tag{7.79}$$

where $G = \begin{bmatrix} 0 & 1 \end{bmatrix}^T$. Therefore, the discrete-time process noise covariance is given by

$$\Upsilon\, Q\, \Upsilon^T = q \int_0^{\Delta t} \begin{bmatrix} 1 & \tau \\ 0 & 1 \end{bmatrix} \begin{bmatrix} 0 \\ 1 \end{bmatrix} \begin{bmatrix} 0 & 1 \end{bmatrix} \begin{bmatrix} 1 & 0 \\ \tau & 1 \end{bmatrix} d\tau \tag{7.80}$$

Evaluating the integral in eqn. (7.80) yields

$$\Upsilon\, Q\, \Upsilon^T = q \begin{bmatrix} \Delta t^3/3 & \Delta t^2/2 \\ \Delta t^2/2 & \Delta t \end{bmatrix} \tag{7.81}$$

Notice, unlike the continuous-time process noise term given by $q\, G\, G^T$, the discrete-time process noise has nonzero values in all elements. This is due to the effect of sampling of a continuous-time process. However, if Δt is small, then eqn. (7.81) reduces down to eqn. (5.136).

Substituting the sensitivity and state matrices of eqns. (7.77) and (7.78) into the discrete-time Kalman update and propagation equations shown in Table 5.1 leads to

$$\hat{r}_k^+ = \hat{r}_k^- + \alpha\, [\tilde{y}_k - \hat{r}_k^-] \tag{7.82a}$$

$$\hat{\dot{r}}_k^+ = \hat{\dot{r}}_k^- + \frac{\beta}{\Delta t}\, [\tilde{y}_k - \hat{r}_k^-] \tag{7.82b}$$

$$\hat{r}_{k+1}^- = \hat{r}_k^+ + \hat{\dot{r}}_k^+\, \Delta t \tag{7.82c}$$

$$\hat{\dot{r}}_{k+1}^- = \hat{\dot{r}}_k^+ \tag{7.82d}$$

where the gain matrix in Table 5.1 is given by $K_k = K \equiv \begin{bmatrix} \alpha & \beta/\Delta t \end{bmatrix}^T$. The gains α and β are often treated as tuning parameters to enhance the tracking performance. However, conventional wisdom tells us that tuning these gains individually is incorrect. To understand this concept we must remember that the model in eqn. (7.76) shows a kinematical relationship. If α and β are chosen separately, then this kinematical relationship can be lost. This means the velocity estimate may not truly be the derivative of the position estimate, even though we know that this relationship is exact. A more true-to-physics approach involves tuning the continuous-time process noise parameter q. From eqn. (7.81) changes in the velocity over the sampling interval are of the order $\sqrt{q\,\Delta t}$, which can be used as a guideline in the choice of q.[19] The complete solution involves the determination of the Kalman gain through the steady-state covariance solution shown by its equation in Table 5.2. Fortunately, the α-β filter is just a subset of the Farrenkopf steady-state analysis shown in §7.2.4.

First, define the following the propagated and updated covariances:

$$P^- \equiv \begin{bmatrix} p_{rr}^- & p_{r\dot{r}}^- \\ p_{r\dot{r}}^- & p_{\dot{r}\dot{r}}^- \end{bmatrix}, \quad P^+ \equiv \begin{bmatrix} p_{rr}^+ & p_{r\dot{r}}^+ \\ p_{r\dot{r}}^+ & p_{\dot{r}\dot{r}}^+ \end{bmatrix} \tag{7.83}$$

Also, define the following variable:

$$\boxed{S_q = q^{1/2}\,\Delta t^{3/2} \Big/ \sigma_n} \tag{7.84}$$

Now, determine the following parameter, ξ, which is related to $p_{r\dot{r}}^-$, using

$$\xi = \frac{1}{2}\left[\left(\frac{S_q^2}{2} + \vartheta\right) + \sqrt{\left(\frac{S_q^2}{2} + \vartheta\right)^2 - 4S_q^2}\,\right] \tag{7.85a}$$

$$\vartheta = \left[4S_q^2 + \frac{S_q^4}{12}\right]^{1/2} \tag{7.85b}$$

The pre-update variance parameters are then given by

$$p_{rr}^- = \sigma_n^2\left[\left(\frac{\xi}{S_q}\right)^2 - 1\right] \tag{7.86a}$$

$$p_{\dot{r}\dot{r}}^- = \left(\frac{\sigma_n}{\Delta t}\right)^2\left[S_q^2\left(\frac{1}{2} - \frac{1}{\xi}\right) + \xi\right] \tag{7.86b}$$

$$p_{r\dot{r}}^- = \frac{\sigma_n^2 \xi}{\Delta t} \tag{7.86c}$$

The Kalman gain and thus the parameters α and β can be determined by using the steady-state version of eqn. (5.42), which leads to

$$K \equiv \begin{bmatrix} \alpha \\ \beta/\Delta t \end{bmatrix} = \frac{1}{p_{rr}^- + \sigma_n^2}\begin{bmatrix} p_{rr}^- \\ p_{r\dot{r}}^- \end{bmatrix} \tag{7.87}$$

This clearly shows that α and β are closely related to one another.

To determine the relationship between α and β, we first will determine the relationship between p_{rr}^- and $p_{r\dot{r}}^-$. Substituting $\xi = \Delta t\, p_{r\dot{r}}^-/\sigma_n^2$ into eqn. (7.86a) and solving the resulting equation for $p_{r\dot{r}}^-$ yields

$$p_{r\dot{r}}^- = \frac{\sigma_n S_q}{\Delta t}\sqrt{p_{rr}^- + \sigma_n^2} \tag{7.88}$$

Next, solving for p_{rr}^- from the definition of α in eqn. (7.87) gives

$$p_{rr}^- = \frac{\sigma_n^2 \alpha}{1 - \alpha} \tag{7.89}$$

Likewise, solving for $p_{r\dot{r}}^-$ from the definition of β in eqn. (7.87) gives

$$p_{r\dot{r}}^- = \frac{\beta\,(p_{rr}^- + \sigma_n^2)}{\Delta t} \tag{7.90}$$

Substituting eqn. (7.89) into eqn. (7.90) and simplifying gives

$$p_{r\dot{r}}^- = \frac{\sigma_n^2 \beta}{\Delta t\,(1-\alpha)} \tag{7.91}$$

Substituting eqns. (7.89) and (7.91) into eqn. (7.88), and after some moderate algebra (which is left as an exercise for the reader), yields

$$\boxed{\frac{\beta^2}{1-\alpha} = S_q^2} \tag{7.92}$$

The quantity S_q is known as the *tracking index*,[20] since it is proportional to the ratio of the process noise standard deviation and the measurement noise standard deviation. We should note that Kalata's index of Ref. [20] is slightly different, which is a function of Δt^2, not $\Delta t^{3/2}$ as shown by eqn. (7.84). This is due to the slightly different model chosen by Kalata, which is defined by

$$\mathbf{x}_{k+1} = \begin{bmatrix} 1 & \Delta t \\ 0 & 1 \end{bmatrix} \mathbf{x}_k + \begin{bmatrix} \Delta t^2/2 \\ \Delta t \end{bmatrix} w_k \tag{7.93}$$

This model assumes that the target undergoes a constant acceleration during the sampling interval and that the accelerations from period to period are independent.[19] This model may ignore the kinematical relationship shown by eqn. (7.76), and thus is not totally realistic.

A plot of α and β versus the tracking index S_q in eqn. (7.84) is shown in Figure 7.9. From this figure both α and β asymptotically approach limiting values. These limits will be assessed through a stability analysis. A simple closed-form solution for α and β can now be derived using eqns. (5.47) and (7.92). Using the steady-state version of eqn. (5.47) with $H = \begin{bmatrix} 1 & 0 \end{bmatrix}$ and $R = \sigma_n^2$ yields the following simple form for the gain K:

$$K \equiv \begin{bmatrix} k_1 \\ k_2 \end{bmatrix} = \sigma_n^{-2} \begin{bmatrix} p_{rr}^+ \\ p_{r\dot{r}}^+ \end{bmatrix} \tag{7.94}$$

where $k_1 = \alpha$ and $k_2 = \beta/\Delta t$. The updated variances are given by eqn. (7.72) using the notation in this section:

$$p_{rr}^+ = \sigma_n^2 \left[1 - \left(\frac{S_q}{\xi} \right)^2 \right] \tag{7.95a}$$

$$p_{r\dot{r}}^+ = \left(\frac{\sigma_n}{\Delta t} \right)^2 \left[S_q^2 \left(\frac{1}{\xi} - \frac{1}{2} \right) - \xi \right] \tag{7.95b}$$

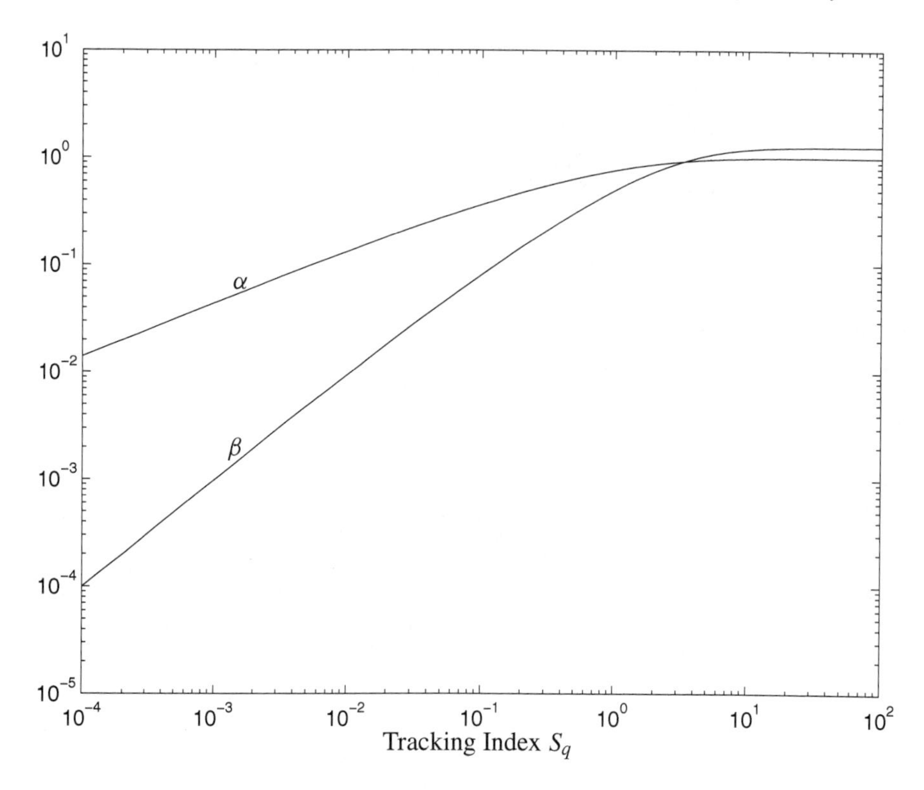

Figure 7.9: α-β Gains versus the Tracking Index

Therefore, from eqn. (7.94) α is simply given by

$$\boxed{\alpha = 1 - \left(\frac{S_q}{\xi}\right)^2} \tag{7.96}$$

Using eqn. (7.92) β is given by

$$\boxed{\beta = S_q\sqrt{1-\alpha}} \tag{7.97}$$

A direct relationship between α and β exists. This relationship is determined by first calculating the steady-state propagated and updated covariance in eqns. (5.35) and (5.44), respectively, with the definitions of Φ and $\Upsilon\, Q\, \Upsilon^T$ in eqns. (7.78) and (7.81), respectively. Substituting $H = \begin{bmatrix}1 & 0\end{bmatrix}$ into eqn. (5.44) gives

$$\begin{bmatrix} p_{rr}^+ & p_{r\dot{r}}^+ \\ p_{r\dot{r}}^+ & p_{\dot{r}\dot{r}}^+ \end{bmatrix} = \begin{bmatrix} p_{rr}^-(1-k_1) & p_{r\dot{r}}^-(1-k_1) \\ p_{r\dot{r}}^- - k_2 p_{rr}^- & p_{\dot{r}\dot{r}}^- - k_2 p_{r\dot{r}}^- \end{bmatrix} \tag{7.98}$$

The matrix in eqn. (7.98) must be symmetric, which gives

$$k_1 = \left(\frac{p_{rr}^-}{p_{r\dot{r}}^-}\right) k_2 \tag{7.99}$$

Substituting eqns. (7.78) and (7.81) into eqn. (5.35) yields

$$\begin{bmatrix} p_{rr}^- & p_{r\dot{r}}^- \\ p_{r\dot{r}}^- & p_{\dot{r}\dot{r}}^- \end{bmatrix} = \begin{bmatrix} p_{rr}^+ + 2p_{r\dot{r}}^+\Delta t + p_{\dot{r}\dot{r}}^+\Delta t^2 & p_{r\dot{r}}^+ + p_{\dot{r}\dot{r}}^+\Delta t \\ p_{r\dot{r}}^+ + p_{\dot{r}\dot{r}}^+\Delta t & p_{\dot{r}\dot{r}}^+ \end{bmatrix} + q \begin{bmatrix} \Delta t^3/3 & \Delta t^2/2 \\ \Delta t^2/2 & \Delta t \end{bmatrix} \tag{7.100}$$

From eqns. (7.98) and (7.100) the 2-2 element gives

$$k_2 = \frac{q\,\Delta t}{p_{r\dot{r}}^-} \tag{7.101}$$

Solving eqn. (7.99) for p_{rr}^- and using eqn. (7.101) gives

$$p_{rr}^- = \frac{k_1 q\,\Delta t}{k_2^2} \tag{7.102}$$

From eqns. (7.98) and (7.100) the 1-2 element gives

$$p_{\dot{r}\dot{r}}^- = p_{r\dot{r}}^- \left(\frac{k_1}{\Delta t} + k_2\right) - \frac{q\,\Delta t}{2} \tag{7.103}$$

From eqns. (7.98) and (7.100) the 1-1 element, with substituting of eqn. (7.103), yields

$$p_{rr}^- k_1 + p_{r\dot{r}}^- \Delta t\,(k_1 - 2) + \frac{q\,\Delta t^3}{6} = 0 \tag{7.104}$$

Solving eqn. (7.101) for $p_{r\dot{r}}^-$, and substituting the resulting equation and eqn. (7.102) into eqn. (7.104) yields

$$k_1^2 \Delta t + k_2 \Delta t^2 (k_1 - 2) + \frac{k_2^2 \Delta t^3}{6} = 0 \tag{7.105}$$

From the definitions of $k_1 \equiv \alpha$ and $k_2 \equiv \beta/\Delta t$, eqn. (7.105) reduces down to

$$\alpha^2 + \beta(\alpha - 2) + \frac{\beta^2}{6} = 0 \tag{7.106}$$

Hence, since β is always positive, which will be proven in the stability analysis, then α and β are related by

$$\boxed{\alpha = -\frac{1}{2}\beta + \frac{1}{2}\sqrt{\beta\,[(\beta/3) + 8]}} \tag{7.107}$$

This equation clearly shows the relationship between α and β, which can be written without S_q directly.

An interesting formula for β can also be derived using its relationship to $p_{r\dot{r}}^{+}$. Substituting eqn. (7.107) into eqn. (7.92) and squaring both sides of the resulting equation yields the following quartic equation:

$$\beta^4 + S_q^2 \beta^3 + S_q^2 \left[(S_q^2/6) - 2 \right] \beta^2 + S_q^4 \beta + S_q^4 = 0 \tag{7.108}$$

Note the similarity to eqn. (7.67)! In fact, the steps leading to eqn. (7.108) can be used to directly derive eqn. (7.67). The only solution that makes β valid in eqn. (7.92) is given by

$$\beta = \frac{1}{2}\left[\left(\frac{S_q^2}{2} + \vartheta \right) - \sqrt{\left(\frac{S_q^2}{2} + \vartheta \right)^2 - 4S_q^2} \right] \tag{7.109}$$

where

$$\vartheta = \left[4S_q^2 + S_q^4/12 \right]^{1/2} \tag{7.110}$$

Also, from eqn. (7.92) α is given by

$$\alpha = \frac{S_q^2 - \beta^2}{S_q^2} \tag{7.111}$$

Both forms for α and β, eqns. (7.96) and (7.97), and eqns. (7.109) and (7.111), are acceptable.

The stability conditions for the α-β filter are now shown. From §5.3.2 the matrix $\Phi_k[I - K_k H_k]$ defines the stability of the Kalman filter. Since this matrix is now constant, its eigenvalues can be evaluated to develop a set of stability conditions for α and β. The eigenvalues of $\Phi_k[I - K_k H_k]$ are given by solving the following equation:

$$|zI - \Phi[I - K\,H]| = \det \begin{bmatrix} z + \alpha + \beta - 1 & -\Delta t \\ \beta/\Delta t & z - 1 \end{bmatrix} = 0 \tag{7.112}$$

Evaluating this determinant leads to the following characteristic equation:

$$z^2 + (\alpha + \beta - 2)z + (1 - \alpha) = 0 \tag{7.113}$$

As mentioned in §3.5 all eigenvalues must lie within the unit circle for a stable system. Even though the characteristic equation is second-order in nature, using the unit circle condition directly to prove stability is arduous. However, Jury's test[21] can be used to easily derive the stability conditions for α and β. Consider the following second-order polynomial:

$$z^2 + a_1 z + a_2 = 0 \tag{7.114}$$

where $a_1 \equiv \alpha + \beta - 2$ and $a_2 \equiv 1 - \alpha$. Jury's test for stability for this second-order equation involves satisfying the following three conditions:

$$a_2 < 1 \tag{7.115a}$$

$$a_2 > a_1 - 1 \tag{7.115b}$$

$$a_2 > -(a_1 + 1) \tag{7.115c}$$

From the definitions of a_1 and a_2, these conditions give $\alpha > 0$, $\beta > 0$, and $2\alpha + \beta < 4$. However, from eqn. (7.96), since $\alpha > 0$ and $(S_q/\xi)^2 > 0$ then the following conditions must be satisfied for stability:

$$0 < \alpha \leq 1 \tag{7.116a}$$

$$0 < \beta < 2 \tag{7.116b}$$

These conditions will always be met since §5.3.2 shows that the Kalman filter is stable as long as $q \geq 0$ and $\sigma_n^2 > 0$.

7.4.2 The α-β-γ Filter

In this section the α-β filter of §7.4.1 is expanded to include an acceleration state. This approach in theory provides better estimates since a higher-order filter is used, but the computational requirements will certainly be greater than the α-β filter. To derive this new filter we begin with the following simple truth model in continuous-time:

$$\dot{\mathbf{x}}(t) = \begin{bmatrix} 0 & 1 & 0 \\ 0 & 0 & 1 \\ 0 & 0 & 0 \end{bmatrix} \mathbf{x}(t) + \begin{bmatrix} 0 \\ 0 \\ 1 \end{bmatrix} w(t) \tag{7.117}$$

where $w(t)$ is the process noise with variance q, and the states $\mathbf{x} \equiv \begin{bmatrix} x_1 & x_2 & x_3 \end{bmatrix}^T$ are position velocity and acceleration denoted by r, $\dot{r}$, and $\ddot{r}$, respectively. Note that the first two states do not contain any process noise, since these are kinematical relationships. Discrete-time measurements of position are assumed, so that

$$\tilde{y}_k = \begin{bmatrix} 1 & 0 & 0 \end{bmatrix} \mathbf{x}_k + v_k \equiv H\mathbf{x}_k + v_k \tag{7.118}$$

where v_k is the measurement noise, which is assumed to be modelled by a zero-mean Gaussian white-noise process with variance σ_n^2. The state transition matrix for the discrete-time model can be computed using eqn. (3.25). Since $F^3 = 0$ for the model in eqn. (7.117), then the discrete-time state matrix is given by

$$\Phi = I + \Delta t\, F + \frac{\Delta t^2}{2} F^2 = \begin{bmatrix} 1 & \Delta t & \Delta t^2/2 \\ 0 & 1 & \Delta t \\ 0 & 0 & 1 \end{bmatrix} \tag{7.119}$$

where Δt is the sampling interval. The discrete-time process noise can be computed using eqn. (7.79), which yields

$$\Upsilon Q \Upsilon^T = q \begin{bmatrix} \Delta t^5/20 & \Delta t^4/8 & \Delta t^3/6 \\ \Delta t^4/8 & \Delta t^3/3 & \Delta t^2/2 \\ \Delta t^3/6 & \Delta t^2/2 & \Delta t \end{bmatrix} \tag{7.120}$$

Note that the lower left 2×2 sub-matrix of eqn. (7.120) is equivalent to the matrix in eqn. (7.81).

Substituting the sensitivity and state matrices of eqns. (7.118) and (7.119) into the discrete-time Kalman update and propagation equations shown in Table 5.1 leads to

$$\hat{r}_k^+ = \hat{r}_k^- + \alpha\,[\tilde{y}_k - \hat{r}_k^-] \tag{7.121a}$$

$$\dot{\hat{r}}_k^+ = \dot{\hat{r}}_k^- + \frac{\beta}{\Delta t}\,[\tilde{y}_k - \hat{r}_k^-] \tag{7.121b}$$

$$\ddot{\hat{r}}_k^+ = \ddot{\hat{r}}_k^- + \frac{\gamma}{2\Delta t^2}\,[\tilde{y}_k - \hat{r}_k^-] \tag{7.121c}$$

$$\hat{r}_{k+1}^- = \hat{r}_k^+ + \dot{\hat{r}}_k^+ \Delta t + \frac{1}{2}\ddot{\hat{r}}_k^+ \Delta t^2 \tag{7.121d}$$

$$\dot{\hat{r}}_{k+1}^- = \dot{\hat{r}}_k^+ + \ddot{\hat{r}}_k^+ \Delta t \tag{7.121e}$$

$$\ddot{\hat{r}}_{k+1}^- = \ddot{\hat{r}}_k^+ \tag{7.121f}$$

where the gain matrix in Table 5.1 is given by $K_k = K \equiv \left[\alpha \;\; \beta/\Delta t \;\; \gamma/(2\Delta t^2)\right]^T$.

As with the α-β filter, the gains of the α-β-γ filter are related to each other. The filter should be designed by tuning q only, where changes in the acceleration over the sampling interval are of the order $\sqrt{q\Delta t}$. However, unlike the α-β filter, a closed-form solution showing a direct relationship of q to the gains is not straightforward. The tracking index in eqn. (7.84) is still useful though. A plot of α, β, and γ versus the tracking index S_q is shown in Figure 7.10. From this figure α, β, and γ asymptotically approach limiting values. These limits will be assessed through a stability analysis, which has been presented in Ref. [22]. Consistent with the analysis shown in §7.4.1, the eigenvalues of $\Phi_k[I - K_k H_k]$ are given by solving the following equation:

$$|zI - \Phi[I - K\,H]| = \det \begin{bmatrix} z + \alpha + \beta + \frac{1}{4}\gamma - 1 & -\Delta t & -\frac{1}{2}\Delta t^2 \\ \frac{1}{2\,\Delta t}(2\beta + \gamma) & z - 1 & -\Delta t \\ \frac{1}{2\,\Delta t^2}\gamma & 0 & z - 1 \end{bmatrix} = 0 \tag{7.122}$$

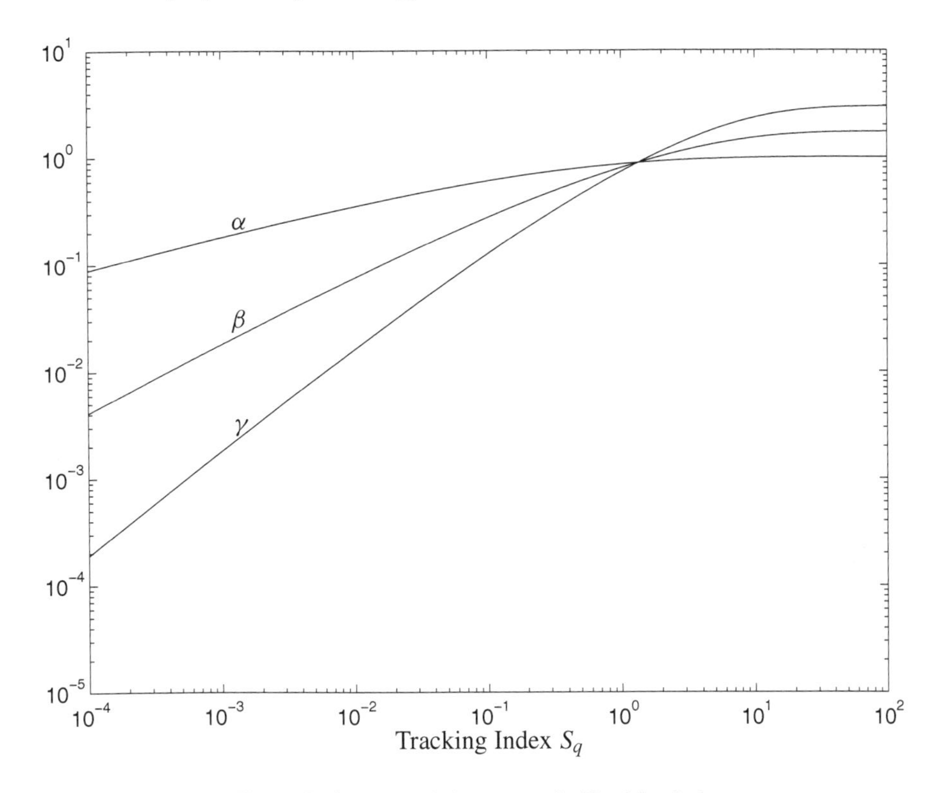

Figure 7.10: α-β-γ Gains versus the Tracking Index

Evaluating this determinant leads to the following characteristic equation:

$$z^3 + (\alpha + \beta + \frac{1}{4}\gamma - 3)z^2 + (3 - 2\alpha - \beta + \frac{1}{4}\gamma)z + (\alpha - 1) = 0 \tag{7.123}$$

Tenne and Singh[22] have evaluated the stability of this characteristic equation using Jury's test.[21] The conditions for stability are given by α and β greater than zero, and

$$2\alpha + \beta < 4 \tag{7.124a}$$

$$0 < \gamma < \frac{4\alpha\beta}{2 - \alpha} \tag{7.124b}$$

From Figure 7.10 these conditions are clearly met for all positive values of q, as expected. Furthermore, if we assume $0 < \alpha \leq 1$, then the stability conditions in eqn. (7.124) reduce down to

$$0 < \alpha \leq 1 \tag{7.125a}$$

$$0 < \beta < 2 \tag{7.125b}$$

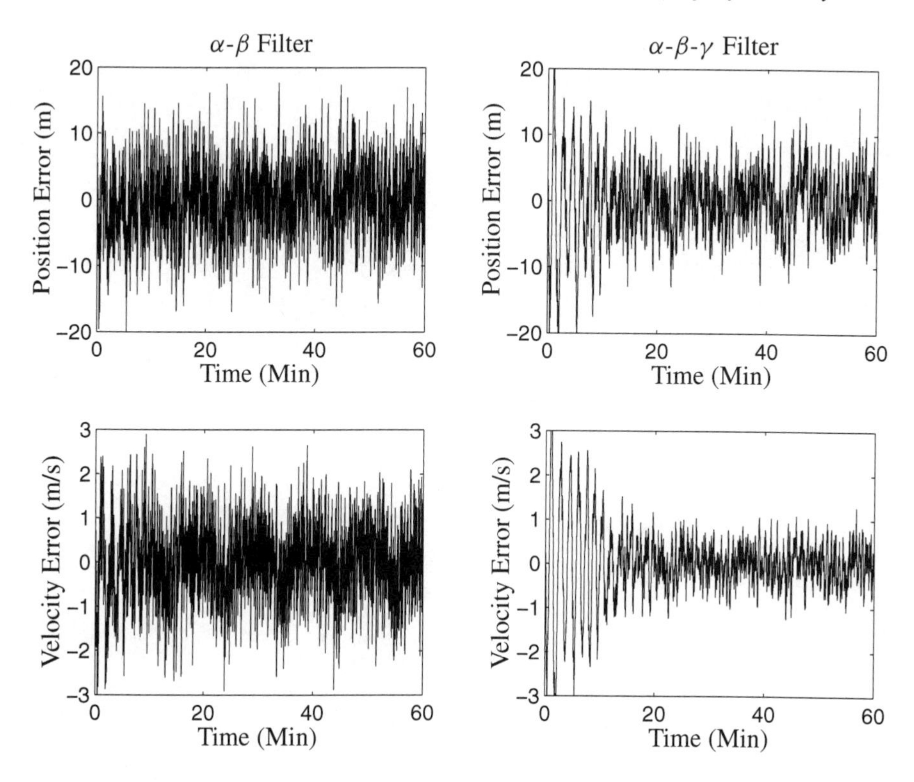

Figure 7.11: Position and Velocity Tracking Error Results Using Both Filters

$$0 < \gamma < \frac{4\alpha\beta}{2-\alpha} \tag{7.125c}$$

Reference [22] also derives metrics to gauge the transient response and steady-state tracking error, and also shows the relationships between the gain parameters for specific maneuvers. These relationships can be used to provide an initial estimate for α, β, and γ, although tuning q is preferred, which exploits the kinematically relationship in the assumed model.

Example 7.4: A simulation involving tracking the vertical position of a 747 aircraft using both the α-β and α-β-γ filters is shown. The longitudinal equations of motion are shown in example 4.4. Using the aircraft flight parameters shown in example 4.4 the equations of motion are integrated over a 60-minute simulation. The thrust is set equal to the computed drag, and the elevator is set to 1 degree down from the trim value for the entire simulation interval. The vertical position, z, has standard deviation of 10 m for the measurement error. Measurements are sampled at 1 sec intervals.

Since we know the truth, then the variance parameter q in both the α-β and α-β-γ filters is tuned to ensure the best possible performance. This parameter is varied until transients begin to appear in the position errors. For the α-β filter the optimal parameter is given by $q = 0.5$. From eqns. (7.96) and (7.97) this value of q gives $\alpha = 0.31344$ and $\beta = 0.05859$. For the α-β-γ filter the optimal parameter is given by $q = 0.0001$. Note this value is much smaller than the value used in the α-β filter. This is due to the fact that q now affects changes in acceleration, which is smaller in magnitude than changes in velocity. Solving the steady-state discrete-time covariance equation in Table 5.2 using the method outlined in §5.3.4 gives $\alpha = 0.18127$, $\beta = 0.01811$, and $\gamma = 0.00181$. A plot of the tracking error results for vertical position and velocity using both filters is shown in Figure 7.11. The 3σ bounds computed from the steady-state error-covariance are 20.27 m (position) and 5.14 m/s (velocity) for the α-β filter, and 14.12 m (position), 1.70 m/s (velocity), and 0.136 m/s^2 (acceleration) for the α-β-γ filter. Clearly, the α-β-γ filter outperforms the α-β filter, but comes at a higher computational cost.

More details on α-β-γ filtering can be found in the references cited in §7.4.1 and §7.4.2. The α-β and α-β-γ filters described here have been widely used in a number of applications, which is mainly due to the simplicity of the filtering mechanisms. For aircraft applications a filter with a more rigorous flight dynamics-based model can significantly improve the tracking accuracy, as shown in Ref. [23]. Also, a simple dynamics-based filter for application to automatic landings on an aircraft carrier is shown in Ref. [24], which gives superior results to the standard α-β-γ filter for control purposes. The reader is highly encouraged to pursue actual applications in the references cited here and in the open literature.

7.4.3 Aircraft Parameter Estimation

In §4.4 parameter identification using a batch set of flight measurement data has been shown. In this section parameter estimation is considered using the extended Kalman filter. This allows for the implementation of real-time estimation, which can be used to update an aircraft model for adaptive control purposes. In this section the focus is only on the longitudinal equations of motions, but this formulation can easily be extended to the general case involving coupled motion. The EKF approach for aircraft parameter estimation involves appending the state vector to include the unknown parameters. The derivative of these parameters is zero, which can easily be put into a state-space form. In this section we present this approach to estimate C_{D_0}, C_{L_0} and C_{m_0} using measurements of angle of attack, velocity, angular rate, and pitch angle. The longitudinal equations of motion are shown in example 4.4. The state vector, $\mathbf{x}$, consists of v_1, v_2, ω_2, θ, C_{D_0}, C_{L_0}, and C_{m_0}. Note that the horizontal and vertical positions, x and z, are not required in this formulation. See §3.9 for a full description of the equations of motion for an aircraft.

Several partial derivatives are required in the EKF. These may be computed numerically using the method described in example 4.4, but we instead choose to derive analytical expressions here. The partial derivatives of α with respect to v_1 and v_3 are given by

$$\frac{\partial \alpha}{\partial v_1} = -\frac{v_3}{v_1^2 + v_3^2} \tag{7.126a}$$

$$\frac{\partial \alpha}{\partial v_3} = \frac{v_1}{v_1^2 + v_3^2} \tag{7.126b}$$

where

$$\alpha = \tan^{-1}\frac{v_3}{v_1} \tag{7.127}$$

The partial derivatives of the drag force, D, with respect to v_1 and v_3 are given by

$$\frac{\partial D}{\partial v_1} = C_D \rho\, v_1 S - \frac{\rho\, C_{D_\alpha} v_3}{2(1+\alpha^2)v_1^2}||\mathbf{v}||^2 S \tag{7.128a}$$

$$\frac{\partial D}{\partial v_3} = C_D \rho\, v_3 S + \frac{\rho\, C_{D_\alpha}}{2(1+\alpha^2)v_1}||\mathbf{v}||^2 S \tag{7.128b}$$

where $||\mathbf{v}||^2 = v_1^2 + v_3^2$ and

$$C_D = C_{D_0} + C_{D_\alpha}\alpha + C_{D_{\delta_E}}\delta_E \tag{7.129}$$

The partial derivatives of the lift force, L, with respect to v_1 and v_3 are given by

$$\frac{\partial L}{\partial v_1} = C_L \rho\, v_1 S - \frac{\rho\, C_{L_\alpha} v_3}{2(1+\alpha^2)v_1^2}||\mathbf{v}||^2 S \tag{7.130a}$$

$$\frac{\partial L}{\partial v_3} = C_L \rho\, v_3 S + \frac{\rho\, C_{L_\alpha}}{2(1+\alpha^2)v_1}||\mathbf{v}||^2 S \tag{7.130b}$$

where

$$C_L = C_{L_0} + C_{L_\alpha}\alpha + C_{L_{\delta_E}}\delta_E \tag{7.131}$$

These partial derivatives will be used in the derivation of the matrix $F(\hat{\mathbf{x}}(t), t)$ for the EKF shown in Table 5.9.

The partial derivative components of $\dot{v}_1$ with respect to the state vector, which give the first row of $F(\mathbf{x}(t), t)$, are given by

$$\frac{\partial \dot{v}_1}{\partial v_1} = \frac{1}{m}\left\{\left[\frac{\partial L}{\partial v_1} + D\frac{\partial \alpha}{\partial v_1}\right]\sin\alpha + \left[L\frac{\partial \alpha}{\partial v_1} - \frac{\partial D}{\partial v_1}\right]\cos\alpha\right\} \tag{7.132a}$$

$$\frac{\partial \dot{v}_1}{\partial v_3} = \frac{1}{m}\left\{\left[\frac{\partial L}{\partial v_3} + D\frac{\partial \alpha}{\partial v_3}\right]\sin\alpha + \left[L\frac{\partial \alpha}{\partial v_3} - \frac{\partial D}{\partial v_3}\right]\cos\alpha\right\} - \omega_2 \tag{7.132b}$$

$$\frac{\partial \dot{v}_1}{\partial \omega_2} = -v_3 \tag{7.132c}$$

$$\frac{\partial \dot{v}_1}{\partial \theta} = -g\cos\theta \tag{7.132d}$$

$$\frac{\partial \dot{v}_1}{\partial C_{D_0}} = -\frac{1}{2m}\rho\,||\mathbf{v}||^2 S\cos\alpha \tag{7.132e}$$

$$\frac{\partial \dot{v}_1}{\partial C_{L_0}} = \frac{1}{2m}\rho\,||\mathbf{v}||^2 S \sin\alpha \tag{7.132f}$$

$$\frac{\partial \dot{v}_1}{\partial C_{m_0}} = 0 \tag{7.132g}$$

The partial derivative components of $\dot{v}_3$ with respect to the state vector, which give the second row of $F(\mathbf{x}(t), t)$, are given by

$$\frac{\partial \dot{v}_3}{\partial v_1} = \frac{1}{m}\left\{\left[-D\frac{\partial \alpha}{\partial v_1} - \frac{\partial L}{\partial v_1}\right]\cos\alpha + \left[L\frac{\partial \alpha}{\partial v_1} - \frac{\partial D}{\partial v_1}\right]\sin\alpha\right\} + \omega_2 \tag{7.133a}$$

$$\frac{\partial \dot{v}_3}{\partial v_3} = \frac{1}{m}\left\{\left[-D\frac{\partial \alpha}{\partial v_3} - \frac{\partial L}{\partial v_3}\right]\cos\alpha + \left[L\frac{\partial \alpha}{\partial v_3} - \frac{\partial D}{\partial v_3}\right]\sin\alpha\right\} \tag{7.133b}$$

$$\frac{\partial \dot{v}_3}{\partial \omega_2} = v_1 \tag{7.133c}$$

$$\frac{\partial \dot{v}_3}{\partial \theta} = -g\,\sin\theta \tag{7.133d}$$

$$\frac{\partial \dot{v}_3}{\partial C_{D_0}} = -\frac{1}{2m}\rho\,||\mathbf{v}||^2 S \sin\alpha \tag{7.133e}$$

$$\frac{\partial \dot{v}_3}{\partial C_{L_0}} = -\frac{1}{2m}\rho\,||\mathbf{v}||^2 S \cos\alpha \tag{7.133f}$$

$$\frac{\partial \dot{v}_3}{\partial C_{m_0}} = 0 \tag{7.133g}$$

The partial derivative components of $\dot{\omega}_2$ with respect to the state vector, which give the third row of $F(\mathbf{x}(t), t)$, are given by

$$\frac{\partial \dot{\omega}_2}{\partial v_1} = \frac{\rho S \bar{c}}{J_{22}}\left[\left(C_{m_0} + C_{m_\alpha}\alpha + C_{m_{\delta_E}}\delta_E\right)v_1 - \frac{C_{m_\alpha} v_3}{2(1+\alpha^2)v_1^2}||\mathbf{v}||^2\right] \tag{7.134a}$$

$$\frac{\partial \dot{\omega}_2}{\partial v_3} = \frac{\rho S \bar{c}}{J_{22}}\left[\left(C_{m_0} + C_{m_\alpha}\alpha + C_{m_{\delta_E}}\delta_E\right)v_3 + \frac{C_{m_\alpha}}{2(1+\alpha^2)v_1}||\mathbf{v}||^2\right] \tag{7.134b}$$

$$\frac{\partial \dot{\omega}_2}{\partial \omega_2} = \frac{1}{4J_{22}}\rho S \bar{c}^2 C_{m_q} \tag{7.134c}$$

$$\frac{\partial \dot{\omega}_2}{\partial \theta} = 0 \tag{7.134d}$$

$$\frac{\partial \dot{\omega}_2}{\partial C_{D_0}} = 0 \tag{7.134e}$$

$$\frac{\partial \dot{\omega}_2}{\partial C_{L_0}} = 0 \tag{7.134f}$$

$$\frac{\partial \dot{\omega}_2}{\partial C_{m_0}} = \frac{1}{2J_{22}}\rho\,||\mathbf{v}||^2 S \bar{c} \tag{7.134g}$$

The 4-3 element of $F(\mathbf{x}(t), t)$ is given by 1, which is derived from the kinematical equation $\dot{\theta} = \omega_2$. All other entries of $F(\mathbf{x}(t), t)$ are zero since C_{D_0}, C_{L_0}, and C_{m_0}

are constants. The output vector is given

$$\mathbf{y} = \begin{bmatrix} \alpha \\ ||\mathbf{v}|| \\ \omega_2 \\ \theta \end{bmatrix} \tag{7.135}$$

The matrix sensitivity matrix H is given by

$$H(\mathbf{x}) = \begin{bmatrix} \frac{\partial \alpha}{\partial v_1} & \frac{\partial \alpha}{\partial v_3} & 0\,0\,0\,0\,0 \\ \frac{\partial ||\mathbf{v}||}{\partial v_1} & \frac{\partial ||\mathbf{v}||}{\partial v_3} & 0\,0\,0\,0\,0 \\ 0 & 0 & 1\,0\,0\,0\,0 \\ 0 & 0 & 0\,1\,0\,0\,0 \end{bmatrix} \tag{7.136}$$

where

$$\frac{\partial ||\mathbf{v}||}{\partial v_1} = \frac{v_1}{||\mathbf{v}||} \tag{7.137a}$$

$$\frac{\partial ||\mathbf{v}||}{\partial v_3} = \frac{v_3}{||\mathbf{v}||} \tag{7.137b}$$

The continuous-discrete extended Kalman filter in Table 5.9 can now be implemented with $F(\hat{\mathbf{x}}(t), t)$ and $H_k(\hat{\mathbf{x}}_k)$ evaluated at the current state estimates.

Example 7.5: To illustrate the power of using the extended Kalman filter for real-time parameter applications, we show an example of identifying the longitudinal parameters of a simulated 747 aircraft. The longitudinal equations of motion are shown in example 4.4. Using the aircraft flight parameters shown in example 4.4 the equations of motion are integrated over a 30-second simulation. The thrust is set equal to the computed drag, and the elevator is set to 1 degree down from the trim value for the first 10 seconds and then returned to the trimmed value thereafter. Measurements of angle of attack, α, velocity, $||\mathbf{v}||$, angular velocity, ω_2, and pitch angle, θ, are assumed with standard deviations of the measurement errors given by $\sigma_\alpha = 0.5$ degrees, $\sigma_{||\mathbf{v}||} = 1$ m/s, $\sigma_{\omega_2} = 0.01$ deg/sec, and $\sigma_\theta = 0.1$ degrees, respectively. Since real-time estimates are required the measurements are sampled at 0.1-second intervals. The continuous-time model and error-covariance are integrated using a time step of 0.01 seconds, which is needed to ensure adequate performance in the EKF propagation.

The initial conditions for $\hat{v}_1$, $\hat{v}_2$, $\hat{\omega}_2$, and $\hat{\theta}$ are set to their true values. The initial conditions for the parameters to be estimated are given by $C_{D_0} = 0.01$, $C_{L_0} = 0.1$, and $C_{m_0} = 0.01$. The initial error-covariance is given by

$$P_0 = \text{diag}\left[1\times 10^{-5}\ 1\times 10^{-5}\ 1\times 10^{-5}\ 1\times 10^{-6}\ 1\ 1\ 1\right]$$

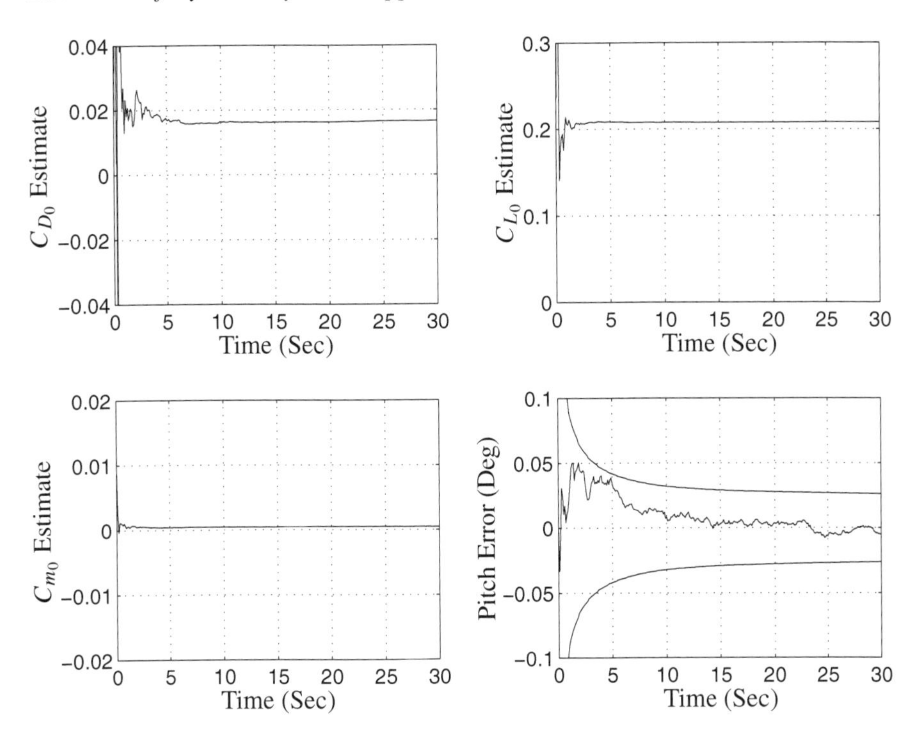

Figure 7.12: Parameter Estimates and Pitch Angle Error

A plot of the parameter estimates is shown in Figure 7.12. The final values at the end of the simulation run are given by $C_{D_0} = 0.0164$, $C_{L_0} = 0.2082$, and $C_{m_0} = 0.0003$, which are close to the batch solutions shown in example 4.4. A plot of the pitch angle errors and associated 3σ bounds is also shown in Figure 7.12. The errors are within their respective 3σ bounds, which indicates that the EKF is performing in an optimal manner. This example clearly shows the usefulness of the extended Kalman filter for real-time parameter estimations. The example shown herein can also be implemented as a real-time dynamics-based filter, without updating the aircraft parameters in the model.[23]

7.5 Smoothing with the Eigensystem Realization Algorithm

The Eigensystem Realization Algorithm (ERA) of §4.5 is fairly accurate for measurements that contain small measurement noise levels. However, significant errors can be produced with high measurement noise, which will be shown in example 7.6. This problem can be overcome by using frequency domain-based filtering methods, which use frequency-response function averaging. But this requires more data sets and computational effort. The approach presented in this section involves first smoothing the measurements using the discrete-time fixed-interval smoothing algorithm of §6.1.1. Since the ERA approach is in essence a batch least squares estimator, it seems natural to use a batch-type estimator to smooth the effects of the large measurement errors. This approach can be shown to be superior over standard band-pass or low-pass filtering of the data.[25]

The theoretical developments of the combined smoother/ERA approach begins with the state-space form of the vibratory system shown in §3.10:

$$\boxed{\begin{aligned}\dot{\mathbf{x}} &= \begin{bmatrix} 0 & I \\ -M^{-1}K & -M^{-1}C \end{bmatrix}\mathbf{x} + \begin{bmatrix} 0 \\ M^{-1} \end{bmatrix}\mathbf{u} + \begin{bmatrix} 0 \\ I \end{bmatrix}\mathbf{w} \\ &\equiv F\mathbf{x} + B\mathbf{u} + G\mathbf{w} \qquad (7.138\text{a}) \\ \tilde{\mathbf{y}}_k &= H\mathbf{x}_k + \mathbf{v}_k \qquad (7.138\text{b})\end{aligned}}$$

where $\mathbf{x}$ now denotes a $2n$ vector of n position states and n velocity states. In this model the process noise is only added to the velocity states since, as discussed in §7.4.1, the first n states of eqn. (7.138a) represent a kinematical relationship. Typically, an *a priori* model of a particular vibratory system is predetermined using a finite element analysis, which was later demonstrated to be a Rayleigh-Ritz method.[26] Exploitation of the second-order block-structure of the model in eqn. (7.138) allows one to use a reduced-order Kalman filter and smoother form.[27, 28] However, since a steady-state gain in the forward-time Kalman filter and backward-time smoother will be used here, which can be determined off-line, we choose to retain the full-order form.

The first step in the Rauch, Tung, and Striebel (RTS) smoother involves executing the Kalman filter forward in time. A method to determine the process noise covariance involves an off-line computation to satisfy the autocorrelation test in eqns. (5.250) and (5.251). Since the state matrices are constant and the measurements are assumed to be sampled frequently, then the steady-state discrete-time Kalman filter shown in Table 5.2 can be used. The discrete-time state matrices, Φ and Γ, can be numerically determined using eqns. (3.112) and (3.113). An analytical solution for the discrete-time process noise covariance is difficult to determine for high-order models. Therefore, eqn. (5.140) will be used to determine this covariance matrix. The steady-state error-covariance matrix computed from the discrete-time algebraic Riccati equation in Table 5.2 is now denoted by P_f^- to reflect the fact that this matrix is the propagated steady-state solution of the forward Kalman filter. The RTS smoother steady-state gain in Table 6.2 is given by

$$\boxed{\mathcal{K} = P_f^+ \Phi^T (P_f^-)^{-1}} \tag{7.139}$$

where P_f^+ is given in Table 6.2 as well:

$$P_f^+ = [I - K_f H] P_f^- \tag{7.140a}$$

$$K_f = P_f^- H^T [H P_f^- H^T + R]^{-1} \tag{7.140b}$$

where R is the covariance of $\mathbf{v}_k$, shown in eqn. (7.138b). From Table 6.2 the steady-state RTS smoother covariance, denoted by P, can be computed by solving the following discrete-time Lyapunov equation:

$$\boxed{P = \mathcal{K} P \mathcal{K}^T + [P_f^+ - \mathcal{K} P_f^- \mathcal{K}^T]} \tag{7.141}$$

This covariance can be used to determine the performance characteristics of the RTS smoothing algorithm.

The procedure to determine the state-space system matrices is as follows. First, determine an initial model of the system at hand. If one is not given, then the ERA algorithm can be employed to determine this model from the noisy measurement sets. Next, implement the discrete-time Kalman filter to determine filtered state estimates. Then, use the discrete-time RTS smoother to determine smoothed output estimates. Finally, use the ERA algorithm with the smoothed output estimates to determine the system matrices. The Modal Amplitude Coherence (MAC) in eqn. (4.91) can be used to compare the performance of the combined smoother/ERA approach with the ERA approach alone. If the smoother is working properly, then an identified mode should have a higher MAC value than the mode identified by ERA alone.

Example 7.6: In this example we will use the ERA to identify the mass, stiffness, and damping matrices of a 4-mode system from simulated high-noise mass-position measurements. The description of the model and the assumed mass, stiffness, and damping matrices are shown in example 4.5. With the exact solution known, Gaussian white-noise of approximately 5% the size of the signal amplitude is added to simulate the output measurements. A 50-second simulation is performed, with measurements sampled every 0.1 seconds. Using all available measurements, the Hankel matrix in the ERA was chosen to be a 400×1600 dimension matrix. After computing the discrete-time state matrices using eqn. (4.85), a conversion to continuous-time state matrices is performed, and the mass, stiffness, and damping matrices are computed using eqn. (4.99). The results of this computation are

$$M = \begin{bmatrix} -0.7376 & 2.0831 & -1.5368 & 0.8198 \\ 2.3310 & -1.7600 & 1.8760 & -0.8917 \\ -1.5544 & 1.9296 & -0.4804 & 0.8381 \\ 0.7807 & -0.8992 & 0.6590 & 0.6519 \end{bmatrix}$$

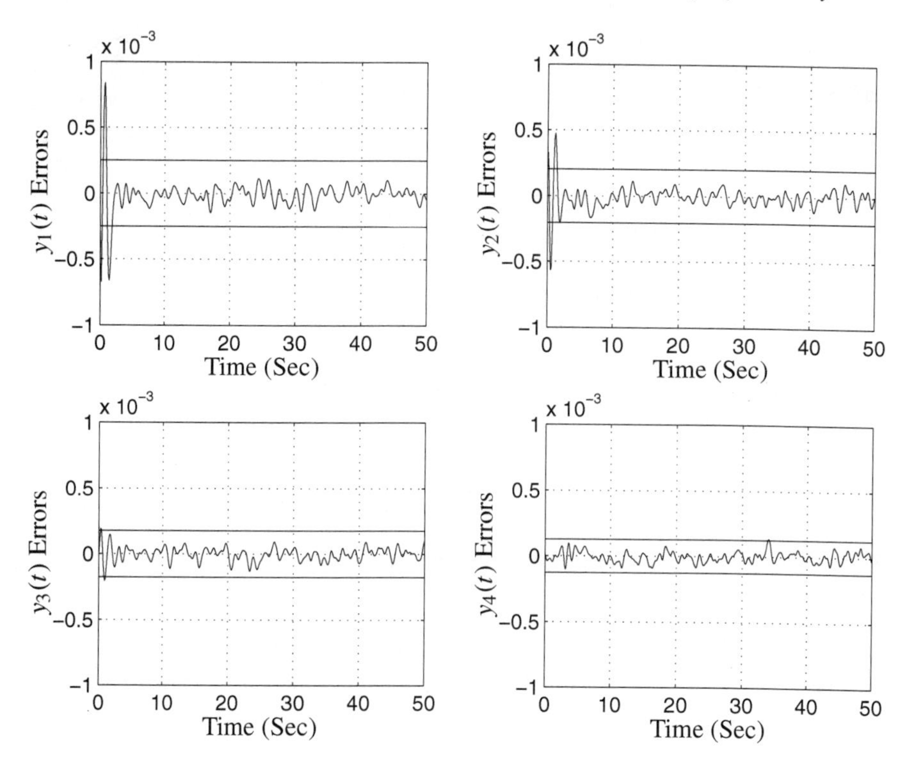

Figure 7.13: Position Errors with 3σ Bounds

$$K = \begin{bmatrix} 6.2382 & -0.2996 & -3.6281 & 1.8429 \\ 1.3916 & 2.4294 & -0.1185 & -1.9367 \\ -9.3119 & 5.9243 & 3.3579 & -2.6469 \\ 6.7156 & -7.4445 & -1.0620 & 9.0596 \end{bmatrix}$$

$$C = \begin{bmatrix} 0.3355 & 1.6663 & -2.9785 & 2.0538 \\ 2.9750 & -2.9882 & 2.4879 & -1.4765 \\ -6.9475 & 6.7105 & -1.6973 & -0.4730 \\ 5.5428 & -5.6182 & 0.5978 & 2.8823 \end{bmatrix}$$

These realized matrices are not close to the true matrices, shown in example 4.5, which is due to the large measurement errors used in the current simulation. Note that some of the diagonal elements are not even positive!

The RTS smoother is implemented to provided smoothed estimates, which are used in the ERA. For the RTS state model we assume that the mass matrix is given by the true mass matrix, but the stiffness matrix is given by 0.9 times the true stiffness matrix. Also, the damping matrix is given by the true *stiffness* matrix divided by 10, which introduces a large error in the state model. This error approach in the assumed model provides a typical scenario where the mass and stiffness matrices are well known, but the damping matrix is not well known. The continuous-time covariance is determined by trial and error. A value of $1 \times 10^{-6} I_{4\times4}$ is found to produce accurate results, which can be verified by the 3σ bounds computed from the diagonal elements of eqn. (7.141). A plot of the position errors with 3σ bounds for an impulse input to the first mass is shown in Figure 7.13. The initial transients are due to the fact that a steady-state gain is used in the RTS smoother. Clearly, the RTS smoother is performing in an optimal fashion. Using the smoothed estimates in the ERA the mass, stiffness, and damping matrices are now computed to be

$$M = \begin{bmatrix} 1.0170 & 0.0023 & 0.0043 & 0.0093 \\ -0.0050 & 1.0093 & -0.0071 & 0.0005 \\ 0.0123 & -0.0027 & 1.0031 & 0.0084 \\ 0.0173 & 0.0145 & -0.0141 & 1.0203 \end{bmatrix}$$

$$K = \begin{bmatrix} 9.4631 & -4.4972 & -0.1975 & -0.0529 \\ -4.4832 & 9.2467 & -4.4816 & -0.1554 \\ -0.0814 & -4.4870 & 9.1694 & -4.3763 \\ 0.0065 & -0.0894 & -4.5658 & 9.5069 \end{bmatrix}$$

$$C = \begin{bmatrix} 1.2389 & -0.4058 & -0.1324 & -0.0259 \\ -0.4370 & 1.1933 & -0.4592 & -0.1316 \\ -0.1377 & -0.4456 & 1.1852 & -0.4273 \\ -0.0232 & -0.1346 & -0.3893 & 1.2430 \end{bmatrix}$$

These matrices are now much closer to the true values than the ones computed using the ERA with the raw measurements. A better comparison involves looking at the identified natural frequencies and damping ratios, which are given by

True		ERA		RTS/ERA	
ω_n	ζ	ω_n	ζ	ω_n	ζ
1.3820	0.1382	1.3814	0.1376	1.3786	0.1354
2.6287	0.2629	2.6563	0.2658	2.6016	0.2155
3.6180	0.3618	0.2545, 1.5778	1.0000	3.4694	0.2231
4.2533	0.4253	3.5146, 4.6940	1.0000	4.0181	0.2184

The modes with a damping ratio of 1 correspond to real-valued modes (i.e., with no complex parts). The MAC factors are given by

ERA		RTS/ERA	
ω_n	MAC	ω_n	MAC
1.3814	1.0000	1.3786	1.0000
2.6563	0.9957	2.6016	0.9990
0.2545, 1.5778	0.7148, 0.7658	3.4694	0.9976
3.5146, 4.6940	0.7841, 0.7398	4.0181	0.9983

Clearly, using the ERA with the raw measurements did not properly identify the high frequency modes, which is due to the fact that the noisy levels make these modes nearly unobservable. The combined RTS/ERA approach did manage to provide a significant improvement in the obtained results. The results are reinforced by the MAC factors, where the higher modes have a MAC close to one using the combined RTS/ERA approach.

7.6 Summary

In this chapter several applications of the linear and extended Kalman filter have been presented for Global Positioning System navigation, spacecraft attitude estimation and gyro bias determination from various sensor devices, orbit determination from ground-based sensors, aircraft tracking from radar measurements and parameter identification using on-board measurements, and robust modal identification of vibratory systems using the RTS smoother to provide optimal estimates. As with Chapter 4, we anticipate that most readers will profit greatly by a careful study of these applications in this chapter. Once again, the constraints imposed by the length of this text did not, however, permit an entirely self-contained and satisfactory development of the concepts introduced in the applications of this chapter. It will likely prove useful for the interested reader to pursue these important subjects in the cited literature. For example, the integration of GPS and Inertial Navigation Systems represents an extremely useful tool in modern-day navigation. However, due to constraints imposed by the length of this text, a full treatise is not possible here. Several texts dedicated just to this subject have been written, (e.g., see Refs. [6], [7], and [29]), which we highly recommend to the interested reader.

A summary of the key formulas presented in this chapter is given below.

- GPS Coordinate Transformations

$$\begin{bmatrix} e_1 \\ e_2 \\ e_3 \end{bmatrix} = \begin{bmatrix} \cos\theta & \sin\theta & 0 \\ -\sin\theta & \cos\theta & 0 \\ 0 & 0 & 1 \end{bmatrix} \begin{bmatrix} i_1 \\ i_2 \\ i_3 \end{bmatrix}$$

$$\begin{aligned} d_{2000} =& 367y - \text{INT}\left\{\frac{7\{y + \text{INT}[(m+9)/12]\}}{4}\right\} \\ &+ \text{INT}\left\{\frac{275m}{9}\right\} + \frac{h + min/60 + s/3600}{24} + d - 730531.5 \end{aligned}$$

$$\theta = 280.46061837 + 360.98564736628 \times d_{2000}$$

- Geodetic to ECEF Conversion

$$N = \frac{a}{\sqrt{1 - e^2 \sin^2 \lambda}}$$

$$\begin{aligned}
x &= (N + h)\cos\lambda \cos\phi \\
y &= (N + h)\cos\lambda \sin\phi \\
z &= [N(1 - e^2) + h]\sin\lambda
\end{aligned}$$

- ECEF to Geodetic Conversion

$$\begin{aligned}
p &= \sqrt{x^2 + y^2} \\
\psi &= \text{atan}\left(\frac{z\,a}{p\,b}\right) \\
\bar{e}^2 &= \frac{a^2 - b^2}{b^2} \\
\lambda &= \text{atan}\left(\frac{z + \bar{e}^2 b\, \sin^3\psi}{p - e^2 a\, \cos^3\psi}\right) \\
\phi &= \text{atan2}(y,\, x) \\
h &= \frac{p}{\cos\lambda} - N
\end{aligned}$$

- GPS Satellite Elevation

$$\begin{aligned}
\mathbf{u} &= \begin{bmatrix} \cos\lambda \cos\phi \\ \cos\lambda \sin\phi \\ \sin\lambda \end{bmatrix} \\
\cos\xi_i &= \boldsymbol{\rho}_i^T \mathbf{u} \\
\boldsymbol{\rho}_i &= \frac{\mathbf{e}_i - \mathbf{r}}{||\mathbf{e}_i - \mathbf{r}||} \\
E_i &= 90^\circ - \xi_i
\end{aligned}$$

- Attitude Estimation

$$\begin{aligned}
\dot{\hat{\mathbf{q}}}(t) &= \frac{1}{2}\Xi\left(\hat{\mathbf{q}}(t)\right)\hat{\boldsymbol{\omega}}(t) \\
\Delta\tilde{\mathbf{x}}(t) &\equiv \begin{bmatrix} \boldsymbol{\delta\alpha} \\ \Delta\boldsymbol{\beta} \end{bmatrix} \\
F(\hat{\mathbf{x}}(t),\, t) &= \begin{bmatrix} -[\hat{\boldsymbol{\omega}}(t)\times] & -I_{3\times3} \\ 0_{3\times3} & 0_{3\times3} \end{bmatrix} \\
G(t) &= \begin{bmatrix} -I_{3\times3} & 0_{3\times3} \\ 0_{3\times3} & I_{3\times3} \end{bmatrix} \\
Q(t) &= \begin{bmatrix} \sigma_v^2 I_{3\times3} & 0_{3\times3} \\ 0_{3\times3} & \sigma_u^2 I_{3\times3} \end{bmatrix}
\end{aligned}$$

$$H_k(\hat{\mathbf{x}}_k^-) = \begin{bmatrix} \left[A(\hat{\mathbf{q}}^-)\mathbf{r}_1\times\right] & 0_{3\times3} \\ \left[A(\hat{\mathbf{q}}^-)\mathbf{r}_2\times\right] & 0_{3\times3} \\ \vdots & \vdots \\ \left[A(\hat{\mathbf{q}}^-)\mathbf{r}_n\times\right] & 0_{3\times3} \end{bmatrix}\Bigg|_{t_k} \tag{7.142}$$

$$\mathbf{h}_k(\hat{\mathbf{x}}_k^-) = \begin{bmatrix} A(\hat{\mathbf{q}}^-)\mathbf{r}_1 \\ A(\hat{\mathbf{q}}^-)\mathbf{r}_2 \\ \vdots \\ A(\hat{\mathbf{q}}^-)\mathbf{r}_n \end{bmatrix}\Bigg|_{t_k} \tag{7.143}$$

$$\Delta\hat{\tilde{\mathbf{x}}}_k^+ = K_k[\tilde{\mathbf{y}}_k - \mathbf{h}_k(\hat{\mathbf{x}}_k^-)]$$

$$\hat{\mathbf{q}}_k^+ = \hat{\mathbf{q}}_k^- + \frac{1}{2}\Xi(\hat{\mathbf{q}}_k^-)\delta\hat{\boldsymbol{\alpha}}_k^+$$
$$\hat{\boldsymbol{\beta}}_k^+ = \hat{\boldsymbol{\beta}}_k^- + \Delta\hat{\boldsymbol{\beta}}_k^+$$

- Discrete-Time Quaternion Propagation

$$\hat{\mathbf{q}}_{k+1}^- = \bar{\Omega}(\hat{\boldsymbol{\omega}}_k^+)\hat{\mathbf{q}}_k^+$$

$$\bar{\Omega}(\hat{\boldsymbol{\omega}}_k^+) \equiv \begin{bmatrix} \cos\left(\frac{1}{2}||\hat{\boldsymbol{\omega}}_k^+||\,\Delta t\right) I_{3\times3} - \left[\hat{\boldsymbol{\psi}}_k^+\times\right] & \hat{\boldsymbol{\psi}}_k^+ \\ -\hat{\boldsymbol{\psi}}_k^{+T} & \cos\left(\frac{1}{2}||\hat{\boldsymbol{\omega}}_k^+||\,\Delta t\right) \end{bmatrix}$$

$$\hat{\boldsymbol{\psi}}_k^+ \equiv \frac{\sin\left(\frac{1}{2}||\hat{\boldsymbol{\omega}}_k^+||\,\Delta t\right)\hat{\boldsymbol{\omega}}_k^+}{||\hat{\boldsymbol{\omega}}_k^+|}$$

- Farrenkopf's Steady-State Analysis

$$\dot{\theta} = \tilde{\omega} - \beta - \eta_v$$
$$\dot{\beta} = \eta_u$$

$$S_u \equiv \sigma_u \Delta t^{3/2} \Big/ \sigma_n$$
$$S_v \equiv \sigma_v \Delta t^{1/2} \Big/ \sigma_n$$

$$\vartheta = \left[S_u^2(4+S_v^2) + S_u^4/12\right]^{1/2}$$
$$\xi = -\frac{1}{2}\left[\left(\frac{S_u^2}{2} + \vartheta\right) + \sqrt{\left(\frac{S_u^2}{2} + \vartheta\right)^2 - 4S_u^2}\,\right]$$

$$p_{\theta\theta}^{-} = \sigma_n^2 \left[\left(\frac{\xi}{S_u} \right)^2 - 1 \right]$$
$$p_{\beta\beta}^{-} = \left(\frac{\sigma_n}{\Delta t} \right)^2 \left[S_u^2 \left(\frac{1}{\xi} + \frac{1}{2} \right) - \xi \right]$$

$$p_{\theta\theta}^{+} = \sigma_n^2 \left[1 - \left(\frac{S_u}{\xi} \right)^2 \right]$$
$$p_{\beta\beta}^{+} = \left(\frac{\sigma_n}{\Delta t} \right)^2 \left[S_u^2 \left(\frac{1}{\xi} - \frac{1}{2} \right) - \xi \right]$$

- Orbit Estimation

$$\ddot{\mathbf{r}}(t) = -\frac{\mu}{||\mathbf{r}(t)||^3}\mathbf{r}(t) + \mathbf{w}(t)$$

- The α-β Filter

$$\hat{r}_k^{+} = \hat{r}_k^{-} + \alpha\,[\tilde{y}_k - \hat{r}_k^{-}]$$
$$\hat{\dot{r}}_k^{+} = \hat{\dot{r}}_k^{-} + \frac{\beta}{\Delta t}\,[\tilde{y}_k - \hat{r}_k^{-}]$$
$$\hat{r}_{k+1}^{-} = \hat{r}_k^{+} + \hat{\dot{r}}_k^{+}\,\Delta t$$
$$\hat{\dot{r}}_{k+1}^{-} = \hat{\dot{r}}_k^{+}$$

$$S_q = q^{1/2}\,\Delta t^{3/2} \Big/ \sigma_n$$

$$\xi = \frac{1}{2}\left[\left(\frac{S_q^2}{2} + \vartheta \right) + \sqrt{\left(\frac{S_q^2}{2} + \vartheta \right)^2 - 4S_q^2} \right]$$
$$\vartheta = \left[4S_q^2 + \frac{S_q^4}{12} \right]^{1/2}$$

$$p_{rr}^{-} = \sigma_n^2 \left[\left(\frac{\xi}{S_q} \right)^2 - 1 \right]$$
$$p_{\dot{r}\dot{r}}^{-} = \left(\frac{\sigma_n}{\Delta t} \right)^2 \left[S_q^2 \left(\frac{1}{2} - \frac{1}{\xi} \right) + \xi \right]$$
$$p_{r\dot{r}}^{-} = \frac{\sigma_n^2 \xi}{\Delta t}$$

$$\frac{\beta^2}{1-\alpha} = S_q^2$$

$$\alpha = 1 - \left(\frac{S_q}{\xi}\right)^2$$
$$\beta = S_q\sqrt{1-\alpha}$$

$$\alpha = -\frac{1}{2}\beta + \frac{1}{2}\sqrt{\beta\,[(\beta/3)+8]}$$

- The α-β-γ Filter

$$\begin{aligned}
\hat{r}_k^+ &= \hat{r}_k^- + \alpha\,[\tilde{y}_k - \hat{r}_k^-] \\
\dot{\hat{r}}_k^+ &= \dot{\hat{r}}_k^- + \frac{\beta}{\Delta t}\,[\tilde{y}_k - \hat{r}_k^-] \\
\ddot{\hat{r}}_k^+ &= \ddot{\hat{r}}_k^- + \frac{\gamma}{2\Delta t^2}\,[\tilde{y}_k - \hat{r}_k^-] \\
\hat{r}_{k+1}^- &= \hat{r}_k^+ + \dot{\hat{r}}_k^+\,\Delta t + \frac{1}{2}\ddot{\hat{r}}_k^+\,\Delta t^2 \\
\dot{\hat{r}}_{k+1}^- &= \dot{\hat{r}}_k^+ + \ddot{\hat{r}}_k^+\,\Delta t \\
\ddot{\hat{r}}_{k+1}^- &= \ddot{\hat{r}}_k^+
\end{aligned}$$

- Smoothing with the Eigensystem Realization Algorithm

$$\begin{aligned}
\dot{\mathbf{x}} &= \begin{bmatrix} 0 & I \\ -M^{-1}K & -M^{-1}C \end{bmatrix}\mathbf{x} + \begin{bmatrix} 0 \\ M^{-1} \end{bmatrix}\mathbf{u} + \begin{bmatrix} 0 \\ I \end{bmatrix}\mathbf{w} \\
&\equiv F\mathbf{x} + B\mathbf{u} + G\mathbf{w}
\end{aligned}$$
$$\tilde{\mathbf{y}}_k = H\mathbf{x}_k + \mathbf{v}_k$$

$$\mathcal{K} = P_f^+\,\Phi^T\,(P_f^-)^{-1}$$
$$P = \mathcal{K}\,P\,\mathcal{K}^T + [P_f^+ - \mathcal{K}\,P_f^-\,\mathcal{K}^T]$$

Exercises

7.1 Write a general computer subroutine that converts a user's position from a known height, longitude, and latitude on the Earth to ECEF position coordinates. Also, write a general computer subroutine that converts ECEF position coordinates to height, longitude, and latitude.

7.2 Find a website that contains the orbital elements for every GPS satellite (as of this writing this website is given by http://www.navcen.uscg.gov/). Pick some user location of longitude and latitude on the Earth's surface. Convert the GPS orbital elements for each satellite into ECEF coordinates using the methods of §3.8.2 and §7.1.1 Assuming that the current time is given by the time of applicability for the GPS satellite, use eqn. (7.10) to determine which GPS satellites are available at that time.

7.3 Using the initial ECI coordinates of each GPS satellite determined from the orbital elements obtained in exercise 7.2, propagate the position of each GPS satellite using eqn. (3.198). Also, at each time-step convert the ECI position vector for each GPS satellite into ECEF coordinates using eqn. (7.1).

7.4 Reproduce the results of the Kalman filter application to GPS in example 7.1. Adjust the process noise variance parameter and discuss its importance on the filtering performance. Try various trajectory motions and speed of the vehicles (e.g., circular motion instead of due southerly motion as shown in the example). Also, artificially reduce down the available number of GPS satellites to three for some time period. How does your EKF design work during this period?

7.5 A simple approach for GPS position and velocity estimation involves using a linear Kalman filter with position "measurements" given from the nonlinear least squares estimates of §4.1. The Kalman filter model now involves only the first six states of the model shown in eqn. (7.11a), since the clock bias is estimated in the least squares solution. The output is now linear in this formulation with $H = \begin{bmatrix} I & 0 \end{bmatrix}$. The measurement covariance in the Kalman filter is given by the first three rows and three columns of error-covariance, P, in eqn. (4.5). Using the simulation parameters shown in example 7.1, test the performance of this simple linear Kalman filter. Compare your results to the results obtained by the full EKF formulation. Discuss the disadvantages of the linear Kalman filter (if any) over the EKF formulation.

7.6 Starting with eqn. (7.25) prove that eqn. (7.27) is indeed correct.

7.7 Show that the second-order errors in eqn. (7.47) are small only if $\delta\hat{\alpha}_k^+$ is small.

7.8 Show that following estimated error angle, defined in §7.2.1, propagation equation is valid up to second order:

$$\delta\dot{\alpha} = -[\hat{\omega}\times]\delta\alpha + \delta\omega - \frac{1}{2}\delta\omega \times \delta\alpha$$

7.9 Reproduce the results of example 7.2. Use the discrete-time propagation for the quaternion in eqn. (7.53) and covariance in eqn. (7.57). Try various values for σ_u and σ_v to generate synthetic gyro measurements, and discuss the performance of the extended Kalman filter under these variations. What parameter, σ_u or σ_v, seems to have the largest effect on the filter's performance?

7.10 Using the same procedure used to derive the eqn. (7.52), fully derive the state transition matrix in eqn. (7.59).

7.11 Fully derive the expressions shown in eqns. (7.71) and (7.72).

7.12 Use Murrell's version shown in Figure 7.5 on the simulated measurements developed in exercise 7.9. Discuss the performance in terms of accuracy and computational savings of Murrell's approach over the standard extended Kalman filter.

7.13 Write a general computer subroutine that solves Farrenkopf's equations in §7.2.4. Discuss how Farrenkopf's equations can be used to provide an initial hardware design from a spacecraft's attitude knowledge requirements. Also, use eqns. (7.71) and (7.72) to assess the expected extended Kalman filter performance for variations in σ_u and σ_v as discussed in exercise 7.9.

7.14 ♣ The extended Kalman filter for attitude estimation in Table 7.1 uses vector observations as measurements. Modify this algorithm to handle the case of quaternion measurements directly (hint: define an error quaternion between the measured quaternion and estimated quaternion).

7.15 Consider the problem of GPS spacecraft attitude estimation using phase difference measurements, as discussed in exercise 4.14. Pick a known position of a low-Earth orbiting spacecraft and simulate the availability of the GPS satellites at that position. Assume that a suitable elevation angle cutoff for the GPS availability in low-Earth orbit is 0 degrees. Generate an Earth-pointing motion in the spacecraft with a true attitude motion given by a constant angular velocity about the y-axis with $\omega = \begin{bmatrix} 0 & -0.0011 & 0 \end{bmatrix}^T$ rad /sec. Assume that the inertia matrix of the spacecraft is given by

$$J = \begin{bmatrix} 100 & 0 & 0 \\ 0 & 120 & 0 \\ 0 & 0 & 90 \end{bmatrix} \quad \text{Nms}$$

Using the dynamics model in eqn. (3.184b) an "open-loop" control input is given by

$$\mathbf{L} = -[\omega \times] J \omega$$

Pick a set of three baseline vectors and generate synthetic phase measurements using a standard deviation of $\sigma = 0.001$ for each measurements. Rederive the extended Kalman filter for attitude estimation, shown in §7.2.1, using the dynamics-based model instead of gyros. Use this filter to estimate the attitude of the vehicle from the GPS measurements and known control-torque input. Simulate process noise errors by varying the true value of J slightly, and tune the process noise covariance until reasonable results are obtained.

7.16 Consider the problem of determining the position and orientation of a vehicle using line-of-sight measurements from a vision-based beacon system based on Position Sensing Diode (PSD) technology, as shown in exercise 4.3. Develop an extended Kalman filter for this problem using the following state model:

$$\dot{\mathbf{q}} = \frac{1}{2}\Xi(\mathbf{q})\boldsymbol{\omega}$$
$$\dot{\boldsymbol{\omega}} = \mathbf{w}_{\omega}$$
$$\dot{\mathbf{p}} = \mathbf{v}$$
$$\dot{\mathbf{v}} = \mathbf{w}_{\upsilon}$$

where $\mathbf{q}$ is the quaternion, $\boldsymbol{\omega}$ is the angular velocity, $\mathbf{p} = \begin{bmatrix} X_c & Y_c & Z_c \end{bmatrix}^T$ is the position vector of the unknown object, and $\mathbf{v}$ is the velocity vector. The variables $\mathbf{w}_{\omega}$ and $\mathbf{w}_{\upsilon}$ are process noise vectors. Use the multiplicative error-quaternion approach of §7.2.1 to develop a 12th-order reduced state vector. Use the simulation parameters discussed in exercise 4.3 to test the performance of your EKF algorithm. Tune your filter design by varying the process noise covariance associated with the vectors $\mathbf{w}_{\omega}$ and $\mathbf{w}_{\upsilon}$. Once the filter is properly tuned, reduce the number of beacons seen by the sensor to 2 beacons. For example, from time period 300 to 500 seconds use measurements from only the first two beacons to update the state in the EKF. Assess and discuss the performance of the estimated quantities during this period.

7.17 Consider the problem of estimating the state (position, $\mathbf{r}$, and velocity, $\dot{\mathbf{r}}$) and drag parameter of a vehicle at launch, as shown in exercise 4.17. Develop a 7-state extended Kalman filter for this problem using the following state model:

$$\ddot{x} = -p\,\dot{x}\,V + w_x$$
$$\ddot{y} = -p\,\dot{y}\,V + w_y$$
$$\ddot{z} = -g - p\,\dot{z}\,V + w_z$$
$$\dot{p} = w_p$$

where w_x, w_y, w_z, and w_p are process noise terms. Use the simulation parameters discussed in exercise 4.17 to test the performance of your EKF algorithm. Tune your filter design by varying the process noise covariance parameters associated with w_x, w_y, w_z, and w_p. Also, use a fully discrete-time version of your filter (i.e., use a discrete-time propagation of the state model and error-covariance). Also, re-derive your algorithm using the following simplified model in the EKF:

$$\ddot{x} = w_x$$
$$\ddot{y} = w_y$$
$$\ddot{z} = -g + w_z$$

Can you achieve reasonable results using this approximate model that ignores the effect of drag on the system?

7.18 Reformulate the parameter identification problem of the coupled weakly nonlinear oscillators shown in exercise 4.20 using the Kalman filter approach discussed in §7.3. Compare the performance of the EKF versus the nonlinear least squares approach developed for exercise 4.20.

7.19 Reproduce the results of example 7.3. Compare your results to the Gaussian Least Squares Differential Correction (GLSDC) of §4.3 for various initial conditions errors. Does the EKF approach always converge in less iterations than the GLSDC?

7.20 ♣ Instead of the extended Kalman filter formulation for orbit estimation shown in §7.3, use the Unscented filter (UF) of §5.7.6 to perform the iterations. Can you achieve better performance capabilities using the UF over the EKF for various initial conditions?

7.21 The orbit navigation problem involves estimating the position and velocity of the spacecraft in real time using an extended Kalman filter. Program a navigation filter where the true orbit trajectory is determined with a nonzero process noise in eqn. (7.75). Use GPS pseudorange measurements sampled at 1-second intervals from the to-be-determined spacecraft to the GPS satellites (assume that the spacecraft is in low-Earth orbit). Assume that a suitable elevation angle cutoff for the GPS availability in low-Earth orbit is 0 degrees. Discuss the performance of the navigation filter as the measurement sampling interval increases.

7.22 Fully derive the relationship shown in eqn. (7.92).

7.23 Using the model in eqn. (7.93) derive analytical expressions for the tracking index and error-covariance matrix. Also, derive a similar expression as shown in eqn. (7.107) for the relationship between α and β. How does this model simplify the analysis?

7.24 Assume that no process noise is given in the model described in eqn. (7.76). Therefore, the discrete-time model is simply given by

$$\mathbf{x}_{k+1} = \begin{bmatrix} 1 & \Delta t \\ 0 & 1 \end{bmatrix} \mathbf{x}_k \tag{7.144}$$

$$\tilde{y}_k = \begin{bmatrix} 1 & 0 \end{bmatrix} \mathbf{x}_k + v_k \tag{7.145}$$

Assuming that no *a priori* information exists, so that $P_0 = \infty$, which corresponds to maximum likelihood estimation, show that the filter gains are given by the following expressions:

$$\alpha_k = \frac{2(2k-1)}{k(k+1)} \tag{7.146}$$

$$\beta_k = \frac{6}{k(k+1)} \tag{7.147}$$

Discuss the significance of these gains as k increases.

7.25 Prove that the only solution that makes β valid in eqn. (7.92) is given by eqn. (7.109).

7.26 ♣ Analytically prove the stability bounds for α, β, and γ shown in eqn. (7.124) are correct.

7.27 Reproduce the results of example 7.4. Try various values for the process noise parameter in each filter, and discuss the robustness of the estimated results to variations in this parameter. Also, perform an assessment on the computation complexity (e.g., the number of Floating Point Operations) of the α-β-γ filter versus the α-β filter.

7.28 From the simulation performed in exercise 7.27, suppose we ignore the relationship between α and β in the α-β filter. Try tuning them separately. We know that this approach ignores the kinematical relationship inherent in the assumed model, but can you achieve better results than the results shown in example 7.4? Also, try varying α, β, and γ independently in the α-β-γ filter.

7.29 Suppose that an acceleration measurement is also available for the system described in example 7.4. Use an acceleration measurement with a standard deviation of 0.1 m/sec^2 in an acceleration-based Kalman filter. The state model is still given by eqn. (7.117), but the observation vector is now given by

$$\mathbf{y}_k = \begin{bmatrix} 1 & 0 & 0 \\ 0 & 0 & 1 \end{bmatrix} \mathbf{x}_k + \mathbf{v}_k \equiv H\mathbf{x}_k$$

Derive a linear Kalman filter with this new observation model. Using the same value for q as in example 7.4, compare the performance of the α-β-γ filter versus this new filter. Also, try increasing the standard deviation of the acceleration measurement error and re-evaluate the performance of the new filter. At what value of this standard deviation does the acceleration measurement become practically useless?

7.30 Consider the nonlinear equations of motion for a highly maneuverable aircraft, as shown in exercise 4.22. Using a known "rich" input for δ_E, create synthetic measurements of the angle of attack α and pitch angle θ with zero initial conditions, as discussed in exercise 4.22. Use the extended Kalman filter to perform two tasks:

(A) Filter the measurements in the system by varying some of the coefficients in the assumed EKF model, using process noise to compensate for this error.

(B) Perform real-time estimation of some of the parametric values associated with the dynamic model. For example, try to estimate the true value (-4.208) associated with α in the differential equation for the pitch angle. Use the methods of §7.4.3 to develop your estimation algorithm. Try estimating other parameters as well.

7.31 Reproduce the results of example 7.5. How sensitive is this filter to variations in the initial state conditions and the initial error-covariance? Try estimating other parameters such as C_{D_α}, C_{L_α}, and C_{m_α}. Derive analytical expressions for the partial derivatives for these new parameters. Compare your EKF results to the results obtained in the nonlinear least squares approach, as shown in example 4.4.

7.32 Implement a nonlinear RTS smoother, shown in Table 6.5, to the simulation performed in exercise 7.31. Discuss the performance enhancement capabilities of the smoother over the EKF.

7.33 Suppose that the model shown in §7.4.3 is used strictly to filter the noisy measurement and for real-time navigation purposes. Use only an 8-state EKF design with states given by v_1, v_3, ω_2, θ, x, and z. The position components follow:

$$\begin{bmatrix} \dot{x} \\ \dot{z} \end{bmatrix} = \begin{bmatrix} \cos\theta & \sin\theta \\ -\sin\theta & \cos\theta \end{bmatrix} \begin{bmatrix} v_1 \\ v_3 \end{bmatrix}$$

The measurement model is now given by

$$\tilde{\mathbf{y}} = \begin{bmatrix} \alpha \\ ||\mathbf{v}|| \\ \omega_2 \\ \theta \\ ||\mathbf{r}|| \end{bmatrix} + \mathbf{v}$$

where $\mathbf{r} = \begin{bmatrix} x & z \end{bmatrix}^T$. Assume that the standard deviation of the measurement error associated with $||\mathbf{r}||$ is given by 10 m. Design an extended Kalman filter to track the position of the aircraft using the simulation parameters shown in example 7.5. Vary some of the coefficients in the assumed EKF dynamics model, and use process noise to compensate for this error. Also, implement an α-β-γ filter with measurements of $||\mathbf{r}||$ only. How do the results using a full dynamics-based model in an EKF compare to the results obtained by the simple α-β-γ filter?

7.34 ♣ Instead of the extended Kalman filter formulation for aircraft parameter estimation shown in §7.4.3, use the Unscented filter (UF) of §5.7.6 to perform the parameter estimation. Can you achieve better performance capabilities using the UF over the EKF for various initial condition and error-covariance errors?

7.35 Reproduce the results of the combined RTS/ERA results shown in example 7.6. Try various noise levels in the synthetic measurements and assess the value of using an RTS smoother as a "pre-filter" to the ERA.

7.36 Using the same simulation parameters shown in example 7.6, implement only the forward-time Kalman filter estimates in the ERA to realize the state model. How do the Kalman filter estimates combined with the ERA compare with the results obtained by the combined RTS/ERA approach? Try various noise levels in the synthetic measurements.

7.37 Instead of using the ERA approach to realize a state model, suppose we use the ARMA model instead, shown in exercise 1.13. Choose some simple second-order model with a significantly "rich" input and use a sequential version of the ARMA model to estimate the parameters of your chosen model. Add a significant amount of noise to the y_k and check the performance of your sequential estimator. Implement a simple linear Kalman filter with some assumed model to pre-filter the measurements before they are used in the sequential ARMA estimator. Finally, ignore the ARMA model estimator approach altogether and use the Kalman filter to directly estimate the coefficients by appending the state vector. Discuss the accuracy and computational requirements of each approach for various noise levels in the synthetic measurements.

References

[1] Hofmann-Wellenhof, B., Lichtenegger, H., and Collins, J., *GPS: Theory and Practice*, Springer Wien, New York, NY, 5th ed., 2001.

[2] Bate, R.R., Mueller, D.D., and White, J.E., *Fundamentals of Astrodynamics*, Dover Publications, New York, NY, 1971.

[3] Wertz, J.R., "Space-Based Orbit, Attitude and Timing Systems," *Mission Geometry: Orbit and Constellation Design and Management*, chap. 4, Microcosm Press, El Segundo, CA and Kluwer Academic Publishers, The Netherlands, 2001.

[4] Meeus, J., *Astronomical Algorithms*, Willman-Bell, Inc., Richmond, VA, 2nd ed., 1999.

[5] Spilker, J.J., "GPS Navigation Data," *Global Positioning System: Theory and Applications*, edited by B. Parkinson and J. Spilker, Vol. 64 of *Progress in Astronautics and Aeronautics*, chap. 4, American Institute of Aeronautics and Astronautics, Washington, DC, 1996.

[6] Farrell, J. and Barth, M., *The Global Positioning System & Inertial Navigation*, McGraw-Hill, New York, NY, 1998.

[7] Grewal, M.S., Weill, L.R., and Andrews, A.P., *Global Positioning Systems, Inertial Navigation, and Integration*, John Wiley & Sons, New York, NY, 2001.

[8] Farrell, J.L., "Attitude Determination by Kalman Filter," *Automatica*, Vol. 6, No. 5, 1970, pp. 419–430.

[9] Lefferts, E.J., Markley, F.L., and Shuster, M.D., "Kalman Filtering for Spacecraft Attitude Estimation," *Journal of Guidance, Control, and Dynamics*, Vol. 5, No. 5, Sept.-Oct. 1982, pp. 417–429.

[10] Crassidis, J.L. and Markley, F.L., "Attitude Estimation Using Modified Rodrigues Parameters," *Proceedings of the Flight Mechanics/Estimation Theory Symposium*, NASA-Goddard Space Flight Center, Greenbelt, MD, May 1996, pp. 71–83.

[11] Pittelkau, M.E., "Spacecraft Attitude Determination Using the Bortz Equation," *AAS/AIAA Astrodynamics Specialist Conference*, Quebec City, Quebec, Aug. 2001, AAS 01-310.

[12] Farrenkopf, R.L., "Analytic Steady-State Accuracy Solutions for Two Common Spacecraft Attitude Estimators," *Journal of Guidance and Control*, Vol. 1, No. 4, July-Aug. 1978, pp. 282–284.

[13] Markley, F.L., "Matrix and Vector Algebra," *Spacecraft Attitude Determination and Control*, edited by J.R. Wertz, appendix C, Kluwer Academic Publishers, The Netherlands, 1978.

[14] Murrell, J.W., "Precision Attitude Determination for Multimission Spacecraft," *Proceedings of the AIAA Guidance, Navigation, and Control Conference*, Palo Alto, CA, Aug. 1978, pp. 70–87.

[15] Andrews, S. and Bilanow, S., "Recent Flight Results of the TRMM Kalman Filter," *AIAA Guidance, Navigation, and Control Conference*, Monterey, CA, Aug. 2002, AIAA-2002-5047.

[16] Crassidis, J.L. and Markley, F.L., "Unscented Filtering for Spacecraft Attitude Estimation," *Journal of Guidance, Control, and Dynamics*, Vol. 26, No. 4, July-Aug. 2003, pp. 536–542.

[17] Yunck, T.P., "Orbit Determination," *Global Positioning System: Theory and Applications*, edited by B. Parkinson and J. Spilker, Vol. 164 of *Progress in Astronautics and Aeronautics*, chap. 21, American Institute of Aeronautics and Astronautics, Washington, DC, 1996.

[18] Brookner, E., *Tracking and Kalman Filtering Made Easy*, John Wiley & Sons, New York, NY, 1998.

[19] Bar-Shalom, Y. and Fortmann, T.E., *Tracking and Data Association*, Academic Press, Boston, MA, 1988.

[20] Kalata, P.R., "The Tracking Index: A Generalized Parameter for α-β and α-β-γ Target Trackers," *IEEE Transactions on Aerospace and Electronic Systems*, Vol. AES-20, No. 2, March 1984, pp. 174–182.

[21] Åström, K.J. and Wittenmark, B., *Computer-Controlled Systems*, Prentice Hall, Upper Saddle River, NJ, 3rd ed., 1997.

[22] Tenne, D. and Singh, T., "Characterizing Performance of α-β-γ Filters," *IEEE Transactions on Aerospace and Electronic Systems*, Vol. AES-38, No. 3, July 2002, pp. 1072–1087.

[23] Mook, D.J. and Shyu, I.M., "Nonlinear Aircraft Tracking Filter Utilizing Control Variable Estimation," *Journal of Guidance, Control, and Dynamics*, Vol. 15, No. 1, Jan.-Feb. 1992, pp. 228–237.

[24] Crassidis, J.L., Mook, D.J., and McGrath, J.M., "Automatic Carrier Landing System Utilizing Aircraft Sensors," *Journal of Guidance, Control, and Dynamics*, Vol. 16, No. 5, Sept.-Oct. 1993, pp. 914–921.

[25] Roemer, M.J. and Mook, D.J., "Enhanced Realization/Identification of Physical Modes," *Journal of Aerospace Engineering*, Vol. 3, No. 2, April 1990, pp. 128–139.

[26] Meirovitch, L., *Principles and Techniques of Vibrations*, Prentice Hall, Upper Saddle River, NJ, 1997.

[27] Hashemipour, H.R. and Laub, A.J., "Kalman Filtering for Second-Order Models," *Journal of Guidance, Control, and Dynamics*, Vol. 11, No. 2, March-April 1988, pp. 181–186.

[28] Crassidis, J.L. and Mook, D.J., "Integrated Estimation/Identification Using Second-Order Dynamic Models," *Journal of Vibration and Acoustics*, Vol. 119, No. 1, Jan. 1997, pp. 1–8.

[29] Rogers, R.M., *Applied Mathematics in Integrated Navigation Systems*, American Institute of Aeronautics and Astronautics, Inc., Reston, VA, 2000.

8

Optimal Control and Estimation Theory

Technology makes it possible for people to gain control over everything, except over technology. Tudor, John

THE optimal estimation foundations and applications of Chapters 2 through 7 are rooted in probability theory. Although the optimal algorithms derived in these chapters can be implemented solely for estimation and filtering applications, they are oftentimes used in control applications as well. For example, the Kalman filter is typically used to provide optimal estimates of state variables that are implemented in a control algorithm to guide a dynamic system along a desired trajectory. A practical scenario of this concept involves using the α-β filter to provide optimal position and rate estimates from position measurements only, which are required for a proportional-derivative controller. If the rate estimates are adequate then a rate hardware sensor may not be needed, which may produce significant cost savings.

The overall pointing error of a dynamic system inherently encompasses both estimation *and* control errors, which can occur from either hardware or algorithmic inaccuracies (or even both). Estimation errors typically arise from measurement errors (hardware), but may include errors associated with tuning parameters (algorithmic), as discussed in 7.4.1. Control errors typically arise from actuation constraints (hardware), as well as modelling errors (algorithmic). Estimation errors can be quantified using probability theory, but control errors usually cannot. When considering the overall pointing error one must keep in mind a dynamic system can only be controlled to within the accuracy of the estimation algorithm, which exemplifies the need for optimal estimation theory discussed in this book.

It seems natural to assume that control theory and estimation theory are two vastly different notions. However, as surmised in §6.4.1.3, the relationship between control and estimation is not a vague facet at all. In particular, §6.4.1.3 shows a derivation of fixed-interval smoother directly from optimal control theory, which proves the existence of a duality between control and estimation. The present chapter serves to provide the necessary foundations and tools of optimal control theory, which can be used to control a dynamic system to a desired point, and to follow a derived trajectory. Also, this theory can be used to fully comprehend the duality between control and estimation.

We begin by showing the most fundamental foundation in optimal control theory, called the *calculus of variations*. Then, Pontryagin's necessary conditions are presented, which can be used for non-smooth control inputs. The linear quadratic regulator is next shown, which provides an algorithm for an optimal controller of a system by minimizing a quadratic loss function using full state knowledge. We follow this theory with the linear quadratic-Gaussian controller, which incorporates the Kalman filter for state estimation. Finally, an example involving spacecraft attitude control is shown to demonstrate the practical aspects of the combined control and estimation theory.

8.1 Calculus of Variations

Modern optimal control theory has its roots in the calculus of variations, a subject placed upon the solid foundations during the 1800s by the monumental works of Lagrange, Hamilton, and Jacobi. Variational calculus was motivated directly by the apparent existence of minimum principles and other variational laws (e.g., Hamilton's principle) in analytical dynamics. In this section we develop the fundamental concepts of calculus of variations and optimal control in a fashion that encompasses a very large class of dynamic systems.

A fundamental class of variational problems seeks an optimum space-time path $\mathbf{x}(t)$ that minimizes (or maximizes) the following loss function:

$$\boxed{J \equiv J(\mathbf{x}(t), t_0, t_f) = \int_{t_0}^{t_f} \vartheta(\mathbf{x}(t), \dot{\mathbf{x}}(t), t)\, dt} \tag{8.1}$$

with $\mathbf{x}(t) = \left[x_1(t)\ x_2(t)\ \cdots\ x_n(t)\right]^T$. Without loss in generality, we assume our task is to minimize eqn. (8.1). It is evident that a simple change of sign converts a maximization problem to a minimization problem.

To obtain the most fundamental classical results, we restrict initial attention to ϑ and $\mathbf{x}$ of class C_2 (smooth, continuous functions having two continuous derivatives with respect to all arguments). Let $\mathbf{x}(t)$, t_0, and t_f represent the unknown path, and start and stop times, respectively, for which J of eqn. (8.1) has a local minimum value. Let an arbitrary neighboring, generally suboptimal, path be denoted by $\bar{\mathbf{x}}(t)$, with neighboring terminal times $\bar{t}_0$ and $\bar{t}_f$. We restrict the *varied path* $\bar{\mathbf{x}}(t)$ to be of class C_2 and to be near $\mathbf{x}(t)$ in the sense that the path variation

$$\delta\mathbf{x}(t) = \bar{\mathbf{x}}(t) - \mathbf{x}(t) \tag{8.2}$$

is of differential size for $\bar{t}_0 \le t \le \bar{t}_f$. We can consider $\bar{\mathbf{x}}(t)$ and $\dot{\bar{\mathbf{x}}}(t)$ to be generated by small arbitrary variations $\delta\mathbf{x}(t)$ of class C_2 as

$$\bar{\mathbf{x}}(t) = \mathbf{x}(t) + \delta\mathbf{x}(t) \tag{8.3a}$$

$$\dot{\bar{\mathbf{x}}}(t) = \dot{\mathbf{x}}(t) + \delta\dot{\mathbf{x}}(t) \tag{8.3b}$$

Clearly $\delta\dot{\mathbf{x}}(t) = \dot{\bar{\mathbf{x}}}(t) - \dot{\mathbf{x}}(t)$ is continuous, since both $\mathbf{x}(t)$ and $\bar{\mathbf{x}}(t)$ are continuous.

Along the varied path $\bar{\mathbf{x}}(t)$ initiating at time $\bar{t}_0 = t_0 + \delta t_0$ and terminating at $\bar{t}_f = t_f + \delta t_f$, the loss function of eqn. (8.1) has neighboring value

$$\bar{J} \equiv J(\bar{\mathbf{x}}(t), \bar{t}_0, \bar{t}_f) = \int_{\bar{t}_0}^{\bar{t}_f} \vartheta(\mathbf{x}(t) + \boldsymbol{\delta}\mathbf{x}(t), \dot{\mathbf{x}}(t) + \boldsymbol{\delta}\dot{\mathbf{x}}(t), t)\, dt \tag{8.4}$$

We define, for the case of finite $\boldsymbol{\delta}\mathbf{x}(t)$, the *finite variation* of J by differencing eqns. (8.4) and (8.1) as

$$\begin{aligned} \Delta J \equiv \bar{J} - J = &\int_{\bar{t}_0}^{\bar{t}_f} \vartheta(\mathbf{x}(t) + \boldsymbol{\delta}\mathbf{x}(t), \dot{\mathbf{x}}(t) + \boldsymbol{\delta}\dot{\mathbf{x}}(t), t)\, dt \\ &- \int_{t_0}^{t_f} \vartheta(\mathbf{x}(t), \dot{\mathbf{x}}(t), t)\, dt \end{aligned} \tag{8.5}$$

We restrict our attention to infinitesimal variations $\boldsymbol{\delta}\mathbf{x}(t_f)$ and δt_f only, since the initial state, $\mathbf{x}(t_0)$, and t_0 are usually defined *a priori*. Therefore, eqn. (8.5) reduces down to

$$\begin{aligned} \Delta J = &\int_{t_0}^{t_f} [\vartheta(\mathbf{x}(t) + \boldsymbol{\delta}\mathbf{x}(t), \dot{\mathbf{x}}(t) + \boldsymbol{\delta}\dot{\mathbf{x}}(t), t) - \vartheta(\mathbf{x}(t), \dot{\mathbf{x}}(t), t)]\, dt \\ &+ \int_{t_f}^{t_f + \delta t_f} \vartheta(\bar{\mathbf{x}}(t), \dot{\bar{\mathbf{x}}}(t), t)\, dt \end{aligned} \tag{8.6}$$

where $\bar{\mathbf{x}}(t) = \mathbf{x}(t) + \boldsymbol{\delta}\mathbf{x}(t)$ and its derivative have been used in eqn. (8.6). Now define the differential *first variation* δJ as the linear part of ΔJ. We find δJ by expanding the first integral of eqn. (8.6) in a Taylor series in $\boldsymbol{\delta}\mathbf{x}(t)$, $\boldsymbol{\delta}\dot{\mathbf{x}}(t)$, and δt_f to be

$$\begin{aligned} \delta J = &\int_{t_0}^{t_f} \left[\frac{\partial \vartheta(\mathbf{x}(t), \dot{\mathbf{x}}(t), t)}{\partial \mathbf{x}^T(t)} \boldsymbol{\delta}\mathbf{x}(t) + \frac{\partial \vartheta(\mathbf{x}(t), \dot{\mathbf{x}}(t), t)}{\partial \dot{\mathbf{x}}^T(t)} \boldsymbol{\delta}\dot{\mathbf{x}}(t) \right] dt \\ &+ \vartheta(\mathbf{x}(t_f), \dot{\mathbf{x}}(t_f), t_f)\, \delta t_f \end{aligned} \tag{8.7}$$

where $\partial\vartheta/\partial\mathbf{x}^T(t)$ and $\partial\vartheta/\partial\dot{\mathbf{x}}^T(t)$ denote row vectors. The second term on the right-hand side of eqn. (8.7) is derived by expanding $\vartheta(\bar{\mathbf{x}}(t_f), \dot{\bar{\mathbf{x}}}(t_f), t_f)$ in a Taylor series as follows

$$\begin{aligned} \vartheta(\bar{\mathbf{x}}(t_f), \dot{\bar{\mathbf{x}}}(t_f), t_f) = &\vartheta(\mathbf{x}(t_f), \dot{\mathbf{x}}(t_f), t_f) \\ &+ \left. \frac{\partial \vartheta(\mathbf{x}(t), \dot{\mathbf{x}}(t), t)}{\partial \mathbf{x}^T(t)} \right|_{t_f} \boldsymbol{\delta}\mathbf{x}(t_f) \\ &+ \left. \frac{\partial \vartheta(\mathbf{x}(t), \dot{\mathbf{x}}(t), t)}{\partial \dot{\mathbf{x}}^T(t)} \right|_{t_f} \boldsymbol{\delta}\dot{\mathbf{x}}(t_f) \end{aligned} \tag{8.8}$$

Substituting eqn. (8.8) into (8.6) yields eqn. (8.7) since $\boldsymbol{\delta}\mathbf{x}(t_f)\delta t_f$ and $\boldsymbol{\delta}\dot{\mathbf{x}}(t_f)\delta t_f$ represent higher-order terms, which vanish in the first variation.

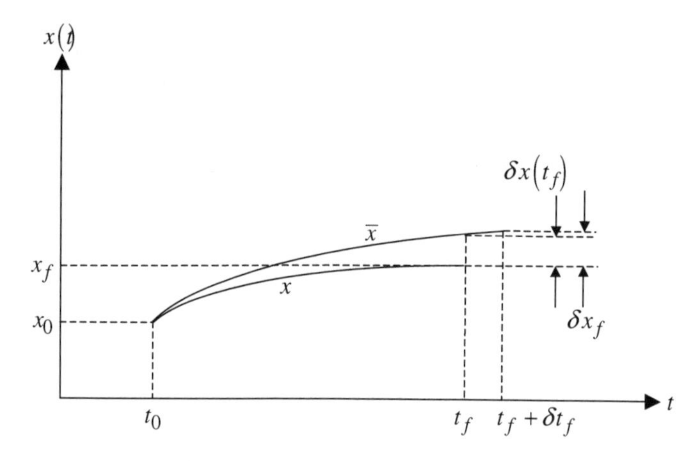

Figure 8.1: An Extremal and an Arbitrary Neighboring Path

In preparation for making arguments on the arbitrariness of $\boldsymbol{\delta}\mathbf{x}(t)$ and δt_f, we seek to eliminate the $\boldsymbol{\delta}\dot{\mathbf{x}}(t)$ term in eqn. (8.7). This is accomplished by using the integration by parts:

$$\int_{t_0}^{t_f} \frac{\partial \vartheta}{\partial \dot{\mathbf{x}}^T(t)} \boldsymbol{\delta}\dot{\mathbf{x}}(t)\, dt = \left. \frac{\partial \vartheta}{\partial \dot{\mathbf{x}}^T(t)} \boldsymbol{\delta}\mathbf{x}(t) \right|_{t_0}^{t_f} - \int_{t_0}^{t_f} \frac{d}{dt}\left[\frac{\partial \vartheta}{\partial \dot{\mathbf{x}}^T(t)}\right] \boldsymbol{\delta}\mathbf{x}(t)\, dt \tag{8.9}$$

Using eqn. (8.9) to replace the second term in the integrand of eqn. (8.7) yields

$$\begin{aligned} \delta J = &\int_{t_0}^{t_f} \left\{ \frac{\partial \vartheta(\mathbf{x}(t), \dot{\mathbf{x}}(t), t)}{\partial \mathbf{x}^T(t)} - \frac{d}{dt}\left[\frac{\partial \vartheta(\mathbf{x}(t), \dot{\mathbf{x}}(t), t)}{\partial \dot{\mathbf{x}}^T(t)}\right]\right\} \boldsymbol{\delta}\mathbf{x}(t)\, dt \\ &+ \left. \frac{\partial \vartheta(\mathbf{x}(t), \dot{\mathbf{x}}(t), t)}{\partial \dot{\mathbf{x}}^T(t)} \right|_{t_f} \boldsymbol{\delta}\mathbf{x}(t_f) + \vartheta(\mathbf{x}(t_f), \dot{\mathbf{x}}(t_f), t_f)\,\delta t_f = 0 \end{aligned} \tag{8.10}$$

Note $\delta t_0 = 0$ since $\mathbf{x}(t_0)$ is assumed to be known. Equation (8.10) is set to zero as a *necessary condition* for J to have a minimum, i.e., we require δJ to vanish for all admissible variations $\boldsymbol{\delta}\mathbf{x}(t)$ and δt_f. As a result the trajectories $\mathbf{x}(t)$ and terminal time t_f satisfying eqn. (8.10) yield a *stationary* value for $J(\mathbf{x}(t), t_0, t_f)$. If both t_f and $\mathbf{x}(t_f)$ are free, a relationship between them still exists. A scalar version of this relationship is demonstrated in Figure 8.1,[1] where δx_f is the difference between the ordinates at the end points. The first-order multidimensional approximation for this relationship is given by

$$\boldsymbol{\delta}\mathbf{x}(t_f) = \boldsymbol{\delta}\mathbf{x}_f - \dot{\mathbf{x}}(t_f)\,\delta t_f \tag{8.11}$$

Substituting eqn. (8.11) into eqn. (8.10) gives

$$\delta J = \int_{t_0}^{t_f} \left\{ \frac{\partial \vartheta(\mathbf{x}(t), \dot{\mathbf{x}}(t), t)}{\partial \mathbf{x}^T(t)} - \frac{d}{dt}\left[\frac{\partial \vartheta(\mathbf{x}(t), \dot{\mathbf{x}}(t), t)}{\partial \dot{\mathbf{x}}^T(t)} \right] \right\} \boldsymbol{\delta}\mathbf{x}(t)\, dt$$
$$+ \left. \frac{\partial \vartheta(\mathbf{x}(t), \dot{\mathbf{x}}(t), t)}{\partial \dot{\mathbf{x}}^T(t)} \right|_{t_f} \boldsymbol{\delta}\mathbf{x}_f \tag{8.12}$$
$$+ \left[\vartheta(\mathbf{x}(t_f), \dot{\mathbf{x}}(t_f), t_f) - \left. \frac{\partial \vartheta(\mathbf{x}(t), \dot{\mathbf{x}}(t), t)}{\partial \dot{\mathbf{x}}^T(t)} \right|_{t_f} \dot{\mathbf{x}}(t_f) \right] \delta t_f = 0$$

Since $\boldsymbol{\delta}\mathbf{x}(t)$ can assume an infinity of functional values, irrespective of the boundary conditions, we see that the integrand of the first term of eqn. (8.12) must vanish identically. Furthermore, since the boundary variations are generally independent of $\boldsymbol{\delta}\mathbf{x}(t)$, the boundary terms must also vanish independently. Thus eqn. (8.12) leads immediately to the *Euler-Lagrange necessary conditions*:

Euler-Lagrange Equations

$$\boxed{\frac{\partial \vartheta(\mathbf{x}(t), \dot{\mathbf{x}}(t), t)}{\partial \mathbf{x}(t)} - \frac{d}{dt}\left[\frac{\partial \vartheta(\mathbf{x}(t), \dot{\mathbf{x}}(t), t)}{\partial \dot{\mathbf{x}}(t)} \right] = \mathbf{0}} \tag{8.13}$$

Transversality Conditions

$$\left. \frac{\partial \vartheta(\mathbf{x}(t), \dot{\mathbf{x}}(t), t)}{\partial \dot{\mathbf{x}}^T(t)} \right|_{t_f} \boldsymbol{\delta}\mathbf{x}_f = 0 \tag{8.14a}$$

$$\left[\vartheta(\mathbf{x}(t_f), \dot{\mathbf{x}}(t_f), t_f) - \left. \frac{\partial \vartheta(\mathbf{x}(t), \dot{\mathbf{x}}(t), t)}{\partial \dot{\mathbf{x}}^T(t)} \right|_{t_f} \dot{\mathbf{x}}(t_f) \right] \delta t_f = 0 \tag{8.14b}$$

For example, if the initial and final times are fixed constants, and if the initial and final states are fully prescribed as $\mathbf{x}(t_0) = \mathbf{x}_0$ and $\mathbf{x}(t_f) = \mathbf{x}_f$, then the admissible path variations $\boldsymbol{\delta}\mathbf{x}(t)$ must vanish at t_0 and t_f, and δt_0 and δt_f must vanish as well. Thus for the *fixed time and fixed end point problem*, we find that the transversality conditions of eqn. (8.14) are trivially satisfied and the necessary conditions reduce to the Euler-Lagrange equations of eqn. (8.13) subject to the $2n$ boundary conditions $\mathbf{x}(t_0) = \mathbf{x}_0$ and $\mathbf{x}(t_f) = \mathbf{x}_f$.

For more general boundary condition specifications, the transversality conditions provide replacement or "natural" boundary conditions for terminal variables not constrained to prescribed values. In the simplest such case, a single variable may be totally "free." For example, if the final time t_f is not constrained (and unknown) and $\mathbf{x}(t_f)$ is specified, we must admit δt_f as nonzero and arbitrary. As a result, it is apparent by inspection of the transversality condition on eqn. (8.14b) that the unknown "free" final time is implicitly determined from the generally nonlinear *stopping con-*

dition

$$\mathbf{x}(t_0) = \mathbf{x}_0 \tag{8.15a}$$

$$\mathbf{x}(t_f) = \mathbf{x}_f \tag{8.15b}$$

$$\vartheta(\mathbf{x}(t_f), \dot{\mathbf{x}}(t_f), t_f) - \left. \frac{\partial \vartheta(\mathbf{x}(t), \dot{\mathbf{x}}(t), t)}{\partial \dot{\mathbf{x}}^T(t)} \right|_{t_f} \dot{\mathbf{x}}(t_f) = 0 \tag{8.15c}$$

If, on the other hand, t_f and $\mathbf{x}(t_f)$ are free and independent, the stopping conditions are given by

$$\mathbf{x}(t_0) = \mathbf{x}_0 \tag{8.16a}$$

$$\left. \frac{\partial \vartheta(\mathbf{x}(t), \dot{\mathbf{x}}(t), t)}{\partial \dot{\mathbf{x}}(t)} \right|_{t_f} = \mathbf{0} \tag{8.16b}$$

$$\vartheta(\mathbf{x}(t_f), \dot{\mathbf{x}}(t_f), t_f) = 0 \tag{8.16c}$$

In §8.2 we will subsequently consider the more general case that the terminal states and time are frequently constrained to lie in a generally nonlinear constraint manifold of the form given by

$$\boldsymbol{\psi}(\mathbf{x}(t_f), t_f) = \mathbf{0} \tag{8.17}$$

where the ψ_j are a set of independent functions of the class C_2.

Notice, in any event, that typically n boundary conditions (i.e., specified boundary conditions and transversality replacement boundary conditions) will be available at time t_0, while the remaining conditions are associated with time t_f. Thus, the terminal boundary conditions on eqn. (8.13) are split, and as a result we have a *two-point-boundary-value-problem* (TPBVP). Equation (8.13) generally provides n second-order nonlinear, stiff differential equations that can usually be solved for the second derivatives in the functional form

$$\ddot{\mathbf{x}}(t) = \mathbf{g}(\mathbf{x}(t), \dot{\mathbf{x}}(t), t) \tag{8.18}$$

Typically, numerical methods are required to solve eqn. (8.18), even if we have an *initial-value problem* in which $\mathbf{x}(t_0)$ and $\dot{\mathbf{x}}(t_0)$ are fully prescribed.[2, 3] Nonlinear TP-BVPs are inherently more difficult to solve than nonlinear initial-value problems. In general, iterative numerical methods must be employed in some fashion to solve TPBVPs, where convergence is usually difficult to guarantee *a priori*.

Given a solution, $\mathbf{x}(t)$, of the Euler-Lagrange equations in eqn. (8.18) satisfying the appropriate terminal boundary conditions in eqn. (8.14) and/or $\mathbf{x}(t_0) = \mathbf{x}_0$ and $\mathbf{x}(t_f) = \mathbf{x}_f$, we have a *stationary trajectory*. If this stationary trajectory in fact minimizes (or maximizes) J, we have a local *extremal trajectory*. Analogous to minima-maxima theory in ordinary calculus, a curvature test is required to establish sufficiency for a local minimum (or maximum). Functional curvature of $J[\mathbf{x}(t) + \delta\mathbf{x}(t)]$ is tested using the *second variation*.[4] Since formal sufficiency tests and the second variation play a relatively restricted role in practical applications, we elect not to treat these concepts here. Fortunately, a resourceful analyst can often achieve a high degree of confidence that a candidate trajectory is at least a local minimum, even if a formal sufficiency test proves intractable.

8.2 Optimization with Differential Equation Constraints

We now turn our attention to development of the fundamental results needed for optimal control of nonlinear systems. Suppose we have a system whose behavior is described by solving ordinary differential equations. It is usually possible to arrange the system of differential equations in the standard first-order form

$$\boxed{\dot{\mathbf{x}}(t) = \mathbf{f}(\mathbf{x}(t), \mathbf{u}(t), t)} \tag{8.19}$$

The $u_i(t)$ are p control functions of class C_2 that are to be chosen to maneuver the system described by eqn. (8.19) from the prescribed initial state

$$\boxed{\mathbf{x}(t_0) = \mathbf{x}_0, \quad t_0 \text{ fixed}} \tag{8.20}$$

to a generally unspecified final time t_f and final state $\mathbf{x}(t_f)$ satisfying a nonlinear *manifold* system of q algebraic equations of the form given by

$$\boxed{\psi(\mathbf{x}(t_f), t_f) = \mathbf{0}} \tag{8.21}$$

The loss function to be minimized has the form given by

$$\boxed{J = \phi(\mathbf{x}(t_f), t_f) + \int_{t_0}^{t_f} \vartheta(\mathbf{x}(t), \mathbf{u}(t), t)\, dt} \tag{8.22}$$

Introducing the two vector of Lagrange multipliers[4, 5] $\boldsymbol{\lambda}(t)$ and $\boldsymbol{\alpha}$ of dimension $n \times 1$ and $q \times 1$, respectively, we form the *augmented functional*

$$\begin{aligned} J &= \phi(\mathbf{x}(t_f), t_f) + \boldsymbol{\alpha}^T \psi(\mathbf{x}(t_f), t_f) \\ &+ \int_{t_0}^{t_f} \left\{ \vartheta(\mathbf{x}(t), \mathbf{u}(t), t) + \boldsymbol{\lambda}^T(t)[\mathbf{f}(\mathbf{x}(t), \mathbf{u}(t), t) - \dot{\mathbf{x}}(t)] \right\} dt \end{aligned} \tag{8.23}$$

Considering the neighboring trajectory associated with the variations $\bar{\mathbf{x}}(t) = \mathbf{x}(t) + \boldsymbol{\delta}\mathbf{x}(t)$, $\bar{\mathbf{u}}(t) = \mathbf{u}(t) + \boldsymbol{\delta}\mathbf{u}(t)$, $\bar{t}_f = t_f + \delta t_f$, we find from the linear part of $\Delta J = \bar{J} - J$ that the first variation of J is

$$\begin{aligned} \delta J &= \int_{t_0}^{t_f} \left[\frac{\partial H}{\partial \mathbf{x}(t)} + \dot{\boldsymbol{\lambda}}(t) \right]^T \boldsymbol{\delta}\mathbf{x}(t)\, dt \\ &+ \int_{t_0}^{t_f} [\mathbf{f}(\mathbf{x}(t), \mathbf{u}(t), t) - \dot{\mathbf{x}}(t)]^T \boldsymbol{\delta\lambda}(t)\, dt + \int_{t_0}^{t_f} \frac{\partial H}{\partial \mathbf{u}^T(t)} \boldsymbol{\delta}\mathbf{u}(t)\, dt \\ &+ \left[H + \frac{\partial \Phi(\mathbf{x}(t), t)}{\partial t} \right] \Bigg|_{t_f} \delta t_f + \left[\frac{\partial \Phi(\mathbf{x}(t), t)}{\partial \mathbf{x}(t)} - \boldsymbol{\lambda}(t) \right]^T \Bigg|_{t_f} \boldsymbol{\delta}\mathbf{x}(t_f) = 0 \end{aligned} \tag{8.24}$$

where the auxiliary definition of the *Hamiltonian* is

$$\boxed{H \equiv \vartheta(\mathbf{x}(t), \mathbf{u}(t), t) + \boldsymbol{\lambda}^T(t)\mathbf{f}(\mathbf{x}(t), \mathbf{u}(t), t)} \tag{8.25}$$

and the augmented terminal function

$$\boxed{\Phi(\mathbf{x}(t_f), t_f) \equiv \phi(\mathbf{x}(t_f), t_f) + \boldsymbol{\alpha}^T \boldsymbol{\psi}(\mathbf{x}(t_f), t_f)} \tag{8.26}$$

It follows, by inspection of the variational statement of eqn. (8.24), that the following necessary conditions hold:

$$\dot{\mathbf{x}}(t) = \frac{\partial H}{\partial \boldsymbol{\lambda}(t)} \equiv \mathbf{f}(\mathbf{x}(t), \mathbf{u}(t), t) \tag{8.27a}$$

$$\dot{\boldsymbol{\lambda}}(t) = -\frac{\partial H}{\partial \mathbf{x}(t)} \equiv -\frac{\partial \vartheta(\mathbf{x}(t), \mathbf{u}(t), t)}{\partial \mathbf{x}(t)} - \left[\frac{\partial \mathbf{f}(\mathbf{x}(t), \mathbf{u}(t), t)}{\partial \mathbf{x}(t)}\right]^T \boldsymbol{\lambda}(t) \tag{8.27b}$$

$$\frac{\partial H}{\partial \mathbf{u}(t)} = \mathbf{0} \tag{8.27c}$$

$$\left[\frac{\partial \Phi(\mathbf{x}(t), t)}{\partial t} + H\right]\Bigg|_{t_f} \delta t_f = 0 \tag{8.27d}$$

$$\left[\frac{\partial \Phi(\mathbf{x}(t), t)}{\partial \mathbf{x}(t)} - \boldsymbol{\lambda}(t)\right]^T\Bigg|_{t_f} \delta \mathbf{x}(t_f) = 0 \tag{8.27e}$$

and, of course, the boundary conditions of eqns. (8.20) and (8.21). If the final time is fixed, then $\delta t_f = 0$ and eqn. (8.27d) becomes trivially satisfied. If none of the $\mathbf{x}(t_f)$ are directly specified and the final time is free, conditions of eqns. (8.27d) and (8.27e) provide the transversality conditions

$$\left[\frac{\partial \phi(\mathbf{x}(t), t)}{\partial t} + \boldsymbol{\alpha}^T \frac{\partial \boldsymbol{\psi}(\mathbf{x}(t), t)}{\partial t} + H\right]\Bigg|_{t_f} = 0 \tag{8.28a}$$

$$\boldsymbol{\lambda}(t_f) = \left\{\frac{\partial \phi(\mathbf{x}(t), t)}{\partial \mathbf{x}(t)} + \left[\frac{\partial \boldsymbol{\psi}(\mathbf{x}(t), t)}{\partial \mathbf{x}(t)}\right]^T \boldsymbol{\alpha}\right\}\Bigg|_{t_f} \tag{8.28b}$$

Equation (8.28a) is the "stopping condition" used to implicitly determine the optimal final time. Notice eqn. (8.28b) determines a *final* boundary condition on the costate $\boldsymbol{\lambda}(t_f)$, which must be considered simultaneously with eqn. (8.21) to determine $\boldsymbol{\alpha}$, whereas eqn. (8.20) provides the *initial* condition on the state $\mathbf{x}(t_0)$. Thus the boundary conditions on eqns. (8.27a) and (8.27b) are *split* and we generally have a TPBVP. The algebraic equation provided by eqn. (8.27c) is usually simple enough to solve for $\mathbf{u}(t)$ as a function of $\mathbf{x}(t)$ and $\boldsymbol{\lambda}(t)$, and thereby eliminate $\mathbf{u}(t)$ from eqns. (8.27a) and (8.27b).

8.3 Pontryagin's Optimal Control Necessary Conditions

In many control applications, the above formulation suffers a serious shortcoming; the requirement (limitation!) that the admissible controls $\mathbf{u}(t)$ be smooth functions with two continuous derivatives immediately precludes on/off controls and the (often necessary) imposition of inequality bounds on the control input's magnitude and its derivatives. Several important generalizations of optimal control formulations have made it possible to routinely solve problems with inequality constraints on both the control and state variables.[1, 5]

If we allow admissible controls which are bounded and only piecewise continuous (in lieu of restricting them to belong to class C_2), the necessary conditions generalize in such a way that the only change from the conditions in eqn. (8.27) is the replacement of eqn. (8.27c) by Pontryagin's Principle:[6] *The optimal control* $\mathbf{u}(t)$ *is determined at each instant to render the Hamiltonian a minimum over all admissible control functions.* For example, Pontryagin's Principle requires for controls of class C_2 that eqn. (8.27c) is true *and* $\partial^2 H/\partial \mathbf{u}^2(t)$ must be positive definite. Thus Pontryagin's Principle is consistent with the developments of §8.2, but with the additional constraint that $\partial^2 H/\partial \mathbf{u}^2(t)$ be positive definite.

The most significant utility of Pontryagin's Principle, however, lies in finding optimal controls when the admissible controls *do not* belong to class C_2. For example, suppose we have an optimal maneuver problem of the form given by

$$\boxed{\dot{\mathbf{x}}(t) = \mathbf{f}(\mathbf{x}(t), t) + \mathbf{u}(t), \quad \mathbf{x}(t_0) = \mathbf{x}_0, \quad \mathbf{x}(t_f) = \mathbf{x}_f} \tag{8.29}$$

The loss function to be minimized is given by

$$\boxed{J = \frac{1}{2}\int_{t_0}^{t_f} \mathbf{x}^T(t)\, \mathcal{Q}\,\mathbf{x}(t)\, dt} \tag{8.30}$$

where $\mathcal{Q}$ is an $n \times n$ positive definite or positive semi-definite matrix. The Hamiltonian for this system is given by

$$\boxed{H = \frac{1}{2}\mathbf{x}^T(t)\, \mathcal{Q}\,\mathbf{x}(t) + \boldsymbol{\lambda}^T(t)[\mathbf{f}(\mathbf{x}(t), t) + \mathbf{u}(t)]} \tag{8.31}$$

If $\mathbf{u}(t)$ is of class C_2 then the solution for the optimal control input simply follows the conditions given in eqn. (8.27). However, we are also given that the admissible control inputs must satisfy the constraints

$$\boxed{|u_j(t)| \leq u_{\max_j}, \quad j = 1, 2, \ldots, p} \tag{8.32}$$

The necessary conditions of eqns. (8.27a) and (8.27b) are still valid, which give

$$\dot{\mathbf{x}}(t) = \mathbf{f}(\mathbf{x}(t), t) + \mathbf{u}(t) \tag{8.33a}$$

$$\dot{\boldsymbol{\lambda}}(t) = -\left[\frac{\partial \mathbf{f}(\mathbf{x}(t), t)}{\partial \mathbf{x}(t)}\right]^T \boldsymbol{\lambda}(t) - \mathcal{Q}\mathbf{x}(t) \tag{8.33b}$$

and Pontryagin's Principle requires the Hamiltonian of eqn. (8.31) to be minimized with respect to $\mathbf{u}(t)$ over all admissible control inputs satisfying eqn. (8.32). Since the Hamiltonian contains $\mathbf{u}(t)$ linearly, we know that the extreme of H with respect to $\mathbf{u}(t)$ must lie on the boundary of the region defined by eqn. (8.32). Thus we find that the $\lambda_i(t)$ are *switching functions* for the element $u_i(t)$ of the control input vector $\mathbf{u}(t)$:

$$\mathbf{u}(t) = \begin{bmatrix} s_1 u_{\max_1} \\ s_2 u_{\max_2} \\ \vdots \\ s_p u_{\max_p} \end{bmatrix} \tag{8.34}$$

where

$$s_i = \text{sign}[\lambda_i(t)] \tag{8.35}$$

Equation (8.34) is not valid, however, for the unusual event that one or more of the elements of $\boldsymbol{\lambda}(t)$ vanishes identically for a finite time interval. This latter case of problems is known as *singular* optimal control problems.[3] While the singular optimal control problem is of significant theoretical and some practical interest, we elect not to treat this subject formally here.

Example 8.1: In this example we consider the case of a rigid body constrained to rotate about a fixed axis, where the equation of motion is given by the single axis version of eqn. (3.180):

$$\ddot{\theta}(t) = \frac{1}{J}L(t) \equiv u(t)$$

where $\dot{\theta} \equiv \omega$ from eqn. (3.180) and J is the inertia (see §3.7.2). Suppose we seek a $u(t)$ of class C_2 that maneuvers the body frame from the prescribed initial conditions

$$\theta(t_0) = \theta_0$$
$$\dot{\theta}(t_0) = \dot{\theta}_0$$

to the desired final conditions

$$\theta(t_f) = \theta_f$$
$$\dot{\theta}(t_f) = \dot{\theta}_f$$

The loss function to be minimized is given by

$$J = \frac{1}{2}\int_{t_0}^{t_f} u^2(t)\, dt$$

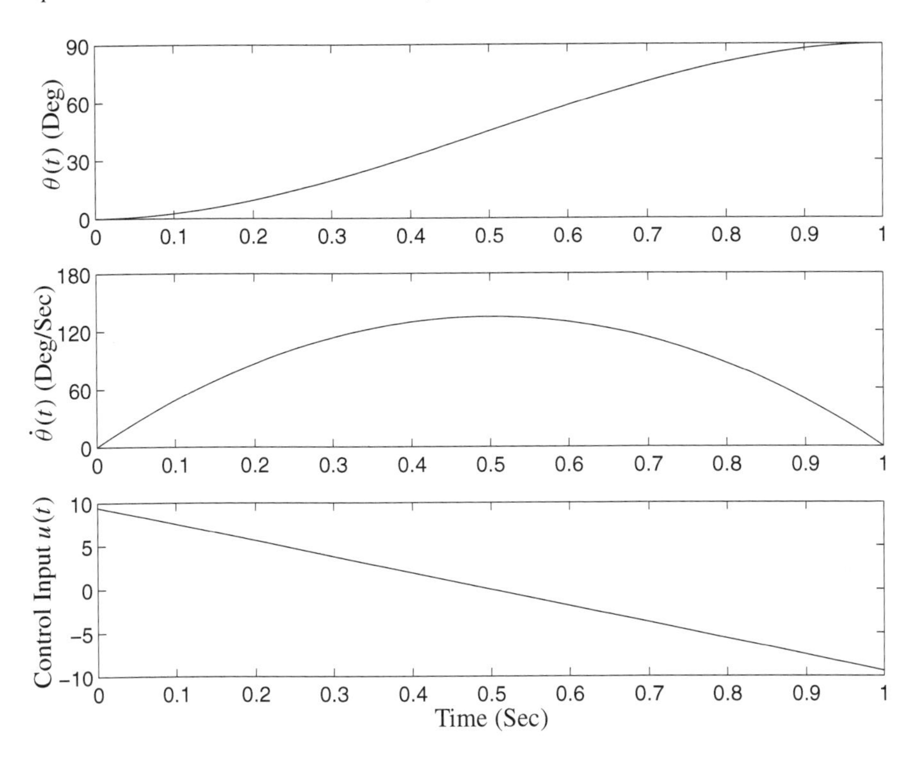

Figure 8.2: Optimal Rest-to-Rest Maneuver for $\ddot{\theta}(t) = u(t)$

where this J is not to be confused with the inertia. We restrict attention to the case that $t_0 = 0$ and $t_f = T$ are fixed. Two methods are considered to derive the optimal maneuver. First we note that direct substitution of the dynamics equation into the loss function yields an equation of the form given by

$$\vartheta(\theta, \dot{\theta}, \ddot{\theta}, t) = \frac{1}{2}\ddot{\theta}^2(t)$$

This form is not identical to the form presented in eqn. (8.1); however, the extension of the Euler-Lagrange equations to higher-order derivatives is straightforward (which is left as an exercise for the reader). For this specific case the Euler-Lagrange equation is given as

$$\frac{d^4\theta(t)}{dt^4} = 0$$

This equation is trivially integrated to obtain the cubic polynomial

$$\theta(t) = a_1 + a_2 t + a_3 t^2 + a_4 t^3$$

as the extremal trajectory.

The four integration constants can be determined as a function of the boundary conditions and the maneuver time T by simply enforcing the boundary conditions on the cubic polynomial equation and its time derivative. The solution of the resulting four algebraic equations gives

$$
\begin{aligned}
a_1 &= \theta_0 \\
a_2 &= \dot{\theta}_0 \\
a_3 &= \frac{3(\theta_f - \theta_0)}{T^2} - \frac{2\dot{\theta}_0 + \dot{\theta}_f}{T} \\
a_4 &= -\frac{2(\theta_f - \theta_0)}{T^3} + \frac{\dot{\theta}_0 + \dot{\theta}_f}{T^2}
\end{aligned}
$$

Furthermore, it is obvious that taking a second time derivative of the cubic polynomial gives the optimal control torque, which is a linear function of time:

$$u(t) = 2a_3 + 6a_4 t$$

As a specific example, consider the following numerical values with $t_0 = 0$ and $T = 1$:

$$
\begin{aligned}
\theta(0) &= 0, \quad \dot{\theta}(0) = 0 \\
\theta(1) &= \pi/2, \quad \dot{\theta}(1) = 0
\end{aligned}
$$

These boundary conditions will yield a *rest-to-rest* maneuver. Using these conditions gives the following control torque:

$$u(t) = \ddot{\theta}(t) = 3\pi(1 - 2t)$$

Also, the maneuver angle, $\theta(t)$, and angular velocity, $\dot{\theta}(t)$, are given by

$$
\begin{aligned}
\theta(t) &= 3\pi\left(t^2/2 - t^3/3\right) \\
\dot{\theta}(t) &= 3\pi\left(t - t^2\right)
\end{aligned}
$$

A plot of the maneuver angle, angular velocity, and control torque is shown in Figure 8.2. Clearly, the initial and final boundary conditions are satisfied with this control torque.

Notice, since we admitted only controls of class C_2, we were able to use the generalized version of Euler-Lagrange's equations in lieu of the Pontryagin-form necessary conditions of §8.2. The constraint in this simple example is enforced by simply substituting it into the loss function directly. In the approach of §8.2, we enforce the differential equation constraints by using the Lagrange multiplier rule. To illustrate the equivalence in the present transparent example, we resolve for the optimal maneuvering using the approach and notations of §8.2.

Before we proceed, it is necessary to convert the dynamics equations $\ddot{\theta}(t) = u(t)$ to the first-order form of eqn. (8.19). This is accomplished by using the change of variables introduced in eqn. (3.3). For the present example the following state variables are introduced:

$$x_1(t) = \theta(t)$$
$$x_2(t) = \dot{\theta}(t)$$

Then the desired equivalent first-order equations follow as

$$\dot{x}_1(t) = x_2(t)$$
$$\dot{x}_2(t) = u(t)$$

The Hamiltonian described in eqn. (8.25) is given by

$$H = \frac{1}{2}u^2(t) + \lambda_1(t)\,x_2(t) + \lambda_2(t)\,u(t)$$

The necessary conditions for the optimal maneuver then follow from eqns. (8.27a) to (8.27c) as

$$\dot{x}_1(t) = x_2(t)$$
$$\dot{x}_2(t) = u(t)$$
$$\dot{\lambda}_1(t) = 0$$
$$\dot{\lambda}_2(t) = -\lambda_1(t)$$
$$u(t) + \lambda_2(t) = 0$$

The solutions for the costate variables $\lambda_1(t)$ and $\lambda_2(t)$ follow as

$$\lambda_1(t) = b_1 = \text{constant}$$
$$\lambda_2(t) = -b_1 t + b_2$$

Also, the control input follows $u(t) = -\lambda_2(t)$:

$$u(t) = b_1 t - b_2$$

Having $u(t)$ then $x_1(t)$ and $x_2(t)$ are trivially solved to be

$$x_1(t) \equiv \theta(t) = b_4 + b_3 t - b_2 t^2/2 + b_1 t^3/6$$
$$x_2(t) \equiv \dot{\theta}(t) = b_3 - b_2 t + b_1 t^2/2$$

This solution is identical to the previous solution using the Euler-Lagrange approach, with the obvious relationship of the integration constants $b_4 = a_1$, $b_3 = a_2$, $b_2 = -2a_3$, and $b_1 = 6a_4$. For the case of one constraint, i.e., one state variable, and controls of class C_2, it appears that the multiplier rule slightly increased the algebra. For the cases in which constraints can be eliminated by direct substitution and for controls of class C_2 this pattern is typical. However, such ideal circumstances represent the minority of applications. Implicit, nonlinear constraints, nonlinear differential

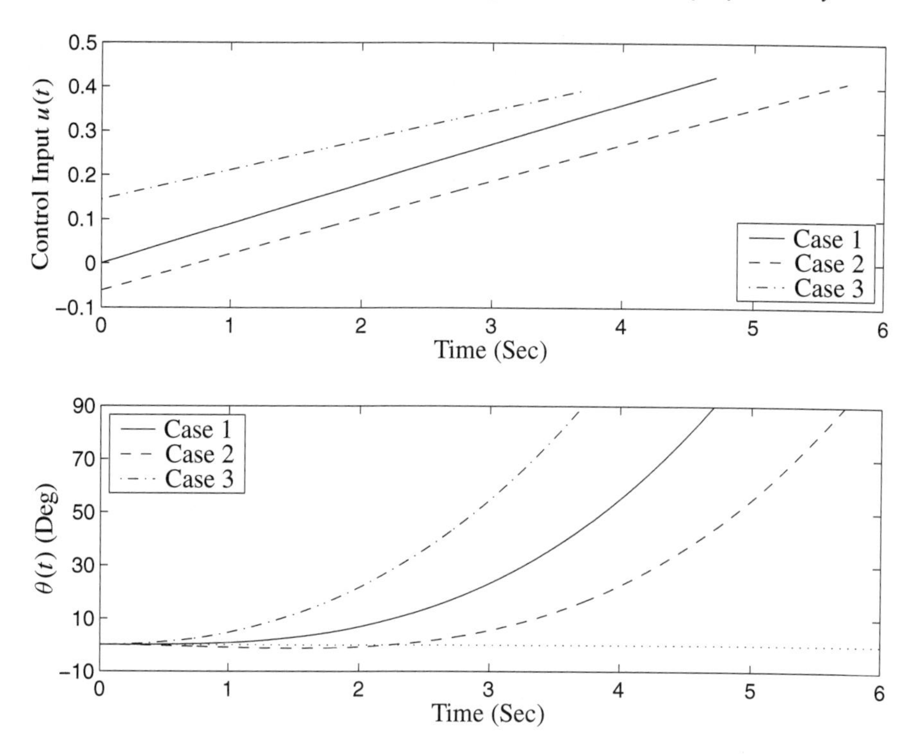

Figure 8.3: Spinup Maneuver: Effect of Final Time Variation

equations, and discontinuous controls abound in modern-day applications. For these cases, the introduction of Lagrange multipliers and the use of Pontryagin-form necessary conditions have been found to be advantageous.

For the case that the final time T is free, we have from eqn. (8.27d) the stopping condition $H(T) = 0$, which leads to

$$H(T) = -\frac{2}{T^4}(aT^2 + bT + c) = 0$$

where

$$a = \dot{\theta}_0^2 + \dot{\theta}_0\dot{\theta}_f + \dot{\theta}_f^2$$

$$b = 6(\theta_0 - \theta_f)(\dot{\theta}_0 + \dot{\theta}_f)$$

$$c = 9(\theta_f - \theta_0)^2$$

Thus, there are three final times for which $H(T) = 0$:

$$T_1^* = \infty, \quad T_{2,3}^* = \frac{3(\theta_f - \theta_0)\left[\dot{\theta}_0 + \dot{\theta}_f \pm \sqrt{\dot{\theta}_0\dot{\theta}_f}\right]}{\dot{\theta}_0^2 + \dot{\theta}_0\dot{\theta}_f + \dot{\theta}_f^2}$$

The value of $T_1^* = \infty$ corresponds to the global optimal free time, whereas T_2^* and T_3^*, when real, are local maxima or minima of J, at finite times; these have some significance in practical applications. It is obvious by inspection of the final time conditions that for the rest-to-rest case ($\dot{\theta}_0 = \dot{\theta}_f = 0$) the only zero of $H(T)$ is $T = \infty$. Thus, the optimum rest-to-rest maneuvers are carried out very slowly. Furthermore, consider the special cases of maneuvers for which $\dot{\theta}_0 = 0$, which cause the discriminant in the solution for $T_{2,3}^*$ to vanish, and we have a double root:

$$T^* = T_2^* = T_3^* = \frac{3(\theta_f - \theta_0)}{\dot{\theta}_f}$$

This causes an inflection at $J(T)$. For $\theta_f = \pi/2$, $\theta_0 = 0$ and $\dot{\theta}_f = 1$, i.e., a spinup maneuver, we show in Figure 8.3 trajectories for the following three cases:

Case 1: $T = T^* = 3\pi/2 = 4.7124$
Case 2: $T = T^* - 1 = 3.7124$ $(T < T^*)$
Case 3: $T = T^* + 1 = 5.7124$ $(T > T^*)$

From Figure 8.3, it is evident that fixing the final time greater than T^* has the undesirable consequence that θ initially counter rotates (e.g., Case 3). The performance, as measured by J, is actually slightly less for Case 3 than for Case 1. This example illustrates that counterintuitive and undesirable results sometimes stem from "optimal" control developments.

If *both* initial and final rates ($\dot{\theta}_0$ and $\dot{\theta}_f$) are zero, the inflection of J disappears, and the only zero of $H(T)$ occurs at $T = \infty$. The global minimum of J is zero and is approached as the maneuver time approaches infinity. The optimal control, angular velocity, and angle of rotation profiles (for this rest-to-rest class of maneuvers) are all completely analogous to the maneuver shown in Figure 8.2.

We should note that the *open-loop* approaches for the solution of optimal control problems shown in §8.1 and §8.2 are not generally robust to parametric variations, unlike *feedback control* methods. This is easily illustrated by multiplying the control torque $u(t)$ in example 8.1 by some scalar, which simulates an error in the inertia J, and using this control input with the identical boundary conditions shown in the example. This will yield suboptimal results for various scalar multiplication factors (which is left as an exercise for the reader to investigate).

8.4 Discrete-Time Control

The importance of discrete-time systems, described in §3.5, is well known with the reliance on digital computers, which are used to process sampled-data systems for estimation and control purposes. As discussed in §8.3, the Lagrange multiplier approach with the use of Pontryagin-form necessary conditions is better suited for modern-day problems. Hence, we only present this approach for the optimal control theory involving discrete-time systems. A more thorough treatise involving the discrete-time Euler-Lagrange equations and associated transversality conditions can be found in Refs. [2] and [3]. Consider finding a control sequence $\mathbf{u}_0, \ldots, \mathbf{u}_{N-1}$ and final time t_f that minimizes the following loss function:

$$\boxed{J = \phi(\mathbf{x}_N, t_f) + \sum_{k=0}^{N-1} \vartheta_k(\mathbf{x}_k, \mathbf{u}_k, k)} \tag{8.36}$$

subject to the constraints

$$\mathbf{x}_{k+1} = \mathbf{f}_k(\mathbf{x}_k, \mathbf{u}_k, k) \tag{8.37a}$$

$$\boldsymbol{\psi}(\mathbf{x}_N, t_f) = \mathbf{0} \tag{8.37b}$$

with $t_f = N\Delta t$, where N is the total number of steps and Δt is the time-step. As in §8.2 we assume that the initial state and time are fixed and known, so that $\mathbf{x}(t_0) = \mathbf{x}_0$ and t_0 is fixed. The augmented functional is formed by introducing two Lagrange multipliers, $\boldsymbol{\lambda}_{k+1}$ and $\boldsymbol{\alpha}$, of dimension $n \times 1$ and $q \times 1$, respectively:

$$\begin{aligned} J &= \phi(\mathbf{x}_N, t_f) + \boldsymbol{\alpha}^T \boldsymbol{\psi}(\mathbf{x}_N, t_f) \\ &+ \sum_{k=0}^{N-1} \vartheta_k(\mathbf{x}_k, \mathbf{u}_k, k) + \boldsymbol{\lambda}_{k+1}^T \left[\mathbf{f}_k(\mathbf{x}_k, \mathbf{u}_k, k) - \mathbf{x}_{k+1}\right] + \boldsymbol{\lambda}_0^T [\mathbf{x}_0 - \mathbf{x}(t_0)] \end{aligned} \tag{8.38}$$

As with the continuous-time development we introduce the following Hamiltonian and augmented terminal function:

$$H_k \equiv \vartheta_k(\mathbf{x}_k, \mathbf{u}_k, k) + \boldsymbol{\lambda}_{k+1}^T \mathbf{f}_k(\mathbf{x}_k, \mathbf{u}_k, k) \tag{8.39a}$$

$$\Phi(\mathbf{x}_N, t_f) \equiv \phi(\mathbf{x}_N, t_f) + \boldsymbol{\alpha}^T \boldsymbol{\psi}(\mathbf{x}_N, t_f) \tag{8.39b}$$

Changing indices of summation on the last term in eqn. (8.38) yields[3, 5]

$$J = \Phi(\mathbf{x}_N, t_f) - \boldsymbol{\lambda}_N^T \mathbf{x}_N + \sum_{k=0}^{N-1} \left[H_k - \boldsymbol{\lambda}_k^T \mathbf{x}_k\right] + \boldsymbol{\lambda}_0^T \mathbf{x}_0 \tag{8.40}$$

Similar to the steps leading to eqn. (8.27), taking the first variation of eqn. (8.40) leads to the following conditions:

$$\mathbf{x}_{k+1} = \frac{\partial H_k}{\partial \boldsymbol{\lambda}_{k+1}} \equiv \mathbf{f}_k(\mathbf{x}_k, \mathbf{u}_k, k) \tag{8.41a}$$

$$\boldsymbol{\lambda}_k = \frac{\partial H_k}{\partial \mathbf{x}_k} \equiv \frac{\partial \vartheta_k(\mathbf{x}_k, \mathbf{u}_k, k)}{\partial \mathbf{x}_k} + \left[\frac{\partial \mathbf{f}_k(\mathbf{x}_k, \mathbf{u}_k, k)}{\partial \mathbf{x}_k}\right]^T \boldsymbol{\lambda}_{k+1} \tag{8.41b}$$

$$\frac{\partial H_k}{\partial \mathbf{u}_k} = \mathbf{0} \tag{8.41c}$$

$$\left[\frac{\partial \Phi(\mathbf{x}_k, t_f)}{\partial \Delta t} + \sum_{k=0}^{N-1} \frac{\partial H_k}{\partial \Delta t}\right] \delta \Delta t = 0 \tag{8.41d}$$

$$\left[\frac{\partial \Phi(\mathbf{x}_k, t_f)}{\partial \mathbf{x}_k} - \boldsymbol{\lambda}_k\right]^T \Bigg|_N \delta \mathbf{x}_N = 0 \tag{8.41e}$$

and, of course, the boundary conditions of $\mathbf{x}(t_0) = \mathbf{x}_0$ and eqn. (8.37b). If none of the $\mathbf{x}_N$ are directly specified and the final time is free, conditions of eqns. (8.41d) and (8.41e) provide the transversality conditions

$$\frac{\partial \Phi(\mathbf{x}_k, t_f)}{\partial \Delta t} + \sum_{k=0}^{N-1} \frac{\partial H_k}{\partial \Delta t} = 0 \tag{8.42a}$$

$$\boldsymbol{\lambda}_N = \left\{\frac{\partial \phi(\mathbf{x}_k, t_f)}{\partial \mathbf{x}_k} + \left[\frac{\partial \psi(\mathbf{x}_k, t_f)}{\partial \mathbf{x}_k}\right]^T \alpha\right\}\Bigg|_N \tag{8.42b}$$

As with the continuous-time formulation, eqn. (8.42a) is the stopping condition used to implicitly determine the optimal final time through the determination of the optimal time step Δt.

8.5 Linear Regulator Problems

The formulations of the foregoing developments naturally lead to *open-loop* optimal controls that are designed to calculate an optimal trajectory from a prescribed initial state to a prescribed final state. Such controls can be pre-computed, under the assumption of perfectly known initial conditions. However, upon application of open-loop controls to a real system, even small modelling errors and initial state errors result in usually unacceptable divergence of the actual system's behavior from the optimal trajectory. In many cases *perturbation feedback controls* need to be superimposed (à la "guidance" in rocket flight path control) to continually correct for model errors and other disturbances.

In some cases, we will see that it is possible to formulate optimal controls so that they can be calculated directly in a *terminal controller* feedback form:

$$\mathbf{u}(t) = \mathbf{f}[\mathbf{x}(t) - \mathbf{x}(t_f), t_f - t] \tag{8.43}$$

in which the optimal control is a function of instantaneous displacement from the desired final state and the "time-to-go" $\tau = t_f - t$. Such controls are of enormous practical impact, since we are, in essence, continuously re-initializing the control calculations with current best estimates of $\mathbf{x}(t)$, from a Kalman filter for example, which can be updated continuously based upon measurements (and thereby counteract the accumulation of ever-present errors due to an erroneous model and other disturbances). In this section we develop one such case for linear time-invariant models belonging to the class of linear regulator problems.

8.5.1 Continuous-Time Formulation

In this section the continuous-time linear quadratic regulator (LQR) problem is solved using Bellman's *Principle of Optimality*[7] and directly from the Hamiltonian formulation of §8.2. If we initiate at an arbitrary start point $[\mathbf{x}(t), t]$, the cost-to-go for an arbitrary control $\mathbf{u}(t)$ is given by

$$J = \phi(\mathbf{x}(t_f), t_f) + \int_t^{t_f} \vartheta(\mathbf{x}(\tau), \mathbf{u}(\tau), \tau)\, d\tau \tag{8.44}$$

Note that unlike eqn. (8.22), the integration is over the interval t to t_f. We are concerned only with trajectories that satisfy the differential equation

$$\dot{\mathbf{x}}(t) = \mathbf{f}(\mathbf{x}(t), \mathbf{u}(t), t) \tag{8.45}$$

and satisfy the terminal constraints

$$\boldsymbol{\psi}(\mathbf{x}(t_f), t_f) = \mathbf{0} \tag{8.46}$$

In §8.2 we developed the necessary conditions for minimizing eqn. (8.44) subject to $\mathbf{x}(t)$ being on a trajectory of eqn. (8.45) satisfying the prescribed boundary conditions. The principle of optimality is concerned with the instantaneous time-to-go $t_f - t$ rather than the fixed $t_f - t_0$ interval. The principle of optimality states that J must be a minimum over every subinterval of the time Δt, satisfying $t_f \geq t + \Delta t \geq t_0$, along an optimal trajectory. Having stated this principle, it seems obviously true that we do not concern ourselves with a formal proof. Clearly, if an optimal control had been employed everywhere *except* during the interval from t to $t + \Delta t$ the only way to minimize J of eqn. (8.44) is to choose $\mathbf{u}(t)$ to minimize J over the interval Δt in question.

The optimal control is implicitly defined by the requirement that it yields the minimum cost-to-go which we denote by

$$J^*(\mathbf{x}(t), t) = \min_{\mathbf{u}(t)} \left\{ \phi(\mathbf{x}(t_f), t_f) + \int_t^{t_f} \vartheta(\mathbf{x}(\tau), \mathbf{u}(\tau), \tau)\, d\tau \right\} \tag{8.47}$$

Notice that $J = J(\mathbf{x}(t), \mathbf{u}(t), t)$ in eqn. (8.44), along with a non-optimal trajectory, but $J^* = J^*(\mathbf{x}(t), t)$ upon carrying out the minimization of eqn. (8.44) over all admissible controls $\mathbf{u}(t)$.

In order to develop an important partial differential equation, we now investigate eqn. (8.47) locally. Suppose optimal control is used everywhere on the interval (t, t_f) *except* during the initial Δt where a non-optimal $\mathbf{u}(t)$ is employed. For Δt sufficiently small, the system will be displaced from $[\mathbf{x}(t), t]$ to a neighboring point $[\mathbf{x}(t)+\mathbf{f}(\mathbf{x}(t), \mathbf{u}(t), t)\,\Delta t, t+\Delta t]$. Now suppose from these perturbed initial conditions an optimal control is employed; it is apparent that the perturbed cost-to-go is

$$\tilde{J}^*(\mathbf{x}(t), t) = J^*[\mathbf{x}(t)+\mathbf{f}(\mathbf{x}(t), \mathbf{u}(t), t)\,\Delta t, t+\Delta t]+\vartheta(\mathbf{x}(t), \mathbf{u}(t), t)\,\Delta t \tag{8.48}$$

Since $\mathbf{u}(t)$ over the interval Δt is generally non-optimal it is clear that

$$\tilde{J}^*(\mathbf{x}(t), t) \geq J^*(\mathbf{x}(t), t) \tag{8.49}$$

The equality holds only if we choose $\mathbf{u}(t)$ to minimize eqn. (8.48). Thus

$$J^*(\mathbf{x}(t), t) = \min_{\mathbf{u}(t)}\left\{J^*[\mathbf{x}(t)+\mathbf{f}(\mathbf{x}(t), \mathbf{u}(t), t)\,\Delta t, t+\Delta t]+\vartheta(\mathbf{x}(t), \mathbf{u}(t), t)\,\Delta t\right\} \tag{8.50}$$

Upon expanding in Taylor's series and taking the limit as $\Delta t \rightarrow 0$,[1] eqn. (8.50) leads directly to the partial differential equation

$$\frac{\partial J^*(\mathbf{x}(t), t)}{\partial t} + \min_{\mathbf{u}(t)}\left\{\vartheta(\mathbf{x}(t), \mathbf{u}(t), t) + \frac{\partial J^*(\mathbf{x}(t), \mathbf{u}(t), t)}{\partial \mathbf{x}^T(t)}\mathbf{f}(\mathbf{x}(t), \mathbf{u}(t), t)\right\} = 0 \tag{8.51}$$

Comparison of eqn. (8.51) with eqn. (8.25) reveals that eqn. (8.51) can be written as the *Hamilton-Jacobi-Bellman* (HJB) equation:

$$\frac{\partial J^*(\mathbf{x}(t), t)}{\partial t} + \min_{\mathbf{u}(t)}\left\{H\left(\mathbf{x}(t), \frac{\partial J^*(\mathbf{x}(t), \mathbf{u}(t), t)}{\partial \mathbf{x}(t)}, \mathbf{u}(t), t\right)\right\} = 0 \tag{8.52}$$

where the costate is defined by

$$\boldsymbol{\lambda}(t) = \frac{\partial J^*(\mathbf{x}(t), \mathbf{u}(t), t)}{\partial \mathbf{x}(t)} \tag{8.53}$$

The significance of finding a globally valid analytical solution of the HJB equation for $J^* = J^*(\mathbf{x}(t), t)$ is that the solution of the Lagrange multiplier $\boldsymbol{\lambda}(t)$ is reduced to taking the gradient of J^*. This immediately allows determination of the corresponding optimal control from Pontryagin's Principle, *in feedback form*.

Unfortunately obtaining such global analytical solutions of the HJB equation can only be accomplished for special cases. The most important special case for which the HJB equation is solvable is the *linear quadratic regulator* for which we seek to minimize

$$J = \frac{1}{2}\mathbf{x}^T(t_f)S_f\mathbf{x}(t_f) + \frac{1}{2}\int_{t_0}^{t_f} \mathbf{x}^T(t)\,\mathcal{Q}(t)\,\mathbf{x}(t) + \mathbf{u}^T(t)\,\mathcal{R}(t)\,\mathbf{u}(t)\,dt \tag{8.54}$$

where S_f, $\mathcal{Q}(t)$ and $\mathcal{R}(t)$ are symmetric, non-negative weight matrices, subject to the constraint

$$\boxed{\dot{\mathbf{x}}(t) = F(t)\,\mathbf{x}(t) + B(t)\,\mathbf{u}(t), \quad \mathbf{x}(t_0) = \mathbf{x}_0} \tag{8.55}$$

The HJB equation of eqn. (8.52) for this case becomes

$$\begin{aligned} \frac{\partial J^*}{\partial t} &+ \min_{\mathbf{u}(t)} \left\{ \frac{1}{2} \left[\mathbf{x}^T(t)\,\mathcal{Q}(t)\,\mathbf{x}(t) + \mathbf{u}^T(t)\,\mathcal{R}(t)\,\mathbf{u}(t) \right] \right. \\ &\left. + \frac{\partial J^*}{\partial \mathbf{x}^T(t)} \left[F(t)\,\mathbf{x}(t) + B(t)\,\mathbf{u}(t) \right] \right\} = 0 \end{aligned} \tag{8.56}$$

Carrying out the minimization over $\mathbf{u}(t)$ of eqn. (8.56) yields

$$\mathbf{u}(t) = -\mathcal{R}^{-1}(t)\,B^T(t)\,\frac{\partial J^*}{\partial \mathbf{x}(t)} \tag{8.57}$$

Thus the HJB equation of eqn. (8.56) becomes

$$\begin{aligned} &\frac{\partial J^*}{\partial t} + \frac{1}{2}\frac{\partial J^*}{\partial \mathbf{x}^T(t)} F(t)\,\mathbf{x}(t) + \frac{1}{2}\mathbf{x}^T(t)\,F^T(t)\,\frac{\partial J^*}{\partial \mathbf{x}(t)} \\ &+ \frac{1}{2}\mathbf{x}^T(t)\,\mathcal{Q}(t)\,\mathbf{x}(t) - \frac{1}{2}\frac{\partial J^*}{\partial \mathbf{x}^T(t)} B(t)\,\mathcal{R}^{-1}(t)\,B^T(t)\,\frac{\partial J^*}{\partial \mathbf{x}(t)} = 0 \end{aligned} \tag{8.58}$$

It can be verified by direct substitution (which is left as an exercise for the reader) that the general solution of the HJB equation of eqn. (8.58) is the quadratic form

$$J^*(\mathbf{x}(t), t) = \frac{1}{2}\mathbf{x}^T(t)\,S(t)\,\mathbf{x}(t) \tag{8.59a}$$

$$\frac{\partial J^*}{\partial \mathbf{x}(t)} = S(t)\,\mathbf{x}(t) \tag{8.59b}$$

$$\frac{\partial J^*}{\partial t} = \frac{1}{2}\mathbf{x}^T(t)\,\dot{S}(t)\,\mathbf{x}(t) \tag{8.59c}$$

where $S(t)$ is a positive definite matrix satisfying the *matrix Riccati equation*

$$\boxed{\dot{S}(t) = -S(t)\,F(t) - F^T(t)\,S(t) + S(t)\,B(t)\,\mathcal{R}^{-1}(t)\,B^T(t)\,S(t) - \mathcal{Q}(t)} \tag{8.60}$$

with the terminal boundary condition

$$S(t_f) = S_f \tag{8.61}$$

Since we gave eqns. (8.53) and (8.57), the optimal control is thus obtained globally in the *time-varying linear feedback* form

$$\boxed{\mathbf{u}(t) = -L(t)\,\mathbf{x}(t)} \tag{8.62}$$

where the *optimal gain matrix* is

$$\boxed{L(t) = \mathcal{R}^{-1}(t)\,B^T(t)\,S(t)} \tag{8.63}$$

Table 8.1: Continuous-Time Linear Quadratic Regulator

Model	$\dot{\mathbf{x}}(t) = F(t)\,\mathbf{x}(t) + B(t)\,\mathbf{u}(t), \quad \mathbf{x}(t_0) = \mathbf{x}_0$
Gain	$L(t) = \mathcal{R}^{-1}(t)\,B^T(t)\,S(t)$
Riccati Equation	$\dot{S}(t) = -S(t)\,F(t) - F^T(t)\,S(t)$ $+S(t)\,B(t)\,\mathcal{R}^{-1}(t)\,B^T(t)\,S(t) - \mathcal{Q}(t), \quad S(t_f) = S_f$
Control Input	$\mathbf{u}(t) = -L(t)\,\mathbf{x}(t)$

Note the similarity between the formulation presented here and the continuous-time Kalman filter in Table 5.4, which leads to the duality results of §6.4.1. A summary of the continuous-time LQR is shown in Table 8.1. Once the gain matrices $\mathcal{R}(t)$ and $\mathcal{Q}(t)$ are chosen, the matrix Riccati solution in eqn. (8.60) is integrated backward in time with boundary conditions given by eqn. (8.61). Storing the entire matrix $S(t)$ over all time, the gain matrix in eqn. (8.63) is then calculated. Finally, eqn. (8.55) is integrated forward in time with the known initial state condition.

The stability of the LQR controller can be proved by using Lyapunov's direct method, which is discussed for continuous-time systems in §3.6. The closed-loop dynamics are given by substituting eqn. (8.62) into eqn. (8.55), which leads to

$$\dot{\mathbf{x}}(t) = \left[F(t) - B(t)\,\mathcal{R}^{-1}(t)\,B^T(t)\,S(t)\right]\mathbf{x}(t) \tag{8.64}$$

We consider the following candidate Lyapunov function:

$$V[\mathbf{x}(t)] = \mathbf{x}^T(t)\,S(t)\,\mathbf{x}(t) \tag{8.65}$$

Taking the time derivative of eqn. (8.65) yields

$$\dot{V}[\mathbf{x}(t)] = \dot{\mathbf{x}}^T(t)\,S(t)\,\mathbf{x}(t) + \mathbf{x}^T(t)\,S(t)\,\dot{\mathbf{x}}(t) + \mathbf{x}^T(t)\,\dot{S}(t)\,\mathbf{x}(t) \tag{8.66}$$

Substituting eqns. (8.60) and (8.64) into eqn. (8.66), and simplifying leads to

$$\dot{V}[\mathbf{x}(t)] = -\mathbf{x}^T(t)\left[S(t)\,B(t)\,\mathcal{R}^{-1}(t)\,B^T(t)\,S(t) + \mathcal{Q}(t)\right]\mathbf{x}(t) \tag{8.67}$$

Clearly if $\mathcal{R}(t)$ is positive definite and $\mathcal{Q}(t)$ is at least positive semi-definite then the Lyapunov condition is satisfied and LQR controller is stable.

In order to implement the control input given by eqn. (8.62), we first must integrate eqn. (8.60) *backward* in time and store matrix $S(t)$ at all times. For the case that all system and weight matrices are constant, and $t_f \to \infty$ in eqn. (8.54), it can be shown (for a controllable system[5, 8]) that $S(t)$ approaches the constant positive semi-definite solution of the algebraic Riccati equation (ARE) given by

$$S\,F + F^T S - S\,B\,\mathcal{R}^{-1}B^T S + \mathcal{Q} = 0 \tag{8.68}$$

Thus, eqns. (8.62) and (8.63) provide a constant gain feedback control that can be implemented in *real time*. The solution of the ARE in eqn. (8.68) can be found by employing the methods of §5.4.4. First, we define the following Hamiltonian matrix:

$$\mathcal{H} \equiv \begin{bmatrix} F & -B\mathcal{R}^{-1}B^T \\ -\mathcal{Q} & -F^T \end{bmatrix} \tag{8.69}$$

The eigenvalues of $\mathcal{H}$ can be arranged in a diagonal matrix given by

$$\mathcal{H}_\Lambda = \begin{bmatrix} \Lambda & 0 \\ 0 & -\Lambda \end{bmatrix} \tag{8.70}$$

where Λ is a diagonal matrix of the n eigenvalues in the right half-plane. Assuming that the eigenvalues are distinct, we can perform a linear state transformation, as shown in §3.1.4, such that

$$\mathcal{H}_\Lambda = W^{-1}\mathcal{H}W \tag{8.71}$$

where W is the matrix of eigenvectors, which can be represented in block form as

$$W = \begin{bmatrix} W_{11} & W_{12} \\ W_{21} & W_{22} \end{bmatrix} \tag{8.72}$$

Going backward in time the stable eigenvalues dominate, which leads to the following solution for S at steady-state:

$$\boxed{S = W_{22}W_{12}^{-1}} \tag{8.73}$$

It is important to note that all states must be observed in order to implement the LQR controller in real time. Unfortunately, this is rarely the case in practice. However, an estimator, such as the Kalman filter, is often employed to provide state estimates for the unmeasured states, which will be discussed in §8.6.

The Riccati solution for the LQR problem can be derived another way. The Hamiltonian of eqn. (8.25) for the minimization problem shown by eqns. (8.54) and (8.55) is given by

$$H = \frac{1}{2}\left[\mathbf{x}^T(t)\,\mathcal{Q}(t)\,\mathbf{x}(t) + \mathbf{u}^T(t)\,\mathcal{R}(t)\,\mathbf{u}(t)\right] + \boldsymbol{\lambda}^T(t)\,[F(t)\,\mathbf{x}(t) + B(t)\,\mathbf{u}(t)] \tag{8.74}$$

From the necessary conditions of eqn. (8.27) the following equations must be satisfied:

$$\dot{\mathbf{x}}(t) = F(t)\,\mathbf{x}(t) + B(t)\,\mathbf{u}(t), \quad \mathbf{x}(t_0) = \mathbf{x}_0 \tag{8.75a}$$

$$\dot{\boldsymbol{\lambda}}(t) = -F^T(t)\,\boldsymbol{\lambda}(t) - \mathcal{Q}(t)\,\mathbf{x}(t) \tag{8.75b}$$

$$\mathbf{u}(t) = -\mathcal{R}^{-1}(t)\,B^T(t)\,\boldsymbol{\lambda}(t) \tag{8.75c}$$

$$\boldsymbol{\lambda}(t_f) = S_f\,\mathbf{x}(t_f) \tag{8.75d}$$

where eqn. (8.27e) has been used to derive eqn. (8.75d). Suppose we assume that the solution for the costate $\boldsymbol{\lambda}(t)$ follows the form of eqn. (8.75d) for all time, which seems to be a reasonable assumption due to the linearity of the system. Hence, we assume

$$\boldsymbol{\lambda}(t) = S(t)\,\mathbf{x}(t) \tag{8.76}$$

Taking the time derivative of eqn. (8.76) gives

$$\dot{\boldsymbol{\lambda}}(t) = \dot{S}(t)\,\mathbf{x}(t) + S(t)\,\dot{\mathbf{x}}(t) = -F^T(t)\,\boldsymbol{\lambda}(t) - \mathcal{Q}(t)\,\mathbf{x}(t) \tag{8.77}$$

where eqn. (8.75b) has been used in eqn. (8.77). Substituting eqn. (8.75c) into eqn. (8.75a) gives

$$\dot{\mathbf{x}}(t) = F(t)\,\mathbf{x}(t) - B(t)\,\mathcal{R}^{-1}(t)\,B^T(t)\,\boldsymbol{\lambda}(t) \tag{8.78}$$

Now, substituting eqn. (8.76) into eqn. (8.78) gives

$$\dot{\mathbf{x}}(t) = F(t)\,\mathbf{x}(t) - B(t)\,\mathcal{R}^{-1}(t)\,B^T(t)\,S(t)\,\mathbf{x}(t) \tag{8.79}$$

Finally, substituting eqns. (8.76) and (8.79) into eqn. (8.77) and collecting terms yields

$$\left[\dot{S}(t) + S(t)\,F(t) + F^T(t)\,S(t) - S(t)\,B(t)\,\mathcal{R}^{-1}(t)\,B^T(t)\,S(t) + \mathcal{Q}(t)\right]\mathbf{x}(t) = \mathbf{0} \tag{8.80}$$

Since eqn. (8.80) must hold for all nonzero $\mathbf{x}(t)$, then the term within the brackets pre-multiplying $\mathbf{x}(t)$ must be zero, which leads directly to eqn. (8.60). Also, substituting eqn. (8.76) into eqn. (8.75c) leads directly to eqn. (8.62).

Example 8.2: In this example we wish to apply the LQR approach to asymptotically control the following linear time-invariant system:

$$\dot{\mathbf{x}}(t) = \begin{bmatrix} 0 & 1 \\ -2 & 2 \end{bmatrix}\mathbf{x}(t) + \begin{bmatrix} 0 \\ 1 \end{bmatrix}\mathbf{u}(t)$$

Note that this system is unstable, with eigenvalues given by $\lambda_{12} = 1 \pm j$. The weighting matrices for the control design are chosen to be $\mathcal{R} = 0.1$ and $\mathcal{Q} = I_{2\times 2}$. Since this system is time-invariant we choose to employ the steady-state feedback gain approach, which allows for real-time implementation. Solving the steady-state ARE in eqn. (8.68) and the steady-state gain in eqn. (8.63) gives

$$S = \begin{bmatrix} 1.9645 & 0.1742 \\ 0.1742 & 0.6181 \end{bmatrix}, \quad L = \begin{bmatrix} 1.7417 & 6.1813 \end{bmatrix}$$

The eigenvalues of the closed-loop system, $F - B\,L$, are given by $\lambda_1 = -1.2974$ and $\lambda_2 = -2.8839$, which yield a stable closed-loop response as expected. A plot of the closed-loop response is shown in Figure 8.4. Clearly, the states approach zero. The weighting matrices dictate the characteristics of the closed-loop response. In general

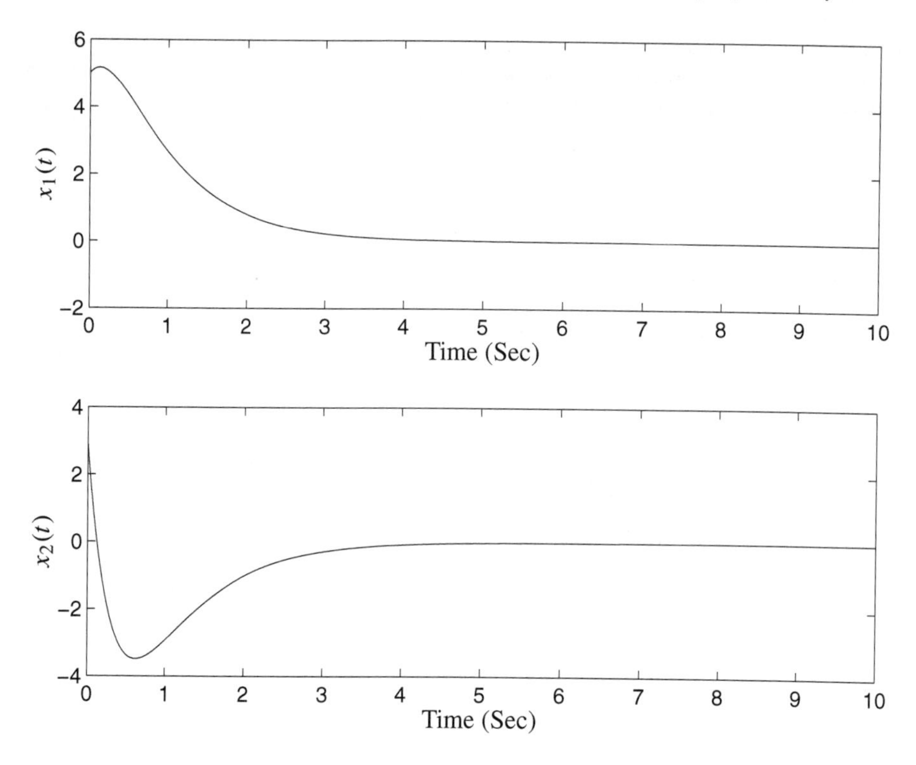

Figure 8.4: Linear Quadratic Regulator Control Example

as $\mathcal{Q}$ is increased, the faster the response time of the closed-loop system, but this comes at the price of a larger control gain. This also occurs as $\mathcal{R}$ is decreased. In a scalar sense it is the ratio of $\mathcal{Q}$ and $\mathcal{R}$ that is important in the final LQR design.

8.5.2 Discrete-Time Formulation

In this section the discrete-time linear quadratic regulator problem is solved using the Hamiltonian formulation of §8.4. The HJB equation can be extended to discrete-time systems, but this is beyond the scope of the present text. Here, we will focus our attentions only on the final discrete-time LQR solution form obtained through a Riccati transformation. Consider the minimization of the following loss function:

$$J = \frac{1}{2}\mathbf{x}_N^T S_f \mathbf{x}_N + \sum_{k=0}^{N-1} \mathbf{x}_k^T \mathcal{Q}_k \mathbf{x}_k + \mathbf{u}_k^T \mathcal{R}_k \mathbf{u}_k \tag{8.81}$$

subject to the constraint

$$\boxed{\mathbf{x}_{k+1} = \Phi_k \mathbf{x}_k + \Gamma_k \mathbf{u}_k, \quad \mathbf{x}(t_0) = \mathbf{x}_0} \tag{8.82}$$

The Hamiltonian of eqn. (8.39a) for the minimization problem shown by eqns. (8.81) and (8.82) is given by

$$H_k = \frac{1}{2}\left[\mathbf{x}_k^T \mathcal{Q}_k \mathbf{x}_k + \mathbf{u}_k^T \mathcal{R}_k \mathbf{u}_k\right] + \boldsymbol{\lambda}_{k+1}^T \left[\Phi_k \mathbf{x}_k + \Gamma_k \mathbf{u}_k\right] \tag{8.83}$$

From the necessary conditions of eqn. (8.41) the following equations must be satisfied:

$$\mathbf{x}_{k+1} = \Phi_k \mathbf{x}_k + \Gamma_k \mathbf{u}_k, \quad \mathbf{x}(t_0) = \mathbf{x}_0 \tag{8.84a}$$

$$\boldsymbol{\lambda}_k = \Phi_k^T \boldsymbol{\lambda}_{k+1} + \mathcal{Q}_k \mathbf{x}_k \tag{8.84b}$$

$$\mathbf{u}_k = -\mathcal{R}_k^{-1} \Gamma_k^T \boldsymbol{\lambda}_{k+1} \tag{8.84c}$$

$$\boldsymbol{\lambda}_N = S_f \mathbf{x}_N \tag{8.84d}$$

where eqn. (8.41e) has been used to derive eqn. (8.84d). Suppose we assume that the solution for the costate $\boldsymbol{\lambda}_k$ follows the form of eqn. (8.84d) for all time, which seems to be a reasonable assumption due to the linearity of the system. Hence, we assume

$$\boldsymbol{\lambda}_k = S_k \mathbf{x}_k \tag{8.85}$$

Taking one time-step ahead of eqn. (8.85) gives

$$\boldsymbol{\lambda}_{k+1} = S_{k+1} \mathbf{x}_{k+1} \tag{8.86}$$

Substituting eqns. (8.85) and (8.86) into eqn. (8.84b), and collecting terms yields

$$\Phi_k^T S_{k+1} \mathbf{x}_{k+1} + (\mathcal{Q}_k - S_k)\mathbf{x}_k = \mathbf{0} \tag{8.87}$$

Substituting eqn. (8.84c) into eqn. (8.84a) gives

$$\mathbf{x}_{k+1} = \Phi_k \mathbf{x}_k - \Gamma_k \mathcal{R}_k^{-1} \Gamma_k^T \boldsymbol{\lambda}_{k+1} \tag{8.88}$$

Now, substituting eqn. (8.86) into eqn. (8.88) gives

$$\mathbf{x}_{k+1} = \Phi_k \mathbf{x}_k - \Gamma_k \mathcal{R}_k^{-1} \Gamma_k^T S_{k+1} \mathbf{x}_{k+1} \tag{8.89}$$

Solving eqn. (8.89) for $\mathbf{x}_{k+1}$ gives

$$\mathbf{x}_{k+1} = \left[I + \Gamma_k \mathcal{R}_k^{-1} \Gamma_k^T S_{k+1}\right]^{-1} \Phi_k \mathbf{x}_k \tag{8.90}$$

Substituting eqn. (8.90) into eqn. (8.87) and collecting terms yields

$$\left\{\Phi_k^T S_{k+1} \left[I + \Gamma_k \mathcal{R}_k^{-1} \Gamma_k^T S_{k+1}\right]^{-1} \Phi_k + \mathcal{Q}_k - S_k\right\} \mathbf{x}_k = \mathbf{0} \tag{8.91}$$

Since eqn. (8.91) must hold for all nonzero $\mathbf{x}_k$, then the term within the brackets pre-multiplying $\mathbf{x}_k$ must be zero, which leads directly to

$$S_k = \Phi_k^T S_{k+1}\left[I + \Gamma_k \mathcal{R}_k^{-1}\Gamma_k^T S_{k+1}\right]^{-1}\Phi_k + \mathcal{Q}_k \tag{8.92}$$

Since S_{k+1} is assumed to have an inverse, then eqn. (8.92) can be rewritten as

$$S_k = \Phi_k^T\left[S_{k+1}^{-1} + \Gamma_k \mathcal{R}_k^{-1}\Gamma_k^T\right]^{-1}\Phi_k + \mathcal{Q}_k \tag{8.93}$$

Using the matrix inversion lemma in eqn. (1.70) with $A = S_{k+1}^{-1}$, $B = \Gamma_k$, $C = \mathcal{R}_k^{-1}$, and $D = \Gamma_k^T$ gives

$$\boxed{S_k = \Phi_k^T S_{k+1}\Phi_k - \Phi_k^T S_{k+1}\Gamma_k\left[\Gamma_k^T S_{k+1}\Gamma_k + \mathcal{R}_k\right]^{-1}\Gamma_k^T S_{k+1}\Phi_k + \mathcal{Q}_k} \tag{8.94}$$

with terminal boundary condition

$$S_N = S_f \tag{8.95}$$

Equation (8.94) represents the discrete-time matrix Riccati equation, which is propagated backward in time. The discrete-time LQR gain for the time-varying linear feedback form is more complicated than the continuous-time case. We first substitute eqn. (8.86) into eqn. (8.84c) to yield

$$\mathcal{R}_k\mathbf{u}_k = -\Gamma_k^T S_{k+1}\mathbf{x}_{k+1} \tag{8.96}$$

Substituting eqn. (8.82) into eqn. (8.96) and solving the resulting equation for $\mathbf{u}_k$ gives

$$\boxed{\mathbf{u}_k = -L_k\mathbf{x}_k} \tag{8.97}$$

where the *optimal gain matrix* is

$$\boxed{L_k = \left[\Gamma_k^T S_{k+1}\Gamma_k + \mathcal{R}_k\right]^{-1}\Gamma_k^T S_{k+1}\Phi_k} \tag{8.98}$$

Note the similarity between the formulation presented here and the discrete-time Kalman filter in Table 5.1, which leads to the duality results of §6.4.1. A summary of the discrete-time LQR is shown in Table 8.2. Once the gain matrices $\mathcal{R}_k$ and $\mathcal{Q}_k$ are chosen, the matrix Riccati solution in eqn. (8.94) is executed backward in time with a boundary condition given by eqn. (8.95). Storing the entire matrix S_k over all time, the gain matrix in eqn. (8.98) is then calculated. Finally, eqn. (8.82) is executed forward in time with the known initial state condition.

The stability of the discrete-time LQR controller can be proved by using Lyapunov's direct method, which is discussed for discrete-time systems in §3.6. The closed-loop dynamics are given by substituting eqn. (8.97) into eqn. (8.82), which leads to

$$\mathbf{x}_{k+1} = [\Phi_k - \Gamma_k L_k]\mathbf{x}_k \tag{8.99}$$

Table 8.2: Discrete-Time Linear Quadratic Regulator

Model	$\mathbf{x}_{k+1} = \Phi_k \mathbf{x}_k + \Gamma_k \mathbf{u}_k, \quad \mathbf{x}(t_0) = \mathbf{x}_0$
Gain	$L_k = \left[\Gamma_k^T S_{k+1} \Gamma_k + \mathcal{R}_k\right]^{-1} \Gamma_k^T S_{k+1} \Phi_k$
Riccati Equation	$S_k = \Phi_k^T S_{k+1} \Phi_k + \mathcal{Q}_k$ $-\Phi_k^T S_{k+1} \Gamma_k \left[\Gamma_k^T S_{k+1} \Gamma_k + \mathcal{R}_k\right]^{-1} \Gamma_k^T S_{k+1} \Phi_k, \; S_N = S_f$
Control Input	$\mathbf{u}_k = -L_k \mathbf{x}_k$

We consider the following candidate Lyapunov function:

$$V(\mathbf{x}) = \mathbf{x}_k^T S_k \mathbf{x}_k \tag{8.100}$$

The increment of $V(\mathbf{x}_k)$ is given by

$$\Delta V(\mathbf{x}) = \mathbf{x}_{k+1}^T S_{k+1} \mathbf{x}_{k+1} - \mathbf{x}_k^T S_k \mathbf{x}_k \tag{8.101}$$

Using the definition of the gain in eqn. (8.98), the Riccati equation in eqn. (8.94) can be rewritten as

$$S_k = \Phi_k^T S_{k+1} \Phi_k - \Phi_k^T S_{k+1} \Gamma_k L_k + \mathcal{Q}_k \tag{8.102}$$

Equation (8.94) can be rewritten as (which is left as an exercise for the reader)

$$S_k = [\Phi_k - \Gamma_k L_k]^T S_{k+1} [\Phi_k - \Gamma_k L_k] + L_k^T \mathcal{R}_k L_k + \mathcal{Q}_k \tag{8.103}$$

Substituting eqns. (8.99) and (8.103) into eqn. (8.101), and simplifying yields

$$\Delta V(\mathbf{x}) = -\mathbf{x}_k^T \left[L_k^T \mathcal{R}_k L_k + \mathcal{Q}_k\right] \mathbf{x}_k \tag{8.104}$$

Clearly if $\mathcal{R}_k$ is positive definite and $\mathcal{Q}_k$ is at least positive semi-definite then the Lyapunov condition is satisfied and the discrete-time LQR controller is stable.

As with the continuous-time case a steady-state discrete-time LQR can be derived if all weighting and system matrices in the Riccati equation of eqn. (8.94) are constant. This leads to the following discrete-time algebraic Riccati equation:

$$S = \Phi^T S \Phi - \Phi^T S \Gamma \left[\Gamma^T S \Gamma + \mathcal{R}\right]^{-1} \Gamma^T S \Phi + \mathcal{Q} \tag{8.105}$$

In order to solve eqn. (8.105) using the method shown in §5.3.4, we must first derive the discrete-time Hamiltonian matrix. Assuming constant system matrices, then solving eqn. (8.84b) for $\boldsymbol{\lambda}_{k+1}$ gives

$$\boldsymbol{\lambda}_{k+1} = \Phi^{-T} \boldsymbol{\lambda}_k - \Phi^{-T} \mathcal{Q} \mathbf{x}_k \tag{8.106}$$

Substituting eqn. (8.106) into eqn. (8.88) gives

$$\mathbf{x}_{k+1} = \left[\Phi + \Gamma \mathcal{R}^{-1} \Gamma^T \Phi^{-T} \mathcal{Q}\right] \mathbf{x}_k - \Gamma \mathcal{R}^{-1} \Gamma^T \Phi^{-T} \boldsymbol{\lambda}_k \tag{8.107}$$

Combining eqns. (8.106) and (8.107) leads to

$$\begin{bmatrix} \mathbf{x}_{k+1} \\ \boldsymbol{\lambda}_{k+1} \end{bmatrix} = \mathcal{H} \begin{bmatrix} \mathbf{x}_k \\ \boldsymbol{\lambda}_k \end{bmatrix} \tag{8.108}$$

where the Hamiltonian matrix is defined by[9]

$$\mathcal{H} \equiv \begin{bmatrix} \Phi + \Gamma \mathcal{R}^{-1} \Gamma^T \Phi^{-T} \mathcal{Q} & -\Gamma \mathcal{R}^{-1} \Gamma^T \Phi^{-T} \\ -\Phi^{-T} \mathcal{Q} & \Phi^{-T} \end{bmatrix} \tag{8.109}$$

The eigenvalues of $\mathcal{H}$ can be arranged in a diagonal matrix given by

$$\mathcal{H}_\Lambda = \begin{bmatrix} \Lambda & 0 \\ 0 & \Lambda^{-1} \end{bmatrix} \tag{8.110}$$

where Λ is a diagonal matrix of the n eigenvalues outside of the unit circle. Assuming that the eigenvalues are distinct, we can perform a linear state transformation, as shown in §3.1.4, such that

$$\mathcal{H}_\Lambda = W^{-1} \mathcal{H} W \tag{8.111}$$

where W is the matrix of eigenvectors, which can be represented in block form as

$$W = \begin{bmatrix} W_{11} & W_{12} \\ W_{21} & W_{22} \end{bmatrix} \tag{8.112}$$

Going backward in time the stable eigenvalues dominate, which leads to the following solution for S at steady-state:

$$\boxed{S = W_{22} W_{12}^{-1}} \tag{8.113}$$

Note that the inverse of Φ must exist for a valid solution. This usually poses no problems though, since Φ does not usually have a zero eigenvalue in practice.

8.6 Linear Quadratic-Gaussian Controllers

The LQR feedback control laws of eqns. (8.62) and (8.97) clearly require full state knowledge, which is not always possible or even practical in real-world systems. It seems natural to use the Kalman filter to provide state estimates, which can be used in place of the "true" states in the LQR feedback control law. In actuality this seemingly ad hoc approach turns out to be the optimal approach, which leads to the so-called *linear quadratic-Gaussian* (LQG) controller.[10] In this section combining the LQR feedback control law with the standard estimator form of the Kalman filter is proven to be optimal using the *Separation Theorem*, which is also known as the *Certainty Equivalence Principle*.[11–13] This theorem states that the solution of overall optimal control problem with incomplete state knowledge is given by the solution of two separate sub-problems: 1) the estimation problem used to provide optimal state estimates, which is solved using the Kalman filter, and 2) the control problem using the optimal states estimates, which is derived from the standard LQR results. Another way to show this separation of the overall control design involves the eigenvalue separation property,[14] which states that the eigenvalues of the overall closed-loop system are given by the eigenvalues of the LQR system together with those of the state estimator system.

8.6.1 Continuous-Time Formulation

In the continuous-time LQG problem we assume that the state model is given by eqn. (5.117):

$$\dot{\mathbf{x}}(t) = F(t)\mathbf{x}(t) + B(t)\mathbf{u}(t) + G(t)\mathbf{w}(t) \tag{8.114a}$$

$$\tilde{\mathbf{y}}(t) = H(t)\mathbf{x}(t) + \mathbf{v}(t) \tag{8.114b}$$

where $\mathbf{w}(t)$ and $\mathbf{v}(t)$ are zero-mean Gaussian noise processes with covariances given by eqn. (5.118). Note that unlike eqn. (8.55), the state model in eqn. (8.114) is random. Therefore, we must take the expected value of the loss function in eqn. (8.54), which leads to the LQG loss function to be minimized:

$$J = E\left\{\int_{t_0}^{t_f} \mathbf{x}^T(t)\,\mathcal{Q}(t)\,\mathbf{x}(t) + \mathbf{u}^T(t)\,\mathcal{R}(t)\,\mathbf{u}(t)\,dt\right\} \tag{8.115}$$

Note that the terminal condition is omitted here for brevity since the results of the Separation Theorem extended easily for this case (also the factor of one half is not needed to prove the theorem). There are many ways to prove the Separation Theorem (e.g., see Refs. [2] and [13]), but we choose to use the approach presented in Ref. [14], which is fairly straightforward without requiring rigorous stochastic optimal control theory. Let us first concentrate on the expression $E\left\{\mathbf{x}^T(t)\,\mathcal{Q}(t)\,\mathbf{x}(t)\right\}$. Adding and subtracting the state estimate $\hat{\mathbf{x}}(t)$ to $\mathbf{x}(t)$ gives

$$E\left\{\mathbf{x}^T(t)\,\mathcal{Q}(t)\,\mathbf{x}(t)\right\} = E\left\{\left[\hat{\mathbf{x}}(t) - \tilde{\mathbf{x}}(t)\right]^T \mathcal{Q}(t)\left[\hat{\mathbf{x}}(t) - \tilde{\mathbf{x}}(t)\right]\right\} \tag{8.116}$$

where the estimation error is defined as $\tilde{\mathbf{x}}(t) \equiv \hat{\mathbf{x}}(t) - \mathbf{x}(t)$. Expanding eqn. (8.116) and using the trace property $\text{Tr}(A\,\mathbf{z}\mathbf{z}^T) = \mathbf{z}^T A\,\mathbf{z}$ (see Appendix A) leads to

$$\begin{aligned} E\left\{\mathbf{x}^T(t)\,\mathcal{Q}(t)\,\mathbf{x}(t)\right\} = {} & E\left\{\hat{\mathbf{x}}^T(t)\,\mathcal{Q}(t)\,\hat{\mathbf{x}}(t)\right\} - 2E\left\{\text{Tr}\left[\mathcal{Q}(t)\tilde{\mathbf{x}}(t)\hat{\mathbf{x}}^T(t)\right]\right\} \\ & + E\left\{\text{Tr}\left[\mathcal{Q}(t)\,\tilde{\mathbf{x}}(t)\,\tilde{\mathbf{x}}^T(t)\right]\right\} \end{aligned} \tag{8.117}$$

The orthogonality principle of the Kalman filter, which is shown for discrete-time systems in §5.3.6 and exercise 5.22, states that the estimation error is orthogonal to the state estimate. This is obviously also true for continuous-time systems, which gives $E\left\{\tilde{\mathbf{x}}(t)\,\hat{\mathbf{x}}^T(t)\right\}=0$. Therefore, eqn. (8.117) reduces down to

$$E\left\{\mathbf{x}^T(t)\,\mathcal{Q}(t)\,\mathbf{x}(t)\right\}=E\left\{\hat{\mathbf{x}}^T(t)\,\mathcal{Q}(t)\,\hat{\mathbf{x}}(t)\right\}+E\left\{\mathrm{Tr}\left[\mathcal{Q}(t)\,\tilde{\mathbf{x}}(t)\,\tilde{\mathbf{x}}^T(t)\right]\right\} \tag{8.118}$$

Using the definition of the covariance $P(t)$ in eqn. (5.125), eqn. (8.118) can be rewritten as

$$E\left\{\mathbf{x}^T(t)\,\mathcal{Q}(t)\,\mathbf{x}(t)\right\}=E\left\{\hat{\mathbf{x}}^T(t)\,\mathcal{Q}(t)\,\hat{\mathbf{x}}(t)\right\}+\mathrm{Tr}\left[\mathcal{Q}(t)\,P(t)\right] \tag{8.119}$$

Substituting eqn. (8.119) into eqn. (8.115) leads to the following equivalent minimization problem:

$$J=E\left\{\int_{t_0}^{t_f}\hat{\mathbf{x}}^T(t)\,\mathcal{Q}(t)\,\hat{\mathbf{x}}(t)+\mathbf{u}^T(t)\,\mathcal{R}(t)\,\mathbf{u}(t)\,dt\right\}+\int_{t_0}^{t_f}\mathrm{Tr}\left[\mathcal{Q}(t)\,P(t)\right]\,dt \tag{8.120}$$

subject to the new dynamic constraint

$$\dot{\hat{\mathbf{x}}}(t)=F(t)\,\hat{\mathbf{x}}(t)+B(t)\,\mathbf{u}(t)+K(t)[\tilde{\mathbf{y}}(t)-H(t)\,\hat{\mathbf{x}}(t)] \tag{8.121}$$

which is the linear continuous estimator for $\mathbf{x}(t)$.

The goal of our overall process is to convert the constrained minimization problem given by eqns. (8.120) and (8.121) into an unconstrained problem (thus avoiding the use of Lagrange multipliers). For the subsequent developments we will need an expression for $W(t)\equiv E\left\{\hat{\mathbf{x}}(t)\,\hat{\mathbf{x}}^T(t)\right\}$. Using the methods of §5.4.1 and the definition of the innovations process in §6.4.2.2, this expression can be shown to follow (which is left as an exercise for the reader)

$$\begin{aligned}\dot{W}(t)&=F(t)\,W(t)+W(t)\,F^T(t)+K(t)\,R(t)\,K^T(t)\\&\quad+E\left\{B(t)\,\mathbf{u}(t)\,\hat{\mathbf{x}}^T(t)+\hat{\mathbf{x}}(t)\,\mathbf{u}^T(t)\,B^T(t)\right\}\end{aligned} \tag{8.122}$$

with $W(t_0)=E\left\{\hat{\mathbf{x}}(t_0)\,\hat{\mathbf{x}}^T(t_0)\right\}$. Also, we need an expression for

$$\frac{d}{dt}[S(t)\,W(t)]=\dot{S}(t)\,W(t)+S(t)\,\dot{W}(t) \tag{8.123}$$

Substituting eqns. (8.60) and (8.122) into eqn. (8.123), taking the trace of the resulting equation, and using the definition of $L(t)$ in eqn. (8.63) leads to

$$\begin{aligned}\mathrm{Tr}\left\{\frac{d}{dt}[S(t)\,W(t)]\right\}&=\mathrm{Tr}\Big[L^T(t)\,\mathcal{R}(t)\,L(t)\,W(t)-\mathcal{Q}(t)\,W(t)\\&\quad+S(t)\,K(t)\,R(t)\,K^T(t)+2E\left\{\hat{\mathbf{x}}^T(t)\,L^T(t)\,\mathcal{R}(t)\,\mathbf{u}(t)\right\}\Big]\end{aligned} \tag{8.124}$$

Using $\text{Tr}[\mathcal{Q}(t)\,W(t)] = \hat{\mathbf{x}}^T(t)\,\mathcal{Q}(t)\,\hat{\mathbf{x}}(t)$ in eqn. (8.124), and solving for the quantity $\hat{\mathbf{x}}^T(t)\,\mathcal{Q}(t)\,\hat{\mathbf{x}}(t)$ yields

$$\begin{aligned}\hat{\mathbf{x}}^T(t)\,\mathcal{Q}(t)\,\hat{\mathbf{x}}(t) &= 2E\left\{\hat{\mathbf{x}}^T(t)\,L^T(t)\,\mathcal{R}(t)\,\mathbf{u}(t)\right\} \\ &+ \text{Tr}\left\{-\frac{d}{dt}[S(t)\,W(t)] + S(t)\,K(t)\,R(t)\,K^T(t) + L^T(t)\,\mathcal{R}(t)\,L(t)\,W(t)\right\}\end{aligned} \tag{8.125}$$

We now find an expression for $\mathbf{u}^T(t)\,\mathcal{R}(t)\,\mathbf{u}(t)$. This is accomplished by first expanding the following expression:

$$\begin{aligned}E\left\{\left[\mathbf{u}(t) + L(t)\,\hat{\mathbf{x}}(t)\right]^T \mathcal{R}(t)\left[\mathbf{u}(t) + L(t)\,\hat{\mathbf{x}}(t)\right]\right\} &= E\left\{\mathbf{u}^T(t)\,\mathcal{R}(t)\,\mathbf{u}(t)\right\} \\ + 2E\left\{\hat{\mathbf{x}}^T(t)\,L^T(t)\,\mathcal{R}(t)\,\mathbf{u}(t)\right\} &+ E\left\{\hat{\mathbf{x}}^T(t)\,L^T(t)\,\mathcal{R}(t)\,L(t)\,\hat{\mathbf{x}}(t)\right\}\end{aligned} \tag{8.126}$$

Using the trace property $\text{Tr}(A\,\mathbf{z}\mathbf{z}^T) = \mathbf{z}^T A\,\mathbf{z}$, and solving eqn. (8.126) for the desired expression, $\mathbf{u}^T(t)\,\mathcal{R}(t)\,\mathbf{u}(t)$, yields

$$\begin{aligned}\mathbf{u}^T(t)\,\mathcal{R}(t)\,\mathbf{u}(t) &= E\left\{\left[\mathbf{u}(t) + L(t)\,\hat{\mathbf{x}}(t)\right]^T \mathcal{R}(t)\left[\mathbf{u}(t) + L(t)\,\hat{\mathbf{x}}(t)\right]\right\} \\ &- 2E\left\{\hat{\mathbf{x}}^T(t)\,L^T(t)\,\mathcal{R}(t)\,\mathbf{u}(t)\right\} - \text{Tr}\left[L^T(t)\,\mathcal{R}(t)\,L(t)\,W(t)\right]\end{aligned} \tag{8.127}$$

Substituting eqns. (8.125) and (8.127) into eqn. (8.120) leads to

$$\begin{aligned}J = \int_{t_0}^{t_f} E\Big\{&\left[\mathbf{u}(t) + L(t)\,\hat{\mathbf{x}}(t)\right]^T \mathcal{R}(t)\left[\mathbf{u}(t) + L(t)\,\hat{\mathbf{x}}(t)\right] + \text{Tr}[\mathcal{Q}(t)\,P(t)] \\ &+ \text{Tr}\left[S(t)\,K(t)\,R(t)\,K^T(t)\right]\Big\}\,dt - \{\text{Tr}[S(t)\,W(t)]\}\big|_{t_0}^{t_f}\end{aligned} \tag{8.128}$$

Minimizing eqn. (8.128) with respect to $\mathbf{u}(t)$ gives

$$\mathbf{u}(t) = -L(t)\,\hat{\mathbf{x}}(t) \tag{8.129}$$

Equation (8.129) is identical to eqn. (8.62) with the exception that the true state $\mathbf{x}(t)$ is replaced with the estimated state $\hat{\mathbf{x}}(t)$! This clearly shows that the optimal solution with partial state information is given by using the linear estimator of the form in eqn. (8.114a). Any estimator with this form is valid; however, the Kalman filter is most widely used in practical applications. This attests to the separation of the estimator design with the control design.

A block diagram of the LQG controller is shown in Figure 8.5. The control input has been generalized in this diagram to be given by

$$\mathbf{u}(t) = -L(t)\,\hat{\mathbf{x}}(t) + \mathbf{u}_{\text{ext}}(t) \tag{8.130}$$

where $\mathbf{u}_{\text{ext}}(t)$ denotes an external input, which may include a term $-L(t)\,\mathbf{x}_d(t)$, where $\mathbf{x}_d(t)$ is some desired state trajectory; or a feedforward term $L_r(t)\,\mathbf{r}(t)$, where

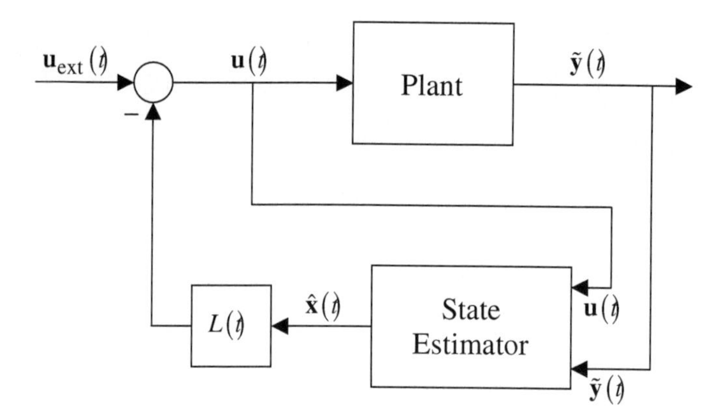

Figure 8.5: The Linear Quadratic-Gaussian Controller

$\mathbf{r}(t)$ is a reference trajectory.[14] Reference [14] also shows other possible arrangements, e.g., where the external input is combined with the output prior to entering the state estimator.

Another, much easier way to show the separation of the estimator and controller involves the investigation of the closed-loop LQG system. We only consider the time-invariant case with constant system matrices to illustrate this concept. Substituting eqn. (8.129) into eqn. (8.114a) gives

$$\dot{\mathbf{x}}(t) = F\,\mathbf{x}(t) - B\,L\,\hat{\mathbf{x}}(t) + G\,\mathbf{w}(t) \tag{8.131}$$

Substituting eqns. (8.114b) and (8.129) into eqn. (8.121) yields

$$\dot{\hat{\mathbf{x}}}(t) = [F - B\,L - K\,H]\hat{\mathbf{x}}(t) + K\,H\mathbf{x}(t) + K\,\mathbf{v}(t) \tag{8.132}$$

Combining eqns. (8.131) and (8.132) leads to

$$\begin{bmatrix} \dot{\mathbf{x}}(t) \\ \dot{\hat{\mathbf{x}}}(t) \end{bmatrix} = \begin{bmatrix} F & -B\,L \\ K\,H & F - B\,L - K\,H \end{bmatrix} \begin{bmatrix} \mathbf{x}(t) \\ \hat{\mathbf{x}}(t) \end{bmatrix} + \begin{bmatrix} G & 0 \\ 0 & K \end{bmatrix} \begin{bmatrix} \mathbf{w}(t) \\ \mathbf{v}(t) \end{bmatrix} \tag{8.133}$$

Unfortunately the stability of this system is not obvious at first glance. To overcome this difficulty we use the definition of the error state from §5.4.1: $\tilde{\mathbf{x}}(t) \equiv \hat{\mathbf{x}}(t) - \mathbf{x}(t)$. Taking the time derivative of this quantity and substituting eqns. (8.131) and (8.132) into the resulting expression yields

$$\dot{\tilde{\mathbf{x}}}(t) = [F - K\,H]\tilde{\mathbf{x}}(t) + K\,\mathbf{v}(t) - G\,\mathbf{w}(t) \tag{8.134}$$

Combining eqns. (8.131) and (8.134) leads to

$$\begin{bmatrix} \dot{\mathbf{x}}(t) \\ \dot{\tilde{\mathbf{x}}}(t) \end{bmatrix} = \begin{bmatrix} F - B\,L & -B\,L \\ 0 & F - K\,H \end{bmatrix} \begin{bmatrix} \mathbf{x}(t) \\ \tilde{\mathbf{x}}(t) \end{bmatrix} + \begin{bmatrix} G & 0 \\ -G & K \end{bmatrix} \begin{bmatrix} \mathbf{w}(t) \\ \mathbf{v}(t) \end{bmatrix} \tag{8.135}$$

The eigenvalues of the state matrix in eqns. (8.133) and (8.135) can be shown to be equivalent (which is left as an exercise for the reader). The form in eqn. (8.135) is much easier to visualize than the one of eqn. (8.133), since the eigenvalues of the block diagonal structure are given by (see Appendix A)

$$\det(\lambda I - F + B\,L)\,\det(\lambda I - F + K\,H) = 0 \tag{8.136}$$

Equation (8.136) clearly shows that the eigenvalues of the controller and estimator are separate from each other in the overall LQG closed-loop system. This again shows the Separation Principle. The obvious advantage of having a time-invariant system is the application of real-time control/estimation in a feedback system. The optimality of the time-invariant closed-loop system is proven by Tse.[15]

Yet another way to prove the Separation Theorem involves using the *Stochastic Hamilton-Jacobi-Bellman* (SHJB) equation,[2] given by

$$\boxed{\begin{aligned}\frac{\partial J^*(\mathbf{x}(t),t)}{\partial t} + \min_{\mathbf{u}(t)}\Bigg\{&\vartheta(\mathbf{x}(t),\mathbf{u}(t),t) + \frac{\partial J^*(\mathbf{x}(t),\mathbf{u}(t),t)}{\partial \mathbf{x}^T(t)}\mathbf{f}(\mathbf{x}(t),\mathbf{u}(t),t)\\ &+\frac{1}{2}\mathrm{Tr}\left[G(t)\,Q(t)\,G(t)\frac{\partial^2 J^*(\mathbf{x}(t),t)}{\partial \mathbf{x}^2(t)}\right]\Bigg\} = 0\end{aligned}} \tag{8.137}$$

with terminal condition

$$J^*(\mathbf{x}(t_f),t_f) = \phi(\mathbf{x}(t_f),t_f) \tag{8.138}$$

The cost-to-go function for the stochastic problem is given by

$$J^*(\mathbf{x}(t),t) = \min_{\mathbf{u}(t)} E\left\{\phi(\mathbf{x}(t_f),t_f) + \int_t^{t_f}\vartheta(\mathbf{x}(\tau),\mathbf{u}(\tau),\tau)\,d\tau|\mathbf{x}(t)\right\} \tag{8.139}$$

subject to the dynamic constraint

$$\dot{\mathbf{x}}(t) = \mathbf{f}(\mathbf{x}(t),\mathbf{u}(t),t) + G(t)\,\mathbf{w}(t) \tag{8.140}$$

For the LQG problem the control input that satisfies the SHJB equation can be shown to be given by eqn. (8.129).

8.6.2 Discrete-Time Formulation

The results of the previous section can be extended to discrete-time systems. Rather than repeating the steps here, we choose to only show the steps required to prove the Separation Theorem for discrete-time systems (see Refs. [2] and [11] for more details). In the discrete-time LQG problem we assume that the state model is given by eqn. (5.27):

$$\mathbf{x}_{k+1} = \Phi_k\mathbf{x}_k + \Gamma_k\mathbf{u}_k + \Upsilon_k\mathbf{w}_k \tag{8.141a}$$

$$\tilde{\mathbf{y}}_k = H_k\mathbf{x}_k + \mathbf{v}_k \tag{8.141b}$$

where $\mathbf{v}_k$ and $\mathbf{w}_k$ are assumed to be zero-mean Gaussian white-noise processes with covariances given by eqns. (5.28) and (5.29), respectively. The discrete-time version of the loss function in eqn. (8.115) is given by

$$J = E\left\{\sum_{k=0}^{N-1} \mathbf{x}_k^T \mathcal{Q}_k \mathbf{x}_k + \mathbf{u}_k^T \mathcal{R}_k \mathbf{u}_k\right\} \tag{8.142}$$

The discrete-time problem involves finding a control input to minimize eqn. (8.142) given a set of measurements $\mathbf{Y}_{k-1} = [\tilde{\mathbf{y}}_0, \ldots, \tilde{\mathbf{y}}_{k-1}]$. Equation (8.142) can be rewritten as

$$J = E\left\{\sum_{s=0}^{k-1} \mathbf{x}_s^T \mathcal{Q}_s \mathbf{x}_s + \mathbf{u}_s^T \mathcal{R}_s \mathbf{u}_s\right\} + E\left\{\sum_{s=k}^{N-1} \mathbf{x}_s^T \mathcal{Q}_s \mathbf{x}_s + \mathbf{u}_s^T \mathcal{R}_s \mathbf{u}_s\right\} \tag{8.143}$$

Note that the second term in the loss function of eqn. (8.143) depends on $\mathbf{u}_k$. Hence, we seek to minimize the following cost-to-go function:

$$J^* = E\left\{\sum_{s=k}^{N-1} \mathbf{x}_s^T \mathcal{Q}_s \mathbf{x}_s + \mathbf{u}_s^T \mathcal{R}_s \mathbf{u}_s\right\} \tag{8.144}$$

It is more convenient to express eqn. (8.144) in terms of a conditional probability, similar to the approach shown in §2.6. This leads to the following equivalent minimizing function:

$$J^* = E\left[\min_{\mathbf{u}_k} E\left\{\sum_{s=k}^{N-1} \mathbf{x}_s^T \mathcal{Q}_s \mathbf{x}_s + \mathbf{u}_s^T \mathcal{R}_s \mathbf{u}_s \middle| \mathbf{Y}_{k-1}\right\}\right] \tag{8.145}$$

where the first expectation in eqn. (8.143) denotes the expectation with respect to the distribution $\mathbf{Y}_{k-1}$, and the minimum is taken with respect to all strategies that express $\mathbf{u}_k$ and a function of $\mathbf{Y}_{k-1}$.[11] Repeating the arguments for $k = N-1,\ N-2, \ldots$ leads to

$$\min_{\mathbf{u}_k, \ldots, \mathbf{u}_{N-1}} E\left\{\sum_{s=k}^{N-1} \mathbf{x}_s^T \mathcal{Q}_s \mathbf{x}_s + \mathbf{u}_s^T \mathcal{R}_s \mathbf{u}_s \middle| \mathbf{Y}_{k-1}\right\} \equiv \tilde{V}_k(\mathbf{Y}_{k-1}) \tag{8.146}$$

Since $\mathbf{x}_k$ and $\mathbf{u}_k$ are not causally affected by $\mathbf{u}_{k+1}, \ldots, \mathbf{u}_{N-1}$, then eqn. (8.146) can be written as[2]

$$\begin{aligned} \tilde{V}_k(\mathbf{Y}_{k-1}) = \min_{\mathbf{u}_k} E\Bigg\{ & \mathbf{x}_k^T \mathcal{Q}_k \mathbf{x}_k + \mathbf{u}_k^T \mathcal{R}_k \mathbf{u}_k \\ & + \min_{\mathbf{u}_{k+1}, \ldots, \mathbf{u}_{N-1}} \left[\sum_{s=k+1}^{N-1} \mathbf{x}_s^T \mathcal{Q}_s \mathbf{x}_s + \mathbf{u}_s^T \mathcal{R}_s \mathbf{u}_s\right] \Bigg| \mathbf{Y}_{k-1}\Bigg\} \end{aligned} \tag{8.147}$$

Equation (8.147) is equivalent to

$$\begin{aligned}\tilde{V}_k(\mathbf{Y}_{k-1}) = \min_{\mathbf{u}_k}\Bigg[& E\left\{\mathbf{x}_k^T \mathcal{Q}_k \mathbf{x}_k + \mathbf{u}_k^T \mathcal{R}_k \mathbf{u}_k \,\middle|\, \mathbf{Y}_{k-1}\right\} \\ & + E\left\{\min_{\mathbf{u}_{k+1},\ldots,\mathbf{u}_{N-1}} E\left\{\sum_{s=k+1}^{N-1} \mathbf{x}_s^T \mathcal{Q}_s \mathbf{x}_s + \mathbf{u}_s^T \mathcal{R}_s \mathbf{u}_s \,\middle|\, \mathbf{Y}_k\right\} \,\middle|\, \mathbf{Y}_{k-1}\right\}\Bigg]\end{aligned} \tag{8.148}$$

Finally, using the definition of the conditional expectation (see Appendix B) allows us to notionally simplify eqn. (8.148) to

$$\begin{aligned}\tilde{V}_k(\mathbf{Y}_{k-1}) = \min_{\mathbf{u}_k}\Big[& E\left\{\mathbf{x}_k^T \mathcal{Q}_k \mathbf{x}_k + \mathbf{u}_k^T \mathcal{R}_k \mathbf{u}_k \,\middle|\, \mathbf{Y}_{k-1}\right\} \\ & + E\left\{E\left\{\tilde{V}_{k+1}(\mathbf{Y}_k) \,\middle|\, \mathbf{Y}_k\right\} \,\middle|\, \mathbf{Y}_{k-1}\right\}\Big]\end{aligned} \tag{8.149}$$

Note that eqn. (8.149) does not include a summation anymore.

In order to prove the Separation Theorem for discrete-time systems, we need to show that for each $k = N, N-1, \ldots, 0$, there exists a function V_k dependent on $\hat{\mathbf{x}}_k$, a matrix S_k, and a scalar s_k such that $\tilde{V}_k(\mathbf{Y}_{k-1}) = V_k(\hat{\mathbf{x}}_k)$. Let us assume that this relationship is of the form given by

$$V(\hat{\mathbf{x}}_k, k) = \hat{\mathbf{x}}_k^T S_k \hat{\mathbf{x}}_k + s_k \tag{8.150}$$

We first concentrate our efforts on the expression $E\left\{\mathbf{x}_k^T \mathcal{Q}_k \mathbf{x}_k | \mathbf{Y}_{k-1}\right\}$. This expression can be given directly from the discrete-time version of eqn. (8.119):

$$E\left\{\mathbf{x}_k^T \mathcal{Q}_k \mathbf{x}_k | \mathbf{Y}_{k-1}\right\} = \hat{\mathbf{x}}_k^T \mathcal{Q}_k \hat{\mathbf{x}}_k + \mathrm{Tr}(\mathcal{Q}_k P_k) \tag{8.151}$$

Starting with the Kalman filter equations of eqn. (5.54), the mean and covariance of $\hat{\mathbf{x}}_{k+1}$ can be shown to be given by (which is left as an exercise for the reader)

$$E\left\{\hat{\mathbf{x}}_{k+1} | \mathbf{Y}_{k-1}\right\} = \Phi_k \hat{\mathbf{x}}_k + \Gamma_k \mathbf{u}_k \tag{8.152a}$$

$$\mathrm{cov}\left\{\hat{\mathbf{x}}_{k+1} | \mathbf{Y}_{k-1}\right\} = \Phi_k K_k \left[H_k P_k H_k^T + R_k\right] K_k^T \Phi_k^T \tag{8.152b}$$

Summing up we find that $V(\hat{\mathbf{x}}_k)$ is given by

$$\begin{aligned}V(\hat{\mathbf{x}}_k) = \min_{\mathbf{u}_k}\Big\{ & \hat{\mathbf{x}}_k^T \mathcal{Q}_k \hat{\mathbf{x}}_k + \mathrm{Tr}(\mathcal{Q}_k P_k) + \mathbf{u}_k^T \mathcal{R}_k \mathbf{u}_k \\ & + \left[\Phi_k \hat{\mathbf{x}}_k + \Gamma_k \mathbf{u}_k\right]^T S_{k+1} \left[\Phi_k \hat{\mathbf{x}}_k + \Gamma_k \mathbf{u}_k\right] \\ & + \mathrm{Tr}\left(S_{k+1} \Phi_k K_k \left[H_k P_k H_k^T + R_k\right] K_k^T \Phi_k^T\right) + s_{k+1}\Big\}\end{aligned} \tag{8.153}$$

Equation (8.153) is equivalent to

$$\begin{aligned}V(\hat{\mathbf{x}}_k) = \min_{\mathbf{u}_k}\Big\{ & \hat{\mathbf{x}}_k^T \left(\Phi_k^T S_{k+1} \Phi_k + \mathcal{Q}_k - L_k^T \left[\Gamma_k^T S_{k+1} \Gamma_k + \mathcal{R}_k\right] L_k\right) \hat{\mathbf{x}}_k \\ & + (\mathbf{u}_k + L_k \hat{\mathbf{x}}_k)^T \left[\Gamma_k^T S_{k+1} \Gamma_k + \mathcal{R}_k\right] (\mathbf{u}_k + L_k \hat{\mathbf{x}}_k) + \mathrm{Tr}(\mathcal{Q}_k P_k) \\ & + \mathrm{Tr}\left(S_{k+1} \Phi_k K_k \left[H_k P_k H_k^T + R_k\right] K_k^T \Phi_k^T\right) + s_{k+1}\Big\}\end{aligned} \tag{8.154}$$

where L_k is given by eqn. (8.98). Also, comparing eqn. (8.150) to eqn. (8.154) gives

$$s_k = s_{k+1} + \text{Tr}\,(Q_k P_k) + \text{Tr}\left(S_{k+1}\Phi_k K_k\left[H_k P_k H_k^T + R_k\right]K_k^T \Phi_k^T\right) \tag{8.155}$$

and

$$S_k = \Phi_k^T S_{k+1}\Phi_k + \mathcal{Q}_k - L_k^T\left[\Gamma_k^T S_{k+1}\Gamma_k + \mathcal{R}_k\right]L_k \tag{8.156}$$

Note that eqn. (8.156) is equivalent to eqn. (8.102) with the gain matrix L_k given by eqn. (8.98)!

The optimal $\mathbf{u}_k$ that satisfies eqn. (8.154) is clearly given by

$$\mathbf{u}_k = -L_k\hat{\mathbf{x}}_k \tag{8.157}$$

Equation (8.157) is identical to eqn. (8.97) with the exception that the true state $\mathbf{x}(t)$ is replaced with the estimated state $\hat{\mathbf{x}}(t)$! In order for eqn. (8.157) to truly achieve the minimum of the LQG loss function, the matrix $[\Gamma_k^T S_{k+1}\Gamma_k + \mathcal{R}_k]$ must be positive definite. This condition will obviously always be met since S_{k+1} is always positive definite, which is shown by the form in eqn. (8.103). For autonomous systems, the discrete-time Separation Theorem can be proved using an eigenvalue separation of the controller and estimator (see exercise 8.30), similar to the steps leading to eqn. (8.135).

8.7 Loop Transfer Recovery

As discussed in example 5.5, the Kalman filter estimates are usually derived by "tuning" the process noise covariance matrix until desired estimation characteristics are obtained. The difficulties of this usually "ad hoc" approach are often mitigated through intimate experience of the dynamical system. However, the process is further complicated when we wish to investigate the robustness properties of the combined estimator/controller in the overall LQG design. As discussed in the introduction section of this chapter, the overall pointing error is a function of both the estimation *and* control errors. Problems with LQG designs may arise from two possible undesirable characteristics: 1) poor stability margins and 2) poor performance of the overall LQG dynamics. One might expect that since the Kalman filter and linear regulator have nice properties, then the LQG controller would exhibit nice properties as well. But, Doyle[16] has shown that LQG designs can exhibit poor stability margins, which leads to the first problem in LQG designs. Also, a natural and seemingly logical assumption in the LQG design involves setting the Kalman gain so that the estimator errors have converged well before the controller errors, which should provide well-behaved feedback properties. However, Doyle and Stein[17] show that stability margins can actually be degraded by making the estimator dynamics faster in some cases, which leads to the second problem in LQG designs.

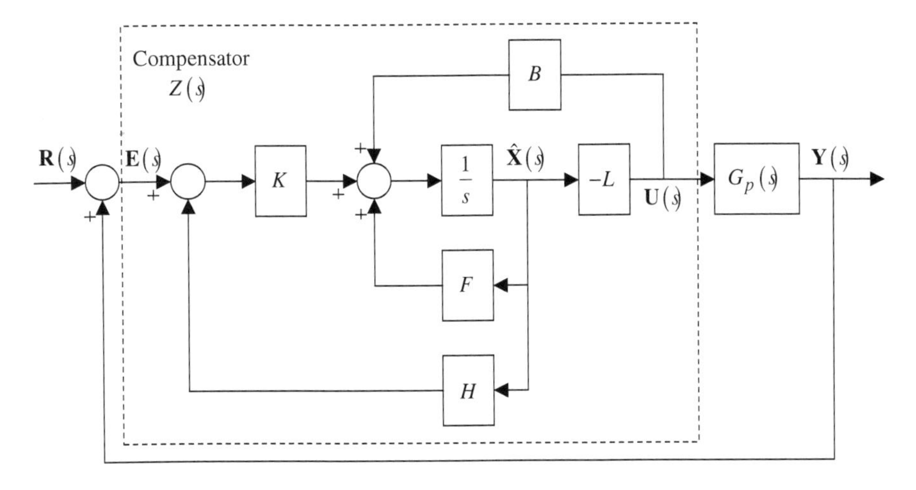

Figure 8.6: The Linear Quadratic-Gaussian Controller with Reference Input

The Loop Transfer Recovery (LTR) approach[17, 18] overcomes these problems by tuning the Kalman filter so that the original (true state) regulator dynamics are "recovered" at the control input. A block diagram of the LQG controller with a reference input is shown in Figure 8.6. Notice that unlike Figure 8.5, we now incorporate unity feedback into the control structure, which provides a more likely plant/controller arrangement for use in practice.[14] Also, this arrangement allows us to compute useful stability/robustness parameters, such as phase and/or gain margins (see Refs. [19]-[24] for more details on these tools).

To help motivate the LTR concept, we begin by considering the closed-loop dynamics of the LQR system. It is assumed that the number of inputs is equal to the number of outputs. Taking the Laplace transform of both sides of eqn. (8.55) leads to

$$\mathbf{X}(s) = (sI - F)^{-1} B\, \mathbf{U}(s) \tag{8.158}$$

Multiplying both sides of eqn. (8.158) by $-L$ and using the definition of the LQR control law in eqn. (8.62) gives

$$\mathbf{U}(s) = -L\,(sI - F)^{-1} B\, \mathbf{U}(s) \tag{8.159}$$

The matrix $-L\,(sI - F)^{-1} B$ represents the desired return ratio. Referring to Figure 8.6, the transfer function from the error signal $\mathbf{E}(s)$ to the state estimate $\hat{\mathbf{X}}(s)$ is given by

$$\hat{\mathbf{X}}(s) = (sI - F + B\,L + K\,H)^{-1} K\, \mathbf{E}(s) \tag{8.160}$$

Taking the Laplace transform of eqn. (8.129) and substituting eqn. (8.160) into the resulting expression gives

$$\mathbf{U}(s) \equiv Z(s)\,\mathbf{E}(s) = -L\,(sI - F + B\,L + K\,H)^{-1} K\, \mathbf{E}(s) \tag{8.161}$$

where $Z(s) \equiv -L\,(s\,I - F + B\,L + K\,H)^{-1} K$ is the LQG compensator matrix. Using the transfer function model of eqn. (3.14), with D assumed to be zero, for the plant $G_p(s)$ gives the following *loop-gain transfer function* matrix:

$$Z(s)\,G_p(s) = -L\,(s\,I - F + B\,L + K\,H)^{-1} K\,H\,(s\,I - F)^{-1} B \tag{8.162}$$

This is the return ratio of the overall LQG system. Our goal is to tune the Kalman filter gain so that $Z(s)\,G_p(s)$ approaches the matrix $-L\,(s\,I - F)^{-1} B$ shown in eqn. (8.159). Define the following quantities:

$$\Phi(s) \equiv (s\,I - F)^{-1} \tag{8.163a}$$

$$\Psi(s) \equiv (s\,I - F + B\,L)^{-1} \tag{8.163b}$$

With these definitions eqn. (8.162) can be rewritten as

$$Z(s)\,G_p(s) = -L\,[\Psi^{-1}(s) + K\,H]^{-1} K\,H\,\Phi(s)\,B \tag{8.164}$$

Equation (8.164) can be shown to be equivalent to (which is left as an exercise for the reader)

$$Z(s)\,G_p(s) = -L\,\Psi(s)\,K[I + H\,\Psi(s)\,K]^{-1} K\,H\,\Phi(s)\,B \tag{8.165}$$

where the matrix inversion lemma of eqn. (1.69) can be used to prove eqn. (8.165).

In the LTR approach it is assumed that the process noise covariance matrix $G\,Q\,G^T$ is replaced by

$$\boxed{G\,Q\,G^T = G\,Q_0 G^T + q^2 B\,B^T} \tag{8.166}$$

where Q_0 is some initial guess for Q, and q is a real and positive tuning parameter. Using eqn. (8.166), the new algebraic Riccati equation for the Kalman filter covariance, shown in Table 5.5, can be written as

$$\boxed{F\left(\frac{P}{q^2}\right) + \left(\frac{P}{q^2}\right) F^T - q^2 \left(\frac{P}{q^2}\right) H^T R^{-1} H \left(\frac{P}{q^2}\right) + \frac{G\,Q_0 G^T}{q^2} + B\,B^T = 0} \tag{8.167}$$

Kwakernaak and Sivan[25] show that if the plant has no transmission zeros in the right half-plane, then

$$\lim_{q \to \infty} \frac{P}{q^2} = 0 \tag{8.168}$$

Equation (8.168) indicates that as q increases, the covariance matrix P is increasing more slowly than the process noise covariance (if the stated assumptions hold).[18] Consequently, from eqn. (8.167) we have

$$q^2 \left(\frac{P}{q^2}\right) H^T R^{-1} H \left(\frac{P}{q^2}\right) \to B\,B^T \tag{8.169}$$

Since the Kalman gain is given by $K = P\,H^T R^{-1}$, then from eqn. (8.169) we now have

$$K \to q\,B\,R^{-1/2} \text{ as } q \to \infty \tag{8.170}$$

Substituting eqn. (8.170) into eqn. (8.165), with the assumption that $H\,\Psi(s)\,B$ is square (i.e., the number of inputs is equal to the number of outputs) yields

$$Z(s)\,G_p(s) \to -L\,\Psi(s)\,R^{-1/2}\left[H\,\Psi(s)\,B\,R^{-1/2}\right]^{-1} H\,\Phi(s)\,B \text{ as } q \to \infty \quad (8.171)$$

Equation (8.171) can be further simplified since R is a square matrix:

$$Z(s)\,G_p(s) \to -L\,\Psi(s)\,[H\,\Psi(s)\,B]^{-1}\,H\,\Phi(s)\,B \text{ as } q \to \infty \quad (8.172)$$

Now, consider the following identity:[18]

$$\Psi(s) = \Phi(s)\,[I + B\,L\,\Phi(s)]^{-1} \quad (8.173)$$

Substituting eqn. (8.173) into eqn. (8.172), and performing some simple algebraic manipulations yields

$$\begin{aligned}\lim_{q\to\infty} Z(s)\,G_p(s) &= -L\,\Phi(s)\,B\,[I + L\,\Phi(s)\,B]^{-1} \\ &\quad \times \left\{H\,\Phi(s)\,B\,[I + L\,\Phi(s)\,B]^{-1}\right\}^{-1} H\,\Phi(s)\,B\end{aligned} \quad (8.174)$$

Since $H\,\Phi(s)\,B$ is assumed to be a square matrix, then eqn. (8.174) reduces down to

$$\lim_{q\to\infty} Z(s)\,G_p(s) = -L\,\Phi(s)\,B = -L\,(sI - F)^{-1}B \quad (8.175)$$

where the definition of $\Phi(s)$ from eqn. (8.163a) has been used. Hence the desired return ratio in eqn. (8.159) is achieved. It can be shown that the LTR approach drives the filter eigenvalues to the plant's zeros as q is increased.[18] Therefore, in order to maintain stability, the plant must have no transmission zeros in the right half-plane, which is also required by eqn. (8.168).

The design procedure for the LTR approach is as follows. First, design a Kalman filter and LQR control law to meet the desired estimation and control characteristics, treating them as separate design issues. The usual checks in the Kalman filter, trading off performance versus noisy estimates, should be employed to tune the initial process noise covariance. Once the initial estimator and control designs are completed, check the characteristics of the overall LQG system. If the stability margins are poor, then use eqn. (8.166) with some value for q to adjust the Kalman filter gain. Increase q until reasonable stability margins are given. It is imperative to make q only as large as possible because, in general, larger values for the process noise covariance introduce more high frequency noise into the filter state estimates.

Example 8.3: In this example we will show the usefulness of the LTR design procedure. We consider the following continuous-time system:[17]

$$\begin{aligned}\dot{\mathbf{x}}(t) &= \begin{bmatrix} 0 & 1 \\ -3 & -4 \end{bmatrix}\mathbf{x}(t) + \begin{bmatrix} 0 \\ 1 \end{bmatrix}u(t) + \begin{bmatrix} 35 \\ -61 \end{bmatrix}w(t) \\ \tilde{y}(t) &= \begin{bmatrix} 2 & 1 \end{bmatrix}\mathbf{x}(t) + v(t)\end{aligned}$$

Table 8.3: Summary of LTR Example Results

q^2	Filter Gain K	Phase Margin	Covariance P
0	$\begin{bmatrix} 30.00 \\ -49.96 \end{bmatrix}$	14.85°	$\begin{bmatrix} 96.23 & -162.46 \\ -162.46 & 274.95 \end{bmatrix}$
100	$\begin{bmatrix} 26.83 \\ -40.21 \end{bmatrix}$	19.39°	$\begin{bmatrix} 139.70 & -252.57 \\ -252.57 & 464.93 \end{bmatrix}$
500	$\begin{bmatrix} 20.38 \\ -17.75 \end{bmatrix}$	32.37°	$\begin{bmatrix} 212.59 & -404.80 \\ -404.80 & 791.85 \end{bmatrix}$
1,000	$\begin{bmatrix} 16.69 \\ -1.93 \end{bmatrix}$	42.50°	$\begin{bmatrix} 244.98 & -473.27 \\ -473.27 & 944.61 \end{bmatrix}$
10,000	$\begin{bmatrix} 6.94 \\ 84.62 \end{bmatrix}$	74.44°	$\begin{bmatrix} 297.68 & -588.43 \\ -588.43 & 1261.47 \end{bmatrix}$

with process noise and measurement noise variances given by $Q_0 = 1$ and $R = 1$, respectively. The resulting Kalman filter gain is given by $K = \begin{bmatrix} 30.00 & -49.96 \end{bmatrix}^T$. The estimator poles are placed at $s = -7.02 \pm 1.95j$ with this gain matrix. Suppose we now design an LQR control law with weighting matrices $\mathcal{R} = 1$ and $\mathcal{Q} = M^T M$, with $M = 4\sqrt{5}\begin{bmatrix} \sqrt{35} & 1 \end{bmatrix}$. Solving the LQR problem with these weighting matrices gives $L = \begin{bmatrix} 50 & 10 \end{bmatrix}$. The closed-loop LQR poles are placed at $s = -7 \pm 2j$ with this gain matrix, which are nearly identical to the estimator poles. The phase margin for the LQR system, which can be derived from the loop $-L(sI - F)^{-1}B$, is 85.94° (in general, the larger the phase margin the better the closed-loop characteristics). This indicates that LQR controller gives a well behaved closed-loop system.

Suppose we now use the Kalman filter in an LQG design with the predetermined Kalman and LQR gain matrices. The phase margin for the LQG system, which can be derived from the loop-gain transfer function matrix in eqn. (8.162), is now only 14.85°. This clearly has decreased the performance of the overall LQG controller design, compared with the original LQR design. Since the estimator poles are nearly identical to the LQR control poles, a natural assumption to make, before ever learning about the LTR approach, might be to place the estimator poles further down the left half-plane. Suppose we use Ackermann's formula, given by eqn. (5.19), to place the estimator's poles at $s = -22 \pm 17.86j$, which gives a gain matrix of $K = \begin{bmatrix} 720 & -1400 \end{bmatrix}^T$. The phase margin for this LQG designed system is now 4.17°, which is even worse than the original design! It fact, the margins go asymptotically to zero for large gains, which is clearly undesirable.

We now employ the LTR design approach, using eqn. (8.166) to recover the desired performance characteristics, with q^2 = 100, 500, 1,000, and 10,000. A summary of the results is shown in Table 8.3. Clearly, as q^2 is increased the phase margin also increases, which provides better closed-loop characteristics. Note when $q^2 = 10,000$ the Kalman gain approaches its limit, given by eqn. (8.170), of $K \rightarrow \begin{bmatrix} 0 & 100 \end{bmatrix}^T$. The improved closed-loop performance comes at a price though, since the filter covariance also increases as expected. The second state corresponds to the rate estimate, and the noise associated with this state substantially increases from the original design of $q^2 = 0$. In practice, hopefully, a satisfactory compromise between closed-loop stability margins and high frequency noise rejection can be found.

8.8 Spacecraft Control Design

In this section an LQG-based control system is designed to optimally orientate a spacecraft along a desired reference trajectory. The control of spacecraft for large angle slewing maneuvers poses a difficult problem. Some of these difficulties include: the highly nonlinear characteristics of the governing equations, control rate and saturation constraints and limits, and incomplete state knowledge due to sensor failure or omission. The control of spacecraft with large angle slews can be accomplished by either open-loop or closed-loop schemes. Open-loop schemes usually require a pre-determined pointing maneuver and are typically determined using optimal control techniques, which involve the solution of a TPBVP (e.g., the time optimal maneuver problem[26]). Also, open-loop schemes are sensitive to spacecraft parameter uncertainties and unexpected disturbances.[27, 28] Closed-loop systems can account for parameter uncertainties and disturbances, and as shown in this chapter, provide a more robust design methodology.

Several spacecraft attitude controllers have been developed that are devoted to the closed-loop design of spacecraft with large angle slews. An exhaustive history of this problem is beyond the present text; a starting reference point for many of these controllers can be found in Refs. [29] and [30]. In fact, a plethora of nonlinear and robust controllers have been developed, each with their own advantages and disadvantages. Our goal in the present text involves first using an LQR approach with *linear* dynamics. Paielli and Bach[31] present an optimal control design that provides linear closed-loop error dynamics for tracking a desired trajectory . However, this approach is singular for $\pm 180°$ error-rotations about any axis. Schaub et al.[32] derive an optimal controller using the modified Rodrigues parameters[33] (MRPs), which are singular for +360° rotations. By switching between the original and alternative sets of MRPs (known as the *shadow set*), it is possible to achieve a globally nonsingular attitude parameterization for all possible $\pm 360°$ rotations. An approach using MRPs is beyond the scope of this text, so we only will present the approach of Ref. [31]. Our derivation is slightly different than the one shown in Ref. [31], but the end result is the same. First, recall the kinematics and dynamics equations of motion given in §3.7:

$$\dot{\mathbf{q}} = \frac{1}{2}\Xi(\mathbf{q})\boldsymbol{\omega} = \frac{1}{2}\Omega(\boldsymbol{\omega})\mathbf{q} \tag{8.176a}$$

$$\dot{\boldsymbol{\omega}} = -J^{-1}[\boldsymbol{\omega}\times]J\boldsymbol{\omega} + J^{-1}\mathbf{L} \tag{8.176b}$$

where $\mathbf{q}$ is the quaternion, $\boldsymbol{\omega}$ is the angular velocity vector, J is the inertia matrix, and $\mathbf{L}$ is the applied torque. Also, the quantities $\Xi(\mathbf{q})$ and $\Omega(\boldsymbol{\omega})$ are defined by eqns. (3.155a) and (3.162), respectively. Suppose that a desired quaternion, $\mathbf{q}_d$, is given that also follows the following kinematics equation:

$$\dot{\mathbf{q}}_d = \frac{1}{2}\Xi(\mathbf{q}_d)\boldsymbol{\omega}_d \tag{8.177}$$

where $\boldsymbol{\omega}_d$ is the desired angular velocity vector. We now define the following error quaternion:

$$\delta\mathbf{q} = \mathbf{q}\otimes\mathbf{q}_d^{-1} \tag{8.178}$$

with $\delta\mathbf{q} \equiv \left[\delta\boldsymbol{\varrho}^T \ \delta q_4\right]^T$. Also, the quaternion inverse is defined by eqn. (3.169). Using the rules of quaternion multiplication, discussed in §3.7.1, $\delta\boldsymbol{\varrho}$ and δq_4 can be shown to be given by

$$\delta\boldsymbol{\varrho} = \Xi^T(\mathbf{q}_d)\mathbf{q} \tag{8.179a}$$

$$\delta q_4 = \mathbf{q}_d^T\mathbf{q} \tag{8.179b}$$

Note that as $\delta\boldsymbol{\varrho}$ approaches zero, then the actual quaternion approaches the desired quaternion. Let us assume that the closed-loop dynamics are desired to have the following prescribed *linear* form:

$$\delta\ddot{\boldsymbol{\varrho}} + L_2\delta\dot{\boldsymbol{\varrho}} + L_1\delta\boldsymbol{\varrho} = \mathbf{0} \tag{8.180}$$

where L_1 and L_2 are 3×3 gain matrices. These matrices can be determined using an LQR approach:

$$\delta\ddot{\boldsymbol{\varrho}} = \mathbf{u} \tag{8.181}$$

with

$$\mathbf{u} = -L\begin{bmatrix}\delta\boldsymbol{\varrho}\\ \delta\dot{\boldsymbol{\varrho}}\end{bmatrix} \tag{8.182}$$

where $L \equiv \left[L_1 \ L_2\right]$. The state-space formulation of eqn. (8.181) is given by

$$\dot{\mathbf{x}} = \begin{bmatrix}0_{3\times 3} & I_{3\times 3}\\ 0_{3\times 3} & 0_{3\times 3}\end{bmatrix}\mathbf{x} + \begin{bmatrix}0_{3\times 3}\\ I_{3\times 3}\end{bmatrix}\mathbf{u} \tag{8.183}$$

where $\mathbf{x} \equiv \left[\delta\boldsymbol{\varrho}^T \;\; \delta\dot{\boldsymbol{\varrho}}^T\right]^T$. If L_1 and L_2 are assumed to be scalars, then these gains can be directly designed to yield the desired closed-loop dynamics without solving the LQR problem.

Our goal is to find a control torque input, $\mathbf{L}$, that achieves the desired closed-loop dynamics given by eqn. (8.180). Toward this end goal, two time derivatives of eqn. (8.179a) are first taken and then substituted into eqn. (8.180), which yields

$$\Xi^T(\mathbf{q}_d)\ddot{\mathbf{q}} + \left[2\Xi^T(\dot{\mathbf{q}}_d) + L_2\Xi^T(\mathbf{q}_d)\right]\dot{\mathbf{q}} + \left[\Xi^T(\ddot{\mathbf{q}}_d) + L_2\Xi^T(\dot{\mathbf{q}}_d) + L_1\Xi^T(\mathbf{q}_d)\right]\mathbf{q} = \mathbf{0} \tag{8.184}$$

Taking the time derivative of eqn. (8.176a) leads to

$$\begin{aligned}\ddot{\mathbf{q}} &= \frac{1}{2}\Xi(\mathbf{q})\dot{\boldsymbol{\omega}} + \frac{1}{2}\Omega(\boldsymbol{\omega})\dot{\mathbf{q}} \\ &= \frac{1}{2}\Xi(\mathbf{q})\dot{\boldsymbol{\omega}} - \frac{1}{4}(\boldsymbol{\omega}^T\boldsymbol{\omega})\mathbf{q}\end{aligned} \tag{8.185}$$

where the identity $\Omega^2(\boldsymbol{\omega}) = -(\boldsymbol{\omega}^T\boldsymbol{\omega})I_{4\times 4}$ has been used. An identical expression for the desired quaternion is also given:

$$\ddot{\mathbf{q}}_d = \frac{1}{2}\Xi(\mathbf{q}_d)\dot{\boldsymbol{\omega}}_d - \frac{1}{4}(\boldsymbol{\omega}_d^T\boldsymbol{\omega}_d)\mathbf{q}_d \tag{8.186}$$

where $\dot{\boldsymbol{\omega}}_d$ can be derived from a desired dynamics equation, using eqn. (8.176b), or it can be pre-specified from the known desired dynamical motion. Substituting eqn. (8.176b) into eqn. (8.185) gives

$$\ddot{\mathbf{q}} = -\frac{1}{2}\Xi(\mathbf{q})J^{-1}[\boldsymbol{\omega}\times]J\boldsymbol{\omega} - \frac{1}{4}(\boldsymbol{\omega}^T\boldsymbol{\omega})\mathbf{q} + \frac{1}{2}\Xi(\mathbf{q})J^{-1}\mathbf{L} \tag{8.187}$$

Substituting eqns. (8.176a) and (8.187) into eqn. (8.184), and solving for $\mathbf{L}$ yields

$$\boxed{\begin{aligned}\mathbf{L} = [\boldsymbol{\omega}\times]J\boldsymbol{\omega} + 2J\left[\Xi^T(\mathbf{q}_d)\Xi(\mathbf{q})\right]^{-1}&\left\{\frac{1}{4}(\boldsymbol{\omega}^T\boldsymbol{\omega})\Xi^T(\mathbf{q}_d) - \Xi^T(\dot{\mathbf{q}}_d)\Omega(\boldsymbol{\omega})\right.\\ &\left. - \Xi^T(\ddot{\mathbf{q}}_d) - L_1\Xi^T(\mathbf{q}_d) - L_2\left[\frac{1}{2}\Xi^T(\mathbf{q}_d)\Omega(\boldsymbol{\omega}) + \Xi^T(\dot{\mathbf{q}}_d)\right]\right\}\mathbf{q}\end{aligned}} \tag{8.188}$$

Note that the inverse of $\Xi^T(\mathbf{q}_d)\Xi(\mathbf{q})$ always exists as long as $\delta q_4 = \mathbf{q}_d^T\mathbf{q}$ is nonzero. This can easily be shown by the following identities:

$$\Xi^T(\mathbf{q}_d)\Xi(\mathbf{q}) = \delta q_4 I_{3\times 3} + [\delta\boldsymbol{\varrho}\times] \tag{8.189a}$$

$$\left[\Xi^T(\mathbf{q}_d)\Xi(\mathbf{q})\right]^{-1} = \delta q_4 I_{3\times 3} - [\delta\boldsymbol{\varrho}\times] + \frac{\delta\boldsymbol{\varrho}\,\delta\boldsymbol{\varrho}^T}{\delta q_4} \tag{8.189b}$$

From the definition of the scalar part of the quaternion in eqn. (3.153b) and from eqn. (8.179b), δq_4 is zero for $\pm 180°$ rotations in the tracking error, which is analogous to the approach shown in Ref. [31]. Hence, care must be exercised when the tracking errors approach $\pm 180°$. If L_1 and L_2 are scalars, with $L_1 = l_1$ and $L_2 = l_2$, then eqn. (8.188) simplifies to

$$\mathbf{L} = [\boldsymbol{\omega}\times]J\boldsymbol{\omega} + J\left\{\delta A\,\dot{\boldsymbol{\omega}}_d - [\boldsymbol{\omega}\times]\delta A\,\boldsymbol{\omega}_d - l_2\delta\boldsymbol{\omega} - 2\left[\frac{4l_1 - (\delta\boldsymbol{\omega}^T\delta\boldsymbol{\omega})}{4\delta q_4}\right]\delta\boldsymbol{\varrho}\right\} \tag{8.190}$$

with

$$\delta A = A(\mathbf{q})A^T(\mathbf{q}_d) \tag{8.191a}$$

$$\delta\boldsymbol{\omega} = \boldsymbol{\omega} - \delta A\,\boldsymbol{\omega}_d \tag{8.191b}$$

where the attitude matrix is defined by eqn. (3.154). Equation (8.190) can be proven using the following identities:

$$\left[\Xi^T(\mathbf{q}_d)\Xi(\mathbf{q})\right]^{-1}\delta\boldsymbol{\varrho} = \frac{\delta\boldsymbol{\varrho}}{\delta q_4} \tag{8.192a}$$

$$2\left[\Xi^T(\mathbf{q}_d)\Xi(\mathbf{q})\right]^{-1}\Xi^T(\dot{\mathbf{q}}_d)\Omega(\boldsymbol{\omega})\mathbf{q} = [\boldsymbol{\omega}\times]\delta A\,\boldsymbol{\omega}_d + \frac{\boldsymbol{\omega}^T\delta A\,\boldsymbol{\omega}_d}{\delta q_4}\delta\boldsymbol{\varrho} \tag{8.192b}$$

$$2\left[\Xi^T(\mathbf{q}_d)\Xi(\mathbf{q})\right]^{-1}\Xi^T(\ddot{\mathbf{q}}_d)\mathbf{q} = -\left[\delta A\,\dot{\boldsymbol{\omega}}_d + \frac{\boldsymbol{\omega}_d^T\boldsymbol{\omega}_d}{2\delta q_4}\delta\boldsymbol{\varrho}\right] \tag{8.192c}$$

$$2\left[\Xi^T(\mathbf{q}_d)\Xi(\mathbf{q})\right]^{-1}\Xi^T(\dot{\mathbf{q}}_d)\mathbf{q} = -\delta A\,\boldsymbol{\omega}_d \tag{8.192d}$$

Equation (8.190) is equivalent to the control law given in Ref. [31]. Note that eqn. (8.190) does not explicitly involve $\dot{\mathbf{q}}_d$ and $\ddot{\mathbf{q}}_d$.

The procedure for the spacecraft attitude controller proceeds as follows. First, given a desired quaternion, $\mathbf{q}_d$, angular velocity vector, $\boldsymbol{\omega}_d$, and angular acceleration vectors, $\dot{\boldsymbol{\omega}}_d$, compute the desired quaternion rate and acceleration using eqns. (8.177) and (8.186). Then, design an LQR feedback gain to achieve the desired closed-loop tracking dynamics shown by eqns. (8.180) and (8.183), which gives the matrices L_1 and L_2. Finally, use the control law given by eqn. (8.188), to drive the spacecraft's attitude and angular velocity to the desired trajectories. This controller can be combined with an extended Kalman filter to filter noisy measurements and to estimate gyro biases, as shown in §7.2.1, which leads to an LQG-type control system.

Example 8.4: In this example the control law given by eqn. (8.188) is combined with the EKF of §7.2.1 to maneuver a spacecraft along a desired trajectory. The assumed sensors include "quaternion-out" star trackers (see exercise 7.14) and three-axis gyros. The noise parameters for the gyro measurements are given by $\sigma_u = \sqrt{10}\times 10^{-10}$ rad/sec$^{3/2}$ and $\sigma_v = \sqrt{10}\times 10^{-7}$ rad/sec$^{1/2}$. The initial bias for each axis is given by 0.1 deg/hr. A combined quaternion from two trackers is assumed for the measurement. In order to generate synthetic measurements the following model is used:

$$\tilde{\mathbf{q}} = \begin{bmatrix} 0.5\mathbf{v} \\ 1 \end{bmatrix} \otimes \mathbf{q}$$

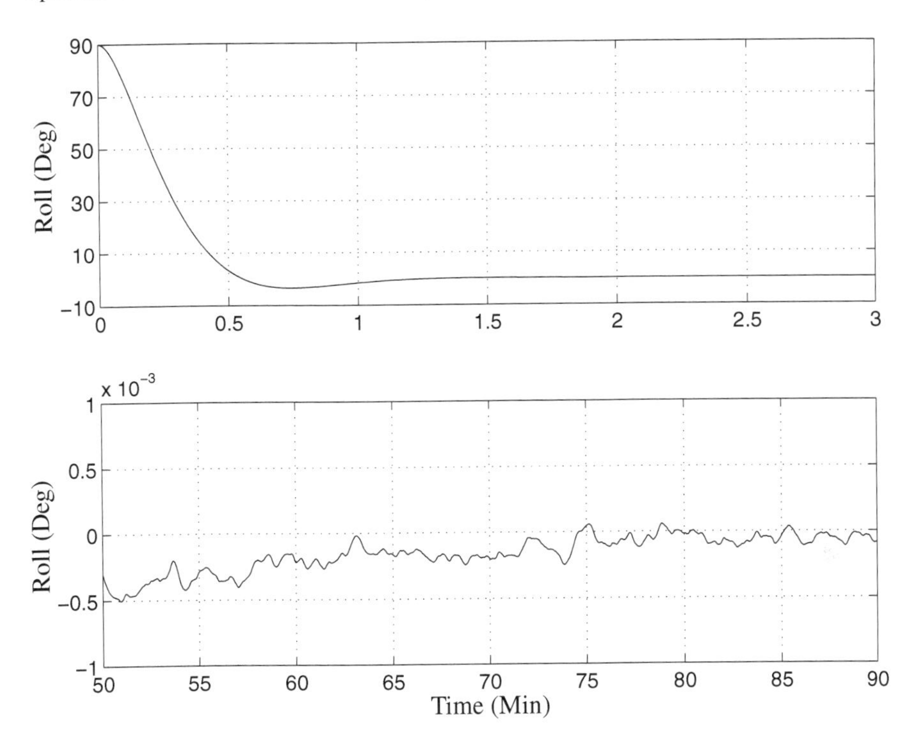

Figure 8.7: Roll Pointing Errors

where $\tilde{\mathbf{q}}$ is the quaternion measurement, $\mathbf{q}$ is the truth, and $\mathbf{v}$ is the measurement noise, which is assumed to be a zero-mean Gaussian noise process with covariance given by $0.001 I_{3\times 3}$ deg^2. Note, the measured quaternion is normalized to within first-order, but a brute-force normalization is still taken to ensure a normalized measurement. All quaternion and gyro measurements are sampled at 10 Hz. The initial covariances for the attitude error and gyro drift are taken exactly from example 7.2.

The spacecraft desired motion includes a constant angular velocity vector given by $\omega_d = \begin{bmatrix} 0 & 0.0011 & 0 \end{bmatrix}^T$ rad/sec, which corresponds to an Earth-pointing spacecraft in low-Earth orbit. The actual initial angular velocity of the spacecraft is given by $\omega(t_0) = \mathbf{0}$. The initial desired and actual quaternions are given by

$$\mathbf{q}_d(t_0) = \begin{bmatrix} \sqrt{2}/2 \\ 0 \\ 0 \\ \sqrt{2}/2 \end{bmatrix}, \quad \mathbf{q}(t_0) = \begin{bmatrix} 0 \\ 0 \\ 0 \\ 1 \end{bmatrix}$$

Equation (8.177) is used to propagate the desired quaternion over time. The space-

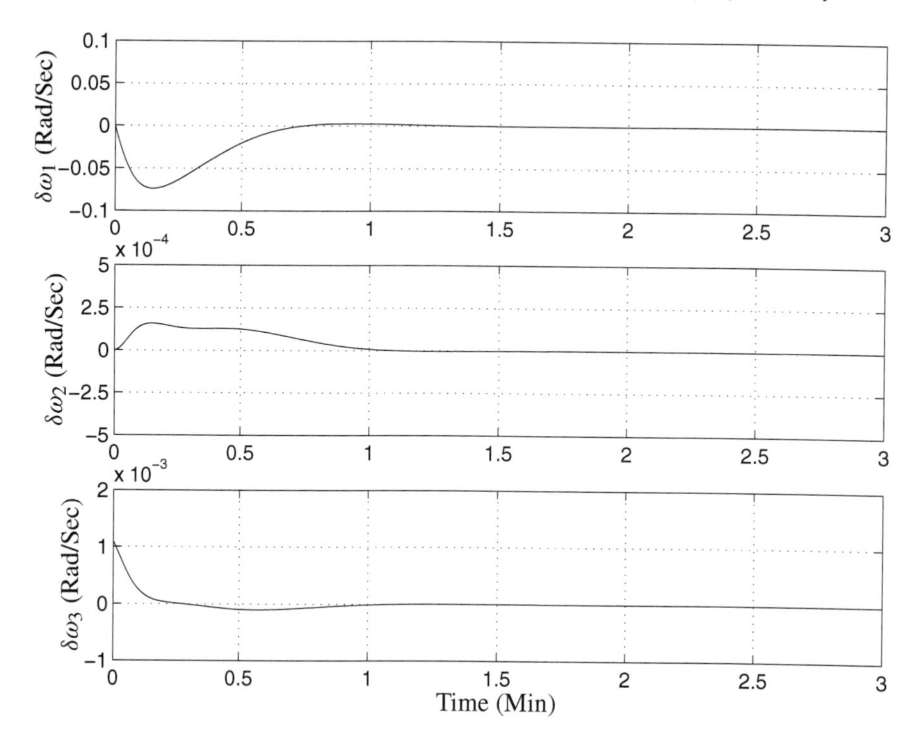

Figure 8.8: Angular Velocity Errors

craft inertia matrix is given by[32]

$$J = \begin{bmatrix} 30 & 10 & 5 \\ 10 & 20 & 3 \\ 5 & 3 & 15 \end{bmatrix} \text{ kg-m}^2$$

An LQR is designed with the model in eqn. (8.183), using the method outlined in §8.5.1. The steady-state Riccati equation in eqn. (8.68) is solved to determine L_1 and L_2. The weighting matrices are given by $\mathcal{Q} = 1 \times 10^{-4} I_{6\times 6}$ and $\mathcal{R} = I_{3\times 3}$. Using these weights gives $L_1 = 0.01 I_{3\times 3}$ and $L_2 = 0.14177 I_{3\times 3}$. The closed-loop natural frequencies and damping ratios (see §3.10) are given by $\omega_n = 0.1$ rad/sec and $\zeta = 0.709$. These gains are used in eqn. (8.188), with the estimated quaternion and angular velocities determined for the EKF (i.e., an LQG-type design), which provides the control torque input into the spacecraft.

A plot of the roll pointing-error trajectory is shown in Figure 8.7. The time required for the damped oscillations to reach and stay within $\pm 2\%$ of the steady-state value is given by $4/(\zeta\omega_n)$.[23] For the LQR design this formula gives a settling time of 56.4175 seconds, which agrees with the result shown in Figure 8.7. This result can also be checked by integrating eqn. (8.180). The bottom plot of Figure 8.7 shows the roll error in finer detail. Clearly, fine pointing can be achieved with this control

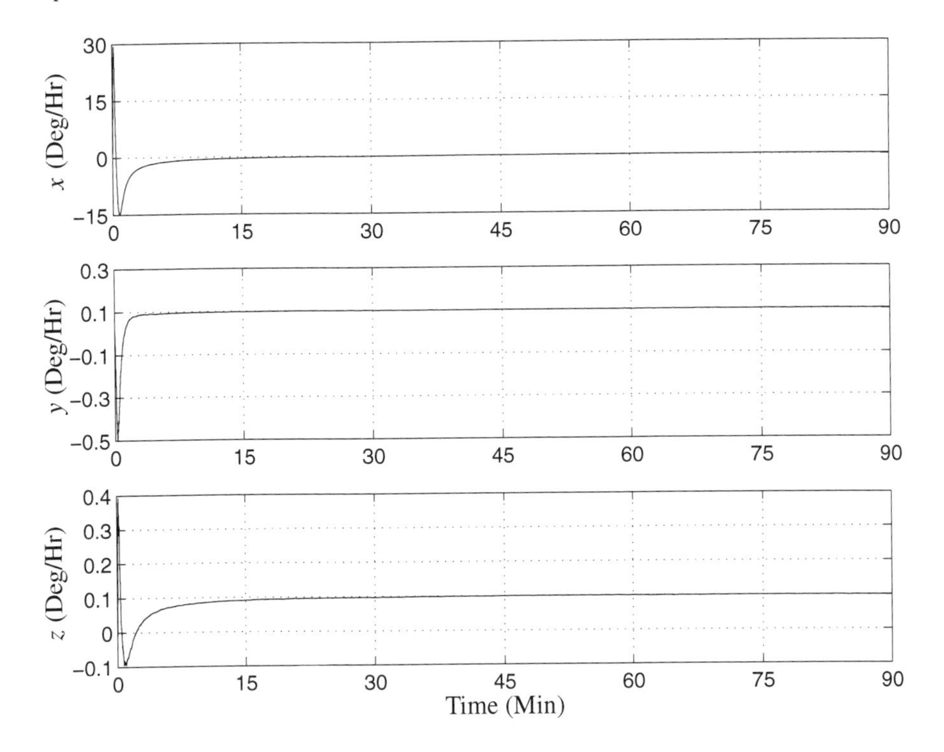

Figure 8.9: Gyro Drift Estimates

law and assumed sensors. A plot of the angular velocity errors is shown in Figure 8.8. Clearly, the desired angular velocity motion is achieved. A plot of the gyro drift estimates using the EKF is shown in Figure 8.9. The x axis has a large response due to the roll maneuver. All axes still converge to the actual bias of 0.1 deg/hr. Note that in a practical setting, the gyro biases are normally allowed to converge before a significant maneuver takes place. Still, this example clearly shows how an EKF can be combined with a control law to achieve effective overall pointing of a very practical system involving large-angle spacecraft maneuvers.

8.9 Summary

This chapter provided only a brief introduction to the theory of optimal control. Several texts and books have been written that provide much more depth in the subject area that can be covered here (e.g., see the references used in this chapter). Optimal control theory has uses well beyond the control of dynamic systems (e.g., optimal path planning for shipping routes), and we encourage the interested reader to pursue other topics where this theory can be used. The main results of this chapter involve the LQR control law and Separation Theorem used in the LQG controller. Although from a practical point of view, the theory used in the Separation Theorem is masked behind the actual control implementation, we believe that the reader will benefit from the derivation and understanding of this elegant theory. Also, the LTR approach of §8.7 is especially useful to recover the originally designed regulator dynamics. A general "rule-of-thumb" is to only use LTR when needed, because increasing the process noise covariance may lead to too much high frequency noise in the output estimates.

A summary of the key formulas presented in this chapter is given below.

- Euler-Lagrange Equations and Transversality Conditions

$$J \equiv J(\mathbf{x}(t), t_0, t_f) = \int_{t_0}^{t_f} \vartheta(\mathbf{x}(t), \dot{\mathbf{x}}(t), t)\, dt$$

$$\frac{\partial \vartheta(\mathbf{x}(t), \dot{\mathbf{x}}(t), t)}{\partial \mathbf{x}(t)} - \frac{d}{dt}\left[\frac{\partial \vartheta(\mathbf{x}(t), \dot{\mathbf{x}}(t), t)}{\partial \dot{\mathbf{x}}(t)}\right] = \mathbf{0}$$

$$\left.\frac{\partial \vartheta(\mathbf{x}(t), \dot{\mathbf{x}}(t), t)}{\partial \dot{\mathbf{x}}^T(t)}\right|_{t_f} \delta\mathbf{x}_f = 0$$

$$\left[\vartheta(\mathbf{x}(t_f), \dot{\mathbf{x}}(t_f), t_f) - \left.\frac{\partial \vartheta(\mathbf{x}(t), \dot{\mathbf{x}}(t), t)}{\partial \dot{\mathbf{x}}^T(t)}\right|_{t_f} \dot{\mathbf{x}}(t_f)\right] \delta t_f = 0$$

- Optimization with Differential Equation Constraints

$$J = \phi(\mathbf{x}(t_f), t_f) + \int_{t_0}^{t_f} \vartheta(\mathbf{x}(t), \mathbf{u}(t), t)\, dt$$

$$\dot{\mathbf{x}}(t) = \mathbf{f}(\mathbf{x}(t), \mathbf{u}(t), t)$$

$$H \equiv \vartheta(\mathbf{x}(t), \mathbf{u}(t), t) + \boldsymbol{\lambda}^T(t)\mathbf{f}(\mathbf{x}(t), \mathbf{u}(t), t)$$

$$\dot{\mathbf{x}}(t) = \frac{\partial H}{\partial \boldsymbol{\lambda}(t)} \equiv \mathbf{f}(\mathbf{x}(t), \mathbf{u}(t), t)$$

$$\dot{\boldsymbol{\lambda}}(t) = -\frac{\partial H}{\partial \mathbf{x}(t)} \equiv -\frac{\partial \vartheta(\mathbf{x}(t), \mathbf{u}(t), t)}{\partial \mathbf{x}(t)} - \left[\frac{\partial \mathbf{f}(\mathbf{x}(t), \mathbf{u}(t), t)}{\partial \mathbf{x}(t)}\right]^T \boldsymbol{\lambda}(t)$$

$$\frac{\partial H}{\partial \mathbf{u}(t)} = \mathbf{0}$$

$$\left.\left[\frac{\partial \phi(\mathbf{x}(t), t)}{\partial t} + \boldsymbol{\alpha}^T \frac{\partial \boldsymbol{\psi}(\mathbf{x}(t), t)}{\partial t} + H\right]\right|_{t_f} = 0$$

$$\boldsymbol{\lambda}(t_f) = \left\{ \frac{\partial \phi(\mathbf{x}(t), t)}{\partial \mathbf{x}(t)} + \left[\frac{\partial \boldsymbol{\psi}(\mathbf{x}(t), t)}{\partial \mathbf{x}(t)} \right]^T \boldsymbol{\alpha} \right\} \Bigg|_{t_f}$$

- Pontryagin's Optimal Control Necessary Conditions

$$J = \frac{1}{2} \int_{t_0}^{t_f} \mathbf{x}^T(t)\, \mathcal{Q}\, \mathbf{x}(t)\, dt$$

$$\dot{\mathbf{x}}(t) = \mathbf{f}(\mathbf{x}(t), t) + \mathbf{u}(t), \quad \mathbf{x}(t_0) = \mathbf{x}_0, \quad \mathbf{x}(t_f) = \mathbf{x}_f$$

$$\boldsymbol{\psi}(\mathbf{x}(t_f), t_f) = \mathbf{0}$$

$$H = \frac{1}{2} \mathbf{x}^T(t)\, \mathcal{Q}\, \mathbf{x}(t) + \boldsymbol{\lambda}^T(t)[\mathbf{f}(\mathbf{x}(t), t) + \mathbf{u}(t)]$$

$$\Phi(\mathbf{x}(t_f), t_f) \equiv \phi(\mathbf{x}(t_f), t_f) + \boldsymbol{\alpha}^T \boldsymbol{\psi}(\mathbf{x}(t_f), t_f)$$

$$|u_j(t)| \le u_{\max_j}, \quad j = 1, 2, \ldots, p$$

$$\dot{\mathbf{x}}(t) = \mathbf{f}(\mathbf{x}(t), t) + \mathbf{u}(t)$$

$$\dot{\boldsymbol{\lambda}}(t) = -\left[\frac{\partial \mathbf{f}(\mathbf{x}(t), t)}{\partial \mathbf{x}(t)} \right]^T \boldsymbol{\lambda}(t) - \mathcal{Q}\, \mathbf{x}(t)$$

$$\mathbf{u}(t) = \begin{bmatrix} s_1 u_{\max_1} \\ s_2 u_{\max_2} \\ \vdots \\ s_p u_{\max_p} \end{bmatrix}, \quad s_i = \text{sign}[\lambda_i(t)]$$

- Discrete-Time Control

$$J = \phi(\mathbf{x}_N, t_f) + \sum_{k=0}^{N-1} \vartheta_k(\mathbf{x}_k, \mathbf{u}_k, k)$$

$$\mathbf{x}_{k+1} = \mathbf{f}_k(\mathbf{x}_k, \mathbf{u}_k, k)$$

$$\boldsymbol{\psi}(\mathbf{x}_N, t_f) = \mathbf{0}$$

$$H_k \equiv \vartheta_k(\mathbf{x}_k, \mathbf{u}_k, k) + \boldsymbol{\lambda}_{k+1}^T \mathbf{f}_k(\mathbf{x}_k, \mathbf{u}_k, k)$$

$$\Phi(\mathbf{x}_N, t_f) \equiv \phi(\mathbf{x}_N, t_f) + \boldsymbol{\alpha}^T \boldsymbol{\psi}(\mathbf{x}_N, t_f)$$

$$\mathbf{x}_{k+1} = \frac{\partial H_k}{\partial \boldsymbol{\lambda}_{k+1}} \equiv \mathbf{f}_k(\mathbf{x}_k, \mathbf{u}_k, k)$$

$$\boldsymbol{\lambda}_k = \frac{\partial H_k}{\partial \mathbf{x}_k} \equiv \frac{\partial \vartheta_k(\mathbf{x}_k, \mathbf{u}_k, k)}{\partial \mathbf{x}_k} + \left[\frac{\partial \mathbf{f}_k(\mathbf{x}_k, \mathbf{u}_k, k)}{\partial \mathbf{x}_k}\right]^T \boldsymbol{\lambda}_{k+1}$$

$$\frac{\partial H_k}{\partial \mathbf{u}_k} = \mathbf{0}$$

$$\left[\frac{\partial \Phi(\mathbf{x}_k, t_f)}{\partial \Delta t} + \sum_{k=0}^{N-1} \frac{\partial H_k}{\partial \Delta t}\right] \delta \Delta t = 0$$

$$\frac{\partial \Phi(\mathbf{x}_k, t_f)}{\partial \Delta t} + \sum_{k=0}^{N-1} \frac{\partial H_k}{\partial \Delta t} = 0$$

$$\boldsymbol{\lambda}_N = \left\{\frac{\partial \phi(\mathbf{x}_k, t_f)}{\partial \mathbf{x}_k} + \left[\frac{\partial \psi(\mathbf{x}_k, t_f)}{\partial \mathbf{x}_k}\right]^T \alpha\right\}\Bigg|_N$$

- Linear Quadratic Regulator (Continuous-Time)

$$\dot{\mathbf{x}}(t) = F(t)\,\mathbf{x}(t) + B(t)\,\mathbf{u}(t), \quad \mathbf{x}(t_0) = \mathbf{x}_0$$

$$\mathbf{u}(t) = -L(t)\,\mathbf{x}(t)$$

$$\dot{S}(t) = -S(t)\,F(t) - F^T(t)\,S(t) + S(t)\,B(t)\,\mathcal{R}^{-1}(t)\,B^T(t)\,S(t) - \mathcal{Q}(t)$$

$$L(t) = \mathcal{R}^{-1}(t)\,B^T(t)\,S(t)$$

- Linear Quadratic Regulator (Discrete-Time)

$$\mathbf{x}_{k+1} = \Phi_k \mathbf{x}_k + \Gamma_k \mathbf{u}_k, \quad \mathbf{x}(t_0) = \mathbf{x}_0$$

$$\mathbf{u}_k = -L_k \mathbf{x}_k$$

$$S_k = \Phi_k^T S_{k+1} \Phi_k - \Phi_k^T S_{k+1} \Gamma_k \left[\Gamma_k^T S_{k+1} \Gamma_k + \mathcal{R}_k\right]^{-1} \Gamma_k^T S_{k+1} \Phi_k + \mathcal{Q}_k$$

$$L_k = \left[\Gamma_k^T S_{k+1} \Gamma_k + \mathcal{R}_k\right]^{-1} \Gamma_k^T S_{k+1} \Phi_k$$

- Stochastic Hamilton-Jacobi-Bellman Equation

$$\frac{\partial J^*(\mathbf{x}(t), t)}{\partial t} + \min_{\mathbf{u}(t)} \left\{ \vartheta(\mathbf{x}(t), \mathbf{u}(t), t) + \frac{\partial J^*(\mathbf{x}(t), \mathbf{u}(t), t)}{\partial \mathbf{x}^T(t)} \mathbf{f}(\mathbf{x}(t), \mathbf{u}(t), t) \right.$$
$$\left. + \frac{1}{2} \mathrm{Tr}\left[G(t)\,Q(t)\,G(t) \frac{\partial^2 J^*(\mathbf{x}(t), t)}{\partial \mathbf{x}^2(t)} \right] \right\} = 0$$

- Loop Transfer Recovery

$$G\,Q\,G^T = G\,Q_0 G^T + q^2 B\,B^T$$

$$F\left(\frac{P}{q^2}\right) + \left(\frac{P}{q^2}\right) F^T - q^2 \left(\frac{P}{q^2}\right) H^T R^{-1} H \left(\frac{P}{q^2}\right) + \frac{G\,Q_0 G^T}{q^2} + B\,B^T = 0$$

- Spacecraft Control

$$\dot{\mathbf{q}} = \frac{1}{2}\Xi(\mathbf{q})\boldsymbol{\omega} = \frac{1}{2}\Omega(\boldsymbol{\omega})\mathbf{q}$$
$$\dot{\boldsymbol{\omega}} = -J^{-1}[\boldsymbol{\omega}\times]J\boldsymbol{\omega} + J^{-1}\mathbf{L}$$

$$\mathbf{L} = [\boldsymbol{\omega}\times]J\boldsymbol{\omega} + 2J\left[\Xi^T(\mathbf{q}_d)\Xi(\mathbf{q})\right]^{-1}\left\{\frac{1}{4}(\boldsymbol{\omega}^T\boldsymbol{\omega})\Xi^T(\mathbf{q}_d) - \Xi^T(\dot{\mathbf{q}}_d)\Omega(\boldsymbol{\omega})\right.$$
$$\left. - \Xi^T(\ddot{\mathbf{q}}_d) - L_1\Xi^T(\mathbf{q}_d) - L_2\left[\frac{1}{2}\Xi^T(\mathbf{q}_d)\Omega(\boldsymbol{\omega}) + \Xi^T(\dot{\mathbf{q}}_d)\right]\right\}\mathbf{q}$$

Exercises

8.1 Suppose that both $\mathbf{x}(t_f)$ and t_f are free, but related by $\mathbf{x}(t_f) = \boldsymbol{\theta}(t_f)$. Derive the transversality condition to determine the final time for this constraint.

8.2 Consider minimizing the loss function in eqn. (8.22), with $\phi(\mathbf{x}(t_f), t_f) = 0$, equality constraint given by eqn. (8.19), and final time fixed. The continuous-time solution is given by the TPBVP shown in eqn. (8.27), with $\boldsymbol{\lambda}(t_f) = \mathbf{0}$. Using first-order finite difference approximations for the state and costate derivatives, develop simple discrete-time approximations to the TPBVP equations involving a constant sampling interval Δt. An alternative approach to this approximation is given by discretizing the loss function and equality constraint:

$$J = \Delta t \sum_{k=0}^{N-1} \vartheta(\mathbf{x}_k, \mathbf{u}_k, k)$$
$$\mathbf{x}_{k+1} = \mathbf{x}_k + \Delta t\, \mathbf{f}(\mathbf{x}_k, \mathbf{u}_k, k)$$

Now, with this discretization derive the associated TPVBP using the methods of §8.4. You will see that the equations associated with this TPBVP are not equivalent to the ones obtained by discretizing the continuous-time TBPVP equations. Under what conditions do both sets of equations give nearly identical solutions?

8.3 Take a Taylor series expansion of eqn. (8.50) to prove the expression given in eqn. (8.51).

8.4 The minimum energy required to charge a capacitor for a portable defibrillator using an RC circuit can be achieved by minimizing the following loss function:

$$J = \int_0^1 \left[\dot{v}(t) + \frac{1}{RC}v(t)\right]^2 dt, \quad v(0) = 0, \quad v(1) = 400 \text{ volts}$$

where R and C are constants, and $v(t)$ is the voltage. Using the methods of §8.1 show the associated Euler-Lagrange equations for this problem. Find the optimal trajectory for $v(t)$ that minimizes J in closed form. How does the trajectory change for increasing RC?

8.5 Consider the minimization of the following loss function:

$$J = \int_0^1 \left[x^2(t) + \dot{x}^2(t)\right] dt, \quad x(0) = 1, \quad x(1) = 0$$

Using the methods of §8.1 show the associated Euler-Lagrange equations for this problem. Find the optimal trajectory for $x(t)$ that minimizes J in closed form. Show that δJ is zero for all admissible perturbations $\delta x(t)$.

8.6 Consider the minimization of the following loss function:

$$J = \frac{1}{2}\int_0^{t_f} \dot{\mathbf{x}}^T(t)\dot{\mathbf{x}}(t)\, dt + \frac{1}{2}x_1^2(t_f), \quad (t_f \text{ free})$$

$$\mathbf{x}(0) = \begin{bmatrix}0 & 1\end{bmatrix}^T, \quad x_2(t_f) = -t_f$$

where $\mathbf{x} = \begin{bmatrix}x_1 & x_2\end{bmatrix}^T$. Using the variational methods of §8.1 derive the following transversality conditions for this problem:

$$\left.\frac{\partial\phi(\mathbf{x}(t),t)}{\partial\dot{\mathbf{x}}^T(t)}\right|_{t_f}\delta\mathbf{x}(t_f) + \left.\frac{\partial\vartheta(\mathbf{x}(t),\dot{\mathbf{x}}(t),t)}{\partial\dot{\mathbf{x}}^T(t)}\right|_{t_f}\delta\mathbf{x}(t_f)$$
$$+\left.\left[\vartheta(\mathbf{x}(t),\dot{\mathbf{x}}(t),t) - \frac{\partial\vartheta(\mathbf{x}(t),\dot{\mathbf{x}}(t),t)}{\partial\dot{\mathbf{x}}^T(t)}\dot{\mathbf{x}}(t)\right]\right|_{t_f}\delta t_f = 0$$

where $\phi(\mathbf{x}(t_f),t_f) \equiv x_1^2(t_f)/2$ and $\vartheta(\mathbf{x}(t),\dot{\mathbf{x}}(t),t) \equiv \dot{\mathbf{x}}^T(t)\dot{\mathbf{x}}(t)/2$. Using the Euler-Lagrange equations with these transversality conditions, determine the solutions for the optimal $\mathbf{x}(t)$ and final time t_f. If t_f is fixed rather than free, how would the optimal trajectory differ?

8.7 Consider the minimization of the following loss function:

$$J = \frac{1}{2}\int_0^1 \left[x^2(t) + \dot{x}^2(t)\right] dt, \quad x(0) = 1, \quad x(1) = \text{free}$$

Using the methods of §8.1 show the associated Euler-Lagrange equations and transversality conditions for this problem. Find the optimal trajectory for $x(t)$ that minimizes J in closed form.

8.8 Consider the minimization of the following loss function:

$$J = \frac{1}{2}\int_{-1}^1 \left[u^2(t) + 5x^2(t)\right] dt$$

$$\dot{x}(t) = -2x(t) + u(t), \quad x(0) = 1$$

Using the methods of §8.1 show the associated Euler-Lagrange equations and transversality conditions for this problem (note: a boundary condition is given at $t = 0$, but the integral is given from $t = -1$ to $t = +1$). Find the optimal trajectory for $x(t)$ that minimizes J in closed form.

8.9 Consider the minimization of the following loss function:

$$J = \frac{1}{2}\int_0^{t_f} \left[1 + u^2(t)\right] dt, \quad (t_f \text{ free})$$
$$\dot{x}(t) = -a\,x(t) + u(t), \quad x(0) = 1, \quad x(t_f) = 1$$

where a is a positive constant. Using the methods of §8.1 show the associated Euler-Lagrange equations and transversality conditions for this problem. Find the optimal trajectory for $x(t)$ that minimizes J in closed form.

8.10 Consider the minimization of the following loss function:

$$J = \frac{1}{2}\int_0^{t_f} \sqrt{1 + \dot{x}^2(t)}\, dt, \quad (t_f \text{ free})$$
$$x(0) = 0, \quad x(t_f) = -5t_f + 15$$

Using the methods of §8.1 and the results from exercise 8.6 show the associated Euler-Lagrange equations and transversality conditions for this problem. Find the optimal trajectory for $x(t)$ and the final time t_f that minimize J in closed form.

8.11 ♣ Consider the following functional:

$$J = \int_{t_0}^{t_f} \vartheta(\mathbf{x}(t), \dot{\mathbf{x}}(t), \ddot{\mathbf{x}}(t), t)\, dt + \phi(\mathbf{x}(t_f), \dot{\mathbf{x}}(t_f), t_f)$$

where t_f is fixed. Express δJ in terms of $\delta\mathbf{x}(t)$ and endpoint perturbations to derive the Euler-Lagrange equations and transversality conditions (hint: integrate by parts twice). Note that this is a generalized extension of the first problem, not a first problem.

8.12 Consider the minimization of the following loss function:

$$J = \phi(\mathbf{x}(t_f), t_f) + \int_{t_0}^{t_f} \vartheta(\mathbf{x}(t), \mathbf{u}(t), t)\, dt$$

Suppose that instead of a differential constraint given by eqn. (8.19) we have the general (possibly nonlinear) constraint given by

$$\mathbf{g}(\mathbf{x}(t), \dot{\mathbf{x}}(t), t) = 0$$

Using a set of Lagrange multipliers derive the associated Euler-Lagrange equations and transversality conditions for this problem. First assume that t_f is fixed, then allow it to be free.

8.13 Consider the minimization of the following loss function:

$$J = \frac{1}{2}\int_0^{t_f} u^2(t)\, dt + \frac{1}{2}x_1^2(t_f), \quad (t_f \text{ free})$$
$$\dot{\mathbf{x}}(t) = \begin{bmatrix} 0 & 1 \\ 1 & 0 \end{bmatrix}\mathbf{x}(t) + \begin{bmatrix} 0 \\ 1 \end{bmatrix} u(t)$$
$$x_1(0) = 0, \quad x_2(0) = \text{free}$$
$$x_2(t_f) = x_1^2(t_f) - 1$$

where $\mathbf{x} = [x_1 \; x_2]^T$. Using the Hamiltonian approach of §8.2 derive the state and costate equations. Also, specify the appropriate boundary conditions. The optimal input $u(t)$ is a function of what two time functions? How would the answer to the solution of this minimization problem change if the constraint $0 \leq |u(t)| \leq 1$ were added to the problem statement?

8.14 In this exercise you will test the robustness of the open-loop control law developed in example 8.1. First, reproduce the results shown in Figure 8.2. Then, multiply the control torque $u(t)$ in example 8.1 by some scalar, which simulates an error in the inertia J, and use this control input with the identical boundary conditions shown in the example. How do the state and control input trajectories change for various scalar multiplication factors?

8.15 In example 8.1 a rigid body constrained to rotate about a fixed axis is considered with the final time fixed at $t_f = T$. Consider the following boundary conditions: $\theta(0) = \theta_0$, $\dot{\theta}(0) = \dot{\theta}_0$, $\theta(T) = 0$, and $\dot{\theta}(T) = 0$. Also, consider only piecewise continuous controls satisfying $|u(t)| \leq 1$. We seek to minimize the maneuver "time-to-go"

$$J = c \int_0^T dt$$

where c is a positive scale factor whose arbitrary value will be chosen to accomplish a useful normalization of the costate variables. Using the Hamiltonian approach with Pontryagin's Principle of §8.3 derive the state and costate equations. Show that only one sign change (at most) can occur in $u(t)$.

8.16 ♣ Using the equations derived in exercise 8.15, show that if $\lambda_1(t)$ and $\lambda_2(t)$ are solutions to the costate differential equations, then $\alpha\lambda_1(t)$ and $\alpha\lambda_2(t)$ are also solutions for α = an arbitrary positive constant. Deduce that the α-scaling on λ_i dictates a specific c value:

$$c = -[\lambda_1(T)\, x_2(T) + \lambda_2(T)\, u(T)]$$

Since an infinity of linearly scaled costates generate the same control, we take advantage of this truth to scale initial conditions on the λ's so that the initial costates lie on the unit circle

$$\lambda_1^2(0) + \lambda_2^2(0) = 1$$

or, alternatively, we can define the complete family of trajectories by introducing an initial phase γ such that $\lambda_1(0) = \cos\gamma$ and $\lambda_2(0) = \sin\gamma$, where $0 \leq \gamma \leq 360°$. Show that the optimal control is given by $u(t) = -\text{sign}(\sin\gamma - t\cos\gamma)$, and that the switch times, t_s, are related to the γ-values as $t_s = \tan\gamma$. Construct an analytical solution for the $u = +1$ and $u = -1$ trajectories. Show that the control in the second quadrant of a $[x_1(t), x_2(t)]$ phase plot switches from positive to negative when the positive torque trajectories intersect the switching curve $x_1(t) = -x_2^2(t)/2$, whereas the control in the fourth quadrant of a $[x_1(t), x_2(t)]$ phase plot switches from negative to positive when the initially negative torque trajectories intersect the switching curve $x_1(t) = +x_2^2(t)/2$. Construct a global portrait of the time optimal "bang-bang" trajectories.

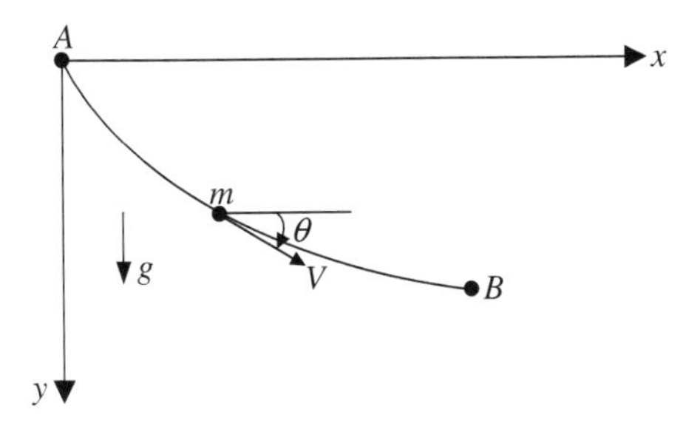

Figure 8.10: The Brachistochrone Problem

8.17 Consider the following second-order dynamical system:

$$\dot{x}_1(t) = x_2(t)$$
$$\dot{x}_2(t) = u(t)$$

where $x_1(t)$ is equivalent to $\theta(t)$ from example 8.1. We now wish to develop a control-rate penalty technique that minimizes the loss function

$$J = \frac{1}{2}\int_0^{t_f} \left[w^2 u^2(t) + \dot{u}^2(t)\right] dt$$

where $u(t)$ is assumed to have two continuous derivatives, and w is a positive constant weight. We can easily convert this loss function into a standard form by simply introducing a new "state variable" $x_3(t) = u(t)$ and defining a new control variable $v(t) \equiv \dot{u}(t)$. Thus we seek to minimize

$$J = \frac{1}{2}\int_0^{t_f} \left[w^2 x_3^2(t) + v^2(t)\right] dt$$

subject to

$$\dot{x}_1(t) = x_2(t)$$
$$\dot{x}_2(t) = x_3(t)$$
$$\dot{x}_3(t) = v(t)$$

Using the Hamiltonian approach of §8.2 derive the state and costate equations. Assume that the state boundary conditions are given by $x_1(0) = \theta_0$, $x_2(0) = \dot{\theta}_0$, $x_1(t_f) = \theta_f$, and $x_2(t_f) = \dot{\theta}_f$. Also, since we require that the control be zero initially and vanish upon completion, we have $u(0) = 0$ and $u(t_f) = 0$. Find analytical expressions for $x_1(t)$ and $u(t)$ that minimize J in closed form. How does the solution change for the special case where $w = 0$?

8.18 Consider the classical *brachistochrone problem* shown in Figure 8.10. Given two points, A and B, in space with A higher than B, but not vertically above B, what shape of wire connecting A to B will have the property that a bead sliding along it under gravity gets from A to B in the shortest time? Using the principle of energy we can write

$$\frac{1}{2}m[\dot{x}^2(t)+\dot{y}^2(t)] = mg\,y(t)$$

where m is the mass of the bead and g is the gravity constant. This equation can be written as

$$\frac{dx(t)}{dt}\left[\sqrt{\frac{\dot{y}(t)}{\dot{x}(t)}}+1\right] = \sqrt{2g\,y(t)}$$

We wish to minimize the time taken, so

$$J = \int_0^T dt = c\int_0^1 \left[\frac{1+\dot{y}^2(t)/\dot{x}^2(t)}{y(t)}\right]^{1/2} dx$$

where $c = 1/\sqrt{2g}$. But since $\dot{y}(t)/\dot{x}(t) = dy/dx$, the minimization problem can be stated as

$$J = c\int_0^1 f\left(y(x),\, \frac{dy}{dx}\right) dx$$

$$y(0) = 0, \quad y(1) = 1$$

where

$$f(y,s) = \left[\frac{1+s^2}{y}\right]^{1/2}$$

The velocity components can be written as[5]

$$\dot{x}(t) = V(y)\cos\theta(t) \tag{8.193}$$

$$\dot{y}(t) = V(y)\sin\theta(t) \tag{8.194}$$

where the velocity is given by $V(y) = \sqrt{V_0^2 + 2g\,y(t)}$, and V_0 is the initial velocity at point A. Solve the Euler-Lagrange equations to show that the paths for $x(t)$ and $y(t)$ are cycloids, i.e., paths generated by a point on a circle rolling without slipping in a horizontal directions, and that $\dot{\theta}$ is constant.

8.19 Verify by direct substitution that the solution of HJB equation in eqn. (8.58) is indeed given by eqn. (8.59).

8.20 Take the second variation of eqn. (8.54). What are the sufficient conditions on $\mathcal{Q}$, $\mathcal{R}$, and S_f to guarantee a minimum?

8.21 In the LQR loss function of eqn. (8.54) no weighting between the cross-correlation of the state $\mathbf{x}(t)$ and input $\mathbf{u}(t)$ is given. Suppose that we now wish to minimize the following loss function, which includes this cross-weighting:

$$J = \frac{1}{2}\mathbf{x}^T(t_f)S_f\mathbf{x}(t_f) + \frac{1}{2}\int_{t_0}^{t_f} \left[\mathbf{x}^T(t)\ \mathbf{u}^T(t)\right]\begin{bmatrix}\mathcal{Q}(t) & \mathcal{N}(t)\\ \mathcal{N}^T(t) & \mathcal{R}(t)\end{bmatrix}\begin{bmatrix}\mathbf{x}(t)\\ \mathbf{u}(t)\end{bmatrix} dt$$

where $\mathcal{N}(t)$ is the cross-weighting matrix. Using the methods of §8.5.1, derive new LQR results using a Riccati transformation that minimizes this loss function.

8.22 A similar loss function to the one shown in exercise 8.21 can be derived for the discrete-time case:

$$J = \frac{1}{2}\mathbf{x}_N^T S_f \mathbf{x}_N + \sum_{k=0}^{N-1} \begin{bmatrix}\mathbf{x}_k^T & \mathbf{u}_k^T\end{bmatrix} \begin{bmatrix} \mathcal{Q}_k & \mathcal{N}_k \\ \mathcal{N}_k^T & \mathcal{R}_k \end{bmatrix} \begin{bmatrix}\mathbf{x}_k \\ \mathbf{u}_k\end{bmatrix}$$

Using the methods of §8.5.2, derive new LQR results using a Riccati transformation that minimizes this loss function.

8.23 Consider the minimization of the following discrete-time loss function:[2]

$$J = \frac{1}{2}\sum_{k=0}^{9} u_k^2$$

$$x_{k+1} = x_k + \gamma\, u_k$$

$$x_0 = 1, \quad x_{10} = 0$$

where γ is a constant. Determine a closed-form solution for x_k that minimizes this loss function and meets the desired boundary conditions.

8.24 Prove that the discrete-time Riccati equation in eqn. (8.103) is equivalent to eqn. (8.102).

8.25 Starting with the expression given in eqn. (8.121), prove the expression given in eqn. (8.122) using the methods of §5.4.1 and the definition of the innovations process in §6.4.2.2.

8.26 Prove that the eigenvalues of the system matrices in eqns. (8.133) and (8.135) are equivalent to each other.

8.27 ♣ Using the Stochastic Hamilton-Jacobi-Bellman equation of eqn. (8.137) to prove the Separation Theorem for continuous-time systems.

8.28 Starting with the Kalman filter equations of eqn. (5.54), show that the mean and covariance of $\hat{\mathbf{x}}_{k+1}$ are given by the expressions in eqn. (8.152).

8.29 Substituting the gain L_k, given by eqn. (8.98), show that eqn. (8.154) is equivalent to eqn. (8.153). Also, show that the matrix $[\Gamma_k S_{k+1} \Gamma_k + \mathcal{R}_k]$ is always positive definite.

8.30 Starting with the Kalman filter estimator form in eqn. (5.54a) and truth model in eqn. (8.141), prove the eigenvalue separation of the combined estimator and controller system by showing that the closed-loop LQG dynamics are given by

$$\begin{bmatrix}\mathbf{x}_{k+1} \\ \tilde{\mathbf{x}}_{k+1}\end{bmatrix} = \begin{bmatrix} \Phi - \Gamma L & -\Gamma L \\ 0 & \Phi(I - K H) \end{bmatrix} \begin{bmatrix}\mathbf{x}_k \\ \tilde{\mathbf{x}}_k\end{bmatrix} + \begin{bmatrix} \Upsilon & 0 \\ -\Upsilon & \Phi K \end{bmatrix} \begin{bmatrix}\mathbf{w}_k \\ \mathbf{v}_k\end{bmatrix}$$

where $\tilde{\mathbf{x}}_k \equiv \hat{\mathbf{x}}_k - \mathbf{x}_k$, and all system and covariance matrices are assumed to be constants.

8.31 Show that eqn. (8.164) is equivalent to eqn. (8.165) by using the matrix inversion lemma.

8.32 Reproduce the results shown for the LTR system in example 8.3. Create synthetic measurements using various standard deviations for the measurement noise with the linear system described in the example. Using the LTR filter gains test the performance of the overall system executing various simulated runs. At what noise levels in the measurements can you find a satisfactory compromise between closed-loop stability margins and high-frequency noise rejection? Discuss the metrics used to qualify this compromise in your simulations.

8.33 In this exercise you will design an optimal controller involving a terminal guidance system for satellite rendezvous. Although the relative equations of the motion for two spacecraft flying in formation are highly nonlinear, if the spacecraft are close to each other, then a linearized solution works well for short periods. A commonly used set of linearized equations is given by the Clohessy-Wiltshire equations or Hill's equations:[34]

$$\begin{aligned}\ddot{r}(t) - 2n\,\dot{s}(t) - 3n^2 r(t) &= F_r(t)\\ \ddot{s}(t) + 2n\,\dot{r}(t) &= F_s(t)\\ \ddot{z}(t) + n^2 z(t) &= F_z(t)\end{aligned}$$

where $r(t)$ is the radial direction, $s(t)$ is the cross-track direction, $z(t)$ is perpendicular to the reference orbit plane, n is the mean motion (see §3.8.2) of the leader spacecraft, and $F_r(t)$, $F_s(t)$, and $F_z(t)$ are control variables. Assuming a low-Earth orbit (with n = 0.0011 rad/sec), design a steady-state LQR controller for this system. For your design assume that the position states in all directions are initially about 1 km with zero velocity errors, and bring the errors to zero within 20 minutes (set $\mathcal{Q}$ to be the identity matrix and adjust $\mathcal{R}$ to meet the design specifications). Use your LQR steady-state control input on the full nonlinear equations of motion, given by

$$\begin{aligned}\ddot{r}(t) - 2n\,\dot{s}(t) - n^2[a + r(t)][1 - g(t)] &= F_r(t)\\ \ddot{s}(t) + 2n\,\dot{r}(t) - n^2 s(t)[1 - g(t)] &= F_s(t)\\ \ddot{z}(t) + n^2 z(t)\,g(t) &= F_z(t)\end{aligned}$$

where

$$g(t) \equiv \frac{a^3}{\{[a + r(t)]^2 + s^2(t) + z^2(t)\}^{3/2}}$$

and a is the semimajor axis given by $a = 6,906.4$ km. How well does your linear controller work for other (larger) initial conditions?

8.34 ♣ Prove the identities in eqns. (8.189) and (8.192). Show that the determinant of the matrix $\Xi^T(\mathbf{q}_d)\Xi(\mathbf{q})$, which is used in the control law given by eqn. (8.188), is given by $\mathbf{q}_d^T\mathbf{q}$. Finally, prove that eqn. (8.188) reduces down to eqn. (8.190) when L_1 and L_2 are scalars.

8.35 A spacecraft equipped with reaction wheels[35] can also be used for attitude maneuvering purposes. Although the spacecraft can no longer be considered a rigid body with the internal wheels, Euler's rotational equations of §3.7.2 can still be used to describe the overall system. The equations of motion using reaction wheels can be written as

$$(J-\bar{J})\dot{\omega} = -[\omega\times](J\,\omega+\bar{J}\,\bar{\omega}) - \bar{\mathbf{u}}$$
$$\bar{J}(\dot{\bar{\omega}}+\dot{\omega}) = \bar{\mathbf{u}}$$

where J is the inertia of the spacecraft which now includes the wheels, $\bar{J}$ is the inertia of the wheels, $\bar{\omega}$ is the wheel angular velocity vector relative to the spacecraft, and $\bar{\mathbf{u}}$ is the wheel torque vector. Derive a wheel control law that provides the linear error-dynamics given by eqn. (8.180). Consider a wheel inertia matrix given by

$$\bar{J} = \begin{bmatrix} 1 & 0 & 0 \\ 0 & 1 & 0 \\ 0 & 0 & 1 \end{bmatrix} \text{ kg-m}^2$$

Assuming that the wheels initially begin at rest, $\bar{\omega}(t_0) = \mathbf{0}$, use the derived wheel control law to maneuver the spacecraft along a desired trajectory, with the simulation parameters shown in example 8.4. Also, test the robustness of the wheel control law by using a different spacecraft inertia matrix in the assumed model. How robust is this control law to parameter variations?

8.36 Consider the nonlinear equations of motion for a highly maneuverable aircraft given in exercise 4.22. Neglecting higher-order terms we can write the equations of motion in linear form as

$$\begin{bmatrix} \dot{\alpha}(t) \\ \dot{\theta}(t) \\ \ddot{\theta}(t) \end{bmatrix} = \begin{bmatrix} -0.88 & 0 & 1 \\ 0 & 0 & 1 \\ -4.208 & 0 & -0.396 \end{bmatrix} \begin{bmatrix} \alpha(t) \\ \theta(t) \\ \dot{\theta}(t) \end{bmatrix} + \begin{bmatrix} -0.22 \\ 0 \\ -20.967 \end{bmatrix} \delta_E(t) \tag{8.195}$$

Design a steady-state LQR controller to bring the states with initial conditions of $\alpha_0 = 1$ deg, $\theta = 10$ deg, and $\dot{\theta}_0 = 0$ deg/sec to zero within 15 to 20 seconds. Use your LQR steady-state control input on the full nonlinear equations of motion. How well does your linear controller work for other (larger) initial conditions?

8.37 Consider using a linear Kalman filter to estimate the states for the system described in exercises 4.22 and 8.36. Design a filter with the linear model shown in exercise 8.36 using measurements of angle of attack, $\alpha(t)$, and pitch angle, $\theta(t)$. Assume standard deviations of the measurement errors to be the same as the ones given in exercise 4.4. Tune the process noise covariance matrix, Q, to yield sufficiently filtered estimates with adequate filter convergence properties. Use the designed estimator in an LQG design to control the aircraft with the gain developed in exercise 8.36. Try various initial conditions in the actual system as well as the Kalman filter to test the overall robustness of your design. Also, use measurements of only the pitch angle and compare the results with those obtained using measurements of both angle of attack and pitch in the Kalman filter.

8.38 Example 4.5 shows mass, stiffness, and damping matrices of a 4-mode system. Convert the continuous-time model in discrete-time using the methods of §3.5 with a sampling interval of 0.1 seconds. Assuming initial conditions of one for the position states and zero for the velocity states, design an LQR controller to bring all states to zero within 10 seconds. Then, use a Kalman filter to estimate all states from position measurements only. Assume that the standard deviation of the measurement noise is given by $\sqrt{1\times 10^{-5}}$ for all measurements. Add discrete-time process noise into the velocity states only (i.e., assume that the kinematically relationships are exact, as discussed in §7.4.1, so do not add process noise to these states). Assume that the discrete-time standard deviation for the process noise is given by 0.1 for all velocity states. Implement an LQG controller using the Kalman filter estimates in the control law. Try various values for the process noise and measurement noise covariances to generate the true states. Discuss the performance of the overall controller to these variations.

References

[1] Kirk, D.E., *Optimal Control Theory: An Introduction*, Prentice Hall, Englewood Cliffs, NJ, 1970.

[2] Sage, A.P. and White, C.C., *Optimum Systems Control*, Prentice Hall, Englewood Cliffs, NJ, 2nd ed., 1977.

[3] Bryson, A.E., *Dynamic Optimization*, Addison Wesley Longman, Menlo Park, CA, 1999.

[4] Gelfand, I.M. and Fomin, S.V., *Calculus of Variations*, Prentice Hall, Englewood Cliffs, NJ, 1963.

[5] Bryson, A.E. and Ho, Y.C., *Applied Optimal Control*, Taylor & Francis, London, England, 1975.

[6] Pontryagin, L.S., Boltyanskii, V.G., Gamkrelidze, R.V., and Mishchenko, E.F., *The Mathematical Theory of Optimal Processes*, John Wiley Interscience, New York, NY, 1962.

[7] Bellman, R., *Dynamic Programming*, Princeton University Press, Princeton, NJ, 1957.

[8] Bryson, A.E., *Applied Linear Optimal Control: Examples and Algorithms*, Cambridge University Press, Cambridge, MA, 2002.

[9] Franklin, G.F., Powell, J.D., and Workman, M., *Digital Control of Dynamic Systems*, Addison Wesley Longman, Menlo Park, CA, 3rd ed., 1998.

[10] Athans, M., "The Role and Use of the Stochastic Linear-Quadratic-Gaussian Problem in Control System Design," *IEEE Transactions on Automatic Control*, Vol. AC-16, No. 6, Dec. 1971, pp. 529–552.

[11] Åström, K.J., *Introduction to Stochastic Control Theory*, Academic Press, New York, NY, 1970.

[12] Davis, M., *Linear Estimation and Stochastic Control*, Chapman and Hall, London, England, 1977.

[13] Stengle, R.F., *Optimal Control and Estimation*, Dover Publications, New York, NY, 1994.

[14] Anderson, B.D.O. and Moore, J.B., *Optimal Control: Linear Quadratic Methods*, Prentice Hall, Englewood Cliffs, NJ, 1990.

[15] Tse, E., "On the Optimal Control of Stochastic Linear Systems," *IEEE Transactions on Automatic Control*, Vol. AC-16, No. 6, Dec. 1971, pp. 776–785.

[16] Doyle, J.C., "Guaranteed Margins in LQG Regulators," *IEEE Transactions on Automatic Control*, Vol. AC-23, No. 4, Aug. 1978, pp. 664–665.

[17] Doyle, J.C. and Stein, G., "Robustness with Observers," *IEEE Transactions on Automatic Control*, Vol. AC-24, No. 4, Aug. 1979, pp. 607–611.

[18] Maciejowski, J.M., *Multivariable Feedback Design*, Addison-Wesley Publishing Company, Wokingham, UK, 1989.

[19] Phillips, C.L. and Harbor, R.D., *Feedback Control Systems*, Prentice Hall, Englewood Cliffs, NJ, 1996.

[20] Kuo, B.C., *Automatic Control Systems*, Prentice Hall, Englewood Cliffs, NJ, 6th ed., 1991.

[21] Nise, N.S., *Control Systems Engineering*, Addison-Wesley Publishing, Menlo Park, CA, 2nd ed., 1995.

[22] Ogata, K., *Modern Control Engineering*, Prentice Hall, Upper Saddle River, NJ, 1997.

[23] Palm, W.J., *Modeling, Analysis, and Control of Dynamic Systems*, John Wiley & Sons, New York, NY, 2nd ed., 1999.

[24] Dorf, R.C. and Bishop, R.H., *Modern Control Systems*, Addison Wesley Longman, Menlo Park, CA, 1998.

[25] Kwakernaak, H. and Sivan, S., *Linear Optimal Control Systems*, Wiley Interscience, New York, NY, 1972.

[26] Scrivener, S.L. and Thompson, R.C., "Survey of Time-Optimal Attitude Maneuvers," *Journal of Guidance, Control, and Dynamics*, Vol. 17, No. 2, March-April 1994, pp. 225–233.

[27] Vadali, S.R. and Junkins, J.L., "Optimal Open-Loop and Stable Feedback Control of Rigid Spacecraft Maneuvers," *The Journal of the Astronautical Sciences*, Vol. 32, No. 2, April-June 1984, pp. 105–122.

[28] Junkins, J.L. and Turner, J.D., *Optimal Spacecraft Rotational Maneuvers*, Elsevier, New York, NY, 1986.

[29] Schaub, H. and Junkins, J.L., *Analytical Mechanics of Aerospace Systems*, American Institute of Aeronautics and Astronautics, Inc., New York, NY, 2003.

[30] Wie, B., *Space Vehicle Dynamics and Control*, American Institute of Aeronautics and Astronautics, Inc., New York, NY, 1998.

[31] Paielli, R.A. and Bach, R.E., "Attitude Control with Realization of Linear Error Dynamics," *Journal of Guidance, Control, and Dynamics*, Vol. 16, No. 1, Jan.-Feb. 1993, pp. 182–189.

[32] Schaub, H., Akella, M.R., and Junkins, J.L., "Adaptive Control of Nonlinear Attitude Motions Realizing Linear Closed Loop Dynamics," *Journal of Guidance, Control, and Dynamics*, Vol. 24, No. 1, Jan.-Feb. 2001, pp. 95–100.

[33] Shuster, M.D., "A Survey of Attitude Representations," *Journal of the Astronautical Sciences*, Vol. 41, No. 4, Oct.-Dec. 1993, pp. 439–517.

[34] Wertz, J.R., "Satellite Relative Motion," *Mission Geometry: Orbit and Constellation Design and Management*, chap. 10, Microcosm Press, El Segundo, CA and Kluwer Academic Publishers, The Netherlands, 2001.

[35] Markley, F.L., "Attitude Dynamics," *Spacecraft Attitude Determination and Control*, edited by J.R. Wertz, chap. 16, Kluwer Academic Publishers, The Netherlands, 1978.

A

Matrix Properties

THIS appendix provides a reasonably comprehensive account of matrix properties, which are used in the linear algebra of estimation and control theory. Several theorems are shown, but are not proven here; those proofs given are *constructive* (i.e., suggest an algorithm). The account here is thus not satisfactorily self-contained, but references are provided where rigorous proofs may be found.

A.1 Basic Definitions of Matrices

The system of m linear equations

$$\begin{aligned} y_1 &= a_{11}x_1 + a_{12}x_2 + \cdots + a_{1n}x_n \\ y_2 &= a_{21}x_1 + a_{22}x_2 + \cdots + a_{2n}x_n \\ &\vdots \\ y_m &= a_{m1}x_1 + a_{m2}x_2 + \cdots + a_{mn}x_n \end{aligned} \tag{A.1}$$

can be written in matrix form as

$$\mathbf{y} = A\mathbf{x} \tag{A.2}$$

where $\mathbf{y}$ is an $m \times 1$ vector, $\mathbf{x}$ is an $n \times 1$ vector (see §A.2 for a definition of a vector) and A is an $m \times n$ matrix, with

$$\mathbf{y} = \begin{bmatrix} y_1 \\ y_2 \\ \vdots \\ y_m \end{bmatrix}, \quad \mathbf{x} = \begin{bmatrix} x_1 \\ x_2 \\ \vdots \\ x_n \end{bmatrix}, \quad A = \begin{bmatrix} a_{11} & a_{12} & \cdots & a_{1n} \\ a_{21} & a_{22} & \cdots & a_{2n} \\ \vdots & \vdots & \ddots & \vdots \\ a_{m1} & a_{m2} & \cdots & a_{mn} \end{bmatrix} \tag{A.3}$$

If $m = n$, then the matrix A is *square*.

Matrix Addition, Subtraction, and Multiplication

Matrices can be added, subtracted, or multiplied. For addition and subtraction, all matrices must of the same dimension. Suppose we wish to add/substract two matrices A and B:

$$C = A \pm B \tag{A.4}$$

Then each element of C is given by $c_{ij} = a_{ij} \pm b_{ij}$. Matrix addition and subtraction are both commutative, $A \pm B = B \pm A$, and associative, $(A \pm B) \pm C = A \pm (B \pm C)$. Matrix multiplication is much more complicated though. Suppose we wish to multiply two matrices A and B:

$$C = A\,B \tag{A.5}$$

This operation is valid only when the number of columns of A is equal to the number of rows of B (i.e., A and B must be *conformable*). The resulting matrix C will have rows equal to the number of rows of A and columns equal to the number of columns of B. Thus, if A has dimension $m \times n$ and B has dimension $n \times p$, then C will have dimension $m \times p$. The c_{ij} element of C can be determined by

$$c_{ij} = \sum_{k=1}^{n} a_{ik} b_{kj} \tag{A.6}$$

for all $i = 1, 2, \ldots, m$ and $j = 1, 2, \ldots, p$. Matrix multiplication is associative, $A\,(B\,C) = (A\,B)\,C$, and distributive, $A\,(B + C) = A\,B + A\,C$, but not commutative in general, $A\,B \neq B\,A$. In some cases though if $A\,B = B\,A$, then A and B are said to *commute*.

The transpose of a matrix, denoted A^T, has rows that are the columns of A and columns that are the rows of A. The transpose operator has the following properties:

$$(\alpha A)^T = \alpha A^T, \text{ where } \alpha \text{ is a scalar} \tag{A.7a}$$

$$(A + B)^T = A^T + B^T \tag{A.7b}$$

$$(A\,B)^T = B^T A^T \tag{A.7c}$$

If $A = A^T$, then A is said to be a *symmetric* matrix. Also, if $A = -A^T$, then A is said to be a *skew symmetric* matrix.

Matrix Inverse

We now discuss the properties of the matrix inverse. Suppose we are given both $\mathbf{y}$ and A in eqn. (A.2), and we want to determine $\mathbf{x}$. The following terminology should be noted carefully: if $m > n$, the system in eqn. (A.2) is said to be *overdetermined* (there are more equations than unknowns). Under typical circumstances we will find that the exact solution for $\mathbf{x}$ does not exist; therefore algorithms for *approximate* solutions for $\mathbf{x}$ are usually characterized by some measure of *how well* the linear equations are satisfied. If $m < n$, the system in eqn. (A.2) is said to be *underdetermined* (there are fewer equations than unknowns). Under typical circumstances, an infinity of exact solutions for $\mathbf{x}$ exist; therefore solution algorithms have implicit some criterion for selecting a particular solution from the infinity of possible or feasible $\mathbf{x}$ solutions. If $m = n$ the system is said to be *determined*, under typical (but certainly not universal) circumstances, a unique exact solution for $\mathbf{x}$ exists. To determine $\mathbf{x}$ for this case, the matrix inverse of A, denoted by A^{-1}, is used. Let A be an $n \times n$ matrix. The following statements are equivalent:

- A has linearly independent columns.
- A has linearly independent rows.
- The inverse satisfies $A^{-1}A = A\,A^{-1} = I$

where I is an $n \times n$ identity matrix:

$$I = \begin{bmatrix} 1 & 0 & \cdots & 0 \\ 0 & 1 & \cdots & 0 \\ \vdots & \vdots & \ddots & \vdots \\ 0 & 0 & \cdots & 1 \end{bmatrix} \tag{A.8}$$

A *nonsingular* matrix is a matrix whose inverse exists (likewise A^T is nonsingular):

$$(A^{-1})^{-1} = A \tag{A.9a}$$

$$(A^T)^{-1} = (A^{-1})^T \equiv A^{-T} \tag{A.9b}$$

Furthermore, let A and B be $n \times n$ matrices. The matrix product $A\,B$ is nonsingular if and only if A and B are nonsingular. If this condition is met, then

$$(A\,B)^{-1} = B^{-1}A^{-1} \tag{A.10}$$

Formal proof of this relationship and other relationships are given in Ref. [1]. The inverse of a square matrix A can be computed by

$$A^{-1} = \frac{\text{adj}(A)}{\det(A)} \tag{A.11}$$

where adj(A) is the *adjoint* of A and det(A) is the *determinant* of A. The adjoint and determinant of a matrix with large dimension can ultimately be broken down to a series of 2×2 matrix cases, where the adjoint and determinant are given by

$$\text{adj}(A_{2\times 2}) = \begin{bmatrix} a_{22} & -a_{12} \\ -a_{21} & a_{11} \end{bmatrix} \tag{A.12a}$$

$$\det(A_{2\times 2}) = a_{11}a_{22} - a_{12}a_{21} \tag{A.12b}$$

Other determinant identities are given by

$$\det(I) = 1 \tag{A.13a}$$

$$\det(A\,B) = \det(A)\det(B) \tag{A.13b}$$

$$\det(A\,B) = \det(B\,A) \tag{A.13c}$$

$$\det(A\,B + I) = \det(B\,A + I) \tag{A.13d}$$

$$\det(A + \mathbf{x}\mathbf{y}^T) = \det(A)\,(1 + \mathbf{y}^T A^{-1}\mathbf{x}) \tag{A.13e}$$

$$\det(A)\det(D + C\,A^{-1}B) = \det(D)\det(A + B\,D^{-1}C) \tag{A.13f}$$

$$\det(A^\alpha) = [\det(A)]^\alpha,\ \alpha \text{ must be positive if } \det(A) = 0 \tag{A.13g}$$

$$\det(\alpha\,A) = \alpha^n \det(A) \tag{A.13h}$$

$$\det(A_{3\times 3}) \equiv \det\left(\left[\mathbf{a}\ \mathbf{b}\ \mathbf{c}\right]\right) = \mathbf{a}^T[\mathbf{b}\times]\mathbf{c} = \mathbf{b}^T[\mathbf{c}\times]\mathbf{a} = \mathbf{c}^T[\mathbf{a}\times]\mathbf{b} \tag{A.13i}$$

where the matrices $[\mathbf{a}\times]$, $[\mathbf{b}\times]$, and $[\mathbf{a}\times]$ are defined in eqn. (A.38). The adjoint is given by the transpose of the *cofactor* matrix:

$$\text{adj}(A) = [\text{cof}(A)]^T \tag{A.14}$$

The cofactor is given by

$$C_{ij} = (-1)^{i+j} M_{ij} \tag{A.15}$$

where M_{ij} is the *minor*, which is the determinant of the resulting matrix given by crossing out the row and column of the element a_{ij}. The determinant can be computed using an expansion about row i or column j:

$$\det(A) = \sum_{k=1}^{n} a_{ik} C_{ik} = \sum_{k=1}^{n} a_{kj} C_{kj} \tag{A.16}$$

From eqn. (A.11) A^{-1} exists if and only if the determinant of A is nonzero. Matrix inverses are usually complicated to compute numerically; however, a special case is when the inverse is given by the transpose of the matrix itself. This matrix is then said to be *orthogonal* with the property

$$A^T A = A A^T = I \tag{A.17}$$

Also, the determinant of an orthogonal matrix can be shown to be ± 1. An orthogonal matrix preserves the length (norm) of a vector (see eqn. (A.27) for a definition of the norm of a vector). Hence, if A is an orthogonal matrix, then $||A\,\mathbf{x}|| = ||\mathbf{x}||$.

Block Structures and Other Identities

Matrices can also be analyzed using block structures. Assume that A is an $n \times n$ matrix and that C is an $m \times m$ matrix. Then, we have

$$\det\begin{bmatrix} A & B \\ 0 & C \end{bmatrix} = \det\begin{bmatrix} A & 0 \\ B & C \end{bmatrix} = \det(A)\det(C) \tag{A.18a}$$

$$\det\begin{bmatrix} A & B \\ C & D \end{bmatrix} = \det(A)\det(P) = \det(D)\det(Q) \tag{A.18b}$$

$$\begin{aligned} \begin{bmatrix} A & B \\ C & D \end{bmatrix}^{-1} &= \begin{bmatrix} Q^{-1} & -Q^{-1}B\,D^{-1} \\ -D^{-1}C\,Q^{-1} & D^{-1}(I + C\,Q^{-1}B\,D^{-1}) \end{bmatrix} \\ &= \begin{bmatrix} A^{-1}(I + B\,P^{-1}C\,A^{-1}) & -A^{-1}B\,P^{-1} \\ -P^{-1}C\,A^{-1} & P^{-1} \end{bmatrix} \end{aligned} \tag{A.18c}$$

where P and Q are *Schur complements* of A and D:

$$P \equiv D - C\,A^{-1}B \tag{A.19a}$$

$$Q \equiv A - B\,D^{-1}C \tag{A.19b}$$

Other useful matrix identities involve the *Sherman-Morrison lemma*, given by

$$(I + A\,B)^{-1} = I - A\,(I + B\,A)^{-1}B \tag{A.20}$$

and the *matrix inversion lemma*, given by

$$(A + B\,C\,D)^{-1} = A^{-1} - A^{-1}B\left(D\,A^{-1}B + C^{-1}\right)^{-1}D\,A^{-1} \tag{A.21}$$

where A is an arbitrary $n \times n$ matrix and C is an arbitrary $m \times m$ matrix. A proof of the matrix inversion lemma is given in §1.3.

Matrix Trace

Another useful quantity often used in estimation theory is the *trace* of a matrix, which is defined only for square matrices:

$$\text{Tr}(A) = \sum_{i=1}^{n} a_{ii} \tag{A.22}$$

Some useful identities involving the matrix trace are given by

$$\text{Tr}(\alpha A) = \alpha \text{Tr}(A) \tag{A.23a}$$

$$\text{Tr}(A + B) = \text{Tr}(A) + \text{Tr}(B) \tag{A.23b}$$

$$\text{Tr}(A\,B) = \text{Tr}(B\,A) \tag{A.23c}$$

$$\text{Tr}(\mathbf{x}\mathbf{y}^T) = \mathbf{x}^T\mathbf{y} \tag{A.23d}$$

$$\text{Tr}(A\,\mathbf{y}\mathbf{x}^T) = \mathbf{x}^T A\,\mathbf{y} \tag{A.23e}$$

$$\text{Tr}(A\,B\,C\,D) = \text{Tr}(B\,C\,D\,A) = \text{Tr}(C\,D\,A\,B) = \text{Tr}(D\,A\,B\,C) \tag{A.23f}$$

Equation (A.23f) shows the cyclic invariance of the trace. The operation $\mathbf{y}\mathbf{x}^T$ is known as the *outer product* (also $\mathbf{y}\mathbf{x}^T \neq \mathbf{x}\mathbf{y}^T$ in general).

Solution of Triangular Systems

An *upper triangular system* of linear equations has the form

$$\begin{aligned}
t_{11}x_1 + t_{12}x_2 + t_{13}x_3 + \cdots + t_{1n}x_n &= y_1 \\
t_{22}x_2 + t_{23}x_3 + \cdots + t_{2n}x_n &= y_2 \\
t_{33}x_3 + \cdots + t_{3n}x_n &= y_3 \\
&\vdots \\
t_{nn}x_n &= y_n
\end{aligned} \tag{A.24}$$

or

$$T\mathbf{x} = \mathbf{y} \tag{A.25}$$

where

$$T = \begin{bmatrix} t_{11} & t_{12} & t_{13} & \cdots & t_{1n} \\ 0 & t_{22} & t_{23} & \cdots & t_{2n} \\ 0 & 0 & t_{33} & \cdots & t_{3n} \\ \vdots & \vdots & & \ddots & \\ 0 & 0 & \cdots & \cdots & t_{nn} \end{bmatrix} \tag{A.26}$$

The matrix T can be shown to be nonsingular if and only if its diagonal elements are nonzero.[1] Clearly, x_n can be easily determined using the upper triangular form. The x_i coefficients can be determined by a *back substitution algorithm*:

for $i = n, n-1, \ldots, 1$

$$x_i = t_{ii}^{-1}\left(y_i - \sum_{j=i+1}^{n} t_{ij} x_j\right)$$

next i

This algorithm will fail only if $t_{ii} \to 0$. But, this can occur only if T is singular (or nearly singular). Experience indicates that the algorithm is well-behaved for most applications though.

The back substitution algorithm can be modified to compute the inverse, $S = T^{-1}$, of an upper triangular matrix T. We now summarize an algorithm for calculating $S = T^{-1}$ and overwriting T by T^{-1}:

for $k = n, n-1, \ldots, 1$

$$t_{kk} \leftarrow S_{kk} = t_{kk}^{-1}$$

$$t_{ik} \leftarrow S_{ik} = -t_{ii}^{-1} \sum_{j=i+1}^{k} t_{ij} s_{jk}, \quad i = k-1, k-2, \ldots, 1$$

next k

where $\leftarrow$ denotes replacement.* This algorithm requires about $n^3/6$ calculations (note: if only the solution of $\mathbf{x}$ is required and not the explicit form for T^{-1}, then the back substitution algorithm should be solely employed since only $n^2/2$ calculations are required for this algorithm).

A.2 Vectors

The quantities $\mathbf{x}$ and $\mathbf{y}$ in eqn. (A.2) are known as *vectors*, which are a special case of a matrix. Vectors can consist of one row, known as a *row vector*, or one column, known as a *column vector*.

Vector Norm and Dot Product

A measure of the length of a vector is given by the norm:

$$||\mathbf{x}|| \equiv \sqrt{\mathbf{x}^T \mathbf{x}} = \left[\sum_{i=1}^{n} x_i^2\right]^{1/2} \tag{A.27}$$

Also, $||\alpha\,\mathbf{x}|| = |\alpha|\,||\mathbf{x}||$. A vector with norm one is said to be a *unit vector*. Any

*The symbol $x \leftarrow y$ means "overwrite" x by the current y-value. This notation is employed to indicate how storage may be conserved by overwriting quantities no longer needed.

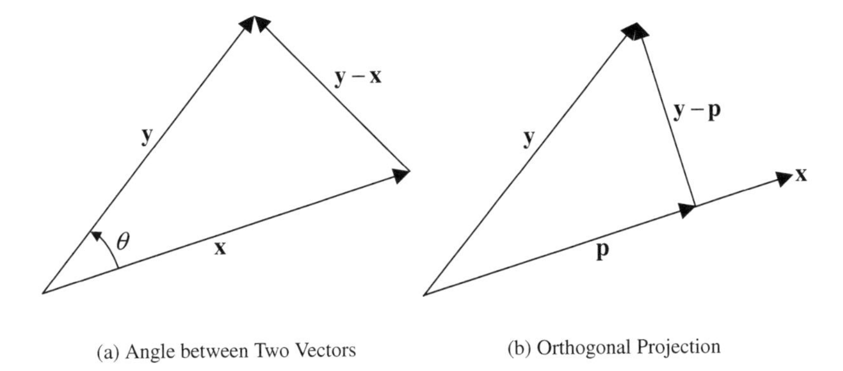

(a) Angle between Two Vectors (b) Orthogonal Projection

Figure A.1: Depiction of the Angle between Two Vectors and an Orthogonal Projection

nonzero vector can be made into a unit vector by dividing it by its norm:

$$\hat{\mathbf{x}} \equiv \frac{\mathbf{x}}{||\mathbf{x}||} \tag{A.28}$$

Note that the carat is also used to denote estimate in this text. The *dot product* or *inner product* of two vectors of equal dimension, $n \times 1$, is given by

$$\mathbf{x}^T\mathbf{y} = \mathbf{y}^T\mathbf{x} = \sum_{i=1}^{n} x_i y_i \tag{A.29}$$

If the dot product is zero, then the vectors are said to be *orthogonal*. Suppose that a set of vectors $\mathbf{x}_i$ $(i = 1, 2, \ldots, m)$ follows

$$\mathbf{x}_i^T\mathbf{x}_j = \delta_{ij} \tag{A.30}$$

where the Kronecker delta δ_{ij} is defined as

$$\begin{aligned} \delta_{ij} &= 0 \quad \text{if } i \neq j \\ &= 1 \quad \text{if } i = j \end{aligned} \tag{A.31}$$

Then, this set is said to be *orthonormal*. The column and row vectors of an orthogonal matrix, defined by the property shown in eqn. (A.17), form an orthonormal set.

Angle Between Two Vectors and the Orthogonal Projection

Figure A.1(a) shows two vectors, $\mathbf{x}$ and $\mathbf{y}$, and the angle θ which is the angle between them. This angle can be computed from the cosine law:

$$\cos(\theta) = \frac{\mathbf{x}^T\mathbf{y}}{||\mathbf{x}||\,||\mathbf{y}||} \tag{A.32}$$

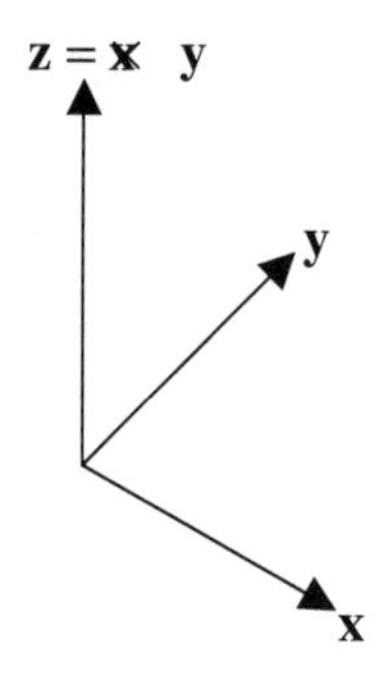

Figure A.2: Cross Product and the Right Hand Rule

Figure A.1(b) shows the *orthogonal projection* of a vector $\mathbf{y}$ to a vector $\mathbf{x}$. The orthogonal projection of $\mathbf{y}$ to $\mathbf{x}$ is given by

$$\mathbf{p} = \frac{\mathbf{x}^T \mathbf{y}}{||\mathbf{x}||^2} \mathbf{x} \tag{A.33}$$

This projection yields $(\mathbf{y} - \mathbf{p})^T \mathbf{x} = 0$.

Triangle and Schwartz Inequalities

Some important inequalities are given by the *triangle inequality*:

$$||\mathbf{x} + \mathbf{y}|| \leq ||\mathbf{x}|| + ||\mathbf{y}|| \tag{A.34}$$

and the *Schwartz inequality*:

$$|\mathbf{x}^T \mathbf{y}| \leq ||\mathbf{x}|| \, ||\mathbf{y}|| \tag{A.35}$$

Note that the Schwartz inequality implies the triangle inequality.

Cross Product

The cross product of two vectors yields a vector that is perpendicular to both vectors. The cross product of $\mathbf{x}$ and $\mathbf{y}$ is given by

$$\mathbf{z} = \mathbf{x} \times \mathbf{y} = \begin{bmatrix} x_2 y_3 - x_3 y_2 \\ x_3 y_1 - x_1 y_3 \\ x_1 y_2 - x_2 y_1 \end{bmatrix} \tag{A.36}$$

The cross product follows the *right hand rule*, which states that the orientation of $\mathbf{z}$ is determined by placing $\mathbf{x}$ and $\mathbf{y}$ tail-to-tail, flattening the right hand, extending it in the direction of $\mathbf{x}$, and then curling the fingers in the direction that the angle $\mathbf{y}$ makes with $\mathbf{x}$. The thumb then points in the direction of $\mathbf{z}$, as shown in Figure A.2. The cross product can also be obtained using matrix multiplication:

$$\mathbf{z} = [\mathbf{x}\times]\mathbf{y} \tag{A.37}$$

where $[\mathbf{x}\times]$ is the *cross product matrix*, defined by

$$[\mathbf{x}\times] \equiv \begin{bmatrix} 0 & -x_3 & x_2 \\ x_3 & 0 & -x_1 \\ -x_2 & x_1 & 0 \end{bmatrix} \tag{A.38}$$

Note that $[\mathbf{x}\times]$ is a skew symmetric matrix.

The cross product has the following properties:

$$[\mathbf{x}\times]^T = -[\mathbf{x}\times] \tag{A.39a}$$

$$[\mathbf{x}\times]\mathbf{y} = -[\mathbf{y}\times]\mathbf{x} \tag{A.39b}$$

$$[\mathbf{x}\times][\mathbf{y}\times] = -\left(\mathbf{x}^T\mathbf{y}\right)I + \mathbf{y}\mathbf{x}^T \tag{A.39c}$$

$$[\mathbf{x}\times]^3 = -\left(\mathbf{x}^T\mathbf{x}\right)[\mathbf{x}\times] \tag{A.39d}$$

$$[\mathbf{x}\times][\mathbf{y}\times] - [\mathbf{y}\times][\mathbf{x}\times] = \mathbf{y}\mathbf{x}^T - \mathbf{x}\mathbf{y}^T = [(\mathbf{x}\times\mathbf{y})\times] \tag{A.39e}$$

$$\mathbf{x}\mathbf{y}^T[\mathbf{w}\times] + [\mathbf{w}\times]\mathbf{y}\mathbf{x}^T = -[\{\mathbf{x}\times(\mathbf{y}\times\mathbf{w})\}\times] \tag{A.39f}$$

$$(I - [\mathbf{x}\times])(I + [\mathbf{x}\times])^{-1} = \frac{1}{1+\mathbf{x}^T\mathbf{x}}\left\{(1-\mathbf{x}^T\mathbf{x})I + 2\mathbf{x}\mathbf{x}^T - 2[\mathbf{x}\times]\right\} \tag{A.39g}$$

Other useful properties involving an arbitrary 3×3 square matrix M are given by[2]

$$M[\mathbf{x}\times] + [\mathbf{x}\times]M^T + [(M^T\mathbf{x})\times] = \mathrm{Tr}(M)[\mathbf{x}\times] \tag{A.40a}$$

$$M[\mathbf{x}\times]M^T = [\{\mathrm{adj}(M^T)\mathbf{x}\}\times] \tag{A.40b}$$

$$(M\mathbf{x})\times(M\mathbf{y}) = \mathrm{adj}(M^T)(\mathbf{x}\times\mathbf{y}) \tag{A.40c}$$

$$[\{(M\mathbf{x})\times(M\mathbf{y})\}\times] = M[(\mathbf{x}\times\mathbf{y})\times]M^T \tag{A.40d}$$

If we write M in terms of its columns

$$M = \begin{bmatrix}\mathbf{x}_1 & \mathbf{x}_2 & \mathbf{x}_3\end{bmatrix} \tag{A.41}$$

then

$$\det(M) = \mathbf{x}_1^T(\mathbf{x}_2\times\mathbf{x}_3) \tag{A.42}$$

Also, if A is an orthogonal matrix with determinant 1, then from eqn. (A.40b) we have

$$A[\mathbf{x}\times]A^T = [(A\mathbf{x})\times] \tag{A.43}$$

Another important quantity using eqn. (A.39c) is given by

$$[\mathbf{x}\times]^2 = -\left(\mathbf{x}^T\mathbf{x}\right)I + \mathbf{x}\mathbf{x}^T \tag{A.44}$$

This matrix is the projection operator onto the space perpendicular to $\mathbf{x}$. Many other interesting relations involving the cross product are given in Ref. [3].

Table A.1: Matrix and Vector Norms

Norm	**Vector**	**Matrix**
One-norm	$\|\|\mathbf{x}\|\|_1 = \sum_{i=1}^n \|x_i\|$	$\|\|A\|\|_1 = \max_j \sum_{i=1}^n \|a_{ij}\|$
Two-norm	$\|\|\mathbf{x}\|\|_2 = \left[\sum_{i=1}^n x_i^2\right]^{1/2}$	$\|\|A\|\|_2 =$ max singular value of A
Frobenius norm	$\|\|\mathbf{x}\|\|_F = \|\|\mathbf{x}\|\|_2$	$\|\|A\|\|_F = \sqrt{\text{Tr}(A^* A)}$
Infinity-norm	$\|\|\mathbf{x}\|\|_\infty = \max_i \|x_i\|$	$\|\|\|A\|\|_\infty = \max_i \sum_{j=1}^n \|a_{ij}\|$

The angle θ in Figure A.1(a) can be computed from

$$\sin(\theta) = \frac{||\mathbf{x} \times \mathbf{y}||}{||\mathbf{x}||\,||\mathbf{y}||} \tag{A.45}$$

Using $\sin^2(\theta) + \cos^2(\theta) = 1$, eqns. (A.32) and (A.45) also give

$$||\mathbf{x} \times \mathbf{y}|| = \sqrt{(\mathbf{x}^T\mathbf{x})(\mathbf{y}^T\mathbf{y}) - (\mathbf{x}^T\mathbf{y})^2} \tag{A.46}$$

From the Schwartz inequality in eqn. (A.35), the quantity within the square root in eqn. (A.46) is always positive.

A.3 Matrix Norms and Definiteness

Norms for matrices are slightly more difficult to define than for vectors. Also, the definition of a "positive" or "negative" matrix is more complicated for matrices than scalars. Before showing these quantities, we first define the following quantities for any complex matrix A:

- *Conjugate transpose*: defined as the transpose of the conjugate of each element; denoted by A^*.

- *Hermitian*: has the property $A = A^*$ (note: any real symmetric matrix is Hermitian).

- *Normal*: has the property $A^* A = A\,A^*$.

- *Unitary*: inverse is equal to its Hermitian transpose, so that $A^* A = A\,A^* = I$ (note: a real unitary matrix is an orthogonal matrix).

Matrix Norms

Several possible matrix norms can be defined. Table A.1 lists the most commonly used norms for both vectors and matrices. The one-norm is the largest column sum. The two-norm is the maximum singular value (see §A.4). Also, unless otherwise stated, the norm defined without showing a subscript is the two-norm, as shown by eqn. (A.27). The Frobenius norm is defined as the square root of the sum of the absolute squares of its elements. The infinity-norm is the largest row sum. The matrix norms described in Table A.1 have the following properties:

$$||\alpha A|| = |\alpha|\,||A|| \tag{A.47a}$$

$$||A+B|| \le ||A|| + ||B|| \tag{A.47b}$$

$$||A\,B|| \le ||A||\,||B|| \tag{A.47c}$$

Not all norms follow eqn. (A.47c) though (e.g., the maximum absolute matrix element). More matrix norm properties can be found in Refs. [4] and [5].

Definiteness

Sufficiency tests in least squares and the minimization of functions with multiple variables often require that one determine the *definiteness* of the matrix of second partial derivatives. A real and square matrix A is

- *Positive definite* if $\mathbf{x}^T A\,\mathbf{x} > 0$ for all nonzero $\mathbf{x}$.
- *Positive semi-definite* if $\mathbf{x}^T A\,\mathbf{x} \ge 0$ for all nonzero $\mathbf{x}$.
- *Negative definite* if $\mathbf{x}^T A\,\mathbf{x} < 0$ for all nonzero $\mathbf{x}$.
- *Negative semi-definite* if $\mathbf{x}^T A\,\mathbf{x} \le 0$ for all nonzero $\mathbf{x}$.
- *Indefinite* when no definiteness can be asserted.

A simple test for a symmetric real matrix is to check its eigenvalues (see §A.4). This matrix is positive definite if and only if all its eigenvalues are greater than 0. Unfortunately, this condition is only necessary but not sufficient for a non-symmetric real matrix. A real matrix is positive definite if and only if its symmetric part, given by

$$B = \frac{A + A^T}{2} \tag{A.48}$$

is positive definite. Another way to state that a matrix is positive definite is the requirement that all the *leading* principal minors of A are positive.[6] If A is positive definite, then A^{-1} exists and is also positive definite. If A is positive semi-definite, then for any integer $\alpha > 0$ there exists a unique positive semi-definite matrix such that $A = B^\alpha$ (note: A and B commute, so that $A\,B = B\,A$). The following relationship:

$$B > A \tag{A.49}$$

implies $(B - A) > 0$, which states that the matrix $(B - A)$ is positive definite. Also,

$$B \ge A \tag{A.50}$$

implies $(B - A) \geq 0$, which states that the matrix $(B - A)$ is positive semi-definite. The conditions for negative definite and negative semi-definite are obvious from the definitions stated for positive definite and positive semi-definite.

A.4 Matrix Decompositions

Several matrix decompositions are given in the open literature. Many of these decompositions are used in place of a matrix inverse either to simplify the calculations or to provide more numerically robust approaches. In this section we present several useful matrix decompositions that are widely used in estimation and control theory. The methods to compute these decompositions is beyond the scope of the present text. Reference [4] provides all the necessary algorithms and proofs for the interested reader. Before we proceed a short description of the *rank* of a matrix is provided. Several definitions are possible. We will state that the rank of a matrix is given by the dimension of the range of the matrix corresponding to the number of linearly independent rows or columns. An $m \times n$ matrix is *rank deficient* if the rank of A is less than the minimum (m, n). Suppose that the rank of an $n \times n$ matrix A is given by $\text{rank}(A) = r$. Then, a set of $(n - r)$ nonzero unit vectors, $\hat{\mathbf{x}}_i$, can always be found that have the following property for a singular square matrix A:

$$A\hat{\mathbf{x}}_i = \mathbf{0}, \quad i = 1, 2, \ldots, n - r \tag{A.51}$$

The value of $(n - r)$ is known as the *nullity*, which is the maximum number of linearly independent null vectors of A. These vectors can form an orthonormal basis (which is how they are commonly shown) for the *null space* of A, and can be computed from the singular value decomposition. If A is nonsingular then no nonzero vector $\hat{\mathbf{x}}_i$ can be found to satisfy eqn. (A.51). For more details on the rank of a matrix see Refs. [4] and [6].

Eigenvalue/Eigenvector Decomposition and the Cayley-Hamilton Theorem

One of the most widely used decompositions for a square $n \times n$ matrix A in the study of dynamical systems is the eigenvalue/eigenvector decomposition. A real or complex number λ is an *eigenvalue* of A if there exists a nonzero (right) *eigenvector* $\mathbf{p}$ such that

$$A\mathbf{p} = \lambda\mathbf{p} \tag{A.52}$$

The solution for $\mathbf{p}$ is not unique in general, so usually $\mathbf{p}$ is given as a unit vector. In order for eqn. (A.52) to have a nonzero solution for $\mathbf{p}$, from eqn. (A.51), the matrix $(\lambda I - A)$ must be singular. Therefore, from eqn. (A.11) we have

$$\det(\lambda I - A) = \lambda^n + \alpha_1\lambda^{n-1} + \cdots + \alpha_{n-1}\lambda + \alpha_n = 0 \tag{A.53}$$

Equation (A.53) leads to a polynomial of degree n, which is called the *characteristic equation* of A. For example, the characteristic equation for a 3×3 matrix is given by

$$\lambda^3 - \lambda^2 \mathrm{Tr}(A) + \lambda\,\mathrm{Tr}[\mathrm{adj}(A)] - \det(A) = 0 \tag{A.54}$$

If all eigenvalues of A are distinct, then the set of eigenvectors is linearly independent. Therefore, the matrix A can be *diagonalized* as

$$\Lambda = P^{-1} A\, P \tag{A.55}$$

where $\Lambda = \mathrm{diag}\begin{bmatrix}\lambda_1 & \lambda_2 & \cdots & \lambda_n\end{bmatrix}$ and $P = \begin{bmatrix}\mathbf{p}_1 & \mathbf{p}_2 & \cdots & \mathbf{p}_n\end{bmatrix}$. If A has repeated eigenvalues, then a block diagonal and triangular-form representation must be used, called a *Jordan block*.[6] The eigenvalue/eigenvector decomposition can be used for linear state variable transformations (see §3.1.4). Eigenvalues and eigenvectors can either be real or complex. This decomposition is very useful when A is symmetric, since Λ is always diagonal (even for repeated eigenvalues) and P is orthogonal for this case. A proof is given in Ref. [4]. Also, Ref. [4] provides many algorithms to compute the eigenvalue/eigenvector decomposition.

One of the most useful properties used in linear algebra is the *Cayley-Hamilton theorem*, which states that a matrix satisfies its own characteristic equation, so that

$$A^n + \alpha_1 A^{n-1} + \cdots + \alpha_{n-1} A + \alpha_n I = 0 \tag{A.56}$$

This theorem is useful for computing powers of A that are larger than n, since A^{n+1} can be written as a linear combination of $(A,\, A^2,\, \ldots,\, A^n)$.[6]

QR Decomposition

The QR decomposition is especially useful in least squares (see §1.6.1) and the Square Root Information Filter (SRIF) (see §5.7.1). The QR decomposition of an $m \times n$ matrix A, with $m \geq n$, is given by

$$A = \mathcal{Q}\mathcal{R} \tag{A.57}$$

where $\mathcal{Q}$ is an $m \times m$ orthogonal matrix, and $\mathcal{R}$ is an upper triangular $m \times n$ matrix with all elements $\mathcal{R}_{ij} = 0$ for $i > j$. If A has full column rank, then the first n columns of $\mathcal{Q}$ form an orthonormal basis for the range of A.[4] Therefore, the "thin" QR decomposition is often used:

$$A = Q\, R \tag{A.58}$$

where Q is an $m \times n$ matrix with orthonormal columns and R is an upper triangular $n \times n$ matrix. Since the QR decomposition is widely used throughout the present text, we present a numerical algorithm to compute this decomposition by the *modified Gram-Schmidt* method.[4] Let A and Q be partitioned by columns $\begin{bmatrix}\mathbf{a}_1 & \mathbf{a}_2 & \cdots & \mathbf{a}_n\end{bmatrix}$ and $\begin{bmatrix}\mathbf{q}_1 & \mathbf{q}_2 & \cdots & \mathbf{q}_n\end{bmatrix}$, respectively. To begin the algorithm we set $Q = A$, and then

for $k = 1, 2, \ldots, n$

$r_{kk} = ||\mathbf{q}_k||_2$

$\mathbf{q}_k \leftarrow \mathbf{q}_k / r_{kk}$

$r_{kj} = \mathbf{q}_k^T \mathbf{q}_j, \quad j = k+1, \ldots, n$

$$\mathbf{q}_j \leftarrow \mathbf{q}_j - r_{kj}\mathbf{q}_k, \quad j = k+1, \ldots, n$$

next k

where $\leftarrow$ denotes replacement (note: r_{kk} and r_{kj} are elements of the matrix R). This algorithm works even when A is complex. The QR decomposition is useful to invert an $n \times n$ matrix A, which is given by $A^{-1} = R^{-1}Q^T$ (note: the inverse of an upper triangular matrix is also a triangular matrix). Other methods, based on the Householder transformation and Givens rotations, can be used for the QR decomposition.[4]

Singular Value Decomposition

Another decomposition of an $m \times n$ matrix A is the *singular-value decomposition*,[4, 7] which decomposes a matrix into a diagonal matrix and two orthogonal matrices:

$$A = \mathcal{U}\mathcal{S}\mathcal{V}^* \tag{A.59}$$

where $\mathcal{U}$ is an $m \times m$ unitary matrix, $\mathcal{S}$ is an $m \times n$ diagonal matrix such that $\mathcal{S}_{ij} = 0$ for $i \neq j$, and $\mathcal{V}$ is an $n \times n$ unitary matrix. Many efficient algorithms can be used to determine the singular value decomposition.[4] Note that the zeros below the diagonal in $\mathcal{S}$ (with $m > n$) imply that the elements of columns $(n+1), (n+2), \ldots, m$ of $\mathcal{U}$ are arbitrary. So, we can define the following reduced singular value decomposition:

$$A = U\,S\,V^* \tag{A.60}$$

where U is the $m \times n$ subset matrix of $\mathcal{U}$ (with the $(n+1), (n+2), \ldots, m$ columns eliminated), S is the upper $n \times n$ matrix of $\mathcal{S}$, and $V = \mathcal{V}$. Note that $U^*U = I$, but it is no longer possible to make the same statement for $U\,U^*$. The elements of $S = \text{diag}\begin{bmatrix} s_1 & \cdots & s_n \end{bmatrix}$ are known as the *singular values* of A, which are ordered from the smallest singular value to the largest singular value. These values are extremely important since they can give an indication of "how well" we can invert a matrix.[8] A common measure of the invertability of a matrix is the *condition number*, which is usually defined as the ratio of its largest singular value to its smallest singular value:

$$\text{Condition Number} = \frac{s_n}{s_1} \tag{A.61}$$

Large condition numbers may indicate a near singular matrix, and the minimum value of the condition number is unity (which occurs when the matrix is orthogonal). The rank of A is given by the number of nonzero singular values. Also, the singular value decomposition is useful to determine various norms (e.g., $||A||_F^2 = s_1^2 + \cdots + s_p^2$, where $p = \min(m, n)$, and the two-norm as shown in Table A.1).

Gaussian Elimination

Gaussian elimination is a classical reduction procedure by which a matrix A can be reduced to upper triangular form. This procedure involves pre-multiplications of a square matrix A by a sequence of *elementary lower triangular* matrices, each chosen to introduce a column with zeros below the diagonal (this process is often called "annihilation"). Several possible variations of Gaussian elimination can be derived.

We present a very robust algorithm called *Gaussian elimination with complete pivoting*. This approach requires data movements such as the interchange of two matrix rows. These interchanges can be tracked by using "permutation matrices," which are just identity matrices with rows or columns reordered. For example, consider the following matrix:

$$P = \begin{bmatrix} 0 & 0 & 0 & 1 \\ 1 & 0 & 0 & 0 \\ 0 & 0 & 1 & 0 \\ 0 & 1 & 0 & 0 \end{bmatrix} \tag{A.62}$$

So PA is a row permutated version of A and AP is a column permutated version of A. Permutation matrices are orthogonal. This algorithm computes the complete pivoting factorization

$$A = P\,L\,U\,Q^T \tag{A.63}$$

where P and Q are permutation matrices, L is a *unit* (with ones along the diagonal) lower triangular matrix, and U is an upper triangular matrix. The algorithm begins by setting $P = Q = I$, which are partitioned into column vectors as

$$P = \begin{bmatrix} \mathbf{p}_1 & \mathbf{p}_2 & \cdots & \mathbf{p}_n \end{bmatrix}, \quad Q = \begin{bmatrix} \mathbf{q}_1 & \mathbf{q}_2 & \cdots & \mathbf{q}_n \end{bmatrix} \tag{A.64}$$

The algorithm for Gaussian elimination with complete pivoting is given by overwriting the A matrix:

for $k = 1, 2, \ldots, n-1$
  Determine μ with $k \le \mu \le n$ and λ with $k \le \lambda \le n$ so
    $|a_{\mu\lambda}| = \max\{|a_{ij}|,\ i = 1, 2, \ldots, n,\ j = 1, 2, \ldots, n\}$
  if $\mu \neq k$
    $\mathbf{p}_k \leftrightarrow \mathbf{p}_\mu$
    $a_{kj} \leftrightarrow a_{\mu j}, \quad j = 1, 2, \ldots, n$
  end if
  if $\lambda \neq k$
    $\mathbf{q}_k \leftrightarrow \mathbf{q}_\lambda$
    $a_{jk} \leftrightarrow a_{j\lambda}, \quad j = 1, 2, \ldots, n$
  end if
  if $a_{kk} \neq 0$
    $a_{jk} \leftarrow a_{kj}/a_{kk}, \quad j = k+1, \ldots, n$
    $a_{jj} \leftarrow a_{jj} - a_{jk}a_{kj}, \quad j = k+1, \ldots, n$
  end if
next k

where $\leftarrow$ denotes replacement and $\leftrightarrow$ denotes "interchange the value assigned to." The matrix U is given by the upper triangular part (including the diagonal elements) of the overwritten A matrix, and the matrix L is given by the lower triangular part (replacing the diagonal elements with ones) of the overwritten A matrix. More details on Gaussian elimination can be found in Ref. [4].

LU and Cholesky Decompositions

The LU decomposition factors an $n \times n$ matrix A into a product of a lower triangular matrix L and an upper triangular matrix U, so that

$$A = LU \tag{A.65}$$

Gaussian elimination is a foremost example of LU decompositions. In general, the LU decomposition is not unique. This can be seen by observing that for an arbitrary nonsingular diagonal matrix D, setting $L' = LD$ and $U' = D^{-1}U$ yield new upper and lower triangular matrices that satisfy $L'U' = LDD^{-1}U = LU = A$. The fact that the decomposition is not unique suggests the possible wisdom of forming the *normalized* decomposition

$$A = LDU \tag{A.66}$$

in which L and U are unit lower and upper triangular matrices and D is a diagonal matrix. The question of existence and uniqueness is addressed by Stewart[1] who proves that the $A = LDU$ decomposition is unique, provided the leading diagonal sub-matrices of A are nonsingular.

There are three important variants of the LDU decomposition; the first associates D with the lower triangular part to give the factorization

$$A = \mathcal{L}U \tag{A.67}$$

where $\mathcal{L} \equiv LD$. This is known as the *Crout reduction*. The second variant associates D with the upper triangular factor as

$$A = L\mathcal{U} \tag{A.68}$$

where $\mathcal{U} \equiv DU$. This reduction is exactly that obtained by Gaussian elimination.

The third variation is possible only for symmetric positive definite matrices, in which case

$$A = LDL^T \tag{A.69}$$

Thus A can be written as

$$A = \mathcal{L}\mathcal{L}^T \tag{A.70}$$

where now $\mathcal{L} \equiv LD^{1/2}$ is known as the *matrix square root*, and the factorization in eqn. (A.69) is known as the *Cholesky decomposition*. Efficient algorithms to compute the LU and Cholesky decompositions can be found in Ref. [4].

A.5 Matrix Calculus

In this section several relations are given for taking partial or time derivatives of matrices.[†] Before providing a list of matrix calculus identities, we first will define

[†]Most of these relations can be found in a website given by Mike Brooks, Imperial College, London, UK. As of this writing this website is given by http://www.ee.ic.ac.uk/hp/staff/dmb/matrix/calculus.html.

the *Jacobian* and *Hessian* of a scalar function $f(\mathbf{x})$, where $\mathbf{x}$ is an $n \times 1$ vector. The Jacobian of $f(\mathbf{x})$ is an $n \times 1$ vector given by

$$\nabla_{\mathbf{x}} f \equiv \frac{\partial f}{\partial \mathbf{x}} = \begin{bmatrix} \frac{\partial f}{\partial x_1} \\ \frac{\partial f}{\partial x_2} \\ \vdots \\ \frac{\partial f}{\partial x_n} \end{bmatrix} \tag{A.71}$$

The Hessian of $f(\mathbf{x})$ is an $n \times n$ matrix given by

$$\nabla^2_{\mathbf{x}} f \equiv \frac{\partial^2 f}{\partial \mathbf{x}\, \partial \mathbf{x}^T} = \begin{bmatrix} \frac{\partial f}{\partial x_1 \partial x_1} & \frac{\partial f}{\partial x_1 \partial x_2} & \cdots & \frac{\partial f}{\partial x_1 \partial x_n} \\ \frac{\partial f}{\partial x_2 \partial x_1} & \frac{\partial f}{\partial x_2 \partial x_2} & \cdots & \frac{\partial f}{\partial x_2 \partial x_n} \\ \vdots & \vdots & \ddots & \vdots \\ \frac{\partial f}{\partial x_n \partial x_1} & \frac{\partial f}{\partial x_n \partial x_2} & \cdots & \frac{\partial f}{\partial x_n \partial x_n} \end{bmatrix} \tag{A.72}$$

Note that the Hessian of a scalar is a symmetric matrix. If $\mathbf{f}(\mathbf{x})$ is an $m \times 1$ vector and $\mathbf{x}$ is an $n \times 1$ vector, then the Jacobian matrix is given by

$$\nabla_{\mathbf{x}} \mathbf{f} \equiv \frac{\partial \mathbf{f}}{\partial \mathbf{x}} = \begin{bmatrix} \frac{\partial f_1}{\partial x_1} & \frac{\partial f_1}{\partial x_2} & \cdots & \frac{\partial f_1}{\partial x_n} \\ \frac{\partial f_2}{\partial x_1} & \frac{\partial f_2}{\partial x_2} & \cdots & \frac{\partial f_2}{\partial x_n} \\ \vdots & \vdots & \ddots & \vdots \\ \frac{\partial f_m}{\partial x_1} & \frac{\partial f_m}{\partial x_2} & \cdots & \frac{\partial f_m}{\partial x_n} \end{bmatrix} \tag{A.73}$$

Note that the Jacobian matrix is an $m \times n$ matrix. Also, there is a slight inconsistency between eqn. (A.71) and eqn. (A.73) when $m = 1$, since eqn. (A.71) gives an $n \times 1$ vector, while eqn. (A.73) gives a $1 \times n$ vector. This should pose no problems for the reader though since the context of this notation is clear for the particular system shown in this text.

A list of derivatives involving linear products is given by

$$\frac{\partial}{\partial \mathbf{x}}(A\mathbf{x}) = A \tag{A.74a}$$

$$\frac{\partial}{\partial A}(\mathbf{a}^T A\,\mathbf{b}) = \mathbf{a}\,\mathbf{b}^T \tag{A.74b}$$

$$\frac{\partial}{\partial A}(\mathbf{a}^T A^T\mathbf{b}) = \mathbf{b}\,\mathbf{a}^T \tag{A.74c}$$

$$\frac{d}{dt}(A\,B) = A\left[\frac{d}{dt}(B)\right] + \left[\frac{d}{dt}(A)\right]B \tag{A.74d}$$

A list of derivatives involving quadratic and cubic products is given by

$$\frac{\partial}{\partial \mathbf{x}}(A\mathbf{x}+\mathbf{b})^T C\,(D\mathbf{x}+\mathbf{e}) = A^T C\,(D\mathbf{x}+\mathbf{e}) + D^T C^T (A\mathbf{x}+\mathbf{b}) \tag{A.75a}$$

$$\frac{\partial}{\partial \mathbf{x}}(\mathbf{x}^T C\,\mathbf{x}) = (C + C^T)\,\mathbf{x} \tag{A.75b}$$

$$\frac{\partial}{\partial A}(\mathbf{a}^T A^T A\,\mathbf{b}) = A\,(\mathbf{a}\,\mathbf{b}^T + \mathbf{b}\,\mathbf{a}^T) \tag{A.75c}$$

$$\frac{\partial}{\partial A}(\mathbf{a}^T A^T C\,A\,\mathbf{b}) = C^T A\,\mathbf{a}\,\mathbf{b}^T + C\,A\,\mathbf{b}\,\mathbf{a}^T \tag{A.75d}$$

$$\frac{\partial}{\partial A}(A\,\mathbf{a}+\mathbf{b})^T C\,(A\,\mathbf{a}+\mathbf{b}) = (C + C^T)\,(A\,\mathbf{a}+\mathbf{b})\,\mathbf{a}^T \tag{A.75e}$$

$$\frac{\partial}{\partial \mathbf{x}}(\mathbf{x}^T A\,\mathbf{x}\mathbf{x}^T) = (A + A^T)\,\mathbf{x}\mathbf{x}^T + (\mathbf{x}^T A\,\mathbf{x})\,I \tag{A.75f}$$

A list of derivatives involving the inverse of a matrix is given by

$$\frac{d}{dt}(A^{-1}) = -A^{-1}\left[\frac{d}{dt}(A)\right]A^{-1} \tag{A.76a}$$

$$\frac{\partial}{\partial A}(\mathbf{a}^T A^{-1}\mathbf{b}) = -A^{-T}\,\mathbf{a}\,\mathbf{b}^T A^{-T} \tag{A.76b}$$

A list of derivatives involving the trace of a matrix is given by

$$\frac{\partial}{\partial A}\mathrm{Tr}(A) = \frac{\partial}{\partial A}\mathrm{Tr}(A^T) = I \tag{A.77a}$$

$$\frac{\partial}{\partial A}\mathrm{Tr}(A^\alpha) = \alpha\left(A^{\alpha-1}\right)^T \tag{A.77b}$$

$$\frac{\partial}{\partial A}\mathrm{Tr}(C\,A^{-1}B) = -A^{-T}\,C\,B\,A^{-T} \tag{A.77c}$$

$$\frac{\partial}{\partial A}\mathrm{Tr}(C^T A\,B^T) = \frac{\partial}{\partial A}\mathrm{Tr}(B\,A^T C) = C\,B \tag{A.77d}$$

$$\frac{\partial}{\partial A}\mathrm{Tr}(C\,A\,B\,A^T D) = C^T D^T A\,B^T + D\,C\,A\,B \tag{A.77e}$$

$$\frac{\partial}{\partial A}\mathrm{Tr}(C\,A\,B\,A) = C^T A^T B^T + B^T A^T C^T \tag{A.77f}$$

A list of derivatives involving the determinant of a matrix is given by

$$\frac{\partial}{\partial A}\det(A) = \frac{\partial}{\partial A}\det(A^T) = [\mathrm{adj}(A)]^T \tag{A.78a}$$

$$\frac{\partial}{\partial A}\det(C\,A\,B) = \det(C\,A\,B)A^{-T} \tag{A.78b}$$

$$\frac{\partial}{\partial A}\ln[\det(C\,A\,B)] = A^{-T} \tag{A.78c}$$

$$\frac{\partial}{\partial A}\det(A^\alpha) = \alpha\det(A^\alpha)\,A^{-T} \tag{A.78d}$$

$$\frac{\partial}{\partial A}\ln[\det(A^\alpha)] = \alpha A^{-T} \tag{A.78e}$$

$$\frac{\partial}{\partial A}\det(A^T C\,A) = \det(A^T C\,A)\,(C + C^T)\,A\,(A^T C\,A)^{-1} \tag{A.78f}$$

$$\frac{\partial}{\partial A}\ln[\det(A^T C\,A)] = (C + C^T)\,A\,(A^T C\,A)^{-1} \tag{A.78g}$$

Relations involving the Hessian matrix are given by

$$\frac{\partial^2}{\partial \mathbf{x}\,\partial \mathbf{x}^T}(A\,\mathbf{x} + \mathbf{b})^T C\,(D\,\mathbf{x} + \mathbf{e}) = A^T C\,D + D^T C^T A \tag{A.79a}$$

$$\frac{\partial^2}{\partial \mathbf{x}\,\partial \mathbf{x}^T}(\mathbf{x}^T C\,\mathbf{x}) = C + C^T \tag{A.79b}$$

References

[1] Stewart, G.W., *Introduction to Matrix Computations*, Academic Press, New York, NY, 1973.

[2] Shuster, M.D., "A Survey of Attitude Representations," *Journal of the Astronautical Sciences*, Vol. 41, No. 4, Oct.-Dec. 1993, pp. 439–517.

[3] Tempelman, W., "The Linear Algebra of Cross Product Operations," *Journal of the Astronautical Sciences*, Vol. 36, No. 4, Oct.-Dec. 1988, pp. 447–461.

[4] Golub, G.H. and Van Loan, C.F., *Matrix Computations*, The Johns Hopkins University Press, Baltimore, MD, 3rd ed., 1996.

[5] Zhang, F., *Linear Algebra: Challenging Problems for Students*, The Johns Hopkins University Press, Baltimore, MD, 1996.

[6] Chen, C.T., *Linear System Theory and Design*, Holt, Rinehart and Winston, New York, NY, 1984.

[7] Horn, R.A. and Johnson, C.R., *Matrix Analysis*, Cambridge University Press, Cambridge, MA, 1985.

[8] Nash, J.C., *Compact Numerical Methods for Computers: linear algebra and function minimization*, Adam Hilger Ltd., Bristol, 1979.

B

Basic Probability Concepts

THIS appendix serves as an overview of the probability concepts that are most important in the present text's approach to estimation theory. These developments are patterned after the excellent survey provided by Bryson and Ho.[1] Still, the interested student is strongly encouraged to study probability theory formally from conventional texts such as Refs. [2]-[5].

B.1 Functions of a Single Discrete-Valued Random Variable

To appeal to the intuitive feel that we have for random variables and elementary probability concepts, attention is first directed to a simple experiment. Consider a single throw of a "true" die; the *probability* of the occurrence of each of the *events* 1, 2, 3, 4, 5, or 6 is exactly the same on a given throw. For a "loaded" die, the probability of certain of the events would be greater than others. If a given discrete-values experiment is conducted N times and N_j is the number of times that the jth event $x(j)$ occurred, then it is intuitively reasonable to define the probability of the occurrence of $x(j)$ as

$$p(x(j)) \equiv \lim_{N\to\infty} \frac{N_j}{N} \tag{B.1}$$

For example, for a throw of a single die the probability of obtaining a value of 3 is given by $p(3) = 1/6$.

A *discrete-valued random variable*, x, is defined as a function having finite number of possible values $x(j)$; with the associated probability of $x(j)$ occurring being denoted by $p(x(j))$. To compact notation, $x(j)$ and $p(x(j))$ are hereafter called x and $p(x)$, whenever this substitution does not cause ambiguity.

Let us expand the die concept for the case of a single throw of two dice. We now have 36 possible outcomes over the entire set. Table B.1 shows the sum of the two dice, the number of times that sum can occur and the probability of that event. Clearly, obtaining a 7 has the highest probability. When multiple dice, $n > 2$, are used this table is much more difficult to produce. Fortunately, a simple mathematical approach known as a *generating function* can be used for this case:

$$f(x) = \left(x + x^2 + x^3 + x^4 + x^5 + x^6\right)^n \tag{B.2}$$

Table B.1: Probabilities for a Single Throw of Two Dice

Sum	Count	$p(x)$
2	1	1/36
3	2	2/36
4	3	3/36
5	4	4/36
6	5	5/36
7	6	6/36
8	5	5/36
9	4	4/36
10	3	3/36
11	2	2/36
12	1	1/36

The coefficients of the powers of x can be used to form the "count" column. The probability of each event is given by the count divided by 6^n.

Let us consider another experiment involving four flips of a coin. We want to look at the number of ways a heads appears for the 16 total number of outcomes. This is presented as a histogram in Figure B.1. Mathematically, the number of ways to obtain x heads in n flips is spoken as the "number of combinations of n things taken x at a time." The number of ways can be computed by

$$\text{Number of Ways} \equiv \binom{n}{x} = \frac{n!}{x!\,(n-x)!} \tag{B.3}$$

For example if $n = 4$ and $x = 2$, then the number of ways is computed to be 6. The probability of obtaining a heads is given by the number of ways divided by the total number of outcomes (16 in our case). This probability can be generalized by noting that the number of outcomes is given by 2^n:

$$p(x) = \frac{\binom{n}{x}}{2^n} = \frac{n!}{x!\,(n-x)!\,2^n} \tag{B.4}$$

For example if $n = 4$ and $x = 2$, then $p(2) = 0.375$.

A compound event can be defined as the occurrence of "either $x(j)$ or $x(k)$"; the probability of a compound event is defined as

$$p(x(j) \cup x(k)) = p(x(j)) + p(x(k)) - p(x(j) \cap x(k)) \tag{B.5}$$

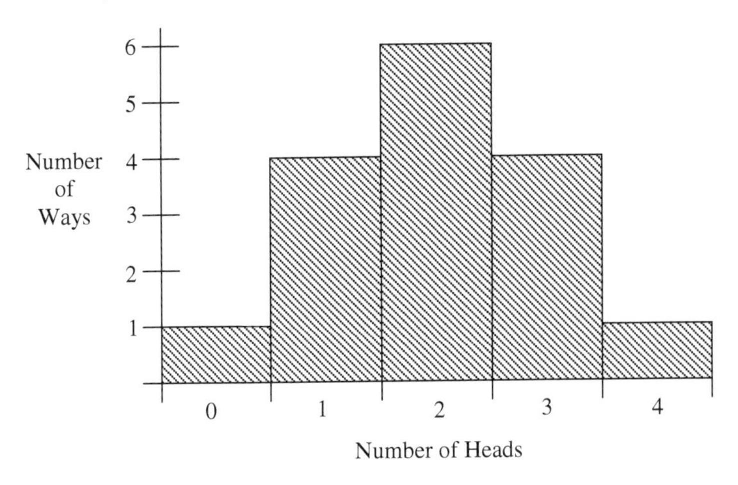

Figure B.1: Histogram of the Number of Ways a Heads Appears

where $x(j) \cup x(k)$ denotes "$x(j)$ or $x(k)$" and $x(j) \cap x(k)$ denotes "$x(j)$ and $x(k)$." The probability of obtaining one event *and* another event is known as the *joint* probability of $x(j)$ and $x(k)$. If $p(x(j) \cap x(k)) = 0$, then the individual probabilities are summed to determine the overall probability. For example, the probability of obtaining less than 3 heads in 4 flips is given by $1/16 + 4/16 + 6/16 = 0.6875$. Note that calculating the probability of obtaining 4 or less heads gives a value of 1! It is clear that a *probability mass function* $p(x(j))$ has the following properties:

$$0 \leq p(x(j)) \leq 1 \tag{B.6a}$$

$$\sum_j p(x(j)) = 1 \tag{B.6b}$$

If events $x(j)$ and $x(k)$ are *independent*, then we have

$$p(x(j) \cap x(k)) = p(x(j))\, p(x(k)) \tag{B.7}$$

For example, the probability of obtaining one heads in two successive trials is given by $(1/4) \times (1/4) = 1/16$.

We now define the *conditional probability* of $x(j)$ given $x(k)$, which is denoted by $p(x(j)|x(k))$. Suppose we know that an event $x(k)$ has occurred. Then $x(j)$ occurs if and only if $x(j)$ and $x(k)$ occur. Therefore, the probability of $x(j)$, given that we know $x(k)$ has occurred, should intuitively be proportional to $p(x(j) \cap x(k))$. However, the conditional probability must satisfy the properties of probability shown by eqn. (B.6). This forces a proportionality constant of $1/p(x(k))$, so that

$$p(x(j)|x(k)) = \frac{p(x(j) \cap x(k))}{p(x(k))} \tag{B.8}$$

In a similar fashion the conditional probability of $x(k)$ given $x(j)$ is

$$p(x(k)|x(j)) = \frac{p(x(k) \cap x(j))}{p(x(j))} \tag{B.9}$$

Combining eqns. (B.8) and (B.9) leads to *Bayes rule*:

$$p(x(j)|x(k)) = \frac{p(x(k)|x(j))\, p(x(j))}{p(x(k))} \tag{B.10}$$

This rule is widely used in estimation theory (e.g., see §2.6). Bayes rule can be used to show some counterintuitive results. For example, say 1 out 1,000 people have a rare disease. Tests show that 99% are positive when they have a disease and 2% are positive when they don't. Bayes rule can be used to show the probability that they actually have a disease when the test is positive is only 0.047! At first glance this seems counterintuitive, but in actuality the result is correct (note: if a 25% incidence rate is given, then the probability is 0.94, which is line with our intuition).

The random variable x is usually described in terms of its *moments*. The first two moments of x are given by the *mean* (μ) of x:

$$\mu \equiv \sum_j x(j)\, p(x(j)) \tag{B.11}$$

and the variance (σ^2) of x:

$$\sigma^2 \equiv \sum_j (x(j) - \mu)^2 p(x(j)) \tag{B.12}$$

The quantity σ is often called the *standard deviation* of x. If $p(x)$ is considered to be a function defining the mass of several discrete masses located along a straight line, then μ locates the center of mass and σ^2 is the moment of inertia of the system of masses about their centroid.

The *expected value* or "average value" of a function $f(x)$ of a discrete random variable x is defined as

$$E\{f(x)\} = \sum_j f(x(j))\, p(x(j)) \tag{B.13}$$

Clearly from eqns. (B.11) and (B.12), the mean and variance are the expected values of x and $(x - \mu)^2$, respectively. Notice that the expected value operator is linear so that

$$E\{a\, f(x) + b\, g(x)\} = a\, E\{f(x)\} + b\, E\{g(x)\} \tag{B.14}$$

for a and b arbitrary deterministic scalars, and $f(x)$ and $g(x)$ arbitrary functions of the random variable x.

B.2 Functions of Discrete-Valued Random Variables

A random vector $\mathbf{x}$ is an $n \times 1$ matrix whose elements x_i are scalar random variables as discussed in §B.1. If each scalar element x_i of $\mathbf{x}$ can take on a finite number, m_i, of discrete values $x_i(j_i)$, for $j_i = 1, 2, \ldots, m_i$, then there are $m_1 m_2 \cdots m_n$ possible vectors. For a complete probabilistic characterization of $\mathbf{x}$, its *joint probability function* $p(j_1, j_2, \ldots, j_n)$ is the probability that x_1 has its j_1^{th} value, x_2 has its j_2^{th} value, ..., x_n has its j_n^{th} value. The function $p(j_1, j_2, \ldots, j_n)$ is often written $p(x_1, x_2, \ldots, x_n)$ when no ambiguity results. On some occasions, one is interested in the *marginal probability mass function* given by

$$p(j_1) = \sum_{j_2=1}^{m_2} \sum_{j_3=1}^{m_3} \cdots \sum_{j_n=1}^{m_n} p(j_1, j_2, \ldots, j_n) \tag{B.15}$$

Note that $p(j_1)$ is the probability of a compound event; that x_1 takes on its j_1^{th} value while $x_2, x_3, \ldots, x_n$ take on arbitrary possible values. Thus, a scalar random variable may represent an elementary or compound event, depending upon the dimension of the underlying space of events.

The marginal probability functions in eqn. (B.15) are sufficient to fully probabilistically characterize the components of $\mathbf{x}$, but to fully characterize $\mathbf{x}$, it is necessary to specify $p(x_1, x_2, \ldots, x_n)$. As in the scalar case, it is customary to describe $p(x_1, x_2, \ldots, x_n)$ and $\mathbf{x}$ in terms of the moments of $\mathbf{x}$. The first two moments are the *mean* ($\boldsymbol{\mu}$) of $\mathbf{x}$:

$$\boldsymbol{\mu} \equiv E\{\mathbf{x}\} = \sum_{j_1=1}^{m_1} \cdots \sum_{j_n=1}^{m_n} \begin{bmatrix} x_1(j_1) \\ \vdots \\ x_n(j_n) \end{bmatrix} p(j_1, j_2, \ldots, j_n) \tag{B.16}$$

and the *covariance* (R) of $\mathbf{x}$:

$$\begin{aligned} R &\equiv E\left\{(\mathbf{x}-\boldsymbol{\mu})(\mathbf{x}-\boldsymbol{\mu})^T\right\} \\ &= E\left\{\begin{bmatrix} (x_1-\mu_1)^2 & (x_1-\mu_1)(x_2-\mu_2) & \cdots & (x_1-\mu_1)(x_n-\mu_n) \\ (x_2-\mu_2)(x_1-\mu_1) & (x_2-\mu_2)^2 & \cdots & (x_2-\mu_2)(x_n-\mu_n) \\ \vdots & \vdots & \ddots & \vdots \\ (x_n-\mu_n)(x_1-\mu_1) & (x_n-\mu_n)(x_2-\mu_2) & \cdots & (x_n-\mu_n)^2 \end{bmatrix}\right\} \end{aligned} \tag{B.17}$$

where the expectation operator $E\{\ \}$ when "operating" upon a matrix, operates upon each individual element. Notice that the covariance matrix R is symmetric. We adopt the following notations:

$$\sigma_i^2 \equiv E\left\{(x_i-\mu_i)^2\right\} = \text{ variance of } x_i \tag{B.18a}$$

$$\sigma_{ij} \equiv E\left\{(x_i-\mu_i)(x_j-\mu_j)\right\} = \text{ covariance of } x_i \text{ and } x_j \tag{B.18b}$$

The covariance matrix is commonly written as

$$R \equiv \begin{bmatrix} \sigma_1^2 & \rho_{12}\sigma_1\sigma_2 & \cdots & \rho_{1n}\sigma_1\sigma_n \\ \rho_{21}\sigma_2\sigma_1 & \sigma_2^2 & \cdots & \rho_{2n}\sigma_2\sigma_n \\ \vdots & \vdots & \ddots & \vdots \\ \rho_{n1}\sigma_n\sigma_1 & \rho_{n2}\sigma_n\sigma_2 & \cdots & \sigma_n^2 \end{bmatrix} \tag{B.19}$$

where ρ_{ij} is the *correlation* of x_i and x_j, defined by

$$\rho_{ij} \equiv \frac{\sigma_{ij}}{\sigma_i\sigma_j} \tag{B.20}$$

This coefficient gives a measure of the degree of linear dependence between x_i and x_j. If x_i is linear in x_j, then $\rho_{ij} = \pm 1$; however, if x_i and x_j are independent of each other, then $\rho_{ij} = 0$. If

$$p(x_1, x_2, \ldots, x_n) = p(x_1)\, p(x_2) \cdots p(x_n) \tag{B.21}$$

for all possible values of $\{x_1, x_2, \ldots, x_n\}$, then the random variables are independent, as discussed in §B.1. Note that while pairwise independence is sufficient to ensure zero correlation of $\{x_1, x_2, \ldots, x_n\}$, it is not sufficient to ensure independence of $\{x_1, x_2, \ldots, x_n\}$.[6]

Example B.1: Consider a vector with two components $\mathbf{x} = \begin{bmatrix} x_1 & x_2 \end{bmatrix}^T$. Suppose that the first component has two possible values:

$$x_1(1) = 0$$
$$x_1(2) = 10$$

Suppose that the second component has three possible values:

$$x_2(1) = -10$$
$$x_2(2) = 0$$
$$x_2(3) = 10$$

Suppose, further, that the six possible events have the following probabilities:

$$p(0,-10) = 0.1, \quad p(0,0) = 0.4, \quad p(0,10) = 0.1$$
$$p(10,-10) = 0.1, \quad p(10,0) = 0.1, \quad p(10,10) = 0.2$$

The expected value (mean) of $\mathbf{x}$ then follows from eqn. (B.16) as

$$\mu = 0.1\begin{bmatrix} 0 \\ -10 \end{bmatrix} + 0.4\begin{bmatrix} 0 \\ 0 \end{bmatrix} + 0.1\begin{bmatrix} 0 \\ 10 \end{bmatrix} + 0.1\begin{bmatrix} 10 \\ -10 \end{bmatrix} + 0.1\begin{bmatrix} 10 \\ 0 \end{bmatrix} + 0.2\begin{bmatrix} 10 \\ 10 \end{bmatrix}$$

which reduces to

$$\mu = \begin{bmatrix} 4 \\ 1 \end{bmatrix}$$

Similarly, the covariance matrix follows from eqn. (B.17) as

$$R = E\begin{bmatrix} (x_1-\mu_1)^2 & (x_1-\mu_1)(x_2-\mu_2) \\ (x_2-\mu_2)(x_1-\mu_1) & (x_2-\mu_2)^2 \end{bmatrix}$$
$$= 0.1\begin{bmatrix} -4 \\ -11 \end{bmatrix}[-4\ -11] + 0.4\begin{bmatrix} -4 \\ -1 \end{bmatrix}[-4\ -1] + 0.1\begin{bmatrix} -4 \\ 9 \end{bmatrix}[-4\ 9]$$
$$+ 0.1\begin{bmatrix} 6 \\ -11 \end{bmatrix}[6\ -11] + 0.1\begin{bmatrix} 6 \\ -1 \end{bmatrix}[6\ -1] + 0.2\begin{bmatrix} 6 \\ 9 \end{bmatrix}[6\ 9]$$

which reduces to

$$R = \begin{bmatrix} 24 & 6 \\ 6 & 49 \end{bmatrix}$$

It may be verified from the results of Appendix A that this covariance matrix is positive definite.

To investigate the definiteness of R in general, let

$$\mu = E\{\mathbf{x}\} \tag{B.22a}$$

$$z = \mathbf{c}^T(\mathbf{x}-\mu) \tag{B.22b}$$

where $\mathbf{c}$ is an $n \times 1$ vector of arbitrary constraints. Investigating the moments of z, we find

$$\mu_z \equiv E\{z\} = E\left\{\mathbf{c}^T(\mathbf{x}-\mu)\right\} = \mathbf{c}^T(\mu-\mu) = 0 \tag{B.23}$$

and

$$\begin{aligned} \sigma_z^2 &\equiv E\left\{(z-\mu_z)^2\right\} = E\left\{\mathbf{c}^T(\mathbf{x}-\mu)(\mathbf{x}-\mu)^T\mathbf{c}\right\} \\ &= \mathbf{c}^T E\left\{(\mathbf{x}-\mu)(\mathbf{x}-\mu)^T\right\}\mathbf{c} \\ &= \mathbf{c}^T R\,\mathbf{c} \end{aligned} \tag{B.24}$$

Since $\sigma_z^2 \geq 0$ and since $\mathbf{c}$ is an arbitrary vector, then R is always *at least* positive semi-definite. For diagonal R, the positive semi-definiteness of R agrees with our intuitive interpretation of σ_i^2; since $\sigma_i^2 < 0$ implies "better than perfect knowledge" or "less than zero uncertainty" in x_i, which is impossible!

B.3 Functions of Continuous Random Variables

For our purposes, the discrete variable concepts of §B.1 and §B.2 can be extended in a natural manner.* By letting $N \to \infty$ with the probability mass function

*There are various theoretical details that must be focused in a rigorous extension of the discrete results to the continuous results (see Ref. [4], for example).

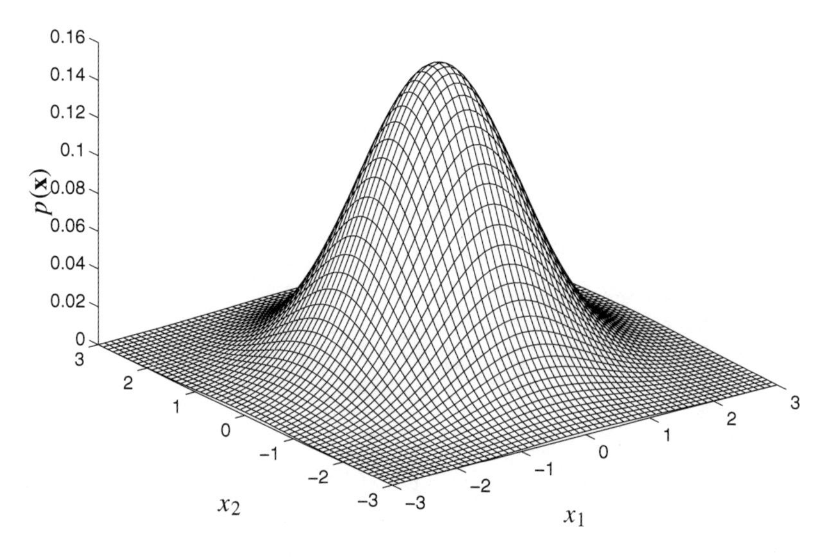

Figure B.2: Two-Dimensional Gaussian Distribution

$p(x_1(j_1), \ldots, x_n(j_n))$ being replaced by a *probability density function* $p(x_1, \ldots, x_n)$; then

$$p(x_1, x_2, \ldots, x_n)\, dx_1\, dx_2 \cdots dx_n \tag{B.25}$$

is the probability that the components of **x** lie within the differential volume given by $dx_1\, dx_2 \cdots dx_n$ centered at $x_1, x_2, \ldots, x_n$. Since all possible **x**-vectors are located in the infinite sphere, it follows that

$$\int_{-\infty}^{\infty} \cdots \int_{-\infty}^{\infty} p(x_1, x_2, \ldots, x_n)\, dx_1\, dx_2 \cdots dx_n = 1 \tag{B.26}$$

The expected value of an arbitrary function $g(x_1, \ldots, x_n)$ is defined in terms of the density function as

$$E\{g(x_1, \ldots, x_n)\} = \int_{-\infty}^{\infty} \cdots \int_{-\infty}^{\infty} g(x_1, \ldots, x_n)\, p(x_1, \ldots, x_n)\, dx_1 \cdots dx_n \tag{B.27}$$

Thus the summation signs of the discrete results of §B.2 are replaced by integral signs to obtain the corresponding continuous results.

B.4 Gaussian Random Variables

The most widely used distribution for state estimation involves the Gaussian random process. Taking the limit as the number of coin flips, used to produce the histogram shown in Figure B.1, approaches infinity leads to the *Gaussian* or *normal* density function for x:

$$p(x) = \frac{1}{\sigma\sqrt{2\pi}} \exp\left[-\frac{(x-\mu)^2}{2\sigma^2}\right] \tag{B.28}$$

with mean given by μ and variance given by σ^2. This function can also be expanded to the multidimensional case for a vector $\mathbf{x}$:

$$p(\mathbf{x}) = \frac{1}{[\det(2\pi R)]^{1/2}} \exp\left[-\frac{1}{2}(\mathbf{x}-\boldsymbol{\mu})^T R^{-1}(\mathbf{x}-\boldsymbol{\mu})\right] \tag{B.29}$$

A plot of this function for two variables, with $\boldsymbol{\mu} = \mathbf{0}$ and $R = I_{2\times 2}$, is shown in Figure B.2. The mean and standard deviation are sufficient enough to define this distribution. Therefore, a simple notation for this distribution is given by

$$p(\mathbf{x}) \sim N(\boldsymbol{\mu}, R) \tag{B.30}$$

The Gaussian distribution is important because of a very useful property that involves any distribution. The *central limit theorem* states that given a distribution with mean μ and variance σ^2, the sampling distribution (no matter what the shape of the original distribution) approaches a Gaussian distribution with mean μ and variance σ^2/N as N, the sample size, increases. This can be clearly seen in Figure B.1, where even for a relatively small sample size the histogram looks like the classic "bell shape" form of the Gaussian distribution. For a formal proof of the central limit theorem see Ref. [7].

A stochastic process is simply a collection of random vectors defined on the same probability space.[8] A *zero-mean Gaussian white-noise* process has the following properties:

$$E\{\mathbf{x}\} = \mathbf{0} \tag{B.31a}$$

$$E\left\{\mathbf{x}(\tau)\mathbf{x}^T(\tau')\right\} = R\,\delta(\tau'-\tau) \tag{B.31b}$$

where $\delta(\tau'-\tau)$ is the delta function. The standard deviation for this process gives a level of confidence that a particular sample lies within the distribution. Also, a process is said to be *stationary* if its random variable statistics do not vary in time (i.e., the probability statistics at time τ have the same mean and covariance as the probability statistics at time τ').

One is often interested in the probability that $\mathbf{x}$ lies inside the quadratic hypersurface

$$(\mathbf{x}-\boldsymbol{\mu})^T R^{-1}(\mathbf{x}-\boldsymbol{\mu}) < G^2 \tag{B.32}$$

where G is a constant. Using an eigenvalue/eigenvector decomposition (see Appendix A) of R, leads to the appropriate orthogonal transformation

$$\mathbf{x} = T\mathbf{y} \tag{B.33a}$$

$$S \equiv \text{diag}\left[\sigma_1^2 \ \sigma_2^2 \ \cdots \ \sigma_n^2\right] = T^T R\, T \tag{B.33b}$$

Therefore, it is always possible to transform coordinates to a principal system in which eqn. (B.32) is reduced to

$$\frac{y_1^2}{\sigma_1^2} + \frac{y_2^2}{\sigma_2^2} + \cdots + \frac{y_n^2}{\sigma_n^2} < G^2 \tag{B.34}$$

We now define another set of change of variables:

$$z_i = \frac{y_i}{\sigma_i}, \quad i = 1, 2, \ldots, n \tag{B.35}$$

so that eqn. (B.34) reduces down to

$$z_1^2 + z_2^2 + \cdots + z_n^2 < G^2 \tag{B.36}$$

The probability of finding z inside this hypersurface is obtained by integrating the Gaussian density function over the volume of the sphere in eqn. (B.36) as

$$p(g^2 \le G^2) = \int_V p(z)\, dV \tag{B.37}$$

where

$$g^2 \equiv \sum_{i=1}^{n} z_i^2 \tag{B.38}$$

Using the element volume $dz_1 dz_2 \cdots dz_n$, eqn. (B.37) can be written as

$$p(g^2 \le G^2) = \int \cdots \int_V \frac{1}{(2\pi)^{n/2}} \exp\left(-\frac{1}{2} g^2\right)\, dz_1 dz_2 \cdots dz_n \tag{B.39}$$

Using an n-dimensional spherically volume element $f(g)\, dg$, eqn. (B.39) can be written as

$$p(g^2 \le G^2) = \frac{1}{(2\pi)^{n/2}} \int_0^G \exp\left(-\frac{1}{2} g^2\right) f(g)\, dg \tag{B.40}$$

For $n = 1, 2, 3$, eqn. (B.40) is explicitly:

- $n = 1$, $f(g)\, dg = 2dg$:

$$\begin{aligned} p(g \le G) &= \sqrt{2/\pi} \int_0^G \exp\left(-\frac{1}{2} g^2\right) dg \\ &= \text{erf}\left(\frac{G}{\sqrt{2}}\right) \end{aligned} \tag{B.41}$$

Table B.2: Probability Values for $g \le G$

	$G=1$	$G=2$	$G=3$
$n=1$	0.683	0.995	0.997
$n=2$	0.394	0.865	0.989
$n=3$	0.200	0.739	0.971

- $n=2$, $f(g)\,dg = 2\pi g\,dg$:

$$\begin{aligned} p(g \le G) &= \int_0^G \exp\left(-\frac{1}{2}g^2\right) g\,dg \\ &= 1 - \exp\left(\frac{-G^2}{2}\right) \end{aligned} \tag{B.42}$$

- $n=3$, $f(g)\,dg = 4\pi g^2\,dg$:

$$\begin{aligned} p(g \le G) &= \sqrt{2/\pi}\int_0^G \exp\left(-\frac{1}{2}g^2\right) g^2\,dg \\ &= \text{erf}\left(\frac{G}{\sqrt{2}}\right) - G\sqrt{2/\pi}\exp\left(\frac{-G^2}{2}\right) \end{aligned} \tag{B.43}$$

where erf is the error function. The numerical value of $p(g < G)$ is often of particular interest in error analysis. Table B.2 displays the "curse of dimensionality" for the probability of g being within 1, 2, and 3 "sigma ellipsoids" for 1, 2, and 3 dimensional spaces.

B.5 Chi-Square Random Variables

The chi-square distribution is often used to provide a consistency test in estimators (see §5.7.3), which is useful to determine whether or not reasonable state estimates are provided. Assuming a Gaussian distribution for the $n \times 1$ vector $\mathbf{x}$, with mean $\boldsymbol{\mu}$ and covariance R, the following variable is said to have a chi-square distribution with n degrees of freedom (DOF):

$$q = (\mathbf{x} - \boldsymbol{\mu})^T R^{-1} (\mathbf{x} - \boldsymbol{\mu}) \tag{B.44}$$

The variable q is the sum of squares of n independent zero-mean variables with variance equal to one. This can be shown by defining the following variable:[9]

$$\mathbf{u} \equiv R^{-1/2}(\mathbf{x} - \boldsymbol{\mu}) \tag{B.45}$$

Then, $\mathbf{u}$ is clearly Gaussian with $E\{\mathbf{u}\} = \mathbf{0}$ and $E\left\{\mathbf{u}\mathbf{u}^T\right\} = I$. The chi-square distribution is written as

$$q \sim \chi_n^2 \tag{B.46}$$

The mean and variance are given by

$$E\{q\} = \sum_{i=1}^{n} E\left\{u_i^2\right\} = n \tag{B.47a}$$

$$E\left\{(q-n)^2\right\} = \sum_{i=1}^{n} E\left\{(u_i^2-1)^2\right\} = \sum_{i=1}^{n}(3-2+1) = 2n \tag{B.47b}$$

where the relationship $E\left\{x^4\right\} = 3\sigma^4$ has been used for the term involving u_i^4. This relationship is given from the scalar version of[9]

$$E\left\{\mathbf{x}^T A\mathbf{x}\mathbf{x}^T B\mathbf{x}\right\} = \text{Tr}(A\,R)\,\text{Tr}(B\,R) + 2\text{Tr}(A\,R\,B\,R) \tag{B.48}$$

where A and B are $n \times n$ matrices.

The chi-square density function with n DOF is given by

$$p(q) = \frac{1}{2^{n/2}\Gamma(n/2)}\, q^{\frac{n-2}{2}} e^{-\frac{q}{2}} \tag{B.49}$$

where the gamma function Γ is defined as

$$\Gamma\left(\frac{1}{2}\right) = \sqrt{\pi} \tag{B.50a}$$

$$\Gamma(1) = 1 \tag{B.50b}$$

$$\Gamma(m+1) = m\,\Gamma(m) \tag{B.50c}$$

Tables of points on the chi-square distribution can be found in Refs. [4] and [9]. For DOF's above 100, the following approximation can be used:[9]

$$\chi_n^2(1-Q) = \frac{1}{2}\left[\mathcal{G}(1-Q) + \sqrt{2n-1}\right]^2 \tag{B.51}$$

where $\chi_n^2(1-Q)$ indicates that to the left of a specific point, the probability mass is $1-Q$. An important quantity used in consistency tests is the 95% *two-sided probability region* for an $N(0,1)$ random variable:

$$[\mathcal{G}(0.025),\, \mathcal{G}(0.975)] = [-1.96,\, 1.96] \tag{B.52}$$

Other values for $\mathcal{G}$ can be found in Ref. [9]. Then, specific values can be calculated for $\chi_n^2(1-Q)$ using eqn. (B.51); e.g., $\chi_{400}^2(0.025) = 346$ and $\chi_{400}^2(0.975) = 457$.

B.6 Propagation of Functions through Various Models

In this section the basic concepts for the propagation of functions through linear and nonlinear models is shown. We shall see that for linear models the original assumed density function is maintained (e.g., a Gaussian input into a linear system produces a Gaussian output), but for nonlinear models this concept does not hold in general.

B.6.1 Linear Matrix Models

We consider the following linear matrix equation:

$$\mathbf{y} = A\,\mathbf{x} + \mathbf{b} \tag{B.53}$$

where A and $\mathbf{b}$ are arbitrary constant matrices with deterministic elements, and $\mathbf{x}$ is a random vector whose first two moments are assumed known:

$$\mu = E\{\mathbf{x}\} \tag{B.54a}$$

$$R = E\left\{(\mathbf{x}-\mu)(\mathbf{x}-\mu)^T\right\} \tag{B.54b}$$

It is desired to determine the first and second moments of $\mathbf{y}$. The mean follows

$$\mu_y \equiv E\{\mathbf{y}\} = E\{A\,\mathbf{x}+\mathbf{b}\} = A\,E\{\mathbf{x}\}+\mathbf{b} \tag{B.55}$$

or

$$\mu_y = A\,\mu + \mathbf{b} \tag{B.56}$$

The covariance matrix is then obtained from the definition

$$R_{yy} \equiv E\left\{(\mathbf{y}-\mu_y)(\mathbf{y}-\mu_y)^T\right\} \tag{B.57}$$

Substituting eqns. (B.53) and (B.56) into eqn. (B.57) gives

$$R_{yy} = E\left\{A\,(\mathbf{x}-\mu)(\mathbf{x}-\mu)^T A^T\right\} = A\,E\left\{(\mathbf{x}-\mu)(\mathbf{x}-\mu)^T\right\} A^T \tag{B.58}$$

or

$$R_{yy} = A\,R\,A^T \tag{B.59}$$

which is a commonly used result for "swapping" covariance matrices through linear systems.

B.6.2 Nonlinear Models

If $\mathbf{x}$ is a random vector whose density function $p(\mathbf{x})$ is known, and if $\mathbf{y} = \mathbf{f}(\mathbf{x})$ is an arbitrary (generally nonlinear) one-to-one transformation, then it can be shown that the density function of $\mathbf{y}$ is given by[1]

$$p(\mathbf{y}) = p(\mathbf{x}) \left| \det\left(\frac{\partial \mathbf{f}}{\partial \mathbf{x}}\right) \right|^{-1} \tag{B.60}$$

with $\mathbf{x}$ on the right-hand side of eqn. (B.60) given by

$$\mathbf{x} = \mathbf{f}^{-1}(\mathbf{y}) \tag{B.61}$$

where $\mathbf{f}^{-1}(\mathbf{y})$ denotes the "reverse" relationship. Thus to convert the density function of $\mathbf{x}$ to the density function of $\mathbf{y}$, simply write the density of $\mathbf{x}$ in terms of $\mathbf{y}$ and multiply by the inverse determinant of the Jacobian matrix.

Example B.2: We will now employ the preceding results using the linear scalar model

$$y = a\,x \tag{B.62}$$

and the following assumed Gaussian density function for x:

$$p(x) = \frac{1}{\sigma\sqrt{2\pi}} \exp\left(\frac{-x^2}{2\sigma^2}\right) \tag{B.63}$$

Then, from eqn. (B.60) we find

$$p(y) = \frac{1}{a\,\sigma\sqrt{2\pi}} \exp\left(\frac{-y^2}{2a^2\sigma^2}\right) \tag{B.64}$$

Note further

$$\begin{aligned} \mu_y \equiv E\{y\} &= \int_{-\infty}^{\infty} y\, p(y)\, dy \\ &= \frac{1}{a\,\sigma\sqrt{2\pi}} \int_{-\infty}^{\infty} y \exp\left(\frac{-y^2}{2a^2\sigma^2}\right) dy \end{aligned} \tag{B.65}$$

Integrating by parts leads to

$$\mu_y = 0 \tag{B.66}$$

which is equivalent to the expected value of x. Similarly, we find from the definition of variance that

$$\begin{aligned} \sigma_y^2 \equiv E\left\{(y-\mu_y)^2\right\} &= E\left\{y^2\right\} \\ &= \int_{-\infty}^{\infty} y^2 p(y)\, dy \\ &= \frac{1}{a\,\sigma\sqrt{2\pi}} \int_{-\infty}^{\infty} y^2 \exp\left(\frac{-y^2}{2a^2\sigma^2}\right) dy \end{aligned} \tag{B.67}$$

which integrates to

$$\sigma_y^2 = a^2\sigma^2 \tag{B.68}$$

This mean and variance of y computed here confirms the previous results shown in eqns. (B.56) and (B.59). Also, we see that y itself is clearly a Gaussian random variable, which confirms that a transformation through a linear model does not alter the form of the distribution.

Example B.3: Assume the following quadratic model:

$$y = a\,x^2 \tag{B.69}$$

Note that for each value of y there are two x-values. Assume that x has the following Gaussian density function:

$$p(x) = \frac{1}{\sigma\sqrt{2\pi}} \exp\left(\frac{-x^2}{2\sigma^2}\right) \tag{B.70}$$

It follows from eqn. (B.60) that

$$p(y) = \frac{1}{2\sigma\sqrt{2\pi\,a\,y}} \exp\left(\frac{-y}{2a\,\sigma^2}\right), \quad \text{for } y > 0 \tag{B.71}$$

and

$$p(y) = 0, \quad \text{for } y < 0 \tag{B.72}$$

It also follows that

$$\mu_y \equiv E\{y\} = a\,\sigma^2 \tag{B.73}$$

and

$$\sigma_y^2 \equiv E\left\{(y-\mu_y)^2\right\} = 2a^2\sigma^4 \tag{B.74}$$

Note that y is no longer a Gaussian variable. Hence, unlike the linear case, a non-linear transformation of a Gaussian variable does not necessarily produce another Gaussian variable.

References

[1] Bryson, A.E. and Ho, Y.C., *Applied Optimal Control*, Taylor & Francis, London, England, 1975.

[2] Cox, D.R. and Hinkley, D.V., *Problems and Solutions in Theoretical Statistics*, John Wiley & Sons, New York, NY, 1978.

[3] Keeping, E., *Introduction to Statistical Inference*, Dover Publications, New York, NY, 1995.

[4] Freund, J.E. and Walpole, R.E., *Mathematical Statistics*, Prentice Hall, Englewood Cliffs, NJ, 4th ed., 1987.

[5] Devore, J.L., *Probability and Statistics for Engineering and Sciences*, Duxbury Press, Pacific Grove, CA, 1995.

[6] Feller, W., *Introduction to Probability Theory and Its Applications*, John Wiley & Sons, New York, NY, 3rd ed., 1966.

[7] Kallenberg, O., *Foundations of Modern Probability*, Springer-Verlag, New York, NY, 1997.

[8] Sage, A.P. and White, C.C., *Optimum Systems Control*, Prentice Hall, Englewood Cliffs, NJ, 2nd ed., 1977.

[9] Bar-Shalom, Y., Li, X.R., and Kirubarajan, T., *Estimation with Applications to Tracking and Navigation*, John Wiley & Sons, New York, NY, 2001.

C

Parameter Optimization Methods

In this appendix classical necessary and sufficient conditions for solution of unconstrained and equality-constrained parameter optimization problems are summarized. We also summarize two iterative techniques for unconstrained minimization, and discuss the relative merits of these approaches.

C.1 Unconstrained Extrema

Suppose we wish to determine a vector $\mathbf{x}$ that minimizes (or maximizes) the following *loss function*:

$$\vartheta \equiv \vartheta(\mathbf{x}) \tag{C.1}$$

with $\mathbf{x} = \begin{bmatrix} x_1 \; x_2 \; \cdots \; x_n \end{bmatrix}^T$. Without loss in generality, we assume our task is to minimize eqn. (C.1). It is evident that a simple change of sign converts a maximization problem to a minimization problem. To obtain the most fundamental classical results, we restrict initial attention to ϑ and $\mathbf{x}$ of class C_2 (smooth, continuous functions having two continuous derivatives with respect to all arguments). Using the matrix calculus differentiation rules developed in §A.5, it follows that a "stationary" or "critical" point can be determined by solving the following necessary condition:

$$\nabla_{\mathbf{x}} \vartheta \equiv \frac{\partial \vartheta}{\partial \mathbf{x}} = \mathbf{0} \tag{C.2}$$

where $\nabla_{\mathbf{x}}$ is the Jacobian (see Appendix A). Unfortunately, satisfying the condition in eqn. (C.2) does not guarantee a *local minimum* in general. If $\mathbf{x}$ is scalar, then the classic test for a local minimum is to check the second derivative of ϑ, which must be positive. This concept can be expanded to a vector of unknown variables by using a matrix check.[1, 2] The sufficiency condition requires that one determine the definiteness of the matrix of partial derivatives, known as the Hessian matrix (see Appendix A). Suppose we have a stationary point, denoted by $\mathbf{x}^*$. This point is a local minimum if the following sufficient condition is satisfied:

$$\nabla^2_{\mathbf{x}} \vartheta \equiv \left. \frac{\partial^2 \vartheta}{\partial \mathbf{x}\, \partial \mathbf{x}^T} \right|_{\mathbf{x}^*} \quad \text{must be positive definite} \tag{C.3}$$

where $\nabla^2_{\mathbf{x}} \vartheta$ is the Hessian (see Appendix A). If this matrix is negative definite, then the point is a maximum. If the matrix is indefinite, then a *saddle point* exists, which corresponds to a relative minimum or maximum with respect to the individual components of $\mathbf{x}^*$. A global minimum is much more difficult to establish though. Consider the minimization of the following function (known as Himmelblau's function):[3]

$$\vartheta(\mathbf{x}) = (x_1^2 + x_2 - 11)^2 + (x_1 + x_2^2 - 7)^2 \tag{C.4}$$

A plot of the contours (lines of constant ϑ) is shown in Figure C.1. Also shown in this plot are the numerical iteration points for the method of gradients (see §C.3.2) from various starting guesses. There is a set of four stationary points which provide local minimums, each of approximately the same importance:

- $\mathbf{x}_1^* = \begin{bmatrix} 3 & 2 \end{bmatrix}^T$, with $\vartheta(\mathbf{x}_1^*) = 0$.
- $\mathbf{x}_2^* = \begin{bmatrix} -3.7792 & -3.2831 \end{bmatrix}^T$, with $\vartheta(\mathbf{x}_2^*) = 0.0054$.
- $\mathbf{x}_3^* = \begin{bmatrix} -2.8051 & 3.1313 \end{bmatrix}^T$, with $\vartheta(\mathbf{x}_3^*) = 0.0085$.
- $\mathbf{x}_4^* = \begin{bmatrix} 3.5843 & -1.8483 \end{bmatrix}^T$, with $\vartheta(\mathbf{x}_4^*) = 0.0011$.

Clearly, a numerical technique such as the method of gradients can converge to any one of these four points from various starting guesses. Fortunately a resourceful analyst can often achieve a high degree of confidence that a stationary point is a global minimum through intimate knowledge of the loss function (e.g., the Hessian matrix for a *quadratic loss function* is constant).

Example C.1: In this example we consider finding the extreme points of the following loss function:[1]

$$\vartheta(\mathbf{x}) = x_1^3 + x_2^3 + 2x_1^2 + 4x_2^2 + 6$$

The necessary conditions for x_1 and x_2, given by eqn. (C.2), are

$$\frac{\partial \vartheta}{\partial x_1} = x_1(3x_1 + 4) = 0$$
$$\frac{\partial \vartheta}{\partial x_2} = x_2(3x_2 + 8) = 0$$

These equations are satisfied at the following stationary points:

$$\mathbf{x}_1^* = \begin{bmatrix} 0 & 0 \end{bmatrix}^T, \quad \mathbf{x}_2^* = \begin{bmatrix} 0 & -\frac{8}{3} \end{bmatrix}^T$$
$$\mathbf{x}_3^* = \begin{bmatrix} -\frac{4}{3} & 0 \end{bmatrix}^T, \quad \mathbf{x}_4^* = \begin{bmatrix} -\frac{4}{3} & -\frac{8}{3} \end{bmatrix}^T$$

The Hessian matrix is given by

$$\nabla^2_{\mathbf{x}} \vartheta = \begin{bmatrix} 6x_1 + 4 & 0 \\ 0 & 6x_2 + 8 \end{bmatrix}$$

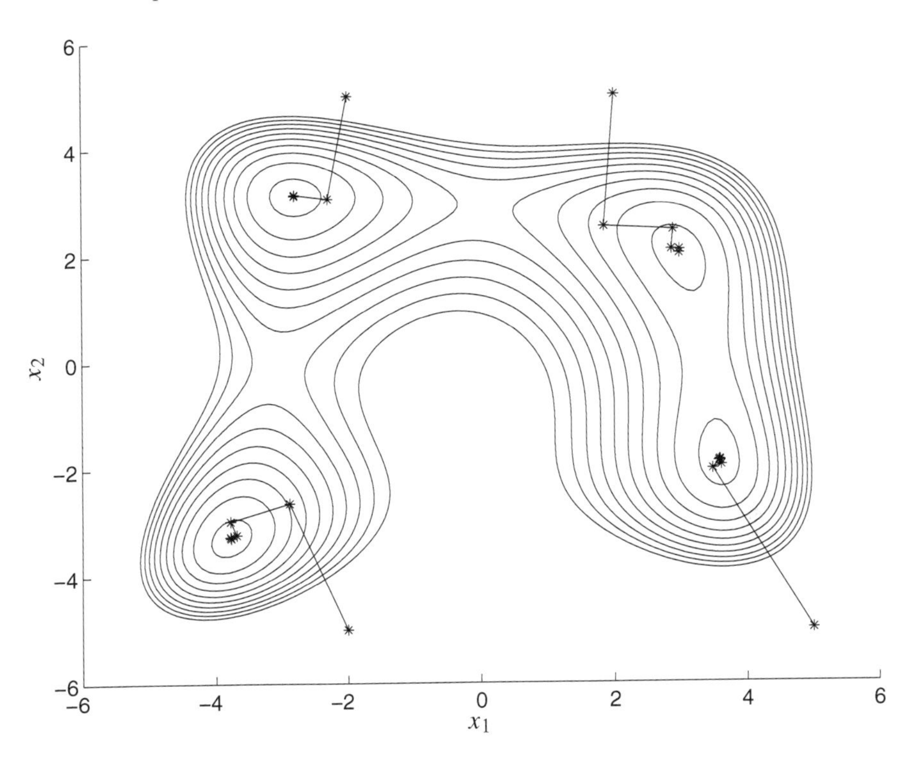

Figure C.1: Himmelblau's Function

Table C.1 gives the nature of the Hessian and the value of the loss function at the stationary points. The first point gives a local minimum, the next two points are saddle points, and the last point gives a local maximum.

C.2 Equality Constrained Extrema

One often encounters problems that must extremize

$$\vartheta \equiv \vartheta(\mathbf{x}) \tag{C.5}$$

subject to the following set of $m \times 1$ *equality constraints*:

$$\boldsymbol{\psi} \equiv \boldsymbol{\psi}(\mathbf{x}) = \mathbf{0} \tag{C.6}$$

Table C.1: Nature of the Hessian and Values for the Loss Function

Point $\mathbf{x}_i^*$	Nature of $\nabla_{\mathbf{x}}^2 \vartheta \vert_{\mathbf{x}_i^*}$	Nature of $\mathbf{x}_i^*$	$\vartheta(\mathbf{x}_i^*)$
$\mathbf{x}_1^* = [0\ 0]^T$	Positive Definite	Relative Minimum	6
$\mathbf{x}_2^* = [0\ -\frac{8}{3}]^T$	Indefinite	Saddle Point	418/27
$\mathbf{x}_3^* = [-\frac{4}{3}\ 0]^T$	Indefinite	Saddle Point	194/27
$\mathbf{x}_4^* = [-\frac{4}{3}\ -\frac{8}{3}]^T$	Negative Definite	Relative Maximum	50/3

where $m < n$. Let us consider the case where $n = 2$ and $m = 1$. Suppose (x_1^*, x_2^*) locally minimizes eqn. (C.5) while satisfying eqn. (C.6). If this is true, then arbitrary admissible differential variations $(\delta x_1, \delta x_2)$ in the differential neighborhood of (x_1^*, x_2^*) in the sense $(x_1, x_2) = (x_1^* + \delta x_1, x_2^* + \delta x_2)$ result in a stationary value of ϑ:

$$\delta\vartheta = \frac{\partial\vartheta}{\partial x_1}\delta x_1 + \frac{\partial\vartheta}{\partial x_2}\delta x_2 = 0 \tag{C.7}$$

Since we restrict attention to neighboring points that satisfy the constraint given by eqn. (C.6), we also require the first variation of the constraint to vanish as a condition on the admissibility of $(\delta x_1, \delta x_2)$ as

$$\delta\psi = \frac{\partial\psi}{\partial x_1}\delta x_1 + \frac{\partial\psi}{\partial x_2}\delta x_2 = 0 \tag{C.8}$$

For notational convenience, we suppress the truth that all partials in eqns. (C.7) and (C.8) are evaluated at (x_1^*, x_2^*). Since eqn. (C.8) constrains the admissible variations, we can solve for either variable and eliminate the constraint equation. The two solutions of the constraint equations are obviously

$$\delta x_1 = -\left(\frac{\dfrac{\partial\psi}{\partial x_2}}{\dfrac{\partial\psi}{\partial x_1}}\right)\delta x_2 \quad \text{and} \quad \delta x_2 = -\left(\frac{\dfrac{\partial\psi}{\partial x_1}}{\dfrac{\partial\psi}{\partial x_2}}\right)\delta x_1 \tag{C.9}$$

Substitution of the "differential eliminations" into the differential of the loss function allows us to locally constrain the variations of ϑ and reduce the dimensionality either of two ways. The first way is given by using

$$\delta\vartheta = \left[\frac{\partial\vartheta}{\partial x_2} - \left(\frac{\dfrac{\partial\vartheta}{\partial x_1}}{\dfrac{\partial\psi}{\partial x_1}}\right)\frac{\partial\psi}{\partial x_2}\right]\delta x_2 = 0 \tag{C.10}$$

The second way is given by using

$$\delta\vartheta = \left[\frac{\partial\vartheta}{\partial x_1} - \left(\frac{\dfrac{\partial\vartheta}{\partial x_2}}{\dfrac{\partial\psi}{\partial x_2}}\right)\frac{\partial\psi}{\partial x_1}\right]\delta x_1 = 0 \tag{C.11}$$

It is evident that either of eqns. (C.10) or (C.11) can be used to argue that the local variations are arbitrary and the coefficient within the brackets must vanish as a necessary condition for a local minimum at (x_1^*, x_2^*). The first form of the necessary conditions is given by

$$\frac{\partial\vartheta}{\partial x_1} - \left(\frac{\dfrac{\partial\vartheta}{\partial x_2}}{\dfrac{\partial\psi}{\partial x_2}}\right)\frac{\partial\psi}{\partial x_1} = 0 \tag{C.12a}$$

$$\psi(x_1, x_2) = 0 \tag{C.12b}$$

The second form of the necessary conditions is given by

$$\frac{\partial\vartheta}{\partial x_2} - \left(\frac{\dfrac{\partial\vartheta}{\partial x_1}}{\dfrac{\partial\psi}{\partial x_1}}\right)\frac{\partial\psi}{\partial x_2} = 0 \tag{C.13a}$$

$$\psi(x_1, x_2) = 0 \tag{C.13b}$$

When this approach is carried to higher dimensions, the number of differential elimination possibilities is obviously much greater, and some of these forms of the necessary conditions may be poorly conditioned if the partial derivatives in the denominator approaches zero.

Lagrange noticed a pattern in the above and decided to "automate" all possible differential eliminations by linearly combining eqns. (C.7) and (C.8) with an unspecified scalar *Lagrange multiplier* λ as

$$\delta\vartheta + \lambda\,\delta\psi = \left[\frac{\partial\vartheta}{\partial x_1} + \lambda\frac{\partial\psi}{\partial x_1}\right]\delta x_1 + \left[\frac{\partial\vartheta}{\partial x_2} + \lambda\frac{\partial\psi}{\partial x_2}\right]\delta x_2 = 0 \tag{C.14}$$

While it "isn't legal" to set the two brackets to zero using the argument that $(\delta x_1, \delta x_2)$ are independent, we can set either one of the brackets to zero to determine λ. Notice that setting the first bracket to zero and substituting the resulting equation for $\lambda = -\left(\frac{\partial\vartheta}{\partial x_1}\right)/\left(\frac{\partial\psi}{\partial x_1}\right)$ into the second bracket renders the second bracket equal to eqn. (C.13a), whereas setting the second bracket to zero, solving for λ and substituting renders the first bracket equal to eqn. (C.12a). Thus the following necessary generalized Lagrange form of the necessary conditions captures all possible differential constraint eliminations (only two in this case):

$$\frac{\partial \vartheta}{\partial x_1} + \lambda \frac{\partial \psi}{\partial x_1} = 0 \tag{C.15a}$$

$$\frac{\partial \vartheta}{\partial x_2} + \lambda \frac{\partial \psi}{\partial x_2} = 0 \tag{C.15b}$$

$$\psi(x_1, x_2) = 0 \tag{C.15c}$$

It is apparent by inspection of eqn. (C.15) that these equations are the gradient of the augmented function $\phi \equiv \vartheta + \lambda \psi$ with respect to (x_1, x_2, λ) and thus the Lagrange multiplier rule is validated. The necessary conditions for a constrained minimum of eqn. (C.5) subject to eqn. (C.6) has the form of an unconstrained minimum of the augmented function ϕ:

$$\frac{\partial \phi}{\partial x_1} = \frac{\partial \vartheta}{\partial x_1} + \lambda \frac{\partial \psi}{\partial x_1} = 0 \tag{C.16a}$$

$$\frac{\partial \phi}{\partial x_2} = \frac{\partial \vartheta}{\partial x_2} + \lambda \frac{\partial \psi}{\partial x_2} = 0 \tag{C.16b}$$

$$\psi(x_1, x_2) = 0 \tag{C.16c}$$

Equations (C.16) provide four equations; all points (x_1^*, x_2^*, λ) satisfying these equations are *constrained stationary points.*

Expanding this concept to the general case results in the necessary conditions for a stationary point, which is applied by the unconstrained necessary condition of eqn. (C.2) to the following *augmented function*:

$$\phi \equiv \phi(\mathbf{x}, \boldsymbol{\lambda}) = \vartheta(\mathbf{x}) + \boldsymbol{\lambda}^T \psi(\mathbf{x}) \tag{C.17}$$

The necessary conditions are now given by

$$\nabla_{\mathbf{x}} \phi \equiv \frac{\partial \phi}{\partial \mathbf{x}} = \frac{\partial \vartheta}{\partial \mathbf{x}} + \left[\frac{\partial \psi}{\partial \mathbf{x}}\right]^T \boldsymbol{\lambda} = \mathbf{0} \tag{C.18a}$$

$$\nabla_{\boldsymbol{\lambda}} \phi \equiv \frac{\partial \phi}{\partial \boldsymbol{\lambda}} = \psi(\mathbf{x}) = \mathbf{0} \tag{C.18b}$$

where $\boldsymbol{\lambda}$ is an $m \times 1$ vector of Lagrange multipliers. The $(n+m)$ equations shown in eqn. (C.18), which define the *Lagrange multiplier rule*, are solved for the $(n+m)$ unknowns $\mathbf{x}$ and $\boldsymbol{\lambda}$. Suppose we have a stationary point, denoted by $\mathbf{x}^*$ with a corresponding Lagrange multiplier $\boldsymbol{\lambda}^*$. The point $\mathbf{x}^*$ is a local minimum if the following sufficient condition is satisfied:

$$\nabla_{\mathbf{x}}^2 \phi \equiv \left.\frac{\partial^2 \phi}{\partial \mathbf{x}\, \partial \mathbf{x}^T}\right|_{(\mathbf{x}^*, \boldsymbol{\lambda}^*)} \quad \text{must be positive definite.} \tag{C.19}$$

The sufficient condition can be simplified by checking the positive definiteness of a matrix that is always smaller than the $n \times n$ matrix shown by eqn. (C.19). Let us rewrite the loss function in eqn. (C.5) as

$$\vartheta(x_1, \ldots, x_m, x_{m+1}, \ldots, x_n) \equiv \vartheta(\mathbf{y}, \mathbf{z}) \tag{C.20}$$

where $\mathbf{y}$ is an $m \times 1$ vector and $\mathbf{z}$ is a $p \times 1$ vector (with $p = n - m$). The necessary conditions are still given by eqn. (C.18) with $\mathbf{x} \equiv \left[\mathbf{y}^T\ \mathbf{z}^T\right]^T$. But the sufficient condition can now be determined by checking the definiteness of the following $p \times p$ matrix:[2]

$$
\begin{aligned}
Q \equiv \Big\{ & [\nabla_{\mathbf{z}}\psi]^T \left[\nabla_{\mathbf{y}}\psi\right]^{-T} \left[\nabla_{\mathbf{y}}^2\phi\right] \left[\nabla_{\mathbf{y}}\psi\right]^{-T} [\nabla_{\mathbf{z}}\psi] + \nabla_{\mathbf{z}}^2\phi \\
& - \left[\nabla_{\mathbf{z}}\nabla_{\mathbf{y}}\phi\right] \left[\nabla_{\mathbf{y}}\psi\right]^{-1} [\nabla_{\mathbf{z}}\psi] - [\nabla_{\mathbf{z}}\psi]^T \left[\nabla_{\mathbf{y}}\psi\right]^{-T} \left[\nabla_{\mathbf{y}}\nabla_{\mathbf{z}}\phi\right] \Big\} \Big|_{(\mathbf{y}^*,\,\mathbf{z}^*,\,\boldsymbol{\lambda}^*)}
\end{aligned}
\tag{C.21}
$$

where $\left[\nabla_{\mathbf{z}}\nabla_{\mathbf{y}}\phi\right]$ and $\left[\nabla_{\mathbf{y}}\nabla_{\mathbf{z}}\phi\right]$ are $p \times m$ and $m \times p$ matrices, respectively, made up of the partial derivatives with respect to $\mathbf{y}$ and $\mathbf{z}$. A stationary point is a local minimum (maximum) if Q is positive (negative) definite. Note that the inverse of an $m \times m$ matrix must be taken. Still, the matrix in eqn. (C.21) is usually simpler to check than using the $n \times n$ matrix in eqn. (C.19).

Example C.2: In this example we consider finding the extreme points of the following loss function, which represents a plane:

$$\vartheta = 6 - \frac{y}{2} - \frac{z}{3}$$

subject to a constraint represented by an elliptic cylinder:

$$\psi(\mathbf{x}) = 9(y-4)^2 + 4(z-5)^2 - 36 = 0$$

where $\mathbf{x} \equiv \left[x\ y\right]^T$. The augmented function of eqn. (C.17) for this problem is given by

$$\phi(\mathbf{x}, \lambda) = 6 - \frac{y}{2} - \frac{z}{3} - \lambda\left[9(y-4)^2 + 4(z-5)^2 - 36\right]$$

From the necessary conditions of eqn. (C.18) we have

$$
\begin{aligned}
\frac{\partial\phi}{\partial y} &= -\frac{1}{2} + 18\lambda(y-4) = 0 \\
\frac{\partial\phi}{\partial z} &= -\frac{1}{3} + 8\lambda(y-5) = 0 \\
\psi(\mathbf{x}) &= 9(y-4)^2 + 4(z-5)^2 - 36 = 0
\end{aligned}
$$

Solving these equations for λ gives $\lambda = \pm 1/(36\sqrt{2})$. Therefore, the stationary points are given by

$$
\begin{aligned}
y^* &= 4 + \frac{1}{36\lambda} = 4 \pm \sqrt{2} \\
z^* &= 45 + \frac{1}{24\lambda} = 5 \pm \frac{3}{2}\sqrt{2} \\
\lambda^* &= \pm\frac{1}{36\sqrt{2}}
\end{aligned}
$$

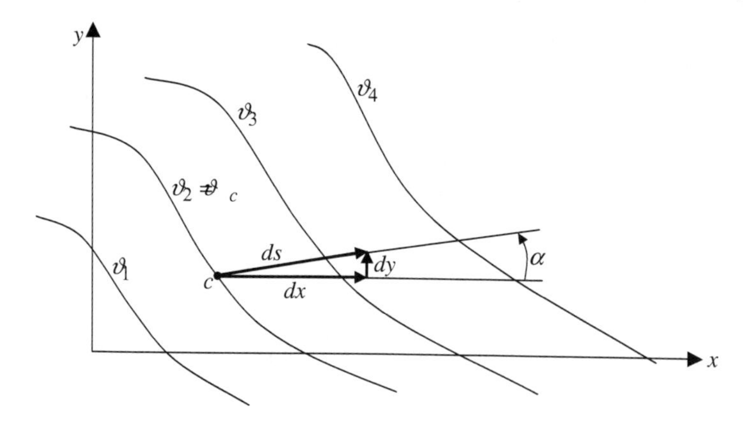

Figure C.2: The Directional Derivative Concept

The sufficient condition of eqn. (C.19) for this problem is given by

$$\nabla_{\mathbf{x}}^2 \phi = \begin{bmatrix} 18\lambda^* & 0 \\ 0 & 8\lambda^* \end{bmatrix}$$

Also, eqn. (C.21) gives

$$Q \equiv q = 8\lambda^* \left[\frac{(z^*-5)^2}{(y^*-4)^2} + 1 \right]$$

Clearly, if $\lambda^* = +1/(36\sqrt{2})$ then the stationary point given by $y^* = 4+\sqrt{2}$ and $z^* = 5+(3/2)\sqrt{2}$ is a local minimum with $\phi = (7/3)-\sqrt{2}$. Likewise, if $\lambda^* = -1/(36\sqrt{2})$ then the stationary point given by $y^* = 4-\sqrt{2}$ and $z^* = 5-(3/2)\sqrt{2}$ is a local maximum with $\phi = (7/3)+\sqrt{2}$.

C.3 Nonlinear Unconstrained Optimization

In this section two iterative methods are shown that can be used to solve nonlinear unconstrained optimization problems. Several approaches can be used to numerically solve these problems, but are beyond the scope of the present text. The interested reader is encouraged to pursue other approaches in the open literature (e.g., see Refs. [1] and [3]).

C.3.1 Some Geometrical Insights

Consider the function $\vartheta(x,y)$ of two variables whose contours are sketched in Figure C.2. From the geometry of Figure C.2 it is evident that

$$\tan\alpha = \frac{dy}{dx} \tag{C.22a}$$

$$\sin\alpha = \frac{dy}{ds} \tag{C.22b}$$

$$\cos\alpha = \frac{dx}{ds} \tag{C.22c}$$

For arbitrary small displacements (dx, dy) away from the "current" point (x_c, y_c), the differential change in ϑ is given by

$$d\vartheta = \left.\frac{\partial\vartheta}{\partial x}\right|_c dx + \left.\frac{\partial\vartheta}{\partial y}\right|_c dy \tag{C.23}$$

If s is the distance measured along an arbitrary line through c, then the rate of change ("differential derivative") of ϑ in the direction of the line is

$$\left.\frac{d\vartheta}{ds}\right|_c = \left.\frac{\partial\vartheta}{\partial x}\right|_c \left.\frac{dx}{ds}\right|_c + \left.\frac{\partial\vartheta}{\partial y}\right|_c \left.\frac{dy}{ds}\right|_c \tag{C.24}$$

Making use of eqns. (C.22b) and (C.22c), we have

$$\left.\frac{d\vartheta}{ds}\right|_c = \left.\frac{\partial\vartheta}{\partial x}\right|_c \cos\alpha + \left.\frac{\partial\vartheta}{\partial y}\right|_c \sin\alpha \tag{C.25}$$

Now, let's look at a couple of particularly interesting cases. Suppose we wish to select the particular line for which $\left.\frac{d\vartheta}{ds}\right|_c = 0$. Equation (C.25) tells us that the angle $\alpha_1 = \alpha$ orienting this line is given by

$$\tan\alpha_1 = \frac{-\left.\dfrac{\partial\vartheta}{\partial x}\right|_c}{\left.\dfrac{\partial\vartheta}{\partial y}\right|_c} \tag{C.26}$$

which gives the "contour direction." Now let's also find the particular direction of which results in the minimum or maximum $\left.\frac{d\vartheta}{ds}\right|_c$. The necessary condition for the extremum of $\left.\frac{d\vartheta}{ds}\right|_c$ requires

$$\frac{d}{d\alpha}\left(\left.\frac{d\vartheta}{ds}\right|_c\right) = -\left.\frac{\partial\vartheta}{\partial x}\right|_c \sin\alpha + \left.\frac{\partial\vartheta}{\partial y}\right|_c \cos\alpha = 0 \tag{C.27}$$

From eqn. (C.27) the angle $\alpha_2 = \alpha$ which orients the direction of "steepest descent" or "steepest ascent" is given by

$$\tan\alpha_2 = \frac{\left.\dfrac{\partial\vartheta}{\partial y}\right|_c}{\left.\dfrac{\partial\vartheta}{\partial x}\right|_c} \tag{C.28}$$

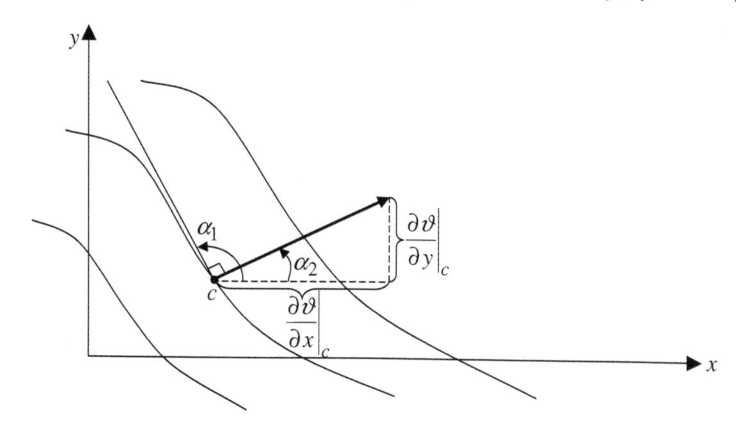

Figure C.3: Geometrical Interpretation of the Gradient Line

which gives the "gradient direction." Notice that $(\tan\alpha_1)(\tan\alpha_2) = -1$. Therefore, α_1 and α_2 orient lines that are perpendicular. So, the contour line is perpendicular to the gradient line, as shown in Figure C.3. These geometrical concepts are difficult to conceptualize rigorously in higher dimensional spaces, but fortunately, the mathematics does generalize rigorously and in a straightforward fashion.

C.3.2 Methods of Gradients

One immediate conclusion of the foregoing is that (based only upon the first derivative information), the most favorable direction to take a small step toward minimizing (or maximizing) the function ϑ is down (or up) the locally evaluated gradient of ϑ. The "method of gradients" (also known as the "method of steepest descent" for minimizing ϑ or the "method of steepest ascent" for maximizing ϑ) is a sequence of one-dimensional searches along the lines established by successively evaluated local gradients of ϑ. Consider ϑ to be a function of n variables which are the elements of $\mathbf{x}$. Let the local evaluations be denoted by superscripts. For example,

$$\vartheta^{(k)} = \vartheta\left(\mathbf{x}^{(k)}\right) \tag{C.29}$$

denotes $\vartheta(\mathbf{x})$ evaluated at the kth set of $\mathbf{x}$-values. The kth one-dimensional search determines a scalar $\alpha^{(k)}$ such that

$$\mathbf{x}^{(k+1)} = \mathbf{x}^{(k)} - \alpha^{(k)}\,[\nabla_{\mathbf{x}}\vartheta]^{(k)} \tag{C.30}$$

results in

$$\vartheta^{(k+1)} = \vartheta\left(\mathbf{x}^{(k+1)}\right) \tag{C.31}$$

being a local minimum or maximum. The one-dimensional search for $\alpha^{(k)}$ can be determined analytically or numerically using various methods (see Refs. [1] and [3]).

It is important to develop a geometrical feel for the method of gradients to understand the circumstances under which it works best, to anticipate failures, and to decide upon remedial action when failure occurs. Sequences of iterations from various starting guesses for Himmelblau's function are shown in Figure C.1. Observe the orthogonality of successive gradients. The successive gradients will be exactly orthogonal only if the one-dimensional minima or maxima are perfectly located. Note, for the case of two unknowns only one gradient calculation may be necessary, since all successive gradients are either parallel or perpendicular to the first. However, this orthogonality condition is obviously insufficient to establish the gradient directions for the case of three or more unknowns (e.g., for three unknowns there exists a *plane* that is perpendicular to the gradient vector).

The convergence of the gradient method is heavily dependent upon the circularity of the contours (see Figure C.5 for a function with nonlinear trenches). As an aside, in 3-space the "contours" most desired are "spherical surfaces"; in n-space the "contours" most desired are "hyperspheres." Also, the gradient method often converges rapidly for the first few iterations (far from the solution), but is usually a very poor algorithm during the final iterations. For any function ϑ with non-spherical contours, the number of iterations to converge exactly is generally unbounded. Satisfactory convergence accuracy often requires an unacceptably large number of one-dimensional searches. This can be overcome by using the Levenberg-Marquardt algorithm shown in §1.6.3, which combines the least squares differential correction process with a gradient search.

Example C.3: In this example the method of gradients is used to determine the minimum of the following quadratic function:

$$\vartheta(\mathbf{x}) = 4x_1^2 + 3x_2^2 - 4x_1x_2 + x_1$$

The starting guess is given by $\mathbf{x}^{(0)} = \begin{bmatrix}-1 & 3\end{bmatrix}^T$. A plot of the iterations superimposed on the contours is shown in Figure C.4. This function has low eccentricity contours with the minimum of $\mathbf{x}^* = \begin{bmatrix}-3/16 & -1/8\end{bmatrix}^T$. The Hessian matrix is constant and symmetric for this function:

$$\nabla_{\mathbf{x}}^2 \vartheta = \begin{bmatrix} 8 & -4 \\ -4 & 6 \end{bmatrix}$$

The eigenvalues of this matrix are all positive, which states that the function is well behaved. The iterations are given by

$$\begin{aligned}
\mathbf{x}^{(1)} &= \begin{bmatrix}0.7576 & 0.9649\end{bmatrix}^T \\
\mathbf{x}^{(2)} &= \begin{bmatrix}-0.2456 & 0.1003\end{bmatrix}^T \\
\mathbf{x}^{(3)} &= \begin{bmatrix}-0.1192 & -0.0462\end{bmatrix}^T \\
\mathbf{x}^{(4)} &= \begin{bmatrix}-0.1917 & -0.1088\end{bmatrix}^T \\
\mathbf{x}^{(5)} &= \begin{bmatrix}-0.1826 & -0.1194\end{bmatrix}^T
\end{aligned}$$

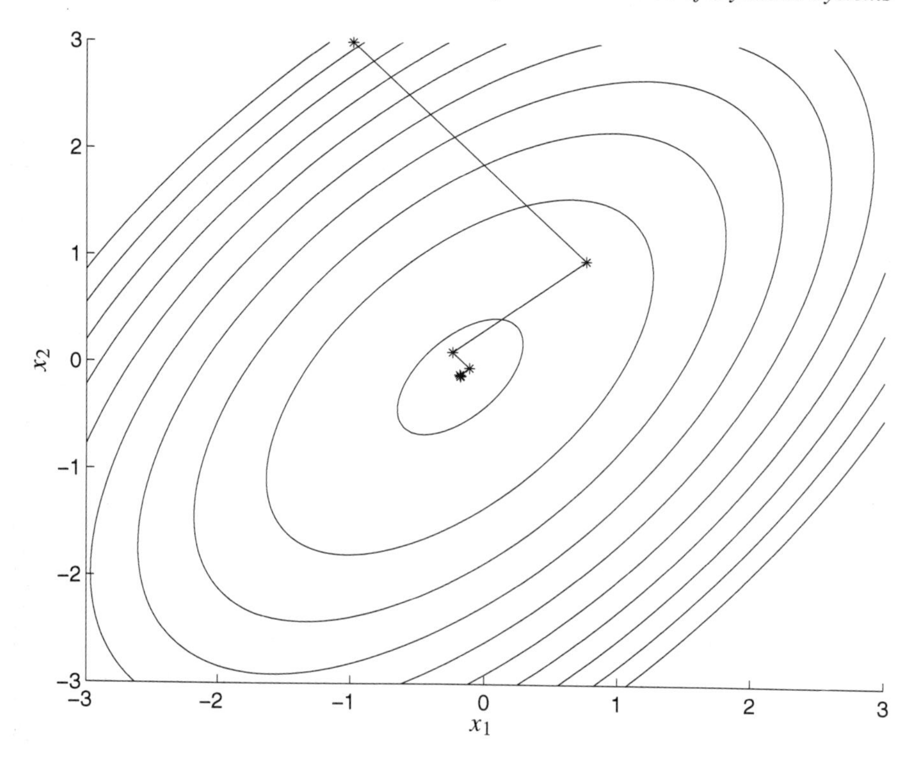

Figure C.4: Minimization of a Quadratic Loss Function

$$\mathbf{x}^{(6)} = \begin{bmatrix} -0.1878 & -0.1238 \end{bmatrix}^T$$
$$\mathbf{x}^{(7)} = \begin{bmatrix} -0.1871 & -0.1246 \end{bmatrix}^T$$
$$\mathbf{x}^{(8)} = \begin{bmatrix} -0.1875 & -0.1250 \end{bmatrix}^T$$

This clearly shows the typical performance of the gradient method, where rapid convergence is given far from the minimum, but slow progress is given near the minimum. Still, the algorithm converges to the true minimum. This behavior is also seen from various other starting guesses.

C.3.3 Second-Order (Gauss-Newton) Algorithm

The Gauss-Newton algorithm is probably the most powerful unconstrained optimization method. We will discuss a "curvature pitfall" that necessitates care in applying this algorithm, however. Say a loss function ϑ is evaluated at a local point

$\mathbf{x}^{(k)}$. It is desired to modify $\mathbf{x}^{(k)}$ by $\Delta\mathbf{x}^{(k)}$ according to

$$\mathbf{x}^{(k+1)} = \mathbf{x}^{(k)} + \Delta\mathbf{x}^{(k)} \tag{C.32}$$

in such a fashion that ϑ is decreased or increased. The behavior of ϑ near $\mathbf{x}^{(k)}$ can be approximated by a second-order Taylor's series:

$$\vartheta \cong \vartheta\left(\mathbf{x}^{(k)}\right) + \Delta\mathbf{x}^T \mathbf{g}^{(k)} + \frac{1}{2}\Delta\mathbf{x}^T H^{(k)} \Delta\mathbf{x} \tag{C.33}$$

where $\mathbf{g}^{(k)} \equiv \nabla_{\mathbf{x}}\vartheta^{(k)}$ (the gradient of ϑ) and $H^{(k)} \equiv \nabla_{\mathbf{x}}^2\vartheta^{(k)}$ (the Hessian of ϑ). The local strategy is to determine the particular correction vector $\Delta\mathbf{x}^{(k)}$ which minimizes (maximizes) the second-order prediction of ϑ. Investigating eqn. (C.33) for an extreme leads to the following:

necessary condition

$$\nabla_{\Delta\mathbf{x}}\vartheta = \mathbf{g}^{(k)} + H^{(k)}\Delta\mathbf{x} = \mathbf{0} \tag{C.34}$$

sufficient condition

$$\nabla_{\Delta\mathbf{x}}^2\vartheta = H^{(k)} \begin{cases} \text{must be positive definite for minimum.} \\ \text{must be negative definite for maximum.} \\ \text{must be indefinite for saddle.} \end{cases} \tag{C.35}$$

From the necessary condition of eqn. (C.34), the local corrections are then given by

$$\Delta\mathbf{x}^{(k)} = -\left[H^{(k)}\right]^{-1} \mathbf{g}^{(k)} \tag{C.36}$$

Substituting eqn. (C.36) into eqn. (C.32) gives the Gauss-Newton second-order optimization algorithm:

$$\mathbf{x}^{(k+1)} = \mathbf{x}^{(k)} - \left[H^{(k)}\right]^{-1} \mathbf{g}^{(k)} \tag{C.37}$$

It is important to note that this algorithm converges in exactly one iteration for a quadratic loss function, regardless of the starting guesses used. For example, the second-order correction for the loss function shown in example C.3 is given by

$$\mathbf{x}^{(1)} = \begin{bmatrix} x_1^{(0)} \\ \\ x_2^{(0)} \end{bmatrix} - \begin{bmatrix} \frac{3}{16} & \frac{1}{8} \\ \\ \frac{1}{8} & \frac{1}{4} \end{bmatrix} \begin{bmatrix} 8x_1^{(0)} - 4x_2^{(0)} + 1 \\ \\ 6x_2^{(0)} - 4x_1^{(0)} \end{bmatrix} = -\begin{bmatrix} \frac{3}{16} \\ \\ \frac{1}{8} \end{bmatrix} \tag{C.38}$$

which gives the optimal solution in one iteration! In many (probably most) solvable unconstrained optimization problems, the second-order approximation underlying eqn. (C.37) becomes valid during the final iterations; the terminal convergence of eqn. (C.37) is usually exceptionally rapid.

There is a pitfall though! If the sufficient condition of eqn. (C.35) is not satisfied, then the correction will be in the wrong direction. It is difficult to attempt minimizing

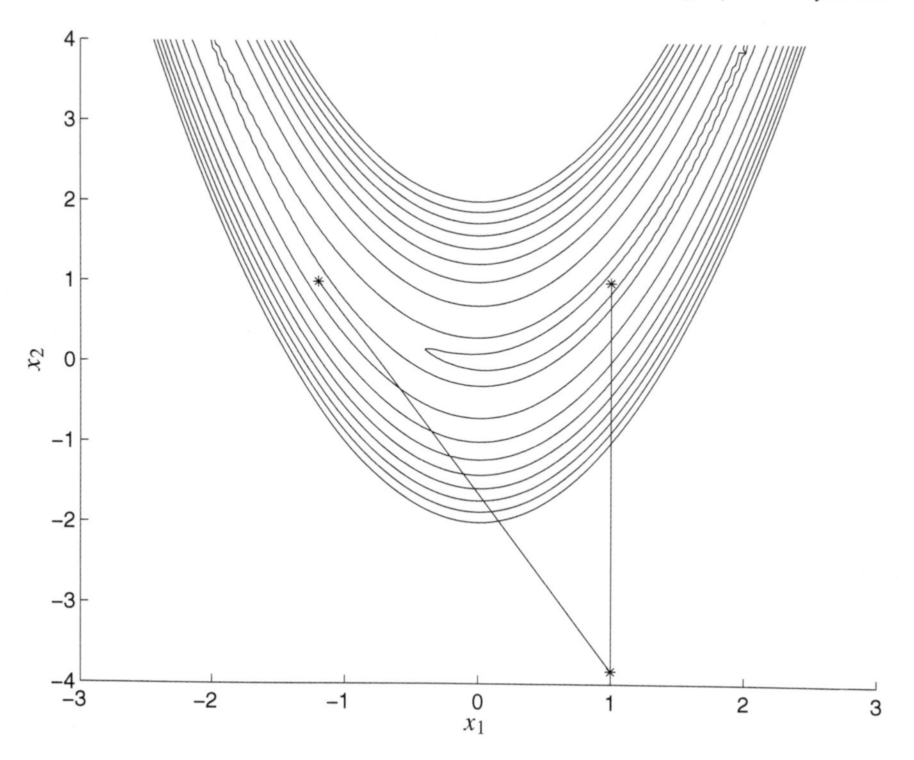

Figure C.5: Minimization of Rosenbrock's Loss Function

a function by solving for local maxima. This pitfall can be circumvented by using a gradient algorithm until the neighborhood of the solution is reached, then testing the sufficient condition of eqn. (C.35) and employing the second-order algorithm if it is satisfied.

Example C.4: In this example the Gauss-Newton algorithm is used to determine the minimum of Rosenbrock's loss function, which has been devised to be a specific challenge to gradient-based approaches:

$$\vartheta(\mathbf{x}) = 100(x_2 - x_1^2)^2 + (1 - x_1)^2$$

A plot of the contours for this function is shown in Figure C.5. Note the highly nonlinear trenches for this function. The starting guess is given by $\mathbf{x}^{(0)} = \begin{bmatrix} -1.2 & 1 \end{bmatrix}^T$. For this particular problem, the gradient method of §C.3.2 does not converge to the true minimum of $\mathbf{x}^* = \begin{bmatrix} 1 & 1 \end{bmatrix}^T$ even after 1,000 iterations. However, the second-order algorithm converges in just two iterations, shown in Figure C.5. The iterations are given by

$$\mathbf{x}^{(1)} = \begin{bmatrix} 1.0000 & -3.8400 \end{bmatrix}^T$$

$$\mathbf{x}^{(2)} = \begin{bmatrix} 1.0000 & 1.0000 \end{bmatrix}^T$$

The Hessian matrix evaluated for this function is given

$$\nabla^2_{\mathbf{x}} \vartheta = \begin{bmatrix} -400(x_2 - x_1^2) + 800x_1^2 + 2 & -400x_1 \\ -400x_1 & 200 \end{bmatrix}$$

which is always positive definite at all the iterations. This example clearly shows the advantages of using a second-order correction in the optimization process.

The overwhelmingly most significant drawback of the second-order correction is the necessity of calculating the matrix of second derivatives. For complicated loss function models, it is usually an expensive consideration to simply determine the n elements of the gradient vector. One is thus motivated to ask the question: "Is it possible to approximate quadratic convergence without the expense of calculating second partial derivatives?" The answer turns out to be yes! Observe that some "second-order information" is contained in the sequence of local function and gradient calculations. Two such techniques have been developed that are in common use today (the Fletcher-Powell[4] and Fletcher-Reeves[5] algorithms). These algorithms are not developed here due to space limitations; the interested reader should see Refs. [1] and [3] for theoretical development and numerical examples of these important algorithms.

It is also significant to note that when the loss function is the sum of squares of a set of functions whose first derivatives are available, that second-order convergence can be approximated by linearizing the functions *before squaring*. The result is a local quadratic approximation of ϑ; this local approximation can be minimized rigorously. The classical example use of this approach is the *Gaussian least squares differential correction*, which is also known as *nonlinear least squares*. This algorithm is developed in §1.4 and is applied to numerous examples in this text.

References

[1] Rao, S.S., *Engineering Optimization: Theory and Practice*, John Wiley & Sons, New York, NY, 3rd ed., 1996.

[2] Bryson, A.E. and Ho, Y.C., *Applied Optimal Control*, Taylor & Francis, London, England, 1975.

[3] Reklaitis, G.V., Ravindran, A., and Ragsdell, K.M., *Engineering Optimization: Methods and Applications*, John Wiley & Sons, New York, NY, 1983.

[4] Fletcher, R. and Powell, M., "A Rapidly Convergent Descent Method for Minimization," *Computer Journal*, Vol. 6, No. 2, July 1963, pp. 163–168.

[5] Fletcher, R. and Reeves, C.M., "Function Minimization by Conjugate Gradients," *Computer Journal*, Vol. 7, No. 2, July 1964, pp. 149–154.

D

Computer Software

ALL of the examples shown in the text have been programmed and simulated using MATLAB®. A website of these programs, listed by chapter, can be found at

http://www.buffalo.edu/~johnc/estim_book.htm

For general information regarding MATLAB or related products, please consult MathWorks, Inc. at

http://www.mathworks.com

After the MATLAB execution file is initiated the following prompt should be present:

```
>>
```

Then to see the program outputs, type "help" and the "filename." For example, for the program example1_1.m, type

```
>> help example1_1
```

This will produce the following output for this example:

> This example illustrates the basic concept of using linear
> least squares to estimate the parameters of a simple dynamic
> system. The program provides a plot of the measurements used
> in least squares and the best fit.

It has been our experience that to thoroughly understand the intricacies of a subject as diverse as estimation theory, one must learn from basic fundamentals first. Although computer routines can provide some insights to the subject, we feel that they may hinder rigorous theoretical studies that are required to properly comprehend the material. Therefore, we strongly encourage students to program their own computer routines, using the codes provided from the website for verification purposes only. We have decided not to include a disk of programs with the text so that up-to-date versions of the computer programs can be maintained on the website. The programs have been written so that anyone with even a terse background in MATLAB should be able to comprehend the relationships between the examples in the text and the coded scripts. We hope that the reader will use these programs in the spirit that they are given; to supplement their reading and understanding of the material in printed text in order to bridge the gap between theoretical studies and practical applications.

Limit of Liability/Disclaimer of Warranty: The computer programs are provided as a service to readers. While the authors have used their best efforts in preparing these programs, they make no representation or warranties with respect to the accuracy or completeness of the programs. The book publisher (CRC Press), the authors, the authors' employers (University at Buffalo and Texas A&M University), or MathWorks, Inc. shall not be liable for any loss of profit or any other commercial or noncommercial damages, including, but not limited to, special, incidental, consequential, or other damages.

Index